T0206014

Removal of Refractory Pollutants from Wastewater Treatment Plants

Removal of Refractory Pollutants from Wastewater Treatment Plants

Edited by
Maulin P. Shah

CRC Press
Taylor & Francis Group
Boca Raton London New York

CRC Press is an imprint of the
Taylor & Francis Group, an **informa** business

First edition published 2022
by CRC Press
6000 Broken Sound Parkway NW, Suite 300, Boca Raton, FL 33487-2742

and by CRC Press
2 Park Square, Milton Park, Abingdon, Oxon, OX14 4RN

CRC Press is an imprint of Taylor & Francis Group, LLC

Library of Congress Cataloging-in-Publication Data

Names: Shah, Maulin P., editor.
Title: Removal of refractory pollutants from wastewater treatment plants /
  edited by Maulin P. Shah.
Description: First edition. | Boca Raton, FL : CRC Press, 2022. | Includes
  index.
Identifiers: LCCN 2021012187 (print) | LCCN 2021012188 (ebook) | ISBN
  9780367758127 (hbk) | ISBN 9781032069081 (pbk) | ISBN 9781003204442
  (ebk)
Subjects: LCSH: Sewage--Purification. | Factory and trade
  waste--Purification. | Refractory materials--Environmental aspects. |
  Water-treatment plants.
Classification: LCC TD745 .R456 2022 (print) | LCC TD745 (ebook) | DDC
  628.3--dc23
LC record available at https://lccn.loc.gov/2021012187
LC ebook record available at https://lccn.loc.gov/2021012188

ISBN: 978-0-367-75812-7 (hbk)
ISBN: 978-1-032-06908-1 (pbk)
ISBN: 978-1-003-20444-2 (ebk)

DOI: 10.1201/9781003204442

Typeset in Times
by Deanta Global Publishing Services, Chennai, India

Contents

Preface

This book describes the state-of-the-art and emerging technologies in environmental bioremediation and reviews their various possible uses together with their related issues and implications. Considering the number of problems that define and concretize the field of environmental microbiology or bioremediation, the role of several bioprocesses and biosystems for environmental protection, control, and health based on the utilization of living organisms are analyzed.

The book aims to provide a comprehensive review of advanced emerging technologies with environmental approaches for wastewater treatment, heavy metal removal, pesticide degradation, dye removal, waste management, microbial transformation of environmental contaminants, and more. With advancements in the area of environmental bioremediation, researchers are looking for new opportunities to improve environmental quality standards. Recent technologies have given impetus to the possibility of using renewable raw materials as potential sources of energy. Cost-effectivet and eco-friendly technologies for producing high quality products and efficient ways to recycle waste to minimize environmental pollution is the need of hour. The use of bioremediation technologies through microbial communities presents another viable option to remediate environmental pollutants, such as heavy metals, pesticides, and dyes.

Since physico-chemical technologies employed in the past have many potential drawbacks, including higher costs and lower sustainability, there is a need to develop efficient biotechnological alternatives to overcome the increasing levels of environmental pollution. Hence, there is a need for environmentally-friendly technologies that can reduce pollutants which have hazardous effects on humans and the surrounding environment.

Environmental remediation, and pollution prevention, detection, and monitoring are evaluated considering the results achieved, as well as the perspectives in the development of biotechnology. Various relevant topics have been chosen to illustrate each of the main areas of environmental biotechnology – wastewater treatment, soil treatment, solid waste treatment, and waste gas treatment – dealing with both the microbiological and process engineering aspects. The distinct role of emerging technologies in environmental bioremediation in the future is emphasized, considering the opportunities to present new solutions and directions in the remediation of contaminated environments, minimizing future waste release, and creating pollution prevention alternatives. To take advantage of these opportunities, innovative strategies, which advance the use of molecular biological methods and genetic engineering technology, are examined. These methods would improve our understanding of existing biological processes in order to increase their efficiency, productivity, and flexibility. Examples of the development and implementation of such strategies are included. Also, the contribution of environmental biotechnology in creating a more sustainable society is revealed.

Editor

Dr. Maulin P. Shah has been an active researcher and scientific writer in the field of microbiology for over 20 years. He received a BSc degree (1999) in Microbiology from Gujarat University, Godhra (Gujarat), India. He also earned his PhD (2005) in Environmental Microbiology from Sardar Patel University, Vallabh Vidyanagar (Gujarat), India. His research interests include biological wastewater treatment; environmental microbiology; and the biodegradation, bioremediation, and phytoremediation of environmental pollutants from industrial wastewaters. He has published more than 240 research papers in several reputable national and international journals on various aspects of the microbial biodegradation and bioremediation of environmental pollutants. He is the editor of 65 books of international repute.

Contributors

Abdul-Wahab Abbew
Jiangsu Key Laboratory of Chemical Pollution
  Control and Resources Reuse
School of Environmental and Biological
  Engineering
Nanjing University of Science and
  Technology
Xiao Ling, Nanjing, PR China

Komal Agrawal
Bioprocess and Bioenergy Laboratory
Department of Microbiology
Central University of Rajasthan
Bandarsindri, India

Azhan Ahmad
Department of Civil Engineering
Indian Institute of Technology
Kharagpur, India

Shatha Al Mandhari
Department of Civil and Environmental
  Engineering
National University of Science and Technology
Muscat, Oman

Ayesha Algade Amadu
Jiangsu Key Laboratory of Chemical Pollution
  Control and Resources Reuse
School of Environmental and Biological
  Engineering
Nanjing University of Science and
  Technology
Xiao Ling, Nanjing, PR China

Vandita Anand
Department of Biotechnology
Motilal Nehru National Institute of
  Technology (MNNIT)
Allahabad, India

Sudipti Arora
Dr. B. Lal Institute of Biotechnology
Jaipur, India

Reyhan Ata
Occupational Health and Safety
Department of Çerkezköy Vocational School,
  Tekirdağ
Namık Kemal University
Tekirdag, Turkey

BF Bakare
Department of Chemical Engineering
Mangosuthu University of Technology
Durban, South Africa

Hafida Bendjama
Laboratory of Environmental Process
  Engineering
Department of Chemical Engineering
University Salah Boubnider
Constantine, Algeria

Pallvi Bhanot
Centre for Fire, Explosives and Environment
  Safety
Delhi, India

Joorie Bhattacharya
Genetic Gains
International Crops Research Institute for the
  Semi-Arid Tropics
Hyderabad, India

JK Bwapwa
Department of Civil Engineering
Mangosuthu University of Technology
Durban, South Africa

S. Mary Celin
Centre for Fire, Explosives and Environment
  Safety
Delhi, India

Indrajit Chakraborty
Department of Civil Engineering
Indian Institute of Technology
Kharagpur, India

Abhijit Chatterjee
Department of Bioengineering
National Institute of Technology
Agartala, India

Ashvini U. Chaudhari
Biotechnology Department
Shivaji University Kolhapur
Kolhapur, India

and

Biochemistry Division
Department of Chemistry
Savitribai Phule Pune University
Pune, India

Parmesh Kumar Chaudhari
Department of Chemical Engineering
National Institute of Technology
Raipur, India

Zhipeng Chen
Jiangsu Key Laboratory of Chemical Pollution
  Control and Resources Reuse
School of Environmental and Biological
  Engineering
Nanjing University of Science and Technology
Xiao Ling, PR China

Praveen Dahiya
Amity Institute of Biotechnology
Amity University Uttar Pradesh (AUUP)
Noida, India

Ananya Das
Department of Civil Engineering
Indian Institute of Technology Delhi
Delhi, India

and

Asian Development Research Institute
Patna, India

Debanko Das
Department of Biotechnology
Haldia Institute of Technology
Haldia, India

Joydeep Das
Department of Chemical Engineering
National Institute of Technology Agartala
Tripura, India

Khushboo Dasauni
Department of Biotechnology
Kumaun University Nainital
Uttarakhand, India

Hemant Dasila
Department of Microbiology
G. B. Pant University of Agriculture and
  Technology
Uttarakhand, India

Prasenjit Debbarma
School of Agriculture
Graphic Era Hill University
Uttarakhand, India

Divya
Department of Biotechnology
Kumaun University Nainital
Uttarakhand, India

Subhasish Dutta
Department of Biotechnology
Haldia Institute of Technology
Haldia, West Bengal

Mostafa M. El-Sheekh
Botany Department
Tanta University
Tanta, Egypt

Nitika Gaurav
Kusuma School of Biological Sciences
Indian Institute of Technology Delhi
Delhi, India

Shijian Ge
Jiangsu Key Laboratory of Chemical Pollution
  Control and Resources Reuse
School of Environmental and Biological
  Engineering
Nanjing University of Science and
  Technology
Xiao Ling, PR China

Sachin Rameshrao Geed
Department of Chemical Engineering and
  Technology
Indian Institute of Technology (BHU)
Varanasi, India

M. M. Ghangrekar
Department of Civil Engineering
Indian Institute of Technology
Kharagpur, India

Damodhar Ghime
Department of Chemical Engineering
National Institute of Technology
Chhattisgarh, India

Prabir Ghosh
Department of Chemical Engineering
National Institute of Technology
Chhattisgarh, India

Sougata Ghosh
Department of Chemical Engineering
Northeastern University
Boston, Massachusetts

and

Department of Microbiology
School of Science
RK University
Gujarat, India

Anupam Guleria
Centre of Biomedical Research
SGPGIMS Campus
Lucknow, India

Ashish Guleria
Department of Applied Sciences
WIT Dehradun, India

Oualid Hamdaoui
Chemical Engineering Department
College of Engineering
King Saud University
Riyadh, Saudi Arabia

Jyoti P. Jadhav
Biochemistry Division
Department of Chemistry
Savitribai Phule Pune University
Pune, India

Divya Joshi
Department of Microbiology
G. B. Pant University of Agriculture and
   Technology
Uttarakhand, India

Rishee K. Kalaria
ASPEE Shakilam Biotechnology Institute
Navsari Agricultural University
Surat, Gujarat

Anchita Kalsi
Centre for Fire, Explosives and Environment
   Safety
Delhi, India

Kisan M. Kodam
Biotechnology Department
Shivaji University Kolhapur
Kolhapur, India

Dinesh Kumar
Centre of Biomedical Research
SGPGIMS Campus
Lucknow, India

Nitesh Kumar
Department of Chemistry
Govt. College Jukhala
Himachal Pradesh, India

Saurabh Kumar
Department of Microbiology
ICAR Research Complex for Eastern
   Region
Bihar, India

Garima Kumari
Department of Biotechnology
Eternal University
Baru Sahib, India

Jaya Lakkakula
Amity Institute of Biotechnology
Amity University
Mumbai

Eder Lima
Institute of Chemistry
Federal University of Rio Grande
Rio Grande, Brazil

Hekmat R. Madian
Petroleum Biotechnology Lab
Processes Design & Develop Department
Egyptian Petroleum Research Institute (EPRI)
Cairo, Egypt

Uttara Mahapatra
Department of Chemical Engineering
National Institute of Technology
Agartala, India

Eman A. Mahmoud
Petroleum Biotechnology Lab
Processes Design & Develop Department
Egyptian Petroleum Research
    Institute (EPRI)
Cairo, Egypt

Damini Maithani
Department of Microbiology
G. B. Pant University of Agriculture and
    Technology
Uttarakhand, India

Ajay Kumar Manna
Department of Chemical Engineering
National Institute of Technology
Agartala, India

Gökçe Faika Merdan
Environmental Engineer in Danone Waters
    Company
İstanbul, Turkey

Slimane Merouani
Laboratory of Environmental Process
    Engineering
Department of Chemical Engineering
University Salah Boubnider
Constantine, Algeria

Sumedha Mohan
Amity Institute of Biotechnology
Amity University Uttar Pradesh (AUUP)
Noida, India

Titikshya Mohapatra
Department of Chemical Engineering
National Institute of Technology
Raipur, India

Aditi Nag
Dr. B. Lal Institute of Biotechnology
Malviya Industrial Area, India

Soma Nag
Department of Chemical Engineering
National Institute of Technology Agartala
Tripura, India

Tapan K. Nailwal
Department of Biotechnology
Kumaun University Nainital
Uttarakhand, India

Sonal Nigam
Amity Institute of Biotechnology
Amity University
Noida, India

Rahul Nitnavare
Division of Plant and Crop Sciences
University of Nottingham
Nottingham, United Kingdom

and

Department of Plant Sciences
Harpenden, United Kingdom

Diana Pacheco
Marine and Environmental Sciences Centre
    (MAREFOZ Laboratory)
University of Coimbra
Figueira da Foz, Portugal

and

Department of Life Sciences
University of Coimbra
Coimbra, Portugal

Anjana Pandey
Department of Biotechnology
Motilal Nehru National Institute of Technology
    (MNNIT)
Allahabad, India

Hiren K. Patel
School of Science
P. P. Savani University
Surat, Gujarat

Sanchita Patwardan
Amity Institute of Biotechnology
Amity University
Mumbai, India

Leonel Pereira
Marine and Environmental Sciences Centre
    (MAREFOZ Laboratory)
University of Coimbra
Coimbra, Portugal

and

Department of Life Sciences
University of Coimbra
Comibra, Portugal

Mohammed Zeeshan Qasim
Jiangsu Key Laboratory of Chemical Pollution
    Control and Resources Reuse
School of Environmental and Biological
    Engineering
Nanjing University of Science and Technology
Xiao Ling, PR China

Shuang Qiu
Jiangsu Key Laboratory of Chemical Pollution
    Control and Resources Reuse
School of Environmental and Biological
    Engineering
Nanjing University of Science and Technology
Xiao Ling, PR China

Rajesh W. Raut
The Institute of Science
Mumbai, India

Ana Cristina Rocha
Marine and Environmental Sciences Centre
    (MAREFOZ Laboratory)
University of Coimbra
Figueira da Foz, Portugal

and

Department of Life Sciences
University of Coimbra
Coimbra, Portugal

Arpita Roy
Department of Biotechnology
School of Engineering & Technology
Sharda University
Greater Noida, India

Bhumika Roy
Delhi Technological University
Delhi, India

Sandeep Kumar Sahai
Centre for Fire, Explosives and Environment
    Safety
Delhi, India

S.M. Sathe
Department of Civil Engineering
Indian Institute of Technology
Kharagpur, India

Devendra Sharma
Dr. B. Lal Institute of Biotechnology
Malviya Industrial Area, India

Yeting Shen
Jiangsu Key Laboratory of Chemical Pollution
    Control and Resources Reuse
School of Environmental and Biological
    Engineering
Nanjing University of Science and Technology
Xiao Ling, PR China

Sudheer Kumar Shukla
Department of Civil and Environmental
    Engineering
College of Engineering
National University of Science and
    Technology
Muscat, Oman

Kulvinder Singh
Department of Chemistry
Maharaja Agrasen University
Baddi, India

Surbhi Sinha
Amity Institute of Biotechnology
Amity University
Noida, India

Ravindra Soni
Department of Agricultural Microbiology
College of Agriculture
Indira Gandhi Krishi Vishwa Vidyalaya
Chhattisgarh, India

Deep Chandra Suyal
Department of Microbiology
Akal College of Basic Sciences
Eternal University
Himachal Pradesh, India

Rajesh Kumar Tanwar
Centre for Fire, Explosives and Environment
    Safety
Delhi, India

Viresh R. Thamke
Biotechnology Department
Shivaji University Kolhapur
Kolhapur, India

and

Biochemistry Division
Savitribai Phule Pune University
Pune, India

Günay Yıldız Töre
Environmental Engineering Department of
    Çorlu Engineering Faculty
Tekirdağ Namık Kemal University
Tekirdag, Turkey

Divyesh K. Vasava
College of Agriculture
Junagadh Agricultural University
Junagadh, Gujarat

Tiago Verdelhos
Marine and Environmental Sciences Centre
    (MAREFOZ Laboratory)
University of Coimbra
Figueira da Foz, Portugal

and

Department of Life Science
University of Coimbra
Coimbra, Portugal

Pradeep Verma
Bioprocess and Bioenergy Laboratory
Department of Microbiology
Central University of Rajasthan
Bandarsindri, India

Nilesh S. Wagh
Amity Institute of Biotechnology
Amity University
Mumbai, India

Lingfeng Wang
Jiangsu Key Laboratory of Chemical Pollution
    Control and Resources Reuse
School of Environmental and Biological
    Engineering
Nanjing University of Science and Technology
Xiao Ling, PR China

Thomas J Webster
Department of Chemical Engineering
Northeastern University
Boston, Massachusetts

Zhengshuai Wu
Jiangsu Key Laboratory of Chemical Pollution
    Control and Resources Reuse
School of Environmental and Biological
    Engineering
Nanjing University of Science and Technology
Xiao Ling, PR China

1 Removal of Pharmaceuticals From Wastewater Using Nanomaterials

Bhumika Roy and Arpita Roy

CONTENTS

1.1 INTRODUCTION

Environmental pollution is one of the biggest challenges our planet has faced in the present century. The increasing world population along with the development of socio-economic, scientific, and technological advancements have directly or inadvertently resulted in ecological deterioration. The presence of pharmaceutical waste in wastewater is a growing threat to the aquatic ecosystem as well as land animals, including human beings. The presence of these contaminants in drinking water is a growing concern; however, proper research about the future impact on the environment is lacking. As a result, these pollutants enter our food chain. Pharmaceuticals are introduced into the environment through various sources like pharmaceutical manufacturing industries, wastewater treatment plants, veterinary use, and leakage in underground sewage systems. Pharmaceutical waste from pharmaceutical manufacturing industries enters the aquatic system due to a lack of properly designed wastewater treatment plants which leads to improper removal of these wastes and finally causes pollution in the aquatic ecosystem (Murray *et al.*, 2010). Animal excreta and its use as manure in agricultural fields introduce these pollutants into the environment. According to available literature, the highest concentration of pharmaceutical waste was found in industrial effluent

DOI: 10.1201/9781003204442-1

followed by hospital and wastewater treatment effluents with surface water, groundwater, and drinking water showing least concentrations (Patel *et al.*, 2019). Figure 1.1 represents how pharmaceutical waste is introduced into the environment.

A global critical evaluation stated that 203 different types of pharmaceutical waste were found in the environment in 41 countries. This data is, however, still incomplete as only a few countries were included in the study and, in some cases, contributors did not have enough resources to cover the true impact of pharmaceutical waste (Patel *et al.*, 2019). Factors such as demography, accessibility of health facilities, size of the manufacturing sector, sewage treatment systems, and effectiveness of regulation guidelines help in determining the ecological footprint of pharmaceuticals. The introduction of pharmaceutical waste into the environment is an important issue since these compounds are synthesized to bind with the receptors that are present in the body. In such a case, consumption by animals that they are not intended for can have grave effects and prove to be toxic to numerous organisms. Various types of pharmaceutical waste are listed in Table 1.1 with their intended functions, which could prove to be toxic to microorganisms. Apart from this, antibiotic resistance has also been observed in bacteria. Observations like alteration in gene expression, abnormal enzymatic activities, and growth deformities in fish, rats, and frogs have made it even more necessary to combat this source of waste in water.

Conventional wastewater treatment systems like activated sludge and biological trickling filters are not adequate enough to eliminate the wide variety of emerging contaminants and thus they remain soluble in the effluent. Various studies have reported the unsuccessful use of physicochemical treatments including coagulation, flocculation, or lime softening for removing pharmaceutical waste. Techniques such as ozonation and photolysis have some shortcomings as well in the form of being expensive, having a short lifetime, and being unable to remove micropollutants. Due to this, these methods need to be improved or replaced in order to successfully remove pollutants in a cost effective and reliable manner. Nanotechnology presents an efficient and advanced treatment process for the successful remediation of persistent water pollutants. Nanomaterials possess promising results in the adsorption of pollutants. Therefore, in this chapter, the application of nanotechnology over conventional methods for the remediation of pharmaceutical waste from wastewater will be discussed.

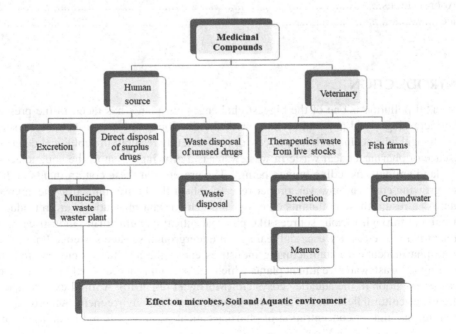

FIGURE 1.1   Entry of pharmaceutical waste into the environment.

TABLE 1.1

Types of Pharmaceutical Wastes, Their Function as Drugs, and Their Use in Specific Treatments

Type of Waste	Function	Use	Example
β-blockers	Block epinephrine and norepinephrine from binding on β receptors on nerves	Treatment of cardiovascular diseases	Atenolol, celiprolol, metoprolol, and propranolol
Hormones	Regulate hormone levels	Treatment of many diseases affected by hormone levels like cancer, growth hormone deficiency, and menopausal symptoms	Estrone, 17-β-estradiol, and 17-α-ethinylestradiol
β-lactam antibiotics	Antimicrobial activity against gram-negative and gram-positive bacteria generally by inhibiting cell wall synthesis	Treating bacterial infections of skin, ear, respiratory and urinary tract	Amoxicillin, cefradine, ceftriaxone, and sultamicillin
Cytostatic drugs	Block mitosis by inhibiting DNA synthesis	Treatment of various types of cancer, autoimmune diseases, and suppress transplant rejections	Cyclophosphamide and ifosfamide
Sulfonamide antibiotics	Inhibit growth and multiplication of bacteria by serving as competitive inhibitors of DHPS in folate synthesis	Treatment of allergies, cough, and antifungal and antimalarial functions	Sulphachlopyridazine, sulfadiazine, sulfadimethoxine, and sulfamerazine
Anticonvulsants and anti-anxiety agents	May block sodium channels or block GABA function	Treatment of mood disorders, depression, and anxiety, as well as to control seizures	Carbamazepine, diazepam, oxazepam, and primidone
Macrolide antibiotics	Inhibit growth of bacteria by blocking 50S ribosomal unit and hindering protein synthesis	Treating common bacterial infections	Azithromycin, clarithromycin, erythromycin, and roxythromycin
Quinolone antibiotics	Hinder the DNA synthesis process	Treating genitourinary infections	Ciprofloxacin, flumequine, norfloxacin and ofloxacin
Analgesic drugs	Affect peripheral and central nervous system in many ways	Treatment of pain and inflammation	Acetaminophen, diclofenac, ibuprofen, and naproxen

1.2 QUANTIFICATION OF PHARMACEUTICAL WASTE

Quantification can be done both before and after the treatment of wastewater to determine remediation effectiveness; however, procuring an accurate estimate is difficult due to low analytic concentration and complex matrix effects. Therefore, performing steps such as sample collection, sample preparation, chromatographic separation, detection, and data analysis is essential.

1.2.1 SAMPLE COLLECTION

A large number of samples are processed to procure reliable results that are as accurate as possible. To further modify the process, the volumes of different samples taken must be in accordance with the concentration of analyte, i.e., samples containing lesser concentrations must be collected in larger volumes. The proper storage of these samples is of utmost importance since errors can lead to the contamination or decomposition of sample analytes. The key players in contamination include microbial activity, unwanted chemical reactions, and exposure to UV radiation. Microbial activity and analyte decomposition can be controlled by adding chemical preservatives or storing in brown amber glass bottles with temperature control (Śliwka-Kaszyńska *et al.*, 2003).

1.2.2 SAMPLE PREPARATION

Samples are prepared by purifying and concentrating them before analysis because of the confounding matrix effect; some pharmaceuticals can show effects in concentrations as low as 1 ng/L. The chemical and physical properties of analyte make this optimization. It must be performed every time the matrix chemicals change. The steps leading up to chromatographic separation include the adjustment of pH, extraction, and elution.

Solid-phase extraction (known for its multiple analyte extraction from complex mixtures) is employed to remove dissolved pharmaceuticals. The mixture is passed through a column where the pharmaceutical separates from the mobile phase to bind to a solid stationary phase. Conditioning of the column is done with a suitable solvent to make the stationary-phase surface wet, followed by loading of the sample onto the column wherein retention of the target analyte is observed. Further washing of the column is done to remove any persistent impurities and later the target pharmaceuticals are eluted and buffered by pH. The solvent is then prepared for analysis by adjusting the volume.

Chromatographic separation is done to further isolate target pharmaceuticals from matrix chemicals through mass spectrometry detection. MS, MS/MS, electrospray ionization, and atmospheric pressure chemical ionization are some techniques used for detection. Because matrix effects can largely reduce analyte signal intensity in environmental pharmaceutical analysis, detection may be hindered. Raman spectroscopy has also been used for pharmaceutical analysis in drug development, drug production, quality control, studies on stability, and drug metabolites analysis. Other analytical techniques include electrophoresis, electrochemical, flow injection analysis, and titrimetric-related processes (Siddiqui *et al.*, 2017).

The identification and quantification of analytes is done by using calibration curves. In this process, the matrix effect plays a vital role. Therefore, a range of matrix chemicals should be included in calibration standards, which helps in the reduction of any errors related to matrix effects. Interference-free quantification in environmental analysis can be ensured by second-order multivariate calibration algorithms, internal standard (IS)-based calibration, standard addition calibration, and matrix-matched calibration.

1.3 ENVIRONMENTAL AND HEALTH ISSUES OF PHARMACEUTICAL WASTE

The environmental exposure of pharmaceuticals generally occurs through, amongst others, manufacturing units, hospital effluents, and land applications. Wastewater treatment plants are not reliably effective at eliminating these pollutants. Therefore, these pollutants enter the aquatic environment where they directly affect aquatic organisms and can be incorporated into food chains. High concentrations of pharmaceuticals can change the structure of microbial communities and ultimately affects food chains. Long-term exposure to even low concentrations of complex pharmaceuticals can result in an acute and chronic damage, behavioral variations, reproductive damage, inhibition of cell proliferation, and more. Several reports suggested that wastewater effluents exhibit reproductive abnormalities in fishes. Furthermore, fish exposed to trace levels of birth control pharmaceuticals showed dramatic reductions in reproductive success, which suggests impacts on population levels.

1.4 CONVENTIONAL TREATMENT METHODS AND THEIR SHORTCOMINGS

Conventional treatment methods include activated carbon adsorption, membrane filtration, biological processes, UV photocatalysis, and ozonation. These methods have their shortcomings in the form of low efficiency or high cost and thus needed to be replaced. Biological processes, for instance, cannot fully degrade recalcitrant pharmaceutically active compounds (PhACs) (like bezafibrate) and carbamazepine. The treatment of wastewater discharged from pharmaceutical industries poses additional difficulties due to the presence of active pharmaceutical ingredients (APIs), organic solvents, reactants, intermediates, raw materials, and catalysts (Chelliapan et al., 2006). Commonly-used adsorbent-like activated charcoal removes hydrophobic substances and charged PhACs from the aqueous phase with the help of non-specific dispersive interactions for removing non-polar PhACs, and interactions of electrostatically activated charcoal and ionic PhACs to remove polar PhACs. This method is mostly used after biological treatment as a post-treatment measure. The drawbacks of this method are in terms of its effectiveness which is governed by the natural dissolved organic matters in the wastewater matrix and its high total energy demand.

Ozonation is used as a secondary treatment method for removing PhACs from wastewater (Dodd et al., 2009). Shortcomings, such as the consumption of high and uncertain effects of oxidation products, necessitate further development of the method. Catalytic ozonation acts as an alternative to the ozonation process and is one of the most advanced oxidation processes. This method's efficiency for removing organic pollutants makes it better than ozonation; however, it has its shortcomings, including a decline in rate due to the limited adsorption and diffusion of PhACs on catalysts steps.

Membrane filtration removes low molecular weight PhACs (Le-Minh et al., 2010) via an adsorption process. In the steady state, the efficiency of removal depends on membrane characteristics (material, surface morphology, and pore size), solution parameters (such as pH and ionic strength), and physicochemical properties of PhACs (such as pK_a, molecular weight, hydrophobicity, or hydrophilicity) (Le-Minh et al., 2010). Membrane fouling causes restrictions in the engineering application of membrane processes and serves as a challenge for effective operation.

UV radiation in association with photocatalysts shows relatively high efficiency in degrading PhACs in wastewater (Murgolo et al., 2015) and has a low energy cost. Conventional photocatalysis methods come with their own drawbacks, such as low solar light activity, high energy consumption, and low quantum yield efficiency. These drawbacks of conventional methods call for a need for new photocatalysts which can improve the removal efficiency of pharmaceutical wastes.

1.5 NANOPARTICLES IN WASTEWATER TREATMENT

Nanoadsorbents like carbon nanotubes and metal oxides have wide applications in wastewater treatment. Their key features such as high surface area, high dispersion area, microporous structure, and being economically viable make their application stronger (Gupta et al., 2015). Nanoparticles used for adsorption include carbon nanotubes, graphene, magnesium oxide, ferric oxides, manganese oxide, zinc oxide, and titanium oxide (Gupta et al., 2015). The factors affecting the process of adsorption include temperature, dose of adsorbent, pH, and contact time. Nanocatalysts are gaining attention for their properties especially the ones obtained from inorganic materials. The different types of catalysts involved in wastewater treatment include photocatalysts, electrocatalysts, and fenton-based catalysts. These catalysts are used for the chemical oxidation of organic pollutants, especially the ones made from noble metals such as gold, platinum, and palladium (Liu et al., 2013). They are advantageous over conventional methods due to lessened treatment times, target recalcitrant compounds, and the transformation of wastes into valuable by-products. Nano-membranes use nanoparticles for membrane filtration technology and are the most effective wastewater treatment method. It is advantageous to use because it provides effective disinfection, quality water treatment, and is efficient, economical, and simple. One-dimensional nanomaterials consisting of organic and inorganic materials such as nanotubes, nanofibers, and nanoribbons are used for constructing these membranes (Liu *et al.*, 2014).

Nanoadsorbents sometimes lead to secondary pollution due to difficulties in the separation of small particles from the aqueous solution, which affect the bioavailability and mobility of pollutants and cause environmental toxicity. Their reuse and regeneration also pose a challenge. Nanocatalysts act as an advancement to catalytic water treatment techniques such as electrocatalysis, photocatalysis, and fenton catalysis, though there are many drawbacks. Zinc oxide and titanium dioxide used in photocatalysis require UV radiation for their activity which raises serious health risks like skin cancer for workers. Nanocatalysts, like AgBr, when introduced in the solution cannot be recycled for reuse. Electrocatalysts widely require Pt, however, its limited availability, poisoning of intermediates, and high cost act as limiting factors (Zhou *et al.*, 2003). Fenton catalysis requires the maintenance of acidic conditions throughout the process and catalyst material is continuously lost. Nanomembrane manufacturing has a large ecological footprint. A study by Khanna *et al.* (2008) showed that carbon nanofiber contributes 100 times more to toxicity, ozone depletion, and global warming than conventional methods. Another disadvantage is membrane fouling, which occurs when organic compounds interact with hydrophobic membranes. This reduces the quality of water treated, and the life and reliability of the membrane equipment (Gu *et al.*, 2013). This problem can be overcome by using biogenic nanomaterials.

1.6   BIOGENIC NANOMATERIAL

Employing the use of biogenic nanomaterials is advantageous over conventional methods. Conventional wastewater treatment methods are, to a certain extent, detrimental to the environment due to their inability to degrade the pollutants into environmental-friendly end products and passing them off from one phase to another. Apart from this, they require large areas to function, high capital, and maintenance costs. Nanotechnology is therefore used as an alternative technique in the form of nano-based adsorbents, catalysts, and membranes (Diallo and Brinker, 2011). They are mass produced using physicochemical methods; however, toxic waste gets introduced into the environment which affects human health and the environment. For instance, sodium borohydride generates hydrogen diborane which is a highly toxic by-product (Li *et al.*, 2006). Biogenic nanoparticles synthesized from bacteria, algae, fungus, and plants are a new alternative. Their attributes makes them suitable for pharmaceutical waste removal from wastewater, including low cost of production, no toxic waste production, greater surface area, stability due to lipid bilayer structure in some, better physiological solubility and manipulation of size and shape by altering pH, contact time, and substrate availability (Li *et al.*, 2011) (see Figure 1.2).

1.7   SYNTHESIS OF BIOGENIC NANOPARTICLES

The synthesis of nanoparticles via microorganisms can take place either intracellularly or extracellularly. The positively charged metal ions diffuse via electrostatic interactions into the negatively charged cell wall in intracellular processes. The uptake is further facilitated by endocytosis, ion channels, and carrier channels. The enzymes convert the toxic metals to non-toxic nanoparticles. Extracellular processes are achieved via enzymes, including nitrate reductase (from fungus), that convert metal ions into metal nanoparticles. Plants are also an alternative source for biogenic synthesis of nanoparticles (see Figure 1.3).

1.7.1   BACTERIAL-FACILITATED SYNTHESIS

Many bacterial strains employ processes like bioleaching, biomineralization, and bioaccumulation to solubilize metal ions and are used as potential biofactories. Their protective and defensive strategy against metal ions aids metallic nanoparticle synthesis, and the ease of manipulation makes

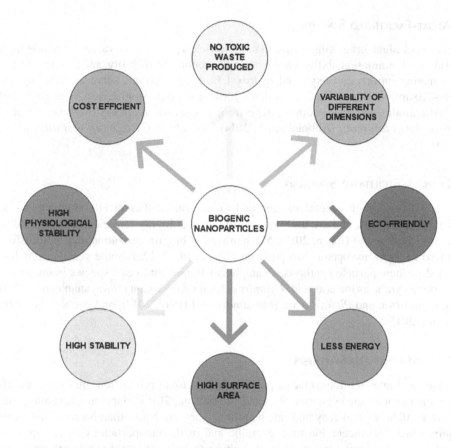

FIGURE 1.2 Advantages of biogenic nanoparticles.

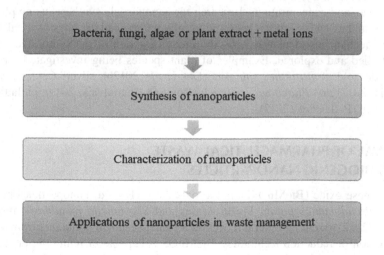

FIGURE 1.3 Biosynthesis of nanoparticles.

them appropriate for both intracellular and extracellular synthesis (Prasad *et al.*, 2016). Examples of bacterial species being investigated for nanoparticle synthesis include gold: *Klebsiella pneumonia* (Malarkodi *et al.*, 2013); iron: *Bacillus subtilis* (Sundaram *et al.*, 2012) and *Klebsiella oxytoca* (Anghel *et al.*, 2012); copper: *Escherichia coli* (Singh *et al.*, 2010) and *Pseudomonas fluorescens* (Shantkriti and Rani, 2014); and silver: *Bacillus cereus* (Sunkar and Nachiyaar, 2012).

1.7.2 ALGAE-FACILITATED SYNTHESIS

The properties of algae promoting its use in nanoparticle synthesis are vast and include high toler-ance, metal bioaccumulation ability, easy handling, economic feasibility, and richness of bioactive molecules having amine, carboxyl, and hydroxyl functional groups serve as reducing and cap-ping agents. Examples of algal species utilized for nanoparticle synthesis include gold: *Chlorella vulgaris* (Annamalai and Nallamuthu, 2015); iron: *Sargassum muticum* (Mahdavi *et al.*, 2013); copper: *Bifurcaria bifurcata* (Abboud *et al.*, 2014); and silver: *Cystophora moniliformis* (Prasad *et al.*, 2013).

1.7.3 FUNGUS-FACILITATED SYNTHESIS

Fungi – being diverse, highly metal tolerant, and a good source of extracellular enzymes – are able to accumulate metal ions, and are used for the synthesis of highly stable and economically-viable nanoparticles (Siddiqi and Husen, 2016). Mechanisms of biomineralization and biotransformation are employed to form mycogenic nanoparticles (Das *et al.*, 2012a). Some yeast strains have also been studied for nanoparticle synthesis. Examples of fungus and yeast species being investigated for nanoparticle synthesis include gold: *Cylindrocladium floridanum* (Narayanan *et al.*, 2013); iron: *Fusarium oxysporum* and *Pleurotus sp.* (Mazumdar and Haloi, 2017); and copper: *Aspergillus sp.* (Cuevas *et al.*, 2015).

1.7.4 PLANT-FACILITATED SYNTHESIS

This involves a single-step biosynthesis process and no toxic production, making it an effective alternative for nanoparticle synthesis (Roy and Bharadvaja, 2019). Plants are advantageous to use due to their availability, and easy and safe handling. They are better than bacteria and fungi since they require much less incubation time. Metallic and oxide nanoparticles can be synthesized on an industrial scale by employing plant tissue culture techniques and downstream processes. The effect of the nanoparticle synthesized varies from species to species and depends on their mode of application, size, and concentrations. Research for nanoparticle synthesis using plants is still in the initial stages and extensive work is required to fully understand the physiological, biochemical, and molecular mechanisms of plants. Also, the mode of action of the synthesized nanoparticles needs to be studied and explored. Examples of plant species being investigated for nanoparticle synthesis include gold: *Amaranthus spinosus* (Das *et al.*, 2012b); silver: *Centella asiatica* (Roy and Bharadvaja, 2017) and *Plumbago zeylanica* (Roy and Bharadvaja, 2019); palladium: *soybean* [*Glycine max* (L.)] (Petla *et al.*, 2012).

1.8 REMOVAL OF PHARMACEUTICAL WASTE USING BIOGENIC NANOPARTICLES

Biogenic manganese oxide (BioMnOx) and bio-palladium (Bio-Pd) nanoparticles have been syn-thesized using *Pseudomonas putida* MnB6 strains and *Shewanella oneidensis*, respectively, by Meerburg *et al.* (2012) to remove several recalcitrant pharmaceutical pollutants from sewage waste-water via oxidation or reduction techniques. BioMnOx is capable of removing organic micropol-lutants such as naproxen, diclofenac, clarithromycin, chlorophene, iohexol, ibuprofen, and steroid hormone estrone at ppb level concentrations (Furgal *et al.*, 2014). Bio-Pd can reportedly remove iomeprol, iopromide, and iohexol via catalytic reduction.

Two biocatalysts, bio-platinum (bio-Pt) and bio-palladium (bio-Pd), derived from *Desulfovibrio vulgaris* were found to be acting against ciprofloxacin, 17b-estradiol, and sulfamethoxazole (Martins *et al.*, 2017) with bio-Pt showing a higher catalytical role than bio-Pd. Ibuprofen was also included in this study; however, both bio-Pt and bio-Pd were unsuccessful in removing it. When whole cells of *D. vulgaris* were introduced, they transformed the ibuprofen, and it was deduced that sulfate-reducing bacteria can completely remove it in an anaerobic environment (Kumari *et al.*, 2019).

Diclofenac is an anti-inflammatory drug that is modestly biodegradable and present in very low concentrations in water. When treated via ozonation, it gave rise to mutagenic by-products. Therefore, bio-Pd nanoparticles were improved to test their activity on it. Bio-Pd nanoparticles synthesized from metal-reducing bacteria when doped with zero-valent Au NPs degrade diclofenac from water. The study concluded that the catalytic activity is determined by the mass ratio between Pd and Au with a 50/L ratio resulting in maximum degradation of the drug in 24 hours (De Corte *et al.*, 2012).

HCl was employed for the regeneration of green composite iron nanoparticles by Ali *et al.* (2016). The regenerated NPs showed a consistent removal rate of 85–92% for ibuprofen.

BioMnOx produced from *Desmodesmus sp.* WR1 algae was similar to bacteria derived BioMnOx in degrading organic pollutants via oxidative degradation and was successful against Bisphenol A (BPA), an endocrine disrupter (Wang *et al.*, 2017).

Magnetic chitosan nanoparticles were used for the removal of tetracycline from wastewater. They showed a maximum adsorption capacity of 78.11 mg/g at pH 5 and a temperature of 25°C. Incorporating magnetic material is advantageous as it promotes easy separation of adsorbent (Raeiatbin and Açıkel, 2017).

Camiré *et al.* (2019) in their study showed the use of novel alkali lignin and poly (vinyl alcohol) (AL: PVA) nanofibrous membrane for the adsorption of pharmaceutical waste. The study showed 90% adsorption of fluoxetine which was in parallel with costly ion-exchange resins (75–80 mg/g). The nanofibers follow the pseudo-first order kinetic model and the Sips isotherm model. This helps in deducing that the nature of the adsorbent was physical and adsorption was taking place at multiple sites (Camiré *et al.*, 2019).

Green-synthesized copper nanoadsorbents were also evaluated for their activity against pharmaceutical waste present in water by Husein *et al.* (2019). Their activity was checked against three pharmaceutical compounds: diclofenac, ibuprofen, and naproxen, and the adsorption capacities were 36, 33.9, 33.9 mg/g, respectively. The nature of the sorption process was stated to be spontaneous, endothermic, and physical.

Silica-based nanoparticles obtained from rice husk were used by Nassar *et al.* (2019) for studying the adsorptive removal of ciprofloxacin from polluted water. The adsorption occurred at pH 7 and the adsorptive capacity of the nanoparticles was 24.1 mg/g. These nanoparticles are cheap to synthesize, are non-toxic to the environment, and are biocompatible. Their advantages such as high surface area, high porosity, and mechanical resistance make their candidature as green nanoparticles strong (Nassar *et al.*, 2019).

Green-synthesized zinc oxide nanoparticles were coated on a ceramic ultrafiltration membrane and were used to remove atenolol and ibuprofen drugs from a synthetic solution, as evaluated by Bhattacharya *et al.* (2020). The membrane's hydrophilicity was enhanced by the coating of zinc oxide nanoparticles and the study concluded that effective removal of the pharmaceutical compounds was observed and can be tested in pharmaceutical wastewater.

1.9 CONCLUSION

A nanotechnology-based approach for the degradation of pharmaceutical waste has gained popularity due to its various advantages. However, toxic and volatile substances being used for the same pose a major concern to manufacturing units. To overcome this challenge, greener ways of producing nanomaterials were explored by various research groups and the best alternative deduced was production with the help of biomolecules from organisms. Nanoparticles from bacteria, algae, fungi, and plants are non-toxic, sustainable, and have a low cost, thus they are gaining popularity. They successfully convert toxic material into non-toxic forms and hence their activity on all types of waste material in water is being explored. Some challenges of biogenic nanoparticles, such as stability and size control, still exist and need to be studied. Research focusing on species that have not been studied yet is underway.

LIST OF ABBREVIATIONS

Ng	Nanogram
L	Liter
MS	Mass spectrometry
MS/MS	Tandem mass spectrometry
PhAC	Pharmaceutically active compound
API	Active pharmaceutical ingredient
Ppb	Parts per billion
Pd	Palladium
Au	Gold
HCl	Hydrochloric acid
NP	Nanoparticle
Mg	Milligram
g	Gram
°C	Celsius

REFERENCES

Abboud Y, Saffaj T, Chagraoui A, El Bouari A, Brouzi K, Tanane O, Ihssane B. (2014) Biosynthesis, characterization and antimicrobial activity of copper oxide nanoparticles (CONPs) produced using brown alga extract (Bifurcaria bifurcata). *Appl Nanosci* 4(5):571–576.

Ali I, Al-Othman ZA, Alwarthan A. (2016) Synthesis of composite iron nano adsorbent and removal of ibuprofen drug residue from water. *J Mol Liq* 219:858–864.

Anghel L, Balasoiu M, Ishchenko LA, Stolyar SV, Kurkin TS, Rogachev AV, Kuklin AI, Kovalev YS, Raikher YL, Iskhakov RS, Duca G. (2012) Characterization of bio-synthesized nanoparticles produced by Klebsiella oxytoca. *J Phys* 351(1):012005.

Annamalai J, Nallamuthu T. (2015) Characterization of biosynthesized gold nanoparticles from aqueous extract of Chlorella vulgaris and their anti-pathogenic properties. *Appl Nanosci* 5(5):603–607.

Bhattacharya P, Mukherjee D, Deb N, Swarnakar S, Banerjee S. (2020) Application of green synthesized ZnO nanoparticle coated ceramic ultrafiltration membrane for remediation of pharmaceutical components from synthetic water: reusability assay of treated water on seed germination. *J Environ Chem Eng* 8(3):103803.

Camiré A, Chabot B, Lajeunesse A. (2019) Sorption capacities of a lignin-based electrospun nanofibrous material for pharmaceutical residues remediation in water. In: *Sorption in 2020s*. Edited by George Kyzas and Nikolaos Lazaridis, IntechOpen, London.

Chelliapan S, Wilby T, Sallis PJ. (2006) Performance of an up-flow anaerobic stage reactor (UASR) in the treatment of pharmaceutical wastewater containing macrolide antibiotics. *Water Res* 40(3):507–516.

Cuevas R, Durán N, Diez MC, Tortella GR, Rubilar O. (2015) Extracellular biosynthesis of copper and copper oxide nanoparticles by Stereum hirsutum, a native white-rot fungus from chilean forests. *J Nanomater* 16(1):57.

Das RK, Gogoi N, Jayasekhar BP, Sharma P, Chandan M, Utpal B. (2012a) The synthesis of gold nanoparticles using *Amaranthus spinosus* leaf extract and study of their optical properties. *Adv Mat Phys Chem* 2(4):275–281.

Das SK, Liang J, Schmidt M, Laffir F, Marsili E. (2012b) Biomineralization mechanism of gold by zygomycete fungi Rhizopous oryzae. *ACS Nano* 6:6165–6173.

De Corte S, Sabbe T, Hennebel T, Vanhaecke L, De Gusseme B, Verstraete W, Boon N. (2012) Doping of biogenic Pd catalysts with Au enables dechlorination of diclofenac at environmental conditions. *Water Res* 46(8):2718–2726.

Diallo M, Brinker CJ. (2011) Nanotechnology for sustainability: environment, water, food, minerals, and climate. In: *Nanotechnology Research Directions for Societal Needs in 2020*. Science Policy Reports, Springer, Dordrecht, 1:221–259.

Dodd MC, Kohler HP, Von Gunten U. (2009) Oxidation of antibacterial compounds by ozone and hydroxyl radical: elimination of biological activity during aqueous ozonation processes. *Environ Sci Technol* 43(7):2498–2504.

Furgal KM, Meyer RL, Bester K. (2014) Removing selected steroid hormones, biocides and pharmaceuticals from water by means of biogenic manganese oxide nanoparticles in situ at ppb levels. *Chemosphere* 136:321–326.

Gu Y, Wang YN, Wei J, Tang CY. (2013) Organic fouling of thin-film composite polyamide and cellulose tri-acetate forward osmosis membranes by oppositely charged macromolecules. *Water Res* 47:1867–1874.

Gupta VK, Tyagi I, Sadegh H, Shahryari-Ghoshekand R, Makhlouf ASH, Maazinejad B. (2015) Nanoparticles as adsorbent: a positive approach for removal of noxious metal ions: a review. *Sci Technol Dev* 34(3):195–214.

Husein DZ, Hassanien R, Al-Hakkani MF. (2019) Green-synthesized copper nano-adsorbent for the removal of pharmaceutical pollutants from real wastewater samples. *Heliyon* 5(8):e02339.

Khanna V, Bakshi B, Lee L. (2008) Carbon nanofiber production: life cycle energy consumption and environmental impact. *J Indust Ecol* 12:394–410.

Kumari S, Tyagi M, Jagadevan S. (2019) Mechanistic removal of environmental contaminants using biogenic nano-materials. *Int J Environ Sci Technol* 16:7591–7606.

Le-Minh N, Khan SJ, Drewes JE, Stuetz RM. (2010) Fate of antibiotics during municipal water recycling treatment processes. *Water Res* 44(15):4295–4323.

Li X, Xu H, Chen ZS, Chen G (2011) Biosynthesis of nanoparticles by microorganisms and their applications. *J Nanomater* 82:489–494.

Li XQ, Elliott DW, Zhang WX. (2006) Zero-valent iron nanoparticles for abatement of environmental pollutants: materials and engineering aspects. *Crit Rev Solid State Mater Sci* 31:111–122.

Liu J, Choe JK, Sasnow Z, Werth CJ, Strathmann TJ. (2013) Application of a Re-Pd bimetallic catalyst for treatment of perchlorate in waste ion exchange regenerant brine. *Water Res* 47(1):91–101.

Liu T, Li B, Hao Y, Yao Z. (2014) MoO_3:–nanowire membrane and $Bi_2Mo_3O_{12}$/ MoO_3 nano-heterostructural photocatalyst for wastewater treatment. *Chem Eng J* 244:382–390.

Mahdavi M, Namvar F, Ahmad MB, Mohamad R. (2013) Green biosynthesis and characterization of magnetic iron oxide ($Fe3O4$) nanoparticles using seaweed (Sargassum muticum) aqueous extract. *Molecules* 18(5):5954–5964.

Malarkodi C, Rajeshkumar S, Vanaja M, Paulkumar K, Gnanajobitha G, Annadurai G. (2013) Eco-friendly synthesis and characterization of gold nanoparticles using Klebsiella pneumoniae. *J Nanostruct Chem* 3(1):30.

Martins M, Mourato C, Sanches S, Noronha JP, Crespo MTB, Pereira IAC. (2017) Biogenic platinum and palladium nanoparticles as new catalysts for the removal of pharmaceutical compounds. *Water Res* 108:160–168.

Mazumdar H, Haloi N. (2017) A study on Biosynthesis of Iron nanoparticles by Pleurotus sp. *J Microbiol Biotechnol Res* 1(3):39–49.

Meerburg F, Hennebel T, Vanhaecke L, Verstraete W, Boon N. (2012) Diclofenac and 2-anilinophenylac-etate degradation by combined activity of biogenic manganese oxides and silver. *Microb. Biotechnol* 5(3):388–395.

Murgolo S, Petronella F, Ciannarella R, Comparelli R, Agostiano A, Curri ML, Mascolo G. (2015) UV and solar-based photocatalytic degradation of organic pollutants by nano-sized TiO2 grown on carbon nanotubes. *Catal Today* 240:114–124.

Murray KE, Thomas SM, Bodour AA. (2010) Prioritizing research for trace pollutants and emerging contaminants in the freshwater environment. *Environ Pollut* 158:3462–3471.

Narayanan KB, Park HH, Sakthivel N. (2013) Extracellular synthesis of mycogenic silver nanoparticles by Cylindrocladium floridanum and its homogeneous catalytic degradation of 4-nitrophenol. *Spectrochim Acta A* 116:485–490.

Nassar MY, Ahmed IS, Raya MA. (2019) A facile and tunable approach for synthesis of pure silica nanostructures from rice husk for the removal of ciprofloxacin drug from polluted aqueous solutions. *J Mol Liq* 282:251–263.

Patel M, Kumar R, Kishor K, Mlsna T, Pittman CUJ, Mohan D. (2019) Pharmaceuticals of emerging concern in aquatic systems: chemistry, occurrence, effects, and removal methods. *Chem Rev* 119(6):3510–3673.

Petla RK, Vivekanandhan S, Misra M, Mohanty AK, Satyanarayana N. (2012) soybean (glycine max) leaf extract based green synthesis of palladium nanoparticles. *J Biomater Nanobiotechnol* 3(1):14–19.

Prasad R, Pandey R, Barman I. (2016) Engineering tailored nanoparticles with microbes: quo vadis? *Wiley Interdiscip Rev Nanomed Nanobiotechnol* 8(2):316–330.

Prasad TN, Kambala VSR, Naidu R. (2013) Phyconanotechnology: synthesis of silver nanoparticles using brown marine algae Cystophora moniliformis and their characterisation. *J Appl Phycol* 25(1):177–182.

Raeiatbin P, Açıkel YS. (2017) Removal of tetracycline by magnetic chitosan nanoparticles from medical wastewaters. *Desalin Water Treat* 73:380–388.

Roy A, Bharadvaja N. (2017) Qualitative analysis of phytocompounds and synthesis of silver nanoparticles from *Centella asiatica*. *Innovative Tech Agric* 1(2):88–95.

Roy A, Bharadvaja N. (2019) Silver nanoparticles synthesis from *Plumbago zeylanica* and its dye degradation activity. *Bioinspired Biomimetic Nanobiomater* 8(2):130–140.

Shantkriti S, Rani P. (2014) Biological synthesis of copper nanoparticles using Pseudomonas fluorescens. *Int J Curr Microbiol Appl Sci* 3(9):374–383.

Siddiqi KS, Husen A. (2016) Fabrication of metal nanoparticles from fungi and metal salts: scope and application. *Nanoscale Res Lett* 11:98.

Siddiqui MR, AlOthman ZA, Rahman N. (2017) Analytical techniques in pharmaceutical analysis: a review. *Arabian J Chem* 10:S1409–S1421.

Singh AV, Patil R, Anand A, Milani P, Gade WN. (2010) Biological synthesis of copper oxide nano particles using Escherichia coli. *Curr Nanosci* 6(4):365–369.

Śliwka-Kaszyńska M, Kot-Wasik A, Namieśnik J. (2003) Preservation and Storage of Water Samples. *Crit Rev Environ Sci Technol* 33:31–44.

Sundaram PA, Augustine R, Kannan M. (2012) Extracellular biosynthesis of iron oxide nanoparticles by Bacillus subtilis strains isolated from rhizosphere soil. *Biotechnol Bioprocess Eng* 17(4):835–840.

Sunkar S, Nachiyar CV. (2012) Microbial synthesis and characterization of silver nanoparticles using the endophytic bacteria *Bacillus cereus*: a novel source in the benign synthesis. *GJMR* 12:43–49.

Wang X, Zhang D, Pan X, Lee DJ, Al-Misned FA, Mortuza MG, Gadd GM. (2017) Aerobic and anaerobic biosynthesis of nano-selenium for remediation of mercury contaminated soil. *Chemosphere* 170:266–273.

Zhou W, Zhou Z, Song S, Li W, Sun G, Tsiakaras P, Xin Q. (2003) Pt based anode catalysts for direct ethanol fuel cells. *Appl Catal B: Environ* 46:273–285.

2 Industrial Wastewater Treatment
Analysis and Main Treatment Approaches

J.K. Bwapwa and B.F Bakare

CONTENTS

2.1 INTRODUCTION

Wastewater can come from five primary sources: municipal sewage, industrial runoff, agricultural runoff, storm water, and urban runoff. It is a complex mixture of natural and man-made organic and inorganic substances. In its broadest sense, wastewaters can be split into domestic (sanitary) wastewater, also known as sewage, industrial (trade) wastewaters, and municipal wastewater, which is a mixture of the two.

As wastewaters receive treatment, increasing emphasis is being placed on their pollution effects and the economics of mitigation against these effects (Schwarzenbach et al. 2010; Rodriguez-Garcia et al. 2011; Preisner et al. 2020; Gaur et al. 2020). Due to industrialization and population growth showing a continuously increasing trend, there is certainly an increase of wastes and pollutants in the environment (Song et al. 2015; Da Silva et al. 2020; Ajibade et al. 2020). The main challenge is to develop or establish effective methods or processes susceptible to remove pollutants from wastewater generated from domestic and industrial activities (Crini and Lichtfouse 2019; Oller et al. 2011; Wang et al. 2014). Industrial wastewaters are sometimes loaded with complex molecules which may not be easily removed by physico-chemical processes. Biological processes are generally seen as a

DOI: 10.1201/9781003204442-2

promising option for industrial wastewater (Guieysse and Norvill 2014; Awaleh and Soubaneh 2014; Ma and Zhang 2008). For instance, the problem of many toxic organic compounds and eutrophication has proven to be treatable using biological treatment processes. Biodegradation is reported as an environmentally friendly and cost-effective treatment option in comparison to chemical processes. Anaerobic or aerobic wastewater treatment is traditionally applied for the treatment of medium- to high-strength wastewaters (Kleerbezem et al. 2005; Van Lier et al. 2015; Song et al. 2020; Meena et al. 2019). With regard to biological treatments, sludge is produced and can be recycled as fertilizer, building material, and biofuel or sent to landfills if the level of toxicity is high (Van Lier 2008; Chojnacka et al. 2019; Lederer and Rechberger 2010). However, biological treatments also have some weaknesses: Transformation rates of insoluble inorganic matter by microorganisms are too low to be of practical importance (Davies 2005). Thus, insoluble inorganic matter is typically removed by preliminary physical unit operations for further treatment and disposal. Besides, biological treatment processes can be effective for industrial wastewater treatment provided that the relevant microorganisms are used under defined operating conditions. Industrial wastewater is made of dissolved or suspended compounds generated from industrial, manufacturing, or transformative processes. The main objective of the treatment is to remove dissolved or suspended substances. The best approach to working out an effective and efficient method of industrial wastewater treatment is to examine those properties of water and of the dissolved or suspended substances that enabled or caused the dissolution or suspension, then to deduce plausible chemical or physical actions that would reverse those processes. Familiarity with the polar characteristics of water is fundamental to being able to make such deductions. This chapter analyzes the approaches to be used for the remediation or recycling of various industrial wastewaters or effluents depending on their nature or sources.

2.2 OVERVIEW OF VARIOUS SOURCES OF INDUSTRIAL WASTEWATER

There are many types of industrial wastewater, based on different industries and their contaminants. Each sector produces its own particular combination of pollutants, as presented in Table 2.1.

Industrial wastewaters are made with physical and chemical characteristics. The main physical and chemical characteristics of industrial wastewaters can be assessed by checking the levels of many parameters including solids content, color, odor, pH, chemical oxygen demand (COD), biochemical oxygen demand (BOD), inorganic and organic substances, volatile organic compounds, and heavy metals. These characteristics result from the combination of raw materials and processes

TABLE 2.1
Effluent Sources and Contaminants

Effluent Source	Major Contaminant Parameter
Steel and iron metallurgical plants	BOD, COD, oil, metals, acids, phenols, and cyanide
Textile industry	BOD, solids, sulfates, and chromium
Pulp and paper industry	BOD, COD, solids, and chlorinated organic compounds
Petrochemical industry	BOD, COD, mineral oils, phenols, and chromium
Chemical manufacturing industry	COD, organic chemicals, heavy metals, suspended solids (SS), and cyanide
Non-ferrous metallurgical plants	Fluorine and SS
Microelectronics	COD and organic chemicals
Mining industry	SS, heavy metals, acids, and salts
Agricultural and pesticide industry	Organics, COD, and BOD
Pharmaceutical industry	Organics, inorganics, BOD, and COD
Food industry	Organics, inorganics, BOD, and COD
Nuclear industry	Radioactive substances, heavy metals, and inorganics
Wood preserving industry	COD, heavy metals, organics, inorganics, and, BOD

used in various industries for manufacturing while discharging industrial wastewaters. This implies that there is a correlation between the contaminants and the raw materials used in an industrial process or manufacturing system to produce a new product. Table 2.2 presents a list of some substances present in industrial wastewaters and their origins.

Generally, industrial wastewater can be categorized in two groups: inorganic industrial wastewater and organic industrial wastewater. The mining industry discharges effluent containing heavy metals such as chromium, nickel, zinc, cadmium, lead, iron, titanium compounds, and many others. The content of the effluent depends on the mineralization of the sites where the mining activities are taking place, the metals being processed, and the chemicals or reactants used in the mineral processing plants or units (Navarro et al. 2008; Favas et al. 2011; Wuana and Okieimen 2011). Their combination with water or surface waters generates an effluent loaded with various heavy metals and inorganic compounds. The pulp and paper industry relies heavily on chlorine-based substances, and as a result, pulp and paper mill effluents contain chloride organics and dioxins, as well as suspended solids and organic wastes (Hanchang 2009; Lindholm-Lehto et al. 2015; Gupta and Gupta 2019). Textile industry effluents contain dyes, and complex organic and inorganic molecules which are not easily removed from conventional wastewater treatment plants. Alternative methods are currently used to effectively remove dyes and complex organic molecules (Carmen and Daniela 2012; Arslan et al. 2016; Wang et al. 2014; Gurses et al. 2020; Mojsov et al. 2016; Al-Degs et al. 2000). The petrochemical industry discharges effluents with phenols, a high number of refractory compounds, and mineral oils that can be harmful to the environment in the event of a poorly-designed effluent strategy. Therefore, appropriate treatment should be established (Hanchang 2009; El-Ashtoukhy et al. 2013; Diya'uddeen et al. 2011).

Effluents from food processing plants are characterized by high amounts of suspended solids and organic material (Cristian 2010; Qasim and Mane 2013; Noukeu et al. 2016). These effluents can cause environmental challenges if they are not handled properly because of the complexity of the contaminants. Nuclear discharges are very toxic and are made up of dangerous wastes that contain radioactive substances (Ojovan et al. 2019; Kyne and Bolin 2016; Grachev 2019).

Effluents require a number of precautions before disposal, and should be treated in a safe zone before being discharged. Radioactive wastes are generally a by-product of nuclear plants or reactors and research facilities dealing with nuclear applications. Nuclear waste can also be generated while taking out or disassembling nuclear reactors and other nuclear facilities (Ojovan et al. 2019; Kyne and Bolin 2016; Grachev 2019). Wastewater from nuclear plants comes from the purification of primary refrigerants and from washing leakage using water. Also, nuclear reactor decontamination operations produce wastewater (Zhang et al. 2019; Grachev 2019). The generated radioactive effluent generally contains soluble and insoluble radioactive and non-radioactive compounds (Ojovan et al. 2019; Kyne and Bolin 2016; Grachev 2019; Natarajan et al. 2020). Standard techniques are normally used to cleanse liquid waste streams. Each treatment method has a specific influence on the radioactive content of the effluent. Pharmaceutical effluents come from processing units that manufacture drugs, vaccines, and other medical consumables. They can also derive from hospitals and health-related units, such as laboratories. The pharmaceutical industry has grown significantly over the last decades and, consequently, the waste related to this type of industry has also increased. Therefore, the treatment of pharmaceutical effluents is crucial. The composition of pharmaceutical effluents is very complex. They contain high concentrations of organic matter, have high levels of microbial toxicity and salinity, and their biodegradability is challenging (Lalwani et al. 2020; Shi et al. 2014). After treatment, the effluent can still contain trace amounts of suspended solids and dissolved organic matter (Lalwani et al. 2020; Bisognin et al. 2020). Agricultural and pesticide effluents come from farming activities including dairy and pig operations. These effluents may contain inorganic and organic compounds from fertilizers, pesticides, and manure wastes.

From this brief overview of the content of effluents, each form of industrial wastewater is based on the type of industry from where it originates. It is therefore important to stress the fact that the treatment process of any industrial wastewater type must be designed specifically for the particular

TABLE 2.2

Substances Present in Industrial Effluents

Contaminating Substance / Compound or Molecule	Origin of the Industrial Wastewater
Acetic acid	Acetate rayon, beetroot manufacturing
Acids	Chemical manufacturing, mines, textiles
Alkalies	Cotton and straw kiering, wool scouring
Ammonia	Gas and coke manufacturing, chemical plants
Arsenic	Sheep dipping
Cadmium	Plating
Chromium	Plating, chrome tanning, alum anodizing
Citric acid	Soft drinks and citrus fruit processing
Copper	Copper plating, copper pickling
Cyanides	Gas manufacturing, plating, metal cleaning
Fats, oils, grease	Wool scouring, laundries, textile industry
Fluorides	Scrubbing of flue gases, glass etching
Formaldehyde	Synthetic resins and penicillin manufacturing
Free chlorine	Laundries, paper mills, textile bleaching
Hydrocarbons	Petrochemical and rubber factories
Free chlorine	Laundries, paper mills, textile bleaching
Mercaptans mills	Oil refining, pulp
Nickel	Plating
Nitro compounds	Explosives and chemical works
Organic acids	Distilleries and fermentation plants
Phenols	Gas and coke manufacturing, chemical plants
Starch	Food processing, textile industries
Sugars	Dairies, breweries, sweet industry
Sulfides	Textile industry, tanneries, gas production
Sulfites	Pulp processing, viscose film manufacturing
Tannic acid	Tanning, sawmills
Tartaric acid	Dyeing, wine, leather, chemical manufacturing
Zinc	Galvanizing zinc plating, rubber process

type of effluent produced. It is possible to use the same type of treatment for various effluents, in cases where the effluent content has the same nature. For example, effluents loaded with heavy metals or inorganic substances may be subjected to the same treatment processes despite the fact that they may originate from different sources. This is because it is possible to precipitate inorganic and heavy metals under operating conditions such as temperature and pH or the use of precipitants, coagulants, or flocculants. The design of the treatment process will mainly rely on the effluent characterization which reveals more about the effluent quality.

2.3 CHARACTERISTICS OF ORGANIC INDUSTRIAL WASTEWATER/EFFLUENT

Organic contaminants have a significant effect on industrial wastewaters or effluents. They can also affect the receiving points such as surface waters, groundwater, and the soil when they come into contact with these receiving points (McCance et al. 2018; Kasonga et al. 2020; Gao et al. 2019; Afrad et al. 2020). The discharged effluents or wastewater from relevant industries may contain various organic compounds including polychlorinated biphenyls (PCBs), formaldehydes, pesticides, nitrobenzenes, herbicides, phenols, polycyclic aromatic hydrocarbons (PAHs), aliphatics, and heterocycles. Agricultural activities also contribute to the discharge of effluents containing some of these contaminants which could compromise the safety of water resources and soil (Wu et al. 2003; Bartolomeu et al. 2018; Gatta et al. 2020). Furthermore, effluents from farming activities may contain high concentration of pesticides or herbicides; effluents from the coke plants may contain various PAHs; wastewater/effluents from the chemical industry may contain various heterogeneity compounds, such as PCBs and polybrominated diphenyl ethers (PBDEs); wastewater/ effluents from the food industry contain microcontaminants or complex organic compounds with high concentrations of suspended solids (SS) and BOD (Zheng et al. 2013; Anjum et al. 2012; Basile et al. 2011; Schwarzenbach et al. 2010; Ratola et al. 2012). Consequently, the effluents in these cases are highly loaded with contaminants and can cause harm to the receiving points, including surface waters and soil. Municipal sewage contains different types of organic contaminants, such as oil, biodegraded food, detergent molecules, dissolved organics, and molecules of surfactants (Bhatt et al. 2020; Shon et al. 2006; Huang et al. 2010; Ruan et al. 2014). These organic pollutants have a harmful effect on the environment and can cause health problems to humans. In terms of environmental impacts, organic contaminants have the capacity to deplete large amounts of soluble oxygen in effluents. Toxicity which is seen as severe and coupled with a higher demand of oxygen can alter the quality of effluents and destroy aquatic life in cases where the effluent is discharged into freshwater resources (Asano 1998; Holeton et al. 2011; Margot et al. 2015). However, their negative impact towards the environment will be brief, because they can be easily and quickly biodegraded by microorganisms. However, the situation is particularly different for persistent organic pollutants (POPs), which have a low water solubility, a high accumulation capacity, and potentially carcinogenic, teratogenic, and neurotoxic properties. Many factors, such as the characteristics of the pollutants, environmental aspects (PH value, temperature, etc.), and the aging process could affect the toxicity of organic industrial wastewater, and their long-term influence to the ecosystem deserve further investigation (Zheng et al. 2013; Luo et al. 2019; Li et al. 2011b). The pollutants from industrial organic wastewaters may come from the textile industry, food processing units and fermentation plants, pharmaceutical and chemical synthesis industries, sugar refineries, and many other. Some of the pollutants from these effluents are biodegradable and others are non-biodegradable. Effluents with biodegradable compounds generally come from domestic sewage, food processing, breweries, etc. These effluents have a high COD and BOD, and are able to break down within a period of time. Most of the available techniques can be used to treat biodegradable organic pollutants, but biological methods are known to the best for their efficiency and cost effectiveness.

2.4   CHARACTERISTICS OF INORGANIC INDUSTRIAL WASTEWATER/EFFLUENT

Quite a few inorganic compounds are common to industrial wastewaters/effluents and are important for the establishment and control of the effluent quality. Inorganic substances in industrial wastewater/effluent may come from precipitates, metal ions, inorganic dyes, and inorganic chemicals from a processing unit or inorganic substances left in the effluent after evaporation (Moore 2012; Awaleh and Soubaneh 2014; Peng and Korshin 2011). Natural waters dissolve rocks and minerals with which they come into contact. It is essential to mention also that some of the inorganic constituents found in natural waters are also found in industrial wastewaters. Many of these constituents are added via a process as mentioned earlier. These inorganic constituents include pH, chlorides, alkalinity, nitrogen, phosphorus, sulfur, toxic inorganic compounds, and heavy metals. The acidic nature of an industrial effluent or any other effluent can intensify toxic contamination issues because sediments release toxic substances in the effluent (Vardham et al. 2019; Rajaram and Das 2008; Carolin et al. 2017).

Common sources of acidity include mine drainage, runoff from mine tailings, and atmospheric deposition (Blowes et al. 2003; Tiwary 2001; Acharya and Kharel 2020). Chloride, in the form of the Cl-ion, is one of the major inorganic characteristics in water and wastewater, including industrial wastewater. Sources of chloride in natural waters include: leaching of chloride from rocks and soils; in coastal areas, saltwater intrusion; agricultural, industrial, domestic, and human wastewater; and infiltration of groundwater into sewers adjacent to saltwater. Another important characteristic of inorganic industrial wastewater quality is alkalinity; it is a measure of the buffering capacity for wastewater to resist changes in pH caused by the addition of acids. Alkalinity is caused by chemical compounds dissolved from soil and geologic formations, and is mainly due to the presence of hydroxyl and bicarbonate ions. These compounds are mostly the carbonates and bicarbonates of calcium, potassium, magnesium, and sodium. The majority of industrial wastewaters loaded with inorganic compounds generally have a high alkalinity (Addy et al. 2004; Xue et al. 2019; Hou et al. 2014). Alkalinity is an essential characteristic for industrial wastewater because, when using anaerobic digestion as a treatment option, alkalinity should be sufficient to keep the pH above 6. In case the alkalinity drops below this level, the methanogenesis phase will not be effective because of the fact that microorganisms activity will be affected or even prevented. Alkalinity is also important when dealing with chemical treatment of wastewater, biological nutrients removal, and whenever ammonia is removed by air stripping. Nitrogen in industrial wastewater can be present in the form of organic nitrogen (N_2), ammonia (NH_3), nitrite (NO_2), or nitrate (NO_3). Ammonia is generally present in many wastewaters, and nitrates are derived from the oxidation of ammonia.

Nitrogen (N_2) data are useful in the assessment of the biological treatment of wastewater, including industrial wastewater. In the event N_2 levels in wastewater are insufficient, it may need to be added to the wastewater in order to make it treatable. After treatment, it is required that the amount of N_2 remaining in the effluent be determined. This will assist in preventing the growth of algae and aquatic plants into receiving water resources.

Phosphorus (P) is a macronutrient needed by all living cells and it is known as an abundant constituent of domestic wastewaters and sometimes in industrial wastewaters. It is generally found in the form of phosphates, the salts of phosphoric acid. Because of harmful algae blooms taking place in surface waters, it is essential to control the amount of phosphorus compounds discharged from domestic and industrial wastewaters into receiving freshwater resources.

Sulfur (S) generally occurs in the form of sulfates in industrial wastewaters that are contaminated by heavy metals. Also, sulfates can be reduced biologically to sulfides, which in turn can combine with hydrogen to form hydrogen sulfide (H_2S). H_2S is toxic to animals and plants, and in certain concentrations, it is considered a deadly toxin. It can also cause corrosion to pipes. Toxic inorganic compounds such as copper (Cu), lead (Pb), silver (Ag), arsenic (As), boron(B), and chromium (Cr) are classified as priority pollutants and are toxic to microorganisms. Their effects should be taken into consideration when using any biological treatment process for industrial wastewater. They can destroy the microorganisms that are useful for the success of the treatment or they may negatively impact the effectiveness of the treatment process.

Heavy metals are generally known as the main contaminants found in industrial wastewaters, especially acid mine water. Heavy metals may adversely affect the biological treatment of wastewater, as mentioned before. Mercury (Hg), lead (Pb), cadmium (Cd), zinc (Zn), chromium (Cr), and plutonium (Pu) are among the most widely known heavy metals. Heavy metals are generally defined as metals with relatively high atomic masses. The presence of any of these metals at high levels may interfere with the beneficial uses of water due to their toxicity. In the industries that discharge heavy metals, chromium is generally reported as the most common metal that is discharged into the environment. However, chromium (Cr) is not the most dangerous heavy metal to living organisms; the most toxic are Cd, Pb, and Hg. They have a remarkable affinity for sulfur and can interrupt enzyme function by developing bonds with sulfur groups in enzymes. Their ions may bind to cell membranes, with the consequence of hampering transport processes through the cell wall. Heavy metals have the capacity to cause the precipitation of phosphate bio-compounds or catalyze their decomposition.

2.5 MAIN TREATMENT OPTIONS FOR INDUSTRIAL WASTEWATERS

2.5.1 PHYSICO-CHEMICAL METHODS

2.5.1.1 Electrocoagulation

Electrocoagulation is a combination of chemical and physical reactions that occur in series in which coagulation, adsorption, precipitation, and flotation take place. This technology allows the coagulation and precipitation of contaminants by a direct electric current through an electrolytic process followed by the separation of flocculent without adding the coagulants. The coagulation takes place while the current is being applied; this allows for the removal of small particles by setting them into motion since direct current is applied. Also, electrocoagulation could reduce residue for waste production and the treatment time (Shammas et al. 2010). In this process, pairs of metal sheets known as electrodes are forming the anode and cathode. From electrochemistry principles, there is a loss of electrons at the cathode which is oxidized while the water is reduced (gains electrons), thereby improving the treatment of industrial wastewater by significantly reducing the contaminant load. The technology presents another advantage of reducing the use of chemicals because the electrodes provide the coagulant. Conversely, many applications involve using chemical coagulants in order to enhance treatment. The process is not effective for all industrial wastewaters, and its effectiveness depends on the contaminant types. The process is suitable on a small scale, but not many trials have been undertaken at pilot and large scales.

2.5.1.2 Advanced Oxidation Process (AOP)

The advanced oxidation process (AOP) is considered as a promising treatment technology for industrial wastewater. AOPs are characterized by a common chemical feature, where they use highly reactive hydroxyl radicals for achieving the complete mineralization of the organic pollutants into carbon dioxide and water. It is a chemical treatment method that has developed significantly in the wastewater treatment industry, and is a highly recommended method for the removal of problematic or very complex organic matter. The basic principle of AOP involves the production of hydroxyl radicals (HO•), which can be generated from hydrogen peroxide (H_2O_2), ozone, photocatalysis, or oxidants in combination with using ultraviolet (UV) radiation. In some cases, two or more radical generators are used in combination. However, it is the HO• that is mainly responsible for the degradation of organic compounds. The HO• is an unselective, strong chemical oxidant. Once produced, it attacks nearly all organic complexes. Therefore, attack by the HO• leads to a complete breakdown of the organic compound, and as a result, AOPs diminish the concentration of the pollutant from a few hundred ppm to less than 5 ppb (Mohajerani et al. 2009) Studies show that with AOP, organic chemicals disintegrate and become smaller and easily biodegradable. The HO• takes away a hydrogen atom from an organic compound (R–H) and causes the formation of an organic radical (•R)

Advanced oxidation processes use strong oxidants, such as hydrogen peroxide, ozone, titanium dioxide, or Fenton's reagent to generate highly reactive intermediates. It is reported that these reagents are used in the presence of UV light to enhance the oxidation process. Ozone, which is a stronger oxidizing agent compared to hydrogen peroxide (H_2O_2), chlorine, bromine, oxides of chlorine, and potassium permanganate are widely used for industrial wastewater treatment. The main advantages of AOPs include fast reaction rates and non-selective oxidation, which allow for the simultaneous removal of multiple contaminants. The expenses related to the use of AOPs generally rely on the quality of the effluent to be treated and the aim of the treatment. Hence, not many studies have been undertaken on the cost estimation of the different types of AOPs. However, it is recommended that pilot testing be carried out to ascertain site-specific costs. It was reported that the main factors influencing costs were the removal efficiency and flow-rate, not owing to the concentration of the influent (Mahamuni and Adewuyi 2010; Canizares et al. 2009; Kommineni et al. 2000; Munter 2001). AOPs have gained the reputation of being effective in the wastewater treatment industry. The main focus of this process is the hydroxyl radical which, once generated, aggressively attacks virtually all organic compounds.

2.5.1.3 Adsorbents/Bio-Adsorbents-Based Treatment

Adsorption is extensively used as a single or hybrid treatment system for the removal of dyes from textiles wastewater (Gürses et al. 2021). Activated carbon, ion exchange membranes, zeolite, and bentonite clay are also reported as some of the adsorption methods used for contaminants removal from wastewater (Li et al. 2019; Mohan et al. 1998, 2007). Also, some remarkable low-cost absorbents including bagasse, polysaccharide, composites, pith, agricultural wastes, mud, and sand are known as effective options for the removal of contaminants. Bio-adsorbents are attractive due to their low-cost and feasibility for large-scale units. Apart from the effluent pH, the main dye adsorption defining parameters include the external mass transfer coefficient and solid diffusivity (Tony 2020). The activity of an adsorbent should be regulated by defined physical and chemical factors. The surface area and pore size are known as the key physical factors that influence the adsorbent accessibility to molecules of dye in order to allow for an effective removal. However, chemical functional groups in the absorbent constitute a decisive factor regarding the type of adsorptive forces as well as desorption ability of the material. Adsorption is also reported in several studies as a combination treatment system for dye removal (Fazal et al. 2020). It is a promising technology because of its flexibility, handling, and feasibility. However, it is a restrictive process because of its limited adsorption to a specific class of dyes. Also, the proper disposal of consumed absorbents is an additional concern. Table 2.3 presents several low cost adsorbents with their removal efficiencies for heavy metals in wastewater.

2.5.1.4 Membrane Treatment Technology: Microfiltration (MF), Ultrafiltration (UF), Nanofiltration (NF), and Reverse Osmosis (RO)

A membrane is a barrier that separates two phases from each other by limiting the movement of components through it in a selective mode (Ravanchi et al. 2009). Membrane treatment technology makes use of a pressure-driven membrane system in which the removal of contaminants takes place from separation processes such as reverse osmosis, nanofiltration, ultrafiltration, and microfiltration (Jhaveri and Murthy 2016; Peters 2010; Dharupaneedi et al. 2019; Li et al. 2011a). Pressure is exerted on the wastewater at one side of the membrane and serves as a driving force to separate both the clean water, named permeate, and a concentrate, named retentate, as indicated in Figure 2.1. Membranes may be a made from polymeric, an organo-mineral, a ceramic, or metallic material. The separation processes differ in pore size, from dense to porous membranes. Depending on the type of technique, salts, small organic molecules, macromolecules, or particles can be retained, and the applied pressure will differ (Mallevialle et al. 1996; Baker 2002; Fane et al. 2011).

TABLE 2.3
Low-Cost Adsorbents and Their Removal Efficiencies

Adsorbent	Heavy Metal Removal Efficiencies [%]					
	Cr	Ni	Cu	Zn	Cd	Pb
Rice husk carbon			100	100	100	100
Sugarcane bagasse activated carbon	99.7					
Untreated tree sawdust		91	86	75.6		
Soybean hulls	98.1	95.6	99.7	96.40		
Cottonseed hulls		47.6	58.8	59. 50		
Defatted rice bran		29.2	71.5	38.40		

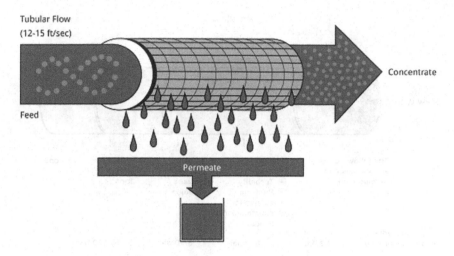

Tubular Flow
(12-15 ft/sec)

Concentrate

Feed

Permeate

FIGURE 2.1 Membrane separation process. (Source: Porex filtration.)

Microfiltration (MF), as with any membrane technology technique, is based on static pressure as the driving force, and the separation process is performed by the action of sieve separation of the membrane (Nath 2017; Baker 2002, 2012, Fane et al. 2011; Fröhlich et al. 2012). This operation is similar to conventional filtration. The small difference resides in the fact that the pore size of MF is smaller than in conventional methods. MF can effectively remove suspended solids (SS), macromolecules, and microorganisms in wastewater. Ultrafiltration (UF) driving force is the pressure difference between the membranes on both sides, and the filter medium is the ultrafiltration membrane. Under certain pressure, when water passes through the membrane surface, water, inorganic salts, and small molecules penetrate, and other macromolecules are trapped (Shi et al. 2014; Pendergast and Hoek 2011; Liang et al. 2008; Moslehyani et al. 2019). UF is mainly used for the removal of macromolecules and colloids in wastewater. In the application of this method, it should be ensured that the membrane has adequate membrane flux and is easily disassembled, replaced, and cleaned. Nanofiltration (NF) is an interesting application for effluent treatment owing to its ability of selectively separating dies, heavy metals, and organic and inorganic compounds (Barakat 2011; Zhao et al. 2016; Abdel-Fatah 2018, 2019). With regard to reverse osmosis (RO), there are two categories of RO membranes: cellulose ester and aromatic polyamide. Its component form includes tube, plate and frame, roll, and hollow fiber types. Reverse osmosis has the ability to remove various types of dissolved and suspended chemical and biological species, including bacteria, from water, and is used in both industrial processes and the production of potable water (Anis et al. 2019; Qasim et al. 2019; Jiang et al 2017). RO differs from filtration in that the mechanism of fluid flow is by osmosis across a membrane. The major removal mechanism in membrane filtration is straining or size exclusion; RO instead involves solvent diffusion across a membrane that is either nonporous or uses pores of 0.001 μm in size, as indicated in Figure 2.2. Membrane technology has gained significant interest over the last two decades not only for the desalination of seawater but also for other types of wastewater, with the aim to produce potable or pure water. Overall, membrane technology is a very effective option. It used to have some weaknesses related to membrane fouling and being energy intensive. However, currently, low-pressure membranes are being developed together with anti-fouling membranes for the effectiveness of the separation process between the permeate and the concentrate.

2.5.1.5 Electrodialysis

Electrodialysis is a mixture of electrolytic and dialysis diffusion processes. Under the action of the electric field created by direct current, anions and cations of the dissolved salts in the wastewater

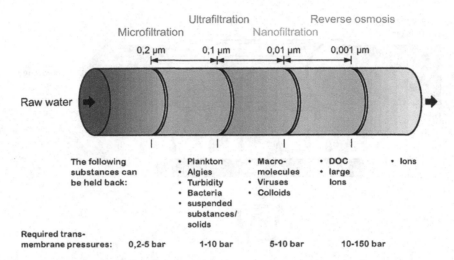

FIGURE 2.2 Different membrane technology operations, particle sizes, and transmembrane pressures.(Source: The HydroGroup, German Company.)

can respectively move to the anode and the cathode. Therefore, the amount of anions and cations in the intermediate section is progressively reduced, and the separation and recovery may be completed. Electrodialysis operates under ambient conditions and its membranes can be replaced with ease. It has a tolerance for high pH and chlorine concentrations and can recover the highest amount of water from the effluent with less electricity consumption. The process is a promising one when it comes to industrial effluent treatment and the recovery of acids, nutrients, metals ions, bases, and valuable organic compounds, generating zero or near zero discharge. It is an energy-efficient, flexible and automated process with very low environmental pollution. However, it has the ability to only remove salts in water, and its desalination efficiency is inferior compared to that of reverse osmosis (RO). It is a sustainable technology that can be used industrially for the treatment of wastewater containing dyes with high salinity using an ion-exchange membrane-based isothermal separation technique involving charge transfer driven by an electric field (Majewska-Nowak 2013). The major weakness of electrodialysis systems is that dye moieties generally ionize to anions and to cations (Lafi et al. 2019). These ionized dyes display strong affinity with the ion-exchange membranes and lead to a drop in the flux and membrane fouling desalting process. This further results in longer process times, the need for early membrane replacement, and high energy investment (Lafi et al. 2019). To overcome these limitations, electrodialysis is usually used as part of combined treatment systems such as hybrid-ultrafiltration and bipolar-membrane electrodialysis (Lin et al. 2019).

2.5.2 BIOLOGICAL PROCESSES

2.5.2.1 Biodegradation of Organic Contaminants

The biological methods used for industrial wastewater treatment start with the biodegradation of contaminants in which microorganisms and their enzymes remove the contaminants from wastewater or make them harmless. Biodegradation is considered as one of the sustainable ways in which wastewater is treated effectively, economically, and safely. Biological treatments can be undertaken for industrial wastewater contaminated by organic substances with the use of appropriate microorganisms. The rate and extent of biodegradation for a contaminant in wastewater depend on the nature of the contaminant (Leahy and Colwell 1990; Tufail et al. 2020; Haritash and Kaushik 2009). There are organic pollutants like organophosphorus pesticide, which have relativity high water solubility and low acute toxicity, are bioavailable, and easy to be degraded. However, some organic substances with higher bioaccumulation, bioadsorption, biomagnification, and biotoxicity properties are reluctant to biodegradation. The biodegradation of organic contaminants can be achieved by aerobic digestion with the use of oxygen or by anaerobic digestion which is done in the absence of oxygen.

2.5.2.1.1 Aerobic Treatment of Industrial Wastewater

In this process, oxygen is needed by microorganisms to achieve the biodegradation at two metabolic sites, at the initial attack of the substrate and at the end of respiratory chain. During this process, oxygenases are produced, and carbon and nutrients are released. Aerobic treatment is used to remove organic compounds (BOD or COD) and to oxidize ammonia to nitrate (Zheng et al. 2013; Chan et al. 2009; Goli et al. 2019; Show and Lee 2017). There are two kinds of relations between microorganisms and organic contaminants present in wastewater: the first is that organic contaminants are used by microorganisms as the only source of energy and carbon; the second is that a growth substrate is used by microorganisms as carbon and an energy source, whereas another organic compound in the organic substrate which is not capable of providing carbon and energy resource is also biodegraded (the process is named co-metabolism). Aerobic treatment is also reported as an effective alternative for removing odor from wastewater. This treatment consumes large amounts of energy and has higher operating and maintenance costs.

Activated sludge is an aerobic process for treating sewage and industrial wastewaters. As a suspended-growth biological treatment process, activated sludge uses a dense culture of microorganisms in suspension allowing the biodegradation of organic material under a steady oxygen supply and forming a biological floc for solid separation. Various aeration methods exist: diffused aeration, surface aeration, or pure oxygen aeration. This technique was developed at the beginning of the last century and is considered as an effective process for big cities as it can handle large amounts of effluent. This process is effective regarding the removal of organic contaminants. However, it is energy intensive and requires regular maintenance. Microorganisms accountable for treatment are maintained in suspension and a solid–liquid separation is completed in a secondary clarifier that follows the activated sludge process. Sludge and/or water is recirculated between the biological reactor and the secondary clarifier. Activated sludge systems are frequently used in combination with other treatment technologies such as membrane technologies. Activated sludge systems are typically located after the preliminary and primary treatment of wastewater and before the advanced tertiary treatment stage. Figure 2.3 presents a summary of the working principles for the activated sludge system.

A membrane bioreactor (MBR) is the mixture of a membrane process, such as microfiltration (MF) or ultrafiltration(UF), with a suspended growth bioreactor. This technique is widely used for municipal and industrial wastewater treatment. MBR processes can generate high quality effluent to

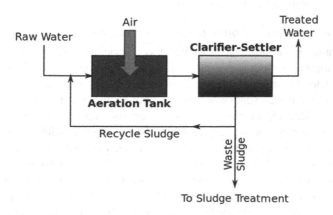

FIGURE 2.3 Activated sludge working principle.

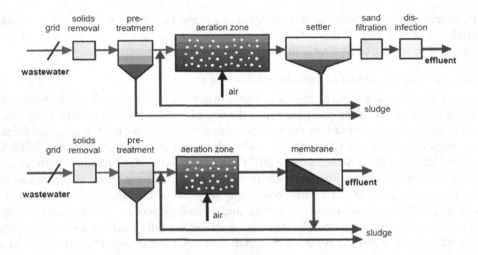

FIGURE 2.4 Schematic representation of the conventional activated sludge process (top) and external membrane bioreactor (bottom).

be discharged to coastal, surface, or brackish waterways or to be reclaimed for urban irrigation. The advantages of MBR over conventional processes include its small footprint, and its ability to retrofit and upgrade old wastewater treatment plants. The principle of this technique is similar to the activated sludge process, except that instead of separating water and sludge through settlement, MBR technology uses the membrane which is more efficient and less dependent on the oxygen concentration of the water. Compared to activated sludge, MBR technology has higher organic contaminant and ammonia removal efficiencies. In addition, MBR technology is effective in treating wastewater with higher mixed liquor suspended solid (MLSS) concentrations in comparison to the activated sludge process. Consequently, this decreases the reactor volume for the achievement of the same loading rate. Fouling is still a common problem, and it reduces the flux and leads to the increase of trans-membrane pressure. However, anti-fouling membranes can be used which could assist in solving this challenging issue. On the other hand, recurrent membrane cleaning and replacement is necessary, which considerably affects the operating costs. Physical cleaning is used for irreversible fouling, such as membrane relaxing, backwashing, a combination of both, and chemical cleaning. In some cases, modification of MBR is needed to improve the performance and to achieve high quality effluent (Mutamim et al. 2013).

Figure 2.4 presents a simplified schematic comparison between a conventional activated sludge process (top) and membrane bioreactor (bottom).

2.5.2.1.2 Anaerobic Biodegradation of Industrial Wastewater

Anaerobic biodegradation of industrial wastewater is a series of reactions in which microorganisms break down organic biodegradable material found in wastewater in the absence of oxygen. The principles of anaerobic biodegradation are described as follows: firstly, the insoluble organic pollutant breaks down the into soluble substance, making them available for other microorganisms; secondly, the acidogenic microorganisms convert the sugars and amino acids into carbon dioxide, hydrogen, ammonia, and organic acid; thirdly, the organic acids convert into acetic acid, ammonia, hydrogen, and carbon dioxide; and finally, the methanogens convert the acetic acid into hydrogen, carbon dioxide, and methane. The effluents from this process will have reduced COD and BOD. Anaerobic digestion can be achieved as a batch process or a continuous process. For a batch process, the addition of biomass to the reactor is done at the start of the process. The reactor is then sealed for the duration of the process. In its simplest form, batch processing needs inoculation with already-processed material to start the anaerobic digestion process. Compared to the aerobic process, the anaerobic process is slow and less effective. However, anaerobic biodegradation not only reduces COD and BOD in the wastewater but also generates renewable energy. Also, nutrient removal is sometimes poor compared to aerobic biodegradation. The final effluent quality is lower compared to the one generated by aerobic biodegradation. However, the process does not require high amounts of energy for it to run. There are two variants of anaerobic reactors: the anaerobic activated sludge process and anaerobic biological membrane process. The anaerobic activated sludge process includes the conventional stirred anaerobic reactor (CSR), upflow anaerobic sludge blanket reactor(UASB), and anaerobic contact tank (ACT). Examples of membrane reactors include anaerobic biological membrane process fluidized bed reactor (FBR), anaerobic rotating biological contactor (ARBC), anaerobic biofilter reactor, and anaerobic baffled reactor membrane bioreactor (ABR-MBR). Furthermore, when industrial wastewater contains inorganic and heavy metals, anaerobic biodegradation can be used with the assistance of sulfate-reducing bacteria (SRB) or sulfate-reducing archaea (SRA). These microorganisms play a key role in the monitoring and bioremediation of acid mine water or any related wastewater. Both SRB and SRA can perform anaerobic respiration using sulfate (SO_4^{2-}) as a terminal electron acceptor, reducing it to hydrogen sulfide (H_2S). Therefore, these sulfidogenic microorganisms "breathe" sulfate rather than molecular oxygen (O_2), which is the terminal electron acceptor reduced to water (H_2O) in aerobic respiration (Ernst-Detlef and Harold 1993; Muyzer and Stams 2008).

SRB may assist in achieving an effective treatment by generating alkalinity and neutralizing the acidic effluent. The design of engineered sulfate-reducing bacteria (SRB) consortia will be an effective tool in optimizing the biodegradation of acid mine wastes or similar effluents in industrial processes. One of the limitations related to SRB-led treatment is the sensitivity of microorganisms to low pH which obviously affects microbial growth and hampers the performance of the bioreactor (Sanchez et al. 2014; Kefeni et al. 2017). Therefore, there is a necessity for making a choice of resistant SRBs to maintain a sustainable operation of the bioreactor while treating wastewater. Overall, the treatment processes for industrial wastewater can combine both anaerobic and aerobic treatments depending on the nature of the waste and the objectives of the treatment.

2.5.2.1.3 Microbial Enzymatic Methods

The need for alternative processes regarding the remediation of wastes has prompted the necessity to introduce and implement strict standards for waste discharged into the environment. Various methods and processes are used, some of which are effective and others that show potential but are still under development. An important number of enzymes from various microorganisms are reported as effective for the treatment of industrial wastewater or any other wastewater. Enzymes have the ability to act specifically on refractory contaminants and remove them by precipitation or conversion to other products. They can also modify the features of a particular waste making it more predisposed to treatment or facilitate the conversion process to other products. The mechanical and thermal stability of the enzymes are improved by immobilization; at the same time the same enzyme immobilization decreases the probability of enzyme leaching into the solution. There are many types of microbial enzymes used in the treatment of wastes; these include oxidoreductases, oxygenases, monooxygenases, dioxygenases, laccases, peroxidases, lipases, cellulases, and proteases.

For instance, horseradish peroxidase has a high activity and stability. It also performs higher phenol conversions when covalently immobilized onto magnetic beads. Nanotubes carrying oxidative enzymes as laccases and peroxidases can be used for the removal of recalcitrant pollutants in industrial wastewater.

When glucose oxidase and chloroperoxidase are immobilized onto carbon nanotubes their functionality is preserved, and there is also an increase in substrate conversion efficiency. The use of enzymes in chemical reactions is particularly important to meet the demands for clean and green technologies for the preservation of our planet.

Table 2.4 presents a summary of some enzymes and their applications in wastewater treatment.

TABLE 2.4

Enzymes and Their Application in Water Treatment (adapted from Pandey et al. 2017)

Enzymes	Applications in Water Treatment
Alkylsulfatase	Surfactant degradation
Chitinase	Bioconversion of shellfish waste to N-acetyl glucosamine
Chloro-peroxidase	Oxidation of phenolic compounds
Cyanidase	Cyanide decomposition
Laccase	Phenols removal, decolorization of Kraft bleaching effluents, binding of phenols, and aromatic amines with humus
Lactases	Treatment of dairy waste and production of value-added products
Lignin peroxidise	Phenols and aromatic compound removals and decolorization of Kraft bleaching effluents
Lipase	Sludge dewatering improvement
Lysozyme	Sludge dewatering improvement
Mn-peroxidase	Oxidation of aromatic dyes
Pectin Lyase	Pectin degradation
Peroxidase	Removal of phenols and aromatic amines, decolorization of Kraft bleaching effluents, and sludge dewatering
Phosphatase	Heavy metals removal
Tyrosinase	Phenols removal

2.6 CONCLUSION

Industrial wastewater is a notable source of environmental pollution. Large amounts of industrial wastewater are being discharged into rivers, lakes, and coastal areas resulting in serious pollution problems characterized by negative effects on the eco-system and human life. Many industrial wastewaters discharge inorganic, organic, and heavy metal contaminants into the environment. The increasing rate of industrial wastewaters that is taking place in many countries results in a growing amount of effluents that require effective treatment. With many industries competing to transform raw materials into high quality finished products, the transformation process will always be accompanied by an effluent discharge that contains contaminants or pollutants that may be complex. The treatment of industrial wastewater should be designed for the specific type of generated effluent. There are two types of industrial wastewaters or effluents: organic industrial effluents and inorganic industrial effluents. Various treatment methods can be used to reduce the contaminant load. The main and popular treatment options are physico-chemical and biological-based. The physico-chemical methods include electrocoagulation; AOP; adsorbent/bio-adsorbent-based treatment; membrane treatment technology such as MF, UF, NF, and RO; and electrodialysis. All these methods are reported as effective in the removal of contaminants despite the fact that they can have some weaknesses related to energy consumption and cost effectiveness. The biological-based treatment methods involve the biodegradation of organic contaminants. This can be achieved by aerobic digestion, anaerobic digestion, and microbial enzymes. These methods are sustainable and are reported to have high removal efficiencies. However, they can be sensitive to environmental parameters due to the use of microorganisms and enzymes, and some methods can be costly. Their removal efficiencies depend on the type of microorganisms, their operating mode, and the complexity of the effluents. There are many kinds of advanced treatment options for industrial wastewaters, each having its own characteristics. Through the rational utilization of various methods, it is possible to improve the quality of effluents from industrial wastewater in order to significantly reduce the impact of pollution when discharged or reused. This is achieved through an appropriate process that can effectively remove contaminants.

REFERENCES

Abdel-Fatah, M.A., Shaarawy, H.H. and Hawash, S.I., 2019. Integrated treatment of municipal wastewater using advanced electro-membrane filtration system. *SN Applied Sciences*, 1(10), pp.1–8.

Acharya, B.S. and Kharel, G., 2020. Acid mine drainage from coal mining in the United States: An overview. *Journal of Hydrology*, vol. 588 p.125061.

Addy, K., Green, L. and Herron, E., 2004. *pH and Alkalinity*. University of Rhode Island.

Afrad, M.S.I., Monir, M.B., Haque, M.E., Barau, A.A. and Haque, M.M., 2020. Impact of industrial effluent on water, soil and rice production in Bangladesh: A case of Turag River Bank. *Journal of Environmental Health Science and Engineering*, 18(2), pp.825–834.

Ajibade, F.O., Adelodun, B., Lasisi, K.H., Fadare, O.O., Ajibade, T.F., Nwogwu, N.A., Sulaymon, I.D., Ugya, A.Y., Wang, H.C. and Wang, A., 2020. Environmental pollution and their socioeconomic impacts. In *Microbe Mediated Remediation of Environmental Contaminants* (pp.321–354). Woodhead Publishing.

Al-Degs, Y., Khraisheh, M.A.M., Allen, S.J. and Ahmad, M.N., 2000. Effect of carbon surface chemistry on the removal of reactive dyes from textile effluent. *Water Research*, 34(3), pp.927–935.

Anis, S.F., Hashaikeh, R. and Hilal, N., 2019. Reverse osmosis pretreatment technologies and future trends: A comprehensive review. *Desalination*, 452, pp.159–195.

Anjum, R., Rahman, M., Masood, F. and Malik, A., 2012. Bioremediation of pesticides from soil and wastewater. In *Environmental Protection Strategies for Sustainable Development* (pp.295–328). Springer.

Arslan, S., Eyvaz, M., Gürbulak, E. and Yüksel, E., 2016. A review of state-of-the-art technologies in dye-containing wastewater treatment: The textile industry case. *Textile Wastewater Treatment*, pp.1–28.

Asano, T. ed., 1998. *Wastewater Reclamation and Reuse: Water Quality Management Library* (Vol. 10). CRC Press.

Awaleh, M.O. and Soubaneh, Y.D., 2014. Waste water treatment in chemical industries: The concept and current technologies. *Hydrology: Current Research*, 5(1), p.1.

Baker, R.W., 2002. *Membrane Technology. Encyclopedia of Polymer Science and Technology*, 3. John Wiley & Sons, Inc.

Baker, R.W., 2012. *Membrane Technology and Applications*. Wiley.

Barakat, M.A., 2011. New trends in removing heavy metals from industrial wastewater. *Arabian Journal of Chemistry*, 4(4), pp.361–377.

Bartolomeu, M., Neves, M.G.P.M.S., Faustino, M.A.F. and Almeida, A., 2018. Wastewater chemical contaminants: Remediation by advanced oxidation processes. *Photochemical & Photobiological Sciences*, 17(11), pp.1573–1598.

Basile, T., Petrella, A., Petrella, M., Boghetich, G., Petruzzelli, V., Colasuonno, S. and Petruzzelli, D., 2011. Review of endocrine-disrupting-compound removal technologies in water and wastewater treatment plants: An EU perspective. *Industrial and Engineering Chemistry Research*, 50(14), pp.8389–8401.

Bhatt, J., Rai, A.K., Gupta, M., Vyas, S., Ameta, R., Ameta, S.C., Chavoshani, A. and Hashemi, M., 2020. Surfactants: An emerging face of pollution. In *Micropollutants and Challenges: Emerging in the Aquatic Environments and Treatment Processes*, p.145. Candice Janco, Elsevier.

Blowes, D.W., Ptacek, C.J., Jambor, J.L., Weisener, C.G., Paktunc, D., Gould, W.D. and Johnson, D.B., 2003. The geochemistry of acid mine drainage. *Environmental Geochemistry*, 9, pp.149–204.

Bisognin, R.P., Wolff, D.B., Carissimi, E., Prestes, O.D., Zanella, R., Storck, T.R. and Clasen, B., 2020. Potential environmental toxicity of sewage effluent with pharmaceuticals. *Ecotoxicology*, 29(9), pp.1315–1326.

Canizares P., Paz R., Sáez C. and Rodrigo M. A., 2009 Costs of the electrochemical oxidation of wastewaters: A comparison with ozonation and fenton oxidation processes *Journal of Environmental Management*, 90(1) 410–420.

Carmen, Z. and Daniela, S., 2012, February. Textile organic dyes–characteristics, polluting effects and separation/elimination procedures from industrial effluents: A critical overview. In *Organic Pollutants Ten Years after the Stockholm Convention-Environmental and Analytical Update* (Vol. 10, p.32373). IntechOpen.

Carolin, C.F., Kumar, P.S., Saravanan, A., Joshiba, G.J. and Naushad, M., 2017. Efficient techniques for the removal of toxic heavy metals from aquatic environment: A review. *Journal of Environmental Chemical Engineering*, 5(3), pp.2782–2799.

Chan, Y.J., Chong, M.F., Law, C.L. and Hassell, D.G., 2009. A review on anaerobic–aerobic treatment of industrial and municipal wastewater. *Chemical Engineering Journal*, 155(1–2), pp.1–18.

Chojnacka, K., Gorazda, K., Witek-Krowiak, A. and Moustakas, K., 2019. Recovery of fertilizer nutrients from materials-Contradictions, mistakes and future trends. *Renewable and Sustainable Energy Reviews*, 110, pp.485–498.

Crini, G. and Lichtfouse, E., 2019. Advantages and disadvantages of techniques used for wastewater treatment. *Environmental Chemistry Letters*, 17(1), pp.145–155.

Cristian, O., 2010. Characteristics of the untreated wastewater produced by food industry. *Analele Universității din Oradea, Fascicula: Protecția Mediului*, 15, pp.709–714.

Da Silva, F.J.G. and Gouveia, R.M., 2020. Global population growth and industrial impact on the environment. In *Cleaner Production* (pp.33–75). Springer.

Davies, P.S., 2005. *The Biological Basis of Wastewater Treatment* (p.3). Strathkelvin Instruments Ltd.

Dharupaneedi, S.P., Nataraj, S.K., Nadagouda, M., Reddy, K.R., Shukla, S.S. and Aminabhavi, T.M., 2019. Membrane-based separation of potential emerging pollutants. *Separation and Purification Technology*, 210, pp.850–866.

Diya'uddeen, B.H., Daud, W.M.A.W. and Aziz, A.A., 2011. Treatment technologies for petroleum refinery effluents: A review. *Process Safety and Environmental Protection*, 89(2), pp.95–105.

El-Ashtoukhy, E.S.Z., El-Taweel, Y.A., Abdelwahab, O. and Nassef, E.M., 2013. Treatment of petrochemical wastewater containing phenolic compounds by electrocoagulation using a fixed bed electrochemical reactor. *International Journal of Electrochemical Science*, 8(1), pp.1534–1550.

Ernst-Detlef S. and Harold A. M., 1993. *Biodiversity and Ecosystem Function* (pp.88–90). Springer, ISBN 9783540581031.

Fane, A.T., Wang, R. and Jia, Y., 2011. Membrane technology: Past, present and future. In *Membrane and Desalination Technologies* (pp.1–45). Humana Press.

Favas, P.J., Pratas, J., Gomes, M.E.P. and Cala, V., 2011. Selective chemical extraction of heavy metals in tailings and soils contaminated by mining activity: Environmental implications. *Journal of Geochemical Exploration*, 111(3), pp.160–171.

Fazal, T., Razzaq, A., Javed, F., Hafeez, A., Rashid, N., Amjad, U.S., Rehman, M.S.U., Faisal, A. and Rehman, F., 2020. Integrating adsorption and photocatalysis: A cost effective strategy for textile wastewater treatment using hybrid biochar-TiO2 composite. *Journal of Hazardous Materials*, 390, p.121623.

Fröhlich, H., Villian, L., Melzner, D. and Strube, J., 2012. Membrane technology in bioprocess science. *Chemie Ingenieur Technik*, 84(6), pp.905–917.

Gao, Q., Xu, J. and Bu, X.H., 2019. Recent advances about metal–organic frameworks in the removal of pollutants from wastewater. *Coordination Chemistry Reviews*, 378, pp.17–31.

Gatta, G., Libutti, A., Gagliardi, A., Disciglio, G., Tarantino, E., Beneduce, L. and Giuliani, M.M., 2020. *Wastewater Reuse in Agriculture: Effects on Soil-Plant System Properties*. Springer, Cham.

Gaur, V.K., Sharma, P., Sirohi, R., Awasthi, M.K., Dussap, C.G. and Pandey, A., 2020. Assessing the impact of industrial waste on environment and mitigation strategies: A comprehensive review. *Journal of Hazardous Materials*, 398, p.123019.

Goli, A., Shamiri, A., Khosroyar, S., Talaiekhozani, A., Sanaye, R. and Azizi, K., 2019. A review on different aerobic and anaerobic treatment methods in dairy industry wastewater. *Journal of Environmental Treatment Techniques*, 6(1), pp.113–141.

Grachev, V.A., 2019. Environmental effectiveness of energy technologies. *International Journal of GEOMATE*, 16(55), pp.228–237.

Guieysse, B. and Norvill, Z.N., 2014. Sequential chemical–biological processes for the treatment of industrial wastewaters: Review of recent progresses and critical assessment. *Journal of Hazardous Materials*, 267, pp.142–152.

Gupta, A. and Gupta, R., 2019. Treatment and recycling of wastewater from pulp and paper mill. In *Advances in Biological Treatment of Industrial Waste Water and Their Recycling for a Sustainable Future* (pp.13–49). Springer.

Gürses, A., Güneş, K. and Şahin, E., 2021. Removal of dyes and pigments from industrial effluents. In *Green Chemistry and Water Remediation: Research and Applications* (pp. 135–187). Elsevier.

Hanchang, S.H.I., 2009. Industrial wastewater-types, amounts and effects. In *Point Sources of Pollution: Local Effects and Their Control*, 2, p.191. Encyclopedia of Life Support Systems.

Haritash, A.K. and Kaushik, C.P., 2009. Biodegradation aspects of polycyclic aromatic hydrocarbons (PAHs): A review. *Journal of Hazardous Materials*, 169(1–3), pp.1–15.

Holeton, C., Chambers, P.A. and Grace, L., 2011. Wastewater release and its impacts on Canadian waters. *Canadian Journal of Fisheries and Aquatic Sciences*, 68(10), pp.1836–1859.

Hou, B., Han, H., Jia, S., Zhuang, H., Zhao, Q. and Xu, P., 2014. Effect of alkalinity on nitrite accumulation in treatment of coal chemical industry wastewater using moving bed biofilm reactor. *Journal of Environmental Sciences*, 26(5), pp.1014–1022.

Huang, M.H., Li, Y.M. and Gu, G.W., 2010. Chemical composition of organic matters in domestic wastewater. *Desalination*, 262(1–3), pp.36–42.

Jhaveri, J.H. and Murthy, Z.V.P., 2016. A comprehensive review on anti-fouling nanocomposite membranes for pressure driven membrane separation processes. *Desalination*, 379, pp.137–154.

Jiang, S., Li, Y. and Ladewig, B.P., 2017. A review of reverse osmosis membrane fouling and control strategies. *Science of the Total Environment*, 595, pp.567–583.

Kasonga, T.K., Coetzee, M.A., Kamika, I., Ngole-Jeme, V.M. and Momba, M.N.B., 2020. Endocrine-disruptive chemicals as contaminants of emerging concern in wastewater and surface water: A review. *Journal of Environmental Management*, 277, p.111485.

Kefeni, K.K., Msagati, T.A.M., and Mamba, B.B., 2017. Acid mine drainage: Prevention, treatment options, and resource recovery: A review. *Journal of Cleaner Production*, 151, pp.475–493.

Kleerbezem R., Beckers, J., Hulshoff Pol, L.W., Lettinga, G., 2005. High rate treatment of terephthalic acid production wastewater in a two stage anaerobic bioreactor‖. *Biotechnology and Bioengineering*, 91(2), pp.169–179.

Kommineni S., Zoeckler J., Stocking A., Liang P.S., Flores A., Rodriguez R., Browne, T., Per R. and Brown A., 2000. *3.0 Advanced Oxidation Processes*. Center for Groundwater Restoration and Protection National Water Research Institute.

Kyne, D. and Bolin, B., 2016. Emerging environmental justice issues in nuclear power and radioactive contamination. *International Journal of Environmental Research and Public Health*, 13(7), p.700.

Lafi, R., Mabrouk, W. and Hafiane, A., 2019. Removal of methylene blue from saline solutions by adsorption and electrodialysis. *Membrane Water Treatment*, 10, pp.139–148.

Lalwani, J., Gupta, A., Thatikonda, S. and Subrahmanyam, C., 2020. An industrial insight on treatment strategies of the pharmaceutical industry effluent with varying qualitative characteristics. *Journal of Environmental Chemical Engineering*, 8(5), p.104190.

Leahy J.G., and Colwell R.R., 1990 Microbial-degradation of hydrocarbons in the environment. *Microbiology and Molecular Biology Reviews*, 54, pp.305–315.

Lederer, J. and Rechberger, H., 2010. Comparative goal-oriented assessment of conventional and alternative sewage sludge treatment options. *Waste Management*, 30(6), pp.1043–1056.

Li, N.N., Fane, A.G., Ho, W.W. and Matsuura, T. eds., 2011a. *Advanced Membrane Technology and Applications*. Wiley.

Li, W., Hua, T., Zhou, Q., Zhang, S. and Rong, W., 2011b. Toxicity identification and high-efficiency treatment of aging chemical industrial wastewater from the Hangu Reservoir, China. *Journal of Environmental Quality*, 40(6), pp.1714–1721.

Li, W., Mu, B. and Yang, Y., 2019. Feasibility of industrial-scale treatment of dye wastewater via bio-adsorption technology. *Bioresource Technology*, 277, pp.157–170.

Liang, H., Gong, W. and Li, G., 2008. Performance evaluation of water treatment ultrafiltration pilot plants treating algae-rich reservoir water. *Desalination*, 221(1–3), pp.345–350.

Lin, J., Lin, F., Chen, X., Ye, W., Li, X., Zeng, H. and Van der Bruggen, B., 2019. Sustainable management of textile wastewater: A hybrid tight ultrafiltration/bipolar-membrane electrodialysis process for resource recovery and zero liquid discharge. *Industrial and Engineering Chemistry Research*, 58(25), pp.11003–11012.

Lindholm-Lehto, P.C., Knuutinen, J.S., Ahkola, H.S. and Herve, S.H., 2015. Refractory organic pollutants and toxicity in pulp and paper mill wastewaters. *Environmental Science and Pollution Research*, 22(9), pp.6473–6499.

Luo, Z., He, Y., Zhi, D., Luo, L., Sun, Y., Khan, E., Wang, L., Peng, Y., Zhou, Y. and Tsang, D.C., 2019. Current progress in treatment techniques of triclosan from wastewater: A review. *Science of the Total Environment*, 696, p.133990.

Ma, L. and Zhang, W.X., 2008. Enhanced biological treatment of industrial wastewater with bimetallic zero-valent iron. *Environmental Science & Technology*, 42(15), pp.5384–5389.

Mahamuni, N.N. and Adewuyi, Y.G., 2010 Advanced oxidation processes (AOPs) involving ultrasound for waste water treatment: A review with emphasis on cost estimation *Ultrasonics Sonochemistry*, 17(6), pp.990–1003.

Majewska-Nowak, K.M., 2013. Treatment of organic dye solutions by electrodialysis. *Membrane Water Treatment*, 4(3), pp.203–214.

Mallevialle, J., Odendaal, P.E. and Wiesner, M.R. eds., 1996. *Water Treatment Membrane Processes.* American Water Works Association.

Margot, J., Rossi, L., Barry, D.A. and Holliger, C., 2015. A review of the fate of micropollutants in wastewater treatment plants. *Wiley Interdisciplinary Reviews: Water*, 2(5), pp.457–487.

McCance, W., Jones, O.A.H., Edwards, M., Surapaneni, A., Chadalavada, S. and Currell, M., 2018. Contaminants of Emerging Concern as novel groundwater tracers for delineating wastewater impacts in urban and peri-urban areas. *Water Research*, 146, pp.118–133.

Meena, R.A.A., Kannah, R.Y., Sindhu, J., Ragavi, J., Kumar, G., Gunasekaran, M. and Banu, J.R., 2019. Trends and resource recovery in biological wastewater treatment system. *Bioresource Technology Reports*, 7, p.100235.

Mohajerani M., Mehrvar M. and Ein-Mozaffari F., 2009 An overview of the integration of advanced oxidation technologies and other processes for water and wastewater treatment. *International Journal of Engineering*, 3(2), pp.120–46.

Mohan, S.V., Bhaskar, Y.V. and Sarma, P.N., 2007. Biohydrogen production from chemical wastewater treatment in biofilm configured reactor operated in periodic discontinuous batch mode by selectively enriched anaerobic mixed consortia. *Water Research*, 41(12), pp.2652–2664.

Mohan, S.V., Sailaja, P., Srimurali, M. and Karthikeyan, J., 1998. Color removal of monoazo acid dye from aqueous solution by adsorption and chemical coagulation. *Environmental Engineering and Policy*, 1(3), pp.149–154.

Mojsov, K.D., Andronikov, D., Janevski, A., Kuzelov, A. and Gaber, S., 2016. The application of enzymes for the removal of dyes from textile effluents. *Advanced Technologies*, 5(1), pp.81–86.

Moore, J.W., 2012. *Inorganic Contaminants of Surface Water: Research and Monitoring Priorities.* Springer.

Moslehyani, A., Ismail, A.F., Matsuura, T., Rahman, M.A. and Goh, P.S., 2019. Recent progresses of ultra-filtration (UF) membranes and processes in water treatment. In *Membrane Separation Principles and Applications* (pp.85–110). Elsevier.

Munter R., 2001 Advanced oxidation processes: Current status and prospects. *Proceedings of the Estonian Academy of Sciences, Chemistry*, 50(2), pp.59–80.

Mutamim, N.S.A., Noor, Z.Z., Hassan, M.A.A., Yuniarto, A. and Olsson, G., 2013. Membrane bioreactor: Applications and limitations in treating high strength industrial wastewater. *Chemical Engineering Journal*, 225, pp.109–119.

Muyzer, G. and Stams, A.J., June 2008. The ecology and biotechnology of sulphate-reducing bacteria. *Nature Reviews Microbiology*, 6(6), pp.441–454.

Natarajan, V., Karunanidhi, M. and Raja, B., 2020. A critical review on radioactive waste management through biological techniques. *Environmental Science and Pollution Research*, 27, pp.29812–29823.

Nath, K., 2017. *Membrane Separation Processes.* PHI Learning Pvt. Ltd.

Navarro, M.C., Pérez-Sirvent, C., Martínez-Sánchez, M.J., Vidal, J., Tovar, P.J. and Bech, J., 2008. Abandoned mine sites as a source of contamination by heavy metals: A case study in a semi-arid zone. *Journal of Geochemical Exploration*, 96(2–3), pp.183–193.

Noukeu, N.A., Gouado, I., Priso, R.J., Ndongo, D., Taffouo, V.D., Dibong, S.D. and Ekodeck, G.E., 2016. Characterization of effluent from food processing industries and stillage treatment trial with Eichhornia crassipes (Mart.) and Panicum maximum (Jacq.). *Water Resources and Industry*, 16, pp.1–18.

Ojovan, M.I., Lee, W.E. and Kalmykov, S.N., 2019. *An Introduction to Nuclear Waste Immobilisation.* Elsevier.

Oller, I., Malato, S. and Sánchez-Pérez, J., 2011. Combination of advanced oxidation processes and biological treatments for wastewater decontamination—A review. *Science of the Total Environment*, 409(20), pp.4141–4166.

Pandey, K., Singh, B., Pandey, A.K., Badruddin, I.J., Pandey, S., Mishra, V.K. and Jain, P.A., 2017. Application of microbial enzymes in industrial waste water treatment. *International Journal of Current Microbiology and Applied Sciences*, 6(8), pp.1243–1254.

Pendergast, M.M. and Hoek, E.M., 2011. A review of water treatment membrane nanotechnologies. *Energy and Environmental Science*, 4(6), pp.1946–1971.

Peng, C.Y. and Korshin, G.V., 2011. Speciation of trace inorganic contaminants in corrosion scales and deposits formed in drinking water distribution systems. *Water Research*, 45(17), pp.5553–5563.

Peters, T., 2010. Membrane technology for water treatment. *Chemical Engineering and Technology*, 33(8), pp.1233–1240.

Preisner, M., Neverova-Dziopak, E. and Kowalewski, Z., 2020. An analytical review of different approaches to wastewater discharge standards with particular emphasis on nutrients. *Environmental Management*, 66(4), pp.694–708.

Qasim, W. and Mane, A.V., 2013. Characterization and treatment of selected food industrial effluents by coagulation and adsorption techniques. *Water Resources and Industry*, 4, pp.1–12.

Qasim, M., Badrelzaman, M., Darwish, N.N., Darwish, N.A. and Hilal, N., 2019. Reverse osmosis desalination: A state-of-the-art review. *Desalination*, 459, pp.59–104.

Rajaram, T. and Das, A., 2008. Water pollution by industrial effluents in India: Discharge scenarios and case for participatory ecosystem specific local regulation. *Futures*, 40(1), pp.56–69.

Ratola, N., Cincinelli, A., Alves, A. and Katsoyiannis, A., 2012. Occurrence of organic microcontaminants in the wastewater treatment process. A mini review. *Journal of Hazardous Materials*, 239, pp.1–18.

Ravanchi, M.T., Kaghazchi, T. and Kargari, A., 2009. Application of membrane separation processes in petrochemical industry: A review. *Desalination*, 235(1–3), pp.199–244.

Ravanchi, M.T., Kaghazchi, T. and Kargari, A., 2010. Supported liquid membrane separation of propylene–propane mixtures using a metal ion carrier. *Desalination*, 250(1), pp.130–135.

Rodriguez-Garcia, G., Molinos-Senante, M., Hospido, A., Hernández-Sancho, F., Moreira, M.T. and Feijoo, G., 2011. Environmental and economic profile of six typologies of wastewater treatment plants. *Water Research*, 45(18), pp.5997–6010.

Ruan, T., Song, S., Wang, T., Liu, R., Lin, Y. and Jiang, G., 2014. Identification and composition of emerging quaternary ammonium compounds in municipal sewage sludge in China. *Environmental Science and Technology*, 48(8), pp.4289–4297.

Sánchez-Andrea, I., Sanz, J.L., Bijmans, M.F., and Stams, A.J., 2014. Sulfate reduction at low pH to remediate acid mine drainage. *Journal of Hazardous Materials*, 269, 98–109.

Schwarzenbach, R.P., Egli, T., Hofstetter, T.B., Von Gunten, U. and Wehrli, B., 2010. Global water pollution and human health. *Annual Review of Environment and Resources*, 35, pp.109–136.

Shammas N.K., Pouet MF., Grasmick A., 2010. Wastewater treatment by electrocoagulation: Flotation. In: Wang L., Shammas N., Selke W., Aulenbach D. (eds) *Flotation Technology. Handbook of Environmental Engineering* (Vol 12). Humana Press.

Shi, X., Tal, G., Hankins, N.P. and Gitis, V., 2014. Fouling and cleaning of ultrafiltration membranes: A review. *Journal of Water Process Engineering*, 1, pp.121–138.

Shon, H.K., Vigneswaran, S. and Snyder, S.A., 2006. Effluent organic matter (EfOM) in wastewater: Constituents, effects, and treatment. *Critical Reviews in Environmental Science and Technology*, 36(4), pp.327–374.

Show, K.Y. and Lee, D.J., 2017. Anaerobic Treatment Versus Aerobic Treatment. In *Current Developments in Biotechnology and Bioengineering* (pp.205–230). Elsevier.

Song, Q., Li, J. and Zeng, X., 2015. Minimizing the increasing solid waste through zero waste strategy. *Journal of Cleaner Production*, 104, pp.199–210.

Song, W., Xie, B., Huang, S., Zhao, F. and Shi, X., 2020. Aerobic membrane bioreactors for industrial wastewater treatment. In *Current Developments in Biotechnology and Bioengineering* (pp.129–145). Elsevier.

Tiwary, R.K., 2001. Environmental impact of coal mining on water regime and its management. *Water, Air, and Soil Pollution*, 132(1–2), pp.185–199.

Tony, M.A., 2020. Zeolite-based adsorbent from alum sludge residue for textile wastewater treatment. *International Journal of Environmental Science and Technology*, 17(5), pp.2485–2498.

Tufail, A., Price, W.E., Mohseni, M., Pramanik, B.K. and Hai, F.I., 2020. A critical review of advanced oxidation processes for emerging trace organic contaminant degradation: Mechanisms, factors, degradation products, and effluent toxicity. *Journal of Water Process Engineering*, 40, p.101778.

Van Lier, J.B., 2008. High-rate anaerobic wastewater treatment: Diversifying from end-of-the-pipe treatment to resource-oriented conversion techniques. *Water Science and Technology*, 57(8), pp.1137–1148.

Van Lier, J.B., Van der Zee, F.P., Frijters, C.T.M.J. and Ersahin, M.E., 2015. Celebrating 40 years anaerobic sludge bed reactors for industrial wastewater treatment. *Reviews in Environmental Science and Bio/Technology*, 14(4), pp.681–702.

Vardhan, K.H., Kumar, P.S. and Panda, R.C., 2019. A review on heavy metal pollution, toxicity and remedial measures: Current trends and future perspectives. *Journal of Molecular Liquids*, 290, p.111197.

Wang, H., Wang, T., Zhang, B., Li, F., Toure, B., Omosa, I.B., Chiramba, T., Abdel-Monem, M. and Pradhan, M., 2014. Water and wastewater treatment in Africa–current practices and challenges. *CLEAN–Soil, Air, Water*, 42(8), pp.1029–1035.

Wu, W.E., Ge, H.G. and Zhang, K.F., 2003. *Wastewater Biological Treatment Technology*. Chemical industry Press.

Wuana, R.A. and Okieimen, F.E., 2011. Heavy metals in contaminated soils: A review of sources, chemistry, risks and best available strategies for remediation. *International Scholarly Research Notices*, 2011. doi:10.5402/2011/402647

Xue, S.G., Wu, Y.J., Li, Y.W., Kong, X.F., Zhu, F., William, H., Li, X.F. and Ye, Y.Z., 2019. Industrial wastes applications for alkalinity regulation in bauxite residue: A comprehensive review. *Journal of Central South University*, 26(2), pp.268–288.

Zhang, X., Gu, P. and Liu, Y., 2019. Decontamination of radioactive wastewater: State of the art and challenges forward. *Chemosphere*, 215, pp.543–553.

Zhao, M., Xu, Y., Zhang, C., Rong, H. and Zeng, G., 2016. New trends in removing heavy metals from wastewater. *Applied Microbiology and Biotechnology*, 100(15), pp.6509–6518.

Zheng, C., Zhao, L., Zhou, X., Fu, Z. and Li, A., 2013. Treatment technologies for organic wastewater. *Water Treatment*, 11, pp.250–86.

3 Prospects for Exploiting Microbes and Plants for Bioremediation of Heavy Metals

Hiren K. Patel, Rishee K. Kalaria, and Divyesh K. Vasava

CONTENTS

DOI: 10.1201/9781003204442-3

3.1 INTRODUCTION

Soil contaminated with heavy metals has become a significant worldwide problem thanks to earth science and phylogenesis activities. Heavy metals are non-degradable and therefore persist indefinitely within the surroundings. Nowadays, varied heavy metals represent a global environmental hazard. The use of microorganisms and plants for the removal of heavy metals has attracted growing attention as a result of their low value and high potency. Microorganisms might be wont to close up metal contamination by removing metals from contaminated water, sequestering metals from soils and sediments, or solubilizing metals to facilitate their extraction (Glick, 2010). In this chapter, we will describe how microorganisms and plants accumulate and detoxify metal ions to boost metal tolerance, accumulation, and detoxification. We will also conjointly describe bioremediation victimization mutualism between plants and microorganisms.

Heavy metals are found in soil and water naturally and are usually discharged from varied natural and phylogeny sources (Khalid et al., 2017). Significant metals disrupt nutrients and water uptake, have an effect on production of reactive oxygen species (ROS), cut back chemical action potency, alter biological process, and alter element metabolism and so have an effect on plant growth. The continuous uptake of significant metals in humans through contaminated foods will cause aerophilic stress by the over production of ROS, higher rates of gastrointestinal cancer, and numerous medical syndromes as well as influence the effects, ontogenesis, and cause of cancer.

Three completely different rectifying processes, i.e., biological, physical, and chemical, are used to clean up heavy metal-contaminated soils (see Figure 3.1). Bioremediation could involve methods during which living organisms are used to take away contamination to re-establish the natural conditions. Usually microorganisms and plants are employed to detoxify significant metals from the soil. Alternative bioremediation techniques like bioventing, bioleaching, bioreaction, bioaugmentation, rhizofiltration, and biostimulation also are useful to remove significant metal toxicity. Bioremediation is one of the most eco-friendly, cost-effective options used to rectify soil contaminants. Bioremediation agents include both microorganisms and plants that detoxify significant metals in soil and water. Phytoremediation is another economical, environmentally-friendly, non-invasive, recently-developed technology used to scrub up or tackle the sites with a low or moderate level of significant metals (Saifullah et al., 2015).

Hyperaccumulator plants have the potential to soak up high levels of metals by up to 50–500 times more than conventional plants. This often involves an organic process adaptation of plants in unfavorable habitats with high concentrations of metal in soil or rocks. However, the benefits of hyperaccumulation of metals in plants remain unknown. Ideally, hyperaccumulator plants have extensive root systems and are able to tolerate a high concentration of metal pollutants. In recent years, different gene-splicing approaches have been utilized to increase the production and productivity of hyperaccumulator plants. Together with standard science practices and gene-splicing, the significant metal absorption by these plants is increased (Bhargava et al., 2012). Thus, there is a pressing need to investigate standard and genetically modified potential in hyperaccumulators,

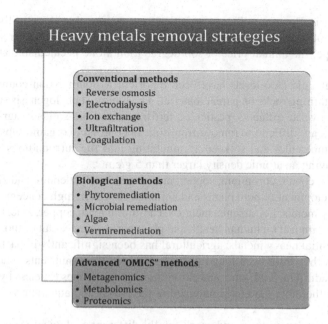

FIGURE 3.1   Different strategies applied to remove heavy metals from wastewater.

which might be planted in contaminated sites to rid the property of significant metals and to maintain the environmental quality of the ecosystem.

3.2   PLANT CLASSIFICATION BASED ON TOXICITY OF HEAVY METAL

Plants are classified based on the toxicity of heavy metals as follows (Baker et al., 1988):

1. **Heavy metal sensitive plants:** In response to increased concentration of heavy metal in soil / water, plants that do not show resistance to heavy metal toxicity and their biochemical machinery shatters.
2. **Heavy metal resistant plants:** Plants that have different mechanisms to prevent heavy metal from accumulating inside their cells either by an active transport mechanism or by restricting heavy metal from soil to root. They are also called excluders.
3. **Hyperaccumulators:** Plants that have a special mechanism for dealing with heavy metal toxicity by deliberately extracting heavy metals from soil/water and accumulating them in the sections above ground. They are the focus of groundbreaking research for new "phytomining" technology.

3.3   HEAVY METALS: SOURCE AND NEGATIVE EFFECTS

Heavy metals accumulate within the soil either from those that naturally occur within the Earth's crust or from phylogenesis sources. Within recent decades, the annual worldwide occurrence of heavy metals reached 22,000 tonnes for metallic elements, 939,000 tonnes for copper, 783,000 tonnes for lead, and 350,000 tonnes for metallic element due to different human activities (Sancenón et al., 2003). There is immense value in efforts to clean up contaminated sites through standard approaches. Within the U.S.A. alone, $6–8 billion is spent annually in correction efforts, with international prices within the vary folks $25–50 billion (Tsao, 2003). The major cost contributor is the cleaning up of various sites contaminated with radionuclides as a result of weapons of mass destruction preparations during a conflict. In line with the European Environment Agency (EEA), it is calculable that, in Europe, polluting activities have occurred at three million sites, of which >8% (or nearly 250 000 sites) are extremely contaminated and projected to increase up to >50% by 2025. Within Europe, the annual price of soil degradation alone is calculable at some \$38 billion (Agency, 2007).

Recently, high arsenic (As) levels have been reported in South Asian countries owing to the presence of As-containing rocks on parent material. The foremost phylogenesis sources are mining, industrial discharge, waste effluents, pesticides, fertilizers, and bio-solids in agriculture (Ali et al., 2013). Heavy metals are difficult to remove from soil; in contrast to organic substances they do not degrade into little molecules and so keep accumulating into the surroundings. Heavy metals are those trace parts having an atomic density larger than 5 g/cm^{-3}.

Heavy metals viz. cobalt, chromium, copper, manganese, molybdenum, and zinc are essential to the growth and development of both plants and animals. However, high concentrations cause toxicity. Poisonous heavy metals, i.e., arsenic, mercury, cadmium, lead, copper, zinc, tin, and chromium, may have an adverse impact on human health and plant performance. Out of those, copper and zinc are considered essential heavy metals; agricultural has been significantly impacted in several areas across the globe as they have been found to be deficient in those nutrients. The toxicity levels of arsenic, mercury, cadmium, lead, zinc, and chromium are sometimes timeless by all organic chemistry pathways and therefore tend to accumulate in terrestrial and aquatic ecosystems (Krämer & Chardonnens, 2001).

Heavy metals enter ecosystems and reach totally different biological process levels resulting in bioaccumulation. Heavy metals are known to have severe health impacts like excretory organ injury, pathology, high blood pressure levels, failure of genital system, and liver disorders (Khalid et al., 2017). Heavy metals (Pb^{+2} and Cd^{+2}) induce carcinogenesis that can cause mortality. In plants, copper is a crucial substance element for growth and development, however, animating thing free Cu^{+2} ions in excess turn out ROS by auto-oxidation and Fenton reaction. Also, radicals react to cause membrane lipid peroxidation, cleavage of the sugar-phosphate backbone of nucleic acids, and protein denaturation. Additionally, Cu^{+2} will displace alternative bivalent cations coordinated with macromolecules, inflicting their inactivation or malfunction. Zn^{+2} is additionally a vital nutrient for plants that acts as a co-factor needed for the structure and performance of various enzyme involved in energy production and the structural integrity of membranes. High levels of Zn^{+2} inhibit several metabolic functions in plants leading to slow growth. Metallic element toxicity limits the expansion of each root and shoot, and produces leaf iron deficiency anemia. Even supposing it is not chemically reactive, higher levels of concentration are poisonous and as a result will displace alternative metals (e.g., iron, manganese, and copper) within the cells. Physical and chemical methods of correction have limitations owing to their high cost, intensive labor, and the production of secondary pollutants. Therefore, bioremediation could be a viable choice to remedy heavy metal contaminated soils in an eco-friendly and cost-effective manner.

3.4 BIOAVAILABILITY OF HEAVY METALS

Although the total concentrations of heavy metal in soil do not offer any specific information regarding metal quality and handiness, it remains necessary to assist in characterizing the precise bio-availability, reactivity, quality, and way plants or different soil small flora take up heavy metals. Metals in soil are categorized into five major geochemical forms: (i) exchangeable, (ii) sure to carbonate part, (iii) sure to metal and manganese oxides, (iv) sure to organic matter, and (v) residual metal. These soil metals will vary greatly in their quality, biological handiness, and chemical behavior in soil, most likely as a result of their ability to react to create organic compounds like low-molecular organic acids, carbohydrates, and enzymes secreted by microorganisms. What is more, the soil-derived microorganism will act firmly with metal ions in soil resolution through their charged surfaces. Microorganism cells have a large capability to modify them to take up and immobilize unhealthy ions from the soil resolution. As an example, Huang et al. (2000) have rumored that once dependent microorganisms like rhizobia are used as inoculants, the sorption of Cu^{+2} and Cd^{+2} in soil is considerably exaggerated. The mechanisms of microorganism effects on the phylogeny and distribution of metals in soils, however, remain poorly understood. Previous studies have used varied strategies to research the doable chemical association of metals in soils, and to assess the metals' quality and bioavailability. The studies illustrated varied strategies like consecutive extraction, single extraction, and soil column-leaching experiments. In single extraction techniques, they used selective chemical extracting like a chelating agent or a gentle neutral salt, and in most cases it did not indicate the bioavailability or quality of significant metals. Moreover, this technique provides valuable information that may predict the bioavailability of metals to plants, movement of metal within the profile, and metal transformation between different forms in soils. There are a number of factors that have an effect on the bioavailability and accumulation of significant metals in soil: (i) soil quality and composition, (ii) physico-chemical characteristics of soils, (iii) plant germplasm and also the photosynthates therein, (iv) soil–plant–microorganism connections, and (v) agronomic methods such as the application of fertilizers, water conservation, and alternating crops.

3.5 HEAVY METAL EFFECT ON HUMAN HEALTH

Heavy metals sometimes enter the body via varying food chains, inhalation, and uptake. Additionally, heavy metals are used by humans for creating metal alloys and pigments for paints, cement, paper, rubber, and alternative materials. The application of heavy metals in some countries is increasing despite their well-known hepatotoxic effects. Once a heavy metal enters the body, it is likely to have negative effects such as nausea, anorexia, vomiting, canal abnormalities, and eczema (Chui et al., 2013). The toxicity of heavy metals may also disrupt or injure the brain and central nervous system, blood composition, lungs, kidneys, liver, and other major organs. The semi-permanent exposures of the human population to heavy metals have additionally shown physical, muscular, and other medical impairments. The chronic processes are the same as with Alzheimer's illness, Parkinson's disease, genetic abnormalities, and sclerosis. Other illnesses, like clogged respiratory organs, have been connected to carcinoma, and injury to humans' metabolic processes have additionally been found to develop following high rates of exposure to metals. Aside from the hepatotoxic effects, certain metals like copper, selenium, and zinc have been observed to play necessary and useful roles in human metabolism. As an example, copper at a lower concentration acts as a co-factor for numerous enzymes of oxidation-reduction in exercise. However, at higher concentrations, it disrupts the human metabolism resulting in anemia, liver and urinary tract injury, and abdominal and enteral irritation.

3.6 HEAVY METALS: IMPORTANCE FOR MICROORGANISMS

The uptake of heavy metals from soil has both direct and indirect effects on microbe composition, metabolism, and differentiation. The interaction of metals and their compounds with soil microbes depends on many factors. These factors embody metal species, interacting organisms and their surroundings, structure, and composition, and microbe functions. Certain metals like copper, zinc, cobalt, and iron are essential for the survival and growth of microbes. However, these metals will exhibit toxicity at higher concentrations and will inactivate protein molecules (Samanovic et al., 2012). Though some metals like aluminum, cadmium, mercury, and lead have unknown biological functions, they nevertheless accumulate in cells and may have the following effects: (i) have an effect on enzyme property, (ii) deactivate cellular functions, and (iii) harm the DNA structure which may result in death. For example, nickel is an essential nutrients and plays an important role in numerous microbe cellular processes. Additionally, several microbes are able to find and absorb metallic elements through permeases or ATP-binding-cassette transport systems. Once the component enters the cell, the metallic element is incorporated into various microbe enzymes like acetyl radical CoA decarbonylase/synthase, urease, methylenediurease, Ni^{+2}- metallic element hydrogenase,

CO dehydrogenase, methyl radical molecule enzyme, superoxide dismutases, and a few glyoxylases. However, metallic elements are harmful to bacterium at higher concentrations. Therefore, bacterium has developed sure ways to manage the level of metallic elements to combat toxicity, as ascertained antecedently in 2 gram-negative bacteria: *E. coli* and *H. pylori*. Additionally, *B. japonica* HypB, that was refined from associate in nursing production of *E. coli* strain, has shown its ability to bind up to 18 metallic element ions per compound, likewise on contain GTPase activity. In general, plant growth-promoting rhizobacteria (PGPR) will produce siderophores in Fe-deficient surroundings. It has been reported that a ferrous Fe-specific substance will increase the plant growth of cucumber (*Cucumis sativus* L.) by facilitating the access of metallic elements in rhizospheric surroundings. In another study, certain strains of bacteria that belong to the genus *Ciceri*, that are associated with chickpeas, have the power to supply phenolate siderophores, like 2-hydroxybenzoic acid and 3-dihydroxybenzoic acid. These compounds are created in response to metallic element deficiency; but their production is considerably influenced by the nutrient elements from the substance. It seems that copper, molybdenum, and manganese ions compete with iron for siderophores, leading to a 34–100% increase in siderophore production. For example, cobalt could be a biologically essential microelement with a broad range of physiological and organic chemistry functions (Okamoto & Eltis, 2011). For dependent association, cobalt is required for N_2 fixation in legumes and root nodules of non-legumes. Curiously, the need for cobalt is much larger for N_2 fixation than for ammonia nutrition, as cobalt deficiency leads to the development of N_2 deficiency symptoms. Previously published studies have ascertained a major increase in leghaemoglobin formation once cobalt is applied, and that it plays a role as an important element of N_2 fixation. It increased the nodule numbers per plant and ultimately multiplied the pod yield of soybeans. Several cobalamin-dependent enzymes of rhizobia are concerned in nodulation and N_2 fixation like essential amino acid synthase, ribonucleotide enzyme, and methylmalonyl coenzyme A mutase. Preliminary studies have shown that the combined inoculations of bacteria genus and cobalt have considerably affected the full uptake of nitrogen, phosphorus, potassium, and cobalt by summer groundnuts. Similarly, molybdenum (Mo) reacts because the chemical process centers on various enzymes. These enzymes are categorized into two classes, supported by their compound composition and chemical process performance: (i) microorganism nitrogenases containing iron-molybdenum-cobalt within the site, and (ii) pterin-based molybdenum enzymes. The second class of enzymes includes sulphite enzymes, organic compound enzymes, and dimethyl-sulfoxide enzymes, each of which plays a specific role and has distinct activities. For example, nitrate enzymes have been reported in *D. desulfuricans* whereas organic compound dehydrogenase in *D. gigas* (Oves et al., 2013) has been reported as chromium-reducing microorganism strains *Stenotrophomonas maltophilia* and *Pseudomonas aeruginosa* from soil samples.

3.7 HEAVY METALS: IMPORTANCE FOR PLANT HEALTH

In general, plants typically want a continuous supply of nutrition so as to stay healthy. Any shortage of a specific nutrient leads to the development of nutrient-deficient symptoms, and extreme conditions of nutrient deficiency could result in early mortality. In comparison, if a nutrient exceeds the required quantities, it will cause injury, and in some cases could lead even to the death of a plant due to high levels of nutrients enriched with heavy metals. However, plants need both macro-nutrients (carbon, hydrogen, nitrogen, oxygen, phosphorous, sulfur, etc.) and micronutrients (boron and chlorine, copper, iron, manganese, molybdenum, nickel, and zinc) for their growth. The nutrients of each class are often found in varied agro-ecological niches. Some plants living in dependent association with nitrogen-fixing microorganisms conjointly need cobalt as a compulsory nutritionary element. Metal can even be considered as a nutrient for plant health, but there are two criteria that categorize metals as essential nutrients for healthy plants: (i) the metal should be needed by the plant to finish its life cycle, and (ii) the metal should form a part of a plant's constituent or substance. Plants are living organisms that use light-weight energy throughout chemical action to convert CO_2 (carbon dioxide) and liquid into energy-rich carbohydrates and O_2 (oxygen). Therefore, plant growth and development normally depend entirely on chemical action that successively depends on a comfortable provider of diverse chemical components together with metals like copper, iron, and manganese. In addition, heavy metals and metalloids can penetrate plants via absorption processes that support entirely different metal transporters (Hossain & Komatsu, 2013). However, in the case of any metal deficiency, plants will regulate that deficiency and increase the metal handiness within the root atmosphere by either: i) lowering the pH through root exudates which can contain organic acids, or ii) through the discharge of metal- complexing agents. After, once the right amount and ample supply are maintained, an indication from the shoot to the basis is transmitted to stop exudation. Many studies have observed that the presence of metals at low rates in plant systems will affect plant growth by influencing the oxidation reduction reaction and typically directly changing it into an integral part of enzymes. Moreover, zinc plays a necessary role in maintaining the integrity of organelle within the formation of carbohydrates in catalyzing the chemical reaction processes and within the synthesis of macromolecules. Similarly, manganese plays a vital role in bound accelerator reactions like malic dehydrogenase and oxalosuccinic enzymes. Additionally, manganese is required for water splitting in photosystem II, as for superoxide dismutase. In microorganisms, cobalt compounds are important for the formation of vitamin B_{12}, and iron is a vital element for several metabolic processes and is indispensable for all organisms.

3.8   HEAVY METAL TOXICITY

3.8.1   SIGNIFICANCE IN MICROBIAL ACTIVITY

Altered structures of microorganism communities in response to metals are a crucial indicator of metal activity and its biological convenience at intervals soil scheme. In this context, the separate application of significant metals like cadmium, chromium, lead, and a mixture of each cadmium sulphate and lead (II) nitrate solutions at totally different rates have shown harmful effects on soil microbes. Thereby, the impure soil with significant metals resulted in a notable decrease of the activities of varied soil enzymes like acid enzyme and enzyme of microbes. Frostegård et al. (1993) have additionally analyzed soil-derived microbes from metal-polluted soils and noted that a gradual amendment within the microorganism community structure supported alterations of the phospholipids and carboxylic acid profiles. However, the response of microorganism communities to varied metals depends on metal solubility, bioavailability, and toxicity in the soil. The metals' characteristics are greatly influenced by the physiochemical methods of the soil like action, precipitation, and complexation ability. Moreover, the interaction of metals with soil depends on the chemical properties of soil, which can vary across the numerous agro-climatic regions of the globe. Broos et al. (2007) has advised that once metal stress is applied, the full quantity of microorganism biomass diminished in parallel with decreasing the potency of substrate utilization of microbes. The decline of the microorganism biomass is an indicator of soil pollution with significant metals, but its quality in environmental observance is restricted, attributable to its high abstraction variability and failure to satisfy bound customary in its measure. Previous studies have additionally shown that the decline of the microorganism biomass was related to changes in the soil microorganism community structure. Another study has ascertained the structural changes of soil microorganism communities, and also that the level of metal tolerance was increased in response to even slight metal contamination of the soil. The ascertained effects of metal toxicity on different soil microorganism communities could also be attributable to the metal-sensitive ability of populations or communities which has been modified. However, no specific threshold for metal toxicity was given; such threshold could also be site specific, as ascertained by Bünemann et al. (2006).

3.8.2   MECHANISM OF HEAVY METAL TOXICITY

Toxic metal species will bind to proteins and thereby have an effect on the biological functions of the target molecule. For example, biomass and soil nutrient production of the microorganisms are adversely riddled with the high metal levels of copper and zinc. Additionally, cyanogenic metal might move powerfully within thiols and disulphides and then cause disruption of the biological activity of bound proteins that contain sensitive S teams. These reactions frequently release ROS that are natural by-products of the traditional metabolism. The destruction of sensitive thiol teams thanks to metal exposure might eventually impair the macromolecule folding or binding of apoenzymes by cofactors, and therefore the traditional biological activity of the proteins is discontinuous. Moreover, bound metals in transition will participate in chemical reactions, referred to as Fenton-type reactions, that turn out ROS. Together, these reactions will put the cell under aerophilous stress and thereby the ROS levels will be considerably exaggerated, which can end in polymer injury, the destruction of lipids and proteins through a variety of organic chemistry routes. Cyanogenic metal species may additionally enter cells through various transporters, or penetrate the cellular membrane and then bind to a lipotropic carrier. The transporter-mediated uptake of cyanogenic metals interferes with the traditional transport of essential substrates and therefore leads to the competitive inhibition of the transport method. Moreover, this transport method acquires energy from the nucleon driving force or nucleotide pool. Some metal oxyanions are reduced by the oxidoreductases that are able to draw electrons from the microorganism transport chain through the chemical compound pool (Su et al., 2011). Explicit cyanogenic metals will cause starvation of the microorganism cells indirectly by siphoning electrons from the metastasis chain. The ROS are created throughout traditional metabolic processes and lead to polymer injury and the destruction of proteins and lipids, but this production is continued throughout the metal pollution process and could result in further cellular damage. Macomber & Hausinger (2011) have advised four mechanisms of metallic element toxicity that involves (a) the replacement of an essential metal by metalloprotein: a macromolecule with a metal particle co-factor, (b) a metallic element that binds to chemical action sites of non-metalloenzymes, (c) a metallic element that binds outside the chemical action site of the associate protein to inhibit its operatation, and (d) a metallic element that indirectly causes aerophilous stress. In another example, the loss of cell viability when chromium (III) exposure wasn't as a result of the membrane injury or catalyst inhibition was, however, in all probability as a result of a chromium (III) cyanogenic result on the cellular morphology of the microorganism. When chromium (VI) was exposed to the cell, the cyanogenic result of chromium (III) seemed to be related to the interaction between living things, which ultimately resulted in misshapen cell morphology.

3.8.3 METAL RESISTANCE STRATEGIES ADOPTED BY BACTERIA

Microscopic organisms have developed a few methods to beat the restrictive effects of harmful metals. Such metal detoxification methods (see Figure 3.2) adopted by microorganism communities are (i) metal exclusion by pervious barriers, (ii) the transportation of metals removed from the cell, (iii) intracellular sequestration of the metal by macromolecule binding, (iv) extracellular sequestration, (v) protein detoxification of metal to a less harmful kind, and (vi) reducing the sensitivity of cellular targets to metal ions. Single or multiple detoxification mechanisms will be applied for someone metal or a bunch of with chemicals connected metals. However, the type of microorganism will have a great effect on the detoxification mechanisms. For example, a bigger range of microbes have specific stress resistance genes that might regulate harmful metal resistance, and such regulative genes are centrally set on plasmids or chromosomes (Shoeb et al., 2012). In general, the regulative system of metal resistance is genes-based, which is a lot more complicated than the cellular inclusion system. On the other hand, plasmid-encoded systems sometimes interact with toxic-ion outflow mechanisms. The determinants of plasmid-encoded resistance systems for harmful metal ions (such as Cu) are inducible. As an example, lead-resistant *E. faecalis* was isolated and known by using organic chemistry approaches and 16S rRNA cistron sequencing. Lead-resistant *E. faecalis* additionally displayed incontestable resistance to different significant metals and antibiotics. Moreover, *E. faecalis* had four plasmids with totally different molecular sizes: 1.58, 3.06, 22.76, and 28.95 kb. The cellular inclusion profile of the cured derivatives has shown that the power of

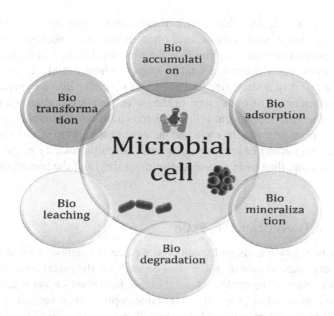

FIGURE 3.2   Different strategies used by microbes for the bioremediation of heavy metals.

lead resistance *E. faecalis* continues even after elimination of all the plasmids (Aktan et al., 2013). What is more, different types of heavy metal resistances are found in microorganism chromosomes, for instance, Hg^{2+} resistance in *Bacillus*, Cd^{2+} outflow in *Bacillus*, and arsenic outflow in *E. coli* (Silver & Phung, 1996). Outflow pumps that were determined by cellular inclusion and body systems are either ATPases or chemiosmotic systems. The mechanisms of outflow pumps show similarity among differing types of bacterium. Cadmium resistance could involve: (i) associate outflow ATPase in gram-positive bacterium, (ii) cation-H^+ antiport in gram-negative bacterium, and (iii) intracellular metallothionein in blue-green algae. Arsenic-resistant gram-negative bacterium have associate arsenite outflow ATPases connected to a salt enzyme which reduces salt [As (V)] to arsenite [As (III)], which comprises the underlying chemical mechanism. A similar system for Hg^{2+} resistance was ascertained on plasmids of gram-positive and gram-negative bacterium with element genes involved in the transportation of Hg^{2+} to the detoxifying protein, during which mercurous enzyme plays a role in the reduction of Hg^{2+} to elemental Hg^0. The protein organomercurial lyase will break the C–Hg^+ bond in organomercurials. In another study, plasmid-encoded resistance to salt seemed to be unconnected to the reduction of salt [chromium (VI)] to chromium (III). Nevertheless, the resistance mechanism of the microbe relies on a reduction in CrO_2^{-4} (chromic acid) absorption (Ramírez-Díaz et al., 2008).

3.8.4   Heavy Metal–Plant Interactions

Exposure to significant metals at higher concentrations results in severe harm to varied metabolic activities, consequently leading to the death of plants. The exposure of excessive metals to plants prevents physiologically processes, inhibiting photosystems and reducing the uptake of nutrients. Sandmann & Böger (1980) in a study disclosed the importance of lipoid peroxidation under metal stress. In various experiments, Janas et al. (2010) analyzed the impact of metallic elements on the development, lipid peroxidation, deposition of phenolic resin compounds on lentil (*Lens Culinaris* Medic.) seedlings. Previous studies have noted that soil dissolved organic substances have important effects on heavy metal transformation by increasing the solubility of the metal, root growth, and plant uptake. Recent studies reveal that maize and soybeans that developed in metal-contaminated (lead) soil had reduced chemical action pigments, a poor stomatal electrical phenomenon, and a considerably lower biomass. On the other hand, cadmium application resulted in a notable decline in the rate of chemical change, stomatal electrical phenomenon, and biomass in cruciferous plants and mustard plants. In another study, cadmium contamination resulted in increased total chlorophyl content in tomatoes and a decreased total biomass (Rehman et al., 2011). Qiu et al. (2011) have shown that as a result of adding zinc and cadmium severally into a hyperaccumulator plant (*P. griffithii*), the accumulation of each metal in the roots, petioles, and leaves was considerably inensified. However, the zinc supplement influenced the cadmium accumulation, within which it decreased cadmium concentration in the roots and increased the buildup of cadmium in the petioles and leaves. The protecting effect of magnesium against cadmium toxicity might partly flow from the upkeep of iron standing, the rise in anti-oxidative capability, and the protection of the chemical action equipment.

3.8.5 HEAVY METAL TRANSLOCATION IN PLANTS

The first interaction of heavy metals with plants takes place throughout a plant's uptake process. The capability of every plant to take up soil metals depends on the metal concentration in the soil and its availableness to plants. The method of metal uptake by a plant's roots depends on: (i) the diffusion of parts on the concentration gradient, (ii) root interception wherever soil volume is displaced by root volume because of root growth, and (iii) mass flow, transport from bulk soil resolution on the water potential gradient.

Some metals will be absorbed by the top sections of a plant whereas others will be absorbed by the entire root surface. Metals are transported mostly into the cells, a number of which are then transferred to the apoplast, and therefore the different half certain to cytomembrane substances. Within the apoplast, metals migrate through the cell wall into the protoplasm where metals might have an effect on the nutrient standing of the plant. For example, the noxious result of metal depends on its phylogenesis, which determines its uptake, translocation, and accumulation. The uptake and accumulation of iron, cadmium, zinc, and nickel by numerous crops is well documented (Tejada-Jiménez et al., 2009). Once the demand for part uptake by a plant root is high, and therefore the nutrient concentration within the soil is low, the weather can move to the plant root by diffusion. Since some metals are essential for plants, the uptake of such metals should be regulated by bound mechanisms. A previous study has confirmed that metal is transported with zinc transporters in a higher abundance with zinc accumulator plant species than in non-accumulator species (Lasat et al., 2000). Metal was actively transported as a free particle across the tonoplast. Other metals like cadmium will simply enter the plant root through plant tissue and be translocated to the lower plant tissues. As cadmium enters the roots, it will reach the vascular tissue through the apoplastic or symplastic pathway and consequently forms complexes with several ligands, like organic acids and/or phytochelatins. Normally, cadmium ions are preserved within the scheme and only small amounts are translocated to the shoot system. Metal ions are most likely absorbed by cells through membrane transport proteins that are designed for nutrient metal acquisition. Another study has found that cadmium and zinc co-exist in aerial elements of *Arabidopsis halleri* plants (Bert et al., 2003). The researchers have confirmed that the uptake of cadmium and zinc is genetically correlated and these metals are absorbed by constant transporters, or their transporters are totally different and may be controlled by common regulators.

3.9 HEAVY METALS TOXICITY IN PLANTS

3.9.1 EFFECT ON CELL WALL AND PLASMA MEMBRANE

The root system of plants, including legumes, is the primary target site for the uptake of any waste material within the soil. Ernst et al. (1990) have postulated that metal interacts well with cytomembranes, however, their findings regarding cytomembranes' binding properties and role in providing resistance to metal were inconsistent. Most heavy metals bind to polygalacturonic acids, to that the affinity of metal ions varies significantly. The absorption of metals among plants varies with the age of plant organs. For example, Nabais et al. (2011) conducted a study to point out the impact of root age on the assimilation of metals, organic compound, and sugars in numerous cellular fractions of perennial bahiagrass (*Paspalum notatum*). The study examined the distribution of metals (aluminum, iron, copper, and nickel), nutritional proteins, and carbohydrates in various *P. notatum* root cells matured in quartz sand at 21 and 120 days old. (Nabais et al., 2011). With these supplements, the scientists reported that young roots seemed to have a better growth rate than the older ones. This variation in metal distribution in roots might be in all probability because of the variations in the metabolic activities and design of roots. However, the metal concentration in older roots was considerably less than the concentration within the younger roots, and this is likely because of the active metabolism of young roots. As an example, aluminum and iron were primarily allotted to plant cytomembranes with cellulose, hemicellulose, and polysaccharide, in both the younger and older roots. However, the older roots showed a large portion of aluminum that was conjointly allotted to the intracellular fraction, suggesting that the older roots were less able to stop the entry of aluminum into the living substance. The proportion of metallic elements was higher in the intracellular elements in both the younger and older root, a clearly indication as a necessary nutrient. Supported this finding, it had been demonstrated that the foundation cells of *P. notatum* suffered from severe changes within the composition of the cytomembrane with aging. In a similar study, the in vitro cadmium application redoubled the wall thickness and caused destruction of the interior organization of the plastid (Bouzon et al., 2012). Another target site for metal toxicity is the semi-permeable membrane, where the metals will bind to the membrane and disrupt the membrane functions. When metal reacts, it can cause changes in the lipid composition of membranes that could lead to distortion to the functioning of the membrane and various essential biological processes. Metal hepatotoxic consequences include the oxidation and cross-linking of macromolecule thiols, the inhibition of key membrane proteins like H^+-ATPase, or changes within the composition and liquidity of membrane lipids. Another study has hypothesized the impact of chromium on the transport activities of semi-permeable membranes (Kabała et al., 2008). Additionally, the inhibition of ATPase activity was urged to flow from to membrane disruption by free radicals generated underneath metal stress. Thereby, the reduction of ATPase activity leads to nucleon extrusion, reduces the transport activities of the foundation semi-permeable membrane, and ultimately limits nutrient uptake by the roots. Moreover, it had been noted that chromium interferes with the mechanism of the dominant intracellular pH. Moreover, chromium might alter the metabolic activities of plants which would result in: (i) the modification of the assembly of chemical action pigments like chlorophyll, and (ii) the increase in the assembly of metabolites like glutathione and antioxidant in response to metal stress which can cause harm to the plants. Among different metals, cadmium and metallic elements have been found to have an adverse effect on the lipid composition of membranes. Moreover, cadmium treatment reduces the ATPase activity of the semi-permeable membrane fraction of roots.

3.9.2 Membrane Lipid Peroxidation by Heavy Metals

High levels of heavy metals induce changes in the carboxylic acid composition of membranes and injury to the membranes may occur once the peroxidation of membrane protein increases because of the action of virulent free radicals. During this process, the metal ions will cause lipid peroxidation to both the semi-permeable membrane and plastid membrane. For example, lipid peroxidation was magnified once plants were fully grown with cadmium, aluminum, copper, chromium, and arsenic. The appliance of Cu^{+2} in the incubation medium and its uptake by the plant tissues resulted in: (i) the reduction of the content of photosynthetic pigments, (ii) the stimulation of lipid peroxidation, and (iii) the increased porosity of the membrane. Specific lipid compositional alterations were attributed to the progressive accumulation of Cu^{+2} in plant cells. For example, a decline in the amount of sulfolipid was discovered in chloroplasts, and a rise in the amount of monogalactosyl diacylg-lycerols, digalactosyl diacylglycerols, and phosphatidyl glycerols in chloroplasts and mitochondria was discovered after one hour exposure to Cu^{+2}. However, after three hours of Cu^{+2} exposure the content of all the lipids accept phosphatidic acids remittent (Meharg, 1993). In general, the metallic elements and Cu^{+2} compounds created many free radicals and magnified the peroxidation. Thus, the alteration in membrane functions caused by metal exposure may well be due to changes in both the structure and peroxidation of membrane lipids. As an example, it is recognized that aluminum causes lipid peroxidation by disorganizing the membrane structure through generating free radicals. The magnified lipid peroxidation additionally alters the membrane properties, like liquidity and porosity, and modulates the activities of membrane-bound ATPases. Indeed, the method of peroxidation may be a chain reaction, within which unsaturated fatty acids are gradually transformed to varied small fragments of organic compounds like malondialdehyde. The ensuing substances from lipid peroxidation processes have shown an injurious impact on the semi-permeable membrane functions resulting in necrobiosis. In a follow-up study, the researchers demonstrated that the appliance of various levels of multiple heavy metal stresses (such as Pb^{2+}, Cd^{2+}, and Hg^{2+}) accelerated the lipid peroxidation of *Bruguiera gymnorrhiza* plants (Zhang et al., 2007).

3.9.3 IMPACT OF HEAVY METALS ON PHOTOSYNTHESIS

Chemical change is one of the most vital physiological options of plant systems. Once plants are fully grown in metal-enriched soils, the method of chemical change is adversely affected by significant metals. The interaction between virulent metals and therefore the chemical change reactions could lead to: (i) the accumulation of metals in leaves and roots, (ii) the alteration of the useful activities of chloroplasts that are teeming with several leaf tissues like stomata, mesophyll, and bundle sheath, (iii) the metal interaction with cytosolic enzymes and organics, (iv) significant effects at the molecular level on growth response and chemical change activity on photosystem I, photosystem II, membrane radical liquids, and carrier proteins in vascular tissues, and (v) the distortion of enzymes concerned with chemical change cycles like photosynthetic carbon reduction (PCR) and xanthophylls. In this context, it has been demonstrated that elevated concentrations of metals like Cu have an injurious impact on the formation of chemical changing pigments. In another study, the photosynthetic process, just like the one concerned within the reduction of carbon once the legume (green gram) was fully grown in heavy metal-contaminated soil, was additionally affected (Maksymiec et al., 1995). Similarly, a previous study has noted that surplus concentrations of Cu^{+2} changed the ultrastructural responses of plastids in runner beans (*Phaseolus coccineus* L.) (Maksymiec et al., 1995). The pigment was reduced once plants were fully grown in metal-treated soils, as observed by others (Wani et al., 2008). Shanker (2003) has considered that the decrease within the pigment a/b quantitative relation following the Cr^{+2} (VI) toxicity in plants was because of the destabilization and degradation of the proteins of the peripheral half. Once significant metal stress is applied, enzymes that mediate the chain reactions of pigment synthesis pathways are inactivated, which may additionally contribute to the final reduction within the pigment content in most of the plants, including legumes.

Furthermore, the impact of Cd^{2+} and Cu^{2+} metal stress on photosystem II activity has been shown in several studies. The Cd^{+2} affects both the PS II reaction center and the light harvesting complex (LHC), which can lead to inefficient energy transfer from the LHC to the reaction center. In general, intensive studies are being conducted to ascertain the Cd^{+2}-mediated inhibition of the reactions of chemical change. Recent studies have revealed that once Cd^{+2} is applied on protoplasts it has a big impact on the Calvin cycle, but it has no impact on Rubisco. In another study, Sheoran et al. (1990) have elicited a notable reduction within the Rubisco activity of pigeon pea plants once treated with Cd^{+2} at an early growth stage. However, Rubisco activity was not affected in older plants.

3.9.4 PHYTOREMEDIATION

Phytoremediation (phyto, which means "plant," and remedium, which means "restoring balance") is a recently-developed cleanup technology that involves the utilization of plants and their associated rhizospheric microbes for treating environmental contaminants like heavy metals, organic compounds, or hot parts in soil, groundwater, or industrial waste. The general advantages of phytoremediation is that it is non-invasive, eco-friendly, energy economical, and cheaper than different strategies like soil excavation, soil laundry, and burning and therefore created an opportunity for metal utilization (Saifullah et al., 2015). A plant's age, its surroundings factors, microorganism organization, the size of the metals, and the translocation to completely different elements are the crucial factors that have an effect on the uptake of heavy metals to the plants. Some tree species, like *Populus* and *Salix* sp., have shown to fulfill all of these uptake requirements, as noted by several researchers (Sebastiani et al, 2004). These trees have a high biomass production, intensive roots, high rates of transpiration, and simple propagation. At the same time, many studies have disclosed noteworthy organism variability in their ability to tolerate heavy metals (Utmazian et al., 2007). Phytoremediation takes advantage of the very fact that a living plant may be thought of as a solar-driven pump, which might extract and accumulate parts from the environment.

3.9.5 PHYTOREMEDIATION: MECHANISM

Plants exhibit completely different physiological mechanisms for dealing with excess metals. These mechanisms control the number of metal uptake by the plants. Metal uptake in roots is regulated by the exudation of organic acid ions, the binding impact of the cytomembrane, and therefore the flux of the ions through plasmalemma metal transporter proteins. In protoplasm, metals are chelate and transported towards organelles by peptidic chelators. Correspondingly, far more antimonial ions are directed to the apoplast by membrane transporters. Metals are mobilized through the vascular tissue from roots to aerial structures in a method driven by transpiration. A coordinated system of trans-membrane proteins and chelators drives metals towards their ultimate destination inside the plant cell. An extra defense line against metal-evoked ROS involves enzymes and reducing metabolites. The response to metal stress includes the release of general defense proteins and signal parts such as calcium and ethylene.

3.9.6 ROOT UPTAKE

The first stage of heavy metals uptake involves particle absorption from the soil and distribution into the foundation cells. There are many compounds which will perform the transportation and accumulation of heavy metals in tissues and alternative locations like metal ligands, e.g., organic acids (OAs) like citrate, malate, and oxalate. International organization conjointly plays an alternate role of excluding metals from plants (Guerra et al., 2011). In wheat, aluminum (Al) uptake is reserved by the exudation of international organization that forms the OA-Al complicated. Similarly, in case of *P. tremula* roots, copper (Cu^{+2}) uptake is reserved by the exudation of formate, malate, and salt, whereas zinc (Zn^{+2}) uptake is reserved by the exudation of formate. Siderophores like mugineic acid and avenic acids are released by some plant species to boost the bioavailability of heavy metals from the soil for root uptake, as reportable in certain grass species.

3.9.7 VASCULAR SEQUESTRATION

After uptake from the root, heavy metals are typically sequestered within the vacuoles of the plant cell. Through ZIP (zinc-regulated, iron-regulated transporter-like proteins) transporters, heavy metal enters the cytoplasm that stimulates phytochelatin synthase (PCS) – the accelerator that catalyzes the synthesis of phytochelatins (PCs) from glutathione. Heavy metal phytochelatin complex is low relative molecular mass complex that are transported to vacuole via tonoplast localized ATP-Binding Cassette (ABC) transporters. Heavy metals are sequestered within the vacuole by tonoplast-localized cation/proton exchanger (CAX) transporters that direct the exchanges of heavy metals with protons. Within the vacuole, low-molecular-weight (LMW) heavy metal phytochelatins complex accumulates into high-molecular-weight (HMW) complex with high levels of heavy metals. Heavy metals might enter the vacuole via an immediate exchange mechanism of various heavy metal-protons money changer transporters like metal tolerance proteins (MTPs) and natural resistance-associated macrophage proteins (NRAMPs). These transporter proteins reside within the tonoplast and mediate the passage of metal ions for compartmentation or remobilization (Yang & Chu, 2011).

3.9.8 METAL UPTAKE ENHANCEMENT

Mycorrhiza is the dependent association between fungi and plants. This association protects a plant from significant metal pollution by binding the metals into the cytomembrane system or by storing high amounts of heavy metals in the cytoplasm. Mycorrhizae also release growth-stimulating substances for plants, thence encouraging the mineral nutrition, augmented growth, and biomass necessary for phytoremediation (Baker, 2000).

Smith & Read (2010) demonstraed that mycorrhiza is found in those plants that are fully grown in natural conditions. Mycorrhiza is shown to play an important role in protecting plant roots from significant metals, as reportable by Galli et al. (1994). However, the level of protection provided varies from species to species and also depends on the kind of heavy metals and the mycorrhizal. The extrametrical plant hyphae will extend deep into the soil and uptake massive amounts of nutrients, as well as significant metals, to the host root spores of *Arbuscular mycorrhizal* plant taxa like *Glomus* and *Gigaspora*. Pawlowska et al. (1997) recovered spores of *Glomusaggregatum, G. fasciculatum,* and *Entrophospora* sp. from the mycorrhizospheres of the plants whereas learning a mineral soil made in cadmium (Cd^{+2}) metal and metallic element at European country. Turnau (1998) studied the localization of significant metals among the plant structures and mycorrhizal roots of *Euphorbia cyparissias* from metallic element-contaminated wastes, and located higher concentrations of metallic element deposited among the plant structure and animal tissue cells of mycorrhizal roots. It was found that *Arbuscular mycorrhiza* flora will transport cadmium (Cd^{+2}) from the soil to subterranean herb plants growing in compartmentalized pots, however, this transfer is restricted because of plant compartmentalization.

3.9.9 METAL COMPARTMENTALIZATION

At the cellular level, cell walls bind to the metal ions helping them reach the protoplasm by ion exchange. Metals will bind to either cellulose or proteins as salt enzymes. Metals will diffuse into the apoplast of some root cells, however, its transport is blocked by the rubber casparian strip within the endodermis layer. At this time, plants have a number of metal transporters involved in the metal absorption and physiological processes that govern their movement towards the symplast and eventual filling up into the vascular tissues (Palmer & Guerinot, 2009).

Now, within the plasmodesmata, metals are transported by heavy metal ATPases (HMAs), ZIP, COPT-type transporters, and cation antiporters. Cadmium (Cd^{+2}) and metallic elements are chemically similar indicating an identical uptake and transport pathway (Obata & Umebayashi, 1993).

3.9.10 GLUTATHIONE (GSH) AND PHYTOCHELATINS (PCS)

In heavy metals toxicity, glutathione acts as a matter sequestration and relieving the aerophilous stress caused by metals. Researchers found an increase within the reduced sort of GSH up to 30 fold against cadmium (Cd^{+2}) toxicity in *Phragmites Australis* (Pietrini et al., 2009). However, in some reports no such incremented GSH synthesis was ascertained, which may conclude that glutathione has no direct role in heavy metal detoxification, and acts through the formation of Phytochelatins. Phytochelatins have a role in the detoxification of heavy metals in plants, as they do in different organisms, where they acts as matter to bind with these heavy metals to create complexes that are signaled for compartmentalization. The term "Phytochelatin" is also related to microorganisms. They're additionally said to be found in roundworm *C. elegans*, slime molds *Dictyostelium*, and aquatic midge *Chironomus oppositus*, as reviewed by Cobbett (2000).

Phytochelatins identified as class III metallothioneins (MTs) are polypeptides accompanied by a Gly terminal with Glu-Cys dipeptide. (-Glu-Cys)-n-Gly, where n>2, is present in many plants and microorganisms and has a similar structural to glutathione (GSH; -Glu-Cys-Gly) and is synthesized by the catalyst phytochelatin synthase that is activated by heavy metal particles and leads to the vacuolar sequestration of heavy metals and even have a role within the physiological state of essential metal ion metabolism (Cobbett, 2000).

3.9.11 METALLOTHIONEINS (MTS)

MTs are low-molecular-mass, cysteine-rich, metal-binding proteins that are used for heavy metal detoxification by intracellular sequestration. These chelators bind to the metals and type a fancy that is transported to the vacuole. As an example Zinc (Zn^{+2}) is transported into the vacuole by MTPs (metal tolerance proteins). MTP1 and MTP3 localize at the vacuolar membrane and are used to form the plant metallic element tolerance (Gustin et al., 2009). Within the case of Cd, AtHMA3 plays a role in its accumulation in the vacuole. For copper (Cu^{+2}), transporters like PAA1 (HMA6) [P-type ATPase of Arabidopsis1 (heavy metal ATPase6)], PAA2 (HMA8) [P-type ATPase of Arabidopsis2 (heavy metal ATPase8)], and HMA1 (heavy metal ATPase1) are important for the transportation of copper (Cu^{+2}) into plastocyanin within the plastid. Copper (Cu^{+2}) may also be transported into the mitochondria once it enters the metastasis negatron transport chain. Then intracellular distribution of metals is completed by chaperons. Metal chaperones along with ATPases assist in the detoxification of heavy metals in roots (Andrés-Colás et al., 2006).

3.9.12 METAL TRANSLOCATION TO SHOOTS AND SHOOT METABOLISM

Heavy metals are transported from a plant's roots to its dermal tissues, then to the pericycle or vascular tissue parenchyma, and eventually loaded into the vascular tissues through transmembranes (Palmer & Guerinot, 2009). In *Dilleniid dicot* genus, ATPases HMA2 (heavy metal ATPase2) and HMA4 (heavy metal ATPase4) are responsible for transporting and accumulating metallic element from the roots to the shoots (Wong & Cobbett, 2009), and ATPase HMA5 (heavy metal ATPase5) is concerned with copper (Cu^{+2}) transportation. The translocation of metals additionally involves amino and organic acids, e.g., Cd^{+2}, Cu^{+2}, and Zn^{+2} need change state, malate, histidine, nicotinamine, etc. In shoots, excess heavy metals will cause aerophilous stress and harm to the exposed cells by the replacement of metal ions in the pigment and other essential molecules related to pigment. Several photo system parts are affected as a result of heavy metal toxicity that disturbs the chemical process. The redox active metals (Cu^{+2}) and non-redox active metals (Cd^{+2} and Zn^{+2}) each will cause aerophilous harm. To forestall the cells from this harm, plants have an inherent antioxidative system that supports reducing metabolites (GSH) and enzymes (peroxidase) that regulate chemical reactions. Glutathione is an elementary molecule that is synthesized from Glu, Cys, and Gly by glutamyl cysteine synthetase and GSH synthetase. Glutathione may be a precursor of phytochelatins (PCs). It will bind to metals and metalloids, and eliminates reactive oxygen radicals elicited by heavy metals in cells and maintains the chemical reaction physiological state for metabolism, signal transduction, and organic phenomenon. The chemical element, strontium, may be a dangerous and inhibits the expansion and development of plant; the response of *Phaseolus mungo* L. to strontium concentration was studied by Meena et al. (2011).

3.9.13   Classification of the Phytoremediation Process

Phytoremediation can be divided into phytofilteration, phytoextraction, phytodegredation, phyto-stabilization, and phytovolatisation (see Figure 3.3). Each will be discussed below.

3.9.13.1   Phytoextraction

Phytoextraction, also referred to as phyto accumulation, is employed by the roots to absorb contamination from soil and water and to translocate contaminants to the shoots and leaves. Hyperaccumulation plants absorb heavy metals up 50–500 times more than non-hyperaccumulation plants with no adverse result on growth and development (Mahar et al., 2016). These plants are usually small and slow growing, and infrequently rare species of restricted population size and restricted distribution in scheme.

Currently, over 450 plant species in 45 families are categorized as hyperaccumulators and represent only 0.2% of all angiosperms (Khalid et al., 2017). These plants form part of the following main families *Brassicaceae, Asteraceae, Fabaceae, Lamiaceae, Poaceae, Euphorbiaceae, Caryophyllaceae,* and *Violaceae*. It has been reported that in contrast to traditional plants, wild hyperaccumulators proliferate roots in patches of high metal convenience. Different species of British genus *Thlaspi caerulescens* plants were used as zinc hyperaccumulators; however, a high uptake of cadmium, cobalt, manganese, and nickel has been observed with the same mechanism of absorption and transportation. Over one metal accumulator conjointly determined in several alternative species of family *Brassicaceae* like leaf mustard, *Brassica napus, Crassulaceaeie,* and *Sedum alfredi.*

Some hyperaccumulators are resilient to higher nickel (Ni^{+2}) concentrations; Ni enters the vascular tissue of roots and transport apace to the shoot vascular tissue alongside simpler transpiration processes. *Brassicacaddii* from Central Africa has a high metallic element tolerance within the soil setting. It effectively absorbs and translocates metals into the shoots, resulting in high concentrations in the leaves (Mahar et al., 2016). Cobalt (Co^{+2}) is additionally accumulated by *Brassicacaddii*

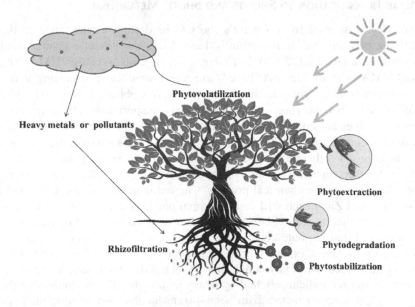

FIGURE 3.3   Different strategies used by plants for the bioremediation of heavy metals.

in both the presence or absence of nickel, but cobalt inhibits nickel absorption. Some tree species, i.e., genus *Populus* species and *Salix* species, are extensively used for zinc and cadmium accumulation from contaminated soils.

Different gene-splicing approaches are used to enhance the ability of hyperaccumulators. Transporter macromolecules like the cation diffusion facilitator (CDF), ZIP, and iron-regulated transporter protein (IRTP) are associated with the accumulation of heavy metals in numerous plant species. Transgenic *Nicotiana tobaccum* increases cadmium (Cd^{+2}) uptake and tolerance, lead (Pb^{+2}) in *Nicotianaglauca*, zinc (Zn^{+2}) in *Lactuca sativa* and *Branicaoleraea*, and arsenic (As) in *Arabidopsis thaliana* (Guo et al., 2012).

3.9.13.2 Phytofiltration

Phytofiltration is a technique used by plants to get rid of impurities from well water and contaminated wastewater. For filtration, different plant elements are used: roots (rhizofiltration), seedlings (blastofiltration), and excised plant shoots (caulofiltration). Phytofilterates may be aquatic, semiaquatic, or terrestrial and have a slow growth and economical metal-binding capability (Singh et al., 2017). Plants grown in husbandry are economical in the use of rhizofiltration to soak up contaminants, as opposed to typical water plants.

Callitrichecophocena effectively treats water contaminated with Thalium (Tl), Cd^{+2}, Zn^{+2}, and Pb^{+2}; *Juncus acutus* is employed for Cr^{+2} contaminated well water (Dimitroula et al., 2015); and *Plectranthus amboinicus* shows tolerance to Pb^{+2}. Two aquatic macrophytes, *Pistia stratiotes* and genus *Azolla*, have been found to remove mercury (Hg^{+2}) pollution from coal mining effluent. *Cladophora algae* has been used treat arsenic-contaminated water. Some terrestrial plants, like sunflowers and *Brassica junecea*, are able to remove heavy metals contaminate from water (Singh et al., 2017).

3.9.13.3 Phytostabilization

Phytostabilization, also known as phytosequestration or phytodeposition, deals with fixing or sequestering pollutants in soil close to the basis, but not in plant tissues that stop heavy metal migration to either well water or in organic phenomenon. Recently, two grasses i.e., genus *Agrostis* species and genus *Festuca* species, were employed in the phytostabilization of Cu^{+2}, Zn^{+2}, and Pb^{+2} contaminated soil from Europe, China, and America. The mixture of grass and tree co-cultivation exhibits photostabilization efficiency. Tree plantations may be useful in scaling back eating away, forestalling water erosion, immobilizing the pollutants by accumulation, and supplying a space around the roots wherever pollutants are present. For instance, spruce (*Picea abien* L.) roots minimize the phytostabilization capability as a result of trace components like Cd^{+2}, Cu^{+2}, Zn^{+2}, and Pb^{+2} being absorbed by their roots. Thus, phytostabilization is different from alternative approaches; in contrast to alternative approaches, it is not a permanent answer and is employed mainly to limit movements of heavy metals. It is a management strategy in the inactivation of toxins.

3.9.13.4 Phytovolatization

Phytovolatization, also referred to as phyto evaporation, is a process where plants uptake volatile organic pollutants and certain heavy metals like mercury, selenium, and arsenic from the soil. It is a contentious technique as it limits the removal of pollutants from the soil; pollutants are transferred from one section (soil) to another (atmosphere) from wherever it will redeposit. Leaf mustard is employed to get rid of selenium from within the soil. It converts the selenium into volatile methyl sclenate and is eliminated. *Astragalus biscularts* is also used for selenium evaporation, where selenium converts into methyl selenocyanate by mistreatment selenocysteine methyltransferase catalyst (Hooda, 2007). In *Pteris vittata* as was effectively gaseous within the kind of arsenite/arsenate. Transgenic *Arabdopsis* and tobacco plants are built with microorganism genes (merA and merB) (mercury reductase) which may volatize mercury nearly 10–100 times more than wild plants.

3.9.13.5   Phytostimulation

Phytostimulation, also known as rhizodegradation, entails microorganisms that are used to break down organic pollutants within the soil. Microorganisms enhance metal convenience and the quality of soil, which helps plants to grow well even under metal strain conditions. Bacteria like *Bacillus mucilaginosus* (K-soluble), *Bacillus megaterium* (P-soluble), and *Azotobacter chroococcum* (Nitrogen-fixing) can facilitate plant growth in many ways, i.e., by lowering the pH scale, and by manufacturing plant growth rules like indoleacetic acid (IAA) and metal chelating compounds like siderophores and biosurfactants (Ullah et al., 2015).

Lead (Pb^{+2}) accumulation is increased by up to 131% with the use of *Pseudomonas* glow, and up to 80% with *Microbacterium* species in rape shoots. Additionally, in secreting extractible metals and organic substrates to facilitate plant growth and development by organisms, plants release certain enzymes to degrade organic contaminants in soils.

3.9.14   LIMITATION OF PHYTOREMEDIATION

Though phytoremediation is non-destructive and uses solar-driven strategies for the aggregation and removal of contaminants from the soil, it does have some drawbacks (Ramamurthy & Memarian, 2012):

1. Time consuming (several years) for cleanup.
2. Slow growth and development of many hyperaccumulators.
3. Biotic factors and disease attack may compromise the accumulation capacity of hyperaccumulators.
4. Climate and weather conditions affect hyperaccumulator plant performance.
5. Only effective in cases of low and moderated levels of contamination.
6. Limited bioavailability, hard to be mobilized by more tightly bound fractions of metal ions from the soil.
7. Risk of food chain contamination in the case of mismanagement.

3.10   CELLULAR MECHANISMS FOR HEAVY METAL DETOXIFICATION AND TOLERANCE

3.10.1   THE CELL WALL AND ROOT EXUDATES

The binding property of a cell wall and its role as a mechanism of metal tolerance has been an arguable one. Though the basis cell membrane is directly in contact with metals within the soil resolution, sorption into the cell membrane is restricted and so has a restricted impact on the metal activity at the surface of the semi-permeable membrane. It is therefore difficult to elucidate metal-specific tolerance by such a mechanism. Nevertheless, Bringezu et al. stated that the heavy metal-resistant *Silene vulgaris* ssp. *humilis* acquired a number of metals inside the epidermal cell walls, either protein-guaranteed or as silicates (Bringezu et al., 1999). One connected method considers the role of root exudates in metal tolerance. Root exudates have a variety of roles as well as that of metal chelators which will enhance the uptake of bound metals. In an investigation into the role of Ni-chelating exudates in Nickel hyperaccumulating plants, it had been ascertained that the Ni-chelating histidine and citrate accumulated within the root exudates of non-hyperaccumulating plants, and so might facilitate in limiting nickel uptake, thereby playing a role in the Ni-detoxification strategy. The variety of compounds exuded is wide, and different exudates play a role in the tolerance of different metals. The clearest example is root secretion intolerance of relevant organic acids and therefore the detoxification of the metal aluminum (Al) (Ma et al., 2001). Buckwheat, for instance, secretes oxalic acid from the roots in response to Al stress, and accumulates non-toxic Al-oxalate within the leaves, so detoxification happens both outwardly and internally. In wheat and maize, there is evidence that such secretion from the roots is mediated by Al-activated ion channels within the semi-permeable membrane (Ma et al., 2001).

3.10.2 PLASMA MEMBRANE

A plant's semi-permeable membrane could be thought to be the primary "living" structure that is a target for heavy metal toxicity. The semi-permeable membrane functions could also be harmed by heavy metals, as seen by an inflated escape from cells within the presence of high concentrations of metals, notably of Cu^{+2}. For instance, it has been shown that Cu^{+2}, but not Zn^{+2}, causes increased K^+ (potassium) efflux from the excised roots of genus *Agrostis capillaries*. Similarly, injury to the cytomembrane, monitored by particle escape, was the first explanation for Cu^{+2} toxicity in roots of *Silene vulgaris, Mimulus guttatus*, and wheat. Such injury might result from varied mechanisms as well as the oxidization and crosslinking of protein thiols, the inhibition of key membrane proteins like the Hq-ATPase, or changes to the composition and thinness of membrane lipids. It has been reported that the direct effects of Cu^{+2} and Cd^{+2} treatments on the lipid composition of membranes can have a direct impact on membrane permeability. Additionally, Cd^{+2} treatments are shown to cut back the ATPase activity of the semi-permeable membrane fraction of wheat and sunflower roots whereas, in *Nitella*, Cu-induced changes in cell permeability are attributed to non-selective electrical phenomena which increase and inhibition of the light-stimulated Hq-ATPase pump. Thus, tolerance involves protecting the integrity of the semi-permeable membrane against heavy metal injury that might result in the increased escape of solutes from the cells. However, there is very little proof to identify how this can be achieved. For instance, metal-tolerant plants do not seem to possess a high tolerance to free radicals or ROS, however, they possess improved mechanisms for metal equilibrium. Once more, these effects on membranes are metal-specific since, in comparison to Cu, zinc protects membranes against oxidization and customarily does not cause membrane escape (Ernst et al., 1992). Another issue concerned with the maintenance of semi-permeable membrane integrity in the presence of heavy metals is increased membrane repair once injury occurs. This might involve heat shock proteins or metallothioneins, and evidence for this will be mentioned in the sections that follow.

Apart from tolerance involving an additional resistant cytomembrane or improved repair mechanisms, the cell wall plays a crucial role in metal equilibrium, either by preventing or reducing entry into the cell or through effluence mechanisms. Several of those cations, of course, are essential to the process of hindering complete exclusion; selective effluence could also be an option. In bacterium, most resistance systems are supported by the energy-dependent effluence of harmful ions (Silver, 1996). It seems that the metabolic penalty for having additional specific uptake mechanisms, and so limiting the entry of harmful ions, is larger than that of getting inducible effluence systems (Silver, 1996).

Examples of exclusion or reduced uptake mechanisms in higher plants is rather restricted. The clearest example of reduced uptake as a custom-made tolerance mechanism is in significant arsenic toxicity. In *Holcus lanatus* roots, phosphate and salt seem to be obsessed by similar systems. However, an arsenate-tolerant genotype showed a far lower rate of uptake for each anion than the non-tolerant genotype, and additionally showed an absence of the high-affinity uptake system. The altered phosphate and salt uptake system were genetically correlated to salt tolerance. Additional work has advised that salt tolerance in *H. lanatus* needs this accommodative suppression of the high-affinity transport system in conjunction with essential phytochelatin (PC) production since arsenate will still accumulate to high levels in intolerant plants (Hartley-Whitaker et al., 2001). PCs are covered in detail in a later section. A cytomembrane transporter in tobacco that confers nickel tolerance and lead hypersensitivity has been represented by Arazi et al. (1999). The transporter, selected not CBP4, could be a calmodulin-binding protein that is structurally similar to bound K^+ and non-selective ion channels. Transgenic plants that over-expressed this transporter showed improved Ni tolerance and hypersensitivity to Pb^{+2} that were related to reduced Ni accumulation and increased lead accumulation. Though the conventional physiological operatation of this transporter remains to be established, it may provide a feasible mechanism for Ni tolerance.

An alternative strategy to combat high metal levels in the cytomembrane involves the active effluence of metal ions, though there is little or no evidence for such a method in plants. However, in bacterium, effluence pumping forms the basis of most harmful-particle-resistance systems, involving transporters like P-type ATPases or cation/H+ antiporters (Silver, 1996). Effluence pumping systems are associated with metals like copper, cadmium, zinc, cobalt, and nickel (Silver, 1996). Effluence transporters play a role in the metal particle equilibrium in animal cells. As an example, a cytomembrane zinc transporter (ZnT-1) was isolated from a rat kidney. Cells remodeled with a mutant ZnT-1 lacking the primary membrane-spanning domain showed zinc sensitivity; normally the ZnT-1 transports zinc out of the cells but in its absence produces hyperbolic sensitivity of the mutant cells to zinc toxicity. It has been thought that zinc effluence involves some form of secondary transport. Another cluster of transporters that seems to be involved in metal equilibrium by a copper export system is the significant metal CPx-ATPases, a branch of the P-type ATPases. Defects in these ATPases are connected to two human disorders, Menkes disease and Wilson disease, which result from defective metal exportation resulting in the build-up of copper in some tissues. In Chinese rodent ovary cells, there is proof that the Menkes P-type ATPase incessantly recycles from histologist to the cytomembrane. Elevated concentrations of metal shift the distribution of the ATPase from the Golgi to the plasma membrane, resulting in the effluence of harmful matter.

Although there is no evidence supporting the role of cytomembrane effluence transporters in significant metal tolerance in plants, the analysis has shown that plants possess many categories of metal transporters that are involved in the metal uptake and equilibrium processes, and so may play a key role in intolerance. These include the heavy metal CPx-ATPases, natural resistance-associated macrophage proteins (Nramps), the CDF family, and the ZIP family. The role of Nramps in iron and cadmium uptake has been studied by Thomine et al. (2000). It was found that the disruption of *AtNramp* caused three gene to be slightly hyperbolic cadmium resistant, whereas over expression resulted in Cd hypersensitivity in *Arabidopsis*. Within the zinc/cadmium hyperaccumulator *Thlaspi caerulescens*, Pence et al. (2000) cloned a ZNT1 transporter which mediates high-affinity metallic element uptake moreover as low-affinity cadmium uptake, and is expressed at high levels within the roots and shoots. Hyperbolic expression ensued from changes within the plant metallic element standing, semiconductor diode to a raised metallic element flow within the roots. However, the transportation, specificity, and cellular location of most of those proteins in plants are unknown. From the proof gathered for microorganisms and class systems, the CPx-ATPases and also the CDF family (which includes the ZnT efflux transporters of humans and rodents) would appear to be the best mechanisms for metal effluence tolerance.

3.10.3 HEAT SHOCK PROTEINS

Heat shock proteins (HSPs) characteristically show augmented expression in response to the expansion of a range of organisms at temperatures higher than their optimum growth temperature. They are grouped together as teams of living organisms, classified in line with molecular size, and are known to react in response to a range of stress conditions together with heavy metals (Lewis et al., 1999). HSPs act as molecular chaperones in traditional folding and assembly, however, they might also operate within the protection and repair of proteins under stressful conditions.

There are many reports of a rise in HSP expression in plants in response to heavy metal stress. Tseng et al. (1993) noted the amount of mRNAs for low-molecular-mass HSPs (16–20 kDa) was increased in every heat and heavy-metal stress in rice, whereas Neumann et al. (1995) indicated that HSP17 is expressed in the roots of *Armeria maritima* adult plants in Cu-rich soils. Tiny heat shock proteins (e.g., HSP17) were additionally shown to extend in the cell cultures of *Silene vulgaris* and *Lycopersicon peruvianum* in response to a spread of heavy metal treatments but no, or very low amounts of, HSPs were found in plants growing in metalliferous soils, suggesting that HSPs do not seem to be responsible for the campion metal tolerance of catchflies.

Working with the cell cultures of *L. peruvianum*, it was shown that a bigger HSP (HSP70) responds to cadmium stress (Neumann et al., 1994). It is of interest to note that protein localization showed that HSP70 was present within the nucleus and living substance, in addition to be present in the cell wall. This implies that HSP70 might be concerned with the protection of membranes against cadmium injury. Expression of HSP70 additionally augmented within the algae *Enteromorpha intestinalis* when exposure to a range of stressors together with metal. Thus, in reference to earlier discussions of tolerance mechanisms involving an additional resistant cell wall or improved repair mechanisms, HSPs might have an important role in this respect. It has been noted that a brief heat stress given before heavy-metal stress induces a tolerance result by preventing membrane injury, as observed in ultrastructural studies (Neumann et al., 1994). Clearly, additional molecular proof is needed to support such an important repairing or protecting role.

3.10.4 PHYTOCHELATINS

Chelation of metals within the cytoplasm by high-affinity ligands is a vital mechanism of heavy metal detoxification and tolerance. Certain ligands embody amino acids and organic acids, and two categories of peptides: the phytochelatins (PCs) and metallothioneins. The PCs are the most widely studied peptides in plants relating to cadmium tolerance (Cobbett, 2000).

PCs are part of a family of metal complexing peptides that have a general structure (c-Glu Cys) n-Gly wherever n¼ 2–11, and are speedily elicited in plants by heavy metal treatments (Cobbett, 2000). Computers are synthesized non-translationally exploitation glutathione as a substrate by PC synthase. A catalyst that has been activated within the presence of metal ions. The genes for phytochelatin have recently been discovered in *Arabidopsis* and yeast.

Evidence has been given for and against the role of PCs in heavy metal tolerance (Cobbett, 2000). However, a clear role in cadmium detoxification has been supported by chemical science and genetic proof. Howden et al. (1995) isolated a series of cadmium-sensitive mutants of genus *Arabidopsis* that varied in their ability to accumulate PCs; the amount of PCs accumulated by the mutants correlated with the degree of sensitivity to cadmium. In Indian mustard, it has been shown that cadmium accumulation is within the middle of a quick induction of PC synthesis that the PC content was on paper comfy to chelate all cadmium taken up. This protects the natural process but does not forestall a decline in the transpiration rate. Again, Inouhe et al. (2000) showed that refined cells of Azuki beans that were hypersensitive to cadmium lacked PC synthase activity. Using genus *Arabidopsis*, Xiang & Oliver (1998) showed that treatment with cadmium and number 29 resulted in the raised transcription of the genes for glutathione synthesis, and thus the response was specific for those metals thought to be detoxified by PCs. Curiously, jasmonic acid treatment activated a constant set of genes, although jasmonic acid production was not stirred by heavy metals in plant cell cultures. Zhu et al. (1999) over-expressed the c-glutamyl cysteine synthetase gene from *E. coli* in Indian mustard resulting in the increased synthesis of glutathione and PCs and a higher tolerance to cadmium. A consistent approach favored *Arabidopsis*; c-glutamylcysteine synthetase was expressed in every sense and antisense orientations resulting in plants with a decent variety of glutathione levels. Plants with low glutathione levels were allergic to cadmium, although elevating the levels in wild-type plants did not increase metal resistance.

Recently, the secret genetic coding for PC synthases in higher plants and yeast has been identified, and it has been shown that the genus *Arabidopsis* cistron could confer a substantial increase in metal tolerance in yeast. The gene for PC synthase (*CAD1*) has been identified in *Dilleniid dicot* genus still as a homologous gene in genus *Schizosaccharomyces pombe*; a mutant of the latter with a targeted deletion of this cistron was PC deficient and cadmium sensitive. To examine the involvement of PCs in metal detoxification, the sensitivity of the cad1–3 mutant was tested for sensitivity to a range of heavy metals in genera *Arabidopsis* and *S. pombe*. PCs seem to play an important role in the detoxification of cadmium and salt, but have no role in the detoxification of zinc, nickel, and selenite ions. In distinction to the *S. pombe* mutant, cad1–3 showed slight sensitivity to copper and mercury. A possible role for PCs in copper tolerance had been suggested from studies on copper-tolerant *Mimulus guttatus*, where exposure to copper in the presence of buthionine sulphoximine (BSO), a potent substance of c-glutamyl-cysteinyl synthetase, caused a considerable reduction in root growth that was not seen in the presence of the substance alone. However, in distinction, once Cu^{+2}-sensitive and Cu^{+2}-tolerant ecotypes of campion were exposed to concentrations of Cu^{+2} giving either no or 50% inhibition of growth for each ecotype, they showed equal PC synthesis at the foundation tips. It fully was over that differential metal tolerance in *S. vulgaris* does not believe differential pc production. The role of PCs in copper tolerance remains to be full understood. A study on the involvement of PCs in salt tolerance has been planned.

Not all studies have supported the role of PCs in metal tolerance. De Knecht et al. (1994) discovered that differential cadmium tolerance in campion plants was not due to a differential production of PCs. While PCs perform a function in *S. vulgaris* cadmium purification, the larger-scale development of PCs is not the method that leads to the increased resistance of cadmiums. Again, treatment with the substance BSO was found to increase the Zn-tolerance of genus *Festuca rubra* roots, distinction of opinion against a key role for PCs at intervals the Zn-tolerance mechanism in these tissues. Thus, although proof for the role for PCs in detoxification is convincing, particularly for cadmium, these peptides may play different very important roles in the cell, furthermore as essential to heavy-metal physiological conditions, sulphur metabolism, or perhaps as anti-oxidants (Cobbett, 2000). Their participation in the detoxification of excess concentrations of some heavy metals might even be a consequence of these different functions. The role of PCs in adaptive tolerance has been questioned. It was suggested that the ultimate lack of proof of CO-tolerance indicates that adaptive tolerance is unlikely to be created by changes in relatively non-specific binding compounds like PCs (or metallothioneins or organic acids).

The final step in cadmium detoxification, definitely within the fission yeast and possibly in higher plants, involves the build-up of cadmium and PCs within the vacuole. This accumulation seems to be mediated by both a Cd^+/H^+ antiporter and an ATP-dependent first principle transporter, situated at the tonoplast. The stabilization of the cadmium-PC advanced within the vacuole involves the incorporation of acid-labile sulfide. Within the fission yeast, a cadmium-sensitive mutant has been isolated that is ready to synthesize PCs, however, is unable to accumulate the cadmium-PC-sulphide advanced; the mutant features a defect in a cistron (hmt 1) that encodes an ABC-type transporter. Similar transporters may possibly be concerned in cadmium compartmentalization in higher plants (Rea et al., 1998).

3.10.5 METALLOTHIONEINS

Higher plants contain two major kinds of cysteine-rich, metal-binding peptides, the metallothioneins (MTs) and phytochelatins (discussed above). MTs are gene encoded polypeptides that are typically classified into two categories. Category one MTs possess cysteine residues that align with a class (equine) excretory organ MT; category two MTs possess similar cysteine clusters, however, these cannot be simply aligned with category one MTs. MT genes have been identified in a wide variety of higher plants together with *Arabidopsis*, and in addition to category one and two of MT genes, MT3 and MT4 types are recognized (Goldsbrough, 2000). Different species also are thought to contain an in depth MT factor family and quite one category of MT factor, whereas expression studies have unconcealed tissue-specific patterns. In plants, there is a scarcity of data regarding the metals that can be removed by MTs, though copper, zinc, and cadmium are the most widely studied (Goldsbrough, 2000).

Although MTs are often elicited by copper treatments and there is proof for a role in heavy metal tolerance in fungi and animals, the role of MTs in heavy metal detoxification in plants remains to be established. However, it has been rumored that MT2 mRNA was powerfully elicited in *Arabidopsis* seedlings by copper, but only slightly by cadmium and zinc. Once genes for MT1 and MT2 from *Arabidopsis* were expressed in an MT-deficient yeast mutant, each gene complemented the mutation and provided a high level of resistance to Cu^{+2}. Van et al. (1995) showed that MT genes are often elicited by Cu^{+2}, where MT2 ribonucleic acid is accumulated in a copper-sensitive mutant of *Arabidopsis* that accumulates high concentrations of Cu^{+2}. Ten ecotypes of *Arabidopsis* were surveyed and a transparent correlation between the copper sensitivity of seedlings and therefore the expression of MT2 ribonucleic acid was shown. Clearly a lot of proof is required to ascertain a relationship between copper sensitivity and MT production. Against this, in a study of the consequences of cadmium exposure on Chinese mustard, it was confirmed that MT2 expression was delayed relative to PC synthesis and that they contended against a role for MT2 in cadmium detoxification. So the role of MTs needs to be clarified. Certainly they may be playing a role in metal metabolic pathways, but their specific output is not evident; they may have distinct roles for different metals. Instead, they can function as antioxidants, although there is no evidence, whereas a work in repairing cytomembranes would be another possibility.

3.10.6 ORGANIC ACIDS AND AMINO ACIDS

Carboxylic acids and amino acids like citric, malic, and histidine are potential ligands for heavy metals and may play a role in tolerance and detoxification; but evidence for tolerance, like a transparent correlation between thr amounts of acid made and exposure to a metal, has not been made to support a widespread role. As an example, a 36-fold increase was reported within the histidine content of the vascular tissue sap on exposure to metal within the nickel-hyperaccumulating plant *Alyssum lesbiacum*. Additionally, provision histidine to a non-accumulating species greatly accumulated both its nickel tolerance and the capability for nickel transport to the shoot. However, the histidine response might not be a widespread mechanism of metal tolerance since it was not determined in another nickel-hyperaccumulator, genus *Thlaspi goesingense* (Persans et al., 1999). A potential role for the histidine found in root exudates as a nickel-detoxifying agent has been mentioned earlier, as has the role of organic acids in aluminum tolerance (see section on cell membrane and root exudates).

3.10.7 VACUOLAR COMPARTMENTALIZATION

The efflux of ions at the cell membrane (discussed above) or transport into the vacuole are two ways of reducing the degree of harmful metals within the cytoplasm and are vital mechanisms for significant metal tolerance. One well-documented example, the build-up of cadmium and PCs within the vacuole involving the first principle transporter, has already been delineated (see section on PCs); however, there is proof that the vacuole could also be vital within the accumulation of different metals involving different tonoplast transport systems.

Earlier studies showed that the vacuole as a site for the build-up of a variety of significant metals together with zinc and cadmium. With the exception of the cadmium–PC accumulation process, zinc provides the most effective evidence for a function of vacuolar aggregation in relation to metal uptake. For instance, meristematic cells of *Festuca rubra* roots show magnified vacuolization on treatment with zinc, whereas uptake analysis victimization zinc-65 with barley leaves suggested that fast compartmentation of zinc into the vacuole was a very important mechanism for handling high levels of Zn^{+2}. Additional studies on barley leaves showed that, although cadmium, zinc, and molybdenum were found principally within the vacuole, nickel was primarily found within the cytoplasm, and this was thought to be associated with the event of leaf injury; however, compartmentalization within the roots was not examined. An analysis of the transport systems at the tonoplast has more support to a vacuolar mechanism of tolerance. Verkleij et al. (1998) isolated tonoplast vesicles from roots of zinc-tolerant and sensitive ecotypes of campion. They showed that at high zinc concentrations, zinc transport was up to five times higher in vesicles from the tolerant lines than from the sensitive ones, suggesting that the tonoplast plays an important role in naturally cho-sen zinc tolerance. In victimization plant crosses, this magnified tonoplast uptake system was shown to correlate genetically with zinc tolerance. Recently, A genus *Arabidopsis* gene *(ZAT)* was isolated which is closely associated with the animal ZnT (Zn transporter) genes (see section on plasma membrane). *ZAT* mRNA appeared to be expressed constitutively throughout the plant and was not elicited by higher zinc concentrations. However, over expression of *ZAT* in transgenic plants led to a big increase in zinc resistance and an increased accumulation within the root experiencing high zinc treatments. So the zinc transporter can be involved in the sequestration of zinc within the vacuole and therefore in zinc tolerance in plants.

Detailed data on different heavy metal transport systems at the tonoplast is limited. Two genes, *CAX1* and *CAX2*, are isolated from genus *Arabidopsis* and are shown to be vacuolar-located high and low potency H^+/Ca^{+2} exchangers. Whilst *CAX1* is believed to be concerned in vacuolar Ca^{+2} accumulation, it is confirmed that *CAX2* can act as a high capability H^+ heavy metal ion transporter. Though there is proof for military installation antiport systems for Ca^+ and Cd^{+2} in oat root tono-plasts, no proof was found for an Ni/H^+ antiporter or a nucleotide-dependent nickel pump, which confirmed that the vacuole is not a significant site for Ni accumulation in this tissue. Brune et al. (1995) came to an analogous conclusion regarding nickel in a study of the heavy metal compart-mentation in barley leaves.

3.11 CONCLUSION

This chapter highlighted the harmful effects of heavy metal contamination caused by certain human activities on the surroundings, the related health hazards, an the varied mechanisms and protein reactions utilized by plants and microbes to effectively remedy impure environments (see Figure 3.4). It revealed the quality of bioremediation as a suitable substitute for the removal of heavy metals from contaminated sites, compared to the physico–chemical ways that are less economical and costly due to the high levels of energy exhausted. Microorganisms and plants possess inherent biological mechanisms that alter them to survive under significant heavy metal stress by removing metals from their surroundings. These microbes use varied processes like precipitation, bioadsorp-tion, protein transformation of metals, complexation, and phytoremedation techniques of which

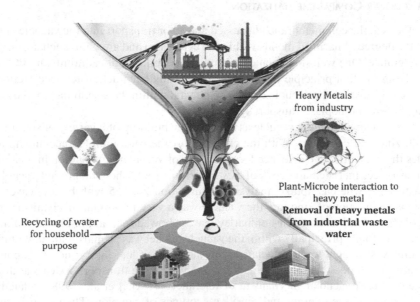

FIGURE 3.4 Effect of plant–microbe interaction on heavy metals.

phytoextraction and phytostabilization are extremely effective. The employment of hyperaccumu-lator plants to remedy contaminated sites depends on the level of metal contamination at that site and the variety of soil. Environmental factors play a crucial role in the success of bioremediation because the microbes used are hampered if specific environmental conditions are not out available.

REFERENCES

Agency, European Environment. (2007). Progress in management of contaminated sites (CSI 015). *Europe Environmental Assessment Agency, Kongan, 6DK-1050, Denmark [Internet].*

Aktan, Yasin, Tan, Sema, & Icgen, Bulent. (2013). Characterization of lead-resistant river isolate Enterococcus faecalis and assessment of its multiple metal and antibiotic resistance. *Environmental Monitoring and Assessment, 185*(6), 5285–5293.

Ali, Hazrat, Khan, Ezzat, & Sajad, Muhammad Anwar. (2013). Phytoremediation of heavy metals: concepts and applications. *Chemosphere, 91*(7), 869–881.

Andrés-Colás, Nuria, Sancenón, Vicente, Rodríguez-Navarro, Susana, Mayo, Sonia, Thiele, Dennis J, Ecker, Joseph R, . . . Peñarrubia, Lola. (2006). The Arabidopsis heavy metal P-type ATPase HMA5 interacts with metallochaperones and functions in copper detoxification of roots. *Plant Journal, 45*(2), 225–236.

Arazi, Tzahi, Sunkar, Ramanjulu, Kaplan, Boaz, & Fromm, Hillel. (1999). A tobacco plasma membrane calmodulin-binding transporter confers Ni2+ tolerance and Pb2+ hypersensitivity in transgenic plants. *Plant Journal, 20*(2), 171–182.

Baker, AJM. (2000). Metal hyperaccumulator plants: a review of the ecology and physiology of a biologi-cal resource for phytoremediation of metal-polluted soils. *Phytoremediation of Contaminated Soil and Water.*

Baker, Alan, Brooks, Robert, & Reeves, Roger. (1988). Growing for gold... and copper... and zinc. *New Scientist, 10*(1603), 44–48.

Bert, Valérie, Meerts, P, Saumitou-Laprade, Pierre, Salis, Pietrino, Gruber, Wolf, & Verbruggen, Nathalie. (2003). Genetic basis of Cadmium tolerance and hyperaccumulation in Arabidopsis halleri. *Plant and Soil, 249*(1), 9–18.

Bhargava, Atul, Carmona, Francisco F, Bhargava, Meenakshi, & Srivastava, Shilpi. (2012). Approaches for enhanced phytoextraction of heavy metals. *Journal of Environmental Management, 105*, 103–120.

Bouzon, Zenilda L, Ferreira, Eduardo C, Dos Santos, Rodrigo, Scherner, Fernando, Horta, Paulo A, Maraschin, Marcelo, & Schmidt, Éder C. (2012). Influences of cadmium on fine structure and metabolism of Hypnea musciformis (Rhodophyta, Gigartinales) cultivated in vitro. *Protoplasma, 249*(3), 637–650.

Bringezu, Kristina, Lichtenberger, Olaf, Leopold, Ines, & Neumann, Dieter. (1999). Heavy metal tolerance of Silene vulgaris. *Journal of Plant Physiology, 154*(4), 536–546.

Broos, Kris, Warne, Michael St J, Heemsbergen, Diane A, Stevens, Daryl, Barnes, Mary B, Correll, Raymond L, & McLaughlin, Mike J. (2007). Soil factors controlling the toxicity of copper and zinc to microbial processes in Australian soils. *Environmental Toxicology and Chemistry: An International Journal, 26*(4), 583–590.

Brune, Andreas, Urbach, Wolfgang, & DIETZ, KARL-JOSEF. (1995). Differential toxicity of heavy metals is partly related to a loss of preferential extraplasmic compartmentation: a comparison of Cd-, Mo-, Ni-and Zn-stress. *New Phytologist, 129*(3), 403–409.

Bünemann, Else K, Schwenke, GD, & Van Zwieten, L. (2006). Impact of agricultural inputs on soil organ-isms: a review. *Soil Research, 44*(4), 379–406.

Chui, SH, Wong, YH, Chio, HI, Fong, MY, Chiu, YM, Szeto, YT, . . . Lam, CWK. (2013). Study of heavy metal poisoning in frequent users of Chinese medicines in Hong Kong and Macau. *Phytotherapy Research, 27*(6), 859–863.

Cobbett, Christopher S. (2000). Phytochelatin biosynthesis and function in heavy-metal detoxification. *Current Opinion in Plant Biology, 3*(3), 211–216.

de Knecht, Joop A, van Dillen, Marjolein, Koevoets, Paul LM, Schat, Henk, Verkleij, Jos AC, & Ernst, Wilfried HO. (1994). Phytochelatins in cadmium-sensitive and cadmium-tolerant Silene vulgaris (chain length distribution and sulfide incorporation). *Plant Physiology, 104*(1), 255–261.

Dimitroula, Helen, Syranidou, Evdokia, Manousaki, Eleni, Nikolaidis, Nikolaos P, Karatzas, George P, & Kalogerakis, Nicolas. (2015). Mitigation measures for chromium-VI contaminated groundwater–the role of endophytic bacteria in rhizofiltration. *Journal of Hazardous Materials, 281*, 114–120.

Ernst, WHO, Schat, H, & Verkleij, JAC. (1990). Evolutionary biology of metal resistance in Silene vulgaris. *Evolutionary Trends in Plants, 4*(1), 45–51.

Ernst, WHO, Verkleij, JAC, & Schat, H. (1992). Metal tolerance in plants. *Acta Botanica Neerlandica, 41*(3), 229–248.

Frostegård, Åsa, Tunlid, Anders, & Bååth, Erland. (1993). Phospholipid fatty acid composition, biomass, and activity of microbial communities from two soil types experimentally exposed to different heavy metals. *Applied and Environmental Microbiology, 59*(11), 3605–3617.

Galli, Ulrich, Schüepp, Hannes, & Brunold, Christian. (1994). Heavy metal binding by mycorrhizal fungi. *Physiologia Plantarum, 92*(2), 364–368.

Glick, Bernard R. (2010). Using soil bacteria to facilitate phytoremediation. *Biotechnology Advances, 28*(3), 367–374.

Goldsbrough, Peter. (2000). 12 Metal tolerance in plants: the role of phytochelatins and metallothioneins. *Norman Terry*.

Guerra, Lina, Guidi, Riccardo, Slot, Ilse, Callegari, Simone, Sompallae, Ramakrishna, Pickett, Carol L, . . . Sjögren, Camilla. (2011). Bacterial genotoxin triggers FEN1-dependent RhoA activation, cytoskeleton remodeling and cell survival. *Journal of Cell Science, 124*(16), 2735–2742.

Guo, Jiangbo, Xu, Wenzhong, & Ma, Mi. (2012). The assembly of metals chelation by thiols and vacuolar compartmentalization conferred increased tolerance to and accumulation of cadmium and arsenic in transgenic Arabidopsis thaliana. *Journal of Hazardous Materials, 199*, 309–313.

Gustin, Jeffery L, Loureiro, Marcello E, Kim, Donggiun, Na, Gunnam, Tikhonova, Marina, & Salt, David E. (2009). MTP1-dependent Zn sequestration into shoot vacuoles suggests dual roles in Zn tolerance and accumulation in Zn-hyperaccumulating plants. *Plant Journal, 57*(6), 1116–1127.

Hartley-Whitaker, Jeanette, Ainsworth, Gillian, Vooijs, Riet, Ten Bookum, Wilma, Schat, Henk, & Meharg, Andrew A. (2001). Phytochelatins are involved in differential arsenate tolerance inholcus lanatus. *Plant Physiology, 126*(1), 299–306.

Hooda, Vinita. (2007). Phytoremediation of toxic metals from soil and waste water. *Journal of Environmental Biology, 28*(2) Supplement, 367.

Hossain, Zahed, & Komatsu, Setsuko. (2013). Contribution of proteomic studies towards understanding plant heavy metal stress response. *Frontiers in Plant Science, 3*, 310.

Howden, Ross, Andersen, Chris R, Goldsbrough, Peter B, & Cobbett, Christopher S. (1995). A cadmium-sensitive, glutathione-deficient mutant of Arabidopsis thaliana. *Plant Physiology, 107*(4), 1067–1073.

Huang, QiaoYun, Wu, JianMei, Chen, Wen, & Li, XueYuan. (2000). Adsorption of cadmium by soil colloids and minerals in presence of rhizobia. *Pedosphere, 10*(4), 299–307.

Inouhe, Masahiro, Ito, Rika, Ito, Shoko, Sasada, Naoki, Tohoyama, Hiroshi, & Joho, Masanori. (2000). Azuki bean cells are hypersensitive to cadmium and do not synthesize phytochelatins. *Plant Physiology, 123*(3), 1029–1036.

Janas, KM, Zielińska-Tomaszewska, J, Rybaczek, D, Maszewski, J, Posmyk, MM, Amarowicz, R, & Kosińska, A. (2010). The impact of copper ions on growth, lipid peroxidation, and phenolic compound accumulation and localization in lentil (Lens culinaris Medic.) seedlings. *Journal of Plant Physiology, 167*(4), 270–276.

Kabała, Katarzyna, Janicka-Russak, Małgorzata, Burzyński, Marek, & Kłobus, Grażyna. (2008). Comparison of heavy metal effect on the proton pumps of plasma membrane and tonoplast in cucumber root cells. *Journal of Plant Physiology, 165*(3), 278–288.

Khalid, Sana, Shahid, Muhammad, Niazi, Nabeel Khan, Murtaza, Behzad, Bibi, Irshad, & Dumat, Camille. (2017). A comparison of technologies for remediation of heavy metal contaminated soils. *Journal of Geochemical Exploration, 182*, 247–268.

Krämer, U, & Chardonnens, A. (2001). The use of transgenic plants in the bioremediation of soils contaminated with trace elements. *Applied Microbiology and Biotechnology, 55*(6), 661–672.

Lasat, Mitch M, Pence, Nicole S, Garvin, David F, Ebbs, Stephen D, & Kochian, Leon V. (2000). Molecular physiology of zinc transport in the Zn hyperaccumulator Thlaspi caerulescens. *Journal of Experimental Botany, 51*(342), 71–79.

Lewis, S, Handy, RD, Cordi, B, Billinghurst, Z, & Depledge, MH. (1999). Stress proteins (HSP's): methods of detection and their use as an environmental biomarker. *Ecotoxicology, 8*(5), 351–368.

Ma, Jian Feng, Ryan, Peter R, & Delhaize, Emmanuel. (2001). Aluminium tolerance in plants and the complexing role of organic acids. *Trends in Plant Science, 6*(6), 273–278.

Macomber, Lee, & Hausinger, Robert P. (2011). Mechanisms of nickel toxicity in microorganisms. *Metallomics, 3*(11), 1153–1162.

Mahar, Amanullah, Wang, Ping, Ali, Amjad, Awasthi, Mukesh Kumar, Lahori, Altaf Hussain, Wang, Quan, . . . Zhang, Zengqiang. (2016). Challenges and opportunities in the phytoremediation of heavy metals contaminated soils: a review. *Ecotoxicology and Environmental Safety, 126*, 111–121.

Maksymiec, W, Baszynski, T, & Bednara, J. (1995). Responses of runner bean plants to excess copper as a function of plant growth stages: Effects on morphology and structure of primary leaves and their chloroplast ultrastructure. *Photosynthetica*.

Meena, Desha, Singh, Hukum, & Chaudhari, SK. (2011). Elucidating strontium response on growth dynamics and biochemical change in Phaseolus mungo L. *International Journal of Agriculture, Environment and Biotechnology, 4*(2), 107–113.

Meharg, Andrew A. (1993). The role of the plasmalemma in metal tolerance in angiosperms. *Physiologia Plantarum, 88*(1), 191–198.

Nabais, Cristina, Labuto, Geórgia, Gonçalves, Susana, Buscardo, Erika, Semensatto, Décio, Nogueira, Ana Rita A, & Freitas, Helena. (2011). Effect of root age on the allocation of metals, amino acids and sugars in different cell fractions of the perennial grass Paspalum notatum (bahiagrass). *Plant Physiology and Biochemistry, 49*(12), 1442–1447.

Neumann, Dieter, Lichtenberger, Olaf, Günther, Detlef, Tschiersch, K, & Nover, Lutz. (1994). Heat-shock proteins induce heavy-metal tolerance in higher plants. *Planta, 194*(3), 360–367.

Neumann, Dieter, Zur Nieden, Uta, Lichtenberger, Olaf, & Leopold, Ines. (1995). How does Armeria maritima tolerate high heavy metal concentrations? *Journal of Plant Physiology, 146*(5–6), 704–717.

Obata, H, & Umebayashi, M. (1993). Production of SH compounds in higher plants of different tolerance to Cd. *Plant and Soil, 155*(1), 533–536.

Okamoto, Sachi, & Eltis, Lindsay D. (2011). The biological occurrence and trafficking of cobalt. *Metallomics, 3*(10), 963–970.

Oves, Mohammad, Khan, Mohammad Saghir, Zaidi, Almas, Ahmed, Arham S, Ahmed, Faheem, Ahmad, Ejaz, . . . Azam, Ameer. (2013). Antibacterial and cytotoxic efficacy of extracellular silver nanoparticles biofabricated from chromium reducing novel OS4 strain of Stenotrophomonas maltophilia. *PloS One, 8*(3), e59140.

Palmer, Christine M, & Guerinot, Mary Lou. (2009). Facing the challenges of Cu, Fe and Zn homeostasis in plants. *Nature Chemical Biology, 5*(5), 333.

Pawlowska, Teresa E, Błaszkowski, Janusz, & Rühling, Åke. (1997). The mycorrhizal status of plants colonizing a calamine spoil mound in southern Poland. *Mycorrhiza, 6*(6), 499–505.

Pence, Nicole S, Larsen, Paul B, Ebbs, Stephen D, Letham, Deborah LD, Lasat, Mitch M, Garvin, David F, . . . Kochian, Leon V. (2000). The molecular physiology of heavy metal transport in the Zn/Cd hyperaccumulator Thlaspi caerulescens. *Proceedings of the National Academy of Sciences, 97*(9), 4956–4960.

Persans, Michael W, Yan, Xiange, Patnoe, Jean-Marc ML, Krämer, Ute, & Salt, David E. (1999). Molecular dissection of the role of histidine in nickel hyperaccumulation in Thlaspi goesingense (Hálácsy). *Plant Physiology, 121*(4), 1117–1126.

Pietrini, Fabrizio, Zacchini, Massimo, Iori, Valentina, Pietrosanti, Lucia, Bianconi, Daniele, & Massacci, Angelo. (2009). Screening of poplar clones for cadmium phytoremediation using photosynthesis, biomass and cadmium content analyses. *International Journal of Phytoremediation, 12*(1), 105–120.

Qiu, Rong-Liang, Thangavel, Palaniswamy, Hu, Peng-Jie, Senthilkumar, Palaninaicker, Ying, Rong-Rong, & Tang, Ye-Tao. (2011). Interaction of cadmium and zinc on accumulation and sub-cellular distribution in leaves of hyperaccumulator Potentilla griffithii. *Journal of Hazardous Materials, 186*(2–3), 1425–1430.

Ramamurthy, Armuthur S, & Memarian, Ramin. (2012). Phytoremediation of mixed soil contaminants. *Water, Air, & Soil Pollution, 223*(2), 511–518.

Ramírez-Díaz, Martha I, Díaz-Pérez, César, Vargas, Eréndira, Riveros-Rosas, Héctor, Campos-García, Jesús, & Cervantes, Carlos. (2008). Mechanisms of bacterial resistance to chromium compounds. *Biometals, 21*(3), 321–332.

Rea, Philip A, Li, Ze-Sheng, Lu, Yu-Ping, Drozdowicz, Yolanda M, & Martinoia, Enrico. (1998). From vacuolar GS-X pumps to multispecific ABC transporters. *Annual Review of Plant Biology, 49*(1), 727–760.

Rehman, F, Khan, FA, Varshney, D, Naushin, F, & Rastogi, J. (2011). Effect of cadmium on the growth of tomato. *Biology and Medicine, 3*(2), 187–190.

Saifullah, Shahid M, Zia-Ur-Rehman, M, Sabir, M, & Ahmad, HR. (2015). Chapter 14: Phytoremediation of Pb-contaminated soils using synthetic chelates. In *Soil Remediation and Plants*. Academic Press, San Diego, 397–414.

Samanovic, Marie I, Ding, Chen, Thiele, Dennis J, & Darwin, K Heran. (2012). Copper in microbial pathogenesis: meddling with the metal. *Cell Host & Microbe, 11*(2), 106–115.

Sancenón, Vicente, Puig, Sergi, Mira, Helena, Thiele, Dennis J, & Peñarrubia, Lola. (2003). Identification of a copper transporter family in Arabidopsis thaliana. *Plant Molecular Biology, 51*(4), 577–587.

Sandmann, Gerhard, & Böger, Peter. (1980). Copper-mediated lipid peroxidation processes in photosynthetic membranes. *Plant Physiology, 66*(5), 797–800.

Sebastiani, Luca, Scebba, Francesca, & Tognetti, Roberto. (2004). Heavy metal accumulation and growth responses in poplar clones Eridano (Populus deltoides× maximowiczii) and I-214 (P.× euramericana) exposed to industrial waste. *Environmental and Experimental Botany*, *52*(1), 79–88.

Shanker, Arun Kumar. (2003). Physiological biochemical and molecular aspects of chromium toxicity and tolerance in selected crops and tree species.

Sheoran, IS, Singal, HR, & Singh, Randhir. (1990). Effect of cadmium and nickel on photosynthesis and the enzymes of the photosynthetic carbon reduction cycle in pigeonpea (Cajanus cajan L.). *Photosynthesis Research*, *23*(3), 345–351.

Shoeb, Erum, Badar, Uzma, Akhter, Jameela, Shams, Hina, Sultana, Maria, & Ansari, Maqsood A. (2012). Horizontal gene transfer of stress resistance genes through plasmid transport. *World Journal of Microbiology and Biotechnology*, *28*(3), 1021–1025.

Silver, Simon. (1996). Bacterial resistances to toxic metal ions-a review. *Gene*, *179*(1), 9–19.

Silver, Simon, & Phung, Le T. (1996). Bacterial heavy metal resistance: new surprises. *Annual Review of Microbiology*, *50*(1), 753–789.

Singh, H, Verma, A, Kumar, M, Sharma, R, Gupta, R, Kaur, M, . . . Sharma, SK. (2017). Phytoremediation: A green technology to clean up the sites with low and moderate level of heavy metals. *Austin Biochem*, *2*(2), 1012.

Smith, Sally E, & Read, David J. (2010). *Mycorrhizal Symbiosis*. Academic Press.

Su, Li, Deng, Yuantao, Zhang, Yingmei, Li, Chengyun, Zhang, Rui, Sun, Yingbiao, . . . Yao, Shixia. (2011). Protective effects of grape seed procyanidin extract against nickel sulfate-induced apoptosis and oxidative stress in rat testes. *Toxicology Mechanisms and Methods*, *21*(6), 487–494.

Tejada-Jiménez, Manuel, Galván, Aurora, Fernández, Emilio, & Llamas, Ángel. (2009). Homeostasis of the micronutrients Ni, Mo and Cl with specific biochemical functions. *Current Opinion in Plant Biology*, *12*(3), 358–363.

Thomine, Sébastien, Wang, Rongchen, Ward, John M, Crawford, Nigel M, & Schroeder, Julian I. (2000). Cadmium and iron transport by members of a plant metal transporter family in Arabidopsis with homology to Nramp genes. *Proceedings of the National Academy of Sciences*, *97*(9), 4991–4996.

Tsao, David T. (2003). *Overview of Phytotechnologies Phytoremediation* (pp. 1–50). Springer.

Tseng, TS, Tzeng, SS, Yeh, KW, Yeh, CH, Chang, FC, Chen, YM, & Lin, CY. (1993). The heat-shock response in rice seedlings: isolation and expression of cDNAs that encode class I low-molecular-weight heat-shock proteins. *Plant and Cell Physiology*, *34*(1), 165–168.

Turnau, Katarzyna. (1998). Heavy metal content and localization in mycorrhizal Euphorbia cyparissias zinc wastes in southern Poland. *Acta Societatis Botanicorum Poloniae*, *67*(1), 105–113.

Ullah, Riaz, Ahmad, Shabir, Atiq, Aimen, Hussain, Hidayat, ur Rehman, Najeeb, Abd Elsalam, Naser M, & Adnan, Muhammad. (2015). Quantification and antibacterial activity of flavonoids in coffee samples. *African Journal of Traditional, Complementary and Alternative Medicines*, *12*(4), 84–86.

Utmazian, Maria Noel Dos Santos, Wieshammer, Gerlinde, Vega, Rosa, & Wenzel, Walter W. (2007). Hydroponic screening for metal resistance and accumulation of cadmium and zinc in twenty clones of willows and poplars. *Environmental Pollution*, *148*(1), 155–165.

van Vliet, Catherine, Andersen, Chris R, & Cobbett, Christopher S. (1995). Copper-sensitive mutant of Arabidopsis thaliana. *Plant Physiology*, *109*(3), 871–878.

Verkleij, Jos AC, Koevoets, Paul LM, Blake-Kalff, Mechteld MA, & Chardonnens, Agnes N. (1998). Evidence for an important role of the tonoplast in the mechanism of naturally selected zinc tolerance in Silene vulgaris. *Journal of Plant Physiology*, *153*(1–2), 188–191.

Wani, Parvaze Ahmad, Khan, Md Saghir, & Zaidi, Almas. (2008). Chromium-reducing and plant growth-promoting Mesorhizobium improves chickpea growth in chromium-amended soil. *Biotechnology Letters*, *30*(1), 159–163.

Wong, Chong Kum Edwin, & Cobbett, Christopher S. (2009). HMA P-type ATPases are the major mechanism for root-to-shoot Cd translocation in Arabidopsis thaliana. *New Phytologist*, *181*(1), 71–78.

Xiang, Chengbin, & Oliver, David J. (1998). Glutathione metabolic genes coordinately respond to heavy metals and jasmonic acid in Arabidopsis. *Plant Cell*, *10*(9), 1539–1550.

Yang, Zhao, & Chu, Chengcai. (2011). Towards understanding plant response to heavy metal stress. *Abiotic Stress in Plants: Mechanisms and Adaptations*, *10*, 24204.

Zhang, Feng-Qin, Wang, You-Shao, Lou, Zhi-Ping, & Dong, Jun-De. (2007). Effect of heavy metal stress on antioxidative enzymes and lipid peroxidation in leaves and roots of two mangrove plant seedlings (Kandelia candel and Bruguiera gymnorrhiza). *Chemosphere*, *67*(1), 44–50.

Zhu, Yong Liang, Pilon-Smits, Elizabeth AH, Tarun, Alice S, Weber, Stefan U, Jouanin, Lise, & Terry, Norman. (1999). Cadmium tolerance and accumulation in Indian mustard is enhanced by overexpressing γ-glutamylcysteine synthetase. *Plant Physiology*, *121*(4), 1169–1177.

4 Application of UV/Periodate Advanced Oxidation Process for the Degradation of Persisting Organic Pollutants
Influence of Water Quality

Slimane Merouani, Hafida Bendjama, and Oualid Hamdaoui

CONTENTS

4.1 INTRODUCTION

Generally, water treatment takes place by several chemical and physical techniques (Hendricks, 2011), but the disadvantages of these techniques outweigh their efficiency (Wang and Xu, 2012). The chemical methods, involving the employment of chlorine, hydrogen peroxide, etc., are restricted by severe mass transfer limitations causing lower disinfection rates (Gogate et al., 2014). Likewise, some of the chemical methods produce non-desirable by-products (Gogate et al., 2014). For example, chlorine, which is extensively used in water treatments, results in the formation of carcinogenic and mutagenic agents, like trihalomethanes, in water and wastewater effluents (Deborde and Gunten, 2008; Allard et al., 2016). Conventional physical methods like adsorption only transfers pollutants from one phase to another, which then requires secondary treatment. The efficiency of certain physical technologies, such as ultraviolet light, is limited by absorbing solutions, light scattering, and the persistence of target contaminants, which are originally refractories (Gogate et al., 2014). Therefore, there is a great necessity for developing alternate techniques for water treatment. Another important requirement of alternative water treatments is that the treatment approach should be able to decrease the scale formation, which possibly will avoid the usage of any external scale inhibitors in the process, giving a superior treatment approach.

DOI: 10.1201/9781003204442-4

Oxyanions like H_2O_2, $S_2O_8^{2-}$, and IO_4^- can be converted into highly reactive species such as hydroxyl (OH), sulfate ($SO_4^{\bullet-}$), and iodyl ($IO_3^\bullet$) radicals via several activations means for advanced oxidation process (AOP) applications (Weavers et al., 1997; Tsitonaki et al., 2010; Wang and Xu, 2012). These radicals offer high-potential oxidation (e.g., $E_0 = 2.6$ V for SO_4^- and 2.8 V for OH) and react with organic compounds with rate constants in the order of $10^6–10^9$ M^{-1} s^{-1} (Neta et al., 1988). For instance, as the one-electron reduction of H_2O_2 by Fe^{2+} leads to the production of OH (Fenton reaction), the high yields of $SO_4^{\bullet-}$ from $S_2O_8^{2-}$ activation via transition metals (e.g., Fe^{2+} and Co^{2+}) have been found effective for the degradation of several recalcitrant organic contaminants (Rodríguez, 2003; Anipsitakis et al., 2006; Long et al., 2014). Alternatively, OH and $SO_4^{\bullet-}$ formed via the homolytic cleavage of peroxide bonds in H_2O_2 and $S_2O_8^{2-}$ under heat, radiolysis, and photolysis efficiently oxidize organic contaminants in groundwater and soil (Parsons, 2004; Petri et al., 2011).

Periodates can also be activated into reactive radicals intermediated under UV irradiation (<300 nm) and has successfully been applied to degrade organic pollutants (Weavers et al., 1997; Chia et al., 2004; Lee and Yoon, 2004; Ghodbane and Hamdaoui, 2016) and lower chemical oxygen demand in industrial wastewater (Tang and Weavers, 2008). The UV/periodate process has been reported to be more effective than UV/H_2O_2 and $UV/S_2O_8^2$ for the degradation of numerous micropollutants, such as C.I. Reactive Red 198 (Wu and Wu, 2011), C.I. Reactive Black 5 (Yu et al., 2010), chlorobiphenyl (Wang and Hong, 1999), methylene blue (Syoufian and Nakashima, 2008), pentafluorobenzoic acid (Ravichandran et al., 2007) and 4-chloro-2-methylphenol (Irmak et al., 2004). However, despite the obvious advantages of the UV/IO_4^- process, the ability of this oxidation process for treating real environmental waters containing mineral and organic compounds was poorly investigated. It is of practical interest to evaluate this process in real matrices, as the several matrix components may meaningfully touch the periodate chemistry and the radical pathways in the UV/IO_4^- system, which may affect the effectiveness of micropollutant abatement. Studies on this matter are very limited.

In this work, the applicability of the UV/IO_4^- process toward the degradation of a textile azo dye, Chlorazol Black (CB), was assessed in different water matrices, namely deionized water, mineral water, and secondary effluents of wastewater treatment (SEWTP). CB is a highly water soluble, acidic azo dye, which is broadly used for the dyeing of leather, fabric, cotton, plastic, and cellulose fibers (Pohanish, 2012). The carcinogenic and mutagenic effects of the dye on humans and animals have been experimentally confirmed (Lewis, 2008). Additionally, laboratory tests confirmed that CB is very persistent to direct oxidation with H_2O_2, $S_2O_8^{2-}$, and IO_4^- (Bendjama et al., 2018). The dependence on water quality impacts various operating conditions including the solution's pH and temperature, initial periodate dosage, and initial dye concentration. As an introduction, we provide an overview of the chemistry of the UV/IO_4^- process and the previous works conducted on this subject.

4.2　CHEMISTRY AND APPLICATION OF THE UV/IO_4^- PROCESS

In aqueous solutions, periodate shows a maximum absorption wavelength of 222 nm with an extinction coefficient of 5.4×10^2 $M^{-1} cm^{-1}$ (Chia et al., 2004). The decomposition of periodate in aqueous solutions has been studied previously using photolysis, flash photolysis, pulse radiolysis, and laser flash photolysis (Barat et al., 1971; Klaning and Sehested, 1978; Klaning et al., 1981; Wagner and Strehlow, 1982; Subhani and Latif, 1992), and a number of reactions have resulted. Iodate ions (IO_3^-), hydrogen peroxide (H_2O_2), oxygen (O_2), and ozone (O_3) were found as products of IO_4^- photolysis. H_2O_2 and O_3 have been detected for pH 3–7 (Barat et al., 1971; Klaning and Sehested, 1978; Klaning et al., 1981; Wagner and Strehlow, 1982; Subhani and Latif, 1992), whereas the yield of products is much lower in basic mediums (Klaning and Sehested, 1978). Intermediates with a short lifetime such as $^\bullet$OH radicals, $O(^3P)$ atoms, $IO_3^\bullet$, as well as unspecified I(VIII) species have been reported

(Barat et al., 1971; Klaning and Sehested, 1978; Klaning et al., 1981; Wagner and Strehlow, 1982). Barat et al. (1971) have also reported the formation of $IO_4^\bullet$. The following mechanism (Equations 4.1–4.12), suggested by Wagner and Strehlow (1982), has been broadly used to express the periodate chemistry under UV irradiation (Weavers et al., 1997; Chia et al., 2004; Lee and Yoon, 2004; Tang and Weavers, 2007, 2008; Ghodbane and Hamdaoui, 2016; Bendjama et al., 2018):

Initiation:

$$IO_4^- + h \rightarrow IO_3^\bullet + O^{\bullet-} \tag{4.1}$$

$$IO_4^- + h \rightarrow IO_3^- + O(^3P) \tag{4.2}$$

Propagation:

$$O^{\bullet-} + H_2O \rightleftharpoons OH + OH^- \tag{4.3}$$

$$O(^3P) + O_2 \rightarrow O_3 \tag{4.4}$$

$$IO_4^- + {}^\bullet OH \rightarrow IO_4^\bullet + OH^- \tag{4.5}$$

$$IO_4^- + IO_3^\bullet \rightarrow IO_4^\bullet + IO_3^- \tag{4.6}$$

$$O_3 + IO_3^\bullet \rightarrow IO_4^\bullet + O_2 \tag{4.7}$$

Termination:

$${}^\bullet OH + {}^\bullet OH \rightarrow H_2O_2 \tag{4.8}$$

$$IO_3^\bullet + IO_3^\bullet \rightarrow I_2O_6 \tag{4.9}$$

$$IO_4^\bullet + IO_4^\bullet \rightarrow I_2O_8 \tag{4.10}$$

$$I_2O_6 + H_2O \rightarrow IO_4^- + IO_3^- + 2H^+ \tag{4.11}$$

$$I_2O_8 + H_2O \rightarrow IO_4^- + IO_3^- + 2H^+ + O_2 \tag{4.12}$$

Based on the above mechanism, photoactivated periodate can result in several reactive radicals and non-radicals species including OH, $IO_3^\bullet$, $IO_4^\bullet$, $O(^3P)$, IO_3, H_2O_2, and O_3. This mechanism also indicates the possible regeneration of IO_4. Tang and Weavers (2008) have quantified the yields of IO_3, O_3, and H_2O_2 during the photo-reduction of periodate (30 mM) in the presence of 1 mM thiodiglycol (TDG) at pH 3. They found that the linear disappearance of IO_4 was accompanied by a linear increase of IO_3 up to 40 minutes; the time when periodate is totally consumed. It was found that iodate ions are liberated with the same rate of periodate depletion (0.612 mMmin in the absence of TDG). On this topic, Chia et al. (2004) showed that O_3 evolution during the degradation of 4-chlorphenol (0.2 mM) in an air-saturated UV/IO_4 system using 266 nm-irradiation light and 5 mM IO_4 increased with time approaching a steady-state at $O_3 \approx 20$ μM. These nonlinear productions of O_3 and H_2O_2 can be explained by the decomposition of these species by direct UV photolysis and/or radical-chain decomposition. Therefore, the UV/IO_4 process may enclose principal reactions occurring in UV/O_3, UV/H_2O_2, and H_2O_2/O_3 advanced oxidation processes. This is perhaps the reason why the UV/IO_4 process was found more efficient than other UV-based AOPs for the degradation of several organic contaminants (Sadik and Nashed, 2008; Wu and Wu, 2011; Saien et al., 2017; Haddad et al., 2019).

Several researchers have reported the degradation of many organic pollutants by UV/IO_4^- process. After exposing a periodate aqueous solution (0.25–1 mM) to UV irradiation at 253.7, Hamdaoui and Ghodbane (2016) and Bendjama et al. (2018) have recorded the fast transformation of periodate, $ca.$ 80% of the pick intensity at 222 nm vanished in less than 10 minutes (Ghodbane and Hamdaoui, 2016; Bendjama et al., 2018). Simultaneously, more than 90% of Acid Blue 25 dye was degraded after only 5 minutes of treatment with 1 mM of periodate. Correspondingly, Bendjama et al. (2018) found that ~70% of Chlorazol Black was removed after only 10 minutes in an acidic seawater solution exposed to 253.7 nm-UV light in the presence of 0.5 mM of periodate. A similar accelerating effect of the UV/IO_4^- process has been reported by Lee and Yoon (2004) for the degradation of Reactive Black 5 (RB5) at a neutral pH. Under low irradiation intensities (23 $\mu W/cm^2$), a wavelength of 266 nm, and a pH of 3, Chia et al. (2004) have observed that periodate photoactivated effectively degraded 4-chlorophenol, following a pseudo-first-order reaction kinetics.

Mineralization by the UV/IO_4 system was also reported for many cases. At pH 7.6, a COD reduction of 90% has been achieved in 3 hours with a xenon lamp, and 30 minutes with a mercury-xenon lamp, during the treatment of triethanolamine (TEA) in deionized water using a $[IO_4]_0/[TEA]_0$ ratio of 30 (Weavers et al., 1997). However, the COD reduction of real wastewater contaminated with TEA varied from 58–82%. A faster rate in total organic carbon (TOC) removal was also found for three hydrolysates of chemical warfare agents, i.e., thiodiglycol (TDG), 3.3-dithiolpropanol (DTP), and 1.4-thioxane (TX) (Tang and Weavers, 2008). More than 80% of TDG was removed in 90 minutes with 60 mM of periodate and slighly less (about 70%) was recored for DTP and TX. Moreover, the TOC elimination varies significantly with pH and the concentration of periodate (Tang and Weavers, 2008). COD abetment of 62.1 and 72.4% after 12 and 24 minutes, respectively, have been reported by Ghodbane and Hamdaoui during the treatment of Acid Blue 25 with the periodate (1 mM) photoactivated system. Saien et al. (2017) showed that the kinetics of quinoline degradation and mineralization increased with time to finally achieve 70% in COD abatement after 70 minutes of reaction.

4.3 EXPERIMENTAL

Wastewater samples were collected from a municipal secondary effluent of wastewater treatment plant (SEWTP), before chlorination, located in the Constantine region of Algeria. The samples were filtrated through a 1 μm GA-100 glass fiber filter before used in the experiments. The filtrates were then stored in a refrigerator in clean bottles. Table 4.1 provides the main characteristics of the SEWTP and mineral water used in this study. Stock solutions of CB (500 mg/L) and periodate (20 mM) were prepared in deionized water. Appropriate dilutions to desirable concentrations were then used by the specific water matrices to examine its effect (deionized water, mineral water, or SEWTP).

The batch experimental setup described by Bendjama et al. (2018 and 2019c) has been adopted herein. The reactor consists on a 500 mL cylindrical glass water-jacketed cell that was wrapped with aluminum foil. A UV lamp (low-pressure mercury, 15 mW cm^{-2}, max. emission at 253.7 nm) mounted in a quartz tube of 1 cm in diameter was placed vertically in the center of the reactor. 250 mL of aqueous solution was used in the runs. The solution pH was adjusted by H_2SO_4 or NaOH (0.1 M). The mixture was vigorously stirred with a magnetic stirrer during the entire reaction process. Timing was started as soon as the UV light was turned on. The dye concentration was determined spectrophotometrically at λ_{max} = 578 nm (Bendjama et al., 2019a, b). Each degradation experiment was performed in triplicate, and the results were presented as averages. Error bars represented the deviation of means.

TABLE 4.1

Characteristics of the Used Matrices:
Mineral Water and the Secondary
Effluent of Wastewater Treatment Plant

	Mineral Water	SEWTP
pH	7.4	7.6
Ca^{2+}	59.0 mg L^{-1}	
Mg^{2+}	45.0 mg L^{-1}	
Na^+	15.0 mg L^{-1}	Salinity=0.8 g/L
K^+	2.0 mg L^{-1}	
Cl	22.0 mg L^{-1}	
SO_4^2	40.0 mg L^{-1}	
HCO_3	378.2 mg L^{-1}	
TOC	0	
COD	0	
BOD_5	0	13 mg/L

*Abbreviations: COD - chemical oxygen demand, TOC - total organic carbon, BOD_5 - biochemical oxygen demand.

4.4 CB DEGRADATION IN DIFFERENT WATER MATRICES

The degradation of CB by direct photolysis and periodate-assisted UV irradiation has been conducted in deionized water (DW), mineral water (MW), and SEWTP using 0.5 mM of periodate at pH 7 and an ambient temperature of 25°C. The assistance of UV irradiation by periodate accelerated the degradation of CB for all the investigated matrices (see Figure 4.1). As compared to UV alone the UV/IO_4 system enhanced the CB removal rate by 8.6 fold for the DW, 11.2 fold for the MW, and 4.4 fold for the SEWTP. A similar enhancement in the degradation rate of CB was previously recorded in seawater; the removal yield being 13.6 fold when UV light was aided by 0.5 mM of IO_4 at pH 3 (Bendjama et al., 2018). Therefore, the UV/IO_4 process showed a potential applicability in degrading CB, whatever the water matrix used during experiments.

The improvement in the CB removal rate with photoactivated periodate was due to the involvement of reactive species in the degradation pathway. As stated early, the photoactivation of periodate at 254 nm can generate several radical and non-radical reactive species, like $IO_3{}^{\bullet}$, $IO_4{}^{\bullet}$, OH, IO_3, $O(^3P)$, H_2O_2, and O_3, with could make the degradation process much fastly (Chia et al., 2004; Tang and Weavers, 2008; Bendjama et al., 2018). Besides, the degradation rate in mineral water is similar to that conducted in pure water, meaning that the low concentration of mineral anions existing in the natural water interact insignificantly with the reactive species generated by the UV/IO_4 system. However, the SEWTP exerted a visual reducing impact, as compared to the mineral water. This may be attributed to the reaction of reactive species with NOM (i.e., natural organic matter) existing in the wastewater matrix (see Table 4.1). Overall, the UV/IO_4 process can be regarded as the proposed technique for the rapid removal of organic pollutants from environmental and real wastewater effluents.

To make sure that hydroxyl radicals participated in the CB degradation by the UV/IO_4 process, tert-butanol (TB, 10 mM) has been added in the reacting medium during the different runs. TB reacts only with $^{\bullet}OH$ ($k_{OH-TB} = 6 \times 10^8$ M^{-1} s^{-1}) and persists for all other reactive species formed in the UV/IO_4 system (Ghodbane and Hamdaoui, 2016). The results of Figure 4.2b clearly shows that the presence of TB in excess (TB to CB ratio equal to 392) appreciably reduces the dye removals, giving evidence that $^{\bullet}OH$ contributes to the degradation process,

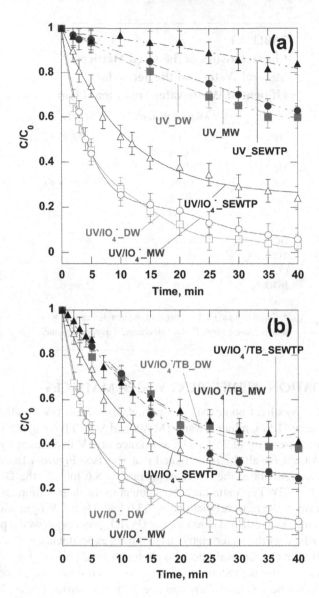

FIGURE 4.1 (a) Effect of water matrices on the removal kinetics of CB upon UV and UV/IO$_4^-$ processes, and (b) the impact of tert-butanol (TB, 10 mM) on the radical chemistry occurring during the CB degradation by the UV/IO$_4^-$ system (C$_0$=20 mg/L, [IO$_4^-$]$_0$=0.5 mM, pH ~7, temp. 25°C). (SEWTP: Secondary effluent of wastewater treatment plant.)

whatever the employed water matrix. Based on TB observations, the contribution of •OH was estimated to be ~70% for deionized water and mineral water and ~45% for SEWTP. Hence, the non-complete quenching of the degradation rate by TB addition confirmed that other reactive species were implicated effectively in the oxidation process of the dye. The same statement has already been noted for the degradation of CB in seawater (Bendjama et al., 2018) and other contaminants in pure water (Weavers et al., 1997; Chia et al., 2004; Tang and Weavers, 2008). Note that there was no recorded decrease in the initial CB concentration with 0.5 mM of periodate or iodate in dark conditions, which confirms the persistence of this pollutant toward direct oxidation by IO$_4^-$ or IO$_3^-$.

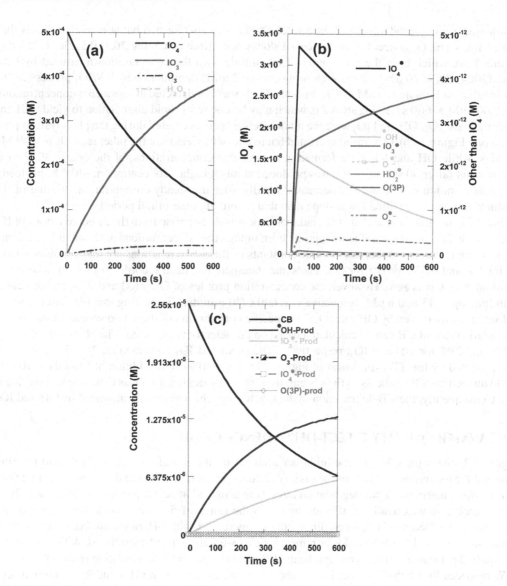

FIGURE 4.2 Concentration profiles of different reactive species during the aqueous oxidation of CB by the UV/periodate process (Djaballah et al., 2020). •OH-prod, $IO_3^•$-prod, O_3-prod, $IO_4^•$-prod, and O(3P)-prod denote the direct products resulted from the CB reaction with •OH, $IO_3^•$, O_3, $IO_4^•$, and O(3P), respectively.

4.5 CONCENTRATION PROFILES OF REACTIVE SPECIES IN THE UV/IO₄⁻ SYSTEM

Djaballah et al. (2020) have recently modeled the reactive species generation during the oxidation of CB in the UV/IO_4 system, for the same experimental conditions of Figure 4.1. The simulations of Figure 4.2 were given for deionized water by adopting a reaction scheme consisting in 45 chemical reactions including, in addition to the initial constitutes (i.e., IO_4^-, CB, H_2O, O_2, H^+, and OH^-), a number of reactive radicals ($IO_3^•$, $IO_4^•$, •OH, $O^{•-}$, $HO_2^•$, $O_2^{•-}$, $O_3^{•-}$) and non-radical intermediates/products [O(3P), O_3, IO_3^-, H_2O_2, HO_2^-, I_2O_6 and I_2O_8].

The consumption of IO_4^- is accompanied by the formation of an appreciable quantity of iodate (see Figure 4.2a), a transient formation of reactive species (see Figure 4.2b), and high levels of

CB-degradation by-products (see Figure 4.2c). Ozone was also formed but at much low levels than that of IO_3^-. The O_3 concentration increased slowly with time achieving 20.7 µM after 600 s (see Figure 4.2a), which is in the same order of magnitude with the concentration measured by Chia et al. (2004) ($O_3 \approx 20$ µM) during 4-chlorphenol (0.2 mM) degradation by UV/IO_4 process at 266 nm-irradiation light and 5 mM of IO_4. Besides, H_2O_2 was not detected at significant concentrations, i.e., 0.002 µM at 600 s (see Figure 4.2a), which may be due to its rapid dissociation to yield •OH and other radicals. IO_4, OH, and IO_3 were the main reactive species created during the photolysis of periodate (see Figure 4.2b). O_2^- was also created but to a lesser extent than the other radicals (~10^{-13} M). Besides, while OH and IO_3 were formed at comparable concentrations, of the order ~10^{-12} M at steady point (after 30 seconds), IO_4 was produced at much higher concentration, ~10^{-8} M at steady. The concentration of IO_4 and OH decreased rapidly after the steady concentration, but that of IO_3 continued to increase with a lower slope than that recorded at the initial period.

From Figure 4.2c, it is seen that degradation products do not arise from the direct reactions of IO_4 and O_3 with CB (i.e., IO_4-prod and O_3-prod), even though the concentrations of O_3 and IO_4 are dominant, which is due to the fact that the rate constants of these two reactions are inappreciable (0 M^{-1} s^{-1} for IO_4 and 0.576 M^{-1} s^{-1} for O_3). The same statement has been recorded for O(3P), whose rate constant with CB is zero. However, the concentration profiles of OH-prod and IO_3-prod increased with time up to 13 and 6 µM, respectively, at 600 s (10 minutes), indicating that the degradation of CB was majorly driven by OH and IO_3. The specific contribution of these two species in the overall degradation rate of CB can be calculated based on the selectivity equation. The obtained values are ~78% and 21% for OH and IO_3, respectively. The predicted IO_3 contribution, 21%, is in line with that provided by the TB experiment of Figure 4.1b (i.e., ~30%). The addition of t-BuOH at 10 mM could quench all •OH radicals in the system; the remaining degradation was therefore mainly due to IO_3. Consequently, the CB degradation in the UV/IO_4 system was mainly mediated by OH and IO_3.

4.6   WATER QUALITY EFFECT-DEPENDENCE OF pH

Figure 4.3 shows the effect of the initial solution pH on the initial CB removal rate and the time required for the removal of half of the initial dye concentration ($t_{1/2}$) for the different water matrices. For the three matrices, faster degradation rates were achieved at acidic conditions, below which the degradation rate was nearly unaffected by pH in the range of 5–7, whereas lowered degradation rates were obtained at pH 9. Accordingly, the $t_{1/2}$ increased with pH elevation from 3 to 5–9 and being highest at pH 9. Overall, lower removal yields were associated with the SEWTP throughout the studied pH range. There is no significant difference between the degradation in the DW and the MW matrices for pH 5–7, but the former showed a rate enhancing at pH 3 and 9. For all matrices, the degradation rate followed the order pH 3>, pH 5~7>, and pH 9.

The speciation diagram of periodate in deionized water indicated that IO_4^- is the predominant form at pH <8, whereas the dimerized form $H_2I_2O_{10}^{4-}$ is predominant at higher pHs (Lee and Yoon, 2004; Chadi et al., 2019a). However, no speciation diagrams were done for mineral water or real wastewater, in which salts and NOM may effectively interact with periodate species, thereby changing the distribution of the resulted species. The lowest process efficiency at pH 9 for the MW and SEWTP was mainly attributed to a precipitation event which resulted from the interaction of $H_2I_2O_{10}^{4-}$ and salts in the basic medium, as captured photographically in Figure 4.3c–e. The same phenomenon has been observed when CB degradation was conducted in seawater (2018). The usefulness of the process at pH 3 has been reported in many studies for different pollutants, i.e., Acid Blue 25 (Ghodbane and Hamdaoui, 2016) and quinoline (Saien et al., 2017). Tang and Weavers (2008) reported that TOC losses during the removal of thiodiglycol (TDG), 3.3-dithiopropanol (TDP), and 1.4-thioxane (TX) by the UV/periodate system decreased in the order pH 3>, pH 7>, and pH 10. Bendjama et al. (2018) have indicated that the degradation rate of Chlorazol Black in seawater decreased progressively with a pH increase in the range 3–7.

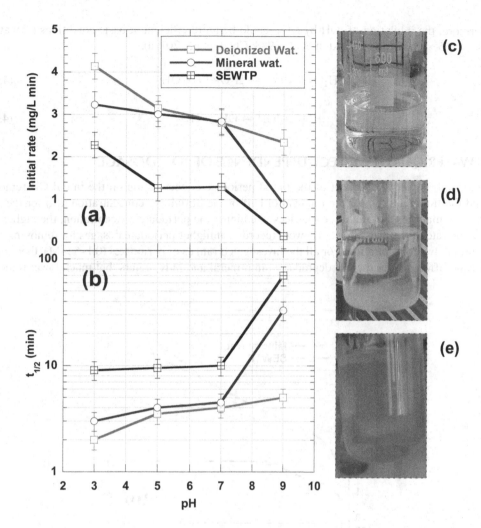

FIGURE 4.3 Effect of pH on (a) initial degradation rate, and (b) the half-life time for CB removal for different water matrices ($C_0 = 20$ mg/l, $[IO_4^-]_0 = 0.5$ mM, temp. 25°C). The y-axis of figure (b) is in logarithmic scale. (SEWTP: Secondary effluent of wastewater treatment plant.) (c), (d) and (e) are photographs of periodate aqueous solutions (0.5 mM) in deionized water, mineral water, and SEWTP, respectively, at pH 9. Precipitation phenomenon has taken place in mineral water and SEWTP at a basic pH.

In acidic solution, H_2O_2 reacts rapidly with IO_4^- to generate free radicals (Equations 4.13 and 4.14). In fact, H_2O_2/IO_4^- has been recently reported as a novel advanced oxidation process that generates several radical and non-radical species (Chadi et al., 2019a, b). The efficiency of the H_2O_2/IO_4^- process toward the degradation of Toluidine Blue decreased with a pH rise in the range of 3–11 (Chadi et al., 2019a).

$$IO_4^- + H_2O_2 \rightarrow IO_3^\bullet + O_2^{\bullet-} + H_2O \tag{4.13}$$

$$H_2O_2 + O_2^{\bullet-} \rightarrow {}^\bullet OH + OH^- + {}^1O_2 \tag{4.14}$$

In parallel, when the pH is basic, H_2O_2 deprotonates into HO_2^- ions. The HO_2^- species react rapidly with H_2O_2 itself (Equation 4.15), leading to dioxygen and water (Merouani et al., 2010a).

Furthermore, the inhibition of $^\bullet$OH by HO_2^- could be more efficient at basic conditions (Equation 4.16, $k_{16} = 7.5 \times 10^9$ M^{-1} s^{-1}) (Buxton et al., 1988; Chadi et al., 2019a).

$$H_2O_2 + HO_2^- \rightarrow H_2O + O_2 + OH^- \qquad (4.15)$$

$$HO_2^- + {}^\bullet OH \rightarrow OH^- + HO_2^\bullet \qquad (4.16)$$

4.7   WATER QUALITY EFFECT-DEPENDENCE OF IO_4^- DOSAGE

Figure 4.4a–b depicts the impact of the initial periodate concentration on the initial CB removal rate and the time required for the removal of half of the initial dye concentration ($t_{1/2}$) for the different water matrices. For the three matrices, the higher the periodate concentration, the higher the destruction rate of the dye. Shorter $t_{1/2}$ were recorded at higher periodate dosages for removing half of the initial dye concentration. For all the investigated ranges of periodates (0.1–5 mM), there is no significant difference for the CB degradation in mineral and pure waters, but much lower removal

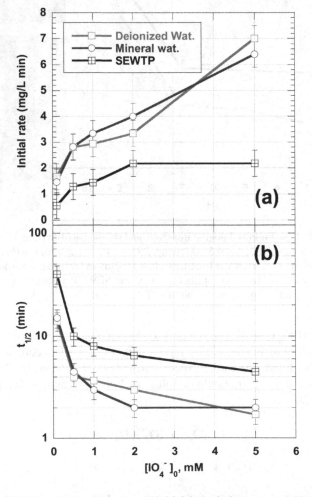

FIGURE 4.4   Effect of initial periodate dosage on (a) initial degradation rate, and (b) the half-life time for CB removal for different water matrices ($C_0 = 20$ mg/L, natural pH$_0$ ~7, temp. 25°C). The y-axis of figure (b) is in logarithmic scale. (SEWTP: Secondary effluent of wastewater treatment plant.)

yields were observed for the SEWTP. It seems that the optimum dosage of periodate, i.e., 2 mM, was attained for the best conversion rate of CB in the SEWTP.

Most studies agree with the fact that increasing the concentration of periodate could increase the number of radicals formed upon IO_4 photolysis (Weavers et al., 1997; Wang and Hong, 1999; Lee and Yoon, 2004; Sadik and Nashed, 2008; Tang and Weavers, 2008; Cao et al., 2010; Yu et al., 2010; Wu and Wu, 2011; Ghodbane and Hamdaoui, 2016; Saien et al., 2017; Bendjama et al., 2018), thereby enhancing the removal rate of contaminates. Some cases registered an optimum IO_4 concentration for the best conversion of pollutants (Weavers et al., 1997; Wang and Hong, 1999; Lee and Yoon, 2004; Sadik and Nashed, 2008; Tang and Weavers, 2008; Cao et al., 2010; Yu et al., 2010; Wu and Wu, 2011; Ghodbane and Hamdaoui, 2016; Saien et al., 2017; Bendjama et al., 2018). The optimal dose of periodate varied from 1 to several mM, depending on the operation l conductions of the works. The surplus of IO_4 in the reacting medium can scavenge radicals. Periodate in that case reacts with hydroxyl and iodyl radicals (Equations 4.5 and 4.6) with high rate constants of $4.5 \times 10^8 \, M^{-1} \, s^{-1}$ and $(2-7) \times 10^8 \, M^{-1} \, s^{-1}$ for $^\bullet OH$ (Barat et al., 1971) and $IO_3^\bullet$ (Chia et al., 2004), respectively. Furthermore, the radical–radical recombination reactions (Equations 4.9 and 4.10) become effective competitors for radical–organic reactions at high radical concentrations. The rate constants of reaction 9 and 10 are both high, of $5.5 \times 10^9 \, M^{-1} \, s^{-1}$ and $7.5 \times 10^8 \, M^{-1} \, s^{-1}$ (Barat et al., 1982), respectively, and this would make them significant parasite reactions that reduce the degradation of the dye at high periodate loading when higher concentrations of radicals are believed to be produced. A similar circumstance has been reported by Hamdaoui and Merouani (2017) for sono-activated periodate in deionized water at high IO_4 concentrations.

4.8 WATER QUALITY EFFECT-DEPENDENCE OF CB CONCENTRATION

The initial pollutant concentration, noted as C_0, is one critical factor that can alter the efficiency of the UV/IO_4 process. Its effect on the initial CB removal rate and the time required for the removal of half of the initial dye concentration ($t_{1/2}$) is shown in Figure 4.5a–b for the different water matrices. The findings of Figure 4.5a reflect that the initial pollutant concentration in the range of 20–50 mg/L could not affect the degradation rate at the initial stage of the treatment. However, the recorded values of $t_{1/2}$ clearly showed that higher time is required for removing half of the initial amount of CB, meaning that the removal efficiency is lower at higher initial dye concentrations. This scenario is largely reported in the literature for several AOPs (Merouani et al., 2010b, 2017; Boutemedjet et al., 2016; Ferkous et al., 2016; Taamallah et al., 2016; Bekkouche et al., 2017a, b; Torres-Palma and Serna-Galvis, 2018; Bendjama et al., 2019a; Chadi et al., 2019a; Belghit et al., 2020). A lower degradation yield in the SEWTP was also maintained for the various investigated values of C_0.

The published results on the UV/IO_4 process indicated that the degradation efficiency was adversely affected with increasing C_0; the trend which was going along with an increase in the initial degradation rate (Ghodbane and Hamdaoui, 2016; Bendjama et al., 2018). Hamdaoui and Ghodbane (2016) have reported that the initial degradation rate of Acid Blue 25 increased with increasing C_0 up to 50 mg/L. However, the degradation extent was found to be inversely proportional to C_0 increase. Similarly, Bendjama et al. (Bendjama et al., 2018) found that the degradation rate of Chlorazol Black in sweater increased with the initial pollutant concentration rise up to a plateau at higher concentration levels, even though the observed degradation efficiency was found to be decreased with C_0 increase. Both studies (Ghodbane and Hamdaoui, 2016; Bendjama et al., 2018) concluded that the degradation kinetics cannot be described by a pseudo-first order kinetics law.

4.9 WATER QUALITY EFFECT-DEPENDENCE OF TEMPERATURE

The temperate of liquids is shown to have an important impact in increasing/inhibiting the effectiveness of various $^\bullet OH$ and $SO_4^{\bullet-}$ based AOPs (Jiang et al., 2006; Son and Zoh, 2012; Merouani et al., 2016, 2017; Bendjama et al., 2019a, c). The effect of liquid temperature on the efficiency of

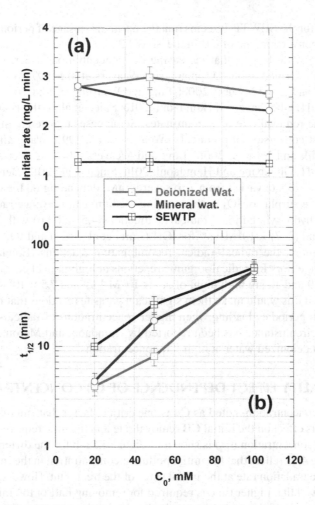

FIGURE 4.5 Effect of initial CB concentration on (a) the initial degradation rate, and (b) the half-life time for CB removal for different water matrices ($[IO_4^-]_0 = 0.5$ mM, natural pH_0 ~7, temp. 25°C). The y-axis of figure (b) is in logarithmic scale. (SEWTP: Secondary effluent of wastewater treatment plant.)

UV/IO_4 for the degradation of CB in seawater has been previously investigated by Bendjama et al. (2018). Their results reflect that the liquid temperature in the range of 25–55°C could not influence the process efficiency. In accordance with these observations, operating with solutions heated to 50°C did not affect the degradation rates of the dye for the three matrices, deionized water, mineral water, and SEWTP, when periodate and pollutant concentrations were initially maintained at 0.5 mM and 20 mg/L, respectively.

4.10 CONCLUSION

Experimental results show that the UV/periodate process is more efficient in the destruction of Chlorazol Black dye in pure, mineral, and wastewaters than UV irradiation alone. The reaction mechanism in the photoactivated process passes through a radical pathway where hydroxyl and iodyl radicals were the predominant reactive species involved in the degradation of Chlorazol Black dye. The natural organic matter of the SEWTP slowed down the degradation process by consuming the reactive species generated upon the photoactivation of periodate. In order to obtain a high removal yield, lower solution pH coupled with high periodate doses have to be applied. In these conditions, periodate could promote the complete degradation of the dye.

ACKNOWLEDGMENTS

The financial support by the Ministry of Higher Education and Scientific Research of Algeria (project No. A16N01UN250320180001) is greatly acknowledged. The authors would like to acknowledge the support provided by King Saud University with the Research Supporting Project No. RSP-2019/37 at King Saud University, Riyadh, Saudi Arabia.

REFERENCES

Allard, S., J. Criquet, A. Prunier, C. Falantin, A. Le Person, J. Y. M. Tang, and J. P. Croué. 2016. Photodecomposition of iodinated contrast media and subsequent formation of toxic iodinated moieties during final disinfection with chlorinated oxidants. *Water Research*, 103:453–461, https://doi.org/10.1016/j.watres.2016.07.050.

Anipsitakis, G. P., D. D. Dionysiuo, and M. A. Gonzalez. 2006. Cobalt-mediated activation of peroxymonosulfate and sulfate radical attack on phenolic compounds . Implications of chloride Ions. *Environmantal Science and Technology*, 40(3):1000–1007.

Barat, F., L. Gilles, B. Hickel, and B. Lesigne. 1971. Transient species in the pulse radiolysis of periodate ion in neutral aqueous solutions. *Chemical Communications*, 847:847–848, https://doi.org/10.1039/C29710000847.

Barat, F., L. Gilles, B. Hickel, and B. Lesigne. 1982. Pulsed radiolysis and flash photolysis of iodates in aqueous solution. *Journal of Physical Chemistry*, 76:302–307, https://doi.org/10.1021/j100647a004.

Bekkouche, S., M. Bouhelassa, A. Ben, S. Baup, N. Gondrexon, C. Pétrier, S. Merouani, and O. Hamdaoui. 2017a. Synergy between solar photocatalysis and high frequency sonolysis toward the degradation of organic pollutants in aqueous phase: Case of phenol. *Desalination and Water Treatment*, 20146:1–8, https://doi.org/10.5004/dwt.2017.20146.

Bekkouche, S., S. Merouani, O. Hamdaoui, and M. Bouhelassa. 2017b. Efficient photocatalytic degradation of Safranin O by integrating solar-UV/TiO2/persulfate treatment: Implication of sulfate radical in the oxidation process and effect of various water matrix components. *Journal of Photochemistry and Photobiology A: Chemistry*, 345:80–91, https://doi.org/10.1016/j.jphotochem.2017.05.028.

Belghit, A. A., S. Merouani, O. Hamdaoui, M. Bouhelassa, A. Alghyamah, and M. Bouhelassa. 2020. Influence of processing conditions on the synergism between UV irradiation and chlorine toward the degradation of refractory organic pollutants in UV/chlorine advanced oxidation system. *Science of the Total Environment*, 736:139623_1–139623_10, https://doi.org/10.1016/j.scitotenv.2020.139623.

Bendjama, H., S. Merouani, O. Hamdaoui, and M. Bouhelassa. 2018. Efficient degradation method of emerging organic pollutants in marine environment using UV/periodate process: Case of chlorazol black. *Marine Pollution Bulletin*, 126(September):557–564, https://doi.org/10.1016/j.marpolbul.2017.09.059.

Bendjama, H., S. Merouani, O. Hamdaoui, and M. Bouhelassa. 2019a. Acetone photoactivated process: Application to the degradation of refractory organic pollutants in very saline waters. *Water and Environment Journal*, 1–8, https://doi.org/10.1111/wej.12507.

Bendjama, H., S. Merouani, O. Hamdaoui, and M. Bouhelassa. 2019b. Using photoactivated acetone for the degradation of Chlorazol Black in aqueous solutions: Impact of mineral and organic additives. *Science of the Total Environment*, 653:833–838, https://doi.org/10.1016/j.scitotenv.2018.11.007.

Bendjama, H., S. Merouani, O. Hamdaoui, and M. Bouhelassa. 2019c. UV-photolysis of Chlorazol Black in aqueous media: Process intensification using acetone and evidence of methyl radical implication in the degradation process. *Journal of Photochemistry and Photobiology A: Chemistry*, 368(August 2018):268–275, https://doi.org/10.1016/j.jphotochem.2018.09.047.

Boutemedjet, S., O. Hamdaoui, and S. Merouani. 2016. Sonochemical degradation of endocrine disruptor propylparaben in pure water, natural water, and seawater. *Desalination and Water Treatment*, 57:19443994, https://doi.org/10.1080/19443994.2016.1177600.

Buxton, G. V., C. L. Greenstock, W. P. Helman, and A. B. Ross. 1988. Critical review of rate constants for reactions of hydrated electrons, hydrogen atoms and hydroxyl radicals (•OH/O-) in aqueous solution. *Journal of Physical and Chemical Reference Data*, 17(2):515–886, https://doi.org/0047-2689/88/020513-374$20.00.

Cao, M. H., B. B. Wang, H. S. Yu, L. L. Wang, S. H. Yuan, and J. Chen. 2010. Photochemical decomposition of perfluorooctanoic acid in aqueous periodate with VUV and UV light irradiation. *Journal of Hazardous Materials*, 179(1–3):1143–1146, https://doi.org/10.1016/j.jhazmat.2010.02.030.

Chadi, N. E., S. Merouani, O. Hamdaoui, M. Bouhelassa, and M. Ashokkumar. 2019a. H2O2/Periodate (IO4–): A novel advanced oxidation technology for the degradation of refractory organic pollutants. *Environmental Science: Water Research and Technology*, 5:1113–1123, https://doi.org/10.1039/c9ew00147f.

Chadi, N. E., S. Merouani, O. Hamdaoui, M. Bouhelassa, and M. Ashokkumar. 2019b. Influence of mineral water constituents, organic matter and water matrices on the performance of the H2O2/IO4: Advanced oxidation process. *Environmental Science: Water Research and Technology*, 5(11):1985–1992, https://doi.org/10.1039/c9ew00329k.

Chia, L. H., X. Tang, and L. K. Weavers. 2004. Kinetics and mechanism of photoactivated periodate reaction with 4-chlorophenol in acidic solution. *Environmental Science and Technology*, 38(24):6875–6880, https://doi.org/10.1021/es049155n.

Deborde, M., and U. Von Gunten. 2008. Reactions of chlorine with inorganic and organic compounds during water treatment: Kinetics and mechanisms : A critical review. *Water Research*, 42:13–51, https://doi.org/10.1016/j.watres.2007.07.025.

Djaballah, M. L., S. Merouani, H. Bendjama, and O. Hamdaoui. 2020. Development of a free radical-based kinetics model for the oxidative degradation of chlorazol black in aqueous solution using periodate photoactivated process. *Journal of Photochemistry and Photobiology A: Chemistry*.

Ferkous, H., S. Merouani, O. Hamdaoui, O. H. H. Ferkous, S. Merouani, H. Ferkous, S. Merouani, and O. Hamdaoui. 2016. Sonolytic degradation of naphtol blue black at 1700 kHz: Effects of salts, complex matrices and persulfate. *Journal of Water Process Engineering*, 9:67–77, https://doi.org/10.1016/j.str.2014.12.012.

Ghodbane, H., and O. Hamdaoui. 2016. Degradation of anthraquinonic dye in water by photoactivated periodate. *Desalination and Water Treatment*, 57(9):4100–4109, https://doi.org/10.1080/19443994.2014.988657.

Gogate, P. R., S. Mededovic-Thagard, D. McGuire, G. Chapas, J. Blackmon, and R. Cathey. 2014. *Hybrid Reactor Based on Combined Cavitation and Ozonation: From Concept to Practical Reality*. Elsevier, Amsterdam, 590–598 pp. https://doi.org/10.1016/j.ultsonch.2013.08.016.

Haddad, A., S. Merouani, C. Hannachi, O. Hamdaoui, and B. Hamrouni. 2019. Intensification of light green SF yellowish (LGSFY) photodegradion in water by iodate ions: Iodine radicals implication in the degradation process and impacts of water matrix components. *Science of the Total Environment*, 652:1219–1227, https://doi.org/10.1016/j.scitotenv.2018.10.183.

Hamdaoui, O., and S. Merouani. 2017. Improvement of sonochemical degradation of brilliant blue R in water using periodate ions: Implication of iodine radicals in the oxidation process. *Ultrasonics Sonochemistry*, 37:344–350, https://doi.org/10.1016/j.ultsonch.2017.01.025.

Hendricks, D. 2011. *Fundamentals of Water Treatment Unit Processes: Physical, Chemical, and Biological*. IWA Publishing, London, https://doi.org/10.1142/p063.

Irmak, S., E. Kusvuran, and O. Erbatur. 2004. Degradation of 4-chloro-2-methylphenol in aqueous solution by UV irradiation in the presence of titanium dioxide. *Applied Catalysis B: Environmental*, 54(2):85–91, https://doi.org/10.1016/j.apcatb.2004.06.003.

Jiang, Y., C. Petrier, and T. D. Waite. 2006. Sonolysis of 4-chlorophenol in aqueous solution: Effects of substrate concentration, aqueous temperature and ultrasonic frequency. *Ultrasonics Sonochemistry*, 13(5):415–22, https://doi.org/10.1016/j.ultsonch.2005.07.003.

Klaning, U. K., and K. Sehested. 1978. Photolysis of periodate and periodic acid in aqueous solution. *Journal of the Chemical Society, Faraday Transactions 1: Physical Chemistry in Condensed Phases*, 74(5):2818–2838,.

Klaning, U. K., K. Sehested, and T. Wolef. 1981. Laser flash photolysis and pulse radiolysis of iodate and periodate in aqueous solution. *Journal of the Chemical Society, Faraday Transactions*, 77(3):1707–1718, https://doi.org/10.1039/F19817701707.

Lee, C., and J. Yoon. 2004. Application of photoactivated periodate to the decolorization of reactive dye: Reaction parameters and mechanism. *Journal of Photochemistry and Photobiology A: Chemistry*, 165(1–3):35–41, https://doi.org/10.1016/j.jphotochem.2004.02.018.

Lewis, R. J. 2008. *Hazardous Chemicals Desk Reference*. Wiley, Hoboken, NJ, https://doi.org/10.1002/9780470335406.

Long, A., Y. Lei, and H. Zhang. 2014. Degradationof toluene by a selective ferrous ion activated persulfate oxidation process. *Industrial & Engineering Chemistry Research*, 53(3):1033–1039.

Merouani, S., O. Hamdaoui, F. Saoudi, and M. Chiha. 2010a. Influence of experimental parameters on sonochemistry dosimetries: KI oxidation, fricke reaction and H2O2 production. *Journal of Hazardous Materials*, 178(1–3):1007–1014, https://doi.org/10.1016/j.jhazmat.2010.02.039.

Merouani, S., O. Hamdaoui, F. Saoudi, and M. Chiha. 2010b. Sonochemical degradation of Rhodamine B in aqueous phase: Effects of additives. *Chemical Engineering Journal*, 158(3):550–557, https://doi.org/10.1016/j.cej.2010.01.048.

Merouani, S., O. Hamdaoui, Z. Boutamine, Y. Rezgui, and M. Guemini. 2016. Experimental and numerical investigation of the effect of liquid temperature on the sonolytic degradation of some organic dyes in water. *Ultrasonics Sonochemistry*, 28:382–392, https://doi.org/10.1016/j.ultsonch.2015.08.015.

Merouani, S., O. Hamdaoui, and M. Bouhelassa. 2017. Degradation of safranin O by thermally activated persulfate in the presence of mineral and organic additives : Impact of environmental matrices. *Desalination and Water Treatment*, 75:202–212, https://doi.org/10.5004/dwt.2017.20404.

Neta, P., R. E. Huie, and A. B. Ross. 1988. Rate constants for reactions of inorganic radicals in aqueous solution. *Journal of Physical and Chemical Reference Data*, 17(3):1027–1284, https://doi.org/0047-2689/88/031027-258/$21.00.

Parsons, S. 2004. *Advanced Oxidation Processes for Water and Wastewater Treatment*. IWA Publishing, London, 356 pp.

Petri, B. G., R. J. Watts, A. Tsitonaki, M. Crimi, T. R.T., and A. L. Teel. 2011. Fundamentals of ISCO using persulfate. In *In Situ Chemical Oxidation for Groundwater Remediation*. R. L. Siegrist,, M. Crimi, and T. J. Simpkin, eds, Springer, New York, 150 pp.

Pohanish, R. 2012. *Sittig's Handbook of Toxic and Hazardous Chemicals and Carcinogens*. Elsevier, Amsterdam.

Ravichandran, L., K. Selvam, and M. Swaminathan. 2007. Effect of oxidants and metal ions on photodefluoridation of pentafluorobenzoic acid with ZnO. *Separation and Purification Technology*, 56(2):192–198, https://doi.org/10.1016/j.seppur.2007.01.034.

Rodríguez, M. 2003. *Fenton and UV-Vis Based Advanced Oxidation Processes in Wastewater Treatment: Degradation, Mineralization and Biodegradability Enhancement*. 296 p.

Sadik, W. A. A., and A. W. Nashed. 2008. UV-induced decolourization of acid alizarine violet N by homogeneous advanced oxidation processes. *Chemical Engineering Journal*, 137(3):525–528, https://doi.org/10.1016/j.cej.2007.05.018.

Saien, J., H. Shafiei, and A. Amisama. 2017. Photo-activated periodate in homogeneous degradation and mineralization of quinoline: Optimization, kinetic, and energy consumption. *Environmental Progress & Sustainable Energy*, 36(6):1621–1627, https://doi.org/10.1002/ep.

Son, H., and K. Zoh. 2012. *Effects of Methanol and Carbon Tetrachloride on Sonolysis of 1,4- Dioxane in Relation to Temperature*.

Subhani, M. S., and R. Latif. 1992. Photolytic decomposition of the periodate ions in aqueous solution. *Pakistan Journal of Scientific and Industrial Research*, 35:484–488.

Syoufian, A., and K. Nakashima. 2008. Degradation of methylene blue in aqueous dispersion of hollow titania photocatalyst: Study of reaction enhancement by various electron scavengers. *Journal of Colloid and Interface Science*, 317(2):507–512, https://doi.org/10.1016/j.jcis.2007.09.092.

Taamallah, A., S. Merouani, and O. Hamdaoui. 2016. Sonochemical degradation of basic fuchsin in water. *Desalination and Water Treatment*, 57(56):27314–27330, https://doi.org/10.1080/19443994.2016.1168320.

Tang, X., and L. K. Weavers. 2007. Decomposition of hydrolysates of chemical warfare agents using photoactivated periodate. *Journal of Photochemistry and Photobiology A: Chemistry*, 187:311–318, https://doi.org/10.1016/j.jphotochem.2006.10.029.

Tang, X., and L. K. Weavers. 2008. Using photoactivated periodate to decompose TOC from hydrolysates of chemical warfare agents. *Journal of Photochemistry and Photobiology A: Chemistry*, 194:212–219, https://doi.org/10.1016/j.jphotochem.2007.08.014.

Torres-Palma, R. A., and E. A. Serna-Galvis. 2018. Sonolysis In *Advanced Oxidation Processes for Wastewater Treatments: Emerging Green Chemical Technology*. Suresh C. Ameta, and R. Ameta, eds, Elsevier, Amsterdam, 177–213 pp, doi: doi.org/10.1016/B978-0-12-810499-6.0.

Tsitonaki, A., B. Petri, M. Crimi, H. Mosbæk, R. L. Siegrist, P. L. Bjerg, H. Mosbk, R. L. Siegrist, and P. L. Bjerg. 2010. In situ chemical oxidation of contaminated soil and groundwater using persulfate : A review. *Critical Reviews in Environmental Science and Technology*, 40(December 2013):55–91, https://doi.org/10.1080/10643380802039303.

Wagner, I., and H. Strehlow. 1982. Flash photolysis in aqueous periodate-solutions. Berichte der Bunsengesellschaft für physikalische Chemie, 86:297–301, https://doi.org/0005-9021/82/0404-0297 S 02.50/0.

Wang, J. L., and L. J. Xu. 2012. Advanced oxidation processes for wastewater treatment: Formation of hydroxyl radical and application. *Critical Reviews in Environmental Science and Technology*, 42(3):251–325, https://doi.org/10.1080/10643389.2010.507698.

Wang, Y., and C. S. Hong. 1999. Effect of hydrogen peroxide, periodate and persulfate on photocatalysis of 2-chlorobiphenyl in aqueous TiO2 suspensions. *Water Research*, 33(9):2031–2036, https://doi.org/10.1016/S0043-1354(98)00436-9.

Weavers, L. K., I. Hua, and M. R. Hoffmann. 1997. Degradation of triethanolamine and chemical oxygen demand reduction in wastewater by photoactivated periodate. *Water Environment Research*, 69(6):1112–1119, https://doi.org/10.2175/106143097X125849.

Wu, M.-C., and C.-H. Wu. 2011. Decolorization of C.I. reactive red 198 in UV/oxidant and UV/TiO2/oxidant systems. *Reaction Kinetics, Mechanisms and Catalysis*, 104(2):281–290, https://doi.org/10.1007/s11144-011-0346-8.

Yu, C. H., C. H. Wu, T. H. Ho, and P. K. Andy Hong. 2010. Decolorization of C.I. Reactive Black 5 in UV/TiO2, UV/oxidant and UV/TiO2/oxidant systems: A comparative study. *Chemical Engineering Journal*, 158(3):578–583, https://doi.org/10.1016/j.cej.2010.02.001.

5 Water Reuse and Recycling
The Great Rejuvenation to the Environment

Khushboo Dasauni, Divya and Tapan K. Nailwal

CONTENTS

DOI: 10.1201/9781003204442-5

5.1   INTRODUCTION

Water covers 70% of our planet so it is easy to think that it will always be plentiful, but only 3% of the world's water is fresh water and two-thirds of that is tucked away in frozen glaciers or otherwise unavailable for our use. As a result, some 1.1 billion people worldwide lack access to water, and a total of 2.7 billion find water scarce for at least one month of the year (Fisherman et al., 2012). Inadequate sanitation is also a problem for 2.4 billion people – they are exposed to diseases, such as cholera and typhoid fever, and other water-borne illnesses. Many water systems, such as rivers, lakes, and aquifers, that keep ecosystems thriving and sustain a growing human population have become stressed. More than half the world's wetlands have disappeared. Agriculture consumes more water than any other source and wastes much of that through inefficiencies. Climate change is altering patterns of weather and water around the world, causing shortages and droughts in some areas and floods in others. At the current consumption rate, this situation will only get worse. By 2025, two-thirds of the world's population may face water shortages. And ecosystems around the world will suffer even more (Fisherman et al., 2012).

Wastewater is 99.9% water; the remaining 0.1% is a cause for concern as it includes nutrients, oil, grease, heavy metals, organic and inorganic chemicals, pesticides, microplastics, disease-causing pathogens, etc., generated by a combination of domestic by-products, surface runoff, and agricultural, industrial, and other commercial activities (see Figure 5.1). Economic and environmental pressures, and the conservation ethic, have led to the widespread and growing application of the recycling of wastewater, including the irrigation of food and non-food crops, green spaces, recovering arid land, fire systems, industrial cooling or industrial processing, sanitation, and even as indirect and possibly direct sources of drinking water.

It is of paramount importance to develop and innovate the wastewater recycling processes and techniques such as desalination technology, bioreactors for wastewater treatment, electrodialysis, ultrafilters and nano-filters (nanotechnology), and biotechnological approaches (see Table 5.1).

Industrial and agricultural work involves the use of many chemicals that can flow into and pollute the water. Metals and solvents in industrial work can pollute rivers and lakes. They are toxic to many forms of aquatic organisms and may slow their development, make them sterile, and even cause death. Pesticides are used in agriculture to control weeds, insects, and fungi. The loss of these pesticides will cause water pollution and aquatic toxicity. As a result, the environment accumulates

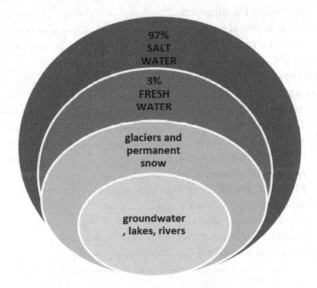

FIGURE 5.1   Percentages of water on Earth.

TABLE 5.1

Different Grades of Water

Water Grade	Definition and Reuse Applications
Wastewater	Combined domestic effluent that contains sewage.
Reclaimed water	Suitable for particular specified purpose, e.g., irrigation, toilet flushing, etc.
Grey water	Water from a potable source that has already been used for bathing, washing, laundry, or washing.
Green water	Reclaimed water treated to a relatively high standard, suitable for general use as a non-potable source in parallel with potable sources.
Drinking water	Very high-quality water assured to be suitable for drinking by humans.

high concentrations of nutrients (eutrophication). Algal blooms resulting from eutrophication can become problematic in marine habitats such as lakes. Petroleum is another form of chemical pollutant, which usually contaminates water through accidental oil spills from marine shipping. Oil spills usually only have a partial impact on wildlife but they can spread for miles. Oil will cause many fish to die and sticks to the feathers of seabirds, causing them to lose their ability to fly. Rising water temperatures may cause the death of many aquatic organisms and destroy marine habitats. For example, rising water temperatures have caused coral bleaching on coral reefs around the world. This is when corals expel the microorganisms they depend on, which results in significant damage to coral reefs and marine life. The increased temperature of the Earth's water is caused by global warming. Global warming is the process of increasing the average global temperature due to the greenhouse effect. The burning of fossil fuels releases greenhouse gases such as carbon dioxide into the atmosphere. This causes the heat from the sun to "stay" in the Earth's atmosphere, so the global temperature rises. Nuclear waste, on the other hand, is produced by industrial, medical, and scientific processes that use radioactive materials. Nuclear waste can have harmful effects on marine habitats. Nuclear waste comes from many sources, including radio waste which is generated by the operations performed by nuclear power plants. Nuclear fuel reprocessing plants in northern Europe are the largest source of man-made nuclear waste in the surrounding ocean. Radioactive traces from these plants were found in Greenland. The mining and refining of uranium is also the cause of marine nuclear waste and is generated in the nuclear fuel cycle used in many industrial, medical, and scientific processes. See Figure 5.2 and Table 5.2 for a breakdown of sources of water pollutants.

5.2 CONSEQUENCES OF WATER POLLUTION AND WATER SCARCITY

Another major source of industrial water pollution is mining. Unless proper precautions are taken, such as the use of sedimentation tanks, the grinding of ore and subsequent water treatment can cause toxic metal fines to be discharged into waterways. Lead and zinc ores usually contain less toxic cadmium. If cadmium is not recycled, it will cause serious water pollution. From 1940–1950, mining was the main source of cadmium poisoning (itai-itai disease) in Japan (Kjellstrom, 1986). Mercury can enter waterways from mining and industrial sites. Another source of environmental pollution by mercury is the incineration of medical waste containing damaged medical equipment. Because metallic mercury is highly volatile, it is also easily transferred into the atmosphere. Sulfate-reducing bacteria and other microorganisms in the underwater sediments of lakes, rivers, or coastal waters can methylate mercury, thereby increasing its toxicity. The accumulation and concentration of methylmercury in the food chain can cause serious neurological diseases or more subtle functional damage to the nervous system (Murata et al., 2004). Due to the local pollution of fish or shellfish, coastal seawater pollution may cause health hazards. For example, in 1956, a water disease broke out in Japan resulting in the mercury pollution of fish (WHO, 1976). Pollution of seawater by

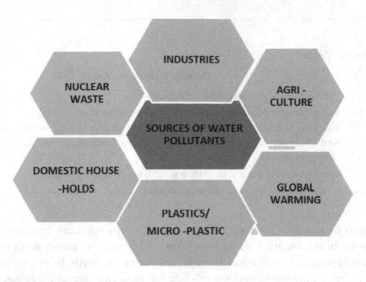

FIGURE 5.2 Sources of water pollutants.

TABLE 5.2
Water Pollutants and Their Sources

Category	Examples	Sources
AFFECTING HEALTH		
Infectious agents	Bacteria, viruses, parasites	Sewage, and human and animal excreta
Organic chemicals	Pesticides, plastics, detergents, oil	Agricultural, and domestic and industrial waste
Inorganic chemical	Acids, caustics, salts, metals	Industrial and domestic effluents
Radioactive materials	Radon, thorium, uranium, etc.	Mining, power plant, and natural resources
AFFECTING ECOSYSTEM		
Plant nutrients	Nitrates, phosphates, etc.	Chemical fertilizers, sewage, and manure
Thermal	Heat	Industries and power plants
Oxygen demanding	Agricultural waste, manure	Sewage and agricultural runoff
Sediments	Silts, soil	Soil erosion

persistent chemicals (such as polychlorinated biphenyls [PCB] and dioxins) can also pose a major health hazard, even at very low concentrations (Yassi et al., 2001).

In addition to soil and sediment that increase turbidity, agricultural runoff also carries nutrients, such as nitrogen and phosphate, which are usually added in the form of animal manure or fertilizer. These chemicals cause eutrophication (excessive levels of nutrients in the water), which increases the growth of algae and plants in the waterway, leading to an increase in cyanobacteria (blue-green algae). The toxic substances released during the decay process are harmful to the human body. In areas where agriculture is increasingly intensive, the use of nitrogen fertilizers may become a problem. These fertilizers increase the concentration of nitrate in the groundwater, resulting in high levels of nitrate in the groundwater source, which can lead to methemoglobinemia, which can cause the life-threatening "blue baby" syndrome. Exposure to nitrate-contaminated water continues to be a common public health problem in rural areas in Eastern Europe (Yassi et al., 2001).

Other examples of high local disease burden include neurological diseases caused by methylmercury poisoning (water am disease), kidney and bone diseases caused by chronic cadmium poisoning (itai-itai disease), and circulatory diseases caused by nitrate exposure (methemoglobin blood) and lead exposure (anemia and hypertension). Acute exposure to pollutants in drinking water can cause irritation or inflammation of the eyes, nose, skin, and gastrointestinal system; however, the most important health effects are due to the long-term exposure to copper, arsenic, or chromium in drinking water (for example, liver toxicity). Chemicals such as cadmium, copper, mercury, and chlorobenzene can be seen. Chemicals excreted through the kidneys target the kidneys (WHO, 2003). Pesticides and other chemical pollutants that enter waterways through agricultural runoff, storm water channels, and industrial wastewater may exist in the environment for a long time and are transported over long distances through the water or air. They may disrupt the function of the endocrine system, leading to reproductive, developmental, and behavioral problems. Endocrine disruptors can reduce fertility and increase the incidence of stillbirths, birth defects, and hormone-dependent cancers such as breast, testicular, and prostate cancer. The effects on the developing nervous system may include impaired mental and psychomotor development, as well as cognitive impairment and behavioral abnormalities (WHO and International Chemical Safety Program, 2002). Examples of endocrine disruptors include organochlorines, polychlorinated biphenyls, alkylphenols, phytoestrogens (natural estrogens in plants), and drugs such as antibiotics and synthetic sex hormones in contraceptives. Chemicals in drinking water can also cause cancer. Disinfection by-products and arsenic have always been of particular concern (International Agency for Research on Cancer, 2004).

5.3 WATER RECYCLING TECHNIQUES

Almost 59% of the water used in developed countries is for industrial purposes and these same countries are responsible for 80% of the world's industrial wastewaters. As for the developing countries, 70% of factory-generated effluent is discharged without receiving any sort of treatment, with consequent contamination of the available water resources. There is thus a clear need for treatment systems suitable for such contaminated effluents (see Figure 5.3).

In order to meet the needs of the ever-increasing population, rapid industrialization, urbanization, and agricultural growth occurred in the past few recessions, which led to the first degradation of our planet. At present and in the future, if appropriate treatment processes are adopted, wastewater will still be a huge resource of available water, enabling wastewater to meet drinking water quality standards. The various techniques that can be employed to make wastewater reusable are discussed below.

5.4 DESALINATION AND OTHER TECHNIQUES

Desalination is a recognized method of removing salt from water to produce process water, ultrapure water, or drinking water. This is done through the use of membranes (reverse osmosis and nanofiltration) and thermal processes (multiple effect distillation, evaporation, and crystallization).

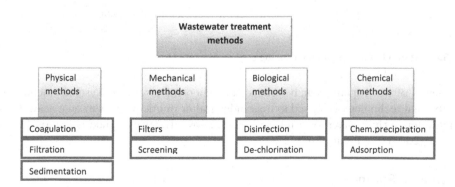

FIGURE 5.3 Wastewater treatment methods.

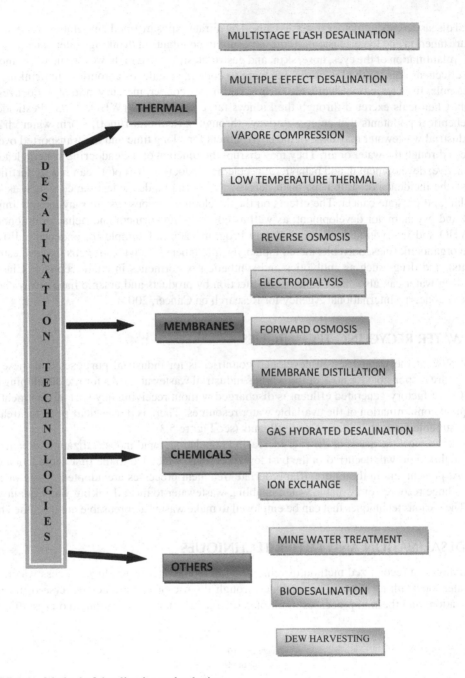

FIGURE 5.4   Method of desalination technologies.

Different levels of water quality can be produced by choosing a series of systems, including membrane or thermal technology designed to treat industrial or municipal wastewater, seawater, brackish groundwater, surface water, or drinking water. A cost-effective method is that of desalination (Miller et al., 2003) (see Figure 5.4).

5.4.1   THERMAL SOLUTIONS

The thermal desalination process is most suitable for water and process streams with a high dissolved salt content. In this case, the applied pressure required by the equivalent membrane system will be too high. If possible, the use of low-cost energy sources, such as those available from thermal power plants, is advantageous in reducing energy costs. Another option is to use a low-temperature, high-efficiency vacuum evaporation system, which can operate using waste heat and low electrical energy (Al-Karaghouli et al., 2013).

5.4.1.1 Multi-Stage Flash Desalination

Multi-stage flash (MSF) desalination processes play a vital role in providing fresh water, and they account for about 34% of the world's desalination capacity. The brine is heated by steam and then sent to a series of containers (effect) where the reduced pressure causes immediate boiling (flashing), and the steam produced is condensed in a series of stages – once-through (no brine recirculation) or brine recirculation (El-Ghonemy et al., 2017).

5.4.1.2 Multiple Effect Desalination

The multiple effect desalination (MED) device is an evaporator in which seawater is evaporated in one or more (up to 14) evaporation stages at low temperature (<70°C) to produce clean distilled water. The MED process is designed to use waste heat from power generation or chemical processes to produce distilled water and produce drinking water. In the MED process, brine is desalinated through evaporation and subsequent condensation. MED has received more attention from heat desalination technology due to its many advantages, such as the low energy consumption compared with MSF, higher total heat transfer coefficient, small specific surface area compared with MSF, low operating steam temperature, and other low-level heat signal sources that can be used to power it. MED processes can be found in various industries, such as sugar, paper and pulp, dairy products, textiles, acid, and desalination (Al-Shammiri et al., 1991 and El-Dessouky et al., 1999).

5.4.1.3 Low Temperature Thermal Desalination

For thermal desalination, if there is no steam, a temperature gradient is required. In this case, the temperature gradient in the ocean can be considered. In the ocean, temperature changes with depth. By rapidly evaporating hot water at low temperatures and condensing the resulting vapor with cold water, the available thermal gradient between warmer surface water and colder deep seawater is utilized. Today, cryogenic distillation is widely regarded as one of the promising methods for desalination. The low temperature thermal desalination (LTTD) process uses the temperature gradient between two bodies of water to evaporate hotter water at low pressure, and uses colder water to condense the resulting steam to obtain high-quality fresh water. The temperature gradient between the different layers of the ocean water column provides huge warm and cold reservoirs, which can be effectively used for power generation and desalination. The main components required by the LTTD plant are the evaporation chamber, condenser, pumps, and pipes to extract warm and cold water, and vacuum pump to maintain the pressure below the atmospheric pressure. One of the advantages of this method is that it can be implemented even when the temperature gradient between two bodies of water is about 8–10°C (Sistla et al., 2009).

5.4.1.4 Vapor Compression

The vapor compression (VC) process is particularly useful for small and medium-sized installations. The VC process relies on decompression operation to drive evaporation. The heat for evaporation is provided by vapor compression using a mechanical compressor (mechanical vapor compression [MVC]) or vapor ejector (thermal vapor compression [TVC]). MVC has proven to be effective in the production capacity of small and medium-sized plants and is very popular in mixed desalination plants. MVC is usually applied to other processing methods to improve energy efficiency (Buro et al., 2000).

5.4.2 MEMBRANE-BASED TECHNIQUES

The membrane filtration process can generally be defined as a separation process involving materials that allow certain molecules to pass through. This material is a semi-permeable membrane with a certain pore size range. The types of membrane filtration include reverse osmosis (RO), nanofiltration (NF), ultrafiltration (UF), and microfiltration (MF). Efficient membranes should have high pollutant retention rates, excellent durability, a high flux, low maintenance costs, and high chemical resistance (Praneeth et al., 2014). Membrane filtration is considered as safe and environmentally friendly (Abid et al., 2012). Membrane technology has been used to treat municipal, industrial, and textile wastewater, as well as beer wastewater. However, the application of membrane filtration has its advantages and disadvantages. The most famous advantage of membrane filtration is the quality of treated wastewater, because the system can effectively remove physical, microbiological, and chemical pollutants compared with other systems. However, due to its high maintenance and operating costs, its disadvantages outweigh its advantages, especially for application in developing countries. The pressure required for membrane filtration depends on the pore size, and the main challenge associated with membrane filtration is membrane fouling (Hosseinzadeh et al., 2013 and Pouet et al., 2014).

5.4.2.1 Ultrafiltration

Ultrafiltration (UF) is a membrane technology that works under low transmembrane pressure to remove dissolved colloidal materials. The use of ultrafiltration technology in wastewater applications is a relatively new concept; although from its initiation it has been widely used in many industrial applications, such as the food or pharmaceutical industries. Since the pore size of the membrane is larger than the dissolved metal ions in the form of hydrated ions or low-molecular-weight complexes, these ions can easily pass through. In order to obtain higher metal ion removal efficiency, micellar-enhanced ultrafiltration (MEUF) and polymer-enhanced ultrafiltration (PEUF) have been proposed (Bade et al., 2011). The MEUF process has been used to remove copper, chromate, zinc, nickel, cadmium, serine, arsenate, and organic matter such as phenol and o-cresol (Ahmad et al., 2007; Aoudia et al., 2003; Adamczak et al., 1999; Bade et al., 2007). By combining MEUF treatment with electrolysis or powdered activated carbon (PAC), metal removal rates can be improved (Bade et al., 2008 and Azoug et al., 1998). The combined treatment of MEUFACF (activated carbon fiber) can also enhance the removal of cetyl ammonium chloride (CPC) and sodium dodecyl sulfate (SDS) surfactants from MEUF (Samper et al., 2009). Surfactants have been recovered from the MEUF retentate by treating the retentate with HNO_3, H_2SO_4, HCl, and NaOH solutions, but the retentate requires further treatment (Syamal et al., 1997). It has been found that electrolysis can better separate metals and surfactants from MEUF retentate (Talens-Alesson et al., 2007; Trivunac et al., 2006; ByhlinH et al., 2003).

5.4.2.2 Reverse Osmosis

Reverse osmosis (RO) involves pushing water under pressure through a membrane that lets the water through but retains the salt and other impurities. RO membrane technology has developed since the later 1970s to a 44% share in world desalting production capacity and an 80% share in the total number of desalination plants installed worldwide (Greenlee et al., 2009). Among the different techniques available, RO has proved to be the most reliable, cost-effective, and energy efficient in producing fresh water. Pressure is used to drive water molecules across the membrane, and the energy needed to drive water molecules across the membrane is directly related to the salt concentration. Therefore, RO has been most often used for brackish waters that are lower in salt concentrations (Buros et al., 2000 and Greenlee et al., 2009).

5.4.2.3 Forward Osmosis

Osmosis, or currently called forward osmosis (FO), has new applications in the separation process of wastewater treatment, food processing, and seawater/brain water desalination. FO is an emerging low-energy desalination technology, which has many advantages compared with other conventional pressure-based RO desalination technologies. Different from the pressure-driven RO desalination process, the driving force of the FO process is the natural permeation of the concentrated extract (DS). FO uses natural osmosis to dilute the seawater feed stream by using an extraction solution with a higher osmotic pressure than the seawater feed, thereby pulling water from the feed solution through the semi-permeable membrane. Then, by using a low heat source (40°C), the extract solute is separated from the diluted extract and recovered. The solute used is usually a mixture of ammonia and carbon dioxide gas. According to reports, the specific energy consumption of the membrane part of the process is lower than 0.25 kWh/m^3 (Cath et al., 2006).

5.4.2.4 Electrodialysis

The electrodialysis (ED) method effectively removes salt from the feed water, in which the feed water cost is relatively high, but lower than the conventional distillation method of the same capacity. ED has been in use for many years and includes an electrochemical method for separating ions on a charged membrane from one solution to another under the influence of a potential difference used as a driving force. This process has been widely used in the production of drinking water and process water from brackish water and seawater, the treatment of industrial wastewater, the recovery of useful substances in wastewater, and the production of salt (Oztekin et al., 2016).

5.4.3 Chemical Solutions

5.4.3.1 Ion-Exchange Technologies

The ion-exchange (IX) technologies for water treatment are often used for water softening. The IX system can best be described as the interchange of ions between a solid phase and a liquid phase surrounding the solid. Partial softening of seawater by means of ion exchange, using the brine concentrated in the evaporators as a regenerant, reduces calcium ion concentrations to such an extent that no deposits occur (Younos et al., 2005). Chemical resins (solid phase) are designed to exchange their ions with liquid phase (feedwater) ions, which purify the water. Resins can be made using naturally occurring inorganic materials (such as zeolites) or synthetic materials. Modern IX materials are prepared from synthetic polymers tailored for different applications. IX can be used in combination with RO processes, such as blending water treated by ion exchange with RO product water to increase water production. The electrical energy used for IX desalination is 1.1 kWh/m^3. The salinity is normally up to 1,500 ppm, and produced water quality is up to 13 ppm. Briefly, saltwater (feedwater) passes over resin beads where salt ions from the saltwater are replaced by other ions. The process removes sodium and chloride ions from the feedwater, thus producing potable water. Generally, it is found that chemical approaches are too expensive to apply to the production of fresh water. IX is an exception in that it is used to soften water and to manufacture high-purity deionized water for specialty applications (Youssef et al., 2014 and Younos et al., 2005).

5.4.3.2 Gas Hydrate Desalination

Gas hydrates are crystalline solids made of the water (host) and the gas molecules (guest), such as methane, carbon dioxide, nitrogen, etc., which are held within the water cavities that are composed of hydrogen-bonded water molecules. The hydrate-based desalination (HBD) process is based on a liquid-to-solid phase change of water, such as a freezing method. In other words, a physical reaction coupled with an exclusion of ions by the hydrogen bonding of water molecules during hydrate formation is the core of the HBD process. It is observed that 1 m^3 of natural hydrate, if dissociated, can produce up to 164 m^3 of gas and 0.8 m^3 of pure water at standard temperature and pressure. Looking at the amount of pure water produced per m^3 of hydrate pellet, the process of using hydrate for desalination seems promising. It is also important to separate the solids (hydrates) from the remaining liquid phase (brines) in the process after hydrate formation. The advantages are contact heat exchange, which means that heat exchangers are not needed, and the fact that high-saline water can also be treated (Bradshaw et al., 2008).

5.4.4 OTHERS

5.4.4.1 Biological Desalination

The bio-desalination process involves the low-salt biological reservoir within seawater which serves as an ion exchanger. Sunlight energizes the desalination process by photosynthetic organisms which offer a potential opportunity to exploit biological processes for this purpose. The use of photosynthetic organisms allows for a cost- and energy-efficient desalination as compared to current alternatives (Minas et al., 2015). Cyanobacteria cultures are used in particular to generate a large biomass in brackish water and seawater, thereby forming a low-salt reservoir within the saline water. It could also be used as an ion exchanger through manipulation of transport proteins in the cell membrane (Amezaga et al., 2014). In many countries, the application of bio-desalination requires a temperature between 20°C and 40°C, and a pH range between 6 and 9 would allow for the utilization of these photosynthetic organisms. When using cyanobacteria for bio-desalination a salinity range of 0–0.5 M (NaCl), or 0–10 g/L of Na+, should be considered to reflect the possible range of salinities encountered. Based on preliminary calculations, a moderate estimate of bacterial-cell densities achieved to date (data not shown) was expected to be in the region of 1,014–1,015 cells/L. One cell could transport approximately 107 of each ion per second across the membrane barrier with ion-transport proteins present at typical densities, Therefore, 1 L of cell volume should be sufficient for the desalination of 2 L of water in 10 minutes (Minas et al., 2015).

5.4.4.2 Mine Water Treatment

Mine water is water that collects in a mine that has to be brought to the surface in order to enable the mine to continue working. In the process of coal mining, huge volumes of mine water are subsequently pumped out to the surface that early was collected in mine sumps and surface. The water that collects in the mine takes in many minerals and metals such as arsenic, cadmium, zinc, copper, mercury, and silver while flowing to the surface (acid mine drainage [AMD]). The mines are dewatered by bringing the water to the surface using wells to prevent mine flooding and groundwater contamination due to mine water runoff. This is later followed by water treatment for reuse. Mine water is becoming an attractive alternative for industrial reuse and for domestic purposes after treatment as water sources are becoming stressed (Younger et al., 2014).

5.4.4.3 Dew Harvesting

Dew harvesting is a method of collecting water vapor present in the atmosphere through condensation to garner potable water. In desert and arid areas where water is scarce, this method has proven effective for agricultural and drinking purposes. Dew harvesting consists of two types: active and passive. Currently, only passive methods for dew collection are being used; active methods still require large-scale testing. Further research and development are required in dew plants to reduce the cost per liter (0.5 INR/L is the current cost) by optimizing the techniques and materials used (Tomaszkiewicz et al., 2015).

5.4.5 TECHNOLOGICAL COMPARISON

Let us consider the LTTD process in comparison to RO and multi-effect distillation. In general, LTTD plants can make use of the large quantity of warm wastewater discharged from power plants to generate fresh water (Sistla et al., 2009). They help to bring down the discharge temperature to an acceptable level of 5°C above ambient temperature. LTTD plants do not require an external source of heat energy as compared to other thermal desalination plants. MSF and MED methods need steam, which LTTD does not require.

Moreover, as it utilizes the discharge water of power plants and as the operation temperature is below 45°C, the requirements for anti-scaling agents and antifouling agents can be eliminated. Materials like high-density polyethylene (HDPE) can be used for the pumping of seawater. RO

TABLE 5.3

Worldwide Share of Desalination Technologies

Classification	Process	Share
Desalination process	Reverse osmosis	53%
	Electro-dialysis	3%
	Multiple effect desalination	8%
	Multi-stage flash	25%
	Others	11%

technology has developed over the last 50 years, and presently holds a nearly 50% share in the world desalination market. RO plants ranging from capacities of 1,000 L/day to 300 million L/day are being manufactured today, and a claim of energy efficiency around 3.8 kW/m^3 is reported (see Table 5.3) (Venkatesan et al., 2014).

5.5 ADVANCED TECHNOLOGIES FOR WATER RECYCLING

5.5.1 BIOTECHNOLOGICAL APPROACHES

The methods based on biotechnology in wastewater treatment are:

- Activated sludge
- Trickling filter
- Anaerobic treatment
- Biofilm technology
- Biological granulation
- Microbial fuel cell (MFC)

5.5.1.1 Activated Sludge

The activated sludge wastewater treatment system has four components: aeration tank, sedimentation tank (clarification tank), return sludge pump, and a system that introduces oxygen into the aeration tank. Sometimes the pretreated and sometimes unpretreated wastewater enters the aeration tank and is mixed with the microbial suspension in the presence of oxygen. This mixture is called "mixed liquid." Microorganisms metabolize organic pollutants in wastewater. After taking an average of time equal to the hydraulic residence time in the aeration tank, the mixed liquid flows into the clarifier where the solid (mixed liquor suspended solids [MLSS]) is separated from the bulk liquid by settling to the bottom. The clarified wastewater then leaves the system. The settled solids from the bottom of the clarifier are collected and a part of the settled solids are recycled to the aeration tank; the rest are discarded. The result is the ability to control the average residence time of microorganisms in the reactor, which is called sludge age, also commonly termed as the solids retention time (SRT), or average cell residence time, also called the mean cell residence time (MCRT). The mixed liquor volatile suspended solids (MLVSS) returned to the aeration tank are microorganisms in a hungry state and have been separated from the untreated wastewater for a long time, so they are called "activated". The return of microorganisms from the clarifier to the aeration tank can increase their concentration to a high level (1,800–10,000 mg/L), and indeed characterizes the activated sludge process itself (Woodard, 2001). The growth of microorganisms in the floc is responsible for the metabolism of liquid and the removal of organic matter. The typical products of this metabolism are carbon dioxide (CO$_2$), nitrate (NO$_3^-$), sulfate (SO$_4^{2-}$), and phosphate (PO$_4^{3-}$). The nature of the floc is important because it determines the degree of separation of the sludge from the treated water and thus the efficiency of the entire process (Barbosa et al., 2007). Usually, aeration units are implemented in parallel, so shutting down a unit will not completely interrupt the plant's operation. With the widespread use of activated sludge plants, gradual aeration, extended aeration, contact stabilization, and oxidation ditch transformation have been developed (Liu and Liptak, 1997).

5.5.1.2 Trickling Filters

The name of this method is misleading because no filtration occurs. Since the 1890s, trickling filters have been used to treat wastewater. Active growths of organisms are formed on rocks, and these organisms obtain food from the waste stream that drips into the rock bed (Weiner and Matthews, 2002). It was found that if the settled wastewater passes through the rock surface, mucus will be produced on the rock and the water will become cleaner. Today, this principle is still used, but in many cases, plastic media is used instead of rocks. In most wastewater treatment systems, the trickling filter is treated by the first stage, including the second settling tank or clarifier. Trickling filters are widely used to treat household and industrial waste. This process is a fixed membrane biological treatment method designed to remove biochemical oxygen demand (BOD) and suspended solids. The trickling filter is composed of a rotating distribution arm that sprays and evenly distributes liquid wastewater on a round bed made of fist-sized rocks, other rough materials, or synthetic media. The space between the media allows air to circulate easily so that aerobic conditions can be maintained. These spaces can also allow wastewater to drip from the media, both around and above. A layer of biological mud that absorbs and consumes the waste that falls on the bed covers the medium material. The organisms decompose solids aerobically, and produce more organisms and stable wastes, which become part of the mucus or are discharged back into the wastewater through the medium. This mucus is mainly composed of bacteria but may also include algae, protozoa, worms, snails, fungi, and insect larvae. The accumulated slime occasionally drops a single medium material which is collected at the bottom of the filter together with the treated wastewater and passed to the secondary sedimentation tank where it is then removed (Spellman, 2003).

5.5.1.3 Anaerobic Treatment

Anaerobic digestion (AD) is a process in which several microbial communities act on complex organic materials where they decompose complex substances into simpler compounds without oxygen (Cioabla et al., 2012). Anaerobic digesters have been used to treat industrial wastewater, including agricultural wastewater and water used in food and beverages. AD is also used to treat solid waste, sludge from sewage treatment plants, and municipal waste. This process is used in the first stage of the treatment process (Lam et al., 2011). Since organic waste is used for energy production, AD can reduce waste disposal (Cioabla et al., 2012). The AD of wastewater is a well-known technology that has been widely used in agricultural industrial processes, such as the treatment of wastewater from palm oil mill effluent and sago factories. In conventional wastewater treatment, the wastewater is treated by AD, and methane and hydrogen sulfide are released, producing an unpleasant smell. By using balloons to cover ponds or pools to control the environment of the AD system, methane can be captured and used for energy recovery or production. The biochemical reactions involved in the anaerobic digestion process include hydrolysis, acid production, acetic acid production, and methane production. In hydrolysis, carbohydrates, proteins, and lipids are hydrolyzed into their respective monomers (Lam et al., 2011). AD is mainly used for waste treatment because it produces biogas, especially methane. Various treatment studies using AD have been reported, involving the treatment of solid waste co-digested with wastewater or using two solid wastes. The factors affecting the bioconversion process include the design of the reactor, the characteristics of the raw materials and the operating conditions. In terms of operating conditions, pH and temperature are important parameters. The optimal temperature is 35°C, and the optimal pH range is about 6.8–7.2, but the pH that the system can withstand is 6.5–8.0 (Cioabla et al., 2012). The integration of AD and microalgae can convert wastewater into beneficial products, such as fertilizer, ammonium sulfate ($[NH_4]_2SO_4$), biogas, crude oil, and other value-added products (Zhou et al., 2002).

5.5.1.4 Biofilm Technology

One of the biological methods that have overcome the problem of bioremediation is biofilm. According to Decho et al. (2000), biofilm-mediated bioremediation provides a capable and safer option for the bioremediation of plankton microorganisms. The definition of biofilm itself is simply defined as the microbial communities or clusters attached to the surface (O'Toole et al., 2000 and Singh et al., 2006). Biofilm is a carrier in which extracellular polymeric substances (EPS) are used to form a thick layer of microbial cells. The formation of biofilm is a natural process in which no chemical substances are used or added, and can be achieved by single or multiple microorganisms that have the ability to form on biological and non-biological surfaces (O'Toole et al., 2000). Generally, there are a few important steps for the development of biofilms. These steps start with the initial attachment and establishment to the surface, followed by maturation, and finally the separation of cells from the surface (O'Toole et al., 2000; Singh et al., 2006; Watnick et al., 2000). EPS are made of polysaccharides, nucleic acids, proteins, or phospholipids. EPS help microbial cells to bind to the surface, thereby protecting microbes from environmental aggressions through a diffusion barrier. This barrier can prevent toxic substances, phagocytes, and fungicides from damaging the cells, but it can also prevent the nutrients needed for cell growth from reaching the cells (Gao et al., 2011).

Biofilm can be used in many types of reactors, including moving bed biofilm reactors (MBBR), air flotation reactors (ALR), continuous stirred tank reactors (CSTR), packed bed reactors (PBR), upflow anaerobic sludge bed (UASB) reactors, fluidized bed reactors (FBR), and expanded granular sludge bed (EGSB) reactors, but they are usually part of MBBRs and fixed bed biofilm reactors. Biofilms can withstand toxic substances because of their high biomass retention capacity, rich microbial species, and better process stability (Wang et al., 2016). However, one of the main disadvantages of using biofilms is bioclogging. This process may occur when the biofilm continues to grow on the solid medium. However, in biological aerated filters, biofilms play a key role in the removal of nitrogen and carbon and help decompose pollutants into less toxic or non-toxic substances (Wang et al., 2016).

5.5.1.5 Aerobic Granulation Technology

In the late 1990s, the invention of a new type of microbial self-immobilization process called biogranulation took place (Morgenroth et al., 1997). In the biological granulation method, granular sludge produces higher biomass retention and reusability, resulting in a more reasonable selection of bio-enhanced bacterial strains and higher microbial density (millions of bacteria per gram of biomass cell) (Liu et al., 2015). Compared with traditional flocculent sludge, granular sludge has excellent settling performance, so it can withstand high-strength organic wastewater and its impact load, and thus has a higher biomass retention capacity and dense microbial structure (Zhu et al., 2013). Biological granulation, which can be processed in a fixed feed, reaction, sedimentation, and can precipitation, is categorized into two types of granular sludge, namely aerobic granular sludge (AGS) and anaerobic granular sludge (AnGS) (Ibrahim et al., 2010). However, AnGS shows some shortcomings, such as long startup time, strict anaerobic environment, relatively high working temperature, which is not suitable for low-concentration organic wastewater, and low efficiency in removing nutrients (nitrogen and phosphate) from wastewater (Liu et al., 2004). On the other hand, AGS 116 (Seow et al., 2016) can overcome all the shortcomings of AnGS mentioned above, thus improving the effectiveness of AGS in treating industrial raw wastewater. Some researchers believe that AGS is a suspended spherical biofilm, including microbial cells, inert particles, degradable particles, and EPS (Liu and Tay, 2004). In the initial stage of biological granulation, the biomass in the reactor will mainly consist of filamentous hypha particles, hyphae deposits formed by fungi and filamentous bacteria, which proved to be very good and can be retained in the reactor. Bacteria do not have this special property and will therefore be almost completely washed away. The mycelial sediment seems to act as a fixed matrix in which bacteria can grow colonies. When the mycelial sediments disperse due to the internal decomposition of the sediment, the bacterial colonies can maintain themselves because they are now large enough to settle. These micro-colonies further grow into denser granular sludge, and with the progress of granulation and eventually bacteria dominate the reactor (Beun et al., 1999, 2002).

5.5.1.6 Microbial Fuel Cell Technology

Recently, the use of power generation to treat wastewater via the application of microbial fuel cell (MFC) technology has been widely reported. MFC is basically a biochemical device to convert the chemical energy present in organic matter (such as glucose) into electrical energy where bacteria are used as biocatalysts (Zhang et al., 2008 and Kim et al., 2007). MFC consists of three components, an anaerobic anode chamber, a cathode chamber, and a proton exchange membrane (PEM) or salt bridge. In MFC technology, the PEM is used to separate the two chambers and only allows protons (H +) to be transferred from the anode chamber to the cathode chamber. The extracellular electron transfer mechanism in MFC is the transfer of electrons from bacteria to the anode, and this can be achieved in three different ways: (1) through direct outer membrane c-type cytochrome transfer, (2) through electron mediators added or produced by the microorganism itself, or (3) through conductive fimbria. Bacteria obtain energy by transferring electrons from their central metabolic system to the anode, which is the ultimate electron acceptor in MFC. Then, through an external circuit the electrons are conducted to the cathode where they combine with oxygen and H + to form water. At present, in MFC technology to generate electricity, both a mixed culture and pure culture of bacteria have been used (Zhang et al., 2008; Hassan et al., 2012; Liu et al., 2010; Rezaei et al., 2009).

5.5.2 NANOTECHNOLOGICAL APPROACHES

Nanomaterials have unique size-dependent properties related to their high specific surface area (fast dissolution, high reactivity, strong adsorption) and discontinuous properties (such as super-paramagnetism, local surface plasmon resonance, and quantum confinement effects). These specific nano-based properties allow for the development of new high-tech materials for more effective water and wastewater treatment processes, namely membranes, adsorbent materials, nano-catalysts, functionalized surfaces, coatings, and reagents.

5.5.2.1 Nano Material-Based Adsorption

Adsorption is the ability of all solid substances to attract gas or solution molecules in close contact to their surface. Solid used to adsorb gases or dissolved substances are called adsorbents, and the adsorbed molecules are usually collectively referred to as adsorbed substances. Due to their high specific surface area, compared with granular or powdered activated carbon, nanoadsorbents have a much higher adsorption rate for organic compounds. They have great potential for novel, more efficient, and faster decontamination processes, designed to remove organic and inorganic pollutants, such as heavy metals and micro pollutants from wastewater or polluted water. Current research activities are mainly focused on the following types of nanoadsorbents:

- Carbon-based nanoadsorbents
- Polymeric nanoadsorbents
- Zeolites
- Metal-based nanoadsorbents

5.5.2.1.1 Carbon-Based Nanoadsorbents

Carbon nanotubes (CNTs) are allotropes of carbon with cylindrical nanostructures. According to their manufacturing process, CNTs are divided into single-wall nanotubes, double wall nanotubes, and multi-wall nanotubes. They can be used to adsorb persistent pollutants and to pre-concentrate and detect pollutants. Metal ions can be absorbed by CNTs through electrostatic attraction and chemical bonding. In addition, CNTs exhibit antibacterial properties by causing oxidative stress in bacteria and destroying cell membranes. Although chemical oxidation occurs, no toxic by-products are produced, which is an important advantage over conventional disinfection processes (such as chlorination and ozonation).

Yan et al. developed plasma-modified ultralong CNTs. CNTs have an extremely high salt-specific adsorption capacity (over 400% by weight). These ultralong CNTs can be realized in a multifunctional membrane, which can not only remove salt, but also organic and metal contaminants. The next-generation drinking water purification equipment equipped with these new CNTs is expected to have excellent desalination, disinfection, and filtration performance.

5.5.2.1.2　Polymeric Nanoadsorbents

Polymer nanoadsorbents, such as dendrimers (repeatingly branched molecules), can be used to remove organics and heavy metals. Organic compounds can be adsorbed by the internal hydrophobic shell, while heavy metals can be adsorbed by custom-made external branches (Hajeh et al., 2013 and Diallo et al., 2005). Integrated dendrimers in ultrafiltration devices are used to remove copper from water. By using the combined dendrimer-ultrafiltration system, almost all copper ions are recovered. The adsorbent can be regenerated only by changing the pH. However, due to the complex multi-stage synthesis of dendrimers, so far, apart from some recently established companies in the People's Republic of China, there are no commercial suppliers. Sadeghi-Kiakhani et al. (2013) produced a highly efficient biosorbent for removing anionic compounds, such as dyes, from textile wastewater by preparing a combined chitosan and side branch nanostructure. The biosorbent is biodegradable, biocompatible, and non-toxic. They can increase the removal rate of certain dyes up to 99%.

5.5.2.1.3　Zeolites

Since the early 1980s, zeolite has been combined with silver atoms. Zeolite has a very porous structure and can be embedded with nanoparticles, such as silver ions (Baker et al., 1985). By exchanging with other cations in the solution, they are released from the zeolite matrix (Egger et al., 2009) compared various materials containing nano-silver, including zeolite. As shown by Agion® (Sciessent LLC, Wakefield, M.A., U.S.A.), when used for sanitary purposes, silver attacks microorganisms and inhibits their growth (Nagy et al., 2011). When in contact with liquid, a small amount of silver ions will be released from the metal surface. Petrik et al. (2014) demonstrated the success of this composition in water disinfection.

5.5.2.1.4　Metal-Based Nanoadsorbents

Nanoscale metal oxides are a promising alternative to activated carbon and are effective adsorbents which can remove heavy metals and radionuclides. In addition to having a high specific surface area they also have a short intra-particle diffusion distance and can be compressed without significantly reducing the surface area. Some of these nano-scale metal oxides (for example, nano-magnetite and nano-magnetite) are superparamagnetic, which facilitates separation and recovery by low-gradient magnetic fields. They can be used in the adsorption media filters and slurry reactors (Qu et al., 2013). Nanometer iron hydroxide (α-FeO[OH]) is a strong, wear-resistant adsorbent with a large specific surface area which can adsorb arsenic from wastewater and drinking water (Aredes et al., 2012). ArsenXnp (SolmeteX Inc., Philadelphia, Pennsylvania, U.S.A.) is a commercially available hybrid ion exchange medium containing iron oxide nanoparticles and polymers. It is very efficient in removing arsenic and requires almost no backwashing (see Table 5.4).

5.5.3　Membranes and Membrane Process

Membrane separation processes are rapidly advancing applications for water and wastewater treatments. Membranes provide a physical barrier for substances, depending on their pore size and

TABLE 5.4

Nano-Metal and Nano-Metal Oxides-Based Adsorbents

Nanometals and Nanometal Oxide	Properties (Positive)	Properties (Negative)	Application	Novel Approaches
Nanosilver and nano-TiO2	Bactericidal, low human toxicity; Nano-TiO$_2$: high chemical stability, very long lifetime	Nanosilver: limited durability; Nano-TiO$_2$ requires UV activation	Water disinfection, antifouling surfaces, decontamination of organic compounds	TiO$_2$ modification for activation by visible light, TiO$_2$ nanotubes
Magnetic nanoparticles	Simple recovery by magnetic field	Stabilization is required	Groundwater remediation	Forward osmosis
Nano zero-valent iron	Highly reactive	Stabilization is required (surface modification)	Groundwater remediation (chlorinated hydrocarbon, perchlorate)	Entrapment in polmeric matrices for stabilization

molecule size. Membrane technology is well established in the water and wastewater area as a reliable and largely automated process. Novel research activities with regard to nanotechnology focus on improving selectivity and flux efficiency by developing antifouling layers. The following sections describe the state-of-the-art field of nanoengineered membrane filtration (see Table 5.5).

5.6 CONCLUSION AND FUTURE PERSPECTIVES

The ultimate goal of wastewater treatment is to protect the environment in a manner commensurate with public health and socio-economic concerns. Understanding the nature of wastewater is essential for designing appropriate treatment technologies to ensure the safety, effectiveness, and quality of the treated wastewater. In addition, it is recommended to improve public education to ensure awareness of technology and its environmental and economic benefits. Also, by comparing the advantages and disadvantages of each method, the traditional and modern methods for wastewater treatment are studied in detail. Industrial and agricultural industrial wastewater is considered to be a highly polluting form of wastewater. However, reusing nutrients by cultivating microalgae in these wastewaters helps to treat wastewater with a high nutrient content. The harvested microalgal biomass can be used as animal feed, fertilizer, and biofuel raw materials. The challenges faced by these wastewater treatment processes are mainly related to internal shading, high suspended solids content, and the harvest and recovery of end-use microalgae. These problems can be overcome by using PBR or applying turbulence in a watercourse pool. The application of microalgae systems in industrial wastewater treatment requires very little energy and reduces operation and maintenance costs. Nano-engineered materials, such as nano-sorbents, nano-metals, nano-films, and photocatalysts, provide the potential for new water technologies that can be easily adapted to customer-specific applications. Most of them are compatible with existing processing technologies and can be simply integrated into conventional modules. Compared with conventional water technology, one of the most important advantages of nanomaterials is their ability to integrate various characteristics, thus forming a multifunctional system, such as nanocomposite membranes, which can retain particles and eliminate pollutants. Desalination technology has been developed to a high-tech preparation level. Many technologies have been adopted and successfully used worldwide. Each technique is site-specific, and the thermal technique differs depending on whether steam is available. Membrane technologies such as RO have become very popular, and because of lower energy requirements compared to thermal systems, they have become cheaper commercially. However, any technology needs to consider

TABLE 5.5
Different Types of Nanomembranes

Nanomembranes	Properties (Positive)	Properties (Negative)	Applications	Novel Approaches	References
Nano-filtration membranes	Charged base repulsion, relative low pressure, high selectivity	Membrane blocking	Reduction of hardness, color, odor, heavy metals	Seawater desalination	(Gehrke et al., 2012, 2015)
Nano-composite membranes	Increased hydrophobicity, water permeability, thermal/mechanical robustness	Resistant bulk material required when using oxidizing nanomaterial, possible release of nanoparticles	Highly dependent on type of composite, e.g., RO, removal of micro-pollutant	Bio nano-composite membranes	(Wegmann et al., 2008; Feng et al., 2013; Kim et al., 2011; Fathizadeh et al., 2011)
Self-assembling membranes	Homogeneous nanopores, tailor-made membranes	Small quantities available (laboratory scale)	ultrafiltration	Process scale up	(Whitesides et al., 2002)
Nanofiber membrane	High porosity, higher permeate efficiency, bactericidal	Pore blocking, possible release of nanofibers	Filter cartridge, ultrafiltration, prefiltration,	Composite Nanofiber membranes, bionanofiber membranes	(Wegmann et al., 2008; Ramakrishna et al., 2006; Feng et al., 2013)
Aquaporin-based membrane	High ionic selectivity and permeability	Mechanical weakness	Low pressure desalination	Stabilization process (surface imprinting, embedding in polymer)	(Tang et al., 2013; Xie et al., 2013)

environmental factors. In the near future, technology based on the enzyme treatment of dyes present in wastewater or industrial wastewater will play a vital role. By using a reactor containing immobilized enzymes, it is also possible to treat wastewater on a large scale. Different types of anaerobic/aerobic bioreactors have been extensively studied, and it is obvious from different combinations of wastewater treatment technologies that regarding the optimal process configuration – that is, the type and coupling of anaerobic/aerobic bioreactors – there is still a lot of research to be done. The number of bioreactors with membrane modules needs to be studied. This includes process conditions, materials of construction, prevention/reduction of membrane fouling, etc. Therefore, there is an urgent need for novel and advanced water technologies, especially through the use of flexibly adjustable water treatment systems to ensure the high quality of drinking water, eliminate micro-pollutants, and enhance industrial production processes.

REFERENCES

Abid, M.F., Zablouk, M.A. and Abid-Alameer, A.M., 2012. Experimental study of dye removal from industrial wastewater by membrane technologies of reverse osmosis and nanofiltration. *Iranian Journal of Environmental Health Science & Engineering*, 9(1), p.17.

Adamczak, H. and Szymanowski, J., 1999. Ultrafiltration of micellar solutions containing phenols and oxyethylated methyl dodecanoates of various hydrophilicity. *Rivista Italiana delle Sostanze Grasse*, 76(12), pp.557–564.

Ahmad, A.L. and Puasa, S.W., 2007. Reactive dyes decolourization from an aqueous solution by combined coagulation/micellar-enhanced ultrafiltration process. *Chemical Engineering Journal*, 132(1–3), pp.257–265.

Al-Karaghouli, A. and Kazmerski, L.L., 2013. Energy consumption and water production cost of conventional and renewable-energy-powered desalination processes. *Renewable and Sustainable Energy Reviews*, 24, pp.343–356.

Al-Shammiri, M. and Safar, M., 1999. Multi-effect distillation plants: state of the art. *Desalination*, 126(1–3), pp.45–59.

Amezaga, J.M., Amtmann, A., Biggs, C.A., Bond, T., Gandy, C.J., Honsbein, A., Karunakaran, E., Lawton, L., Madsen, M.A., Minas, K. and Templeton, M.R., 2014. Biodesalination: a case study for applications of photosynthetic bacteria in water treatment. *Plant Physiology*, 164(4), pp.1661–1676.

Aoudia, M., Allal, N., Djennet, A. and Toumi, L., 2003. Dynamic micellar enhanced ultrafiltration: use of anionic (SDS)–nonionic (NPE) system to remove Cr3+ at low surfactant concentration. *Journal of Membrane Science*, 217(1–2), pp.181–192.

Aredes, S., Klein, B. and Pawlik, M., 2013. The removal of arsenic from water using natural iron oxide minerals. *Journal of Cleaner Production*, 60, pp.71–76.

Azoug, C., Steinchen, A., Charbit, F. and Charbit, G., 1998. Ultrafiltration of sodium dodecylsulfate solutions. *Journal of Membrane Science*, 145(2), pp.185–197.

Bade, R. and Lee, S.H., 2007. Micellar enhanced ultrafiltration and activated carbon fibre hybrid processes for copper removal from wastewater. *Korean Journal of Chemical Engineering*, 24(2), pp.239–245.

Bade, R. and Lee, S.H., 2008. Chromate removal from wastewater using micellar enhanced ultrafiltration and activated carbon fibre processes: validation of experiment with mathematical equations. *Environmental Engineering Research*, 13(2), pp.98–104.

Bade, R. and Lee, S.H., 2011. A review of studies on micellar enhanced ultrafiltration for heavy metals removal from wastewater. *Journal of Water Sustainability*, 1(1), pp.85–102.

Barbosa, V.L., Tandlich, R. and Burgess, J.E., 2007. Bioremediation of trace organic compounds found in precious metals refineries' wastewaters: a review of potential options. *Chemosphere*, 68(7), pp.1195–1203.

Beun, J.J., Hendriks, A., van Loosdrecht, M.C.M., Morgenroth, E., Wilderer, P.A. and Heijnen, J.J., 1999. Aerobic granulation in a sequencing batch reactor. *Water Research*, 33, pp. 2283–2290.

Beun, J.J., Van Loosdrecht, M.C.M. and Heijnen, J.J., 2002. Aerobic granulation in a sequencing batch airlift reactor. *Water Research*, 36(3), pp.702–712.

Buros, O.K., 2000. *The ABCs of Desalting* (p. 30). Topsfield, MA: International Desalination Association.

Byhlin, H. and Jönsson, A.S., 2003. Influence of adsorption and concentration polarisation on membrane performance during ultrafiltration of a non-ionic surfactant. *Desalination*, 151(1), pp.21–31.

Cath, T.Y., Childress, A.E. and Elimelech, M., 2006. Forward osmosis: principles, applications, and recent developments. *Journal of Membrane Science*, 281(1–2), pp.70–87.

Cioabla, A.E., Ionel, I., Dumitrel, G.A. and Popescu, F., 2012. Comparative study on factors affecting anaerobic digestion of agricultural vegetal residues. *Biotechnology for Biofuels*, 5(1), pp.1–9.

Decho, A.W., 2000. Microbial biofilms in intertidal systems: an overview. *Continental Shelf Research*, 20, pp.1257–1273.

Diallo, M.S., Christie, S., Swaminathan, P., Johnson, J.H. and Goddard, W.A., 2005. Dendrimer enhanced ultrafiltration. 1. Recovery of Cu (II) from aqueous solutions using PAMAM dendrimers with ethylene diamine core and terminal NH2 groups. *Environmental Science & Technology*, 39(5), pp.1366–1377.

Egger, S., Lehmann, R.P., Height, M.J., Loessner, M.J. and Schuppler, M., 2009. Antimicrobial properties of a novel silver-silica nanocomposite material. *Applied and Environmental Microbiology*, 75(9), pp.2973–2976.

El-Dessouky, H.T. and Ettouney, H.M., 1999. Multiple-effect evaporation desalination systems. Thermal analysis. *Desalination*, 125(1–3), pp.259–276.

El-Ghonemy, A.M.K., 2018. Performance test of a sea water multi-stage flash distillation plant: case study. *Alexandria Engineering Journal*, 57(4), pp.2401–2413.

Fathizadeh, M., Aroujalian, A. and Raisi, A., 2011. Effect of added NaX nano-zeolite into polyamide as a top thin layer of membrane on water flux and salt rejection in a reverse osmosis process. *Journal of Membrane Science*, 375(1–2), pp.88–95.

Feng, C., Khulbe, K.C., Matsuura, T., Tabe, S. and Ismail, A.F., 2013. Preparation and characterization of electro-spun nanofiber membranes and their possible applications in water treatment. *Separation and Purification Technology*, 102, pp.118–135.

Fishman, C., 2012. *The Big Thirst: The Secret Life and Turbulent Future of Water.* Simon and Schuster.

Gao, D., Liu, L. and Wu, W.M., 2011. Comparison of four enhancement strategies for aerobic granulation in sequencing batch reactors. *Journal of Hazardous Materials*, 186(1), pp.320–327.

Gehrke, I., Geiser, A. and Somborn-Schulz, A., 2015. Innovations in nanotechnology for water treatment. *Nanotechnology, Science and Applications*, 8, p.1.

Gehrke, I., Keuter, V. and Groß, F., 2012. Development of nanocomposite membranes with photocatalytic surfaces. *Journal of Nanoscience and Nanotechnology*, 12(12), pp.9163–9168.

Greenlee, L.F., Lawler, D.F., Freeman, B.D., Marrot, B. and Moulin, P., 2009. Reverse osmosis desalination: water sources, technology, and today's challenges. *Water Research*, 43(9), pp.2317–2348.

Hajeh M., Laurent S. and Dastafkan K., 2013. Nanoadsorbents: classification, preparation, and applications (with emphasis on aqueous media). *Chemical Reviews*, 113:S7728–S7768.

Hassan, S.H., Kim, Y.S. and Oh, S.E., 2012. Power generation from cellulose using mixed and pure cultures of cellulose-degrading bacteria in a microbial fuel cell. *Enzyme and Microbial Technology*, 51(5), pp.269–273.

Hosseinzadeh, M., Bidhendi, G.N., Torabian, A. and Mehrdadi, N., 2013. Evaluation of membrane bioreactor for advanced treatment of industrial wastewater and reverse osmosis pretreatment. *Journal of Environmental Health Science and Engineering*, 11(1), p.34.

Ibrahim, Z., Amin, M.F.M., Yahya, A., Aris, A. and Muda, K., 2010. Characteristics of developed granules containing selected decolourising bacteria for the degradation of textile wastewater. *Water Science and Technology*, 61(5), pp.1279–1288.

Kim, B.H., 2007. Challenges in microbial fuel cell development and operation. *Applied Microbiology and Biotechnology*, 76, pp. 485–494.

Kim, E.S. and Deng, B., 2011. Fabrication of polyamide thin-film nano-composite (PA-TFN) membrane with hydrophilized ordered mesoporous carbon (H-OMC) for water purifications. *Journal of Membrane Science*, 375(1–2), pp.46–54.

Kjellström, T., 1986. Critical organs, critical concentrations, and whole body dose-response relationships. In *Cadmium and Health: A Toxicological and Epidemiological Appraisal, Vol. 2, Effects and Response* (pp.231–246). CRC Press, Boca Raton, FL.

Lam, M.K. and Lee, K.T., 2011. Renewable and sustainable bioenergies production from palm oil mill effluent (POME): win–win strategies toward better environmental protection. *Biotechnology Advances*, 29(1), pp.124–141.

Liu D.H.F. and Liptak B.G., 1997. *Environmental Engineers' Handbook*, Second Edition, CRC, Boca Raton, FL.

Liu, M., Yuan, Y., Zhang, L.-X., Zhuang, L., Zhou, S.-G. and Ni, J.-R., 2010. Bioelectricity generation by a Gram-positive Corynebacterium sp. strain MFC03 under alkaline condition in microbial fuel cells. *Bioresource Technology*, 101, pp. 1807–1811.

Liu, Y. and Tay, J.H., 2004. State of the art of biogranulation technology for wastewater treatment. *Biotechnology Advances*, 22(7), pp.533–563.

Liu, Y., Kang, X., Li, X. and Yuan, Y., 2015. Performance of aerobic granular sludge in a sequencing batch bioreactor for slaughterhouse wastewater treatment. *Bioresource Technology*, *190*, pp.487–491.

Liu, Y.Q., Liu, Y. and Tay, J.H., 2004. The effects of extracellular polymeric substances on the formation and stability of biogranules. *Applied Microbiology and Biotechnology*, *65*(2), pp.143–148.

Miller, J.E., 2003. *Review of Water Resources and Desalination Technologies* (Vol. 49, pp.2003–0800). Sandia National Laboratories, Albuquerque, NM.

Minas, K., Karunakaran, E., Bond, T., Gandy, C., Honsbein, A., Madsen, M., Amezaga, J., Amtmann, A., Templeton, M.R., Biggs, C.A. and Lawton, L., 2015. Biodesalination: an emerging technology for targeted removal of Na+ and Cl– from seawater by cyanobacteria. *Desalination and Water Treatment*, *55*(10), pp.2647–2668.

Morgenroth, E., Sherden, T., Van Loosdrecht, M.C.M., Heijnen, J.J. and Wilderer, P.A., 1997. Aerobic granular sludge in a sequencing batch reactor. *Water Research*, *31*(12), 3191–3194.

Murata, K., Weihe, P., Budtz-Jørgensen, E., Jørgensen, P.J. and Grandjean, P., 2004. Delayed brainstem auditory evoked potential latencies in 14-year-old children exposed to methylmercury. *The Journal of Pediatrics*, *144*(2), pp.177–183.

Nagy, A., Harrison, A., Sabbani, S., Munson Jr, R.S., Dutta, P.K. and Waldman, W.J., 2011. Silver nanoparticles embedded in zeolite membranes: release of silver ions and mechanism of antibacterial action. *International Journal of Nanomedicine*, *6*, p.1833.

O'Toole, G., Kaplan, H.B. and Kolter, R., 2000. Biofilm formation as microbial development. *Annual Review of Microbiology*, *54*, 49–79.

Oztekin, E. and Altin, S., 2016. Wastewater treatment by electrodialysis system and fouling problems. *Turkish Online Journal of Science & Technology*, *6*(1).

Petrik, L., Missengue, R., Fatoba, O., Tuffin, M. and Sachs, J., 2012. Silver/zeolite nano composite-based clay filters for water disinfection. In *Water Research Commission, WRC Report (KV 297/12)*. ISSN: 9781431203062.

Pouet, M.F., Grasmick, A., Homer, F., Nauleau, F. and Cornier, J.C., 1994. Tertiary treatment of urban wastewater by cross flow microfiltration. *Water Science and Technology*, *30*(4), p.133.

Praneeth, K., Manjunath, D., Bhargava, S.K., Tardio, J. and Sridhar, S., 2014. Economical treatment of reverse osmosis rejects of textile industry effluent by electrodialysis–evaporation integrated process. *Desalination*, *333*(1), pp.82–91.

Qu, X., Alvarez, P.J. and Li, Q., 2013. Applications of nanotechnology in water and wastewater treatment. *Water Research*, *47*(12), 3931–3946.

Ramakrishna S., Fujihara K., Teo W.E., Yong T., Ma Z.W. and Ramaseshan R., 2006. Electrospun nanofibers: solving global issues. *Mater Today*, 9: 40–50.

Rezaei, F., Xing, D., Wagner, R., Regan, J.M., Richard, T.L. and Logan, B.E., 2009. Simultaneous cellulose degradation and electricity production by Enterobacter cloacae in a microbial fuel cell. *Applied and Environmental Microbiology*, *75*(11), pp.3673–3678.

Sadeghi-Kiakhani M., Mokhtar Arami M. and Gharanjig K., 2013. Dye removal from colored-textile wastewater using chitosan-PPI dendrimer hybrid as a biopolymer: optimization, kinetic, and isotherm studies. *Journal of Applied Polymer Science*, 127:2607–2619.

Samper, E., Rodríguez, M., De la Rubia, M.A. and Prats, D., 2009. Removal of metal ions at low concentration by micellar-enhanced ultrafiltration (MEUF) using sodium dodecyl sulfate (SDS) and linear alkylbenzene sulfonate (LAS). *Separation and Purification Technology*, *65*(3), pp.337–342.

Seow, T.W., Lim, C.K., Nor, M.H.M., Mubarak, M.F.M., Lam, C.Y., Yahya, A. and Ibrahim, Z., 2016. Review on wastewater treatment technologies. *International Journal of Applied Environmental Sciences*, *11*(1), pp.111–126.

Simmons, B.A., Bradshaw, R.W., Dedrick, D.E., Cygan, R.T., Greathouse, J.A. and Majzoub, E.H., 2008. *Desalination Utilizing Clathrate Hydrates (LDRD Final Report)* (No. SAND2007-6565). Sandia National Laboratories.

Singh, R., Paul, B. and Jain, R.K., 2006. Biofilms: implications in bioremediation. *Trends in Microbiology*, 14, pp.389–397.

Sistla, P.V., Venkatesan, G., Jalihal, P. and Kathiroli, S., 2009, January. Low temperature thermal desalination plants. In Eighth ISOPE Ocean Mining Symposium. International Society of Offshore and Polar Engineers.

Spellman F.R., 2003. *Handbook of Water and Wastewater Treatment Plant Operations*, Lewis Publishers, Ririe, ID.

Syamal M., De S. and Bhattacharya P.K., 1997 Phenol solubilization by cetyl pyridinium chloride micelles in micellar enhanced ultrafiltration. *Journal of Membrane Science*, Elsevier, 137(1–2): 99–107.

Talens-Alesson, F.I., 2007. Behaviour of SDS micelles bound to mixtures of divalent and trivalent cations during ultrafiltration. *Colloids and Surfaces A: Physicochemical and Engineering Aspects*, *299*(1–3), pp.169–179.

Tomaszkiewicz, M., Abou Najm, M., Beysens, D., Alameddine, I. and El-Fadel, M., 2015. Dew as a sustainable non-conventional water resource: a critical review. *Environmental Reviews*, *23*(4), pp.425–442.

Trivunac, K. and Stevanovic, S., 2006. Removal of heavy metal ions from water by complexation-assisted ultrafiltration. *Chemosphere*, *64*(3), pp.486–491.

Venkatesan, R., 2014. Comparison between LTTD and RO process of sea-water desalination: an integrated economic, environmental and ecological framework. *Current Science*, pp.378–386.

Wang, Z., Gao, M., Wei, J., Ma, K., Zhang, J., Yang, Y. and Yu, S., 2016. Extracellular polymeric substances, microbial activity and microbial community of biofilm and suspended sludge at different divalent cadmium concentrations. *Bioresource Technology*, *205*, pp.213–221.

Watnick, P. and Kolter, R., 2000. Biofilm, city of microbes. *Journal of Bacteriology*, 182, pp.2675–2679.

Wegmann, M., Michen, B. and Graule, T., 2008. Nanostructured surface modification of microporous ceramics for efficient virus filtration. *Journal of the European Ceramic Society*, 28(8), pp.1603–1612.

Weiner R.F. and Matthews R.A., 2002. *Environmental Engineering*, Fourth Edition, Butterworth-Heinemann.

Whitesides G.M. and Grzybowski B., 2002. Self-assembly at all scales. *Science*, 295:2418–2421.

Woodard, F., 2001. *Industrial Waste Treatment Handbook*. Elsevier, Amsterdam.

Xie, W., He, F., Wang, B., Chung, T.S., Jeyaseelan, K., Armugam, A. and Tong, Y.W., 2013. An aquaporin-based vesicle-embedded polymeric membrane for low energy water filtration. *Journal of Materials Chemistry A*, *1*(26), pp.7592–7600.

Yassi, A., Kjellström, T., De Kok, T. and Guidotti, T.L., 2001. *Basic Environmental Health*. Oxford University Press, Oxford.

Younger, P.L. and Wolkersdorfer, C., 2004. Mining impacts on the fresh water environment: technical and managerial guidelines for catchment scale management. *Mine Water and the Environment*, 23, p.s2.

Younos, T. and Tulou, K.E., 2005. Overview of desalination techniques. *Journal of Contemporary Water Research & Education*, *132*(1), pp.3–10.

Youssef, P.G., Al-Dadah, R.K. and Mahmoud, S.M., 2014. Comparative analysis of desalination technologies. *Energy Procedia*, *61*, pp.2604–2607.

Zhang, T., Cui, C., Chen, S., Yang, H. and Shen, P., 2008. The direct electrocatalysis of Escherichia coli through electroactivated excretion in microbial fuel cell. *Electrochemistry Communications*, *10*(2), pp.293–297.

Zhou, H. and Smith, D.W., 2002. Advanced technologies in water and wastewater treatment. *Journal of Environmental Engineering and Science*, *1*(4), pp.247–264.

Zhu, L., Dai, X., Lv, M. and Xu, X., 2013. Correlation analysis of major control factors for the formation and stabilization of aerobic granule. *Environmental Science and Pollution Research*, *20*(5), pp.3165–3175.

6 Treatment of Pharmaceutical Compounds Present in Wastewater Using Microbial Fuel Cells

M. M. Ghangrekar, Azhan Ahmad,
S.M. Sathe, and Indrajit Chakraborty

CONTENTS

6.1 INTRODUCTION

Pharmaceuticals such as antibiotics, anti-inflammatories, analgesics, beta-blockers, lipid regulators, and X-ray contrast media are widely used forms of medication. Pharmaceuticals are classified as emerging contaminants (EC) as their existence is not yet regulated by the Environmental Protection Agency (Botero-Coy et al., 2018). The consumption of pharmaceuticals for improved health conditions in developed countries is set to increase in the coming years. For instance, more than 100 new chemical entities were approved in the year 2013 by the U.S. Food and Drug Administration for medication use (Couto et al., 2019). These compounds can enter the water stream through sources such as the pharmaceutical manufacturing industry, hospital wastewater, leakage from landfill leachate, human excretion, etc. (Archer et al., 2017). The presence of numerous pharmaceuticals is well documented in surface water, ground water, and even in treated and untreated wastewater (Taheran et al., 2016). The release of pharmaceuticals into the environment is a matter of concern for researchers because these can cause aquatic toxicity, antibiotic resistance in microbes, endocrine disruption, and genotoxicity. Moreover, these pollutants are persistent and can bioaccumulate in living species (Botero-Coy et al., 2018; Couto et al., 2019).

Pharmaceuticals, being bio-refractory in nature, cannot be efficiently removed from conventional wastewater treatment plants (WWTPs) and are detected in the effluent of WWTPs with a similar concentration as in the influent (Botero-Coy et al., 2018). Moreover, the biotoxicity imposed by these pharmaceuticals reduces the efficiency of the microbes that form the backbone of secondary biological treatment processes. Secondary biological treatment processes are designed for the removal of labile organic matter such as carbohydrates, carboxylic acids, fats, proteins, amino acids, and other forms of nitrogen; whereas, pharmaceutical compounds are resistant to biodegradation (García-Galán et al., 2020). For instance, ketoprofen, an anti-inflammatory drug that has

DOI: 10.1201/9781003204442-6

two aromatic rings, which was detected with a maximum concentration of 1,080 ng L^{-1} in the final effluent of a municipal WWTPs in India (Balakrishna et al., 2017). Diclofenac, an analgesic with an aromatic ring structure, was detected at a level of 49.18 µg L^{-1} in the effluent of municipal WWTPs (Thalla and Vannarath, 2020). Hence, these compounds should be removed from wastewater prior to disposal of treated water in receiving water bodies, as long-term exposure to these pollutants, even at low concentrations, can have adverse effects on ecosystems.

Due to the incapability of secondary biological systems to remove pharmaceuticals, tertiary treatments are adopted in WWTPs for removal thereof. Treatment technologies such as ozonation, membrane filtration, adsorption, UV-H_2O_2 treatment, membrane biological reactor, and electrochemical oxidation are proven to be successful for the removal of pharmaceuticals from wastewater (Khan et al., 2020). However, every technology has some disadvantages, for instance, by-products of parent pharmaceuticals are formed in ozonation treatment, and as a result longer treatment time is required for complete degradation of these pollutants (Alharbi et al., 2016). In addition to this, high operational costs are involved in membrane filtration, and external power is required for electrochemical methods (Ahmad et al., 2020; Zaied et al., 2020). Therefore, an alternate technology is required, which will cater to the need for pharmaceutical removal at a low operating cost without the supplementation of external chemicals.

In this regard, microbial fuel cell (MFC) technology has incredible potential to eliminate pharmaceuticals from wastewater, while simultaneously producing bioelectricity. The use of MFCs is considered as an effective tool for the simultaneous treatment of oxidizable organic and other matters, EC removal, and bio-electricity production. While the anodic chamber allows for the oxidation of these pollutants, the cathodic chamber follows reduction mechanisms (Kumar et al., 2018; Wang et al., 2017b). Thus, by taking advantage of both mechanisms, pharmaceuticals can be effectively eliminated in anodic as well as cathodic chambers of an MFC, which is discussed in detail in this chapter. The principle mechanism will also be discussed along with the possible future direction of research.

6.2 OVERVIEW OF MICROBIAL FUEL CELLS

The MFC is an upcoming technology capable of converting chemical energy of the substrate to electrical energy using anaerobic microorganisms as biocatalysts. The anaerobic microorganisms in the anodic chamber of MFCs consume the substrate, which can be organic matter present in domestic wastewater, industrial wastewater, landfill leachate, etc., to generate electrons, protons, and CO_2 (Eq. 1; Logan, 2008). The protons migrate from the bulk anolyte to the cathodic chamber via the proton exchange membrane (PEM), which separates the anodic and cathodic chambers; whereas, electrons flow towards the cathode from the external circuit. Electricity is harvested during the flow of electrons from the anode to cathode through this external circuit. The electroactive microorganisms present in the anodic biofilm are capable of extracellular electron transfer to the anode. While the conventional anaerobic degradation process is dependent on the availability of terminal electron acceptors (TEA), such as; NO_3^-, SO_4^{2-}, however, in the case of the MFC the anode itself acts as an inexhaustible electron sink. Further, the protons in the cathodic chamber combine with molecular oxygen as a TEA, thus completing the circuit and producing electricity via the redox reaction (Logan et al., 2006; Pham et al., 2006).

$$CH_3COO^- + 4H_2O \rightarrow 2HCO_3^- + 9H^+ + 8e \quad \left(E' = 0.296 \text{ V, vs. SHE at pH} = 7\right) \qquad (6.1)$$

6.3 ANODIC DEGRADATION OF PHARMACEUTICALS IN MICROBIAL FUEL CELLS

The anodic chamber facilitates the growth of electroactive microbiomes that degrade organic matter to produce electrons, protons, and CO_2. The electrons and protons are transferred to the cathodic chamber via an external circuit and PEM that separates both chambers, respectively. The selectivity of the anode over the other dissolved TEAs accelerates the metabolic process, as the potential difference between the anode and the outer membrane of the cell is more stable, unlike the dissolved TEAs, wherein the concentration becomes a rate limiting factor (Morris et al., 2009). This initiates a faster metabolic regime that is again rate limited by substrate availability, among other factors. Higher metabolic rates enable the prolific growth of microbes that enhances the biofilm.

Past investigations have exhibited that matured and thicker biofilms with a dense intercalated network of extra-cellular polymeric substances render resistance to the inner layers of the microbiome in the anodic biofilm (Yang et al., 2019). Additionally, higher metabolic rates also aid in harvesting more adenosine triphosphate (ATP), which can render the energy required to catabolize the complex molecules of the pharmaceutical compounds. Hence, it is evident that the MFC provides a niche environment to the microbes as compared to conventional anaerobic microbiomes, wherein the microbial metabolism is often rate limited by the availability of the TEAs; availability of which is again controlled by diffusion and mass transport rates. This section will provide a detailed discussion on the biodegradation of pharmaceuticals in the anodic chamber of MFCs.

In a study, the performance of MFCs was investigated for the degradation of norfloxacin (NFLX) antibiotic, electricity generation, and antibiotics resistance genes (ARGs) generation (Ondon et al., 2020). It was observed that 65.5% of NFLX and 94.5% of chemical oxygen demand (COD) removals were achieved, with an initial NFLX concentration of 4 mg L^{-1}. Moreover, when the initial concentration of NFLX increased to 128 mg L^{-1}, the removal efficiency decreased to 48.4%. This can be attributed to the inhibition effect of NFLX on anodic electrogenic microorganisms and, as a result, the biodegradation ability of these microbes was partially weakened. On the other hand, when MFC was operated using sodium acetate as a carbon source in the absence of NFLX, a maximum power density of 1,497 mW m^{-2} was achieved. However, after adding 2 mg L^{-1} of NFLX, about a 53% reduction in power density was observed. This is attributed to the fact that the addition of NFLX in the anodic chamber can inhibit the growth of exoelectrogens, thus adversely affecting power generation. Furthermore, it was observed that ARG generation in the MFC was about 10^5 copies of DNA μL^{-1}, which was considerably low compared to the conventional treatment methods (10^6–10^8 copies of DNA μL^{-1}) for the removal of antibiotics from wastewater. In the investigation, the duration of assessment was not mentioned as long-term exposure to pharmaceutical-laden wastewater might develop ARGs in the anodic chamber of the MFC as well. Hence, long-term investigation is required to analyze the impact of pharmaceuticals on the formation of ARGs in MFC. The investigation highlighted that MFCs are efficient in NFLX removal from wastewater along with offering the advantage of harvesting electricity, and have the potential of minimizing ARG production during antibiotic-laden wastewater treatment.

In another demonstration, energy production along with bioremediation of acetaminophen (APAP) and its by-product para-aminophenol from wastewater was observed using MFCs with *Scedosporium dehoogii* fungi as a bio anode (Pontié et al., 2019). During the process of electrodeposition of fungi to catalyze the oxidation of APAP, the optimal potential against the saturated calomel electrode with 3.5 ± 0.2 μm thickness was found to be +0.8 V. It was noticed that APAP is an efficient source of carbon for electricity production in MFCs and after 40 h of contact time, a maximum power density of 50 mW m^{-2} with an electromotive force (EMF) of 550 mV was achieved. The power density obtained during APAP degradation using fungal MFCs in this exploration was 7.5 times higher than the previously reported power density (6.5 mW m^{-2}) using the first MFC based on *S. dehoogii* fungi. Moreover, the optimal resistance of 3 Ω was observed from the polarization curves during the biodegradation of para-aminophenol. In the investigation, the effluent quantification of APAP was measured in the form of total organic carbon (TOC), which was reported to be 22 mg TOC L^{-1} after 7 days of contact time, with an initial APAP concentration of 100 mg TOC L^{-1}. Hence, using fungi on bio-anodes in MFC can offer a better alternative for energy production along with the removal of pharmaceuticals from wastewater.

Song et al. (2013) evaluated the performance of MFC in the biodegradation of metronidazole (MNZ) from synthetic wastewater (Song et al., 2013). Using glucose as a co-substrate (1 g L^{-1}), about 85% removal of MNZ was noticed with an initial MNZ concentration of 10 mg L^{-1} within 24 h of retention time. However, the removal efficiency of MNZ decreased by about 1.2 fold with an increase in initial concentration of MNZ from 10 to 30 mg L^{-1} (Song et al., 2013). This can be attributed by the increasing toxic effect on electrogenic bacteria due to high pharmaceutical concentrations. Moreover, the power density of 99.23 mW m^{-2} was achieved, corresponding to an MNZ concentration of 10 mg L^{-1}.

A dual chamber MFC was used for the treatment of real wastewater collected from a non-steroidal anti-inflammatory drugs production plant along with the degradation of diclofenac sodium (DS) from synthetic wastewater. Moreover, the influence of two different types of cathodes on COD removal and the current characteristics were also investigated (Amari and Boshrouyeh Ghandashtani, 2020). Using Pt-coated Ti as a cathode (Phase 1), 93% and 78% of COD removal efficiencies were achieved, with the real and synthetic wastewater having an initial COD of 7,440 mg L^{-1} and retention time of 30 h, respectively. This was higher than using Pd/Ir-coated Ti as the cathode, as used in Phase 2. The reason is due to the presence of Pt, which can increase the rate of reaction significantly, thereby improving the removal rate in Phase 1. Moreover, the lower degradation of COD in synthetic wastewater can be due to the de-conjugation of DS during the treatment (Zwiener and Frimmel, 2003). Further, the biodegradation of DS is limited due to the hydrophobicity of DS in water bodies. This can be attributed to the presence of sodium ion with carboxyl groups in DS, which can render hydrophobicity and it can also change the mobility and sorption of DS by reducing their net charge (Amari and Boshrouyeh Ghandashtani, 2020).

Additionally, in Phase 1, the maximum power density of 20.5 and 6.5 W m^{-3} was produced for the treatment of real and synthetic wastewater, respectively, with corresponding internal resistance of 54 and 65 Ω. The high internal resistance in the case of synthetic wastewater can be attributed to the presence of aromatic benzene acetic acid groups in DS, which hinders the treatment. Thus, a lower power density was achieved in the case of synthetic wastewater. However, in Phase 2, the Pt coated Ti cathode was replaced by Pd/Ir-coated Ti, and a maximum power density of 17.81 W m^{-3} was produced during the treatment of real wastewater with an internal resistance of 65 Ω. Hence, Phase 1 had 1.15 times more power density while treating real wastewater as compared to Phase 2. The enhanced power recovery in Phase 1 can be attributed to the presence of platinum in the cathodic chamber which has a low charge transfer resistance that in turn reduces the overall internal resistance of the cell. Thus, a dual chamber MFC can be a feasible alternative in treating DS present in wastewater.

In a recent investigation, MFCs were used to degrade four pharmaceuticals, namely estrone (EST), lamivudine (3TC), trimethoprim (TRI), and levofloxacin (LFL) present in source-separated urine (Sharma et al., 2020). Subsequent to 30 h of retention time, 78.58 ng L^{-1} of EST, 70.15 ng L^{-1} of 3TC, 58.39 ng L^{-1} of TRI, and 37.29 ng L^{-1} of LFL concentrations were achieved with an initial concentration for each of the pharmaceutical of 2,000 ng L^{-1}. Thus, 96% of EST, 96.49% of 3TC, 97% of TRI, and 98.1% of LFL reduction was achieved. The higher level of removal of contaminants from the urine is due to the strong mechanism of electron transfer in microbes (ATP level) to degrade the pharmaceuticals using MFCs. Moreover, 93% of COD reduction was obtained with an initial COD of 7,000 mg L^{-1}. The COD reduction is owing to the degradation of pharmaceuticals by microorganisms in the anodic chamber. Hence, MFC can be employed in the biodegradation process for the efficient removal of pharmaceuticals from wastewaters.

In another recent demonstration, the effect of sulfamethoxazole (SMX) antibiotic on electroactive biofilm was explored to enhance the power generation in an MFC (Wu et al., 2020). When 20 mg L^{-1} of SMX was used as a co-substrate along with 1 g L^{-1} of sodium acetate, with an 18% increment in maximum power density it reached to 40 W m^{-3} while treating SMX. Moreover, with the addition of SMX as a co-substrate along with sodium acetate as a substrate, almost three times the enhancement in energy recovery was reported (9.6 kWh kg^{-1} COD) than using sodium acetate as a sole carbon source without adding SMX (3.57 kWh kg^{-1} COD). The linear sweep voltammogram of the MFC fed with SMX (20 mg L^{-1}) exhibited a prominent reduction peak that is characteristic of the cytochrome-C protein. As compared to the biofilm that is not exposed to SMX, the biofilm exposed to SMX exhibited a "thin film behavior." The biofilm exhibiting "thin film behavior" transfers electrons at a faster rate and as a singular entity due to the presence of redox proteins on the surface. Such thin film action also enables faster permeation of solute through the film interstices (Armstrong et al., 1997). Further, after 48 h of retention time, 98% of SMX removal was achieved with an initial concentration of 20 mg L^{-1} (Wu et al., 2020). Another investigation has demonstrated that the pre-treatment of the anodic inoculum with fungal peptide antibiotic (peptaibiotics) in MFC can enhance its power density due to the suppression of methanogens (Ghosh et al., 2017). Similarly, the enhancement of current in MFC after the addition of SMX antibiotic might be due to the suppression of methanogens.

In a separate investigation, the anode of MFC was modified with metal and metal oxides nanoparticles, and its effect in pharmaceuticals removal and power generation was evaluated (Xu et al., 2018). When anodes modified with nanoparticles of MnO_2, Fe_3O_4, and Pd were used individually in an MFC, 82.2%, 83.4%, and 81.6% of carbamazepine (CBZ); 18.8%, 20.1%, and 20.0% of ibuprofen (IBF); and 50.6%, 48.7%, and 52.6% of diclofenac (DCF) removal was achieved, respectively, when an initial concentration of 5 mg L^{-1} was present for each pharmaceutical at the same time in the influent of the MFC. On the other hand, when carbon black was used as an anode in an MFC, 78.3% of CBZ, 14.6% of IBF, and 28.8% of DCF removal was achieved. Thus, nanoparticles-decorated anodes achieved better efficiency in the removal of pharmaceuticals when compared to carbon black anodes. Moreover, MFCs with anodes having catalysts Pd, Fe_3O_4, and MnO_2 generated a maximum power density of 824, 728, and 782 mW m^{-2}, respectively, which was higher than the carbon black control anode (680 mW m^{-2}). Further, it was also observed that anode modification with Fe_3O_4 and MnO_2 has favored the enrichment of exoelectrogenic bacteria (Xu et al., 2018).

In another demonstration, the influence of tobramycin antibiotic on the performance of MFC was evaluated. In the absence of tobramycin, sodium acetate was used as a sole carbon source in the anodic chamber of an MFC. As a result, 1.2–1.4 mA of current was generated with the internal resistance of 300 Ω. The addition of the antibiotic in different concentrations (0.2–4 mM) exhibited no immediate change in the overall current generation (Wu et al., 2014). However, different current profiles were observed at different tobramycin concentrations and these profiles are more notable at higher concentrations. It was observed that inhibition was noticed at the beginning of the batch with concentrations of 0.2, 0.5, and 1 mM, after which a stable current was observed. When the tobramycin concentration was increased from 2 to 4 mM, a significant inhibition of the bacterial metabolism was observed. Further, an exponential increase in the inhibition ratio was also observed with tobramycin concentrations between 0.1 and 1.9 g L^{-1} (Wu et al., 2014). Apart from these, other pharmaceuticals, such as penicillin, roxithromycin, aureomycin, tetracycline, etc., have been successfully degraded in MFCs (see Table 6.1).

It can be concluded from the above section that the anodic degradation of pharmaceuticals as substrates and co-substrates in the MFC was influenced by pharmaceutical concentration in the influent. The power generation in the MFC was also enhanced by the addition of pharmaceuticals such as SMX in the anodic chamber. Whereas, pharmaceuticals like NFLX adversely affect the power generation in the MFC with an increase in concentration in the influent. The removal of pharmaceuticals using MFC technology is still in the experimental phase. Nevertheless, the successful degradation of pharmaceuticals in the anodic chamber of the MFC is being reported in certain laboratory-scale demonstrations, as cited above. However, more investigations are required for optimizing the effect of operating conditions and configurations of the MFC for simultaneous degradation of pharmaceuticals and enhancing electricity production.

TABLE 6.1

Removal of Various Pharmaceuticals Using Microbial Fuel Cells

Pharmaceuticals	Operation Time (h)	Maximum Power Density (W m^{-3})	Removal efficiency (%)	References
Ceftriaxone sodium	24	113	91	(Wen et al., 2011b)
Penicillin	24	101.2	98	(Wen et al., 2011a)
Sulfanilamide	96	0.316 W m^{-2}	90	(Guo et al., 2016b)
Chloramphenicol	48	0.223 W m^{-2}	83.7	(Guo et al., 2016a)
Neomycin sulfate	24	0.251 W m^{-2}	54	(Catal et al., 2018)
Oxytetracycline	192	0.0866 W m^{-2}	91	(Chen et al., 2018)
Tetracycline	168	2.5	80	(Wang et al., 2017a)
Cefazolin sodium	30	30.4	> 70	(Zhang et al., 2018)
Aureomycin, Roxithromycin,	-	-	~ 100	(Zhou et al., 2018)
Norfloxacin, and Sulfadiazine	-	-	~ 100	(Zhou et al., 2018)
	-	-	~ 100	(Zhou et al., 2018)
	-	-	99.9	(Zhou et al., 2018)

6.4 CATHODIC DEGRADATION OF PHARMACEUTICALS IN MICROBIAL FUEL CELLS

At the cathode of an MFC, electrons received from the anode are further transferred to the TEA. Oxygen is considered as the most appropriate TEA due to its availability and lack of harmful by-products formation. Typically, the cathodic reduction reaction in an MFC proceeds based on the use of the cathode material and cathode catalyst. Generally, electrocatalysts reduce oxygen in the cathode via four-electron pathways, whereas the use of graphite-based cathodes results in a two-electron reduction reaction. The four- and two-electron pathway cathodic reactions produce water and hydrogen peroxide at the cathodes, respectively (Equation 6.2 and Equation 6.3). From the aspect of power generation, the four-electron pathway is preferred over the two-electron pathway. However, two-electron pathways leading to H_2O_2 production can still be beneficial as H_2O_2 is a strong oxidizing agent, which is capable of disinfecting and degrading pollutants in wastewater (Asghar et al., 2015; Logan et al., 2006). With in situ H_2O_2 production, the cathodic chamber can be used for the polishing treatment of wastewater without any additional energy or chemical requirements.

$$O_2 + 4H^+ + 4e^- \rightarrow 2H_2O \quad \left(E' = 0.805 \text{ V, vs. SHE at pH} = 7\right) \tag{6.2}$$

$$O_2 + 2H^+ + 2e^- \rightarrow H_2O_2 \quad \left(E' = 0.328 \text{ V, vs. SHE at pH} = 7\right) \tag{6.3}$$

In an electro-Fenton process, cathodic H_2O_2 is produced under the effect of poised potential followed by •OH generation utilizing ferric ions (Equation 6.4 and Equation 6.5). The applicability of Fenton oxidation for the removal of ECs, such as surfactants, pharmaceuticals, and dyes, has been documented in the literature (Liu et al., 2018; Panizza et al., 2013; Panizza and Cerisola, 2009). On the other hand, MFC with the cathodic production of H_2O_2 assisted with ferric ions embedded in the cathode can lead to a Fenton reaction in the cathodic chamber of an MFC (see Figure 6.1). This hybridization of two separate processes (MFC and Fenton oxidation) is termed as bio-electro-Fenton. The resulting •OH is a strong oxidizing agent with an oxidizing potential of 2.8 eV, thereby making it one of the strongest oxidizing agents in water and wastewater treatment applications.

$$Fe^{3+} + e^- \rightarrow Fe^{2+} \tag{6.4}$$

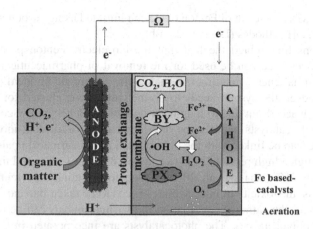

FIGURE 6.1 Process of bio-electro-Fenton. (PX: pharmaceuticals, BY: intermediate by-products.)

$$H_2O_2 + Fe^{2+} \rightarrow OH + OH^- + Fe^{3+} \tag{6.5}$$

This novel approach of bio-electro-Fenton MFC, taking advantage of the low operating costs of Fenton oxidation, was implemented for the first time by Zhu and Ni for the degradation of p-nitrophenol. The cathodic Fenton oxidation contributed to the complete removal of p-nitrophenol in 12 h, simultaneously producing a maximum power density of 143 mW m^{-2}. The 85% TOC removal in 96 h of operation confirms that the majority of the pollutant was mineralized (Zhu and Ni, 2009). Following the successful application of this hybridization, the combined system of Fenton and MFC (bio-electro-Fenton MFC) has been effectively applied for the removal of antibiotics, dyes, endocrine disturbing chemicals, pharmaceuticals, etc. from wastewater (Kahoush et al., 2018; Li et al., 2020a).

As outlined in Section 1, pharmaceuticals are bio-refractory compounds that require advanced treatment processes for their elimination from wastewater; bio-electro-Fenton is one such initiative for this process. Paracetamol, which is the most widely used pharmaceutical drug, was successfully degraded using the bio-electro-Fenton process. The cathodic Fenton process initiated by the addition of FeSO$_4$ resulted in a 70% removal of paracetamol in 9 h at a highly acidic pH. The removal efficiency was directly proportional to the •OH concentration with a maximum quantity of •OH generated at a pH of 2. The paracetamol was successively degraded into p-aminophenol, p-nitrophenol, and maleic acid before forming dicarboxylic and carboxylic acids in the end (Zhang et al., 2015). In another investigation, 90% carbamazepine removal using an Fe-Mn cathode catalyst was achieved in the cathodic chamber of an MFC. The carbamazepine removal rate (0.375 mg L^{-1} h^{-1}) was higher than other methods experimented, such as anodic treatment in the MFC, rotating biological contactor, and bioreactors enriched with selective microorganisms (Wang et al., 2018).

In another investigation, the anodic effluent was circulated in the cathodic chamber as catholyte for further polishing treatment using Fenton oxidation. Combined, anodic followed by cathodic treatment revealed an 88.73% elimination of erythromycin and 1.96 log reduction of ermB – one of the antibiotic-resistant genes, using a dual-chambered MFC with carbon nanotubes, stainless steel mesh, and γ-FeOOH as the composite cathode. The maximum power density of 0.193 W m^{-2} was obtained from the system with 50 μg L^{-1} of erythromycin in the feed. The first stage of the treatment was carried out in the anodic chamber by the microbial consortia, which shows the ability of microbial consortia to tolerate small concentrations of erythromycin (50 μg L^{-1}) without affecting the performance of the MFC (Li et al., 2020c). Following a similar operation of anodic followed by cathodic treatment, 94.66% of sulfamethoxazole was removed in 48 h. The initial anodic treatment contributed to 56.28% removal of sulfamethoxazole, which is due to microbial activity; whereas, the cathodic removal was as a result of Fenton oxidation initiated using carbon nanotubes, stainless steel mesh, and γ-FeOOH cathodes (Li et al., 2020b).

These investigations highlighted the fact that the bio-electro-Fenton process is an effective tertiary treatment process that can be used for the removal of pharmaceuticals from wastewater. Additionally, the minimal energy requirement makes this technique an ideal choice for practical implementation. However, the system needs to be meticulously evaluated for durability as cathode catalysts can get leached out due to long-term operation. Hence, a low-cost approach for the regeneration of cathode catalysts needs to be developed so as to make it more practicable. The quantification of TOC can be linked with the mineralization of pharmaceuticals. It is evident from studies that, even though a high removal of paracetamol (89%) was observed in 24 h, the resulting TOC removal was only 25% (Zhang et al., 2015). Hence, for deciding the retention time of the bio-electro-Fenton process, the toxicity of intermediates needs to be taken into careful consideration.

Another approach for the cathodic degradation of pharmaceuticals can be through the coupled system of MFC and photocatalysts. The photocatalysts are incorporated in the MFC as cathode catalysts and under irradiation act as photo-electrocatalysts. Under irradiation with artificial light or sunlight, photocatalysts can absorb photons to generate photoelectrons, which react with oxygen to produce superoxide radicals (see Figure 6.2). These photoelectrons have reducing capabilities, which can improve the cathodic oxygen reduction reaction (Han et al., 2017). Additionally, the photo-induced holes created in the valence band react with water to produce H_2O_2 and then $\cdot$OH, which is capable of degrading pollutants (Jiang et al., 2016). The use of photocatalysts as cathode catalysts in an MFC not only increases the quantum of electrons available but also produces reactive oxygen species that can be used to degrade a variety of pollutants (Bhowmick et al., 2019). Additionally, photocatalysts when used as electrocatalysts in an MFC are not in suspension but instead immobilized on the cathode surface; hence, a separate mechanism for the separation of photocatalysts from the effluent is not required.

The use of zero valent iron and TiO_2 as cathode catalysts assisted in the cathodic degradation of tetracycline. In the presence of a light source, the cathodic degradation of tetracycline, which was monitored in terms of COD, reached 70% after 2 h of contact time. This was 1.2 times greater than that in the absence of light, indicating the formation of reactive oxygen species from the photocatalytic reaction that aid in antibiotic removal efficiency (Jiang et al., 2016). In a recent investigation, the photocatalytic oxidation of ibuprofen using a $CuInS_2$ catalyst was successfully attained in photo-catalytic MFC. The ibuprofen removal was highest (75.9%) at a pH of 5, which is close to the pK_a range of 4.5–4.9. The ibuprofen degradation rate followed the pseudo first-order kinetic equation with a degradation rate of 0.0419 h^{-1}. With a maximum power density of 0.566 W m^{-2}, the system achieved close to 70% TOC removal in 36 h with an influent ibuprofen concentration of

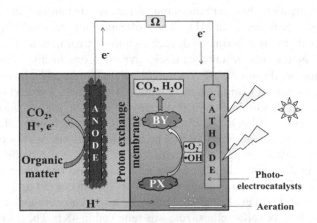

FIGURE 6.2 Process of photo-electrocatalytic MFC. (PX: pharmaceuticals, BY: intermediate by-products.)

10 mg L^{-1} (Xu et al., 2020b). In one of the latest investigations by the same group of scientists, ofloxacin was successfully degraded using LiNbO$_3$ as a photocathode catalyst in the dual-chamber MFC. The system achieved 86.5% ofloxacin removal efficiency, when an initial concentration of 0.1 mM L^{-1} was used, while simultaneously producing a maximum power density of 0.546 W m^{-2} (Xu et al., 2020a). The CuInS$_2$ was irradiated in visible light; whereas, LiNbO$_3$ was irradiated using a 300-W xenon lamp. The photocatalysts that irradiate in visible light can be operated in daylight, reducing the dependency on UV lights, which can further reduce operational costs.

It can be seen from the above section that for the cathodic degradation of pharmaceuticals using an MFC, two strategies have been implemented, namely bio-electro-Fenton and photo-electrocatalysis. In both cases, the degradation of the target pollutant is via the formation of •OH as a result of either Fenton oxidation or photo-catalytic reactions. Unlike conventional advanced oxidation processes, these bio-electrochemically-assisted processes are effective at near neutral pH also; hence, reducing the cost of chemicals and time required for the pH adjustment. The use of Fe-based or photo-electro-catalysts in bio-electro-Fenton and photo-electro-catalysis not only benefits the pollutant degradation process but also reduces the internal resistance, thereby improving the electrical performance of MFCs. Considering the encouraging results of the laboratory-scale investigations, future research in the direction of upscaling of these techniques is required after addressing some of the critical issues, which are elaborated in the subsequent section.

6.5 FUTURE SCOPE

The utilization of MFCs for the remediation of pharmaceuticals present in wastewater is an ambitious proposition as the scale of applicability in past investigations is limited to the bench top level. To drive this technology towards pilot-scale applications, several factors have to be considered, such as maintaining the membrane as well as the electrode surface area ratio with volume (Du et al., 2007), the development of novel anode (Mohamed et al., 2018) and cathode catalysts (Chakraborty et al., 2020; Das et al., 2018), the utilization of efficient air cathodes with and without gas diffusion layers (Li et al., 2020d; You et al., 2011), and low internal resistance.

It has been demonstrated in past investigations that the regulation of anode potential can be instrumental in biofilm formation as well as the proliferation of microbes on the electrode surface (Hou et al., 2020; Nath et al., 2021). A thicker biofilm formation is beneficial for MFCs catering to the degradation of the pharmaceutical compounds that are often bio-toxic in nature. This stems from the fact that the diffusion of the bio-toxic pharmaceutical products to the inner layers can be reduced in the case of thicker biofilms that can protect the inner layers of microbial cells by secretion of extracellular polymeric substances (Obata et al., 2020). Hence, the regulation of anode potential by operating the system in microbial electrolysis cell mode can be a pragmatic approach towards stimulating faster biodegradation and the development of more robust biofilms. The anode potential can also be regulated to some extent by manipulating the connected external load while operated in MFC mode (Kaur et al., 2014). In that case, it is more effective if the cathodic electron drag is stronger, thus promulgating more pumping of electrons from the anode. This enhancement in the cathodic drag can be realized by the utilization of highly efficient cathode catalysts (Chen et al., 2018).

Apart from employing cathode catalysts for the indirect realization of the effective degradation of pharmaceutical compounds via an enhanced rate of oxygen reduction reaction, the application of anode catalysts can directly enhance the degradation as well as power performance (Qiu et al., 2020; Wu et al., 2020). An anode catalyst that is biocompatible, highly conductive with low charge transfer resistance, supports bacterial growth, and has adequate specific surface area would have a profound effect on the anodic removal efficiency of the pharmaceutical compounds. Furthermore, the anodic treatment of pharmaceuticals can be achieved by cathodic treatment by employing a carbon-Fe combination as a cathode modification technique/cathode catalyst for ensuring Fenton reaction in the cathodic chamber (see Figure 6.3). Such two-stage removal of pharmaceuticals

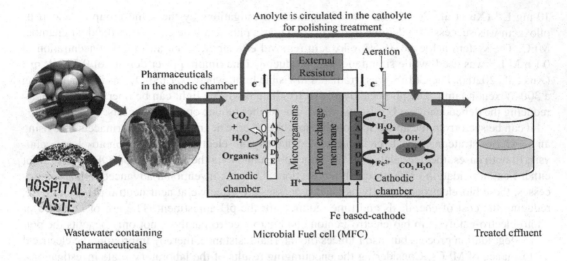

FIGURE 6.3 Degradation of pharmaceuticals in MFC. (PH: pharmaceuticals, BY: by-products.)

from wastewater would ensure a higher removal efficiency of these pharmaceutical compounds from wastewater. Cathodic treatment can also ensure the destruction of the ARGs developed in the anodic biofilm that are released through the lysis of the microbial cells exposed to the antibiotics. However, it is worth mentioning that cathode can be susceptible to biofouling when anodic effluent with high micro-organism content is diverted towards it, which in turn can affect the Fenton oxidation process. Hence, adequate tailoring is required to optimize the hydraulic retention time in both the anodic and cathodic chambers of an MFC.

Based on the above discussions, the future direction of research should focus on the development of efficient anode and cathode catalysts (preferably one facilitating cathodic Fenton reaction as well as power enhancement), flat-plate MFC, or other relevant configurations of anodic and cathodic chambers to reduce diffusion losses and by employing efficient power management for the proper utilization of the electricity harvested. Furthermore, the development of antibiotic-resistant genes and the efficient removal of these from wastewater is another research area that should be focussed upon. Another research topic that needs attention in future is the development of bio-electrochemical sensors for the detection of antibiotic concentrations in effluents. This is a prerequisite for the real-time monitoring of the efficiency of employed systems, as the present detection methods are both cumbersome and involve the usage of costly chemicals and instruments.

6.6 CONCLUSION

This chapter explores the possibility of treating pharmaceutical compounds present in wastewater using MFCs. The findings indicate that MFCs have an excellent scope for removing pharmaceuticals from wastewater, both in anodic as well as cathodic chambers, and simultaneously for producing bioelectricity. The removal of antibiotics and other pharmaceutical products from wastewater has been successfully addressed in MFCs using laboratory-scale investigations, and further research work should be propagated towards the upscaling of this technology for augmenting future wastewater treatment plants. This can be achieved by furthering the knowledge gained in past investigations as a founding base for designing future treatment units. As MFC represents a one-step power recovery technique from waste, the concomitant treatment of such complex wastewater and power generation have the potential to become modern solutions used by wastewater treatment plants in the future.

REFERENCES

Ahmad, A., Das, S., Ghangrekar, M.M., 2020. Removal of xenobiotics from wastewater by electrocoagulation: A mini-review. *J. Indian Chem. Soc.* 97, 493–500.

Alharbi, S.K., Price, W.E., Kang, J., Fujioka, T., Nghiem, L.D., 2016. Ozonation of carbamazepine, diclofenac, sulfamethoxazole and trimethoprim and formation of major oxidation products. *Desalin. Water Treat.* 57, 29340–29351. https://doi.org/10.1080/19443994.2016.1172986

Amari, S., Boshrouyeh Ghandashtani, M., 2020. Non-steroidal anti-inflammatory pharmaceutical wastewater treatment using a two-chambered microbial fuel cell. *Water Environ. J.* 34, 413–419. https://doi.org/10.1111/wej.12476

Archer, E., Petrie, B., Kasprzyk-Hordern, B., Wolfaardt, G.M., 2017. The fate of pharmaceuticals and personal care products (PPCPs), endocrine disrupting contaminants (EDCs), metabolites and illicit drugs in a WWTW and environmental waters. *Chemosphere* 174, 437–446. https://doi.org/10.1016/j.chemosphere.2017.01.101

Armstrong, F.A., Heering, H.A., Hirst, J., 1997. Reaction of complex metalloproteins studied by protein-film voltammetry. *Chem. Soc. Rev.* 26, 169. https://doi.org/10.1039/cs9972600169

Asghar, A., Raman, A.A.A., Daud, W.M.A.W., 2015. Advanced oxidation processes for in-situ production of hydrogen peroxide/hydroxyl radical for textile wastewater treatment: A review. *J. Clean. Prod.* 87, 826–838. https://doi.org/10.1016/j.jclepro.2014.09.010

Balakrishna, K., Rath, A., Praveenkumarreddy, Y., Guruge, K.S., Subedi, B., 2017. A review of the occurrence of pharmaceuticals and personal care products in Indian water bodies. *Ecotoxicol. Environ. Saf.* 137, 113–120. https://doi.org/10.1016/j.ecoenv.2016.11.014

Bhowmick, G.D., Chakraborty, I., Ghangrekar, M.M., Mitra, A., 2019. TiO2/Activated carbon photo cathode catalyst exposed to ultraviolet radiation to enhance the efficacy of integrated microbial fuel cell-membrane bioreactor. *Bioresour. Technol. Rep.* 7, 100303. https://doi.org/10.1016/j.biteb.2019.100303

Botero-Coy, A.M., Martínez-Pachón, D., Boix, C., Rincón, R.J., Castillo, N., Arias-Marín, L.P., Manrique-Losada, L., Torres-Palma, R., Moncayo-Lasso, A., Hernández, F., 2018. An investigation into the occurrence and removal of pharmaceuticals in Colombian wastewater. *Sci. Total Environ.* 642, 842–853. https://doi.org/10.1016/j.scitotenv.2018.06.088

Catal, T., Yavaser, S., Enisoglu-atalay, V., Bermek, H., Ozilhan, S., 2018. Bioresource technology monitoring of neomycin sulfate antibiotic in microbial fuel cells. *Bioresour. Technol.* 268, 116–120. https://doi.org/10.1016/j.biortech.2018.07.122

Chakraborty, I., Bhowmick, G.D., Ghosh, D., Dubey, B.K., Pradhan, D., Ghangrekar, M.M., 2020. Novel low-cost activated algal biochar as a cathode catalyst for improving performance of microbial fuel cell. *Sustain. Energy Technol. Assess.* 42, 100808. https://doi.org/10.1016/j.seta.2020.100808

Chen, J., Hu, Y., Huang, W., Liu, Y., Tang, M., Zhang, L., Sun, J., 2018. Biodegradation of oxytetracycline and electricity generation in microbial fuel cell with in situ dual graphene modified bioelectrode. *Bioresour. Technol.* 270, 482–488. https://doi.org/10.1016/j.biortech.2018.09.060

Couto, C.F., Lange, L.C., Amaral, M.C.S., 2019. Occurrence, fate and removal of pharmaceutically active compounds (PhACs) in water and wastewater treatment plants: A review. *J. Water Process Eng.* 32, 100927. https://doi.org/10.1016/j.jwpe.2019.100927

Das, I., Noori, M.T., Bhowmick, G.D., Ghangrekar, M.M., 2018. Synthesis of bimetallic iron ferrite Co0.5Zn0.5Fe2O4 as a superior catalyst for oxygen reduction reaction to replace noble metal catalysts in microbial fuel cell. *Int. J. Hydrogen Energy* 43, 19196–19205. https://doi.org/10.1016/j.ijhydene.2018.08.113

Du, Z., Li, H., Gu, T., 2007. A state of the art review on microbial fuel cells: A promising technology for wastewater treatment and bioenergy. *Biotechnol. Adv.* 25, 464–482. https://doi.org/10.1016/j.biotechadv.2007.05.004

García-Galán, M.J., Arashiro, L., Santos, L.H.M.L.M., Insa, S., Rodríguez-Mozaz, S., Barceló, D., Ferrer, I., Garfí, M., 2020. Fate of priority pharmaceuticals and their main metabolites and transformation products in microalgae-based wastewater treatment systems. *J. Hazard. Mater.* 390, 121771. https://doi.org/10.1016/j.jhazmat.2019.121771

Ghosh, R., Noori, M.T., Ghangrekar, M.M., 2017. Novel application of peptaibiotics derived from Trichoderma sp. for methanogenic suppression and enhanced power generation in microbial fuel cells. *RSC Adv.* 7, 10707–10717. https://doi.org/10.1039/c6ra27763b

Guo, W., Geng, M., Song, H., Sun, J., 2016a. Removal of chloramphenicol and simultaneous electricity generation by using microbial fuel cell technology. *Int. J. Electrochem. Sci.* 11, 5128–5139. https://doi.org/10.20964/2016.06.42

Guo, W., Song, H., Zhou, L., Sun, J., 2016b. Simultaneous removal of sulfanilamide and bioelectricity generation in two-chambered microbial fuel cells. *Desalin. Water Treat.* 57, 24982–24989. https://doi.org/10.1
080/19443994.2016.1146923

Han, H.X., Shi, C., Yuan, L., Sheng, G.P., 2017. Enhancement of methyl orange degradation and power generation in a photoelectrocatalytic microbial fuel cell. *Appl. Energy* 204, 382–389. https://doi.org/10.1016/j
.apenergy.2017.07.032

Hou, R., Luo, C., Zhou, Shaofeng, Wang, Y., Yuan, Y., Zhou, Shungui, 2020. Anode potential-dependent
protection of electroactive biofilms against metal ion shock via regulating extracellular polymeric substances. *Water Res.* 178, 115845. https://doi.org/10.1016/j.watres.2020.115845

Jiang, C., Liu, L., Crittenden, J.C., 2016. An electrochemical process that uses an Fe0/TiO2 cathode to degrade
typical dyes and antibiotics and a bio-anode that produces electricity. *Front. Environ. Sci. Eng.* 10, 15.
https://doi.org/10.1007/s11783-016-0860-z

Kahoush, M., Behary, N., Cayla, A., Nierstrasz, V., 2018. Bio-Fenton and bio-electro-Fenton as sustainable
methods for degrading organic pollutants in wastewater. *Process Biochem.* 64, 237–247. https://doi.org/
10.1016/j.procbio.2017.10.003

Kaur, A., Boghani, H.C., Michie, I., Dinsdale, R.M., Guwy, A.J., Premier, G.C., 2014. Inhibition of methane
production in microbial fuel cells: Operating strategies which select electrogens over methanogens.
Bioresour. Technol. 173, 75–81. https://doi.org/10.1016/j.biortech.2014.09.091

Khan, N.A., Khan, S.U., Ahmed, S., Farooqi, I.H., Yousefi, M., Mohammadi, A.A., Changani, F., 2020. Recent
trends in disposal and treatment technologies of emerging-pollutants: A critical review. *TrAC: Trends
Anal. Chem.* 122, 115744. https://doi.org/10.1016/j.trac.2019.115744

Kumar, R., Singh, L., Zularisam, A.W., Hai, F.I., 2018. Microbial fuel cell is emerging as a versatile technology: A review on its possible applications, challenges and strategies to improve the performances. *Int.
J. Energy Res.* 42, 369–394. https://doi.org/10.1002/er.3780

Li, S., Hua, T., Li, F., Zhou, Q., 2020a. Bio-electro-Fenton systems for sustainable wastewater treatment:
Mechanisms, novel configurations, recent advances, LCA and challenges. An updated review. *J. Chem.
Technol. Biotechnol.* 95, 2083–2097. https://doi.org/10.1002/jctb.6332

Li, S., Hua, T., Yuan, C.S., Li, B., Zhu, X., Li, F., 2020b. Degradation pathways, microbial community and
electricity properties analysis of antibiotic sulfamethoxazole by bio-electro-Fenton system. *Bioresour.
Technol.* 298, 122501. https://doi.org/10.1016/j.biortech.2019.122501

Li, S., Liu, Y., Ge, R., Yang, S., Zhai, Y., Hua, T., Ondon, B.S., Zhou, Q., Li, F., 2020c. Microbial electro-
Fenton: A promising system for antibiotics resistance genes degradation and energy generation. *Sci.
Total Environ.* 699, 134160. https://doi.org/10.1016/j.scitotenv.2019.134160

Li, Y., Yang, W., Liu, X., Guan, W., Zhang, E., Shi, X., Zhang, X., Wang, X., Mao, X., 2020d. Diffusion-layer-
free air cathode based on ionic conductive hydrogel for microbial fuel cells. *Sci. Total Environ.* 743,
140836. https://doi.org/10.1016/j.scitotenv.2020.140836

Liu, X., Zhou, Y., Zhang, J., Luo, L., Yang, Y., Huang, H., Peng, H., Tang, L., Mu, Y., 2018. Insight into electro-
Fenton and photo-Fenton for the degradation of antibiotics: Mechanism study and research gaps. *Chem.
Eng. J.* 347, 379–397. https://doi.org/10.1016/j.cej.2018.04.142

Logan, B.E., 2008. *Microbial Fuel Cells*. Wiley-Blackwell. https://doi.org/10.1002/9780470258590

Logan, B.E., Hamelers, B., Rozendal, R., Schröder, U., Keller, J., Freguia, S., Aelterman, P., Verstraete, W.,
Rabaey, K., 2006. Microbial fuel cells: Methodology and technology. *Environ. Sci. Technol.* 40, 5181–
5192. https://doi.org/10.1021/es0605016

Mohamed, H.O., Obaid, M., Poo, K.M., Ali Abdelkareem, M., Talas, S.A., Fadali, O.A., Kim, H.Y., Chae,
K.J., 2018. Fe/Fe2O3 nanoparticles as anode catalyst for exclusive power generation and degradation
of organic compounds using microbial fuel cell. *Chem. Eng. J.* 349, 800–807. https://doi.org/10.1016/j.
cej.2018.05.138

Morris, J.M., Jin, S., Crimi, B., Pruden, A., 2009. Microbial fuel cell in enhancing anaerobic biodegradation
of diesel. *Chem. Eng. J.* 146, 161–167. https://doi.org/10.1016/j.cej.2008.05.028

Nath, D., Chakraborty, I., Ghangrekar, M.M., 2021. Methanogenesis inhibitors used in bio-electrochemical
systems: A review revealing reality to decide future direction and applications. *Bioresour. Technol.* 319,
124141. https://doi.org/10.1016/j.biortech.2020.124141

Obata, O., Greenman, J., Kurt, H., Chandran, K., Ieropoulos, I., 2020. Resilience and limitations of MFC
anodic community when exposed to antibacterial agents. *Bioelectrochemistry* 134, 107500. https://doi.
org/10.1016/j.bioelechem.2020.107500

Ondon, B.S., Li, S., Zhou, Q., Li, F., 2020. Simultaneous removal and high tolerance of norfloxacin with
electricity generation in microbial fuel cell and its antibiotic resistance genes quantification. *Bioresour.
Technol.* 304, 122984. https://doi.org/10.1016/j.biortech.2020.122984

Panizza, M., Cerisola, G., 2009. Electro-Fenton degradation of synthetic dyes. *Water Res.* 43, 339–344. https://doi.org/10.1016/j.watres.2008.10.028

Panizza, M., Barbucci, A., Delucchi, M., Carpanese, M.P., Giuliano, A., Cataldo-Hernández, M., Cerisola, G., 2013. Electro-Fenton degradation of anionic surfactants. *Sep. Purif. Technol.* 118, 394–398. https://doi.org/10.1016/j.seppur.2013.07.023

Pham, T.H., Rabaey, K., Aelterman, P., Clauwaert, P., De Schamphelaire, L., Boon, N., Verstraete, W., 2006. Microbial fuel cells in relation to conventional anaerobic digestion technology. *Eng. Life Sci.* 6, 285–292. https://doi.org/10.1002/elsc.200620121

Pontié, M., Jaspard, E., Friant, C., Kilani, J., Fix-Tailler, A., Innocent, C., Chery, D., Mbokou, S.F., Somrani, A., Cagnon, B., Pontalier, P.Y., 2019. A sustainable fungal microbial fuel cell (FMFC) for the bioremediation of acetaminophen (APAP) and its main by-product (PAP) and energy production from biomass. *Biocatal. Agric. Biotechnol.* 22, 101376. https://doi.org/10.1016/j.bcab.2019.101376

Qiu, B., Hu, Y., Liang, C., Wang, L., Shu, Y., Chen, Y., Cheng, J., 2020. Enhanced degradation of diclofenac with Ru/Fe modified anode microbial fuel cell: Kinetics, pathways and mechanisms. *Bioresour. Technol.* 300, 122703. https://doi.org/10.1016/j.biortech.2019.122703

Sharma, P., Kumar, D., Mutnuri, S., 2020. Probing the degradation of pharmaceuticals in urine using MFC and studying their removal efficiency by UPLC-MS/MS. *J. Pharm. Anal.* https://doi.org/10.1016/j.jpha.2020.04.006

Song, H., Guo, W., Liu, M., Sun, J., 2013. Performance of microbial fuel cells on removal of metronidazole. *Water Sci. Technol.* 68, 2599–2604. https://doi.org/10.2166/wst.2013.541

Taheran, M., Brar, S.K., Verma, M., Surampalli, R.Y., Zhang, T.C., Valero, J.R., 2016. Membrane processes for removal of pharmaceutically active compounds (PhACs) from water and wastewaters. *Sci. Total Environ.* 547, 60–77. https://doi.org/10.1016/j.scitotenv.2015.12.139

Thalla, A.K., Vannarath, A.S., 2020. Occurrence and environmental risks of nonsteroidal anti-inflammatory drugs in urban wastewater in the southwest monsoon region of India. *Environ. Monit. Assess.* 192, 1–13. https://doi.org/10.1007/s10661-020-8161-1

Wang, J., He, M.F., Zhang, D., Ren, Z., Song, T.S., Xie, J., 2017a. Simultaneous degradation of tetracycline by a microbial fuel cell and its toxicity evaluation by zebrafish. *RSC Adv.* 7, 44226–44233. https://doi.org/10.1039/c7ra07799h

Wang, W., Lu, Y., Luo, H., Liu, G., Zhang, R., Jin, S., 2018. A microbial electro-fenton cell for removing carbamazepine in wastewater with electricity output. *Water Res.* 139, 58–65. https://doi.org/10.1016/j.watres.2018.03.066

Wang, Y., Feng, C., Li, Y., Gao, J., Yu, C.P., 2017b. Enhancement of emerging contaminants removal using Fenton reaction driven by H2O2-producing microbial fuel cells. *Chem. Eng. J.* 307, 679–686. https://doi.org/10.1016/j.cej.2016.08.094

Wen, Q., Kong, F., Zheng, H., Cao, D., Ren, Y., Yin, J., 2011a. Electricity generation from synthetic penicillin wastewater in an air-cathode single chamber microbial fuel cell. *Chem. Eng. J.* 168, 572–576. https://doi.org/10.1016/j.cej.2011.01.025

Wen, Q., Kong, F., Zheng, H., Yin, J., Cao, D., Ren, Y., Wang, G., 2011b. Simultaneous processes of electricity generation and ceftriaxone sodium degradation in an air-cathode single chamber microbial fuel cell. *J. Power Sources* 196, 2567–2572. https://doi.org/10.1016/j.jpowsour.2010.10.085

Wu, D., Sun, F., Chua, F.J.D., Zhou, Y., 2020. Enhanced power generation in microbial fuel cell by an agonist of electroactive biofilm – Sulfamethoxazole. *Chem. Eng. J.* 384, 123238. https://doi.org/10.1016/j.cej.2019.123238

Wu, W., Lesnik, K.L., Xu, S., Wang, L., Liu, H., 2014. Impact of tobramycin on the performance of microbial fuel cell. *Microb. Cell Fact.* 13, 1–7. https://doi.org/10.1186/s12934-014-0091-6

Xu, H., Quan, X., Xiao, Z., Chen, L., 2018. Effect of anodes decoration with metal and metal oxides nanoparticles on pharmaceutically active compounds removal and power generation in microbial fuel cells. *Chem. Eng. J.* 335, 539–547. https://doi.org/10.1016/j.cej.2017.10.159

Xu, P., Zheng, D., He, Q., Yu, J., 2020a. The feasibility of ofloxacin degradation and electricity generation in photo-assisted microbial fuel cells with LiNbO3/CF photocatalytic cathode. *Sep. Purif. Technol.* 250, 117106. https://doi.org/10.1016/j.seppur.2020.117106

Xu, P., Zheng, D., Xie, Z., He, Q., Yu, J., 2020b. The degradation of ibuprofen in a novel microbial fuel cell with PANi@CNTs/SS bio-anode and CuInS2 photocatalytic cathode: Property, efficiency and mechanism. *J. Clean. Prod.* 265, 121872. https://doi.org/10.1016/j.jclepro.2020.121872

Yang, G., Huang, L., Yu, Z., Liu, X., Chen, S., Zeng, J., Zhou, S., Zhuang, L., 2019. Anode potentials regulate Geobacter biofilms: New insights from the composition and spatial structure of extracellular polymeric substances. *Water Res.* 159, 294–301. https://doi.org/10.1016/j.watres.2019.05.027

You, S.J., Wang, X.H., Zhang, J.N., Wang, J.Y., Ren, N.Q., Gong, X.B., 2011. Fabrication of stainless steel mesh gas diffusion electrode for power generation in microbial fuel cell. *Biosens. Bioelectron.* 26, 2142–2146. https://doi.org/10.1016/j.bios.2010.09.023

Zaied, B.K., Rashid, M., Nasrullah, M., Zularisam, A.W., Pant, D., Singh, L., 2020. A comprehensive review on contaminants removal from pharmaceutical wastewater by electrocoagulation process. *Sci. Total Environ.* 726, 138095. https://doi.org/10.1016/j.scitotenv.2020.138095

Zhang, E., Yu, Q., Zhai, W., Wang, F., Scott, K., 2018. High tolerance of and removal of cefazolin sodium in single-chamber microbial fuel cells operation. *Bioresour. Technol.* 249, 76–81. https://doi.org/10.1016/j.biortech.2017.10.005

Zhang, L., Yin, X., Li, S.F.Y., 2015. Bio-electrochemical degradation of paracetamol in a microbial fuel cell-Fenton system. *Chem. Eng. J.* 276, 185–192. https://doi.org/10.1016/j.cej.2015.04.065

Zhou, Y., Zhu, N., Guo, W., Wang, Y., Huang, X., Wu, P., Dang, Z., Zhang, X., Xian, J., 2018. Simultaneous electricity production and antibiotics removal by microbial fuel cells. *J. Environ. Manage.* 217, 565–572. https://doi.org/10.1016/j.jenvman.2018.04.013

Zhu, X., Ni, J., 2009. Simultaneous processes of electricity generation and p-nitrophenol degradation in a microbial fuel cell. *Electrochem. Commun.* 11, 274–277. https://doi.org/10.1016/j.elecom.2008.11.023

Zwiener, C., Frimmel, F.H., 2003. Short-term tests with a pilot sewage plant and biofilm reactors for the biological degradation of the pharmaceutical compounds clofibric acid, ibuprofen, and diclofenac. *Sci. Total Environ.* 309, 201–211. https://doi.org/10.1016/S0048-9697(03)00002-0

7 Molecular Advances in Bioremediation of Hexavalent Chromium from Soil and Wastewater

Aditi Nag, Devendra Sharma, and Sudipti Arora

CONTENTS

7.1 INTRODUCTION

Industrial wastewater is a major source of heavy metal contamination in our environment. Several inorganic metals like magnesium (Mg), nickel (Ni), chromium (Cr), copper (Cu), calcium (Ca), manganese (Mn), and zinc (Zn) are vital elements needed in small quantities for metabolic and redox functions. Heavy metals are of economic significance in industrial use and are therefore essential within the environment. However, the environmental pollution caused by heavy metals has become a significant threat to living organisms in an ecosystem. Metal toxicity is of great environmental concern due to the bioaccumulation and non-biodegradability nature of metals. Bioremediation has now become a popular approach to convert toxic heavy metals into less harmful states using microorganisms or their enzymes to clean-up pollutants from various contexts of the environment. Bioremediation is thought to be an eco-friendly and cost effective approach in the revitalization of

DOI: 10.1201/9781003204442-7

the environment. As the term suggests, Microorganisms play a central role in the bioremediation process, thus the process is dependent on and limited by several factors. For example, the microorganisms which are actively involved within the bioremediation of a selected pollutant might be inhibited by other pollutants present within the same environment. Further, the speed of degradation of pollutants by microorganisms is usually very slow which limits the feasibility of using them in public practice or wide application processes (Whiteley et al., 2006). In this context, Sutherland et al. (2004) have mentioned several studies which were carried out in previous years that confirmed the utilization of sole enzymes isolated from bacterial species can prove to be more advantageous than using microorganisms as a whole. An added advantage of the enzymatic biotransformation is that it does not usually generate toxic by-products, as often found in the case of chemicals and a few microbiological processes, and therefore possesses less risk for biological contamination on ecosystems. It has been shown that the action of enzymes is restricted to the substrate as compared to using whole microorganisms, and they also are more mobile than microorganisms because of their smaller size (Gianfreda and Bollag, 2002). Although enzymatic treatment processes have tremendous scope for bioremediation, the application of these processes often faces several challenges in terms of low productivity, and the activity and stability of the enzymes which impede the sustainability of their application. Efforts are being made in searching for potential microbes capable of producing enzymes that can transform toxic metal ions to less-toxic or non-toxic forms under various environmental conditions for their practical application in bioremediation processes.

7.2   CHROMIUM: THE ELEMENT

The element chromium (Cr) is a heavy metal. It has an atomic number of 24. It is present as the first element in group number six and as the fourth transition metal on the periodic table. Cr is reported as the 17th most abundant element present in the Earth's mantle, the most abundantly available heavy metal in the lithosphere, along with zinc, and stands 21st on the list of most abundant elements in the Earth's crust ranging from 100–300 µg/g (Cervants et al., 2001). Thus, overall it is reported that Cr is the seventh most abundant element on Earth (Mohanty and Kumar Patra, 2013). Cr occurs naturally in rocks, soil, plants, animals, volcanic dust, and gases, among others It can be found as chromite ($FeCr_2O_4$) or as a complex with other metals like crocoite ($PbCrO_4$), bentorite ($Ca_6[Cr,Al]_2[SO_4]_3$), tarapacaite (K_2CrO_4), and vauquelinite ($CuPb_2CrO_4PO_4OH$). Cr is widely used in the textile, leather, and metal industries for making textile dyes and mordants, pigments, tanning of animal hides as plating, as an alloying agent, in the inhibition of water corrosion, in ceramic glazes and refractory bricks, etc. This wide range of applications has led to a high anthropogenic use of Cr, which has consequently led to environmental contamination. Thus, Cr presence in the environment has become an increasing concern in the last few years (Oliveira, 2012).

7.3   CHEMICAL AND PHYSICAL PROPERTIES

Cr is a lustrous, brittle, hard metal. It has silver-gray color to it and can be highly polished. Cr is known to resist tarnish in air. Upon heating it burns and forms the green-colored chromic oxide. Cr is unstable in the presence of oxygen. It immediately produces a thin oxide layer which is not permeable to oxygen, thus protecting the metal below. Table 7.1 details various physical and chemical properties of this heavy metal.

7.4   OXIDATION STATES OF CHROMIUM

Cr exhibits a wide range of oxidation states. These states may range from −4 to +6; out of these the +3 oxidation state is most energetically stable. The +3 and +6 oxidation states are the two most commonly observed in Cr compounds. The oxidation states +1, +2, +4, and +5 are relatively rare. The oxidation states −4, −3, −2, −1, and 0 are very rare.

TABLE 7.1

Physical and Chemical Properties of Chromium

S.no.	Physical and Chemical Properties	Value
1	Density	7.18 g/cc
2	Melting point	2,130 K
3	Boiling point	2,945 K
4	Appearance	Very hard, crystalline, steel-grayish metal
5	Solubility	Insoluble in water but soluble in acid (except nitric) and strong alkalies
6	Molecular weight	51.99
7	Physical state	Solid
8	Color	Bright green to red
9	Atomic radius	130 pm
10	Atomic volume	7.23 cc/mol
11	Covalent radius	118 pm
12	Specific heat @20°C	0.488 J/g mol
13	Fusion heat	21 KJ/mol
14	Oxidation states	6, 3, 2, 0
15	Lattice structure	Body-centered cubic

7.5   TRIVALENT AND HEXAVALENT STATES OF CHROMIUM

Cr is known to occur in several oxidation states in the environment ranging from Cr(II) to Cr(VI) (Rodriguez et al., 2009). Out of these forms, the most commonly occurring states of Cr are trivalent and hexavalent. Both of these commonly occurring states of Cr are reported to be toxic to animals, humans, and plants to different extents (Mohanty and Kumar Patra, 2013). Cr is usually produced upon the burning of oil, coal, and petroleum from oil well drilling and metal plating, tanneries, other refractory materials like pigment, oxidants, catalysts, steel, fertilizers, etc. Cr(III), or trivalent chromium, resides in the organic matter of soil and aquatic environments in the form of oxides, hydroxides, and sulphates (Cervantes et al., 2001). As mentioned above, it is extensively used in electroplating, metallurgy, the production of paints and pigments, wood preservation, tanning, chemical production, and pulp and paper production. These industries play a major role in Cr pollution with an adverse effect on biological and ecological species (Ghani, 2011). Various industrial and agricultural practices increase toxic levels in the environment and cause concern about the pollution caused by Cr. Pollution of the environment by Cr, particularly hexavalent chromium, has been the greatest concern in recent years (Mitra et al., 2017).

Cr(VI), or hexavalent chromium, is 10–100 times more bioavailable, relatively stable, and more toxic than Cr(III). Due to its strong oxidation potential, higher solubility in water, and rapid permeability through biological membranes, Cr(VI) is considered the most toxic form of chromium. The Agency for Toxic Substance and Disease Registry (ATSDR) classified Cr(VI) as one of the 17 most toxic chemicals, and the United States Environmental Protection Agency (U.S. EPA) listed it as a grade "A" human carcinogen. All Cr(VI) compounds are considered occupational carcinogens by the National Institute for Occupational Safety and Health (NIOSH). The concentration of Cr in drinking water, as per the World Health Organization (WHO) and U.S. EPA, should be less than 0.05 mg/L. In 2012, the Central Pollution Control Board (CPCB) gave the permissible limit of Cr(VI) for industrial discharge water as 0.1 ppm. The principle behind Cr detoxification is conversion of Cr(VI) to Cr(III). Numerous polluting heavy metals and their compounds are discharged into water streams by tanneries (Nath et al., 2008). Due to the presence of excess oxygen in the environment, Cr(III) gets oxidized to Cr(VI), which is extremely toxic and highly soluble in water (Cervantes et al., 2001). In India, the Cr level in underground water has been witnessed to be more than 12 mg/L and 550–1,500 ppm/L. The discharge of industrial wastes and groundwater contamination has drastically increased the concentration of Cr in soil (Bielicka et al., 2005). Cr(VI) is one of the valence states (+6) of the element chromium. It is usually produced by an industrial process. Cr(VI) is known to cause cancer. In addition, it targets the respiratory system, kidneys, liver, skin, and eyes. Cr metal is added to alloy steel to increase hardness and corrosion resistance. A major source of worker exposure to Cr(VI) occurs during "hot work" such as welding on stainless steel and other alloy steels containing Cr metal. Cr(VI) compounds may be used as color pigments in dyes, paints, inks, and plastics. It also may be used as an anticorrosive agent added to paints, primers, and other surface coatings. The chromic acid (Cr[VI] compound) is used to electroplate Cr onto metal parts providing a decorative as well as protective coating.

Cr(VI) is easily soluble and 100 times more toxic than Cr(III) (Saha et al., 2011). Cr(VI) has been recognized as one of the most dangerous environmental pollutants due to its ability to cause mutations, irritations, corrosion of the skin and respiratory tract to most microorganisms, and lung carcinoma in humans. The presence of Cr(VI) in the environment puts selective pressure on microflora, which can possess resistance to high levels while others are sensitive. The bacterial chromate resistance is generally combined to plasmids but it can also be coupled to chromosomal DNA. Cr(VI) is a strong oxidizing agent and is potentially mutagenic, carcinogenic, a potent inducer of tumors in experimental animals, immunotoxic, neurotoxic, reproductive toxic, genotoxic, and can induce a wide spectrum of DNA damage, gene mutations, sister chromatid exchanges, and chromosomal aberrations. Cr(VI) acts as both sensitizer and irritant when it comes in contact with skin. Cr(VI) gets reduced to Cr(III) after entering the organism, and then binds to proteins and creates haptens which trigger an immune system reaction.

7.6   EFFECTS OF CHROMIUM CONTAMINATION AND POLLUTION ON HEALTH

The characteristic clinical presentations of patients with Cr(VI) compound exposure include sinusitis, nasal septum perforation, allergic and irritant dermatitis, skin ulcers, respiratory irritation, bronchitis, asthma, effects on kidney, liver, and lung cancers (Lewis, 2004).

7.6.1   CARCINOGENIC EFFECTS AND THE MOLECULAR MECHANISMS INVOLVED

The mechanism of carcinogenicity of Cr(VI) relies on Cr(III), which is a product of intracellular reduction after penetration of Cr into the cell. Chromium goes into the body through the lungs, gastrointestinal tract, and to a lesser extent through the skin. Inhalation is the most common route for occupational exposure, whereas non-occupational exposure usually occurs via ingestion of chromium-containing food or water. Regardless of the route of exposure, Cr(VI) is more readily absorbed which accounts for its toxicological activity and carcinogenicity. All the ingested Cr(VI) is reduced to Cr(III) before entering the bloodstream. Since the excretion of Cr is usually done via urine (after being processed in the kidneys) and feces (containing Cr via bile), high doses of Cr and long-term exposure to it can give rise to various cytotoxic and genotoxic reactions that affect the body systemically (Hertel et al., 1986; EPA, 1998). The molecular mechanism of the Cr(VI)-induced cytotoxicity is not entirely understood. However, a series of in vitro and in vivo studies have tried to demonstrate that Cr(VI) induces oxidative stress through the enhanced production of reactive oxygen species (ROS) leading to genomic DNA damage and the oxidative deterioration of lipids and proteins. Low levels of DNA damage may eventually lead to genomic instability. Following Cr(VI)-induced oxidative stress, a cascade of cellular events occur, including increased lipid peroxidation and genomic DNA fragmentation, the enhanced production of superoxide anion and hydroxyl radicals, the modulation of intracellular oxidized states, apoptotic cell death, the activation of protein kinase C, and altered gene expressions (Bagchi et al., 2001). The process by which Cr(VI) is reduced to Cr(III) can itself cause many forms of DNA damage: oxidative DNA lesions such as strand breaks, chromium–DNA adducts, DNA–DNA interstrand crosslinks, and DNA–protein crosslinks. Reports published in the March 1994 issue of *Molecular Carcinogenesis* specify that Cr(III)-induced DNA–DNA interstrand crosslinks are the lesions responsible for blocking DNA replication. This observed mutagenicity complements other studies on Cr(III)-dependent DNA lesions, which demonstrate the importance of a Cr(III)-dependent pathway in Cr(VI) carcinogenicity (Branco et al., 2008) .

7.6.2 Reproductive and Developmental Effects and the Molecular Mechanisms Involved

Animal studies provide evidence that Cr(VI), after oral exposure, is a developmental toxicant in rats and mice (ATSDR, 2000). Adverse developmental effects in animals include a greater incidence of post-implantation loss, decreased fetal body weight, reduced ossification, and decreased number of live fetuses. In the case of humans, however, ambiguity still exists. One of the studies shows that the wives of stainless steel welders are at a higher risk of spontaneous abortions (Olsen et al, 1992). A more recent study (Bonde et al., 1992), however, did not corroborate those findings. The mechanism of chromium-induced genotoxicity is not fully understood. In a study it has been shown that Cr(VI) plus glutathione induces DNA damage in vitro, whereas Cr(III) with or without glutathione does not. It seems that Cr exerts its genetic effects by directly binding onto DNA. This may produce DNA strand breaks, stable DNA–chromium complexes, DNA–DNA and DNA–protein crosslinks, etc. The active species of Cr which binds with DNA seems to be the trivalent form (Meditext, 2005). It has been recently reported that strong DNA oxidative damage was observed in the urinary samples of patients who ingested two to three grams of potassium dichromate (Hantson et al., 2005). Moreover, an involvement of the oxidative damage pathway in the mechanism of toxicity of Cr in occupationally exposed individuals has been shown (Goulart et al., 2005). Cr(VI) compounds are clearly mutagenic in the majority of experimental situations (Meditext, 2005). They can cause chromosome aberrations in mammalian cells and have been associated with increased frequencies of chromosome aberrations in lymphocytes in chromate production workers. Additionally, an increase in the sister chromatid exchanges were seen in lymphocytes from workers exposed to chromium, cobalt, and nickel dusts (WHO, 1990; Meditext, 2005).

7.7 BIOREMEDIATION EMERGING AS A SUSTAINABLE SOLUTION

Environmental pollution has already been recognized as one of the major hazards of the modern world due to rapid industrialization and the use of chemicals, including dyes manufactured and used in everyday life (Saharan et al., 2011). Most industries use dyes and pigments to color their products, which include the textile, tannery, food, paper and pulp, printing, carpet, and mineral processing industries (Asamudo et al., 2005). The effluents of these industries are thus highly pigmented, and the disposal of these wastes into receiving waters causes damage to the environment (Saharan et al., 2011). The microbial heavy metal accumulation comprises two phases: an initial rapid phase involving physical adsorption or ion exchange at the cell surface, and a subsequent slower phase involving the active metabolism-dependent transportation of heavy metal into the bacterial cells. During the bioaccumulation process, many features of a living cell-like intracellular sequestration can occur followed by localization within specific organelles, metallothionein binding, particulate metal accumulation, extracellular precipitation, and complex formation (Cabrera et al., 2007). Over the past two decades, there have been advances in bioremediation techniques with the goal being to effectively restore the polluted environments using a low-cost, eco-friendly approach. Researchers have developed different bioremediation techniques modeled for different types of pollutants and contexts. These processes can be undertaken by indigenous microorganisms which might be using and degrading the pollutants for their basic survival processes (Verma and Jaiswal, 2016). Eco-friendly and low-cost features are amongst the main advantages of bioremediation compared to chemical and physical methods of remediation. The process of pollutant removal depends primarily on the nature of the pollutant. The nature, depth, location, and degree of the pollution; the type of environment; and cost are the variables that are considered when choosing a bioremediation technique (Frutos et al., 2012; Smith et al., 2015). Bioremediation processes are dependent on several factors like temperature, pH, oxygen, nutrient concentrations, and other abiotic factors; therefore, before starting a successful bioremediation project it is important to consider these factors. The challenges associated with biodegradation and bioremediation, however, can become limiting factors for these approaches to become practical solutions for pollutant removal.

7.8 UNDERSTANDING THE MECHANISMS AND MOLECULAR INTERVENTIONS AGAINST CHROMIUM POLLUTION

7.8.1 MOLECULAR TARGETS INVOLVED IN CHROMIUM TOXICITY

In recent years, Cr contamination in the environment has become a major concern. Waste from certain industries are used as filling materials which pose considerable health hazards. The toxicity of Cr, due to its ability to easily penetrate cell membranes and cause DNA damage by oxidative stress induced by Cr(VI), has been extensively reported in both eukaryotic and prokaryotic cells. Cr(VI) enters cells using the sulphate transport system of the membrane in the cells of organisms that are able to use sulfate (Wang et al, 2017). Cr(VI) is a strong oxidizing agent and in the presence of organic matter is reduced to Cr(III) by various enzymatic and non-enzymatic processes. During this process, reactive oxygen species (ROS) are formed that exert deleterious effects on cells by interacting with protein and nucleic acid (Cheung and Gu, 2007). Cr produces a wide spectrum of genomic damages such as DNA strand breaks, alkali-labile sites, DNA protein, DNA–DNA cross-links, and Cr(III)–DNA adducts (Reynolds et al., 2009). Cr(VI) can also bind to cellular materials and consequently hinder their normal physiological functions. In contrast, Cr(III) is less toxic and bioavailability is limited (He et al., 2009) as it readily forms insoluble hydroxide/oxides above pH ~ 5.5. The detoxification of Cr(VI) by its reduction to Cr(III) is of great environmental importance.

7.8.2 MICROBES INVOLVED IN THE BIOREMEDIATION OF CHROMIUM TOXICITY

Microbial diversity has gained much interest in bioremediation due to its adaptability, and the search for new capable strains continues, as only a minor portion of the microbial species in the environment have been explored. Microbial populations in metal-polluted environments contain microorganisms that have adapted to toxic concentrations of heavy metals and become metal resistant. The first Cr(VI)-reducing strain isolated from industrial wastewater is *Pseudomonas* sp. Many Cr(VI)-resistant microorganisms have been isolated from different sources by researchers. Bacteria isolated from Cr(VI)-contaminated sites are reported to be highly resistant. Cr-resistant bacteria have been isolated from tannery effluents and waste, industrial sludge, discharge water, electroplating effluents, activated sludge, evaporation ponds, and consortium culture which is isolated from the environment (Narayani and Shetty, 2013; Chang and Kim, 2007; Shakoori et al., 2000; Park et al., 2000). Overwhelming evidence on Cr-resistant bacteria and important findings by various researchers has been seen, which suggests that gram-positive bacteria are predominant over gram-negative bacteria for their resistance against Cr(VI) in which genus *Bacillus* is prominent among all gram-positive bacteria and genus *Pseudomonas* is prominent among all gram-negative bacteria for their Cr(VI) resistance.

There has been considerable interest by researchers in recent years regarding the use of bioremediation and phytoremediation to treat Cr-contaminated soils, sediments, and wastewater (Chandra et al., 1997). Other methods require high energy inputs or large quantities of chemical reagents.

These methods are effective when high concentrations of Cr(VI) present in large volumes of waste-water and generate toxic sludges; whereas biological methods of Cr(VI) are safe, sustainable, and effective at lower concentrations of Cr(VI) (Shakoori et al., 2000).

7.8.3 Metagenomics, Genetic, and Phylogenetic Studies

With advancements in fields of genetics and genomics, many groups have tried to adapt and apply molecular approaches in bioremediation studies. Many studies so far have focussed on the molecular characterization of the newly-discovered microbes in various contexts which showed high rates of bioremediation of refractory metals like Cr.

Due to the inherent ability of a habitat to remediate the presence of toxic heavy metals like Cr, it is crucial to identify the microbes that are capable of processing the bioremedial steps within their physiology. In other words, they should have the genetic makeup and molecular mechanisms to remove or reduce the toxicity within their own habitat. So far, biochemical tests have been done to classify and identify such organisms, but with advancements in molecular genetics and genomics, molecular characterization has now become the gold standard to identify microbes.

16s ribosomal subunit is highly conserved and common in microbial translation processes, with only a few species or strain level variations. Primers can be designed from the conserved regions flanking the variable regions of 16s rDNA so that the molecular characterization can be done by amplifying variable regions in the background of a highly conserved sequence. The variations in the amplified sequences, identified by sequencing, are used to analyze and phylogenetically connect these microorganisms and their strain/s to their nearest relatives (Baldiris et al., 2018; Huang et al., 2016).

Our group obtained six isolates which were able to tolerate and survive on a 0.2% Cr concentration. These isolates were further categorized into three categories as the isolates giving an optimal reduction of Cr on less (0.1% Cr) to more (0.25% Cr) concentrated exposures than the initial concentration of Cr in the growth medium. Genomic DNA was isolated from the isolates of each category and the variable region of 16s rDNA was amplified using two universal primers (see Figure 7.1a). Universal primers are those which are designed against the conserved sequences and therefore are likely to amplify sequences even from the yet unidentified and unsequenced genomes. An alternate method is to prepare a molecular profile of rapidly amplified polymorphic DNA (RAPD), which is a technique in which random primers are used to amplify DNA fragments in unknown isolates. This random amplification gives very specific bands for a particular species as well as various strains of microorganisms. Thus, unique molecular profiles can be created to specifically identify any microorganism even to their strains (see Figure 7.1b).

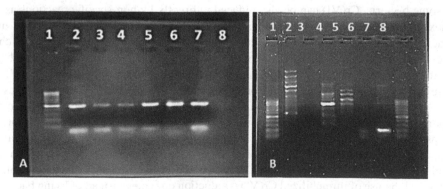

FIGURE 7.1   A. rDNA isolation by amplification of PCR; B. Strain-specific RAPD profiles of the isolates obtained in Cr bioremediation.

The genes involved in the process of reduction have been analyzed for their presence or absence. The presence of such genes can be checked by a simple technique of polymerase chain reaction (PCR), in which a specific sequence of DNA can be amplified and visualized by using the specific primers against such genes. This technique, although simple, requires an understanding of at least a partial sequence of the genes related to the processing of Cr(VI) to Cr(III). In order to identify such toxicity removal genes and their sequences, bioinformatics and high throughput sequencing strategies have been used. In cases where there is no prior information regarding the gene sequence, the gene and genome sequencing of various bioremedial microbial strains is done using the next generation sequencing Illumina platform. The sequences thus obtained are used to predict amino acid sequences. Various software programs, like ClustalW, MEGA, etc., have been used to find the similarity of such genes with already-known proteins and enzymes where the function has been predicted. This prediction, however, remains to be validated in most part; thus, a collaborative investigation into the actual functional relationships and interactions between the predicted molecular players needs to be done.

7.8.4 MOLECULAR MECHANISMS UNEARTHED IN CHROMIUM RESISTANCE AND REMEDIATION

It has been reported previously that microbes use mechanisms like metal efflux, adsorption uptake, DNA methylation, and metal transformation to avoid metal toxicity to survive toxic metal-polluted environments (Camargo et al. 2005; Soni et al., 2012). Various mechanisms of Cr(VI) resistance or detoxification have been described, such as the reduced uptake of Cr(VI); the efflux of chromate ions from the cell cytoplasm; the reduction of extracellular Cr(VI) to Cr(III); ROS scavenging; the activation of enzymes involved in the ROS detoxifying processes, e.g., catalase and superoxide dismutase; the repairing of DNA by SOS response enzymes (RecA, RecG, RuvAB); and the regulation of iron uptake to prevent the production of hydroxyl radicals through fenton reduction and sulfur metabolism (Diaz-Perez et al., 2007; Oljsak et al., 2010; Nies et al., 2006; Branco et al., 2008; Priester et al., 2006; Cheng et al., 2009; Llagostera et al., 1986; Christl et al., 2012). Chromate resistance is an important mechanism from the perspective of bioremediation in the microbial reduction of Cr(VI) to Cr(III) (Cervantes et al., 2001). A variety of Cr-resistant bacteria with a high Cr(VI)-reducing potential have been reported, including *Pseudomonas, Bacillus, Enterobacter, Deinococcus, Shewanella, Agrobacterium, Escherichia, Thermus,* and other species (Ohtake et al., 1987). Cervantes and Campos-Garcia (2007) have summarized the reduction mechanisms. Microbial reduction of Cr(VI) to Cr(III) can be considered a chromate detoxification mechanism and is usually not plasmid associated (Cervantes et al., 2001). Two direct Cr(VI) reduction mechanisms have been explained: firstly Cr(VI) is reduced under aerobic conditions commonly associated with soluble chromate reductases that use NADH or NADPH (Park et al., 2000), and secondly, Cr(VI) can be used as an electron acceptor in the electron transport chain under anaerobic conditions by some bacteria. Cr(VI) can be also reduced indirectly by non-specific reactions associated with redox intermediate organic compounds such as amino acids, nucleotides, sugars, vitamins, organic acids, or glutathione (Myers et al., 2000; Robins et al., 2013). The first analyzed chromate reductase was a membrane-associated enzyme from *enterobacter cloacae* strain HOI that transfers electrons to Cr(VI) via NADH-dependent cytochromes (Ohtake et al., 1990). Cr(VI) reductases were primarily characterized in the context of alternative substrates. These enzymes usually show NADH:flavin oxidoreductase activity and can also act as chromate reductases. Representative examples are the reductase NfsA/NfsB from *Vibrio harveyi* with a nitrofurazone nitroreductase property as a primary activity and chromate reductase as a secondary function (Kwak et al., 2003), and the ferric reductase FerB from *Pseudomonas denitrificans*, which uses Fe(III)–nitrilotriacetate and chromate as substrates (Mazoch et al., 2004). On the basis of results achieved, Robins et al., 2013 suggested the use of immobilized Cr(VI) – reduction enzymes instead of living bacterial cells in the bioremediation processes to overcome the limit of the toxicity of Cr(VI) itself to the remediating bacteria.

7.8.4.1 Anaerobic Cr(VI) Reduction

Anaerobic reduction by bacteria is usually associated with membranes with bound reductases like flavin reductases, cytochromes, and hydrogeneses that can be part of electron transport systems and use chromate as the terminal electron acceptor. Microbial Cr(VI) reduction was first reported using anaerobic bacterium *Pseudomonas dechromaticans*, isolated from sludge. Studies with *E. cloacae* HO1 have implicated the respiratory chain in the transfer of reducing equivalents to Cr(VI) through cytochrome c. The chromate reductase activity of *Shewanella putrefaciens* MR-1 was found to be associated with the cytoplasmic membranes of anaerobically grown cells where formate and NADH served as electron donors for the reductase (Myers et al., 2000).

The overall bio-reduction of Cr(VI) and precipitation of Cr(III) is illustrated in Equation (7.1) and Equation (7.2). Under anaerobic conditions with glucose as an electron donor, microbial Cr(VI) reduction is related to Equation (7.3) (Singh et al., 2011).

$$CrO^{2-}_{4}(aq) + 8H^+(aq) \rightarrow Cr^{3+}(aq) + 4H_2O \tag{7.1}$$

$$Cr_3(aq) + 4H_2O \rightarrow Cr(OH)_3(s) + 3H^+ + H_2O \tag{7.2}$$

$$C_6H_{12}O_6 + 8CrO^{2-}_{4}(aq) + 14H_2O \rightarrow Cr(OH)_3(s) + 100H^-(aq) + 6HCO^-(aq) \tag{7.3}$$

Sulphate-reducing bacteria (SRB) and iron-reducing bacteria (IRB) are important members of anaerobic microbial communities with economic, environmental, and biotechnological interest. Cr(VI) reduction by biogenic iron(II) and sulphides generated by IRB and SRB are ~100 times faster than that by CRB alone. SRB produces H$_2$S, which serves as a Cr(VI) reductant. Anaerobic bacteria mainly use Cr(VI) as a terminal electron acceptor or reduce Cr(VI) in the periplasmic space by hydrogenase or reduced cytochrome.

7.8.4.2 Aerobic Cr(VI) Reduction

Aerobic Cr(VI) is generally associated with soluble proteins and requires NAD(P)H as an electron donor (Shen et al., 1993). These soluble proteins are localized as cytoplasmic proteins, as demonstrated by Puzon et al. (2002) in *E. coli*. Apart from this, the Cr(VI)-reducing activity in aerobes like *Pseudomonas ambigua*, *P. putida*, *E. coli* (Shen and Wang, 1993), and *Bacillus coagulans* (Philip et al., 1998) have been found to be present in the soluble fractions of cells (Ishibashi et al., 1990). Garbisu et al. (1998) have reported that gram-positive aerobe *Bacillus subtilis* reduced Cr(VI) by using cell-free extracts of the bacterium. In the case of gram-negative bacteria, several *Pseudomonas* have also been investigated to assess their Cr(VI) reduction abilities.

The NADH was also the preferred electron donor for the reduction of chromate by the soluble enzyme present in the cytoplasm of *Pseudomonas* sp. CRB5 (McLean and Beveridge, 2001). Previous studies also showed that NADH/NADPH served as electron donors for Cr(VI) reduction by the soluble enzymes present in *P. ambigua* (Suzuki et al., 1992) and *P. putida* (Ishibashi et al., 1990).

In the presence of oxygen, the enzymes from the aerobes responsible for reducing Cr(VI) to Cr(III) require NAD(P)H. Two reaction steps have been suggested to be involved in the reduction reactions: firstly, Cr(VI) accepts one electron from one molecule of NADH to generate Cr(V) as an intermediate (Equation 7.4), and secondly, Cr(V) accepts two electrons to form Cr(III) (Equation 7.5) (Singh et al., 2011).

$$Cr^{6+} + e^- \rightarrow Cr^{5+} \tag{7.4}$$

$$Cr^{5+} + 2e^- \rightarrow Cr^{3+} \tag{7.5}$$

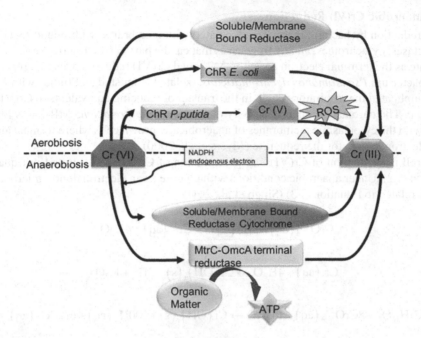

FIGURE 7.2   Mechanisms of enzymatic reduction of Cr(VI).

The ability to reduce Cr(VI) by aerobic bacteria such as *P. aeruginosa* has been extensively studied. It has been found that the resistance to Cr(VI) is proportional to the increased efflux of the heavy metal by cell membrane. In contrast, several species, like *P. fluorescens*, showed that the Cr(VI) resistance does not depend on the capacity of their reduction of heavy metal. Soluble enzymes that catalyze the reduction of Cr(VI) to Cr(III) can detoxify their environment. Once reduced, Cr(III) usually binds to the electronegativity-charged functional groups on the bacterial cell surface, which serve as nucleation sites for further precipitation. Cr(III) also undergoes complexation with a cell envelope and capsule exo-polymer, which in turn inhibits the metals from entering the cytoplasm. By employing this process, chromate can be effectively detoxified and removed from Cr(VI)-enriched medium (McLean and Beveridge., 2001). A summary of the anaerobic and aerobic mechanisms is given in Figure 7.2.

7.9   SIGNIFICANCE OF MOLECULAR INTERVENTIONS IN THE CONTEXT OF CHROMIUM BIOREMEDIATION

Bioremediation is a promising approach to sustainably improve the health of the environment and reduce pollution by exploiting the inherent ability of microflora to process pollutants, for example heavy metals like Cr(VI). Laboratory-scale studies have been successful in identifying some species and strains present within contaminated soils or wastewaters that show potential for bioremediation. These studies, however, are limited by the fact that natural microflora are highly variable, and active remedial species in one habitat might be absent from another. Thus, it is crucial to identify the active members and delineate the molecular mechanisms involved in the bioremedial process so that these findings can be uniformly applied in the different types of real-world scenarios for sustainable environmental solutions.

REFERENCES

Agency for Toxic Substances and Disease Registry. (2000) Toxicological profile for chromium. http://www. atsdr.cdc.gov/toprofiles/tp7.html.

Asamudo NU, Daba AS, Ezeronye OU. (2005) Bioremediation of textile effluent using Phanerochaete chrysosporium. *Afr J Biotechnol* 4(13):1548–1553.

Baldiris R, Acosta-Tapia N, Montes A, Hernández J, Vivas-Reyes R. (2018) Reduction of hexavalent chromium and detection of chromate reductase (ChrR) in Stenotrophomonas maltophilia. *Molecules* 23:406.

Bielicka A, Bojanowska I, Wisniewski A. (2005) Two faces of chromium-pollutant and bioelement. *Pol J Environ Stud* 14(1):5–10.

Bonde, J., et al. (1992) Adverse pregnancy outcome and childhood malignancy with reference to paternal welding exposure. *Scand J Work Environ Health* 18(3):169–77.

Branco, R., Chung, A.P., Johnston, T., Gurel, V., Morais, P., Zhitkovich, A. (2008) The chromate-inducible chrBACF operon from the transposable element TnOtChr confers resistance to chromium(VI) and superoxide. *J Bacteriol* 190(21):6996–7003.

Cabrera G, Viera M, Gomez JM, et al. (2007) Bacterial removal of chromium (VI) and (III) in a continuous system. *Biodegradation* 18:505–13.

Camargo FAO, Okeke BC, Bento FM, Frankenberger WT. (2005) Hexavalent chromium reduction by immobilized cells and cell free extract of Bacillus sp. ES 29. *Biorem J* 8:23–30.

Cervantes C., Campos-García J. (2007) Reduction and efflux of chromate by bacteria. In: Nies D.H., Silver S. (eds.), *Molecular Microbiology of Heavy Metals. Microbiology Monographs*, vol 6. Springer, Berlin, Heidelberg.

Cervantes, C., Campos-Garcia, J., Devars, S., Corona, F.G., Loza-Tavera, H., Guzman, J., Moreno-Sanchez, R. (2001) Interactions of chromium with microorganisms and plants. *FEMS Microbiol Rev* 25(3):335–347.

Chandra P., et al. (1997) *Phytoremediation of Soil and Water Contaminants.* April 8, 1997, 274–282.

Chang IS, Kim BH. (2007) Effect of sulfate reduction activity on biological treatment of hexavalent chromium [Cr(VI)] contaminated electroplating wastewater under sulfate-rich condition. *Chemosphere* 68:218–26.

Cheng Y, et al. (2009) Identification and characterization of the chromium(VI) responding protein from a newly isolated Ochrobactrum anthropi CTS-325. *J Environ Sci* 21(12):1673–1678, ISSN 1001-0742.

Cheung KH, Gu JD. (2007) Mechanism of hexavalent chromium detoxification by microorganisms and bioremediation application potential: a review. *Int Biodeter Biodegrad* 59:8–15.

Christl I, Imseng M, Tatti E, Frommer J, Viti C, Giovannetti L, Kretzschmar R. (2012) Aerobic reduction of chromium (VI) by Pseudomonas corrugata 28: influence of metabolism and fate of reduced chromium. *Geomicrobiol J* 29:173–185.

Diaz-Perez C, Cervantes C, Campos-Garcia J, Julian-Sanchez A, Riveros-Rosas H. (2007) Phylogenetic analysis of the chromate ion transporter (CHR) superfamily. *FEBS J* 274:6215–6227.

EPA. (1998) Toxicological review of trivalent chromium. In: CAS No. 16065-83-1. *In support of Summary Information on the Integrated Risk Information System (IRIS)*. U.S. Environmental Protection Agency, Washington, DC.

Frutos FJG, et al. (2012) Remediation trials for hydrocarbon-contaminated sludge from a soil washing process: evaluation of bioremediation technologies. *J Hazard Mater* 199:262–271.

Garbisu C, Alkorato I, Llama MJ, Serra JL. (1998) Aerobic chromate reduction by Bacillus subtilis. Biodegradation 9:133–41.

Ghani A. (2011) Effect of chromium toxicity on growth, chlorophyll and some mineral nutrients of *Brassica juncea* L. *Egyptian Acad J Biol Sci* 2(1):9–15.

Gianfreda, L., Bollag, J.M. (2002) Isolated enzymes for the transformation and detoxification of organic pollutants. In: Burns, R.G., Dick, R. (eds.), *Enzymes in the Environment: Activity, Ecology and Applications.* MarcelDekker, NewYork, pp. 491–538.

Goulart M., Batoréu M.C., Rodrigues A.S., Laires A., Rueff J. (September 2005) Lipoperoxidation products and thiol antioxidants in chromium exposed workers. *Mutagenesis* 20(5):311–315.

Hantson, P. O., et al. (2005) Hexavalent chromium ingestion: biological markers of nephrotoxicity and genotoxicity. *Clin Toxicol* 43(2):111–2.

He Z, et al. (2009) Isolation and characterization of a Cr(VI)-reduction Ochrobactrum sp. strain CSCr-3 from chromium landfill. *J Hazard Mater* 163:869–73.

Helena Oliveira. (2012) Chromium as an environmental pollutant: insights on induced plant toxicity. *J Bota* 2012: 375843, 8 pages.

Hertel RF. (1986) Sources of exposure and biological effects of chromium. *IARC Scient Publ* 71:63–77.

Huang, H., et al. (2016) A novel Pseudomonas gessardii strain LZ-E simultaneously degrades naphthalene and reduces hexavalent chromium, *Bioresource Technology* 207, 370–378, ISSN 0960-8524.

Ishibashi, Y., et al. (1990) Chromium reduction in Pseudomonas putida. *Appl. Environ. Microbiol.* 56 (7), 2268–2270.

KwakYH, et al. (2003) Vibrio harveyi nitroreductase is also a chromate reductase. *Appl Environ Microbiol* 69:4390–5.

Lewis R. (2004) Occupational exposures: metals. In: LaDou, J. (ed.), *Current Occupational and Environmental Medicine*. 3rd Ed. Lange Medical Books/McGraw-Hill Companies, Inc.: pp. 439–441.

Llagostera M, Garrido S, Guerrero R, Barbe J. (1986) Induction of SOS genes of Escherichia coli by chromium compounds. *Environ Mutagen* 8:571–577.

Mazoch J, Tesar R, Sedlacek V, Kucera I, Turanek J. (2004) Isolation and biochemical characterization of two soluble iron (III) reductases from Paracoccus denitrificans. *Eur J Biochem* 271:553–562.

McLean J, Beveridge TJ. (2001) Chromate reduction by a pseudomonad isolated from a site contaminated with chromated copper arsenate. *Appl Environ Microbiol* 67:1076–84.

Meditext: Medical Management. (2005) Chromium hexavalent salts. In: *TOMES Information System*. Micromedex, Inc., Denver, CO.

Mitra, S., Sarkar, A., Sen, S. (2017) Removal of chromium from industrial effluents using nanotechnology: a review. *Nanotechnol Environ Eng* 2:11.

Mohanty M, Kumar Patra H. (2013) Effect of ionic and chelate assisted hexavalent chromium on mung bean seedlings (Vigna Radiata l. Wilczek. Var k-851) during seedling growth. *J Stress Physiol Biochem* 9(2):232–241.

Myers CR, Carstens BP, Antholine WE, Myers JM. (2000) Chromium (VI) reductase activity is associated with the cytoplasmic membrane of anaerobically grown Shewanella putrefaciens MR-1. *J Appl Microbiol* 88:98–106.

Narayani M., Shetty K. Vidya. (2013) Chromium-resistant bacteria and their environmental condition for hexavalent chromium removal: a Review, *Crit Rev Env Sci Technol*, 43(9):955–1009.

Nath K, et al. (2008) Effect of chromium and tannery effluent toxicity on metabolism and growth in cowpea (Vigna sinensis L. Saviex Hassk) seedling. *Res Environ Life Sci* 1:91–94.

Nies DH, et al. (2006) Paralogs of genes encoding metal resistance proteins in Cupriavidus metallidurans strain CH34. *J Mol Microbiol Biotechnol* 11:82–93.

Ohtake H, Cervantes C, Silver S. (1987) Decreased chromate uptake in Pseudomonas fluorescens carrying a chromate resistance plasmid. *J Bacteriol* 169:3853–6.

Ohtake H, Fujii E, Toda K. (1990) Reduction of toxic chromate in an industrial effluent by use of a chromate-reducing strain of Enterobacter cloacae. *Environ Technol* 11:663–8.

Oliveira H. (2012) Chromium as an environmental pollutant: Insights on induced plant toxicity. *J Bot* 2012:375843.

Olsen S. F., Sørensen J. D., Secher N. J., Hedegaard M., Henriksen T. B., Hansen H. S., Grant A. (1992) Randomised controlled trial of effect of fish-oil supplementation on pregnancy duration, *Lancet (London, England)*, 339(8800):1003–1007.

Park CH, et al. (2000) Purification to homogeneity and characterization of a novel Pseudomonas putida chromate reductase. *Appl Environ Microbiol* 66:1788–95.

Philip L, Iyengar L, Venkobacchar C. (1998) Cr (VI) reduction by Bacillus coagulans isolated from contaminated soils. *J Environ Eng* 124(12):1165–1170.

Priester JH, et al. (2006) Enhanced exopolymer production and chromium stabilization in Pseudomonas putida unsaturated biofilms. *Appl Environ Microbiol* 72:1988–1996.

Puzon GJ, Petersen JN, Roberts AG, Kramer DM, Xun L. (2002). A bacterial flavin reductase system reduces chromates(III)eNADþ complex. *Biochem Biophys Res* 294(1):76e81.

Reynolds MF, Peterson-Roth EC, Bespalov IA, Johnston T, Gurel VM, Menard HL, Zhitkovich A. (2009) Rapid DNA double-strand breaks resulting from processing of Cr-DNA cross-links by both MutS dimers. *Cancer Res* 69(3):1071–1079.

Robins KJ, Hooks DO, Rehm BHA, Ackerley DF. (2013). *Escherichia coli* NemA is an efficient chromate reductase that can be biologically immobilized to provide a cell free system for remediation of hexavalent chromium. *PLoS ONE* 8(3):1–8.

Rodriguez E, Azevedo R, Fernandes P, Santos C. (2011) Cr(VI) induces DNA damage, cell cycle arrest and polyploidization: a flow cytometric and comet assay study in *Pisum sativum*. *Chem Res Toxicol* 24:1040–1047.

Saha R, Nandi R, Saha B. (2011) Sources and toxicity of hexavalent chromium. *J Coord Chem* 64(10):1782–1806.

Saharan, B. S., Ranga, P. (2011) Enhanced decolourization of congo red dye under submerged fermentation (SMF) process by newly isolated Bacillus subtilis SPR42. *J Appl Nat Sci* 3(1):51–53.

Santal AR, Singh NP, Saharan BS. (2011) Biodegradation and detoxification of melanoidin from distillery effluent using an aerobic bacterial strain SAG5 of *Alcaligenes faecalis*. *J Hazard Mater* 193:319–324.

Shakoori AR, Makhdoom M, Haq RU. (2000) Hexavalent chromium reduction by a dichromate-resistant gram-positive bacterium isolated from effluents of tanneries. *Appl Microbiol Biotechnol* 53:348–51.

Shen H, Wang YT. (1993) Characterization of enzymatic reduction of hexavalent chromium by Escherichia coli ATCC 33456. *Appl Environ Microbiol* 53:3771–7.

Singh R, Kumar A, Kirrolia A, et al. (2011) Removal of sulphate, COD and Cr(VI) in simulated and real wastewater by sulphate reducing bacteria enrichment in small bioreactor and FTIR study. *Bioresource Technol* 102:677–82.

Smith E, Thavamani P, Ramadass K, Naidu R, Srivastava P, Megharaj M. (2015) Remediation trials for hydro-carbon-contaminated soils in arid environments: evaluation of bioslurry and biopiling techniques. *Int Biodeterior Biodegrad.*

Soni S.K., Singh R, Awasthi A. et al. (2013) In vitro Cr(VI) reduction by cell-free extracts of chromate-reducing bacteria isolated from tannery effluent irrigated soil. *Environ Sci Pollut Res 20,* 1661–1674.

Stohs SJ, Bagchi D, Hassoun E, Bagchi M. (2001) Oxidative mechanisms in the toxicity of chromium and cadmium ions. *J Environ Pathol Toxicol Oncol* 20(2):77–88.

Sutherland, T.D., Horne, I., Weir, K.M., Coppin, C.W., Williams, M.R., Selleck, M., Russell, R.J., Oakeshott, J.G. (2004) Enzymatic bioremediation: from enzyme discovery to applications. *Clin Exp Pharmacol Physiol* 31(11):817–21.

Suzuki T, Miyata N, Horitsu H, et al. (1992) NAD(P)H-dependent chromium(VI) reductase of Pseudomonas ambigua G-1: a Cr(V) intermediate is formed during the reduction of Cr(VI) to Cr(III). *J Bacteriol* 174:5340–5.

Verma JP, Jaiswal DK. (2016) Book review: advances in biodegradation and bioremediation of industrial waste. *Front Microbiol* 6:1–2.

Wang, Y., et al. (2017) Carcinogenicity of chromium and chemoprevention: a brief update. *OncoTargets Ther* 10:4065–4079. https://doi.org/10.2147/OTT.S139262

Whiteley, C.G., Lee, D.J. (2006) Enzyme technology and biological remediation. *Enzym Microbiol Technol* 38(3e4):291–316.

World Health Organization. (1990) *Chromium (Environmental Health Criteria 61) International Programme on Chemical Safety*, Switzerlands, Geneva.

8 Fenton-Like Degradation of Organic Pollutants Using Rice Husk-Based Catalysts

Damodhar Ghime and Prabir Ghosh

CONTENTS

8.1 INTRODUCTION

Environmental pollution is a major public concern that is increasing in recent decades. As a result, the regulation of water quality is becoming more stringent. The researchers and scientists working in the area of environmental engineering are making great efforts to reduce the toxicity of wastewater released from various anthropological sources (Prousek et al., 2007). The different organic contaminants (azo dyes and harmful chemicals) are known for their presence in industrial effluents. The biological processes are not able to oxidize these water contaminants completely. Therefore, advanced oxidation processes (AOPs) have been projected as a promising solution for the removal of water pollutants. AOPs use the generated hydroxyl ($^{\bullet}OH$) radicals for the degradation of these recalcitrant organic compounds. Fenton and Fenton-like processes are found to be useful for the purification of industrial effluents to the maximum allowable limit.

DOI: 10.1201/9781003204442-8

8.1.1 FENTON AND FENTON-LIKE PROCESSES

These processes have been successfully employed for the mineralization of persistent organic pollutants in wastewater. The hydroxyl radicals play a crucial role in converting toxic compounds to less harmful chemicals. In addition to hydroxyl radicals, sometimes reactive oxygen species (ROS) like per hydroxyl radicals, superoxide radical anions, and oxygen also take part in AOPs (Koltsakidou et al., 2017). The Fenton process uses a mixture of ferrous salt (Fe^{2+}) and hydrogen peroxide (H_2O_2), acting as a strong oxidant, which further generates a highly reactive species, i.e., hydroxyl radicals (see Equation 1). These radicals are capable of degrading recalcitrant organic compounds in industrial effluents (Jiang et al., 2010). The heterogeneous Fenton process is carried out by using ferric (Fe^{3+}) ions directly or other transition metal ions. Such a process is frequently studied as a prototype of the modified Fenton process.

$$Fe^{2+} + H_2O_2 \rightarrow Fe^{3+} + HO^{\bullet} + HO^{-} \qquad (8.1)$$

In a heterogeneous Fenton process, the Fe^{3+}/H_2O_2 Fenton-like reaction forms a complex as Fe(III)-H_2O_2, and followed by the breakdown of this complex yields Fe^{2+} ions and per hydroxide ($H_2O^{\bullet}$) and superoxide ($O_2^{\bullet}$) radicals. De Oliveira et al. (2015) investigated the oxidation of caffeine in wastewater using Fenton and Fenton-like processes. The initial concentration of 1,000 µg/L of caffeine in deionized water was taken to degrade it with Fenton's reagent (Fe[II]/H_2O_2) and Fenton-like reagent (Fe[III]/H_2O_2). Both the processes were compared for their efficiencies towards the degradation of caffeine. About 95% degradation efficiency was obtained in the Fenton process with the molar ratio of 3:10 for Fe(II)/H_2O_2 in just 30 minutes of treatment time. Jiang et al. (2013) studied the intrinsic mechanisms for the interconversion of Fe(III)/Fe(II) in Fenton and Fenton-like processes for the treatment of wastewater. Nitrobenzene and atrazine (aromatic compounds) were chosen as targeted water pollutants with the reaction conditions as pH 3, oxidant H_2O_2 in excess amounts, and Fe in the form of catalytic concentrations. The intermediate species formed during the oxidation of aromatic compounds can influence the interconversion of Fe(III)/Fe(II). Generally, Fenton-like reactions include basic Fenton reactions, while Fenton reactions comprise Fenton-like reaction steps. Fenton-like reactions show autocatalytic behavior for the degradation of aromatic compounds. Hashemian (2013) investigated the oxidative degradation of malachite green dye solutions with a Fenton-like (Fe^{3+}/H_2O_2) process. From the obtained results, it was observed that the Fenton-like process is an economically feasible process for the removal of toxic pollutants in wastewater. For the dye concentration of 3×10^{-5} M, H_2O_2 concentration of 5×10^{-2} M with a ferric ion concentration of 1×10^{-3} M, about 95% percentage decolorization and 70% dye mineralization were obtained using Fenton-like processes. Manaa et al. (2019) used a cost-effective photo-Fenton-like process for the treatment of Patent Blue V dye in wastewater. The Fenton-like process efficiency was enhanced by the use of UV-irradiation and a small dose of reagents at a neutral pH value. The Patent Blue V dye (initial concentration of 10 mg/L) was completely oxidized with 0.98 mg/L Fe^{3+} concentration and oxidant (H_2O_2) dose of 39.1 mg/L at a pH value of 6.4 in just 1 hour of treatment time. The catalyst Fe^{3+} (iron sludge in the Fenton process) can be recycled in the photo-Fenton-like process, which can enhance the degradation efficiency and makes the process cost-effective and environment friendly.

8.1.2 POSSIBLE USE OF RICE HUSKS FOR HETEROGENEOUS DEGRADATION

Silica (SiO_2) is found in its most abundant form in the Earth's crust. Also, it can be prepared by artificial means for industrial application as a valuable inorganic chemical compound with multiple uses. It consists of two surface functional groups: silanol (Si-O-H) and siloxane (Si-O-Si). Out of these two functional groups, siloxane is considered as non-reactive. The manufacturing of silica consists of multiple stages with the use of a tremendous amount of energy and high pressure (Adam et al., 2012). This makes the silica synthesis process costly and not very environmentally friendly. Therefore, researchers and scientists are searching alternative ways for the use of naturally-occurring silica. It was found that agro-waste can offer an alternative source for the replacement of commercial silica. Rice husk sawdust is one of the forms of agro-wastes. Rice is a significant source of food for billions of people and crops cover around 1% of the entire Earth's surface. Annually, approximately 600 million tons of rice is produced around the globe. Nearly 220 kg (22%) of raw rice husk is produced for each 1,000 kg of paddy mill. If this large amount of rice husks is burnt, it can cause serious environmental problems and human health issues. So, it is vital to identify a useful pathway for the complete utilization of rice husk sawdust. Silica can be pyrolyzed at varying temperatures to get rice husk ash, or it can be extracted in the form of a brownish suspension of sodium silicate solution using a solvent extraction method. The amorphous silica in rice husk ash is obtained at 500–800°C, above which crystalline ash is formed. Rice husk ash is the most favorable and useful form in comparison to the direct use of raw rice husks. Because the silica in the ash form undergoes several structural transformations, these transformations are significant for its further use. Sometimes, transition metals like iron (Fe) and copper (Cu) are supported on silica, while using the raw rice husks for application. These metals are easily supported on the silica and have good potential for the use of heterogeneous catalysts. Chen et al. (2013) have successfully used the rice husk ash as a catalyst for the production of biodiesel. The catalyst was prepared with a simple solid-state reaction by using 1.0 gm of rice husk ash with 1.23 gm of lithium carbonate (Li_2CO_3), which was further calcined at 900°C in the presence of air for about 4 hours. The catalyst was highly efficient due to its basic strength. The complete conversion to biodiesel was achieved using optimum conditions: molar ratio of methanol to oil of 24:1; catalyst amount to 4% and at a temperature of 65°C in 3 hours. Kumar et al. (2016) introduced the most economical and eco-benign approach to treating rice mill wastewater with a Fenton-like process. A rice husk ash-based silica-supported iron catalyst was prepared and used for the abatement of rice mill effluent. The complete characterization was carried out with different analyzing techniques like scanning electron microscopy (SEM), X-ray diffraction (XRD), and Fourier transform infrared (FT-IR) spectroscopy. About 75.5% COD reduction was observed for a solid catalyst dose of 4.12 gm/L, H_2O_2 4.2 gm/L, and a pH of 3.2. Thus, it was concluded that one waste (rice husk) could be used for the treatment of severe secondary waste (rice mill wastewater). This proposed scheme offers a sustainable solution to minimize the different wastes in rice mills.

8.1.3 Aim and Objectives of the Chapter

This chapter aims to discuss the preparation of immobilized catalysts by using raw rice husks as a support material. The solid catalyst was prepared by using the sol-gel technique. Further, the prepared catalyst was characterized by different analyzing techniques like SEM, BET surface area, XRD, and FT-IR, respectively. Moreover, the catalyst efficiency was tested for the degradation of various organic pollutants. These contaminants may include azo dyes and different harmful chemicals. The chapter discussion also extends to study the influence of certain operating conditions such as pH, catalyst loading, oxidant (H_2O_2) concentration, and temperature on the process efficiency. Therefore, this chapter will focus on the efficient use of one biogenic and waste material (rice husk sawdust) for the treatment of industrial effluents.

8.2   RICE-HUSK BASED CATALYST PREPARATION AND CHARACTERIZATIONS

The silica-supported iron catalyst can be prepared with a sol-gel technique using rice husk ash or directly with the use of raw rice husks (Adam et al., 2006; Ghime and Ghosh, 2017).

8.2.1 Catalyst Synthesis

Raw rice husks were collected from the nearest rice mill. Further, the husks were washed with distilled water for the removal of dust and other inorganic material. The husks were air dried for about 48 hours. The washed rice husks were then pyrolyzed in a heating furnace at about 700°C for 5–6 hours until a white-colored silica powder was produced. The silica powder was then stirred with a 1.0 molar solution of nitric acid (HNO_3) for 24 hours, and then filtered and washed thoroughly with deionized water until a neutral pH value was obtained. Finally, the rice husk ash (RHA) was obtained this way. 10 gm of RHA was taken for further treatment. It was added and stirred continuously into 500 mL of 5.0 molar solution of sodium hydroxide (NaOH) for 24 hours to produce a sodium silicate solution. The transition metal was then immobilized into the obtained silica. 10wt.% of Fe (3.62 gm of $Fe[NO_3]_3.9H_2O$) was loaded into 200 mL of 3.0 molar solution of HNO_3. Then the acidic solution of Fe was added dropwise from the burette into a measured amount of sodium silicate solution until a pH of 5 was reached. A gel was formed, which can be aged for 2–4 days. It was further filtered using suction pressure and dried in a hot air oven at 110°C for 24 hours. This way, the silica-supported iron catalyst RHA-Fe700 was prepared. The catalyst preparation could also be done with the use of raw rice husks (Ghime and Ghosh, 2017). There is no need to pyrolyze the rice husks. One can directly follow these catalyst preparation steps without pyrolysis (RHA-Fe).

8.2.2 CHARACTERIZATION

The prepared catalyst was characterized for different analyzing methods, including BET surface area, SEM, XRD, and FT-IR analyses, respectively.

8.2.2.1   SEM Analysis of Silica-Supported Iron Catalysts

The SEM images of silica-supported iron catalysts are shown in Figure 8.1(a) and (b). The catalyst without pyrolysis (RHA-Fe) reveals more porosity than the catalyst prepared with pyrolysis (RHA-Fe700) at a temperature of 700°C (Adam et al., 2006). The presence of more porosity in the synthesized catalyst results in the solid consuming high surface area for the oxidation reaction to occur.

8.2.2.2   BET Surface Area Measurement of Prepared Catalysts

The specific surface area of RHA-Fe and RHA-Fe700 solid catalysts was measured with the Brunauer-Emmett-Teller (BET) method, calculated from the isotherms of N_2 adsorption at −196°C. From the obtained results it was found that the value of the specific surface area was higher in the case of the catalyst prepared without pyrolysis (RHA-Fe), 87.40 m²/gm compared to the solid catalyst with pyrolysis phenomenon (55.83 m²/gm) (Adam et al., 2006).

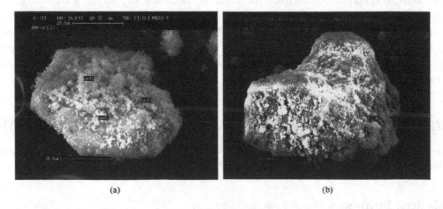

(a)   (b)

FIGURE 8.1   Scanning electron microscope images of the silica-supported iron catalyst: a) RHA-Fe (× 1.50 K) and b) RHA-Fe700 (× 1.50 K). Source: F. Adam et. al, 2006; @Elsevier.

8.2.2.3 XRD Analysis of Silica-Supported Iron Catalysts

Figure 8.2 shows the XRD spectrogram of silica-supported iron catalysts RHA-Fe and RHA-Fe700. Similar types of spectrograms were obtained in both the cases. The XRD analysis of both catalysts shows that RHA-Fe and RHA-Fe700 are present in the amorphous nature. Besides, it was found that the solid catalyst retains its amorphous character upon calcinations at a higher temperature (700°C) (Adam et al., 2006). There was a good dispersion of microcrystalline iron (Fe) particles on the silica support. It means that the active iron particles are too small to diffract X-rays (Gan and Li, 2013). The amorphous nature of the prepared catalyst can also be predicted from the SEM images obtained (see Figure 8.1(a) and 1(b)).

8.2.2.4 FT-IR Spectroscopy of RHA-Fe Catalysts

The FT-IR spectra of the silica-supported iron catalyst RHA-Fe are shown in Figure 8.3. Different peaks were obtained at varying wave numbers, indicating the presence of respective functional groups in the prepared catalyst. The FT-IR spectrum was taken in the range of 400–4,000 cm^{-1} (Ghime and Ghosh, 2017). The peak at wave number 1,048 cm^{-1} is due to the presence of a structural siloxane bond (Si-O-Si), which confirms the silica present in the solid catalyst. The broadband ranging from 3,430–3, cm^{-1} denotes the stretching vibration of the hydroxyl functional group (O–H). The sharp peak at 1,638 cm^{-1} measures the bending mode vibrations of adsorbed water molecules in the silica matrix. The band at 1,385 cm^{-1} confirms the nitrate ion species, while the 826 cm^{-1} band represents the deformation of the Si–O bond in the prepared catalyst. In this way, the obtained peaks are the result of a specific functional group in the RHA-Fe catalyst.

8.3 FENTON-LIKE DEGRADATION OF ORGANIC POLLUTANTS

Gan and Li (2013) evaluated the performance of a silica-supported iron catalyst RHA-Fe for the Fenton-like degradation of a xanthene dye, Rhodamine B. 3wt.% of iron was loaded into the silica matrix. The complete decolorization of xanthene dye was obtained after 10 minutes of treatment time at a pH value of 3 (reaction conditions: dye concentration of 5 mg/L, the dosage of RHA-Si catalyst: 1 gm/L and H$_2$O$_2$ concentration of 0.98 mM). From the obtained results, it is clear that the

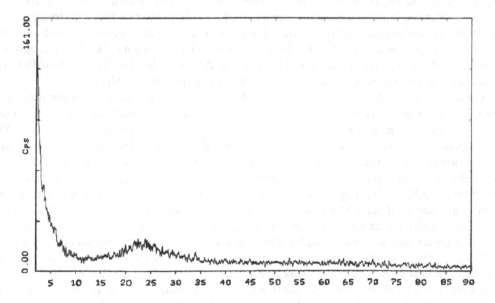

FIGURE 8.2 X-ray diffraction spectrogram of RHA-Fe (similar spectrogram was obtained in the case of RHA-Fe700 catalyst). Source: F. Adam et. al, 2006; @Elsevier.

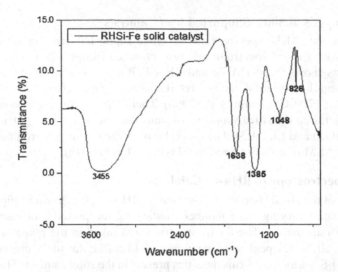

FIGURE 8.3 Fourier transform infrared spectroscopy of silica-supported iron catalyst RHA-Fe. Source: D. Ghime and P. Ghosh, 2017; @Elsevier.

RHA-Fe catalyst possesses a better structural ability and exhibits a lesser amount of iron leaching during application. The prepared catalyst demonstrated great potential for enhancing the treatability of dye-containing effluent. This is a low-cost and green catalyst as rice mill waste was utilized for its preparation.

Moreover, 10wt.% of the iron-loaded catalyst was used for the degradation of oxalic acid in wastewater by Ghime and Ghosh (2017). About 60% degradation efficiency was observed for the oxalic acid degradation in 60 minutes (reaction conditions: oxalic acid concentration of 1,000 mg/L, the dosage of RHA-Si catalyst: 1 gm/L, and H_2O_2 concentration of 0.98 mM). The obtained percentage of degradation in the Fenton-like process was higher while comparing the percentage degradation efficiency in the homogeneous process (20.18%). This shows that the catalyst was efficient for the oxidation of oxalic acid. Olga et al. (2021) reported the Fenton-like degradation of lignin hydrolysates, which was generated in the process of getting a fibrous mass from the rice husks. RHA-Fe catalyst with UV-irradiation was used for this degradation study. The phenolic compound was degraded to 90% for the catalyst dose of 1 gm/L. The bimetallic catalyst in this category also plays a vital role in the oxidation of organic dyes in wastewater. Suligoj et al. (2020) loaded the silica-supported catalyst with copper (Cu) and manganese (Mn) for the decolorization of methylene blue dye.

The use of copper in the catalyst preparation helps in reducing the leaching of manganese from the porous silica support. The adsorption of this textile dye on the surface of the photocatalyst was entirely dependent on the manganese. The bimetallic catalysts provide two active sites (Cu and Mn) for the decolorization of organic dyes in wastewater. The silica-supported metal catalyst can activate hydrogen peroxide (H_2O_2) at a neutral pH level. The stoichiometric efficiency is defined as the number of moles of organic compounds transformed per mole of H_2O_2 consumed during oxidation (Pham et al., 2009). The RHA-Fe catalyst also exhibits stochiometric ability like naturally-occurring iron ores (hematite and goethite), which was 10 to 40 times more than that of iron oxides. If the catalyst is also loaded with alumina and iron on the silica support, the stoichiometric efficiency might be 50 to 80 times higher than that of iron oxides. The interaction of iron with aluminum and silica in the mixed oxides plays a significant role in the enhancement of oxidant production in solution. Ikhlaq et al. (2018) compared the catalytic ozonation process and Fenton-like oxidation by using rice husk ash-supported iron catalysts for the degradation of methylene blue dye in wastewater. Both operations were found efficient for the removal of color in the effluent. The Fenton-like process showed a higher rate of decolorization towards methylene blue dye oxidation at acidic pH values. Sometimes, platinum can also be loaded in place of iron on the silica support from rice husk waste (Sellick et al., 2015). The oxidation of polyaromatic hydrocarbon naphthalene was successfully carried out with the help of silica-supported platinum catalysts. The catalyst activity was found to increase with 2.5wt.% of platinum loading into the prepared catalyst, while the further loading of platinum (beyond 2.5wt.%) into the silica-matrix reduced the catalyst activity. Alemany et al. (1997) replaced the iron with titanium on the silica support and used silica-supported titania catalysts for the degradation of phenol in the treating solution. The species of titanium dioxide (TiO_2) are homogeneously distributed on the catalyst surface as a crystal of small anatase molecules and show titania-like behavior. Those species demonstrate a process performance which might be equivalent to the use of bulk TiO_2 for the oxidation of phenol under UV-irradiation. There are some advantages in the use of distributed species of titanium dioxide in oxidation, more fractions of titania can be used, and the absorption of an excessive amount of titania by UV-irradiation is avoided during oxidation.

8.4 INFLUENCE OF OPERATING CONDITIONS

The influence of different experimental parameters (pH, H_2O_2 dosage, catalyst loading, and reaction temperature) was studied to show the efficiency of silica-supported metal catalysts on degradation efficiency.

8.4.1 EFFECT OF pH

Industrial effluents containing toxic organic compounds are discharged at varying pH values. But the Fenton-like catalyst RHA-Fe shows productive activity at all pH values. The dye wastewater (Rhodamine B of concentration 5 mg/L) was treated for pH values of 3, 5, and 7 using a silica-supported iron catalyst (Gan and Li, 2013). The complete decolorization of Rhodamine B dye in the solution was obtained at pH 3 for the reaction conditions (catalyst dosage: 1 gm/L and H_2O_2 concentration to 0.98 mM). The decolorization rate was found to decrease with an increase in solution pH as an excessive amount of HO^- ions in a solution may form ferric hydroxide complexes, which deactivate the RHA-Si catalyst and reduce the rate of the oxidation reaction. At pH< 3, hydroperoxyl radicals ($^\bullet OOH$) might have been produced instead of hydroxyl radicals. These hydroperoxyl radicals are less reactive than the hydroxyl radicals, thus reducing the rate of the oxidation reaction. The rate of degradation is generally enhanced at an acidic pH of 3 in Fenton and Fenton-like processes (Karale et al., 2014; Youssef et al., 2016). Sometimes, oxonium ion (H_3O^{2+}) is generated at pH <3 from the reaction of H_2O_2 and photons (H^+) which might have reduced the reactivity of ferrous ions (Fe^{2+}) in the treating solution.

8.4.2 EFFECT OF OXIDANT (H_2O_2) CONCENTRATIONS

H_2O_2 plays a vital role as an oxidant in advanced oxidation processes. The selection of an optimal concentration of H_2O_2 is necessary because of its cost. The decolorization of azo dye Acid Red 1 was attempted by taking its concentration in the range of 4–32 mM (Daud and Hameed, 2010). The rate of decolorization was higher when the concentration of oxidant was 8 mM, but at a lower concentration of 4 mM and a higher concentration of 32 mM, the rate of degradation reaction for Acid Red 1 dye was decreased. The reason for this is that at lower concentrations of oxidant, H_2O_2 could not produce enough hydroxyl ($^\bullet OH$) radicals. For the concentration of 8 mM, as predicted, enough hydroxyl radicals were formed to decolorize the dye in the aqueous medium. But the generated hydroxyl radicals show some scavenging action with an increase in concentration of H_2O_2 (32 mM). As a result, less reactive per hydroxyl radical was formed, which lowers the rate of decolorization, as per Equation 8.2.

$$HO^\bullet + H_2O_2 \rightarrow H_2O + HO_2^\bullet \tag{8.2}$$

8.4.3 Effect of Catalyst (RHA-Fe) Dosage and its Stability

It is crucial to evaluate the stability of silica-supported iron catalysts for the practical implementation of heterogeneous catalytic systems. Therefore, the reusability and iron leaching tests should be performed for successive runs, keeping all other operating conditions the same.

The catalyst can be regenerated by washing it with water only. If the catalyst was dried and then used for further runs, it might increase the process cost and can consume more energy. Gan and Li (2013) increased the catalyst doses from 0.5 to 2 gm/L for the degradation of rhodamine B dye. The decolorization efficiency was enhanced for the catalyst doses range of 0.5 to 1 gm/L. This is because of the availability of more active sites of transition metal (Fe) that can speed up the generation of more reactive oxygen species on the surface of the catalyst. The rate of decolorization declined for the catalyst doses ranging from 1 to 2 gm/L. This declination in the decolorization occurs due to the formed agglomeration of particles that decrease the available surface area of solid catalysts. So, the optimal dose of the catalyst was identified as 1 gm/L. This catalyst was tested for reusability for three consecutive runs. Figure 8.4 shows the effect of varying catalyst doses on the decolorization of rhodamine B dye in an aqueous medium. The amount of transition metal leached (Fe) into the solution was measured as 5.0 mg/L (first cycle), 0.8 mg/L (second cycle), and 0.1 mg/L (third cycle). There was no significant decay in the decolorization efficiency observed for the successive runs.

8.4.4 Effect of Reaction Temperature

The reaction temperature also shows its effect on the oxidation of refractory organic pollutants in wastewater. The rate of decolorization is increased with the increase in the temperature by accelerating the reaction rate at 343 K. This phenomenon is anticipated because of the exponential dependency of the rate of oxidation reaction with the temperature (Gan and Li, 2013). About 70% degradation efficiency was recorded for the oxidation of oxalic acid at 343 K in a just 60 minutes of treatment time, which was only 59% at the temperature of 303 K (Ghime and Ghosh, 2017). But the further increase in the value of temperature to 353 K did not enhance the percentage degradation efficiency. It occured due to the thermal decomposition of the oxidant (H_2O_2) used in the heterogeneous catalysis phenomenon. At a temperature of 343 K, more reaction oxygen species were

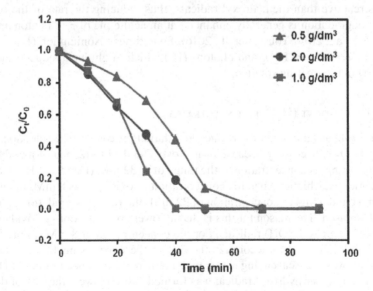

FIGURE 8.4 Effect of catalyst dosage on the decolorization of rhodamine B dye. (Reaction conditions: Dye concentration: 5 mg/L, reaction temperature: 323 K, H_2O_2: 0.98 mM and pH 5.) Source: P. P. Gan and S. F. Y. Li, 2013; @Elsevier.

generated because of the increase in temperature value from 303 to 343 K. This means that the process efficiency can be enhanced by increasing the value of the reaction temperature.

8.5 CONCLUSIONS

The transition of metal-immobilized silica from raw rice husks or rice husk ash has been successfully performed with the sol-gel method of catalyst preparation. The catalyst prepared without pyrolysis holds more surface area than the prepared catalysis with pyrolysis. The use of silica-supported iron catalysts is also feasible for the treatment of textile effluent (removal of rhodamine B in wastewater) and the removal of toxic organic pollutants (oxalic acid present in industrial effluents). The biogenic catalyst RHA-Fe opens a door for the use of one waste material (rice husks) for the treatment of another waste (rice mill wastewater). This work has excellent potential for the utilization of silica-supported iron catalysts as a cost-effective material for the Fenton-like degradation of toxic dyes and refractory organic compounds. The catalyst also retains its stability in iron leaching while using it for the treatment.

8.6 FUTURE PERSPECTIVE

The efficiency of the Fenton-like process can be further improved by combining it with different reactor systems. These reactor designs include the use of fixed and fluidized bed reactors. In both models, the catalytic beds are generally used for degradation purposes. With the beds of silica-supported iron catalysts, the efficacy of the Fenton-like process can be improved and will be higher than the simple heterogeneous processes.

GLOSSARY

Transition metals: These are the d-block elements that have partially filled d-orbitals and can conduct electricity and heat.
Impregnation: The diffusion of an element via a medium or substance.
Catalyst support: Solid material with a higher surface area to which a catalyst particle is immobilized.
Pollutant: A substance that pollutes water or the environment.
Rice husks: The hard-protective coverings of grains of rice. They are also known as rice hulls.
Heterogeneous catalysis: The type of catalysis where the catalyst phase differs from that of the reactants or products.
Textile effluent: Effluent that is highly contaminated with dyes, dissolved solids, suspended solids, and toxic metals.

REFERENCES

Adam, F., Kandasamy, K., & Balakrishnan, S. (2006). Iron incorporated heterogeneous catalyst from rice husk ash. *Journal of Colloid and Interface Science*, 304, 137–143.
Adam, F., Nelson, J., & Iqbal, A. (2012). The utilization of rice husk silica as a catalyst: Review and recent progress. *Catalysis Today*, 190, 2–14.
Alemanyay, L. J., Bafiaresb, M. A., Pardoa, E., Martin, F., Gal, M., & Blascoa, M. (1997). Photodegradation of phenol in water using silica-supported titania catalysts. *Applied Catalysis B: Environmental*, 13, 289–297.
Chen, K., Wang, J., Dai, Y., Wang, P., Liou, C., Nien, C., Wu, J., & Chen, C. (2013). Rice husk ash as a catalyst precursor for biodiesel production. *Journal of the Taiwan Institute of Chemical Engineers*, 44, 622–129.
Daud, N. K., & Hameed, B. H. (2010). Decolorization of acid red 1 by Fenton-like process using rice husk ash-based catalyst. *Journal of Hazardous Materials*, 176, 938–944.

De Oliveira, T. D., Martini, W. S., Santos, M. D. R., Matos, M. A. C., & Rocha, L. L. (2015). Caffeine oxidation in water by Fenton and Fenton-like processes: Effects of inorganic anions and ecotoxicological evaluation on aquatic organisms. *Journal of the Brazilian Chemical Society*, 26, 178–184.

Gan, P. P., & Li, S. F. Y. (2013). Efficient removal of rhodamine B using a rice hull-based silica supported iron catalyst by Fenton-like process. *Chemical Engineering Journal*, 229, 351–363.

Ghime, D., & Ghosh, P. (2017). Heterogeneous Fenton degradation of oxalic acid by using silica supported iron catalysts prepared from raw rice husk. *Journal of Water Process Engineering*, 19, 156–163.

Hashemian, S. (2013). Fenton-like oxidation of malachite green solutions. *Journal of Chemistry*, 2013, 1–7.

Ikhlaq, A., Muhammad, H., Munir, S., Khan, A., Javed, F., & Joya, K. S. (2018). Comparative study of catalytic ozonation and Fenton-like processes using iron-loaded rice husk ash as catalyst for the removal of methylene blue in wastewater. *Ozone: Science and Engineering*, 41, 250–260.

Jiang, C., Pang, S., Ouyang, F., Ma, J., & Jiang J. (2010). A new insight into Fenton and Fenton-like processes for water treatment, *Journal of Hazardous Materials*, 174, 813–817.

Jiang, C., Gao, Z., Qu, H., Li, J., Wang, X., Li, P., & Liu, H. (2013). A new insight into Fenton and Fenton-like processes for water treatment: Part II. Influence of organic compounds on Fe (III)/ Fe (II) interconversion and the course of reactions. *Journal of Hazardous Materials*, 250–251, 76–81.

Karale, R. S., Manu, B., & Shrihari, S. (2014). Fenton and photo-Fenton oxidation processes for degradation of 3-aminopyridine from water. *Procedia: APCBEE*, 9, 25–29.

Koltsakidou, A., Antonopoulou, M., Sykiotou, M., Evgenidou, E., Konstantinou, I., & Lambropoulou, D. A. (2017). Photo-Fenton and Fenton-like processes for the treatment of the antineoplastic drug 5-fluorouracil under simulated solar radiation. *Environmental Science and Pollution Research*, 24, 4791–4800.

Kumar, A., Priyadarshinee, R., Singha, S., Dasgupta, D., & Mandal, T. (2016). Rice husk ash-based silica-supported iron catalyst coupled with Fenton-like process for the abatement of rice mill wastewater. *Clean Technologies and Environmental Policy*, 18, 2565–2577.

Manaa, Z., Chebli, D., Bouguettoucha, A., Atout, H., & Amrane, A. (2019). Lowcost photoFenton like process for the removal of synthetic dye in aqueous solution at circumneutral pH. *Arabian Journal for Science and Engineering*, 44, 9859–9867.

Olga, D. A., Marina, S. V., Liudmila, A. Z., & Anna, S. T. (2020). Heterogeneous photo-Fenton oxidation of lignin of rice husk alkaline hydrolysates using Fe-impregnated silica catalysts. *Environmental Technology*, 2220–2228. doi: 10.1080/09593330.2019.1697376.

Pham, A. L., Lee, C., Doyle, F. M., & Sedlak, D. L. (2009). A silica-supported iron oxide catalyst capable of activating hydrogen peroxide at neutral pH values. *Environmetal Science and Technology*, 43, 8930–8935.

Prousek, J., Palackova, E., Priesolova, S., Markova, L., & Alevova A. (2007). Fenton- and Fenton-like AOPs for wastewater treatment: From laboratory-to-plant-scale application. *Separation Science and Technology*, 42, 1505–1520.

Sellick, D. R., Morgan, D. J., & Taylor, S. H. (2015). Silica supported platinum catalysts for total oxidation of the polyaromatic hydrocarbon naphthalene: An investigation of metal loading and calcination temperature. *Catalysts*, 5, 690–702.

Suligoj, A., Ristic, A., Drazic, G., Pintar, A., Logar, N. Z., & Tusar, N. N. (2020). Bimetal Cu-Mn porous silica-supported catalyst for Fenton-like degradation of organic dyes in wastewater at neutral pH. *Catalysis Today*, 358, 270–277. doi: 10.1016/j.cattod.2020.03.047.

Youssef, N. A., Shaban, S. A., Ibrahim, F. A., & Mahmoud, A. S. (2016). Degradation of methyl orange using Fenton catalytic reaction. *Egyptian Journal of Petroleum*, 25, 317–321.

9 Activated Sludge Process for Refractory Pollutants Removal

Reyhan Ata, Gökçe Faika Merdan, and Günay Yıldız Töre

CONTENTS

Evaluation of the removal of antibiotic residues from the hybrid domestic wastewater treatment plant which was combined with a new generation photocatalytic treatment process containing NFC doped TiO_2 nanoparticles as an efficient technology.

9.1 INTRODUCTION

Industrial facilities produce increasing amounts of refractory pollutant wastewater that is contaminated with toxic and hazardous organic compounds, which cause serious problems for the environment during and as a result of their processes. Refractory pollutants, which are not readily biodegradable in nature, are resistant, difficult to biodegrade, toxic or inhibitory to bacterial growth, and cause increasing problems in water and wastewater treatment systems. These pollutants have different variations in that they degrade poorly and/or exhibit a low value for the ratio of biological oxygen demand (BOD) to chemical oxygen demand (COD). These refractory compounds, originating from anthropogenic or natural sources, can be organic (e.g., halogenated organics, aromatic and aliphatic hydrocarbons, pesticides, and pharmaceutical products) or inorganic (e.g., metals, metal alloys).

Wastewater, including refractory pollutants produced in many industrial processes, may contain organic compounds that are often toxic and are not suitable for biological treatment only. Although conventional biological treatment technologies, including biological, thermal, and physicochemical processes used until recent years, are widely applied in the treatment of wastewater. They require a long time for microorganisms to degrade refractory/recalcitrant pollutants and are not suitable for treating toxic pollutants because of biomass poisoning.

Moreover, it causes a high energy consumption and significant emission of other dangerous compounds. Supporting this situation, scientific studies revealed that the conventional treatment methods, for the treatment of recalcitrant/refractory compounds limiting its biodegradability, have been inadequate in the degradation of complex pollution loads of industrial effluents. In addition to industrial wastewater, domestic/urban/sewage/municipal wastewater also contains refractory/persistent organic pollutants such as pharmaceuticals and personal care products (PPCPs) sourced from residential areas, hospitals, and restaurants.

If these waters containing even trace amounts of refractory, toxic, and metallic pollutants are used in agricultural irrigation after treatment with conventional biological methods, as a result of the intake of transformed refractory pollutant metabolites, through the food chain this would cause harmful effects on the endocrine and reproductive systems of humans. For these reasons, advanced oxidation processes (AOPs), as alternative treatment methods, have gained importance such as ozonation, photocatalytic oxidation, fenton and photo-fenton, membrane filtration, and electrochemical methods which are integrated into conventional biological treatment methods. These applications provided efficient results in terms of the biodegradation of most refractory pollutants.

This chapter firstly discusses the treatability of refractory pollutants from different types of wastewaters and then as a case study, the effect of NFC-doped TiO_2 photocatalysts on the performance of the activated sludge system has been evaluated to remove antibiotic residues from domestic wastewater. This evaluation has been demonstrated in comparison with experimental studies conducted in an aerobic condition using both synthetic and real wastewater.

9.2   REFRACTORY / PERSISTENT SUBSTANCES / POLLUTANTS

Refractory substances are organic or inorganic natural or anthropogenic compounds with certain physical and chemical properties that remain undissociated for a long time when released into the environment due to their resistance to photolytic, chemical, and biological degradation. The word refractory comes from the French *réfractaire*, meaning "high-melting." These refractory compounds can be classified as below:

- Refractory/persistent organic compounds (ROCs/ POPs)
- Refractory inorganic compounds (RIOCs)

9.2.1   REFRACTORY / PERSISTENT ORGANIC COMPOUNDS (ROCS OR POPS), MAIN WASTEWATER SOURCES, TYPICAL CHARACTERISTICS, AND TREATMENT ALTERNATIVES

Refractory organic compounds (ROCs) are poisonous chemicals that do not easily clear up in the environment, accumulate in the food chain, and are easily spread and transported over long distances from their sources through the air, water, and soil. They are also known as persistent organic pollutants (POPs) in the literature. POPs pose global environmental and health problems because they can spread over long distances around the world via air currents, and accumulate in animal and human tissues by entering the food chain. They have a semi-volatile structure with high solubility in oil and low solubility in water.

POPs are carbon-based compounds. The carbon chain is usually surrounded by hydrogen and oxygen atoms and halogens such as chlorine or bromine. Due to the chemical industry's dependence on chlorine with numerous structural possibilities, most POPs are known to belong to the organochlorine chemical group [e.g., dichlorodiphenyltrichloroethane (DDT), aldrin, endrin, and chlorine]. Furthermore, POPs have a structure that cannot be broken or dissolved in the natural environment and therefore they accumulate permanently in the environment over long periods of time. These compounds, which are also biologically persistent, have a long-lasting feature in the adipose tissues of animals as they have a fat-soluble structure and can accumulate in high concentrations, especially in predators such as eagles and humans at the top of the food chain because they can easily enter the food chain. Moreover, because of their bioaccumulative nature, POPs are chronically toxic and cause serious long-term health problems to humans and wildlife. Although evidence of the damage caused by POPs is more common in animals, it also plays a role in liver damage, immune and reproductive system disorders, negative affects to child development and even death in humans.

POPs can travel long distances with water waves, stream cycles, and deposition paths. POPs evaporating at tropical temperatures can reach high altitudes, and at lower temperatures, especially at the Poles, they concentrate and accumulate in these regions (Grasshopper Effect).

These chemicals are the subject of global interest on an international and regional scale. The Stockholm Convention on Persistent Organic Compounds is a global environmental agreement focusing on the protection of humans and the environment against this group of harmful chemicals. Important regional activities have been associated with the United Nations Economic Commission for Europe (UNECE), the Convention on Long-Range Transboundary Air Pollution (CLRTAP), and the POPs Protocol.

These substances are classified as refractory/persistent organic pollutants according to the following properties:

- Being toxic to living organisms (especially by causing deterioration in endocrine functions)
- Having a structure that can accumulate in adipose tissues
- Having permanent features (resistance to photolytic, chemical, and biological reactions due to their stable structure)
- Due to their semi-volatile nature, they cause global environmental problems by showing long-distance transport in the atmosphere

Twenty-three POPs (12 existing and 11 new) can be grouped under three separate categories, taking into account their properties and usage areas:

- Pesticides
- Industrial chemicals
- Unintentionally produced POPs

POPs are anthropogenic chemicals (human made or industrial) manufactured for different usages in many industrial sectors and agricultural activities. They are also a by-product of industrial processes and combustion processes.

9.2.1.1 Pesticides Wastewaters

Pesticides are chemical agents that are generally used in agriculture to inactivate and kill pests, viruses, bacteria, and fungi. They include insecticides that control different types of insects, disinfectants that prevent the spread of bacteria, herbicides that destroy weeds and other unwanted plants, fungicides that prevent mold and mildew growth, and compounds that control mice and rats. Pesticides are grouped according to the types of pests which they kill:

- Insecticides kill insects
- Fungicides kill fungi
- Herbicides kill plants

- Larvicides kill larvae
- Rodenticides kill rodents (rats and mice)
- Bactericides kill bacteria

Pesticides are used for controlling weeds (herbicides), insects (insecticides), and plant diseases. Pesticides promote the growth, harvesting, and marketability of crops, making the pesticide industry a significant economic player in the world market. At the same time, the widespread use of pesticides for agricultural and non-agricultural (veterinary) purposes has resulted in the presence of their residues the environment resulting in various problems. Traces of these products are frequently detected in surface water and in some cases in ground water, which is a major source of drinking water around the world. Toxicological and epidemiological studies have revealed that treated wastewater, including pesticide residual contamination, has potential health risks such as cancer, genetic malformations, neuro-developmental disorders, and immune system disorders. Pesticides included in the banned list of the Stockholm Convention are given in Table 9.1.

Pesticide wastewater contains high levels of chemical oxygen demand (COD), due to compounds such as volatile aromatics, halomethanes, and phenols. They are toxic substances with complex properties and poor biodegradability. Since pesticides are often chemically stable and difficult to mineralize, it makes them resistant to biodegradation and able to inhibit microbial activity due to their toxicity. Physicochemical processes like steam-stripping, activated carbon adsorption, chemical oxidation, or resin adsorption techniques are used before biological treatment processes. A sample wastewater characterization containing pesticides is given in Table 9.2.

Conventional treatment processes are also insufficient for the removal of pesticide-contaminated wastewater. In this context, a study done by Hongchao et al. (20200) on pesticide wastewater from a pesticide production in China. The wastewater – which was high in COD and low in biochemical oxygen demand (BOD_5), the concentrations of which varied between 6,324–13,213 $mg.L^{-1}$ and 317–1,113 $mg.L^{-1}$, respectively – was treated by a system which mainly combined a hybrid anaerobic reactor (HAR), upflow anaerobic sludge blanket (UASB) reactor, and biological contact oxidation

---

TABLE 9.1

Stockholm Convention Banned List (SC. 2010)

Annex A Elimination List	Lists POPs for which parties must take measures to eliminate production and use	POPs listed in Annex A (as of 26 August 2010): aldrin, chlordane, chlordecone, dieldrin, endrin, heptachlor, hexabromobiphenyl, hexabromodiphenyl ether and heptabromodiphenyl ether, hexachlorobenzene, alpha hexachlorocyclohexane, beta hexachlorocyclohexane, lindane, mirex, pentachlorobenzene, PCBs, tetrabromodiphenyl ether and pentabromodiphenyl ether, and toxaphene. Annex A contains specific provisions on the elimination of PCBs, hexabromodiphenyl ether and heptabromodiphenyl ether, and tetrabromodiphenyl ether and pentabromodiphenyl ether.
Annex B Restriction List	Addresses restrictions on production and use	Annex B applies to DDT and perfluorooctane sulfonic acid, its salts, and perfluorooctane sulfonyl fluoride.
Annex C Unintentional Production	Lists POPs for which parties must take measures to reduce their unintentional release, with the objective of continuing to minimize and eventually eliminate them where feasible	POPs listed in Annex C (as of 26 August 2010): hexachlorobenzene, pentachlorobenzene, PCBs, and polychlorinated dibenzo-p-dioxins and polychlorinated dibenzofurans.

TABLE 9.2

Pesticide Wastewater Characteristics from Pesticide Production in China (Hongchao et al. 2020)

Parameter	Sample 1	Sample 2	Sample 3
COD (mg.L^{-1})	18,000–22,000	6,000–8,000	2,000–3,000
Flow (ton.d^{-1})	1.0 ± 0.1	1.7 ± 0.1	2.5 ± 0.1
BOD$_5$/COD	0.05 ± 0.01	0.1 ± 0.02	0.1 ± 0.04

*Reproduced with permission from Ref, (Hongchao et al. 2020). Copyright © 2020, Elsevier.

pool (BCOP). It was difficult to achieve satisfactory performance using this factory's existing original treatment system for the aeration of pesticide wastewater containing refractory material of this type. The study showed that advanced treatment technologies (such as biofilters and coagulation/sedimentation) are still needed to ensure that wastewater quality meets the effluent standard of urban wastewater treatment plants (UWWTPs) (Hongchao et al. 2020). For this purpose, advanced oxidation processes have recently been used in combination with conventional treatment techniques as a first or second treatment step in order to remove refractory pesticide contaminants from wastewater or reuse for irrigational water or for drinking water. These treatments include solar photo-Fenton treatment (Zapata et al. 2009), moving-bed biofilm reactor combined with Fenton-coagulation (Chen et al. 2009), combining large-scale homogeneous solar photocatalysis and biological treatment (Zapata et al. 2010), nanofiltration and photo-Fenton oxidation (Zhang et al. 2010), electrochemical oxidation combined with advanced treatment processes, solar photocatalytic oxidation, UV Fenton and sequencing batch reactor, and advanced oxidation (Klančar et al. 2016).

Although some studies in the literature have shown that Fenton pre-treatment can improve the biodegradability of wastewater, high reagent consumption and higher sludge production are problems that need to be solved in refractory pesticide treatment with Fenton processes (Li et al. 2009). In 2015, Cheng et al. investigated a Fenton reagent enhanced by the use of a microwave electrodeless ultraviolet (MWEUV)/Fenton method for the advanced treatment of non-biodegradable organic substances in pesticide-containing bio-treated wastewater. It has been observed that MWEUV lamps provide higher efficiency than other commercial mercury lamps in eliminating COD with Fenton process. With the MWEUV/Fenton process, the three main pesticides in wastewater, dimethoate, triazophos, and malathion, are completely purified in 120 minutes and a high rate of pesticide decomposition and mineralization has been proven with inorganic anions (Cheng et al. 2015). Studies, have shown that electrochemical and bio-electrochemical processes are more effective and have widespread interest in the treatment of wastewaters containing refractory pollutants with a high salinity and conductivity compared to Fenton processes. In this direction, Beegle et al. (2018) combined that anaerobic digestion and electrochemical systems to provide high energy efficiency and potential economic benefits (Beegle et al., 2018). Garcia-Mancha et al. (2017) confirmed that using an expanded granular sludge bed (EGSB) reactor and aerobic post-treatment is a viable and effective option to treat wastewater containing pesticides (Garcia et al. 2017).

Pesticides, determined by the Stockholm Convention, include aldrin, chlordane, dichlorodiphenyltrichloroethane (DDT), dieldrin, endrin, heptachlor, hexachlorobenzene (HCB), mirex, toxaphene – these are all organochloride that member of chloroalkane in common definition.

Aldrin, with a $C_{12}H_8Cl_6$ molecular formula, is an organochlorine insecticide. Aldrin, which is neurotoxic, adversely affects the central nervous system. Although the usage of aldrin is banned in most of countries it is one of the most widely occurring pesticides in the surface water of most countries. It is used for the control of termites and leads to the localization of the chemical compound in the air, soil, and water (NIOSH 2020).

Chlordane is an organochlorine compound (OC) used as a pesticide. It is a white solid with a molecular formula of $C_{10}H_6Cl_8$ and used for termite control of corn and citrus crops. It is a banned pesticide, except for use in termite control. Chlordane, which has low solubility in water (in hydrophobic structure), can cause type-2 diabetes, lymphoma, prostate cancer, obesity, testicular cancer, and breast cancer in the case of exposure. It can be bioaccumulated in the environment with persistent refractory chemical properties (Tang-Peronard 2011).

Dichlorodiphenyltrichloroethane (DDT), with the molecular formula $C_{14}H_9Cl_5$, is a colorless, tasteless, and almost odorless crystalline chemical compound. It is a banned organochlorine insecticide that causes refractory organic pollutant when adsorbed into soils and sediments. It can bioaccumulate especially in predatory birds because of its lipophilic properties. DDT is toxic to a wide range of living organisms, including marine animals such as crayfish, daphnids, sea shrimp, and many species of fish. It is toxic in food chains and becomes a disruptor [26] and considered to be a human carcinogen (breast cancer) (DDT 2014).

Dieldrin and endrin are a banned organochloride insecticides with a similar molecular formula of $C_{12}H_8Cl_6O$ but at different angles. Both are extremely refractory organic pollutants and do not easily break down. They are reproductive toxic and have harmful effects on fetuses. They are toxic to the environment and aquatic environment (NIOSH 2020).

Heptachlor is an organochlorine compound that was previously used as an insecticide with the molecular structure of $C_{10}H_5Cl_7$. It is a refractory chemical which is lipophilic and poorly soluble in water. It is considered as a possible human carcinogen and is toxic to the environment and aquatic environment (NIOSH 2020).

Hexachlorobenzene is an organochloride with the molecular formula C_6Cl_6. It is a fungicide formerly used as a seed treatment, especially on wheat to control fungal disease bunt. It is an animal carcinogen and is considered to be a probable human carcinogen and toxic to the environment (NIOSH 2020).

Mirex is an organochloride insecticide that was later banned because of its impact on the environment. This white, crystalline, odorless solid is a derivative of cyclopentadiene with the molecular formula of $C_{10}Cl_{12}$. Mirex is highly cumulative and moderately toxic for a range of aquatic organisms. It is a potent endocrine disruptor, interfering with estrogen-mediated functions such as ovulation, pregnancy, and endometrial growth. It also induces liver cancer by interaction with estrogen in female rodents (NIOSH 2020).

Toxaphene is an insecticide, with the molecular formula of C_3H_8Cl, used primarily for cotton. It is a refractory chemical that can cause damage to the lungs, nervous system, liver, kidneys, and in extreme cases, may even cause death. It is thought to be a potential carcinogen in humans and toxic to the environment (NIOSH 2020).

There are many degradation options for general POPs, such as bioremediation, photochemical, and adsorption. However, these treatment processes are costly, have long reaction times, and require high technology. Although aldrin is included in the list of restricted pesticides in Mexico, it is one of the most widely occurring pesticides in the surface water of this country. Photocatalysis of other pesticides from this same family have already been investigated. Degradation was performed using titanium dioxide as a catalyst. Each of the tested solar collectors was capable of similar degradation of aldrin (NIOSH 2020). The photochemical processing of aldrin and dieldrin in frozen aqueous solutions under arctic field conditions have been investigated (Rowland et al. 2011). Biological remediation or bioremediation is an attractive technology that is used to convert organic compounds into less harmful end products such as CO_2 and H_2O, and considered as low cost and environmentally friendly compared to physical or chemical methods for removing contaminants. There are several examples of pesticide remediation using diverse bacterial and fungal strains (Nwankwegu et al. 2018).

Activated carbon (AC) is an excellent adsorbent which can be used to remove pesticides from wastewater. Beside AC, other adsorbents are used such as natural clay, biochar, and zeolite in order to remove the degradation performance of polymerics, and ceramic adsorbents are also used.

Membrane processes were studied for the degradation of pesticides by using a high technology nano-structured membrane. Apart from these techniques operated at ambient temperature and pressure, recently advanced oxidation processes (Cheng et al. 2016), that are more environmental friendly and based on the production of hydroxyl radicals (·OH), are preferred to degrade virtually all types of organic pollutants into harmless products. by using heterogen photocatalysts like TiO_2 and solar/UV irridation. Some of the most effective methods for the oxidation of organic pollutants is Fenton, electro-Fenton or UV/fenton processes. Studies have shown that ultraviolet (UV) radiation is capable of degrading many organic contaminants, including pesticides (Fernandez-Alvarez et al. 2010). Zero valued iron (ZVI) is widely used in the degradation of environmentally harmful organochlorides and is a low-cost option. As a result of the combination of ZVI (Fe0) with magnetite (Fe_3O_4) particles, a successful reagent degrades organochlorine pesticides (OCPs) such as DDT, lindane (γ-HCH), and aldrin in water by using iron-based materials in the water (Shoiful et al. 2016). A study was conduced in India for the presence and distribution of OCPs in paddy fields. Pesticide residue analysis was done using gas chromatography with an electron capture detector (GC-ECD). Twenty-one soil samples were collected for the multiresidual analysis of OCPs. The residues observed were α- BHC; β-BHC; γ-BHC; δ-BHC; α-chlordane; γ-chlordane; heptachlor; 4,4-DDT; 4,4-DDE; 4,4-DDD; α-endosulfan; β-endosulfan; aldrin; dieldrin; endrin aldehyde; and endrin ketone. The findings indicated that the concentrations of OCPs were within the permissible limits of the United States Environmental Protection Agency; thus, the human population in the study area was deemed safe (Sruthi et al. 2017).

Consequently, advanced oxidation processes are preferred as the first or secondary treatment step in order to attain optimum pesticide removal, to reuse or for drinking water treatment (taking into account all the parameters involved, such as pH, type of matrix, temperature, quantity of water, etc.) combined with other treatment techniques such as physico-chemical or conventional biological treatment process.

9.2.1.2 Industrial Chemical Production Wastewaters

Industrial chemicals are anthropogenic (human made or industrial) chemicals produced by the industrial production process. As new products are continuously introduced to the market, the number of anthropogenic substances in the water is becoming more and more threatening with each passing day. The increasing use and discharge of "emerging pollutants" or micropollutants such as medicines, PPCPs, steroid hormones, surfactants, industrial chemicals, and pesticides into the receiving environment is a concern. The discharging of these micropollutants, referred to as anthropogenic refractory compounds (ARCs), usually occurs in the parts-per-trillion (ppt) to parts-per-billion (ppb) range. Almost all chemicals, which are indispensable for modern life, are harmful to humans, animals, and the environment. However, it becomes even more important to know extent of damage of a group of chemicals, defined as refractor or persistent chemicals, in order to take the necessary precautions. The chemical groups listed below are defined as refractor or persistence chemicals by the Stockholm Convention:

- Polybrominated diphenyl ethers (PBDEs)
- Perfluorooctane sulfonic acid (PFOS), its salts, and perfluorooctane sulfonyl fluoride (PFOS-F)
- Polychlorinated biphenyls (PCBs)
- Hexachlorobenzene (HCB)
- Other refractory chemicals such as persistent pharmaceuticals/antibiotics, humic acids, tanning acids

Although they are not clearly specified in this group by the Stockholm Convention, due to their refractory and recalcitrant properties and damaging effects to the environment and living bodies, wastewater treatment will be discussed in this group of chemicals. The Stockholm Convention on

POPs also targets industrial chemicals. Hexabromobiphenyl, tetrabromodiphenyl and pentabromo-diphenyl ethers (components of commercial pentabromodiphenyl ether), and hexa and heptabro-modiphenyl ethers (components of commercial octabromodiphenyl ether) are only produced for industrial purposes. They are member of PBDEs that are a class of recalcitrant and bioaccumulative halogenated compounds emerging as a major environmental pollutant.

Polybrominated diphenyl ethers (PBDEs) have been used since the 1960s. Its application in daily life includes consumer goods such as paints and coatings, plastics, electrical equipment, construction materials, textiles, rugs, polyurethane foam (furniture padding), flame retardants, televisions, building materials, airplanes, and automobiles. When evaluated in terms of environmental effects, PBDEs are refractory compounds that resist degradation in the environment. Low-brominated PBDEs such as tetra, penta, and hexa, which have a high affinity for lipids, can accumulate in the bodies of humans and animals. Studies have shown that the breast milk of North American mothers contains much higher amounts of PBDEs than the milk of Swedish mothers. Scientific studies conducted to date have revealed that tetra- and penta-BDEs are more toxic and bioaccumulating of PBDE compounds than octa- and deca-BDEs. PBDEs are commercially available as a mixture of approximately equal amounts of tetra-BDEs and penta-BDEs (both believed to be the most toxic) under names such as "pentabromodiphenyl ether" and "octabromodiphenyl ether," or "pentabromo" for short. This mixture has been banned by the European Union, but despite the research results mentioned above, it is still used in North America. This is because the United States is the leading producer and user of pentabromo. Although the toxicology of PBDEs is not fully understood, PBDEs have been associated with tumors, neuro-developmental toxicity, and thyroid hormone imbalances. The neurotoxic effects of PBDEs are similar to those observed for PCBs, and children exposed to PBDEs have been observed to be prone to subtle but measurable developmental problems. Although it is assumed that PBDEs are endocrine disruptors, it is a known fact that research in this area is insufficient and requires attention (Siddiqi et al. 2003).

Considering the issue as a source of PBDE pollutants, with the Stockholm Convention decisions and the gradual decline in the production and use of PBDEs with global awareness, it can be assumed that the risks of exposure to PBDE species to ecosystem services and humans are also reduced. However, even if production is reduced, it is still found in domestic and industrial areas even after production: in the air, water, and soil. Therefore, it has become necessary to take into account the release of PBDEs into the environment at various points in their lifecycle. In a study conducted in Taiwan, particularly high levels of PBDEs in particulate matter were noted in sewage treatment works discharged directly into controlled waters, without tertiary treatment methods available to deal with the contamination. This approach causes harmful anthropogenic effects for Taiwan and similar countries where one of the main industires is the chemical industry (O'Driscoll et al. 2016). The stable and persistence character of PBDEs in nature makes them difficult to degrade. There are many studies that have been proposed for de-bromination and degradation of PBDEs such as microbial degradation, electrocatalysis, photocatalysis, and ZVI reduction. However, the results of the studies revealed that these proposed methods can only be useful in the degradation of low amounts of PBDEs in the environment. They need to be treated with combined system of advanced oxidation, biological, and chemical treatment processes together to achieve high removal performance.

Polycyclicaromatic hydrocarbons (PAHs) as well as PBDEs are ubiquitous environmental contaminants. The main sources of anthropogenic PAHs pollutants include the incomplete combustion of fossil fuels and the discharge of petroleum-related materials during their use in homes and factories. Studies revealed that levels of PBDEs pollutants are increasing in e-waste, personal automobiles, cars wastes, and associated dust. The effluent discharged from a conventional municipal sewage treatment plant frequently includes persistent organic pollutants to the receiving water bodies of local aquatic environments surrounding the location of these discharges. Large amounts of PBDE and PAH pollutants that are discharged into receiving waters can potentially threaten drinking water supplies and fishery resources (Fatone et al. 2011).

Organic pollutants, such as PBDE and PAH, are measured with ISO 22032:2006 (Water quality – determination of selected polybrominated diphenyl ethers in sediment and sewage sludge – Method using extraction and gas chromatography/mass spectrometry).

Treatment methods such as sludge treatment processes, adsorption, electrocoagulation, and UV-rays are some of the treatment techniques which were applied for the removal of PAH from wastewater. Apart from these processes, polycyclic aromatic hydrocarbons and phenols were removed from coking wastewater by simultaneously synthesized organobentonite in a one-step process which could improve biological treatment and decrease the environment risk of the effluent. In a study done by Vogelsang et al. in Norway in 2006, it was evident that different types of treatment methods, such as mechanical pre-treatment and biological treatment with simultaneous chemical precipitation, were applied to remove organic micropollutants. Best results were obtained by combining chemical and biological treatments (Vogelsang et al. 2006). In 2016, Mezzanotte et al. studied the distribution and removal of polycyclic aromatic hydrocarbons in two Italian municipal wastewater treatment plants in Italy. They applied the dissolved phase extraction and particulate phase extraction techniques to determine distribution after the collection of wastewater samples in the influent and the effluent points from the activated sludge reactor, and the final effluent after disinfection (ozonation). For both plants, PAHs were mostly removed in the biological section while disinfection had a minor role. It is now necessary to confirm this result by using the same effluent for the two disinfection treatments (Mezzanotte et al. 2016). The conventional activated sludge treatment process was investigated in the wastewater treatment plant in northern Greece in 2004 for the removal of ROCs including PCBs and 19 organochlorine pesticides (such as aldrin, dieldrin, endrin, isodrin, etc.). Removal percentages throughout the whole treatment process ranged from 65% to 91% for individual POP species. As a result of this study, strong hydrophobic characters are principally removed through sorption to sludge particles and transfered to the sludge processing systems (Katsoyiannis and Samara 2004). The major source of PAH and its derivatives (toxic) is wastewater treatment plant (WWTP) effluent. Low molecular weight compounds were mainly removed by mineralization/transformation and adsorption into the anaerobic unit, and by volatilization in the aerobic unit. High molecular weight compounds were mainly removed by adsorption in the anaerobic unit. The percentage of organic PAHs (OPAHs) was higher in summer than in winter. In 2010 in Spain, Rodríguez et al. experimented with the removal of polycyclic aromatic hydrocarbons from organic solvents by ashes wastes (Rodríguez et al. 2016). In 2018, Zhao et al. investigated the removal of PAH, such as the removal of fate of polycyclic aromatic hydrocarbons in a hybrid anaerobic–anoxic–oxic process for highly toxic coke wastewater treatment in China. According to this study, PAHs were significantly removed (99%) by the optimized hybrid A1/A2/O process and activated sludge treatment was crucial for the elimination of PAHs through adsorption (Zhao et al. 2018). In addition to these methods, direct photolysis processes were applied in order to remove polycyclic aromatic hydrocarbons in drinking water. Widely-used low-pressure lamps were tested in terms of their efficiency to degrade PAHs listed as priority pollutants by the European Water Framework Directive. The formation of photolysis by-products was found to be highly dependent on the source waters tested (Sanches et al. 2011).

In recent years, ecological WWTPs (EWWTPs) have been developed as an alternative form of WWTPs, combining microbiology with botany which is efficient for the removal of nitrogen and organic matter, as well as deodorization. There is no data on the occurrence, removal, and fate of the PAHs and soluble PAHs (SPAHs) in EWWTPs. Qiao et al. used an microfiltration and UV disinfection treatment, an advanced treatment process, for EWWTP. In this study, the main removal mechanisms of PAHs and SPAHs in the ecological treatment process were probably biodegradation and adsorption by the sludge and film. The additional contribution of the microfiltration and UV disinfection treatment (0–20%) for the elimination of the PAHs and SPAHs was much less than the ecological treatment (80–100%) (Qiao et al. 2009).

Perfluorooctane sulfonic acid (PFOS), its salts, and perfluorooctane sulfonyl fluoride (PFOS-F) fulfills the criteria of the Stockholm Convention Annex B (Restriction) list as refractory (persistent, bio-accumulative, and toxic) chemicals with extensive industrial application. They are man-made (synthetic), non-biodegradable, and striking in terms of the strength of organic pollution created in the receiving environment where it is discharged without treatment. They do not occur naturally in the environment and are known as surfactants or fluorosurfactants [perfluorooctanoic acid (PFOA)]. Surfactants are compounds that reduce the surface tension (or interfacial tension) between two liquids, between a gas and a liquid, or between a liquid and a solid. Surfactants have the ability to play the role of detergent, wetting agent, emulsifier, foaming agent, and dispersant. The volume of surfactants (through detergents, cleaning agents, and PPCPs) released into the environment at a rising rate are becoming an increasing concern. Linear alkylbenzene sulfonates (LAS) (used for detergent production) and alkyl phenol ethoxylates (APE) (used for detergent-like compounds production), known as the two basic surfactants, break down into nonylphenol, which is considered an endocrine disruptor in sewage treatment plants and soil (Rebello et al. 2014).

Fluorosurfactants are surfactants that contain a perfluoroalkyl group. They can also reduce high surface tension. They prevent dirt from being caught on the surface and can be used in water- or solvent-based products such as varnishes, paints and coatings, non-stick cookware, waterproof fabrics, food packaging, and forming foams. A fluorosurfactant offers long-term dirt resistance and easy oil stain removal. PFOS, is a man-made fluoro-surfactant that has been used as lubricant component of polishes, paints, paper, textile coatings, food packaging, and fire-retardant foams. Products containing PFOS, which are less soluble in water and resistant to biodegradation, can become mobile when released into the environment and released from landfills in water bodies, and bio-accumulates into living (human or animal) bodies thorough the food chain (NCBI 2018).

The strong structure of carbon-fluorine bonds is the factor that makes perfluoroalkyl chains resistant to biodegradation and stabilize over wide temperature, pH, and pressure ranges. The persistence, toxicity, and bioaccumulation potential of certain perfluoroalkyl chain lengths of perfluoroalkyl carboxylic and sulfonic acids (PFCAs and PFSAs, respectively) have resulted in numerous phase-outs and bans worldwide. The introduction of PFOAs into the environment occurs through various mechanisms, such as the active use of fire-fighting surfactants, discharge from chemical production facilities (i.e., air emissions or wastewater), improperly treated wastewater, and the improper disposal of household products produced with PFAs. Studies have shown that PFOS can bioaccumulate in the human body by oral, inhalation, and dermal routes or by consuming contaminated seafood and drinking water. PFOS taken into the human body through diet can cause serious toxic effects such as liver toxicity, developmental toxicity, and immunotoxicity. It may also have negative effects on reproductive organs, cause endocrine disrupting effects, and increase the risk of breast cancer. Consequently, poly- and perfluoroalkyl substances (PFASs), which are not biodegradable due to their high thermal and chemical stability, are widely distributed in the environment, especially in the receiving aquatic environment (Saikat et al. 2013).

The alternative treatment of PFOAs and PFOSs is limited to granular activated carbon (GAC), which is preferred by industry branches today. The GAC treatment option traps contaminants that can leach into the environment if not treated properly. It is necessary to develop a destructive purification method to allow PFOA and PFOS to be broken down into smaller, more easily purifiable substances. Treatment techniques used to remove PFOA from wastewater include reverse osmosis/membrane filtration, anion exchange, GAC, and incineration. Besides these techniques, filtration systems have shown potential for the removal PFOA and PFOS from the respective waste stream, but result in concentrated reject streams that must be sent for further treatment which can be costly and inefficient due to its high energy use. In recent years, treatment techniques that aim to actually destroy PFOA/PFOS molecules include electro-chemical degradation, photochemical degradation, and sonochemical decomposition, collectively known as AOPs. In 2017, Eriksson et al. investigated the contribution precursor compounds make to the release of PFASs from WWTPs. They used influent and effluent sewage water and sludge from three municipal wastewater treatment plants in Sweden for measuring several classes of persistent PFASs: precursors, transformation intermediates, and newly identified PFASs. As a result of their study, they reported that the load of precursors and intermediates in influent water and sludge combined with net mass increase support the hypothesis that the degradation of precursor compounds is a significant contributor to PFAS contamination in the environment. The effect of the precursor compounds is further demonstrated by a significant increase observed for the majority of persistent PFCAs and PFSAs in water after wastewater treatment with mean values of 83%, 28%, 37%, and 58%, respectively (Eriksson et al. 2017). Ebersbach et al. researched an alternative treatment method for fluorosurfactant-containing wastewater by aerosol-mediated separation from electroplating wastewater. In this study, fluorosurfactants were enriched and scavenged by the formation of gas bubbles in the solution and the transfer of gas bubbles to the water surface. The bubbles then collapsed and released an aerosol enriched with florosurfactants. Thereby 99.8% of the initial amount was revocered in the collected aerosols. The concentration of fluorosurfactant in solution decreased by 99.6%, 99.9%, and 99.8%, respectively, simultaneously with PFOA and PFOS over 2 to 6 minutes (Ebersbach et al. 2016). In 2015, Lina et al. performed a photocatalytic treatment of municipal reverse osmosis concentrate (ROC) with relatively high efficiency and low cost. In this study, in which adsorption and coagulation methods and their effectiveness in removing fluorosurfactants were investigated, it was revealed that Ferric clotting (FER) coagulation was quite effective in removing fluorosurfactants, but the powdered activated carbon (PAC) adsorption was insufficient. On the other hand, the FER pretreatment process has been found to outperform the post-treatment in removing fluo-surfactants. FER selectively removed at 62.19 % the bulky fluorosurfactants with long branches but not the slim ones with short or no branches (Lina et al. 2015). These treatment technologies may experience low treatment efficiency and secondary water quality problems due to their long operating times and their tendency to have unexpected realizations with organic materials and naturally occurring geochemical components. For instance, the hydroxyl radical, which is a powerful oxidant, formed as a result of many destructive technologies related to PFAS, may not react effectively with PFAAs. Naturally, the chloride formed in the reaction oxidizes bromide or trivalent chromium to perchlorate (Schaefer et al. 2019) and hexavalent chromium (Zhang et al. 2014), and all these formations have restrictive regulatory criteria. Because of these factors, it complicates the commercial applicability of PFAS-related disruptive technologies to affected water and supports the fact that pre- or post-treatment process components may be required to achieve final water quality goals. Moreover, AOPs have also been widely used in the purification of PFOAs from wastewater and continue to be used.

Zhang et al. conducted a scientific study on PFOA mineralization via γ-irradiation in 2014. In this study, experiments were conducted on 20 mg/L aqueous PFOA solutions at various pH ranges to determine the effectiveness of PFOA degradation when exposed to gamma (γ)-radiation produced by a 60Co source. Below pH 7, PFOA purification was minimally determined, and at a pH of 13.0 PFOA degraded almost 100% after exposure to γ-irradiation for approximately 6 hours. In this study in which the hydroxyl radicals (•OH) and aqueous electrons (eaq-) were successfully produced, it was determined that individual components produced undesirable results, and both components (•OH and eaq-). were required for the degradation process to work effectively. At the end of the study, rapid, complete mineralization and overall deflorination of PFOA were successfully achieved (Zhang et al. 2014).

Polychlorinated biphenyls (PCBs), HCBs, dioxins, and furans are described in the Stockholm Convention in Annex C and classified as "Unintentional Production", which means that they are not specially produced, and pose in the presence of chlorine as a result of industrial activities caused by the exposure of components to high temperatures. Moreover, PCBs, as a group of chemicals, have contaminated rivers and lakes in industrial areas, killing or poisoning fish. PCBs are also unintentional by-products of combustion and industrial processes, along with HCB and pentachlorobenzene, used in the past for industrial and agricultural purposes (as pesticides). PCBs, which are a group of anthropogenic (man-made) organic chemicals (chlorinated hydrocarbons), include carbon, hydrogen, and chlorine atoms. The number and location of chlorine atoms in a PCB molecule determine most of its physical and chemical properties. They were manufactured from 1929 and later banned in 1979. They have a range of toxicity values and vary in consistency from thin, light-colored liquids to yellow or black waxy solids. Their non-flammability, chemical stability, high boiling point, and electrical insulating properties make PCBs feasible for hundreds of industrial and commercial applications such as electrical, heat transfer and hydraulic equipment, plasticizers in paints, plastics and rubber products, pigments, dyes, carbonless copy paper, and other industrial applications. Although no longer commercially produced in the United States, PCBs may be included in products and materials produced before the 1979 PCB ban. Products that may contain PCBs include transformers and capacitors; electrical equipment including voltage regulators, switches, re-closers, bushings, and electromagnets; oil used in motors and hydraulic systems; old electrical devices or appliances containing PCB capacitors; fluorescent light ballasts; cable insulation; thermal insulation material including fiberglass, felt, foam, and cork; adhesives and tapes; oil-based paint; caulking; plastics; carbonless copy paper; and floor finishes. Sources of pollutants can be classified as poorly maintained hazardous waste sites, the illegal or improper dumping of PCB wastes, leaks or releases from electrical transformers containing PCBs, the disposal of PCB-containing consumer products into municipal or other landfills not designed to handle hazardous waste, the burning of wastes in municipal and industrial incinerators, and exposure to PCB-contaminated soils (treated with WWTP sludge) through different intake pathways. PCBs do not readily degrade in the environment and can remain for long periods cycling between air, water, and the soil. PCBs can be carried long distances and have been found in snow and sea water in areas far from where they were released into the environment. As a consequence, they are found all over the world. PCBs can accumulate in the leaves and other above-ground portions of plants and food crops. They are also taken up into the bodies of small organisms and fish. As a result, people who ingest contaminated fish may be exposed to PCBs that have bioaccumulated in the fish. The quantification analysis of PCBs in water is realized by gas chromatography (GC). PCB, which is known commercially as Aroclor, quantification is based on the Aroclor method; on the other hand, the quantification of individual PCB congeners is determined by the congener-specific method (EPA 2020).

PCBs pose high toxicological risks to wildlife, agricultural ecosystems, and consequently the human food chain due to their stable properties and ubiquitous distribution. As it is known, activated carbon adsorption (ACA) is frequently used to remove apolar pollutants from wastewater. Hence, ACA is usef as an effective remover of PCBs which are apolar. As alternative absorbents, Nolleta and Roelsan (2003) revealed the removal of PCBs from wastewater using fly ash for commercially available carbon due to its low cost and high efficiency (Nolleta et al. 2003). Also, a significant linear relationship was observed between removal efficiency and log Kow for PCBs, suggesting that compounds with a strong hydrophobic character are principally removed through sorption to sludge particles and transfered to the sludge processing systems. The total concentrations of PCBs in sewage sludge ranged between 185 and 765 ng.g-1 dw being below the European Union limit for use of sludge in agriculture (Katsoyiannis and Samara 2004). Spatial distribution and partitioning of polychlorinated biphenyl and organochlorine pesticide were provided in water and sediment from the Sarno River and Estuary in southern Italy (Montuori et al. 2014). A hybrid sequential biofiltration system (HSBS) was established for the improvement of nutrient removal and PCB control in municipal wastewater in 2017. The HSBS achieved a significant load of total phosphorous (TP) (0.415 kg), total nitrogen (TN) (3.136 kg), and PCB EQ (0.223 g) per square meter removal per year. The use of low-cost hybrid HSBSs as a post-treatment step for wastewater treatment was found to be an effective ecohydrological biotechnology that may be used for reducing point source pollution and improving water quality in river catchments. This study offered the low-cost HSBS as an additional treatment step and an alternative ecohydrological biotechnology for reducing point source pollution and improving water quality in river catchments (Edyta et al. 2017). The biodegradation of PCBs is provided by using a moving bed biofilm reactor (MBBR). In a 2015 study, it was found that membrane biorecator (MBR) systems were widely preferred for the removal of PCBs; in this context the nano-occurrence and removal of PCBs within Europe's largest petrochemical MBR system were studied. Studies revealed that the MBBR technology could have potential for biodegradation use of PCBcontaminated wastewater (Dong et al. 2015). Fenton ve biological fenton oxidation processes were used for improving the removal rate of pentachlorophenol. Fenton processes provide a tractable alternative to more detailed approaches and a more accurate strategy than simple models. Beside all these processes, organic micropollutants at mechanical, chemical, and advanced wastewater treatment plants were used as a combined system for the improved removal of PCBs, instead of single processes, to get them to benefit from the purification advantages of combined systems. As another alternative method, a combination of the multi-oxidation process and electrolysis process was used for the pretreatment of PCB industry wastewater and recovery of highly-purified copper in China in 2018 (Min et al. 2018). Studies have revealed that combined systems present more efficient results for the removal of PCBs, as is the case with other pollutants.

Hexachlorobenzene (HCB) or PCB, with the molecular formula C_6Cl_6, is a kind of fungicide first introduced in 1945 used as a seed treatment especially on wheat to control the fungal disease bunt. It has been banned globally under the Stockholm Convention on POPs listed in Annex C as "Unintentional Production." As a physical property, HCB is a white crystalline solid that has negligible solubility in water. HCB is banned in Austria, Belgium, the Czech Republic, Denmark, Germany, Hungary, Liechtenstein, Netherlands, Panama, Switzerland, Turkey, the United Kingdom, and the Union of Soviet Socialist Republics. It is severely restricted or has been voluntarily withdrawn in Argentina, New Zealand, Norway, and Sweden. The effects of HCB on humans who ingested the treated seeds include a range of symptoms such as photosensitive skin lesions, hyperpigmentation, hirsutism, colic, severe weakness, porphyrinuria, and debilitation. Children born to these women developed a condition called pembe yara or "pink sore," with a reported mortality rate of approximately 95%. There is inadequate evidence for the carcinogenicity of HCB in humans. It is toxic to animals, highly toxic to aquatic organisms, and may cause long-term adverse effects to the aquatic environment when bioaccumulated in the environment. Sources of HCB water pollution are mainly attributable to the agricultural production process, rainfall, and atmospheric deposition. The extensive use of agricultural pesticides, industrial solid waste incineration, medical waste incineration, sewage sludge incineration, melting scrap metal, waste tires burning, waste oil combustion, cement manufacturing municipal sewage and sludge treatment processes, chlorine bleaching (paper) process, wood processing, sintering plants, etc., cause the release of HCB pollution to the environment and hence to water. Moreover, the melting and incineration processes of copper, aluminum, and magnesium can cause HCP pollutants as well. HCB pollution comes mainly from the extensive use of pesticides in agriculture, industrial solid waste incineration, medical waste incineration, sewage sludge incineration, melting scrap metal, waste tires burning, waste oil combustion, cement manufacturing municipal sewage and sludge treatment process, chlorine bleach (paper) process, wood processing, sintering plants, and so on. Copper, aluminum, magnesium smelting, and the incineration process can also generate the HCB. High performance liquid chromatography (HPLC) is used for the quantative instrumental analysis of HCBs. Since the late 1900s, the main treatment technique for HCB is the dechlorination process in municipal sewage sludge by an anaerobic mixed culture application. According to scientific studies, the factors affecting dechlorinating HCB in sludge are substrate concentration, total solids concentration, pH, and agitation. Recently, there have been many studies on HCB degradation. Activated carbon fiber's adsorption of hexachlorobenzene at different pH levels and at different temperatures can be counted as an ultrasonic instrument occurring with petroleum ether and acetone, as the solvent extraction can be counted among the methods of HCB degradation. As it is known, the redox potential of the environment is low under anaerobic or anoxic conditions. When enzymes act on the bezene ring, having a low electron density due to the reducing agent susceptible to lead nucleophilic attack, chlorine atoms get easily substituted, and therefore become bio-degradable. Consequently, chloride is replaced by hydrogen under an enzymatic reaction known as dechlorination reaction. A type of bacteria, which is isolated from digested sludge, river sediments, ditch sediment, and soil in anaerobic environments, under the premise of glucose, formic acid, acetic acid, and inorganic iron salts and other nutrients, is capable of degrading the HCB. There are many studies on the anaerobic dechlorination of higher chlorinated organic compounds, in particular in Europe and the U.S.A. *Dehalococcoides* spp. is the only bacterium that can completely dechlorinate dechloroethenes and convert other chlorinated compounds as well. Other microorganisms that are able to degrade chlorinated organic contaminants through organohalide respiration are *Dehalobacter* spp., *Desulfomonile* spp., *Desulfuromonas* spp., and *Desulfitobacterium* spp. Again, as it is determined that AOPs combined with conventional treatment techniques provide a more efficient removal/degradation performance of HCB contaminants (Ji et al. 2015, Wang and He 2013).

Dioxins and furans, which are organochlorine compounds, are listed as "Unintentional Production" which are not specially produced, pose in the presence of chlorine as a result of industrial activities caused by the exposure of components to high temperatures, according to the Stockholm Convention. Polychlorinated dibenzo para dioxins (PCDDs) are the chemical name of dioxin: TCDD. The name "dioxins" is often used for the structurally- and chemically-related family PCDDs and PCDFs. Some dioxin-like PCBs with similar toxic properties are also included under the term "dioxins." A total of 419 types of compounds associated with dioxin have been identified, but only 30 of them are considered to have significant toxicity, although TCDD is the most toxic. Dioxins are environmental pollutants. They are a group of hazardous chemicals known as refractory and persistent organic pollutants and are also known as the "dirty dozen." Dioxins are alarming because of their high toxic potential. Experiments have shown that they affect a range of organs and systems. After dioxins enter the body, they predominantly pass into the adipose tissue and are then stored in the body for a long time. Their half-life in the body is estimated to be 7 to 11 years. In the environment, dioxins tend to accumulate in the food chain. The higher an animal is in the food chain, the higher the concentration of dioxins. More than 90% of human exposure passes through food, mostly meat, dairy products, fish, and shellfish. Many national authorities have programs to monitor food supply. Short-term human exposure to high levels of dioxins can cause lesions such as patchy darkening of the skin and changes in liver function. Prolonged exposure causes the immune system, developing nervous system, endocrine system, and reproductive functions to deteriorate. The chronic exposure of animals to dioxins has resulted in various types of cancers, classified as a "known human carcinogen" (group 1 carcinogen) by the International Agency for Research on Cancer (IARC). However, TCDD is not genotoxic (does not affect genetics) but increases cancer risk after a certain amount of exposure (Özdoğan 2020).

Furan or furfurans (PCDFs), with the molecular formula of C_4H_4O, are heterocyclic aromatic organic compounds. Furan is a colorless, volatile, flammable liquid, and is toxic and carcinogenic. It is used as a starting point to other specialty chemicals. Produced by the thermal degradation of natural food ingredients, furan is found in heat-treated commercial foods. It can be measured in roasted coffee, instant coffee, and processed baby foods. Furan is listed as a possible human carcinogen. PCDDs and furans (PCDFs), which are products of the industrial production of chlorinated organic compounds and chemical processes involving combustion, are found almost everywhere in the environment. This situation poses a potential risk for human health. PCDD and PCDF compounds accumulate especially in fatty foods (whole milk and dairy products, meat, and eggs) and transfer to the human body. The structures of PCDD and PCDF compounds; sources of contamination; toxic effects; milk, dairy products, and breast milk as sources of contamination; and precautions to be taken are discussed here. There are 75 different dioxins, 135 different furans, and 209 different PCB types in nature, of which 29 are the compounds with the most toxic effects. These toxic compounds are dioxin derivatives containing four or more chlorine atoms. These compounds include tetra, penta, hexa, and orthodioxins. Dioxins and furans have three rings in their structure, while biphenyls have only two rings (see Figure 9.1) (Hişmioğulları et al. 2012).

Dioxins and furans are not commercially produced compounds and have no known uses. These are often revealed as undesirable by-products in the production of chemical products (Hişmioğulları et al. 2012).

Dioxins are mainly by-products of industrial processes, but can also result from natural processes such as volcanic eruptions and forest fires. Dioxins are undesirable by-products of a wide variety of production processes, including melting, bleaching pulp with chlorine, and the production

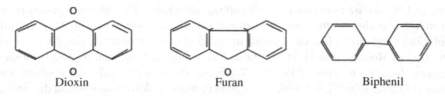

Dioxin   Furan   Biphenil

FIGURE 9.1 Diagrams for dioxin, furan, and biphenils (Hişmioğulları et al. 2012).

of certain herbicides and pesticides. In terms of dioxin release into the environment, uncontrolled waste incineration plants (solid waste and hospital waste) are often the worst culprits due to complete combustion. Technology is available that allows controlled waste incineration with low dioxin emissions. In many cases, dioxin contamination occurs through contaminated animal feed, for example, increased dioxin levels in milk or animal feed are traced to clay, oil, or citrus pulp pellets used in animal feed production (Özdoğan 2020).

The basic sources of dioxin and furan are summarized as burning waste, ferrous and non-ferrous metal production, electricity generation and heating, mineral (lime, cement, ceramic, glass, and asphalt mixtures) production, motor vehicles, uncontrolled combustion processes, the production of chemicals and consumer foods (paper, textile, leather, and chemical industries), landfill and deposit (sludge treatment, composting, and accumulation of waste oil), cigarette smoke, natural phenomena such as forest fires, volcanic eruptions, and animal food. Other important sources can be classified as ferrous and non-ferrous metal production in the production of chemicals and consumer foods, energy production, and heating activities (Hişmioğulları et al. 2012).

Apart from all these, it is known that dioxin is formed during the production of herbicides, fungicides, insecticides, and bactericides. In order to protect the logs used as raw material in paper production, they are treated with chlorophenyls and after the chlorination process in the paper bleaching phase, dioxin is produced. Frequently used pharmaceutical materials (medicine, dentistry, and cosmetic products) also contain TCDD isomers at the level of 2,000–5,000 ppm. In recent years, some products that make our lives easier are claimed to contain dioxin. Plastic cups and plates, plastic bottles, foam materials, chlorine bleached toilet paper, tissue paper, milk and fruit juice cartons, diapers, and napkins are among these materials. It is reported that dioxins pass into water with the effect of heat as a result of keeping hot drinks in plastic materials and the keeping water sold in plastic bottles under the sun for a long time (Hişmioğulları et al. 2012). Instrumental measurement methods for determining the ratio of dioxin and furan are GC or GC-mass spectrometer (MS), which are the only legally accepted analytical techniques.

It was found that the conventionally activated sludge process could perform a good rate of removal of PCDDs/Fs and PCBs, but the relatively low solid retention time applied and the presence of suspended solids in the effluent limited the removal capability of the system. On the other hand, the membrane bioreactor was capable of both the removal of solids from the effluent (permeate) and the application of prolonged solid retention times, which enabled the bioconversion of those compounds as demonstrated by the mass balances. In a study conducted by Bolzonellaa et al., it was found that the conventional activated sludge process could perform efficient removal of PCDDs/Fs and PCBs, but the solid retention time adopted (typically less than 20 days) and the presence of suspended solids in the effluent (some 10 mg.L^{-1}) limited the removal rates of the system if compared to the membrane bioreactor (David et al. 2010).

Adsorption processes were also used effectively for the removal of dioxins and furans from wastewater or fresh water sources. A study was carried out to determine the properties of adsorbent AC to achieve a high removal performance. The results showed that the O + N content of the adsorbent should be less than about 2–3 mmol/g to ensure that activated carbons are hydrophobic enough to effectively remove organic contaminants from aqueous solutions (Li et al. 2002).

Dioxins and furans are also endocrine disrupting chemicals. The reverse osmosis process has shown promising results in removing dioxins and furans. An assessment of the potential health impacts of dioxin and dioxin-like compounds in recycled water for indirect potable reuse was conducted in 2008 by Rodriguez et al. in Australia. The results indicate that reverse osmosis (RO) is able to reduce the concentration of PCDD, PCDF, and dioxin-like PCBs and produce water of a high quality (RQ after RO=0.15). No increased human health risk from dioxin and dioxin-like compounds is anticipated if highly-treated recycled water is used to augment drinking water supplies. Recommendations for a verification monitoring program are offered (Rodriguez et al. 2008).

Regarding the overall treatment processes for the removal of endocrine disrupters (EDCs), coagulation processes using iron or aluminum salts do not allow for any EDC removal and they are an expensive treatment option. On the other hand, polyaluminium chloride (PAC) coagulation could remove a considerable amount of small-sized contaminants such as EDCs including hormones. Filtration processes, such as ultrafiltration (UF), microfiltration (MF), and nanofiltration (NF), used as hybrid process or not, can allow relatively high rates of EDC removal. But these processes are also expensive and require a significant amount of maintenance to avoid membranes clogging. Membrane bioreactors combine the adsorption and biodegradation processes, and thus would be a good option for EDC removal. AOPs (single or combined with other conventional process) allow for high removal rates of recalcitrant compounds; however, many by-products are released and could have an estrogenic activity higher than their precursors. As a conclusion, EDCs (dioxins and furans) are of a general concern and are a significant research subject. Therefore, future research priorities should include wastewater treatment plant optimization to increase EDC (dioxins and furans) removal from wastewaters.

9.2.1.3   Other Refractory Chemicals

Other refractory chemicals that are not listed by the Stockholm Convention, but are in organic refractory nature such as persistent antibiotics, humic acids, and tanning acids in industrial or domestic wastewater, will be investigated in detailed in this section.

Pharmaceuticals/antibiotics are natural, semi-synthetic, or synthetic chemical preparations that make it possible to diagnose, cure, or reduce the symptoms of a disease with the effect it creates on a living cell. They are also used in the prevention of disease, given to living things by different application methods. Pharmaceuticals are generally classified according to their physiological effects such as medicines affecting the nervous system, medicines affecting the cardiovascular system, drugs affecting the digestive system, medicines affecting the respiratory system, chemotherapeutic drugs, vitamins, hormones, disinfectants and antiseptic drugs, and antibiotics.

Antibiotics are organic chemicals produced by microorganisms or obtained by semi/total chemical synthesis. When used in low concentrations, they inhibit the growth of microorganism chemicals. Most of them have a long chain or cyclic chemical structure that is difficult to degrade. Antibiotics are also human-made refractory organic chemicals with persistent occurence in industrial or domestic wastewater.

In the production of sterile medicinal products, minimum world for injection (WFI) and purified water quality, non-sterile medicinal products and purified water quality for granulation and tablet-coating processes are used. All equipment used in pharmaceutical production is washed and rinsed with water of minimum drinking water quality. It is mandatory to comply with good manufacturing practices (GMP) and good laboratory practices (GLP) standards at all stages of the drug production process.

Pharmaceutical wastewater characterization is determined by the designed formulation of drug and bulk, manufacturing methods and processes of bulk and drug, and used equipment. These are important considerations in the design treatment process. Discharging of the wastewater, including pharmaceutical refractory compounds without sufficient treatment, to the receiving environment has become an environmental concern due to the negative effects on aquatic ecosystems which cause an increase of antibiotic-resistant bacteria and therefore an increase in antibiotic tolerance in humans and animals (Ata et al. 2017, 2019). The haracteristics of mixed raw wastewater of pharmaceutical companies is presented in Table 9.3.

Conventional biological treatment methods were generally used in the treatment of wastewater containing refractory pharmaceuticals, however, biological WWTPs have recently played a major role in reducing the damage caused to the environment. Recent scientific studies have revealed that conventional biological treatment methods, in the treatment of pharmaceutical wastewater including refractory compounds limiting its biodegradability, are insufficient in degrading complex pollution loads.

Toxic and recalcitrant pharmaceuticals, at even very low concentrations, especially antibiotics residues, with refractory compounds are becoming a bigger problem in wastewater treatment plants day by day. The toxicity they create on organisms in the ecosystem and biological treatment systems and the emergence of antibiotic-resistant bacteria disrupt the ecological balance in the receiving water environment. If wastewater, including refractory pharmaceuticals/antibiotics, is treated but leaves antibiotic residues (even at nanogram levels) after conventional biological process treatment is used in agricultural irrigation, converted antibiotic residues and metabolites are taken back to the body through the consumption of agricultural products (through the food chain) and results in negative effects to living metabolisms, such as endocrine disruption, toxicity, and reproductive damage. For this reason, the development of alternative treatment methods in the treatment of refractory pharmaceutical/antibiotic residues, which are also called new generation (emerging) pollutants, has become more important. A combination of conventional biological treatment processes (CBTP) and AOPs such as ozonation, photocatalytic methods, fenton and photo-fenton, membrane filtration, and electrochemical methods provide much more efficient results in terms of biodegradation, in the sense of detoxification and reuse of purified water (Ata et al. 2017).

In another study conducted by Pengxiao et al. in 2014, a systematic study was undertaken of nanofiltration combined with ozone-based advanced oxidation processes for the removal of trace antibiotics from wastewater obtaining more than 98% rejection in all sets of NF experiments. This study revealed that nanofiltration could efficiently remove antibiotics from WWTP effluent. Additionally, the UV/O_3 process was able to further eliminate the antibiotics in the NF concentrate, resulting in a synergy between O_3 and UV in the degradation of the selected antibiotics during the UV/O_3 treatment (Pengxiao et al. 2014). Iakovides et al. studied the removal of antibiotics, antibiotic-resistant *Escherichia coli*, and antibiotics in continuous mode at different hydraulic retention times (Hrts) (i.e., 10, 20, 40, and 60 minutes) and specific ozone doses [i.e., 0.125, 0.25, 0.50, and 0.75 g O_3.g dissolved oxygen concentration (DOC)$^{-1}$] from urban wastewater by using ozone. In this study, done in 2019, the efficiency of ozonation was highly ozone dose and contact time dependent, as expected, and the inactivation of total cultivable *E. coli* was achieved under the experimental conditions of HRT 40 minutes and 0.25 g O_3 .g DOC^{-1}, at which all antibiotic compounds were already degraded (Iakovides et al. 2019). Kurt et al. showed that most of the investigated advanced oxidation treatment processes for the oxidation of antibiotics in water are direct and indirect photolysis with the combinations of H_2O_2, TiO_2, ozone, and Fenton's reagent in order to increase ozone degradation performance (Kurt et al. 2017).

A combination of membrane processes is also helpfull for the removal of antibiotics from wastewaters. Many studies have been done (with a succes of 90% for all the refractory antibiotic classes) to remove refractory antibiotic residues from wastewater by using RO, NF, and UF processes. In 2018, Thai-Hoang Le et al. used a combination of CAS and MBR systems to purify 19 antibiotics, 10 antibiotic-resistant bacteria, and 15 ARGs from wastewater. This study concluded that the highest removal of amoxicillin, azithromycin, ciprofloxacin, chloramphenicol, meropenem, minocycline, oxytetracycline, sulfamethazine, and vancomycin was by CAS or MBR systems with a removal efficiency of >70%, while trimethoprim and lincomycin were recalcitrant in the CAS system with a median removal efficiency of <50%. Similarly, antibiotic resistant bacteria (ARB) and antibiotic resistant genes (ARGs) were omnipresent in the raw influent samples, with average concentrations as high as 2.6×10^6 CFU/mL and 2.0×10^7 gene copies/mL, respectively (Thai-Hoang Le et al. 2018).

TABLE 9.3

Characterization of Pharmaceutical Industry Wastewater (Rana et al. 2014)

Parameters	Gome and Upadhyay (2013)	(Choudhary and Parmar 2013)	Wei et al. (2012)	Lokhande et al. (2011)	Saleem (2007)	Idris et al. (2013)	(Imran 2005)
pH	6.9	5.8–7.8	7.2–8.5	3.69–6.77	6.2–7.0	5.65 ± 0.65–6.89 ± 0.12	5.8–6.9
Total suspended solids (TSS) (mg.l^{-1})	370	230–830	48–145	280–1,113	690–930	29.67 ± 4.22–123.03 ± 4.56	761–1,202
Total dissolved solids (TDS) (mg.l^{-1})	1,550	650–1,250	–	1,770–4,009	600–1,300	136.33 ± 5.83–193.05 ± 5.35	1,443–3,788
Total solids (mg.l^{-1})	1,920	880–2,040	–	2,135–4,934	–	–	263–330
BOD$_5$ (mg.l^{-1})	120	20–620	480–1,000	995–1,097	1,300–1,800	–	–
COD (mg.l^{-1})	490	128–960	2,000–3,500	2,268–3,185	2,500–3,200	–	2,565–28,640
Biodegradability (BOD$_5$/COD)	0.259	–	0.20–0.39	–	–	–	–
Alkalinity (mg.l^{-1})	–	130–564	–	–	90–180	–	–
Total nitrogen (mg.l^{-1})	–	–	80–164	–	–	–	–
Ammonium nitrogen (NH$_4^+$-N) (mg.l^{-1})	–	–	74–116	–	–	–	–
Total phosphate-T-P (mg.l^{-1})	–	–	18–47	–	2.2–3.0	17.22 ± 0.78–28.78 ± 1.18	–
Turbidity (NTU)	–	–	76–138	–	–	–	–
Chloride (mg.l^{-1})	–	–	–	205–261	–	–	–
Oil and grease (mg.l^{-1})	–	–	–	0.5–2.9	–	–	1,925–3,964
Phenol (mg.l^{-1})	–	–	–	–	95–125	–	–
Conductivity (IS.cm^{-1})	–	–	–	–	–	157 ± 115.84–1,673 ± 119.36	–
Temperature (°C)	–	–	–	–	–	32 ± 2.23–46 ± 3.41	31–34

Reference

Another powerful tool used to degrade the most resistant refractory pharmaceutical compounds is electrochemical oxidation. In 2005, Hirose et al. applied electrochemical oxidation to various antibiotics: epirubicin (anthracycline), bleomycin (glycopeptides), and mitomycin C (Hirose et al. 2005). Only epirubicin was mostly removed in this study.

Among the AOPs, heterogeneous semiconductor photocatalytic (with TiO_2) processes under UV light are the most effective for the removal of photo-stable refractory pharmaceuticals (antibiotic) from wastewater. In these types of processs, anatase as a form of TiO_2 has excellent photocatalytic activity, toxicity, long-term stability, and affordable production costs, although it has two drawbacks that limit the application of the photocatalytic process. Doping TiO_2 with N, F, B, C, and S is done to extend the photo-response from the UV to the visible light region in order to enhance the removal performance of refractory compounds. Furthermore, the immobilizing process of doping photocatalysts is a promising method to avoid the high costs of particle separation processes after the treatment. Recent studies have focused on the immobilization of doped TiO_2 with various substrates, such as fabrics, reactor walls, hollow glass spheres, and polymers. Polystyrene (PS) is a cheap, inert, non-toxic, and low density thermoplastic polymer, widely used in the food service and retail industries. As a result, studies revealed that conventional biological tretament process should be combined with AOP systems as a secondary treatment in order to treat wastewater containing persistent and resistant refractory pharmaceutical compounds against biodegradation more effectively.

Pharmaceutical and Personal Care Products (PPCPs) consist of drugs, hormones, cosmetics, etc., that are widely used in many fields such as medicine, livestock farming, aquaculture, and also for everyday use by people. They have poor removal properties by conventional biological wastewater treatment plants. Environmental pollution caused by the extensive application of PCPs is becoming more and more serious. The PCPs can be released into the environment by direct and indirect pathways. Sources of PCPs are industries, hospitals, households, and wastewater treatment plants and through biosolids that are spread on agricultural land, which can reach the groundwater by leaching or bank filtration. PCPs can be detected with gas chromatography-tandem mass spectrometry (GC-MS/MS) and liquid chromatography-tandem mass spectrometry (LC-MS/MS) in even trace levels in wastewater. PCPs are recognized as pseudo persistent organic pollutants in the environment due to their continuous introduction into the environment by differents means, such as sewage treatment plants.

The removal of the PCPs from wastewater is a necessity. As a result, many methods including physical, chemical, and biological have been developed to remove the PPCPs in wastewater treatment. The CBTS (activated sludge system) has the capacity to remove the PCPs from the wastewater, but the removal efficiency of PCPs changes according to hydraulic retention time, sludge retention time, and pH. In addition, in conventional treatment processes the removal efficiency of each ibuprofen, naproxen, ketoprofen, diclofenac, bezafibrate, sulfamethoxazole, and trimethoprim range from 60% to 100% (Kasprzyk-Hordern et al. 2008). Consequently, conventional treatment processes are unable to remove the PPCPs from wastewater completely. Several treatment methods have been applied to remove PCPs from wastewater, such as physical adsorption processes with the carbon-based adsorptive materials, including activated carbon, graphene, and carbon nanotubes.

The activated sludge process was also widely used as a biological treatment method in conventional WWTPs. Plosz et al. developed an activated sludge modelling framework for xenobiotic trace chemicals (ASM-X), used to assess the factors that influence the removal of diclofenac and carbamazepine in activated sludge (Plosz et al. 2012).

As mentioned above, the conventional wastewater treatment process cannot remove the PPCPs completely. Therefore, advanced chemical processes are needed to deal with wastewater containing PPCPs. The chemical oxidation processes, such as ozonation and other AOPs, involving O3/UV, UV/H_2O_2, Fenton, and Fenton-like oxidation, as well as electrochemical oxidation were effective for the degradation of toxic or refractory organic pollutants in aqueous solutions. Recently, Fenton-like systems, such as electro-Fenton and photo-Fenton oxidation, have been developed in addition to conventional Fenton oxidation. Feng et al. have summarized the electro-Fenton process (Feng et al. 2013) and Bokare and Choi have reviewed the Fenton-like systems. Another new application used for the removal of PPCPs is UV treatment (Bokare and Choi 2014). In order to overcome the shortcomings of single biological or single AOPs, a combination of AOPs and biological methods has been preferred in recent years. Generally, a chemical method is used as a pre-treatment process, where persistent pollutants could be transformed into biodegradable intermediates. Afterward, a biological method is carried out to degrade the intermediates completely. For example, tetracyclines in a single biodegradation process presents difficult elimination due to their toxicity. Gómez-Ramos et al. employed an integrated process of ozone and biological treatment to remove sulfamethoxazole. The results showed a low degree of mineralization; therefore, biological treatment followed by ozone oxidation is used to enhance the mineralization and reduce the COD (Gómez-Ramos et al. 2011). Vargas revealed that the electro-Fenton method as pre-treatment process could decrease the recalcitrance and toxicity of furosemide and ranitidine, and produce biodegradable intermediates which can be completely degraded by the following aerobic system. Instead of a pre-treatment process, a chemical method can be also employed as a post-treatment process (Vargas 2014). Due to the fact that PCPs are resistant to biodegradation, advanced oxidation techniques are preferred for pre-treatment to decompose the PCPs. In addition to that, bioaugmentation may be a feasible technology for increasing the removal of PPCPs, or advanced oxidation can be combined with conventional treatment processes by using secondary or tertiary treatments in order to break food chain.

Tanning acids and humic acids are also concluded in natural organic refractory compounds. Tannins, a complex family of polyphenolic compounds, are widely distributed where they act as growth inhibitors against many microorganisms, including bacteria, yeasts, and fungi. They are one of the main components of domestic wastewater and cause serious environmental pollution. There are two main classes of tannins: hydrolysable and condensed tannins. Tannins have toxic effects on animals and effects on a wide variety of organisms such as the inhibition of growth of microorganisms. Basically, gallic acid and its oxidative metabolites and multiple esters of D-glucose are hydrolysable tannins. Condensed tannins (proanthocyanidins) consist of flavan-3-ol units linked by a carbon-acid-carbon (CAC) bond. Tannic acid, which is a specific commercial form of tannin, a type of polyphenol, can be used in the pharmaceutical, food, and industrial fields. In the medicinal, pharmaceutical, and cosmetic industries, tannic acid is used as an ingredient in ointments and suppositories for treating hemorrhoids and in burn lotions. In the food industry, it is used in clarifying alcoholic beverages (especially beer and wines). It is also used as a flavor ingredient in most major categories of foods, including alcoholic and non-alcoholic beverages, frozen dairy desserts, candy, baked goods, gelatins and puddings, and meat and meat products. In addition, in the industrial sector, tannic acid is used as a mud treatment agent in petroleum Also, it can be used in the production of leather tanning agents, mordant, rubber curing agents, agent of proteins, alkaloids precipitation agents, mineral inhibitors, wine clarifying agents, the extraction of germanium, and the configuration of the blue-black ink, etc. The main sources of tannins can be categorized as slaughterhouses, leachates, tanneries, dyes, olive mill wastewater (OMW), and pulp/paper wastewater. Tannins inhibit the treatmeant of such wastewaters through biological treatment processes. Tannins are polyphenolic compounds that are widely prevalent in plants and plant residues that have large molecular weights and a strong affinity to proteins, alkaloids, and heavy metals to form complex molecules. In recent years, the tannery industry has become one of the main economic sectors in developing countries, but also the main source of environmental pollution due to the discharge of large volumes of potentially toxic and hazardous wastewater. Scientific studies revealed that tannins have the ability to hinder the treatment process, as the effluent taken from each phase of the tannins treatment process is generally mixed before the treatment begins. The high concentrations of tannins in the treatment process make it difficult to assess the effects of baths, for example it is difficult to predict whether such streams will be treated separately or not. The biological treatment of tannery wastewater and, in a much more general view, the treatment of wastewater containing polyphenolic compounds, such as tannins, are difficult and laborious for the following reasons:

- The presence of aerobic and anaerobic conditions at high concentrations of non-biodegradable COD
- The presence of inhibitory conditions for biomass, especially nitrification biomass
- The requirement to run at a high solids retention time (SRT) to achieve stable nitrification and complete hydrolysis of the particulate organic fraction (Munz et al. 2007)

In the research carried out by Munza et al., the role played by tannins in tannery wastewater treatment was evaluated by establishing a pilot MBR facility and a full-scale conventional activated sludge treatment (CASP) plant operated in parallel. In the study, the proposed methodology has established the preliminary use of respirometry to examine the biodegradability of a selection of commercial products (synthetic and natural tannins). The results show that a consistent percentage of TOC in the effluent of the biological phase of the plants can be attributed to the presence of natural and synthetic [sulfonated naphthalene-formaldehyde condensates (SNFC)] tannins: 17% and 14%, respectively. This study revealed that tannins are not significant in terms of the resulting differences by comparing the two technologies (CASP and MBR) in terms of their role in biodegradability (Munza et al. 2009). He et al., in 2007 investigated the biodegradability of wastewater containing tannins, taking into account the practical applications in wastewater treatment plants. For this, wastewater obtained from vegetable tanning in the leather industry was used. Vegetable tannin extracts were used as tanning agents to convert skins into leather in an aqueous solution. The study showed that the biodegradability of tannin-containing wastewater varies greatly according to the tannin content in the wastewater. It was observed that the treatment efficiency of tannery wastewater with a high tannin content was lower due to the inhibition of tannins in the activated sludge process. The results further show that the biodegradability of the tannin content and its BOD_5, COD, and TS values must be appropriately controlled in order to ensure an effective biodegradation of wastewater containing tannins in wastewater treatment plants (He et al. 2007). The wastewater characterization of tannery wastewater varies from tannery to tannery. Tannery wastewaters are generally dark brown, have an objectionable odor, and contain high levels of pH, COD, BOD_5, TDS, chromium, sulphate, phosphate, and nitrate, and various highly toxic organic substances, chemicals, and heavy metals. The leather tanning process is a heavy industrial process where large amounts of highly toxic chemicals, such as chromium salts, vegetable and synthetic tannins, phenolic compounds, azo dyes, surfactants, pesticides, and sulphonated oils, are used in large quantities in order to convert the raw hide/skins into commercial leather or leather products, causing these chemicals to be disposed of into the environment due to not being fully absorbed by hide/skins during the tanning process. Thus, these chemicals are a major source of environmental (soil and water) pollution and cause serious health threats to humans and the aquatic environment. Tannery wastewater generates a large volume of wastewater characterized by high concentrations of non-biodegradable organic compounds besides inorganic salts, including NH_4-N, sulphide, heavy metals, sulphonated mono-, di-and tri-cyclic aromatic compounds, alicyclic and aliphatic long-chain compounds, volatile organic compounds, diazocompounds, fatty acids, aryl amides, and aliphatic amides. As outlined in Table 9.4, tannery wastewater contains ammonium substances, salts (i.e., chloride and sulphate), as well as high concentrations of organic matter (COD) with a significant percentage of refractory organic compounds such as aliphatic amines, aromatic compounds, surfactants, oils, and dyes at considerable concentrations. Moreover, the presence of highly toxic non-degradable metals, especially chromium (Cr^{6+}), have been reported in many studies. Tannery wastewater containing sulphur and BOD_5 is complex in nature and has placed more emphasis on the physico-chemical and oxidation systems than to biological treatment methods; however, these are comparatively expensive and may cause secondary pollution because additional chemicals are needed (Tore et al. 2011).

Considering tannery wastewater content (poorly biodegradable polyphenolic compounds and very toxic pollutants), chromium precipitation, primary sedimentation, biological oxidation and secondary sedimentation process are additionally applied after conventional treatment in order to meet the required limits for most of the problematic parameters such as COD, salinity, ammonium,

TABLE 9.4

Characteristics and Pollution Profile of the Segregated Wastewater Streams for COD, Ammonia, and Chromium (Daily Production: 6.4 Ton Skin) of Tannery Wastewater (Tore et al. 2011)

Process	Production	Total COD				NH3-N				T-Cr			
	[ton skin/day]	mg.L-1	kg COD.d-1	kg COD.(ton Skin)-1	%	mg.L-1	kg NH3.d-1	kg NH3.(ton Skin)-1	%	mg.L-1	kg T-Cr.d-1	kg T-Cr.(ton Skin)-1	%
					Wastewater stream no. 1 – washing								
Pre-soaking	6.4	4,264	409	64	7.3	101	10	2	8.1	N/A*	N/A*	N/A*	N/A*
Soaking	6.4	5,847	561	88	10	48	5	1	3.9	N/A*	N/A*	N/A*	N/A*
Wool washing	6.4	5,031	724	113	13	44	6	1	5	N/A*	N/A*	N/A*	N/A*
Washing	6.4	4,880	195	31	3.5	22	0.9	0.14	0.75	N/A*	N/A*	N/A*	N/A*
Pickling	6.4	9,466	379	59	6.7	25	1	0.2	0.85	N/A*	N/A*	N/A*	N/A*
Depickling	6.4	7253	406	63	7.2	15	0.9	0.1	0.7	N/A*	N/A*	N/A*	N/A*
Bating	6.4	4,979	279	44	5.0	342	19.2	3.0	16.1	N/A*	N/A*	N/A*	N/A*
Composite no. 1	**6.4**	**5,594**	**2,954**	**462**	**53**	**81**	**43**	**7**	**36**	**N/A***	**N/A***	**N/A***	**N/A***
					Wastewater stream no. 2 - tanning								
Tanning I	6.4	5,897	330	52	5.9	101	6	1	4.8	1,686	94.4	14.7	69.2
Tanning II	6.4	3,276	183	29	3.3	119	7	1	5.6	352	19.7	3.1	14.5
Neutralization I	6.4	2,571	144	22	2.6	84	4.7	0.7	4.0	0	0	0	0
Composite no. 2	**6.4**	**3,914**	**658**	**103**	**12**	**101**	**17**	**3**	**14**	**679**	**114**	**17.8**	**83.7**
					Wastewater stream no. 3 - greasing								
Fatliquoring	6.4	6,941	389	61	6.9	147	8.2	1.3	6.9	N/A*	N/A*	N/A*	N/A*
Retanning	6.4	3,852	308	48	5.5	133	10.6	1.7	9.0	279	22.3	3.5	16.3
Composite no. 3	**6.4**	**5,124**	**697**	**109**	**12**	**139**	**19**	**3**	**16**	**164**	**22.3**	**3.5**	**16.3**
					Wastewater stream no. 4 – dyeing								
Wool dyeing	6.4	3,552	568	89	10.1	77	12.3	1.9	10.4	N/A*	N/A*	N/A*	N/A*
Neutralization II	6.4	1,656	265	41	4.7	77	12.3	1.9	10.4	N/A*	N/A*	N/A*	N/A*
Fur–suede dyeing	6.4	3,016	482	75	8.6	98	15.7	2.5	13.2	N/A*	N/A*	N/A*	N/A*
Composite no. 4	**6.4**	**2,741**	**1,316**	**206**	**23**	**84**	**40**	**6**	**34**	**N/A***	**N/A***	**N/A***	**N/A***
Plant composite	**6.4**	**4,287**	**5,624**	**879**	**100**	**91**	**119**	**19**	**100**	**104**	**136**	**21.3**	**100**

N/A*: Not available

and surfactant. Table 9.4 includes tannery wastewater characterization. In the studies carried out until today different methods have been studied as alternatives to aerobic, anaerobic, biological, and classical physicochemical processes for tannery wastewater treatment, e.g., adsorption and coagulation. Ganesh et al. used sequencing batch reactors (SBRs) to study the periods of satisfactory and unsatisfactory performance in the nitrification and denitrification properties. At the end of the lengthy SBR process, a 14.1 mg/L h maximum nitrification rate was obtained. It has been observed that the possible cause of the instability in the nitrification process is inhibiting in the tannery wastewater (Ganesh et al. 2015). The combined treatment system, for example an anaerobic sludge bed (ASB) reactor with an aerobic post-treatment, helped to increase the wastewater treatment and COD removal performance. Munz et al. used a pilot-scale MBR and CASP to treat the same tannery wastewaters at the same operating conditions, comparing the overall treatment efficiency. The results showed that MBR had a higher COD removal rate (+4%) and a more stable and complete nitrification (Munz et al. 2008). Banu and Kaliappan treated the tannery wastewater by means of a hybrid upflow anaerobic sludge blanket reactors with the fixed film and upflow sludge blanket treatment. The reactor's treatment performance resulted in an organic loading rate of 2.74 kg $COD.m^{-3}.d^{-1}$ and 3.14 kg $COD.m^{-3}.d^{-1}$ at an HRT of 70 and 60, respectively (Banu and Kaliappan 2007). In order to improve the treatment performance of tannery wastewater, the following were applied in addition to the aforementioned techniques: bio-adsorption, electroflotation, electro-chemical filtration, and ion-exchange combined with nano filtration, or membrane filtration only. Murugananthan et al. treated tannery wastewater by using the electroflotation technique. This study revealed that electroflotation treatment process is very effective especially for the elimination of pathogenic bacteria and color of the effluent (Muruganathan et al. 2004). Szpyrkowicz et al. used an undivided electrochemical reactor with parallel plate Ti/Pt–Ir anodes and stainless-steel cathodes to treat tannery wastewater at constant current densities and different stirring rates, in order to improve the removal of the all pollutants, for which direct anodic oxidation was an additional process for their destruction. Gando-Ferreira et al. researched the possibility of integrating both ion-exchange (IX) and nanofiltration (NF) processes for the recovery of Cr(III) salts from a synthetic solution prepared with concentrations of Cr(III) and Cl – in the range of industrial effluents of tanneries. Results showed that with a combination of both IX and NF, the feed solution after treatment with the resin was fed to NF where the ratio led to the best operating conditions for this process [90% of Cr(III) rejection and up to 77% for ions] (Gando-Ferreira et al. 2015).

In addition to the leather tannery industry, the cork processing industry also contains high levels of organic and phenolic compounds, such as tannins, with a low biodegradability and significant toxicity. These refractory compounds are difficult to treat with conventional municipal wastewater treatment methods, which is largely based on primary sedimentation followed by biological treatment. Various membrane separation (UF, NF) processes were used in order to assess its potential for biological treatment and having in view its valorization through tannins recovery via using different cork wastewater fractions. It is determined that the permeated fraction from the membrane had a minimized phenols toxicity that enables it to undergo biological treatment and so, to be treated in a municipal wastewater treatment plant (Bernardo et al. 2011).

Scientific studies demonstrated that if these waters, containing even trace amounts of refractory, toxic, and metallic pollutants, are used in agricultural irrigation after treatment with conventional biological methods, the intake of transformed refractory pollutant metabolites back into the living body through the food chain cause harmful effects on the endocrine and reproductive systems. Therefore the usage of alternative treatment methods individually or combined with or after conventional treatment processes – such as AOPs, ozonation, photocatalytic methods, fentone and photo-phentone, membrane filtration, electrochemical methods – have gained importance as alternatives to traditional biological treatment methods in recent years. The emerging AOP technologies proposed, which are based on the in-situ formation of hydroxyl (•OH) radicals, are combined with the biological treatment of tannery wastewater for the improved elimination of refractory compounds in it. The purpose of using AOPs technologies alone or in combination with conventional WWT technologies is to cause an elimination through the separate decomposition of different types of recalcitrant refractory and toxic organic compounds or even complete mineralization/degradation. Also supporting this, an increasing number of researchers have proposed a sequential AOP or electrochemical process combined with the biological treatment of tannery wastewater for improved elimination. The use of ozone as an oxidizing reagent for the removal of organic matter in tannery wastewater treatment and for treating wastewater has also been studied by several researchers. These studies also provided knowledge and experience on determining appropriate conditions and preventing the use of excessive amounts of ozone, which is costly. Consequently, hybrid systems (such as conventional biological treatment processes with advanced oxidations systems) have the potential to sufficiently treat high-strength refractory-characterized tannery wastewater. Since ancient times, there has been a ceaseless attempt to increase productivity in the agricultural field and to obtain quality products. For this purpose, the use of chemical fertilizers is one of the oldest methods used. With chemical fertilization, plants are able to provide some inorganic nutrients, but chemical fertilizers mixed into the soil cannot be used sufficiently by the plant. The parts that are not used by the plant are removed from the soil through rain, snow, and irrigation water and causes significant ecological problems by mixing into groundwater, seas, and lakes. Pollutants arising in this way are called humic substances (humic acid vs. fulvic acid). Humic substances (HSs), which are an important component of humus and the major organic fraction of soil, peat, and coal (and also a constituent of many upland streams, dystrophic lakes, and ocean water), have large molecular weights. They are created by the microbial degradation of plant and animal tissues and ultimately by the dispersal of biomolecules (lipids, proteins, carbohydrates, and lignin) in the environment after the death of living cells (Turkay et al. 2015). HSs, which are the refractory decomposition products of biological material, can be found in both terrestrial and aquatic ecosystems and play a fundamental role in the environment. The remaining organic substances are a polydisperse and heterogeneous mixture of partially biodegradable matter, and are often also referred to as refractory organic substances (ROS) in scientific literature. Although the structural properties of humic and fulvic substances are similar, there are two important differences between them: molecular weight and functional groups. While the molecular weight of fulvic acids varies between 200 and 1,000 gr, the molecular weight of humic acids increases up to 200,000 gr (see Figure 9.2).

The oxygen content of fulvic acid is higher and it contains more functional groups per unit weight (COOH^{-1}, OH^{-1}, C-O). Despite these differences between humic and fulvic acid, humic acid is generally used to refer to both compounds (Figure 9.3).

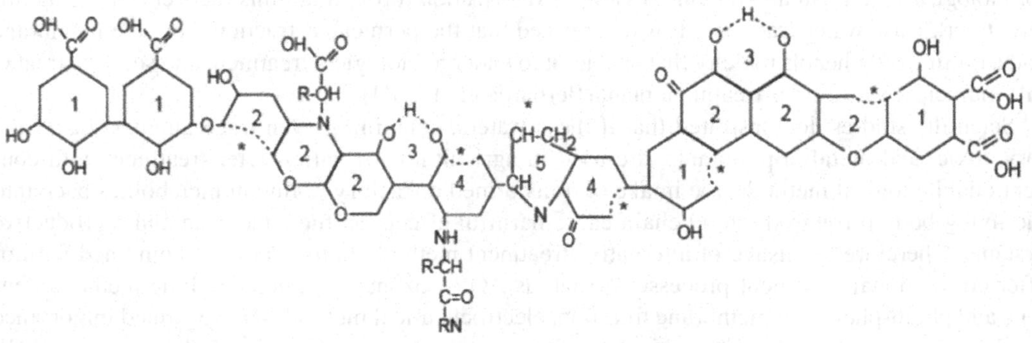

FIGURE 9.2   Scheme of the humic acid model structure (Zularisam et al. 2007). *Reproduced with permission from Ref, (Zularisam et al. 2007). Copyright © 2007, Elsevier.*

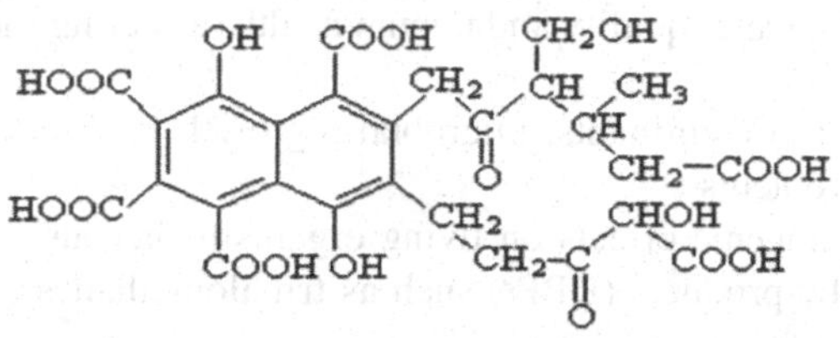

FIGURE 9.3 Buffle model structure of fulvic acids (Ateşli 2006).

Humic acids (HAs) are insoluble in acidic water (pH <2) but soluble at higher pH values. Fulvic acids (FAs) are soluble in water under all pH conditions in wastewater. HSs are known to be inert to biodegradation under the following conditions:

- Both aerobic and anaerobic conditions necessary for HS can be biodegraded
- In biological processes biochar and activated carbon act as redox mediators

HSs represent most of the organic substances found in soil, lignite, peat coal, sewage waters, spring waters, and sediments. Humic substances are divided into three categories: FAs, HAs, and humin. One of the most important parts of humic substances is humic acid. Humic acids and fulvic acids represent humus structures that dissolve in alkaline mediums. HSs can also be of industrial origin such as from the ceramic industry and pulp and paper industry. In the ceramic industry, humic substances are mainly used to increase the mechanical strength of unprocessed ceramics, to make the casting properties of ceramics efficient, to paint mud ceramics, and to prepare pottery. In addition, humic materials are used in plastic production, especially for dyeing Nylon 6 or polyvinyl chloride (PVC) plastics, to harden polyurethane foams, or as plasticizing agents for PVC plastics (Ay 2015). Humic substances are also used in the pulp and paper industry. For example, they are used in the production of paper with high tensile strength and in the recycling of paper.

Other industrial applications using humic substances, in the form of ion exchangers, include using them as a source of synthetic hydrocarbons and fuel oil. Humic materials have great potential in holding transition metals and forming organometallic compounds. It has also been found that it is possible to produce liquid packaging cardboard (LPB) by recycling the cellulose contained in humic acid. Since humic substances in aquatic water systems interact with organic and inorganic water components and thus affect the distribution and transport of pollutants, studies conducted in recent years have focused on the structural and functional properties of these substances. Since the 1970s, many studies have been carried out for structural characterization of HS and/or ROS and the determination of humic substances based on fractionation properties as FAs (acid and base soluble), HAs (base soluble), humin (insoluble in either acid or base, available only in soil samples), and non-humic substances (NHS). When pollutants in the type of humic material are thrown into the environment without sufficient treatment, they not only pollute the receiving surface water environment but also cause the pollution of our freshwater ecosystems (including drinking and potable water). More than 50% of natural organic substances, sourced by the microbiological breakdown of living things or leaves, mixed into drinking water of 40–80% in surface waters constitute humic substances, and is an important fraction of the biodegradation of organic substances in nature, especially plants, and of biologically persistent materials in effluents. It is composed of aromatic blocks and a skeleton consisting of alkyl chains with various functional groups, especially carboxyl, phenol, hydroxyl, and quinone groups. HSs are the major fraction of natural organic matters sourced from microbial degradation of plant/animal tissues. They cause:

- Undesirable effects on water quality and human health, affecting the color, taste, and odor of drinking water
- The transportation of contaminants, microbial regrowth in the distribution system, and membrane fouling problems
- Mutagenic and carcinogenic effects on living organisms because they are major precursors of disinfection by-products (DBPs) such as trihalomethanes (THMs) and haloacetic acids (HAAs)
- Enhanced biological growth in distribution systems
- Complexation sites for heavy metals
- An increase in coagulant and disinfectant dose requirements (Kang et al. 2002)

Therefore, HSs must be removed from water. Since they are in low concentrations, they must be removed from water by special methods. There are many treatment techniques used to remove humic and fulvic acids. Some of these techniques include coagulation and enhanced coagulation, adsorption, nanofiltration, catalytic ozonation, and AOPs, which have been developed for the elimination of HS from water. Conventional coagulation and filtration processes were used to treat humic substances from the wastewater by using aluminum sulphate before the chlorination process. Moreover, electrocoagulation (EC) treatment was shown to be a feasible technology for treating real peat bog drainage waters. At the end of the study, high pollutant removal efficiencies were achieved in Finland in 2015 (Kuokkanen et al. 2015). Since depicted as large polymers, HSs are considered as resistant to biodegradation. Black carbon materials, including char (biochar) and AC, long recognized as effective sorbents, have been recently discovered to provide effective degradation of organic pollutants in a way of coating. In a study, humic acid was removed from water by using pyrolysed coke originated from olive mill wastes and alcohol industry sewage sludge as a raw materials. It has been proved that it has the same removal rate as activated carbon adsorption, and the possibility of reuse within the framework of waste minimization has been investigated. However, only 60–70% of the HS concentration can be treated under optimal treatment conditions. It has been found that conventional coagulation-flocculation is not an effective method to inactivate some pathogenic microorganisms (Gerrity et al. 2009). Therefore, residual organic matter (about 30–40%) that remains after coagulation can lead to the formation of DBPs and will need further treatment (Hua and Reckhow 2007). There are also studies on the use of combined systems to increase treatment efficiency of humic acid removal. In this context, Gündağ Ö. studied the removal of humic acid substances in drinking water using a process of chemical treatment, ozonation, fenton oxidation, and activated carbon adsorption. In optimum pH and laboratory conditions, it obtained more than 90% treatment yield by chemical treatment, approximately 90% treatment yield by ozonation, and approximately 60% treatment yield by activated carbon, by UV light, and by Fenton process (Gündağ 2017). Amano et al. studied ferrate-based technologies in water treatment due to their potential for insitu production and they do not form any harmful by-products in order to compare the oxidative performance of Fe (VI) generated by an electrochemical process, with H_2O_2-UV irradiation for removing natural organic matter from the Suwannee River in the U.S.A. The results suggest that integrating ferrate pre-oxidation and polyaluminum chloride (PACl) coagulation improves the treated water quality by the combined effects of humic substance oxidation and coagulation (Amano et al. 2018). As it is well known, concentrated leachates from membrane treatment processes including non-biodegradable humic substances can cause potential hazards to the ecological environment. In order to treat this concentrated leachate from RO and NF, continuous ozone generating-reaction integrated equipment was used in 2016 by Chinese researchers. The results showed that the humic substances, including HAs and FAs, were effectively removed after 110 minutes of reaction. Because the effluent of the two concentrated leachates did not meet the maximum allowable criterion for leachate direct or indirect discharge standards in China, further biological treatment or other advanced treatments was deemed mandatory (Wang et al. 2016). Beside that, the influence of preozonation on the adsorptivity of HSs onto activated carbon was investigated by Rodríguez et al. in Berlin, in 2016, which are the usual stages in drinking water treatment. Adsorption isotherms show that preozonation improved the adsorptivity of the commercial HASs onto AC, whereas no appreciable effect was observed for the case of the FAs (Rodríguez et al. 2016). In China in 2017, concentrated leachate (CL), which is the by-product of leachate, contains many humic-like substances, was treated by the membrane process after bio-treatment processes. Ozonation processes were beneficial for optimizing the further removal of those refractory substance. MBR-O_3 was more effective in 35% COD removal, more than the normal one, and the reaction rate of MBR-O_3 is three times higher than normal bubbles. Consequently, MBR-O_3 could be offered for the advanced treatment of concentrated leachate (Wang et al. 2017).

9.2.2 Refractory Inorganic Compounds, Main Sources, Typical Characteristics, and Treatment Alternatives

In addition to organic pollutants in wastewater, refractory inorganic pollutants will be discussed in detail in this section. Inorganic pollutants, such as heavy metals/ions, oxoanions, and radioactive materials, etc., may persist longer in the aqueous systems and cause further deterioration of the water quality due to their non-biodegradability. Elements with a density greater than 5 g/cm^3 or an atomic weight of 50 or more are called heavy metals. Examples of heavy metals inclucd copper (Cu), iron (Fe), zinc (Zn), lead (Pb), mercury (Hg), cobalt (Co), chromium (Cr), nickel (Ni), and cadmium (Cd). Refractory inorganic compounds (RICs) in wastewater may include heavy metals such as cadmium (Cd), iron (Fe), nickel (Ni), zinc (Zn) and lead (Pb) [measured by using an atomic absorption spectrophotometer (AAS)], arsenic (As), antimony (Sb), titanium (Ti), and vanadium (V) (measured using a UV-VIS spectrophotometer). Other cations measured were (Na^+, K^+, Mg^{2+}, Ca^{2+}) and anions (Cl^-, HCO_3, NO_3, SO_4^{-2}, and PO_4^{-3}).

9.2.2.1 Metals / Metal Alloys Wastewater

Heavy metals are widely present in many areas of our daily life due to their chemical properties. Heavy metals are generally used in metal cleaning, metal plating, and electroplating. They are found together with other pollutants in wastewater from the following industries: iron and steel, auto parts, paint, printing, textile, petrochemical, and leather. Most of the heavy metals in wastewater are found in treatment sludges. The dissolved parts reach and stay in water resources such as surface waters, the sea, and lakes. From here, heavy metals can move again and reach the food chain and drinking water. Heavy metals that contaminate the food chain are released from the body chemically or biologically. They cannot be degraded and therefore accumulate in the body. Heavy metals in wastewater cannot biodegrade like organic compounds. The widespread use of some heavy metals results in undesirable concentrations in wastewater. Heavy metals, which are found in significant amounts in the wastewater of various industries, are included in the "priority pollutants" lists. Heavy metal concentrations are particularly high in waste and wastewater produced by the coating, mining, and metal industries.

Cadmium(II) pollution occurs at undesirable concentrations in the wastewater of the mining, metal plating, electroplating, metal processing, paint, and battery industries. Cadmium in wastewater containing cadmium(II), Cd(II), or Cd(I) (in very low concentrations) exist as organic complexes with amine, halogen, or cyanide. While cadmium is found in the form of cadmium(II) ions in the low pH range (pH 2.0–4.5) in wastewater, it precipitates as a Cd (OH)$_2$ compound at higher pH levels (pH 8.0–11.0). Cadmium(II) ions are purified from wastewater by processes such as reduction precipitation, ion exchange, recovery by evaporation, and electrolysis. The wastewaters of the plating baths, plastic plating, steel, battery, and paint industries contain high concentrations of nickel(II) ions. While nickel is mostly found in wastewater as nickel(II) ions in the pH range of 2.0–4.5, it can also occur as Ni(I) and Ni(III) ions at low concentrations at the same pH ranges. Classical methods used in the treatment of nickel(II) ions are precipitation, ion exchange, evaporation, and reverse osmosis. High concentrations of cadmium(II)–nickel(II) contamination are encountered in the watstewaters of the mining, metal plating, metal processing, battery, and paint industries. Cadmium in wastewater containing cadmium(II)–nickel(II) pollution is found in the form of cadmium(II) and nickel is found in the form of nickel(II). The pH of wastewater containing cadmium(II)–nickel(II) contamination together is quite low and generally in the range of pH 2.0–4.5. These ions are purified from wastewater by processes such as reduction precipitation, ion exchange, and recovery by evaporation (Gurnham 1965). Classical treatment techniques such as reduction, oxidation, and neutralization used in the removal of heavy metal ions are far from practical and economical due to the high chemical and equipment costs and low treatment efficiency. In addition, it is not possible to recover heavy metal ions precipitated by chemical methods, and the sludge formed is a pollutant in itself. Therefore, more efficient and nano-level heavy metal removal rates can be achieved with techniques such as ion exchange and adsorption. Heavy metals can enter our body by food, drinking water, and through our airways. They can be toxic and cause intoxication in high concentrations and can be bioaccumulated in living bodies. RIOCs can cause serious illnesses (sometimes even cancer, abnormal growth, diabetes, obesity, etc.) in humans. It also reduces microbial activity and negatively affects biological wastewater treatment processes. Heavy metals have also been found to inhibit nitrification and denitrification mechanisms and reduce microbial oxidation of organic compounds. In addition, it has been shown that heavy metal contamination in wastewater depends on factors such as slurry concentration, pH, metal types and their concentrations, and the solubility of metal ions. RIOCs in wastewater were treated by chemically transforming the metal ion into a precipitatable compound, although it is generally dependent on the capacity of the enterprise, wastewater flow and characteristics, process, treatment plant, and the chemicals used. In precipitation, generally reduction-precipitation, oxidation-precipitation, and neutralization-precipitation methods are used (Matlock et al. 2002). Reverse osmosis (Rao et al. 2009), solvent extraction (Makrlik and Vanura 2005), biosorption (Ofomaja et al. 2010), adsorption (Imamoglu and Tekir 2008), and membrane filtration (Bernabé et al. 2018) are other treatment methods used widely for heavy metal removal. These applications are not practical and economical due to reasons such as low treatment efficiency, especially at low metal ion concentrations, high investment and operating costs, and the formation of new pollutants.

Reduction-precipitation method: With this method, the metal ion with high valence is neutralized after it is reduced to a precipitating compound, the excess of the chemical reagent used precipitates the metal. Mixing, flocculation, thickening, and filtering processes are also performed in precipitation.

Oxidation-precipitation method: With this method, the reduced metal is converted into stable, oxidized, and insoluble forms. In this type of waste treatment process, there are three consecutive steps: aeration, sedimentation, and filtration. For metals that are not easily oxidized, it is necessary to add the chemical oxidation step to the process in question. This method is generally used in the treatment of wastes containing iron and manganese ions.

Neutralization-precipitation method: Heavy metal ions such as Cr $(^{+6})$, Cu $(^{+2})$, Zn $(^{+2})$, Ni $(^{+2})$, Fe $(^{+2})$, and Cd $(^{+2})$ are added to the environment with lime, soda, and/or neutralization with the addition of sodium hydroxide, and metals are precipitated as hydroxides and removed from the wastewater.

Ion exchange method: The ion exchange method is based on the principle of keeping metal ions on the solid surface as functional groups with the help of electrostatic forces and replacing them with different types of ions in the environment. For this purpose, mostly ion exchange synthetic resins are used (Deniz 2010).

Adsorption: Treatment with adsorption occurs in the form of the selective attachment of heavy metal ions in wastewater to active groups on a solid adsorbent surface by various physical and chemical bonds (Deniz 2010).

Electrocoagulation method: In wastewater, coagulated particles absorb colloidal microparticles and ions in wastewater by attracting them towards them. The balls formed precipitate and can be lifted to the water surface with the help of the gases formed in electroflotation. This method is used in the removal of color, heavy metal, COD, TOC, and SS from textile wastewater (Can 2002).

The advantages and disadvantages of these methods used in heavy metal removal are explained in detail in Table 9.5. Adsorption, biosorption, and phytoremediation methods are effective applications in which biological molecules are used in metal removal. For the biosorption of metal ions from aqueous media by biomass, dissolved substances to be held on the biosorbent surface must pass through the solvent liquid film surrounding the biomass. In order for the biosorption conditions to be realized, optimal conditions must be formed. The biosorption method is affected by physicochemical factors such as the metal ion type, biomass type and amount, concentration, temperature, and solution pH. Adsorption is the combination of molecules with that surface according to the pulling forces on the surface they come into contact with. Phytoremediation is an environmental improvement technology using plants as one of the biological materials (Rasim 2012).

Adsorption is one of the most favorable methods for the removal of pollutants with its simple, highly effective, and economical properties. A lot of adsorbents were used for the adsorption process such as clay minerals, natural and modified zeolites, metal oxides, agricultural wastes, biomass, polymeric materials, chitin, and chitosan. Moreover, CNTs and graphene are useful in improving metal ions adsorption capacities. Compared with activated carbon, carbon-based nanomaterials have a high adsorption ability and low cost. CNTs have been used to adsorb Cu(II), Pb(II), Cd(II), Co(II), Cu(II), Zn(II), Mn(II), Ni(II), and Hg(II) (Chen and Wang 2006).

Biosorption is the removal of metal ions from aqueous media by biomass. The biosorption (biological adsorption) of metals is generally based on adsorption, ion exchange, complexation, and micro-precipitation, and is a rapid and reversible event. It is known as a cost-effective biotechnological method for the treatment of low concentrations and high volumes of metals in wastewater using biomaterials such as bacteria, crab shells, fungus, and algae. In other words, the accumulation of biological materials in the cell surface or in the cell surface of waste materials in aqueous solutions is called biosorption. Uluözlü et al., in 2008, investigated the biosorption of Pb(II) and Cr(III) from aqueous solutions in their biosorption study with *Parmelina tiliacea* lichen species (Uluözlü et al. 2008). Ekmekyapar et al., in 2006, examined the biosorption of copper(II) from *Cladonia rangiformis* non-living dead biomass (Ekmekyapar et al. 2006). Yalçın et al, in 2010, studied the Cu^{+2} and Zn^{+2} biosorption of *Roccella psychopsis* species. With these features, lichens have been used as a biosorbent in metal removal in recent years (Yalçın et al. 2010).

TABLE 9.5
Conventional Metal Removal Methods and Their Advantages and Disadvantages (Cındık Akıncı et al. 2016)

Method	Advantages	Disadvantages
Precipitation	Simple and affordable	Difficult separation at high concentrations Not active Waste sludge formation
Electrochemical methods	Recovery of metal	Costly Effective only in high concentrations
Chemical oxidation and reduction	Inactivation	Environmental sensitivity
Ion exchange	Efficient treatment and recovery of pure waste metal	Sensitive to particles and expensive resins
Evaporation	Pure waste generation	Excessive energy requirement Costly Waste sludge formation
Reverse osmos	Pure waste for recycling	High pressure membrane size Costly
Adsorbtion	Sorbents activated Carbon usage	Not applicable for all metals

It seems that the biosorption method is one of the most suitable alternative methods for metal removal. Biosorption is also referred to as the removal of metal ions by dead biomass in a solution. Since the surfaces of organisms are negatively charged, they have the ability to adsorb positively-charged metal ions. Microorganisms take up metal both actively (accustomed cell) and passively (biosorption). Studies show that in the biosorption method, passive intake is applied more than active intake. This is because living systems (active intake) require additional nutrients that are frequently added, thus increasing the BOD_5 or COD at the exit channel of the reaction.

Another method, phytoremediation, is a technology to improve the environment based on plants. With this technology, organic and inorganic materials can be eliminated from the area where the plant creates pollution. The properties sought in the plants used in these studies are that they can grow in a healthy way without being damaged by the existing pollutants in polluted areas, and that they can form roots and green parts at sufficient levels. The popularity of phytoremediation technology is increasing day by day. Hyperaccumulator plants contain metal ions in high concentrations and can detoxify them. In this case, plant treatment is an inexpensive option compared to other methods. Green breeding (phytoremediation) is thought to have more economical, technical, and environmental advantages than traditional physical and chemical treatment methods. However, it is stated that the recovery period is long. There are many different technologies under the name of phytoremediation: phytoextraction, phytostabilization, phytodegradation, and phytovolatilization. Benaissa et al. investigated the benefit of dried activated sludge for the removal of Cu^{+2} from synthetic aqueous solutions. Kinetic data and equilibrium biosorption isotherms were measured in bulk conditions. The effects of some parameters such as contact time, initial copper concentration, initial pH of the solution, and the nature of copper salt on copper biosorption kinetics were investigated. Maximum copper biosorption was found at an initial pH of 5. These results showed that activated sludge is a suitable biosorbent for the removal of Cu^{+2} ions in synthetic aqueous solutions (Mcintyre 2003).

In a study, Vanlı investigated the cleaning of soils contaminated with Pb, Cd, and B elements by the phytoremediation method. For this purpose, phytoremediation studies were carried out using corn, sunflower, and canola plants in soils supplemented with Pb, Cd, and B elements. In addition, various doses of complexing chelate were added to the soil to increase the phytoremediation capacity, and changes in the elemental removal performance of the plants were observed. Within the scope of experimental studies, after the addition of Pb $(NO_3)_2$, $CdCl_2$, and H_3BO_3 to the soil, sunflower, corn, and canola, biki seeds were planted and their development was observed by watering them at appropriate intervals as needed. Seven days before the plant harvest, EDTA was added to the soil and then the plants were harvested. In order to determine the elements taken by the root and above-ground organs of the plant, they were cut one by one in pots and solubilized, and the element contents were determined with AAS and inductively coupled plasma (ICP) devices. In light of the obtained results, the amount of elements the plants received per dry weight was determined (Vanlı 2007). A new trends in removing inorganic pollutants is the combination of water-soluble polymers, membranes, and electrocatalysis/photocatalysis. Solar photocatalytic degradation of Zn^{2+} is carried out by using graphene-based TiO_2.

Besides heavy metal ions, arsenides and fluorides are common inorganic toxic species. Although fluoride ingestion, within the permissible limits, is beneficial to the human body, if it is present in excess, it causes mottling of teeth and lesions of the endocrine glands, thyroid, liver, and other organs. The suitable level of fluoride in drinking water specified by the World Health Organization (WHO) is 1.5 mg.L^{-1}. Sources of fluoride wastewater may be photovoltaic (PV) energy manufacturing processes producing hydrofluoric acid (HFA), a major source pollutant of this waste. Fluoride contamination is highly toxic in certain aquatic systems. Treatment technologies for fluoride wastewaters are precipitation, ion exchange, membrane, electro-chemical and adsorption processes, proposed and tested for the removal efficiency of excess fluoride from drinking water as well as industrial effluents. In the case of high fluoride concentrations, lime precipitation is commonly used to form CaF_2 precipitate, which is the mos cost-effective way; however, this process causes hardness of the wastewater.

In recent years, EC is used to effectively treat restaurant wastewater, textile wastewater, electroplating wastewater, and fluoride-containing wastewater. The EC technique uses a direct current source between metal electrodes immersed in polluted water. During EC defloration, no contaminants enter the environment and useful ingredients in raw water may remain, which presents a very attractive method. The EC defloration technique was also used for the treatment of textile wastewater treatment, for wastewater containing Cu^{2+}, Zn^{2+}, and Cr (VI) by electrocoagulation, for the defluoridation of Sahara water by small plant electrocoagulation using bipolar aluminum electrodes and for the removal of fluoride from drinking water.

Arsenic (As) is ubiquitous and presents in seawater and the human body. It is a silver-grey, brittle crystalline solid. It is classified as one of the most toxic and carcinogenic chemical elements. Arsenic is mobilized by natural weathering reactions, biological activity, geochemical reactions, volcanic emissions, and other anthropogenic activities. Sources of As pollutant are mining activities; the combustion of fossil fuels; the use of arsenic pesticides, herbicides, and crop desiccants; and the use of arsenic additives to livestock feed. Arsenic is present in water mainly as oxoanyons: arsenate [As(V)] and arsenite ions [As(III)] related to arsenic acid (H_3AsO_4) and arsenous acid (H_3AsO_3). In order to remove arsenic, water-soluble polymers are combined with UF membranes and used in LPR techniques. Generally, coagulation by ferric (Fe) or aluminum (Al) coagulants has been identified as the most affordable options to treat wastewater with high levels of arsenic, and the removal of As is ascribed to As adsorbing onto Fe or Al hydroxides and to the solid–liquid separation of precipitates with adsorbed As by subsequent sedimentation and filtration. Ferrate [Fe(VI)] may be employed as both an oxidant and a coagulant for As removal. The Fe(II)–$KMnO_4$ process has also been proposed for arsenite [As(III)] removal. Additionally, the electrocoagulation also have good capability to treat high concentrations of arsenic in wastewater (Guan 2009). The Fe–Mn binary oxide combined with coagulation by poly-aluminum chloride was used for the removal of arsenic-containing wastewater using in situ removal (Wang 2003). Recently, several techniques were developed to remove arsenic from wastewater, such as chemical, physicochemical, and biological methods. The chemical methods to remove arsenic, including neutralization precipitation, coagulation sedimentation, and sulfide precipitation, primarily involve adding chemical reagents to the wastewater to form insoluble arsenic compounds. The potential use of magnesite was evaluated as an initial step in the pragmatic removal of arsenic from aqueous solutions. In terms of economic viability, the technology was feasible (Vhahangwele et al. 2016). A novel method was offered to remove arsenic from HAWA through a combination of copper powder and chloride. As(III) was reduced to non-toxic As(0) residues by copper powder in the presence of chloride (An et al. 2020), and also a multi-start distributor, for sulphide feeding, was used for the efficient removal of arsenic from "dirty acid" wastewater recently (Li et al. 2015).

Radioactive wastes, such as cesium, cobalt, strontium, thorium, radium, and uranium, are a threat to people and the environment with their toxic, carcinogenic, mutagenic, and bioaccumulative properties. Sources of radioactive pollutants include nuclear power reactors and the application of radionuclides in medicine, agriculture, industry, and research, all of which have adverse effects on human health and the environment. Water and wastes which have been discharged from schools, residences, and businesses flow through municipal sewer systems and are treated at wastewater treatment plants, also referred to as publicly owned treatment works (POTWs). Cesium is one of the important radioactive pollutants in the water. It is removed by several methods such as potassium cobalt hexacyanoferrate compounds (KCo-HCFe's) impregnated onto a 6-nylon fiber by radiation-induced graft polymerization and subsequent chemical modifications. With this method, nearly 95% of cesium was removed (Okamuraet al. 2014). Karamanis and Assimakopoulos studied aluminum-pillared-layered montmorillonites (PILMs) for their potential application in the removal of copper or cesium from aqueous solutions (Karamanis and Assimakopoulos 2007). Dwivedi et al. developed a novel synthesis method using calcium alginate beads as a template. Equilibrium data were found to be well-fitted with the Langmuir isotherm equation, with a monolayer sorption capacity of 490.2 mg.g^{-1}, and the complete elution of the sorbed cesium ions was also possible (Dwivedi et al. 2012). In another study, PAN-based manganese dioxide composite materials were used for the removal of Cs in aqueous solutions which resulted in the significant removal of Cs^+ ions at the pH range of 4–9 (Nilchi et al. 2012). The proposed technologies to treat radioactive wastes were thermal treatment, precipitation, extraction, membrane, adsorption, and ion exchange. Ion exchange is one of the most promising processes, and many types of inorganic ion exchange materials and adsorbents have been applied such as magnetite, silicotitanate, natural zeolite, and hexacyano ferrate complexes. Another method for low-level radioactive wastewater treatment included testing the effect of ion exchange membranes on the removal efficiency of continuous electrodeionization (CEDI) (Li et al. 2017). For another radioactive pollutant, strontium, various techniques such as solvent extraction, inorganic ion exchange, and organic ion exchange for the removal of strontium have been extensively studied. A 2016 study, in this direction, also involved strontium removal from radioactive contaminated wastewater where silica-based titanate adsorbents were prepared and applied. As a result of this study, the change in flow rate and column diameter influenced the adsorption capacity of Sr^{2+} for the treatment of Fukushima wastewater. Beside all, macrocyclic organic compounds can effectively remove strontium through molecular recognition (Chen 2016).

Thorium is a naturally occurring radioactive element that represents a significant component of the fuel used in nuclear breeder reactors and in the thorium fuel cycle, which can be applied in most of the reactor processes. In a 2019 stude, alginate-immobilized *Aspergillus niger* was prepared for the removal of thorium ions from radioactive wastewater. The study revealed that this biosorbent was environmentally friendly and can be regenerated and used repeatedly (Ding et al. 2019). Bivalent radioactive ions, such as Sr^{2+} and Ra^{2+}, are sourced from the heap-leach residues of the uranium mining industry. Sodium niobate adsorbents were doped with tantalum (TaV) for the removal of bivalent radioactive ions in wastewaters (Blain et al. 2011). Cobalt is one of the toxic radioactive elements. Co^{2+} was removed from radioactive wastewater by a polyvinyl alcohol (PVA)/chitosan magnetic composite. A study showed that the PVA/chitosan magnetic composite is promising adsorbent for removing Co^{2+} radioactive wastewater (Zhu et al. 2014). Uranium mining and hydrometallurgy usually produce large volumes of low-concentration radioactive wastewater with U(VI) concentration of lower than 1 mg.L^{-1}. The conventional methods such as chemical precipitation, membrane separation, solvent extraction, and ion exchange are both ineffective and uneconomical for removing U(VI) from the low-concentration radioactive wastewater. Although some researchers have studied the biosorption methods, results show that the biosorption method based on the ethylenediamine-modified biomass of *A. niger* was an effective and economical one for treating the low-concentration radioactive wastewater (De Xin et al. 2014). Barium is one of the most toxic elements present in radioactive wastes and is a long-lived radionuclide. It is removed by using a combination of maghemite and titania nanoparticles in PVA and alginate beads (Zohreh et al.2015).

9.2.2.2 Asbestos Wastewaters

Asbestos fibers are widely used because of their excellent physical properties such as great tensile strength, poor heat conduction, non-biodegradability, high electrical and alkali, acid attack resistance, and sound absorption capabilities. Although the use of asbestos was banned because of its carcinogenic properties, asbestos-containing wastes are still present in the environment. They are currently landfilled or encapsulated with resins. Asbestos is found as natural fibrous silicate minerals that come in various forms. Because of these properties, asbestos has been widely used, particularly as a building and insulation material acoustic and in thermal sprays, plasters, paints, flooring products, flat sheets, tiles, corrugated sheets for roofing, rainwater, pressure pipes, and several other building materials.

Asbestos pollutant can be removed through three methods:

- *Removal of asbestos materials:* This is the most used and expensive method with high disposal costs, producing toxic and harmful waste.

- **Encapsulation:** The asbestos-containing material is coated with penetrating and buried products to provide a protective film between the environment and the asbestos fiber.
- **Confinement:** The asbestos-containing material is isolated from the environment by creating a structure that separates it from the environment and the fibers are released in this structure. The cost is lower than the other two methods.

Sources of asbestos contamination in the aqueous environment and in drinking water have waterborne effects on natural deposits such as serpentine rocks, airborne effect, release from asbestos-cement pipes, processes associated with mining, and the production of certain types of iron ore. Whether asbestos fibers are airborne or waterborne, or both, they remain in the environment for decades or longer (Valerio et al. 2019).

Ultimately, asbestos contamination has to be reduced and prevented from entering water sources. Some of the water treatment methods include conventional water treatments such as coagulation/filtration. The filtering process of asbestos fibers requires low-pore filters. However, such filters tend to be clogged with organic matter and other suspended inorganic particles. In addition, although asbestos filtration can be applied with success to small and controlled systems, such as drinking water supplies or garbage leachate, it is not readily applicable to larger scenarios, such as the natural release of fibers from asbestos spoil dumps, excavated faces in formerly quarried or mined sites, or the dumping of mining tailings into river or lakes. An innovative method, by means of the synergistic effect of power ultrasound (US) and an acidic metal chelator, oxalic acid (Ox), was proposed in order to cope with such issues. Afterward, researchers tested a commercial sample of chrysotile from a mining site (Turci et al. 2008). Thermal treatments are applied for the modification of the crystal-chemical structure of asbestos fibers transforming asbestos into an inert material using high temperature. The considered thermal treatments include both standard vitrification and thermal treatments with controlled recrystallization. Other applied methods for reducing asbestos contamination in nature are chemical treatments using chemical reactions to convert asbestos into harmless compounds, and mechanochemical treatments where fibers are degraded by mechanical milling (Kusiorowski et al. 2015). The best practices reported for chemical treatments are based on strong acidic or basic solutions, as well as the use of fluorine. The microwave thermal inertization of asbestos is another application for thermal treatments (Leonelli et al. 2006). In a study carried out by Min et al., waste asbestos was melted using a mixture of hydrogen and oxygen produced from water electrolysis in order to provide treatment (Min et al. 2008). A more innovative decomposition technique, of friable asbestos by CHClF 2-decomposed acidic gas, was applied (Yanagisawa et al. 2009). In 2009, an industrial process was established for the direct temperature induced recrystallization of asbestos and/or mineral fibers containing waste products using a tunnel kiln and recycling (Gualtieri and Zanatto 2009). In 2012, the products of thermal inertization of cement-asbestos were recycled in geopolymers (Gualtieri et al. 2012). Moreover, asbestos materials were detoxified by microwave treatment. Today, there is an increasing need for such recycling efforts. In recent years, with the increasing cases of cancer, more scientific and feasible studies are needed to remove asbestos pollution, which is carcinogenic and hardly disappear after being released into nature, or to make asbestos fibers harmless.

9.3   REFRACTORY COMPOUNDS IN DOMESTIC/URBAN/ SEWAGE/MUNICIPAL WASTEWATER AND TREATMENT

In addition to industrial wastewater, domestic/urban/sewage/municipal wastewater also contains refractory/persistent organic pollutants. The potential organic contaminants in domestic/urban/sewage wastewater are:

- Steroid hormones and pharmaceuticals, and their metabolites
- Domestic and industrial detergent residues and degradants

- Antimicrobial agents (e.g., triclosan, TCS)
- Other PPCPs (e.g., musk fragrances, sunscreens)
- Brominated flame retardants (e.g., PBDEs)
- Persistent organic pollutants (e.g., PAH, PCBs)
- Plasticizers (e.g., bisphenol A-BPA)
- Disinfection by-products such as N-nitrosodimethylamine (NDMA)

Sources of these pollutants could be residential: domestic wastewater streams made up of black and grey water; or commercial waste: discharges from small-scale commercial establishments such as medical and veterinary clinics, hospitals, and restaurants or trade wastes and a variety of small-, medium-, and large-scale industrial discharges (e.g., pharmaceuticals, paper and pulp mills, and intensive livestock operations). Human sourced ROCs/POPs include natural and synthetic steroid hormones (e.g., E2 and EE2), household products such as surfactants and their by-products in detergents [e.g., alkylphenol polyethoxylates (APEOs)], linear alkylbenzene sulfonates (LAS), and pharmaceuticals (e.g., analgesics and antibiotic drugs). According to researchers, an estimated 60–75% of antibiotics, which can be partially metabolized after being used by humans and animals, are excreted without being absorbed, and thus they are mixed into the wastewater unchanged or as metabolites via urine and feces (Shareef et al. 2011).

Potential treatment technologies (PTT) include conventional biological treatment processes or AOP to remove organic refractory contaminants from domestic wastewater. In the literature there many studies on the removal or degradation of organic refractory contaminants from domestic/urban/sewage/municipal wastewater. The occurrence of 26 POPs was investigated at different stages of a conventional activated sludge WWTP that receives mainly domestic wastewater and urban runoff in Greece. A total of 22 ROCs were detected such as PCBs, DDTs, dieldrin, aldrin, HCHs, heptachlor, and HCBs in untreated, primarily treated, and secondarily treated wastewater. In a study of the wastewater treatment plant in the city of Thessaloniki, the removal efficiencies of the various POPs during the overall treatment process varied between 65% and 91% during the conventional activated sludge treatment process (Athanasios et al. 2004). In 2017, the efficiency of an innovative HSBS was evaluated for removing phosphorus, nitrogen, and PCBs from the original municipal wastewater produced by the Poland Municipal Wastewater Treatment Plant. The HSBS proved an effective method of ecohydrological biotechnology for the treatment of ROCS from wastewater from WWTPs (Edyta et al. 2017).

As an overall review, domestic wastewater treatment includes several technologies such as adsorption, activated sludge, MBR, biofilm, membrane filtration, coagulation/electrocoagulation, advanced/photocatalytic oxidation, and combined processes. Activated carbon, clay, industrial by-products, zeolites, polymeric materials, biomass, agricultural waste, and natural absorbents like zeolite are used for adsorption processes. There are many treatment techniques applied in domestic wastewater treatment, as in industrial wastewater treatment, and each of these treatment techniques has advantages and disadvantages. According to the type and structure of the pollution to be treated, the highest treatment efficiency is the systems where advanced oxidation processes and other processes are applied in combination.

9.4 ACTIVATED SLUDGE PROCESS FOR REFRACTORY COMPOUNDS/POLLUTANTS REMOVAL

Biological treatment techniques are the most used processes for the treatment of waste materials. Biological treatment has been widely used for the treatment and disposal of wastewater for more than a century. In CBTPs, organic substances are oxidized with the help of microorganisms and converted into mineral forms. For example, organic carbon, carbon dioxide, nitrogen, nitrate, sulfur, and sulphate are converted into phosphorus. Some of these minerals take place in the bacterial cell structure. Therefore, dissolved organic substances in wastewater that cause environmental pollution are partially oxidized by microorganisms and partially enter the microorganism structure formed, and are then purified by separating these microorganisms from the water. The organic carbon contained in the microorganisms is also converted into methane gas microbiologically in the anaerobic system. Biological decomposition of organic substances can occur in an environment with free oxygen (aerobic) or in the absence of free oxygen (anaerobic). Therefore, biological treatment systems are divided into two groups: aerobic and anaerobic. In each group, different processes have been developed according to whether the microorganism culture is suspended or adherent. Again, these processes can be thermophile or mesophyll, separately. Therefore, biological wastewater treatment systems can be of many different types. In practice, aerobic, mesophilic systems are widely used for the purification of biodegradable organic substances. Generally, anaerobic systems are preferred less due to their slow kinetics, and they are applied when concentrated and difficult to aerobically degrade organic substances which need to be separated. Despite their rapid kinetics, thermophilic systems are not common due to the difficulties that can arise in flocculation and precipitation. Since the types of microorganisms in all types of biological treatment systems are different, it is possible to summarize the main types of microorganisms in other important systems, starting from the most widely used biological systems as follows.

The activated sludge system is the most widely used biological wastewater treatment process. The system is based on aerobic, mesophilic, and suspended cultures. There are many different types of microorganisms in it. Depending on the organic substance type, concentration, and other environmental factors (pH, temperature, oxygen concentration, toxic substance, etc.), the types of microorganisms in activated sludge also change. Species such as bacteria, fungi (fungi), protozoa, and sometimes nematodes are the main microorganisms that are widely found in activated sludge.

Activated sludge biomass is a heterogeneous complex ensemble that, in equilibrium, provides the flexibility to operate the system within certain limits. However, compounds that are difficult to biodegrade and have inhibitory properties affect this system negatively and cause a decrease in treatment efficiency and even collapse of the system (Gutiérrez and de las Fuentes 2002).

The best known inhibition tests for aerobic microorganisms are based on the respirometric approach [304]. The respirometric test is an accepted effective method for evaluating the inhibitory effect and/or toxic level of selected compounds on activated sludge microorganisms. For organisms, the oxygen uptake rate (OUR) indicates the rate of oxygen consumed in the process. The respirometer device allows for the measurement of the OUR parameter related to substrate utilization and biomass production, to which the inhibition is directly related. MPV can be interpreted to evaluate COD fractions and the stoichiometry and kinetics of important biochemical reactions.

In the treatment of refractory substances, conventional activated sludge systems cannot work efficiently. These micro-contaminants inhibit activated sludge microorganisms and cause the current treatment efficiency to decrease. However, degradation intermediates are more toxic than the original compounds, the removal of which remains a major challenge in conventional systems. In the case study section, the performance and renewability of active sludge exposed to a nanoparticular photocatalyst in real domestic wastewater biodegradation is evaluated in detail.

9.5 CASE STUDY

The contamination of substances such as PPCPs, endocrine disrupting compounds, antipyretics, pesticides, and artificial sweeteners and their metabolites into nature threatens the environment and human health to a great extent. In the treatment of these substances, conventional activated sludge systems cannot work efficiently. The micro-contaminants inhibit activated sludge microorganisms and cause the current treatment efficiency to decrease. However, degradation intermediates are more toxic than the original compounds, the removal of which remains a major challenge in conventional systems. Since wastewater treatment plants discharge directly into aquatic environments, micropollutants can be found in aquatic environments from ng / L to mg / L concentrations.

Micropollutants entering domestic wastewater treatment plants and photocatalysts used in their removal affect the efficiency of conventional activated sludge systems by inhibiting the activated sludge. In this study, following the evaluation of the acute and chronic effects of NFC-doped TiO_2 nanoparticles on activated sludge in real domestic wastewater and peptone biodegradation, which are known to be efficient in antibiotic removal, their renewability for 20 days after exposure to photocatalysts was evaluated.

9.5.1 MATERIALS AND METHODS

Domestic wastewater used in this study was taken from the balancing unit of the Çorlu Domestic Wastewater Treatment Plant, which is known to have a high non-biodegradable substance such as antibiotic residues. The activated sludge was taken from the aeration unit of this plant. The acclimation of activated sludge into both synthetic wastewater (peptone) and real domestic wastewater was carried out in separate reactors with a sludge age of 10 days and a hydraulic retention time of 1 day. The acute and chronic effects of NFC-doped TiO_2 photocatalysts on the microbial population in both reactors was determined by the OUR profiles measured during the respirometric experiments conducted with the Ra-Combo (Applitek Co., Nazareth, Belgium) respirometer device. After acclimation, respirometric analyses were performed on the microbial population that was fed only wastewater and peptone. An evaluation of the effect of nanoparticular structure photocatalysts on the biodegradation of pepton-containing synthetic wastewater and real domestic wastewater was made by comparing the data obtained from the respirometric analysis carried out before and after the addition of photocatalysts to the same wastewater. The microbial culture used in the evaluation of the acute effect was provided from the reactor after acclimation to the wastewater containing photocatalysts. The microbial culture used in the evaluation of the chronic effect was obtained after feeding of the NFC-doped TiO_2 photocatalyst at a concentration of 2.25 mg.L^{-1} together with the peptone-containing synthetic / real domestic wastewater after the acute effect was determined. The addition of a new generation photocatalyst agent (inhibitor) in the nanoparticular structure to the synthetic / real domestic wastewater was made after the respirometer reactor was operated at a level of endogenous decay for a certain period of time. In chronic effect determination studies, prior to the respirometric study, acclimation reactors were fed with NFC-doped TiO_2 photocatalysts everyday with a dosage of 2.25 mg.L^{-1}, which is stated in the literature as effective in antibiotic removal, in addition to peptone-containing synthetic wastewater/real domestic wastewater. After the 30th day, the photocatalyst feed was discontinued for 20 days and the renewable potential of the activated sludge was determined by feeding only synthetic wastewater / real domestic wastewater. The acute effect was evaluated before photocatalyst feeding. The chronic effect and recovery of biomass were evaluated with the biomass taken from reactors on the 30th day and 50th day, respectively. The experimental design, OUR measurements, and respirometry device are given in Table 9.6. and Figures 9.5 and 9.6, respectively. During the study, all measurements, except COD and COD$_{sol}$, were performed according to standard methods. COD and color measurements were performed according to ISO 6060 and ISO 7887 procedures, respectively. In respirometric experiments, a nitrification inhibitor at a concentration of 1 mg.L^{-1} was added to the respirometer reactor to prevent O_2 consumption from nitrification (Figure 9.7 and Tables 9.7 and 9.8).

9.5.2 RESULTS AND DISCUSSION

When the oxygen consumption rate and COD experiment data obtained as a result of acute respirometric tests for domestic and peptone wastewaters are examined, it was observed that a decrease in the total amount of consumed O_2, 94 mg.L^{-1} and 171 mg.L^{-1} of total consumed oxygen are used in easily degradable COD biodegradation, respectively, and also determined that the amount of consumed O_2 in the endogenous decay process decreased by approximately 3% and 20%, respectively. According to these data, it was understood that a dramatic inhibition of bacterial activity had not been observed but a decrease in the amount of substrate. On the other hand, when the

TABLE 9.6

Experimental Design of Case Study (Merdan 2019)

Experiment	Day	Photocatalyst (mg.L⁻¹)	Species Analysis
Domestic ww-control	0	-	+
Domestic ww-acute	0	2.25	+
Domestic ww-chronic	30	2.25	+
Domestic ww- recovery	50	-	+
Peptone-control	0	-	+
Peptone ww-acute	0	2.25	+
Peptone ww -chronic	30	2.25	+
Peptone ww-recovery	50	-	+

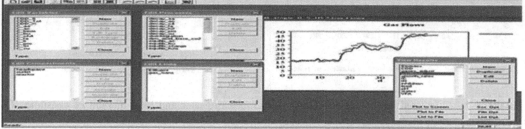

FIGURE 9.4 Respirometry device and OUR measurement examples (Merdan 2019).

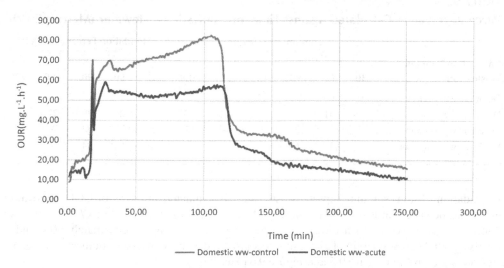

FIGURE 9.5 Domestic wastewater-control and acute analysis OUR data.

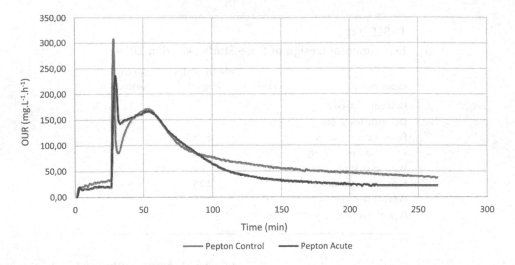

FIGURE 9.6   Pepton wastewater-control and acute analaysis OUR data.

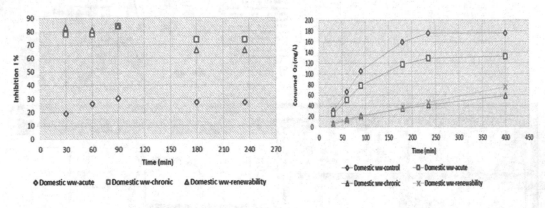

FIGURE 9.7   Calculated percentage values for domestic wastewater and total amount of consumed O_2 against time.

TABLE 9.7
Percentage Values Calculated Against Time for Domestic Wastewater (Merdan 2019)

Calculation Time (min)	Consumed O_2 (mg.L⁻¹)						Inhibition Percentage (%I)				
Wastewater	30	60	90	180	235	400	30	60	90	180	235
Domestic ww-control	31	66	105	159	176	176	-	-	-	-	-
Domestic ww-acute	25	51	78	117	129	132	19	26	30	27	27
Domestic ww-chronic	7	14	21	34	41	59	78	78	84	74	74
Domestic ww-renewability	5	12	18	36	46	75	83	81	84	66	66

biodegradation of domestic and peptone wastewaters is evaluated in terms of COD removal efficiency, it has been determined that the removal efficiency decreased from 87% to 77%, respectively, in domestic wastewater, and wastewater with peptone decreased by approximately 10% and 8%, respectively. OUR graphs obtained as a result of control and acute analysis for domestic and peptone wastewaters are given in Figure 9.5 and Figure 9.6 (Merdan 2019).

TABLE 9.8

Percentage Values Calculated Against Time for Peptone Wastewater (Merdan 2019)

Calculated COD Fraction Wastewater	Consumed O$_2$ (mg.L^{-1})			Calculated COD Fraction	Inhibition Percentage (%I)	
	Readily Biodegradable	Slowly Biodegradable	Total		Readily Biodegradable	Slowly Biodegradable
Peptone ww-control	141	153	294	4,5	-	-
Peptone ww-acute	171	55	226	4	10	42
Peptone ww-chronic	102	99	201	5	40	44
Peptone ww-renewability	105	100	205	7.5	57	63

As can be observed from Figure 9.5 and Figure 9.6, it is understood that the NFC-doped TiO$_2$ photocatalyst inhibits OUR to a small extent, but when the COD removal efficiencies are considered, the performance of the reactors does not change dramatically. When the OUR and COD experiment data are studied, obtained as a result of respirometric experiments performed to determine the chronic inhibition of the biodegradation of domestic and peptone wastewaters are examined, it is observed that a substantial reduction in the total amount of consumed O$_2$, 30 mg.L^{-1} and 102 mg.L^{-1} of total consumed oxygen are used in easily degradable COD biodegradation, respectively, and it was determined that the amount of consumed O$_2$ in the endogenous decay process increased approximately 30% in domestic wastewater, while it remained almost the same in peptone wastewater.

On the other hand, when the biodegradation of domestic and peptone wastewaters is evaluated in terms of COD removal efficiency, it has been determined that the removal efficiency decreased from 87% to 77% in domestic wastewater, and wastewater with peptone decreased by approximately 92% and 67%, respectively. According to the chronic experiment results, in domestic wastewater biodegradation, activated sludge was found to be highly affected by NFC-doped TiO$_2$ photocatalysts than peptone wastewater.

On the other hand, the total amount of consumed O$_2$ and the corresponding inhibition percentages in the selected minutes from the OUR graphs of domestic and peptone wastewater were calculated, and inhibition occurrence was evaluated in this decreased rate of removal efficiency. In this evaluation, the first plateau on the OUR plot reflects easily degradable COD biodegradation, and the curve after the first plateau reflects slow degradable COD biodegradation. So, in the respirometric analysis, the inhibition percentages and consumed O$_2$ amounts for domestic and peptone wastewaters are given in Fig. 5.3 and Fig. 5.4 and Table 5.2 and Table 5.3, respectively (Merdan 2019).

As it can be seen from Fig. 5.3, Table 5.2 and Table 5.3, activated sludge microorganisms were inhibited by increasing photocatalyst exposure time in domestic wastewater biodegradation. Acute inhibition increased by 10% every 30 minutes in the first 90 minutes of respirometric analysis, but remained stable in the following minutes. Inhibition increased significantly after 30 days of chronic exposure. In the first 60 minutes of the respirometric analysis, it remained stable around 80%, then decreased by 10% and continued steadily. After the termination of the photocatalyst feeding, over 20 days, in the first 90 minutes of respirometric analysis, it was observed that the inhibition continued at about a 10% increase compared to the chronic analysis of domestic wastewater. However, in the following minutes, inhibition decreased by approximately 20% and reached constant levels. The reason for the different inhibition values observed in the selected time intervals is thought to be due to the different kinetics of different biochemical reactions corresponding to each time interval. In the first 10 minutes of the respirometric experiment, the first microorganism growth and substrate storage took place. After then, while the biodegradation of readily biodegradable COD fractions occurred in up to 180 minutes, slowly biodegradable COD fractions occurred after then. The total amount of consumed O$_2$ was inhibited by increasing the prolongation of the photocatalyst exposure, and inhibition continued after the end of the photocatalyst feeding. Contrary to the absence of improvement in inhibition, it was thought that the reason for the increase of inhibition may be the other micropollutants found in the domestic wastewater. However, the inhibition of the total amount of consumed O_2 as a result of chronic exposure is significantly higher than the inhibition that occurs after the termination of the photocatalyst feed. From this point of view, it can be said that the inhibition from other micropollutants that were found in domestic wastewater is much lower than the inhibition from photocatalysts. After determining the inhibition effect on the domestic wastewater and peptone biodegradation of the NFC-doped TiO_2 photocatalyst, photocatalyst feeding to the reactors was stopped for 20 days and fed only with the selected substrate source; thus, to what extent the performance of microorganisms improved after the inhibitory effect. Therefore, the microorganism recovery performance was observed. According to this result, the peptone reactor was found to be able to tolerate nanoparticular material at a higher level than the domestic wastewater reactor. There was no improvement in COD removal efficiency in the domestic wastewater reactor, and an increase in inhibition of oxygen uptake rate was observed. On the other hand, in the peptone reactor, the inhibition of oxygen uptake rate continued, but the microorganisms consumed more organic matter over a longer period of time and a 19% improvement in COD removal efficiency was observed (see Table 9.8).

9.6 CONCLUSION

It is important to determine the character of the raw wastewater of a potential conventional activated sludge treatment plants to be supported by the advanced oxidation process where NFC-doped TiO_2 is used as a photocatalyst. Because, compared to peptone wastewater reactors, a high inhibition is observed in domestic wastewater reactors and the unfavorable renewal of activated sludge is due to refractory pollutants found in domestic wastewater. A combined system should be implemented, since the refractory pollutants found in domestic wastewater can be removed by advanced oxidation processes, like the inhibition of the level observed in peptone can be observed in real wastewater. At the same time, when there is a need for treatment only with the conventional activated sludge process, the advanced oxidation process can be disabled and, in this case, the COD removal efficiency of the activated sludge can improve, as observed in the peptone reactor analysis results in this study. Within the framework of this information, if domestic wastewater discharge standards are met with a possible 30% inhibition of COD removal efficiency in a potential combined treatment plant, it may be appropriate to use combined treatment systems. Generally, since domestic wastewater treatment plants' raw wastewater COD load is low, combined systems can be considered as an alternative that can be applied. When needed, for example, when the input COD load increases in certain seasons at the combined treatment plant and high COD removal is required, the advanced oxidation process can be disabled and an increase in COD removal efficiency can be observed over the 20-day period. Thus, discharge standards can be achieved at desired levels by applying combined systems in domestic wastewater, where seasonal changes are observed, and by disabling the advanced oxidation process in the seasons when the input COD load increases because of refractory compounds.

ACKNOWLEDGEMENTS

This research was funded by the Tekirdağ Namık Kemal University Scientific Research Project, Project No: NKUBAP.06.GA.18.183. We would like to thank the technical and administrative staff of the investigated plant for their cooperation and assistance in this study.

ABBREVIATIONS

NFC   Nitrogen-fluoride-carbon
COD   Chemical oxygen demand
BOD_5 Biochemical oxygen demand
PPCPs   Pharmaceutical and personal care products

AOPs	Advanced oxidation processes
ROCs/POPs	Refractory/persistent organic Compounds
RIOCs	Refractory inorganic compounds
DDT	Dichlorodiphenyltrichloroethane
POPs	Persistent organic compounds
CLRTAP	Convention on Long-Range Transboundary Air Pollution
HAR	Hybrid anaerobic reactor
UASB	Upflow anaerobic sludge blanket
BCOP	Biological contact oxidation pool
UWWTPs	Urban wastewater treatment plants
MWEUV	Microwave electrodeless ultraviolet
HARs	Hydrolytic acidification reactors
EGSB	Expanded granular sludge bed
HCB	Heptachlor, Hexachlorobenzene
OC	Organochlorine compound
DDT	Dichlorodiphenyltrichloroethane
AC	Activated carbon
ZVI	Zero valued iron
GC-ECD	Gas chromatograph with electron capture detector
PPCPs	Persistent PCPs
ARCs	Anthropogenic refractory compounds
PBDEs	Polybrominated diphenyl ethers
PFOS	Perfluorooctane sulfonic acid
PFOS-F	Perfluorooctane sulfonyl fluoride
PCBs	Polychlorinated biphenyls
HCBs	Hexachlorobenzene
SC	Stockholm Convention
PAHs	Polycyclicaromatic hydrocarbons
WWTP	Wastewater treatment plant
EWWTP	Ecological WWTP
OPAHs	Organic PAHs
SPAHs	Soluble PAHs
PFOA	Perfluorooctanoic acid
LAS	Linear alkylbenzene sulfonates
APE	Alkyl phenol ethoxylates
PFAS	Poly- and perfluoroalkyl materials
GAC	Granular activated carbon
ROC	Reverse osmosis concentrate
FER	Ferric clotting
PAC	Powdered activated carbon
GC	Gas chromatography
ACA	Activated carbon adsorption
HSBS	Hybrid sequential biofiltration system
T-P	Total phosphate
T-N	Total nitrogen
MBBR	Moving bed biofilm reactor
PVC	Polyvinyl chloride
FAs	Fulvic acids
LPB	Liquid packaging cardboard
NHSs	Non-humic substances
DBPs	Disinfection by-products
THMs	Trihalomethanes

HAAs	Haloacetic acids
EC	Electrocoagulation
PACl	Polyaluminium chloride
CL	Concentrated leachate
AAS	Atomic absorption spectrophotometer
ICP	Inductively coupled plasma
WHO	World Health Organization
PV	Photovoltaic
HFA	Hydrofluoric acid
POTWs	Publicly owned treatment works
PILMs	Pillared-layered montmorillonites
CEDI	Continuous electrodeionization
PVA	Polyvinyl alcohol
US	Ultrasound
PBDEs	Brominated flame retardants
BPA	Bisphenol A
NDMA	N-nitrosodimethylamine
APEOs	Alkylphenol polyethoxylates
LAS	Linear alkylbenzene sulfonates
PTT	Potential treatment technologies
COD_{sol}	Soluble COD
MBR	Membrane biorecator
HPLC	High performance liquid chromatography
PCDDs	Polychlorinated dibenzodioxins
TCDD	2,3,7,8-tetrachlorodibenzo p-dioxin
PCDFs	Polychlorinated dibenzofurans
IARC	International Agency for Research on Cancer
EDCs	Endocrine disrupters
PAC	Polyaluminium chloride
MF	Microfiltration
UF	Ultrafiltration
NF	Nanofiltration
WFI	World for injection
GMP	Good manufacturing practices
GLP	Good laboratory practices
CBTP	Conventional biological treatment process
TOC	Total organic carbon
TSS	Total suspended solids
TDS	Total dissolved solids
T-N	Total nitrogen
T-P	Total phosphate
NH_4^+-N	Ammonium nitrogen
HRTs	Hydraulic retention times
DOC	Dissolved oxygen concentration
SMX	Sulfamethoxazole
ARB	Antibiotic resistant bacteria
ARGs	Antibiotic resistant genes
TiO_2	Titanium dioxide
PS	Polystyrene
LC-MS/MS	Liquid chromatography-tandem mass spectrometry
CAC	Carbon-acid-carbon
OMW	Olive mill wastewater

SRT Solids retention time
CASP Conventional activated sludge treatment
SNFC Sulfonated naphthalene-formaldehyde condensates
SBR Sequencing batch reactor
ASB Anaerobic sludge bed
RLT Residual liquors of tannery
UASB Upflow anaerobic sludge bed
EO Electrochemicaloxidation
BAF Biological aerated filter
PST Primary settling tank
SST Secondary settling tank
HSs Humic substances
HAs Humic acids

REFERENCES

Amano M., Lohwacharin J., Dubechotd A., Takizawa S. 2018. Performance of integrated ferrateepolyaluminum chloridecoagulation as a treatment technology for removing freshwaterhumic substances, *Journal of Environmental Management*, 212, 323–331.

An W., Kanggen Z., Xuekai Z., Dingcan Z., Changhong P., Wei C. 2020. Arsenic removal from highlyacidic wastewater with high arsenic content by copper-chloride synergistic reduction, *Chemosphere*, 238(January), 124675.

Ani I., Muhd Z, Abd M. 2015. Efficiency of barium removal from radioactive waste water using the combination of maghemite and titania nanoparticles in PVA and alginate beads, *Applied Radiation and Isotopes*, 105, 105–113.

Ata R., Yıldız Töre G. 2019. Characterization and removal of antibiotic residues by NFC-doped photocatalytic oxidation from domestic and industrial secondary treated wastewaters in Meric-Ergene Basin and reuse assessment for irrigation, *Journal of Environmental Management*, 233. pp. 673–680.

Ata R., Sacco O., Vaiano V., Rizzo L., Yıldız Töre G., Sannino D. 2017. Visible light active N- doped TiO2 immobilized on polystyrene as efficient system for wastewater treatment, *Journal of Photochemistry and Photobiology A: Chemistry*, 348, pp. 255–262.

Ateşli A. 2006. *Effects of Humic Substances on Drinking Water Disinfection Process*, Master's Thesis, Uludağ University, Institute of Science, Bursa, 5–6.

Athanasios K., Constantini S. 2004. Persistent organic pollutants (POPs) in the sewage treatment plant of Thessaloniki, northern Greece: Occurrence and removal, *Water Research*, 38(11), 2685–2698.

Ay F. 2015. Geological and economical importance of humic acid and humic acid, Sivas, *Cumhuriyet University Faculty of Science. Science Journal*, 36(1). ISSN: 1300-1949.

Banu J.R. and Kaliappan S. 2007 Treatment of tannery wastewater using hybrid upflow anaerobic sludge blanket reactor, *Journal of Environmental Sciences*, 6, 415–421.

Beegle, J.R., Borole, A.P. 2018. Energy production from waste: Evaluation of anaerobic digestion and bioelectrochemical systems based on energy efficiency and economic factors, *Renewable and Sustainable Energy Reviews*, 96, 343–351.

Bernardo M., Santos A., Cantinho P., Miguel Minhalmaa M. 2011. Cork industry wastewater partition by ultra/nanofiltration: Abiodegradation and valorisation study, *Water Research*, 45, 904–912.

Blain P., Dongjiang Y, Wayde N., Ray L.F. 2011. Sodium niobate adsorbents doped with tantalum (TaV) for the removal of bivalent radioactive ions in waste waters, *Journal of Colloid and Interface Science*, 356(1), 240–247.

Bokare A.D., Choi W. 2014. Review of iron-free Fenton-like systems for activating H_2O_2 in advanced oxidation processes, *Journal of Hazardous Materials*, 275, 121–135.

Can O.T. 2002. *Alüminyum Elektrotlar Kullanılarak Tekstil Atıksu ve Boyalarının Elektrokoagülasyon ile Arıtımı*, Yüksek Lisans Tezi, GYTE Fen Bilimleri Enstitüsü, Gebze.

Chen C., and Wang X. 2006. Adsorption of Ni(II) from aqueous solution using oxidized multiwall carbon nanotubes, *Industrial & Engineering Chemistry Research*, 45, 9144–9149. doi: 10.1021/ie060791z.

Chen S., Sun D., Chung J.S. 2007. Treatment of pesticide wastewater by moving-bed biofilm reactor combined with Fenton-coagulation pretreatment, *Journal of Hazardous Materials*, 144, pp. 577–584. doi: 10.1016/j.jhazmat.2006.10.075.

Chen Z., Wu Y., Wei Y. & Mimura H. 2016. Preparation of silica-based titanate adsorbents and application for strontium removal from radioactive contaminated wastewater, *Journal of Radioanalytical and Nuclear Chemistry*, 307, 931–940.

Cheng G., Lu J., Zhao X., Cai Z., and Fu J. 2015. Advanced treatment of pesticide-containingwastewater using fenton reagent enhanced by microwaveelectrodeless ultraviolet Hindawi publishing corporation, *BioMed Research International*, 2015, 205903, 8 pages, http://dx.doi.org/10.1155/2015/205903.

Cheng M., Zeng G., Huang D., Lai C., Xu P., Zhang C., Liu Y. 2016. Hydroxyl radicals based advanced oxidation processes (AOPs) for remediation of soils contaminated with organic compounds: A review, *Chemical Engineering Journal*, 284, 582–598, https://doi.org/10.1016/j.cej.2015.09.001.

Cındık Akıncı Y., Yüksek T., Demirel Ö. 2016. Remediation contaminated soil by heavy metal: Vetiver Grass (Vetiveria zizanioides (Linn.) Nash) and Earthworms, *Journal of Architecture Sciences and Applications*, 1(1), pp.1–11.

David B., Francesco F., Paolo P., Franco C. 2010. Poly-chlorinated dibenzo-p-dioxins, dibenzo-furans and dioxin-like poly-chlorinated biphenyls occurrence and removal in conventional and membrane activated sludge processes, *Bioresource Technology*, 101(24), 9445–9454.

DDT (pdf). 2014. *National Toxicology Program*.

De Xin D., Xin X., Le L., Nan H., Guang Yue L., Yong Dong W. & Ping Kun F. 2014. Removal and recovery of U(VI) from low concentration radioactive wastewater by ethylenediamine-modified biomass of Aspergillus niger, *Water, Air, & Soil Pollution*, 225, 2206.

Deniz C. 2010. *Ağır Metal ve Renk İçeren Atıksuların Gideriminin Adsorpsiyon/Biyosorpsiyon Yöntemleriyle Araştırılması*, Yüksek Lisans Tezi, Cumhuriyet Üniversitesi Fen Bilimleri Enstitüsü, Sivas.

Ding H., Luo X., Zhang X., Yang H. 2019. Alginate-immobilized Aspergillus niger: Characterization and biosorption removal of thorium ions from radioactive wastewater, *Colloids and Surfaces A: Physicochemical and Engineering Aspects*, 562, 186–195, 5 Pages.

Dong B., Chen H-Y., Yang Y., He Q., Dai X-H. 2015. Biodegradation of polychlorinated biphenyls using a moving-bed biofilm reactor, *CLEAN–Soil, Air, Water*, 43(7), 967–1114.

Dwivedi C, Kumar, Ajish Juby K., Kumar M., Kishen P. Wattal, N.B.P. 2012. Preparation and evaluation of alginate-assisted spherical resorcinol–formaldehyde A. resin beads for removal of cesium from alkaline waste, *Chemical Engineering Journal*, 200–202, 491–498.

Ebersbach I., Ludwig S. M., Constapel M., Kling H-W. 2016. An alternative treatment method forfluorosurfactant-containing wastewater by aerosol-mediated separation, *Water Research*, 101, 333–340.

Edyta K., Magdalena U., Marcin K., Adam J., Agnieszka B. 2017. The use of a hybrid sequential biofiltration system for the improvement of nutrient removal and PCB control in municipal wastewater, *Scientific Reports*, 7, 1–14. doi: 10.1038/s41598-017-05555-y.

Ekmekyapar F., Arslan A., Bayhan Y.K., Cakici A. 2006. Biosorption of copper(II) by non living lichen biomass of Cladonia rangiformis Hoffm, *Journal of Hazardous Materials*, 137, 293–298.

EPA, www.epa.go-PCBs, accessed August 2020.

Eriksson U., Haglund P., Kärrman A. 2017. Contribution of precursor compounds to the release of per- and polyfluoroalkyl substances (PFASs) from waste watertreatment plants (WWTPs), *Journal of Environmental Sciences*, 61, 80–90.

Fatone F., Di Fabio S., Bolzonella D., Cecchi F. 2011. Fate of aromatic hydrocarbons in Italian municipal wastewa-ter systems: An overview of wastewater treatment usingconventional activated-sludge processes (CASP) and membrane bioreactors (MBRs), *Water Research*, 45(1), 93–104.

Feng L., van Hullebusch E.D., Rodrigo M.A., Esposito G., Oturan M.A. 2013. Removal of residual anti-inflammatory and analgesic pharmaceuticals from aqueous systems by electrochemical advanced oxidation processes. A review, *Chemical Engineering Journal*, 228, 944–964.

Fernandez-Alvarez, M., Llompart, M., Lores, M., Garcia-Jares, C., Cela, R., Dagnac, T. 2010. The photochemical behaviour of five household pyrethroid insecticides and a synergist as studied by photo solid phase microextraction, *Analytical and Bioanalytical Chemistry*, 388(5–6), 1235–1247.

Gando-Ferreira L. M., Marques J.C. and Quina M.J. 2015. Integration of ion-exchange and nanofiltration processesfor recovering Cr(III) salts from synthetic tannery wastewater, *Environmental Technology*, 36, 2340–2348. doi:10.1080/09593330.2015.1027284.

Ganesh R., P. Sousbie, M. Torrijos, N. Bernet, and R.A. Ramanujam. 2015. Nitrification and denitrification characteristics in a sequencing batch reactor treatingtannery wastewater, *Clean Technologies and Environmental Policy*, 17, 735–745. doi:10.1007/s10098-014-0829-1

Garcia-Mancha N., Monsalvo V.M., Puyol D., Rodriguez J.J., Mohedano A.F. 2017. Enhanced anaerobic degradability of highly polluted pesticides-bearing wastewater under thermophilic conditions, *Journal of Hazardous Materials*, 339, 320–329.

Gerrity D., Mayer B., Ryu H., Crittenden J., Abbaszadegan M. 2009. A comparison of pilot-scale photocatalysis and enhanced coagulation for disinfection byproduct mitigation, *Water Research*, 43, 1597–1610.

Gómez-Ramos Del Mar M., Mezcua M., Agüera A., Fernández-Alba A.R., Gonzalo S., Rodríguez A., Rosal R. 2011. Chemical and toxicological evolution of the antibiotic sulfamethoxazole under ozone treatment in water solution, *Journal of Hazardous Materials*, 192(1), 18–25.

Gualtieri A.F., Zanatto I. 2009. Industrial process for the direct temperature induced recrystallization of asbestos and/or mineral fibres containing waste products using a tunnel kiln and recycling.

Gualtieri A.F., Veratti L., Tucci A., Esposito L. 2012. Recycling of the product of thermal inertisation of cement-asbestos in geopolymers, *Construction and Building Materials*, 31, 47–51.

Guan X.H., Ma J., Dong H.R., Jiang L. 2009. Removal of As from water: Effects of calcium ions on As(III) removal in the KMnO4–Fe(II) process, *Water Research*, 43, 5119–5128.

Gündağ Ö. 2017. *Comparison Of Humic Acid Removal Alternatives in Water*, MSc. Thesis.

Gurnham C.F. 1965. *Industrial Waste Water Control*, Academic Press, Cambridg, MA, 168–199, 221–285, 339–357.

Gutiérrez M. E. J, de las Fuentes L. 2002. Evaluation of wastewater toxicity: Comparative study between Microtox and activated sludge oxygen uptake inhibition. *Water Research*, 36, 919–924.

He Q., Yao K., Sun D., Shi B. 2007. Biodegradability of tannin-containing wastewater from leather industry, *Biodegradation*, 18, 465–472. doi: 10.1007/s10532-006-9079-1.

Hişmioğulları ŞE., Hişmioğulları AA., Kontaş AT. 2012. Dioksin ve dioksin benzeri kimyasalların toksik etkileri, *Balıkesir Sağlık Bilimleri Dergisi*, 1(1), 23–27.

Hirose J., Kondo F., Nakano T., Kobayashi T., Hiro N., Ando Y., Takenaka H., Sano K. 2005. Inactivation of antineoplastics in clinical waste water by electrolysis. *Chemosphere*, 60, 1018–1024.

Hongchao M., Dongxue H., Hongcheng W., Yuanyi Z., Yubo C., Kongyan L., Lufeng Z., Wenyu L., Pan W., Hui G., Aijie W., Ying Z. 2020. Electrochemical-assisted hydrolysis/acidification-based processes as a costeffective and efficient system for pesticide wastewater treatment, *Chemical Engineering Journal*, 397, 125–417.

Hua G, Reckhow DA. 2007. Characterization of disinfection byproduct precursors based on hydrophobicity and molecular size. *Environmental Science and Technology*, 41, 3309–3315.

Iakovides I.C., Michael-Kordatou I., Moreira N.F.F., Ribeirod A.R., Fernandes T., Pereira M.F.R., Nunes O.C., Manaia C.M., Silva A.M.T., Fatta-Kassinosa D. 2019. Continuous ozonation of urban wastewater: Removal of antibiotics, antibiotic-resistant Escherichia coli and antibiotic, *Water Research*, 159, 333–347. doi: https://doi.org/10.1016/j.watres.2019.05.025

Imamoglu M., Tekir O. 2008. Removal of copper(II) and lead(II) ions from aqueous solutions by adsorption on activated carbon from a new precursor hazelnut husks, *Desalination*, 228, 108–113.

Ji X., Lin W., Zhang W., Yin F., Zhao X., Wang C., Liu J., Yang H., Liu S. 2015. A comprehensive review of the process on HCB degradation, http://www.matec-conferences.orgorhttp://dx.doi.org/10.1051/matec-conf/20152204007.

Kang K.-H., Shin H. S., and Park H. 2002. Characterization of humic substances present in landll leachates with different landll ages and its implications, *Water Research*, 36(16), 4023–4032.

Karamanis D., Assimakopoulos P.A. 2007. Efficiency of aluminum-pillared montmorillonite on the removal of cesium and copper from aqueous solutions, *Water Research*, 41, 1897–1906.

Kasprzyk-Hordern B., Dinsdale R.M., Guwy A.J. 2008. The occurrence of pharmaceuticals, personal care products, endocrine disruptors and illicit drugs in surface water in South Wales, UK, *Water Research*, 42(13), 3498–3518.

Katsoyiannis A., Samara C. 2004. Persistent organic pollutants(POPS) in the sewage treatment plant of Thessaloniki, north-ern Greece: Occurrence and removal, *Water Research*, 38(11), 2685–2698.

Klančar A., Trontelj J., Kristl A., Meglič A., Rozina A., Justin M.Z., Roškar R. 2016. An advanced oxidation process for wastewater treatment to reduce the ecological burden from pharmacotherapy and the agricultural use of pesticides, *Ecological Engineering*, 97, 186–195.

Kuokkanen V., Kuokkanen T., , Ramo J., Lassi U. 2015. Electrocoagulation treatment of peat bog drainagewater containing humic substances, *Water Research*, 79, 79e87.

Kurt A., Mert B. K., Özengin N., Sivrioğlu Ö. and Yonar T. 2017. *Treatment of Antibiotics in Wastewater Using Advanced Oxidation Processes (AOPs)*. doi: 10.5772/67538.

Kusiorowski R., Zaremba T., Gerle A., Piotrowski J., Simka W., Adamek J. 2015a. Study on the thermal decomposition of cro-cidolite asbestos, *Journal of Thermal Analysis and Calorimetry*, 120, 1585–1595.

Kusiorowski R., Zaremba T., Piotrowski J., Podworny J. 2015b. Utilisation of cement-asbestos wastes by thermal treatment and the potential possibility use of obtained product for the clinker bricks manufacture, *Journal of Materials Science*, 50, 6757–6767.

Le Thai-Hoang, Ng Charmaine, HanTran Ngoc, Chen Hongjie, Gin Karina Yew-Hoong 2018. Removal of antibiotic residues, antibiotic resistant bacteria and antibiotic resistance genes in municipal wastewater by membrane bioreactor systems, *Water Research*, 145, 498–508.

Leonelli C., Veronesi P., Boccaccini D.N., Rivasi M.R., Barbieri L., Andreola F., Lancellotti I., Rabitti D., Pellacani G.C. 2006. Micro-wave thermal inertisation of asbestos containing waste and its recycling in traditional ceramics, *Journal of Hazardous Materials B*, 135, 149–155.

Li F., Zhang X., Zhao X. 2017. Effect of ion exchange membrane on the removal efficiency of continuous electrodeionization (CEDI) during low level radioactive wastewater treatment, *Nuclear Engineering and Design*, 322(October), 159–164.

Li G., Yaguang D., Qiushi Y., Dunshun L., Longwen C., Dongyun D. 2015. Efficient removal of arsenic from "dirty acid" wastewater by using a novel immersed multi-start distributor for sulphide feeding, *Separation and Purification Technology*, 142, 209–214.

Li L., Quinlivan P.A, Knappe D.R.U. 2002. Effects of activated carbon surface chemistry and pore structure on the adsorption of organic contaminants from aqueous solution, *Carbon*, 40(12), 2085–2100.

Li R., Yang C. , Chen H. , Zeng G. , Yu G., Guo J. 2009. Removal of triazophos pesticide from wastewater with Fenton reagent, *Journal of Hazardous Materials*, 167, 1028–1032.

Lina X. H., Sriramulua D., Lia S. F. Y. 2015. Selective removal of photocatalytic non-degradablefluorosurfactants from reverse osmosisconcentrate, *Water Research*, 68, 831–838.

Makrlik E., Vanura P.2005. Solvent extraction of lead using a nitrobenzene solution of strontium dicarbollylcobaltate in the presence of polyethylene glycol PEG 400, *Journal of Radioanalytical and Nuclear Chemistry*, 267(1), 233–235.

Matlock M.M., Howerton B.S., Atwood D.A. 2002. Chemical precipitation of lead from lead battery recycling plant wastewater, *Industrial & Engineering Chemistry Research*, 41(6), 1579–1582.

Mcintyre T. 2003. Phytoremediation of heavy metals from soils, *Advances in Biochemical Engineering/ Biotechnology*, 78, 97–123.

Merdan F.G. 2019. *Evaluation of the Performance of Microbial Community Induced by NFC -Doped TiO₂ Nanoparticles on Aerobıc Biodegradation of Sewage Wastewater and Its' Recoverability*, MSc. Thesis, Tekirdağ Namık Kemal University, Çorlu Eng.Fac., Env.Eng.Dept., Tekirdağ/Turkey.

Mezzanotte V., Anzano M., Collina E., Marazzi F.A., Lasagni M. 2016. Distribution and removal of polycyclic aromatic hydrocarbons in two Italian municipal wastewater treatment plants in 2011–2013, *Polycyclic Aromatic Compounds*, 36, 213–228.

Min S.Y., Maken S., Park J.W., Gaur A., Hyun J.S. 2008. Melting treatment of waste asbestos using mixture of hydrogen and oxygen produced from water electrolysis, *Korean Journal of Chemical Engineering*, 25, 323–328.

Min X., Luo, X., Deng F., Shao P., Wu X., Dionysios D. 2018. Combination of multi-oxidation process and electrolysis for pretreatment of PCB industry wastewater and recovery of highly-purified copper, *Chemical Engineering Journal*, 354, 228–236.

Montuori P., Cirillo T., Fasano, E., Nardone, A., Esposito, F., Triassi, M. 2014. Spatial distribution and partitioning of polychlorinated biphenyl and organochlorine pesticide in water and sediment from Sarno River and Estuary, Southern Italy, *Environmental Science and Pollution Research*, 21(7), 5023–5035.

Munz G., Gori R., Mori G., Lubello C. 2007. Powdered activated carbon and membrane bioreactors (MBR-PAC) for tannery wastewater treatment: Long term effecton biological and filtration process performances, *Desalination*, 207, 349–360.

Munz G., M. Gualtiero L. Salvadori B. Claudia and L.Claudio. 2008. Process efficiency and microbial monitoring in MBR (membrane bioreactor) and CASP (conventional activated sludge process) treatment of tannery wastewater, *Bioresource Technology*, 99, 8559–64. doi:10.1016/j.biortech.2008.04.006.

Munza G., De Angelis D., Goria R., Moric G., Casarci M., Lubello C. 2009. The role of tannins in conventional and membrane treatment of tannery wastewater, *Journal of Hazardous Materials*, 164, pp. 733–739.

Murugananthan M., Raju B.G. and Prabhakar S. 2004. Separation of pollutants from tannery effluents by electroflotation, *Separation and Purification Technology*, 40, 69–75. doi:10.1016/j.seppur.2004.01.005.

National Center for Biotechnology Information (NCBI). 2018. https://www.ncbi.nlm.nih.gov/pmc/articles/ PMC6323993/ (accessed August 2020).

National Institute for Occupational Safety and Health (NIOSH). 2020. *"Aldrin" Immediately Dangerous to Life and Health Concentrations (IDLH)*. https://www.cdc.gov/niosh/idlh/309002.html (accessed August, 2020).

Nilchi A., Saberi R., Rasouli Garmarodi S., Bagheri A. 2012. Evaluation of PAN-based manganese dioxide composite for the sorptive removal of cesium-137 from aqueous solutions, *Applied Radiation and Isotopes*, 70, 369–374.

Nolleta H., Roelsa M., Lutgen P., der Meeren P. V., Verstraete W. 2003. Removal of PCBs from wastewater using fly ash, *Chemosphere*, 53(6), 655–665.

Nwankwegu, A.S., Onwosi C.O. 2018. Bioremediation of gasoline contaminated agricultural soil by bioaugmentation, *Environmental Technology and Innovation*, 7, 1–11. https://doi.org/10.1016/j.eti.2016.11.003. 2062 Environ Sci Pollut Res., 25, pp.2051–2064.

O'Driscoll K., Robinson J., Chiang W-S., Chen Y-Y., Kao R-C., Doherty R. 2016. The environmental fate of polybrominated diphenyl ethers (PBDEs) in western Taiwan and coastal waters: Evaluation with a fugacity-based model, *Environmental Science and Pollution Research*, 23, 13222–13234. Doi: 10.1007/s11356-016-6428-4.

Ofomaja A.E., Naidoo E.B., Modise S.J. 2010. Biosorption of copper(II) and lead(II) onto potassium hydroxide treated pine cone powder, *Journal of Environmental Management*, 91(8), 1674–1685.

Okamura Y., Fujiwara K., Ishihara R., Sugo T., Kojima T., Umeno D., Saito K. 2014. Cesium removal in freshwater using potassium cobalt hexacyanoferrate-impregnated fibers, *Radiation Physics and Chemistry*, 94, 119–122.

Özdoğan M. 2020. https://www.drozdogan.com/dioksinler-farkindaligin-cok-az-oldugu-ciddi-bir-saglik-tehditi/

Pengxiao L., Hanmin Z., Yujie F., Fenglin Y., Jianpeng Z. 2014. Removal of trace antibiotics from wastewater: A systematic study ofnanofiltration combined with ozone-based advanced oxidationprocesses, *Chemical Engineering Journal*, 240, 211–220.

Plosz B.G., Langford K.H., Thomas K.V. 2012. An activated sludge modeling framework for xenobiotic trace chemicals (ASM-X): Assessment of diclofenac and carbamazepine, *Biotechnology and Bioengineering*, 109, 2757–2769.

Qiao M., Fu L., Cao W., Bai Y., Huang Q., Zhao X. 2009. Occurrence and removal of polycyclic aromatic hydrocarbons and their derivatives in an ecological wastewater treatment plant in South China and effluent impact to the receiving river, *Environmental Science and Pollution Research*, 26, 5638–5644. https://doi.org/10.1007/s11356-018-3839-4.

Rana R. S., Singh P., Kandari V., Singh R., Dobhal R., Gupta S. 2014. A review on characterization and bioremediation of pharmaceutical industries' wastewater: An Indian perspective, *Applied Water Science*, 7, 1–12.

Rao C.N.R., Sood A.K., Subrahmanyam K.S., Govindaraj A. 2009. Graphene: The new two-dimensional nanomaterial. *Angewandte Chemie International Edition*, 48(42), 7752–7777.

Rasim H., Adnan Berk D., Demet D.C., Sümer A. 2012. Biosorption, adsorption, phytoremediation methods and applications, *Turk Hij Den Biyol Derg*, 69(4), 235–53.

Rebello S., Asok Aju K., Mundayoor S., Jisha M. S. 2014. Surfactants: Toxicity, remediation and green surfactants, *Environmental Chemistry Letters*, 12(2), 275–287. doi:10.1007/s10311-014-0466-2.

Rivas B.L., Urbano B.F. and Sánchez J. 2018. Water-soluble and insoluble polymers, nanoparticles, nanocomposites and hybrids with ability to remove hazardous inorganic pollutants in water, *Frontiers in Chemistry*, 31 July 2018 | https://doi.org/10.3389/fchem.2018.00320.

Rodriguez C., Cook A., Devine B., Buynder P.V., Lugg R., Linge K. and Weinstein P. 2008. Dioxins, Furans and PCBs in recycled water for indirect potable reuse.

Rodríguez F.J, García-Valverde M. 2016. Influence of preozonation on the adsorptivity of humic substances onto activated carbon, *Environmental Science and Pollution Research*, 23, 21980–21988. Doi: 10.1007/s11356-016-7414-6.

Rowland G. A., Bausch A. R., Grannas A. M. 2011. Photochemical processing of aldrin and dieldrin in frozen aqueous solutions under arctic field conditions, *Environmental Pollution*, 159(5), 1076–1084.

Saikat S., Kreis I., Davies B., Bridgman S., Kamanyire R. 2013. The impact of PFOS on health in the general population: A review. *Environmental Science: Process and Impacts*, 15(2), 329–335. doi:10.1039/c2em30698k.

Sanches S., Leitao C., Penetra A., Cardoso V.V., Ferreira E., Benoliel M.J., Crespo M.T.B., Pereira V.J. 2011. Direct photolysis of polycyclic aromatic hydrocarbons in drinking water sources, *Journal of Hazardous Materials*, 192, 1458–1465.

SC. 2010. *A Guide To The Stockholm Convention On Persistent Organic Pollutants*.

Schaefer C.E., Andaya C., Maizel A., Higgins C.P. 2019. Assessing continued electrochemical treatment of groundwater impacted by aqueous film-forming foams, *Journal of Environmental Engineering*, 145(12).

Shareef A., Kookana R., Kumar A. and Tjandraatmadja G. 2011. Sources of emerging organic contaminants in domestic wastewater, *Water for a Healthy Country Flagship Report*. ISSN: 1835-095X.

Shoiful A., Ueda Y., Nugroho R., Honda K. 2016. Degradation of organochlorine pesticides (OCPs) in water by iron(Fe)-based material, *Journal of Water Process Engineering*, 11, 110–117.

Siddiqi M. A., Laessig R. H. and Kurt D. R. 2003. Polybrominated diphenyl ethers (PBDEs): New pollutants-old diseases, *Clinical Medicine and Research*, 1(4), 281–290. doi: 10.3121/cmr.1.4.281

Sruthi S. N., Shyleshchandran M. S., Mathew S. P., Ramasamy E. V. 2017. Contamination from organochlo-rine pesticides (OCPs), in agricultural soils of Kuttanad agroecosystem in India, and related potential health risk, *Environmental Science and Pollution Research*, 24, 969–978.

Tang-Peronard, J. L. 2011. Endocrine-disrupting chemicals and obesity development in humans: A review. *Obesity Reviews*, 12(8), 622–636. doi: 10.1111/j.1467-789x.2011.00871.x. PMID 21457182.

Turci F., Tomatis M., Mantegna S., Cravotto G., and Fubini B. 2008. A new approach to the decontami-nation of asbestos-pollutedwaters by treatment with oxalic acid underpower ultrasound, *Ultrasonics Sonochemistry*, 15, 420–427.

Turkay O, Inan H, Dimoglo A. 2015. Experimental study of humic acid degradation and theoretical modelling of catalytic ozonation, *Environmental Science and Pollution Research*, 22(1), 202–210.

Uluözlü O.D., Sarı A., Tuzen M., Soylak M. 2008. Biosorption of Pb(II) and Cr(III) from aqueous solution by lichen (Parmelina tiliaceae) biomass, *Bioresource Technology*, 99, 2972–80.

Valerio P., Laura T., Marco S., Daniele B., Flavia L., Marco T., Francesco P. 2019. Asbestos treatment technologies, *Journal of Material Cycles and Waste Management*, 21, 205–226 https://doi.org/10.1007/s10163-018-0793-7.

Vanlı Ö. 2007. *Pb, Cd ve B elementlerinin topraklardan şelat destekli fitoremediasyon yöntemiyle giderilmesi*, Yüksek Lisans Tezi, İstanbul Teknik Üniversitesi Fen Bilimleri Enstitüsü.

Vargas H.O. 2014. *Study on the Fate of Pharmaceuticals in Aqueous Media: Synthesis, Characterization and Detection of Biotic and Abiotic Transformation Products using Electrochemical Advanced Oxidation Processes and Bioconversions*, Université Paris-Est.

Vhahangwele M., Mugera W. G.. 2016. Removal of arsenic from wastewaters by cryptocrystalline magnesite: Complimenting experimental results with modelling, *Journal of Cleaner Production*, 113, 318–324.

Vogelsang C., Grung M., Jantsch T. G., Tollefsen K. E., Liltved H. 2006. Occurrence and removal of selected organicmicropollutants at mechanical, chemical and advancedwastewater treatment plants in Norway, *Water Research*, 40, 3559–3570.

Wang H., Wang Y., Li X., Sun Y., Wud H., Chen D. 2016. Removal of humic substances from reverse osmosis (RO) and nanofiltration (NF) concentrated leachate using continuously ozone generation-reaction treat-ment equipment, *Waste Management*, 56, 271–279.

Wang H., Wang Y., Lou Z., Zhu N., Yuan H. 2017. The degradation processes of refractory substances in nano-filtration concentrated leachate using micro-ozonation, *Waste Management*, 69(November), 274–280.

Wang J.W., Bejan D., Bunce N.J. 2003. Removal of As from synthetic acid mine drainage by electrochemical pH adjustment and coprecipitation with iron hydroxide, *Environmental Science and Technology*, 37, 4500–4505.

Wang S, He J. 2013. Dechlorination of commercial PCBs and other multiple halogenated compounds by a sediment-free culture containing Dehalococcoides and Dehalobacter, *Environmental Science and Technology*, 47, 10526–10534. doi:10.1021/es4017624.

Yanagisawa K., Kozawa T., Onda A., Kanazawa M., Shinohara J., Takanami T., Shiraishi M. 2009. A novel decomposition technique of friable asbestos by CHClF 2-decomposed acidic gas, *Journal of Hazardous Materials*, 163, 593–599.

Yıldız Töre, G., Insel, G.,,Ubay Çokgör, E., Ferlier, E., Kabdaşlı, I., Orhon, D. 2011. Pollution profile and bio-degradation characteristics of fur-suede processing effluents, *Environmental Technology*, 32, 1151–1162.

Zapata A, Velegraki T., Sánchez-Pérez J.A., Mantzavinos D., Maldonado M.I., Malato S. 2009. Solar photo-Fenton treatment of pesticides in water: Effect of iron concentration on degradation and assessment of ecotoxicity and biodegradability, *Applied Catalysis B: Environmental*, 88, 448–454. doi: 10.1016/j.apcatb.2008.10.024.

Zapata A., Oller I., Sirtori C., Rodríguez A., Sánchez-Pérez J.A., López A., Mezcua M., Malato S. 2010. Decontamination of industrial wastewater containing pesticides by combining large-scale homoge-neous solar photocatalysis and biological treatment, *Chemical Engineering Journal*, 160, 447–456. doi: 10.1016/j.cej.2010.03.042.

Zhang Y., Pagilla K. 2010. Treatment of malathion pesticide wastewater with nanofiltration and photo-Fenton oxidation, *Desalination*, 263, 36–44.

Zhang Z., Chen J.-J., Lyu X.-J., Yin H., Sheng G.-P. 2014. Complete mineralization of perfluorooctanoic acid (PFOA) by γ-irradiation in aqueous solution, *Scientific Reports*, 4, 7418.

Zhao W., Sui Q., Huang X. 2018. Removal and fate of polycyclic aromatic hydrocarbons in a hybrid anaero-bic–anoxic–oxic process for highly toxic coke wastewater treatment, *Science of the Total Environment*, 635, 716–724.

Zhu Y., Hu J., Wang J. 2014. Removal of Co2+ from radioactive wastewater by polyvinyl alcohol (PVA)/chito-san magnetic composite, *Progress in Nuclear Energy*, 71, 172–178.

Zularisam A.W. , Ismail A.F. , Salim M.R. , Sakinah M., Ozaki H. 2007. The effects of natural organic mat-ter (NOM) fractions on fouling characteristics and flux recovery of ultrafiltration membranes, Elsevier, *Desalination*, 212, 191–208.

10 Microbial Removal of Toxic Chromium for Wastewater Treatment

Joorie Bhattacharya, Rahul Nitnavare,
Thomas J Webster, and Sougata Ghosh

CONTENTS

10.1 INTRODUCTION

Chromium (Cr) is present in several forms in the environment, namely chromium (0), chromium (III), and chromium (VI). Out of these, chromium (III) is an essential micronutrient while the other two are found in industrial discharge and waste. Chromium is used in several industries such as steel, leather, textile, and mining. However, the oxidized form, chromium (VI), is known to be highly toxic as well as carcinogenic. While Cr(III) has been found to be insoluble at neutral pH and is an integral part of maintaining human health and metabolism, Cr(VI) is known to cause severe

DOI: 10.1201/9781003204442-10

185

damage to the environment as well as to humans. Cr(VI) is mobile in the soil and aquatic ecosystem and is able to penetrate the cell wall. It causes skin irritations at below threshold levels and kidney damage to death at higher dosages. Cr(III) is hazardous only at extremely high concentrations (Owlad et al. 2009). The maximum permissible concentration of chromium in water is about 0.05 mg/L in accordance with the U.S. Environmental Protection Act (U.S. EPA 1998).

Due to this, it is crucial to either remove the toxic Cr(VI) from the environment or reduce it to its non-lethal form, Cr(III). Conventional methods of chromium removal include adsorption using synthetic and natural materials, ion exchange, filtration, electrolysis, reverse osmosis, foam flotation, surface adsorption, etc. However, such physical and chemical removal techniques are not able to remove chromium completely and thus they pose an additional threat due to the production of toxic by-products. Also, such methods are often rendered comparatively futile due to the high concentrations of heavy metals in water bodies at a range of 1–100 mg/L (Liu et al. 2007). Bioremediation has therefore gained mass popularity owing to its eco-friendly, cost effective, as well as efficient nature (Kulkarni et al. 2018; Luikham et al. 2018; Joutey et al. 2015). Numerous microorganisms like bacteria and yeast strains have been identified as having the capacity to reduce Cr(VI) to Cr(III), popularly known as biotransformation, which mostly involves cellular reductases (Fernández et al. 2018). Another bioremediation method includes biosorption which essentially uses microbes as a biosorbent. This, along with a conventional method as an accessory, would lead to an efficient system. Another mechanism referred to as bioaccumulation is the ability of microbes to incorporate active metals from aqueous solutions into cellular compartments. This has also been observed and reported in various bacteria, fungi, and algae (Ghosh et al. 2020a–e).

Considering global industrial growth and the increasing anthropogenic activities, contamination levels have also seen a rise in the environment. This chapter focuses on the damage caused by hexavalent chromium, the need for bioremediation techniques for their efficient reduction, the cellular mechanism of microorganisms in chromium removal, and the scope of newer technologies, such as genetic modifications in the removal/recovery of chromium.

10.2 SOURCES OF CHROMIUM

Chromium is found naturally in ultramafic rocks. Its ore, chromite ($FeOCr_2O_3$), is found in igneous rocks and to a certain extent in sedimentary and metamorphic rocks. Its other lesser-known ore is crocoite. Dhal et al. (2013) reported the total world reserve is estimated at $>480 \times 10^6$ metric tons of shipping-grade chromite ore with $\sim45\%$ Cr_2O_3. Major resources are located ($\times10^6$ tons) in Kazakhstan (220), Southern Africa (200), India (54), and the United States (0.62), as seen in Table 10.1. Chromium rarely exists as a free metal in the environment. It is also known to be found in oceanic crusts as ore deposits. Its elemental form, Cr(0), is also rarely found in nature. In the oxidized states, chromium can exist in any oxidation form that may range from −2 to +4. The most commonly known forms are Cr(III) and Cr(VI). In groundwater, chromium is found in the above-mentioned oxidized states. However, Cr(VI) is comparatively more soluble than Cr(III). Cr(VI) is also a known toxicant and therefore it is a primary contaminant of groundwater (Zhitkovich 2011). Cr(III) is vital to the human body and actively participates in metabolic processes. In food, Cr(III) is present in cheese, calf liver, and wheat germ. Apart from this, chromium is present in the atmosphere and is usually observed in significant amounts in volcanic residues and windblown sand. The anthropogenic sources of chromium occur mainly due to leaching into soil and water from industrial waste and sludge, agricultural by-products, mining activities, coal, and fly-ash disposals. From an agricultural perspective, chromium is present in phosphate fertilizers in relatively high concentrations which eventually penetrate the soil ecosystem. The chromium-containing effluents are released from industries such as metal plating, dyes, pigments, textiles, tanning, ink manufacturers, leather etc., most of which are dumped into the water as waste sludge. The chromium occurring in such sources is both in the trivalent and hexavalent forms. While 100% of Cr(III) can be removed

TABLE 10.1

World Reserves (Shipping Grade) and Production (Ore and Concentrate) of Chromium (by Principal Countries). Reprinted with permission from Dhal, B.; Thatoi, H.N.; Das, N.N.; and Pandey, B.D., 2013. Chemical and microbial remediation of hexavalent chromium from contaminated coil and mining/metallurgical solid waste: A review. J. Hazard. Mater. 250–251, 272–291. Copyright © 2013 Elsevier B.V.

Country	Reserves	Production (in tons)				
		2007	2008	2009	2010	2011
World: Total (rounded)	>480,000	23,900	23,600	18,700	23,700	24,000
India[a]	54,000	4,873	3,980	3,372	3,800	3,800
Kazakhstan	220,000	3,687	3,552	3,333	3,830	3,900
South Africa	200,000	9,647	9,683	6,865	10,900	11,000
USA	620	–	–	–	–	–
Finland	–	556	614	247		
Brazil	–	628	700(e)	700(e)		
Russia	–	777	913	416		
Turkey	–	1,679	1,886	1,770		
Zimbabwe	–	614	442	194		
Other countries	N/A	1,439	1,829	1,803	5,170	5,300

[a] Production of chromite in India in 2007–2008, 2008–2009, and 2009–2010 was 4.9 million tons, 4.1 million tons, and 3.4 million tons, respectively.

from such sources, only about 26–48% of Cr(VI) is removed from wastewater. This makes Cr(VI) a highly precarious contaminant. Apart from this, mining valleys and smelter wastes are primary contributors of chromium as CrO_4^{2-}, which is highly phytotoxic. Industrial releases from cement and brick factories are additional sources of chromium pollution into the atmosphere. Chromium electroplating industries consist of a procedure wherein bubbles of gases are emitted into the air while plating using chromic acid. This forms an important aspect of chromium contamination in several countries holding such industries (Saha et al. 2011).

10.3 CHROMIUM TOXICITY

In spite of being an essential trace element, even the slightest increase in Cr(VI) level is enough to render it toxic. The mutagenicity and carcinogenicity of chromium to biological systems are greatly attributed to the high oxidizing potential of Cr(VI). Moreover, Cr(VI) is not able to directly interact with the cellular DNA. Thus, its toxicity is believed to be due to the intracellular reduction of Cr(VI) to Cr(III). Additionally, the intracellular reduction of Cr(VI) forms intermediate chromium species, essentially Cr(V) and Cr(IV). This also leads to an increase in the concentration of reactive oxygen species (ROS) which causes oxidative DNA damage (Dhal et al. 2013).

The chromate ion (CrO_4^{2-}) is structurally analogous to sulfate ions. Due to this, it is able to easily utilize the sulfate transport system to cross the cellular membrane and enter into the cell. Under normal conditions, Cr(VI) reacts with intracellular reductants such as ascorbate and glutathione to generate Cr(V) and Cr(IV). After entering the cell, the intermediate species of chromium interact and bind to the cellular materials and cause physiological damage (Ksheminska et al. 2005). On reduction, the Cr(III) interacts with negatively-charged DNA containing phosphate groups and often interferes with nuclear mechanisms such as replication and transcription. Cr(III) also binds with carboxyl and thiol groups of native cellular enzymes and may cause severe alteration in their structural and functional chemistry (Joutey et al. 2015).

Apart from severe cellular disruptions, chromium toxicity is known to cause stark health hazards affecting various systems, resulting in respiratory or nasal irritations, asthma, lung cancer, skin dermatitis and allergic reactions, digestive or intestinal damage, and stomach cancer. It may also cause deleterious effects in a developing fetus and can get accumulated in the placenta (Cheung and Gu 2007).

10.4 CONVENTIONAL METHODS FOR CHROMIUM REMOVAL

Most of the conventional methods used for chromium removal greatly rely on the principle of recovery and reuse. Since the extent of chromium toxicity lies in its oxidation state, it is ideal to formulate treatment strategies accordingly. The following mentioned methods deal with the removal of Cr(VI).

10.4.1 PRECIPITATION / COAGULATION

Precipitation reactions or chemical precipitation is an efficient *ex-situ* treatment used for the purification of contaminated aqueous solutions. Metal coagulants have been used for some time and involve alum and ferric (III) hydroxides for the removal of Cr(III) from contaminated water via precipitation, with a removal efficiency of 85–89%. However, these chemical coagulants were found to be ineffective in the removal of Cr(VI). This may be due to the high solubility of chromate and dichromate (Martín-Domínguez et al. 2018). In addition to this study, Fe(II) sulfate was able to remove Cr(VI) contamination completely. This may be due to the fact that Fe(II) gets oxidized to Fe(III) in the process, in turn reducing Cr(VI) to its non-toxic Cr(III) form. The removal of chromium via precipitation or coagulation is a highly pH-dependent process. The reduction of Cr(VI) to the trivalent form is done in acidic conditions and, consequently, precipitation is obtained by adding a basic agent to it. However, the precipitated metals may form smaller particles that do not settle easily and therefore need an additional filtration unit. Precipitation cannot be achieved if the metals are in complex form or in their anionic form. Also, a copious amount of sludge is produced in this process along with other dendrite-like deposits which would eventually require a separate filtration unit (Sharma et al. 2008).

Electrochemical precipitation (ECP) is another technique that has been taken up to remove Cr(VI) contamination using a cathode and an anode. Electric potential, pH, and the concentration of contamination in the solution are the defining factors. ECP was found to be highly cost effective as well as selective. Additionally, it was seen to produce lesser amounts of dissolved salts and consequently 50% less sludge as compared to chemical precipitation (Martín-Domínguez et al. 2018). However, this method was also seen to produce high amounts of metallic hydroxides which make the recovery of metals from aqueous solutions challenging.

10.4.2 ADSORPTION

Adsorption includes the attachment of molecules onto the surface of another component which is the adsorbent. The adsorption capacity depends on the concentration and solubility of the compound. Chromium contaminants can be removed from water bodies using this technique. A wide range of adsorbents have been identified to date, which are able to efficiently remove chromium to a certain extent. Adsorption is advantageous over other conventional methods of chromium removal due to its simplicity, cost effectiveness, ease of operation, as well as environment-friendly nature. Adsorbents utilized can be synthetic or occur naturally in nature (Owlad et al. 2009). The most common of them is the usage of activated carbon due to its high porosity and availability of ample surface area. Several other adsorbents have been identified, namely ion-exchange resins, chitosan-based materials, and starch-based materials.

Activated carbon is able to remove both trivalent and hexavalent chromium. However, their high cost, loss of carbon during reactivation, and non-selectivity are major drawbacks. Ion-exchange resins, on the other hand, overcome the previous perils of selectivity and adsorbent loss. Cationic resins remove Cr(III) while anionic resins remove Cr(VI). But, these resins are not environmentally friendly as they are derived from petroleum-based raw materials. Also, they are relatively expensive. Chitosan-based adsorbents are versatile, environmentally-friendly, low-cost, non-porous, pH-dependent polymers with a high regeneration capacity. Among other biopolymers, starch-based polymers, with a high-swelling capacity in water, are selective at a wide range of concentrations.

Biosorption is a comparatively newer and inexpensive process used to remove chromium from water using agricultural by-products and wastes. Various biosorbents used to date include wool, sawdust, pine needles, cactus, and almond and walnut shells. Such components provide a high content of lignin, hemicellulose, and cellulose which are responsible for metal chelation and adsorption. Some of the notable limitations of biosorbents are low surface area and high chemical and biological oxygen demand along with high organic carbon leaching. Thus, there is a need to develop improved techniques for chromium removal/recovery (Pakade et al. 2019). Industrial waste products and waste slurry are also used as adsorbents. Iron(III) hydroxide is the most popularly used and extensively studied adsorbent used for the removal of chromate from aqueous systems. It has been reported that Fe(III) is able to absorb 0.47 mg of Cr(VI) effectively at all pH levels. Other such industrial wastes which have been studied to understand their adsorption capacity are blast furnace slag (by-product of steel plants), bagasse fly ash (by-product of the sugar industry), and red mud (by-product of the aluminum industry) (Owlad et al. 2009).

10.4.3 Ion Exchange

Ion exchange is another frequently used method for water treatment that consists of a reversible procedure of solid and liquid phases wherein an insoluble component, such as resin, removes the chromium contamination without altering its own chemistry. The resins can then be extracted and eluted to obtain the concentrated metals with appropriate reagents. Other natural ion exchangers have been identified, such as clinoptilolite and zeolite, which are able to purify wastewater. Synthetic resins have now been developed, such as Dowex 2-X4, which are known to completely remove Cr(III) and Cr(VI). Natural ion exchangers have an upper hand over synthetic ones owing to their cost effectiveness. Ion exchange treatments do not produce any kind of additional sludge as compared to the previous methods. They are also feasible to use due to their simplicity in principle. However, ion exchange treatments require preliminary removal and filtration of solid and other suspended particles in the effluents to be treated. Also, they cannot be used for the removal of a wider range of heavy metals, making them limited to certain metals only (Kurniawan et al. 2006).

10.4.4 Membrane Filtration

Membrane filtration or reverse osmosis has been known to be effective in the removal of Cr(VI). In this technique, a pressure-driven flow is used wherein the solvent passes while the solute/metal remains (Kurniawan et al. 2006). A wide range of membrane filtration techniques have been employed, such as liquid membrane, polymeric membrane, and inorganic membrane for the same. Due to the thermal stability, inorganic membranes have the appropriate pore dimensions for the purpose of membrane filtration. Ceramic membranes containing an ion exchange component, such as zirconium dioxide, have been effectively used for Cr(VI) removal. Polymeric membranes, such as those comprised of polyethyleneimine, pectin, and chitosan, have been known to remove Cr(VI) from aqueous solutions using a polymer-enhanced ultrafiltration process. Filtration through a polymeric membrane is highly pH and ionic strength dependent.

Liquid membrane filtration is made up of two types: emulsion based and immobilized liquid membrane based. Here, both purification and extraction are combined into a single step, resulting in a reduced number of steps involved and making it time and cost effective. Liquid membranes have been used previously for the purification of chromium from tannery wastes using various derivatives as a carrier through the aqueous solutions. Depending on the characteristics of the membrane, this system is able to remove a wide range of chromium under different conditions. Membrane filtrations do not produce any secondary sludge or waste and are therefore advantageous over the other methods mentioned above. They are also highly selective and require low energy. However, they have the major issue of membrane fouling, resulting in the need for the periodic changing of membranes which makes it cost-ineffective. Liquid membranes also have low chemical and thermal stability (Owlad et al. 2009).

The above-mentioned physical and chemical methods for the removal of toxic chromium from wastewater are able to reduce the negative effect of chromium. However, they pose other serious issues in terms of cost, feasibility, durability, as well as from an environmental perspective. It is therefore necessary to adopt solutions and techniques which would be able to overcome these limitations using existing resources. Microbial removal of Cr(VI) is thus a promising avenue as this could be a sustainable alternative strategy, which is discussed in the following section.

10.5 MICROBIAL REMOVAL OF CHROMIUM

Previously, biosorption using plant-based materials has been mentioned as a technique for the removal of hexavalent chromium. The adaptive capacity and versatility of microorganisms in adverse conditions and their applicability in biotechnological tools and techniques have rendered them as excellent candidates for biosorption. Biosorption using microbes is a promising technique as compared to its conventional counterparts. Table 10.2 presents the microorganisms used for this purpose, either in the active or inactive state depending on its purpose. The nature of microbial biomass, specific growth rate, nutritional requirements, cell viability, by-product formation, and regeneration capacity are some of the factors which need to be deliberated while choosing an appropriate microbe for bioremoval (Vendruscolo et al. 2017). The cell and cell wall composition of a microbe play an important role in the biosorption of metal contaminants. For example, microalgae consist of different carbohydrates, proteins, and lipids, and the cell wall of fungi comprised of polysaccharides and glycoproteins provides metal binding sites composed of amide, phosphates, sulfhydryls, carboxyls, amines, and hydroxyls functional groups (Vimala and Das 2011). The cell wall of gram-positive and gram-negative bacteria differs in the thickness of the peptidoglycan layer which contributes to the former's rigidity and the latter's fragility. The other methods through which microorganisms are known to detoxify chromium-contaminated environments are bioaccumulation, biotransformation, and bioreduction (Guillen-Jimenez et al. 2009). In the following section, Cr(VI) removal via various microorganisms is discussed in more detail.

10.5.1 BACTERIA MEDIATED BIOREMOVAL

Many of the bodies of water have low temperatures for a large portion of the year. Thus, the remediation of Cr(VI) by bacteria, known as psychrophiles, existing at these lower temperatures has been explored. A strain of *Arthrobacter aurescens* was identified using 16S rDNA sequencing which was able to enzymatically reduce toxic and soluble Cr(VI) from contaminated aquifers into its non-toxic Cr(III) from temperatures as low as 10–18°C. A significant amount of biomass was also generated at these low temperatures within two days along with enzymatic reduction, which would also indicate efficient cell viability of this bacterial species. Other species of *Arthrobacter* also have the ability for subsurface bioremediation progression at a wide range of temperatures (Horton et al. 2006).

Streptomyces species have extensive use in biotechnological interventions such as antibiotic production. Studies have also reported their role in the tolerance and removal of Cr(VI) and Cr(III) from contaminated soil and water ecosystems. In a study performed by Morales et al. (2007), a species of *Streptomyces thermocarboxidus* NH50 was isolated from wastewater-contaminated soil

TABLE 10.2
Summary of Reported Data on Cr(VI) Reduction by Microorganisms. Reprinted with permission from Morales-Barrera, L. and Cristiani-Urbina, E., 2006. Removal of hexavalent chromium by *Trichoderma viride* in an airlift bioreactor. Enzyme Microb Technol. 40(1):107–113. Copyright © 2006 Elsevier Inc.

Organism	Cr(VI) concentration [mM]	Cr(VI) reduction efficiency [%]
Acinetobacter lwoffii	1	13.2–36.9
Acinetobacter sp.	1	24.7
Activated sludge	0.096	100
Aureobacterium esteroaromaticum	1	22.3–33.1
Bacillus subtilis	0.1–0.5	100
	1–2	0
Bacterial consortium	0.048	84
Candida maltose	1.15	85
Cellulomonas sp.	0.2	99.5
Deinococcus radiodurans R1	0.5	100
Desulfovibrio desulfuricans	0.5	86–96
Enrichment consortium	0.6	98.5
	0.04–0.38	100
	0.96	60
Enterobacter cloacae HO-1	0.25	100
	0.5	75
	1.5	13
Escherichia coli ATCC 33456	0.38	100
	1.5	75
	3.07	38.5
Hydrogenophaga pseudoflava	1	24.8
Microbacterium liquefaciens MP30	0.1	100
Microbacterium sp.	0.1	100
	0.5	40
Ochrobactrum anthropi	1	25.6–34.2
Phanerochaete chrysosporium	0.58	100
	0.96	0
Pseudomonas aeruginosa A2Chr	0.77	60
	0.96	16
	0.19	96
Pseudomonas fluorescens var. *pseudo-iodinum* P-11	0.58	57
Pseudomonas fragi B-184	0.58	83
Pseudomonas mendocina P-13	0.58	50
Pseudomonas putida B-117	0.58	97
LB303	0.02	100
Pseudomonas "rhatonis" P-17	0.58	100
Pseudomonas stutzeri	0.1	88
Pyrobaculum islandicum	0.42	100
Shewanella oneidensis MR-1		
(a) Anaerobic conditions	0.05	29.4
(b) Aerobic conditions	0.4	35
(c) Fumarate-reducing conditions	0.04	94

(*Continued*)

TABLE 10.2 (CONTINUED)
Summary of Reported Data on Cr(VI) Reduction by Microorganisms. Reprinted with permission from Morales-Barrera, L. and Cristiani-Urbina, E., 2006. Removal of hexavalent chromium by *Trichoderma viride* in an airlift bioreactor. Enzyme Microb Technol. 40(1):107–113. Copyright © 2006 Elsevier Inc.

Organism	Cr(VI) concentration [mM]	Cr(VI) reduction efficiency [%]
Streptomyces griseus	0.096	100
	1.15	100
Sulfate-reducing bacteria (isolate TKW)	0.36	94.5
Sulfidogenic and non-sufidogenic microbial consortia	0.85	100
	1.35	100
Thiobacillusferrooxidans	0.19	65
Trichoderma viride	1.3	100
	1.6	100
	1.94	94.3

from the tanning industry, containing about 3,765 parts per million (ppm) of chromium. This contained both Cr(VI) (8%) and Cr(III) (92%). A significant increase in concentrations of Cr(III) was observed in the experiment with a reduction in Cr(VI), which could imply the occurrence of a reduction process. The removal of Cr(VI) was attributed to the enzymatic reduction process and not to other removal processes like bioaccumulation or biosorption. The chromate reductase (*chR*) gene identified in *P. putida* and *E. coli* is known to be responsible for the reduction of Cr(VI). However, the activity and role of this gene could not be proven in *Streptomyces sp.* (Morales et al. 2007).

Another species of bacteria identified along similar lines of study is *Exiguobacterium* sp. GS1, which is a halotolerant, thermotolerant, and alkalitolerant, facultatively anaerobic bacterium capable of Cr(VI) reduction in diverse environmental conditions. This bacterial species removed Cr(VI) at both high and low concentrations (1–200 µg mL^{-1}) within 12 hours. A substantial bioremoval of 50% was observed in 3 hours and then a subsequent 90% was observed in 8 hours. Interestingly, *Exiguobacterium,* was active at a wide range of temperatures from 18–45°C. Bioremoval was also seen at high salt and acidic conditions. However, there was a slight reduction in the removal capacity. Although *Exiguobacterium* exhibited significant bioremediation capacity, the metabolic factors and the exact mechanism of the bioremoval for this species of bacteria are yet to be studied (Okeke 2008).

Bacillus sp. JDM-2-1 and *Staphylococcus capitis* isolated from wastewater were studied for their tolerance to Cr(VI) under varying pH levels and dissolved oxygen. *Bacillus* sp. JDM-2-1 and *S. capitis* were resistant to chromium levels up to 4,500 and 2,800 µg/mL, respectively. A gradual increase in the reduction capacity was observed for both the bacterial species over a span of 96 hours in industrial effluent containing chromium. *Bacillus* sp. JDM-2-1 reduced Cr(VI) by about 40–85% while *S. capitis* was able to reduce it by 29–81%. This enzymatic reduction and removal of Cr(VI) by *Bacillus* sp. JDM-2-1 and *S. capitis* can be exploited further for bioremediation purposes (Zahoor and Rehman 2009). Similarly, *Ochrobactrum anthropic* isolated from activated sludge produced an exopolysaccharide at pH 2.0 that could remove Cr(VI) even with an increase in initial concentrations from 30.0 to 280 mg/L capacity (Ozdemir et al. 2003).

Tanneries are one of the major sources of chromium contamination in water bodies. The waste generated from tanneries consists of hexavalent chromium and another compound pentachlorophenol (PCP), both of which are known toxicants. Bacterial species capable of degrading PCP and reducing Cr(VI) are considered as ideal candidates for bioremediation purposes. In a study performed by Srivastava and Thakur (2007), *Serratia mercascens*, *Pseudomonas fluorescens*, *E. coli*, *Pseudomonas aeruginosa*, and *Acinetobacter* sp. were isolated from tannery wastewater. Out of them, *Acinetobacter* sp. had the potency to remove about 70% chromium under aerobic conditions at pH 7. Additionally, a transmission electron microscopy analysis indicated bacterial bioaccumulation of chromate in the cell in the form of hydroxyl and carboxyl groups.

Another *Bacillus* species studied for its chromium bioremoval ability is *Bacillus cereus*, which was able to reduce Cr(VI) into Cr(III) but only at lower concentrations of 50 mg/L. Beyond this, a gradual decrease in the growth rate of the bacteria was noted. This species showed optimum biosorption of chromium at 37°C (Zhao et al. 2012).

In a study performed by Bennett et al. (2013), bacteria like *Klebsiella pneumonia*, *Bacillus firmus*, and *Mycobacterium* sp. isolated from contaminated water reduced Cr^{6+} by up to 87%, 96%, and 91%, respectively, when the initial concentration of Cr^{6+} was 100 mg/L at 30°C and at pH 7. The bacterial species were capable of the bioaccumulation of Cr^{6+} into the cellular biomass, which was attributed to several functional groups, such as hydroxyls, amines, and carbonyls. Metal binding is followed by the intracellular bioaccumulation and biotransformation of Cr^{6+} to Cr^{3+} (Bennett et al. 2013).

An interesting study carried out by Qu et al. (2014) reported the removal of metal-loaded bacteria from solutions as a potential strategy for bioremediation. Magnetotactic bacteria contain magnetosomes in their cellular structures, which aid in their removal from a solution by applying a magnetic field after the removal of toxic metals such as chromium. Magnetosomes are membranous, nano-sized particles possessing a dipole moment. Magnetotactic bacteria were found to have a high adsorptive capacity of chromium and were able to reduce Cr(VI) significantly in a wide range of environmental conditions, namely pH 5.5–7.5 and a temperature of 20–50°C. Initially, the bacteria adsorbed the metal ions and then the metal-loaded bacteria were isolated by the application of a magnetic field. The application of the magnetic field accelerated the bacterial and ion movement. The application of magnetotactic bacteria in the bioremediation process is promising due to the advantages it offers, such as enhanced reusability, increased biomass loading, and its environmentally-friendly nature (Qu et al. 2014).

Enzymatic reduction is a predominant mechanism adopted by bacteria for chromate reduction. Several reductases have been identified previously in gram-negative bacteria like *E. coli*, *Rhodobacter sphaeroids*, *Pseudomonas* sp., and *Enterobacter* sp. The chromate reductases have been hypothesized to be part of an inherent mechanism of the bacteria for fortification against reactive oxygen species (ROS). Chromate reductase isolated from a novel strain of *Ochrobactrum* sp. Cr-B4 was well-studied and revealed that the enzymatic activity of the reductase was inhibited by certain other metal ions such as Ni^{2+}, Zn^{2+}, and Cd^{2+}, while metal ions like Cu^{2+} and Fe^{3+} enhanced its activity. Understanding the catalytic and inhibitory properties of this enzyme would help to further validate other uncharacterized reductases as well as in exploring its application for Cr^{6+} reduction (Hora and Shetty 2015). The reduction of Cr^{6+} to Cr^{3+} was catalyzed by an enzyme, NADPH (reduced nicotinamide adenine dinucleotide phosphate)-dependent flavin mononucleotide reductase (FMN_red), which is encoded by the *ChrT* gene or the chromate reductase gene. Deng et al. (2014) cloned the *ChrT* gene from *Serratia* sp. CQMUS2 and studied it further for structural analysis. The gene also showed high similarity with the FMN_red gene of *Klebsiella pneumonia*, *E. coli*, and *Raoultella ornithinolytica*. The characterization of reductase genes forms the basis for further studies on *ChrT* using a genomic and proteomic approach to design efficient strategies for bioremediation (Deng et al. 2014). In gram-positive bacteria like *Bacillus subtilis*, the *nfrA* gene from the nitro/flavin reductase family was responsible for reductase activity (Zheng et al. 2015).

Shewanella oneidensis MR-1 also exhibited Cr(VI) reduction under fumarate and nitrate-reduction conditions due to the transition of microorganisms from oxic to anoxic states. This indicated non-enzymatic reactions playing a role in the Cr(VI) reductions. The late log phase anaerobic cells were most suited for Cr(VI) reduction (Viamajala et al. 2002).

Polycyclic aromatic hydrocarbons (PAH) are another form of toxic waste generated along with hexavalent chromium. Naphthalene is the simplest homolog belonging to PAH. It has been observed that the microbial biomass is altered if there is an additional pollutant along with metal ions, or vice versa. Therefore, the possibility of a single strain being able to tackle combined pollutants is much sought after. A novel strain of *Pseudomonas gessardii* LZ-E was isolated from the wastewater of petrochemical corporations containing copious amounts of naphthalene and Cr^{6+}. The bacterial strain was able to degrade naphthalene along with the reduction of chromium. Also, the degradation of naphthalene further aided the reduction process due to the production of catechol and phthalic acid produced during its breakdown. *P. gessardii* strain LZ-E is therefore an excellent candidate for combating combined pollution and researchers are exploring its application in bioremediation (Huang et al. 2016). Along with PAH, chlorobenzenes, which are usually used as raw materials in the dye and chemical industries, are also a matter of concern due to their adverse effects on health and the environment. A lipophilic compound *o*-dichlorobenzene (*o*-DCB) is one such solvent used in dye manufacturing industries which, along with Cr(VI), exhibits extremely high toxicity when released in wastewater. *Serratia marcescens* ZD-9 isolated from dye industry effluent was able to effectively remove both *o*-DCB and Cr(VI) at concentrations of 1.29 mg/L and 20 mg/L, respectively. The optimum conditions required were a pH of 8 and a temperature of 30°C. Biosorption and bioreduction resulted in the removal of Cr(VI) which was confirmed by the presence of Cr(III) in the intracellular spaces of the bacteria. A fourier-transform infrared spectroscopy (FTIR) analysis supported the fact that the presence of functional groups such as in hydroxyls, amides, and polysaccharides may have played an important role in the bioremoval. The *o*-DCB was absorbed by the bacterial cell and degraded subsequently which rationalized the application of *S. marcescens* ZD-9 for the bioremediation of organic/inorganic and metal ion simultaneously in wastewater (Xu et al. 2018).

Metal organic frameworks (MOFs) are hybrid frameworks wherein metal ions and organic ligands are utilized concomitantly for application in separation techniques, ion exchange, and drug delivery. Metals like zirconium have been widely used in MOF with high thermal stability, porosity, and surface area. In a study performed by Sathvika et al. (2019), a novel *Nitrosomonas* modified zirconium MOF was developed to study its capacity in the bioremoval of Cr(VI). The porosity and stability of MOF and functional groups of bacterial strains provided an enhanced and robust framework for chromate removal, as seen in **Figure 10.1**. The biosorbent's affinity was attributed to the electrostatic interactions. The bacterial modified MOF (BMOF) and Zr-MOF showed average particle sizes of 39 nm and 54 nm, respectively. A significant bioremoval of 95% at 23.69 mg/g of Cr(VI) was observed along with an ability for reusability of the said biosorbent via facile desorption

FIGURE 10.1   FESEM images of (a) BMOF (b) BMOF after Cr(VI) adsorption. Reprinted with permission from Sathvika, T.; Balaji, S.; Chandra, M.; Soni, A.; Rajesh, V.; and Rajesh, N., 2019. A co-operative endeavor by nitrifying bacteria *Nitrosomonas* and zirconium-based metal organic framework to remove hexavalent chromium. Chem. Eng. J. 360, 879–889. Copyright © 2018 Elsevier B.V.

using NaOH (sodium hydroxide). Thus, nitrifying bacteria can serve as potential biosorbents for Cr(VI) removal (Sathvika et al. 2019).

Microbial mats are diverse clumps of microbes found attached to a surface or floating on water bodies. Abed et al. (2020) reported a microbial mat which could remove approximately 1 mg/L Cr(VI) in a period of 7 days under aerobic conditions. The microbial mat contained bacterial species, namely *Proteobacteria*, *Verrucomicrobiae*, *Firmicutes*, *Actinobacter*, and *Alphaproteobacteria*. Even though all these microbial strains survived on the mat, the microbial diversity was observed to have shifted to *Verrucomicrobiae* and *Alphaproteobacteria* when exposed to Cr(VI) contamination (Abed et al. 2020). Further research is therefore required in isolation of species from microbial mats and their characterization to explore the in-depth mechanism of aerobic Cr(VI)-reducing microbiota in such mats.

10.5.2 Fungus Mediated Bioremoval

Apart from bacteria, several higher fungi serve as potential biosorbents. *Trichoderma viride* has the ability to remove high dosages of Cr(VI) under aerobic conditions in stirred tank bioreactors and airlift bioreactors. Morales-Barrera and Cristiani-Urbina (2006) reported the removal of Cr(VI) by *T. viride* which was isolated from tannery effluents. Initially, the microbial biomass was prepared in a liquid growth medium containing, per liter of distilled water, 10 g of glucose, 3 g of $(NH_4)_2SO_4$, 1 g of KH_2PO_4, 0.3 g of $MgSO_4 \cdot 7H_2O$, 0.1 g of KCl, 0.1 g of yeast extract, 0.05 g of $CaCl_2$, and 1 mg of $FeCl_3$. In unbaffled flasks, the cell mass increased when incubated at initial Cr(VI) concentrations ranging from 1–2 mM resulting in 97–100% of Cr(VI) removal. However, in the stirred tank bioreactors, the intensive agitation caused mycelial fragmentation and reduced activity of Cr(VI) removal. Airlift bioreactors, on the other hand, were able to considerably remove high concentrations of Cr(VI).

Another fungus studied for the removal of Cr(VI) was *Aspergillus niger*. The bioremediation capacity of *A .niger* was optimized at pH 6 and 30°C. At 250 and 500 ppm of chromate, the fungus was able to remove about 70% of the chromium. However, it was observed that at higher concentrations of chromium (1,000 and 2,000 ppm), the percentage removal reduced at 55% and 45%, respectively. The elevated levels of chromium probably had an inhibitory effect on the growth of the fungus. A 19.8 mg/g uptake was observed at 250 ppm which was the maximum capacity, while only 7.9 mg/g dry wt. of chromium was taken up at 2,000 ppm (Srivastava and Thakur 2006).

Mucor racemosus is another abundantly found fungus in nature and its biomass was exploited for the bioremoval of chromium from wastewater. Cr(VI) removal by this species was dependent on several factors, namely pH, metal concentrations, biomass generated, and the contact time. At pH 0.5–1.0, an optimum metal removal of 94.9% was seen. However, at higher pH levels, the removal was found to be reduced. This could be attributed to the increase in H^+ ions in the solution at lower pH which interacts with the -OH group to form $-OH_2^+$. This led to protonation of the surface of the fungal biomass. At higher pH levels, however, the -OH group was negatively ionized to form $-O^-$ which eventually repelled the negatively-charged Cr(VI) and thus reduced the adsorption capacity. Additionally, at higher biomass dosages the removal capacity also increased. This could be due to the greater surface area for the binding of Cr(VI). The bioreduction of Cr(VI) to Cr(III) and also adsorption of Cr(VI) by several functional groups of *M. racemosus* were the predominant mechanisms of chromium removal (Liu et al. 2007).

Studies performed on bioremoval mechanisms by fungal biomass are associated with biotransformation and biosorption. However, the exact functional groups involved in this have not yet been determined. Inactive/dead fungal biomass is usually easy to handle as it is not deleterious to the environment and is not a source of contamination. Further, dead biomass can be stored for longer periods without putrefaction, thus making their transportation easy as well. Additionally, the activity of dead fungal biomass can be enhanced with the application of physical and chemical treatment. *Termitomyces clypeatus* is an edible fungus and its heat inactivated form along with chemically inactivated form was used to study the mechanism of chromium bioremoval. The fungus was cultured under submerged condition in a complex medium (%, w/v) comprised of the following ingredients: sucrose 5.0, malt extract 1.0, boric acid 0.057, KH_2PO_4 0.15, $CaCl_2 \cdot 2H_2O$ 0.037, $MgSO_4$ 0.05, $MnCl_2 \cdot 4H_2O$ 0.0036, $ZnSO_4 \cdot 7H_2O$ 0.031, $FeSO_4 \cdot 7H_2O$ 0.025, and $(NH_4)_2PO_4$ 2.5 at pH 3.5 at 30°C for 5 days. The biomass was recovered which, on chemical pre-treatment with HCl (0.1 N) and $CaCl_2$ (1%), showed remarkable enhancement in Cr(VI) removal up to 100% and 99.07%, respectively. However, the opposite result was obtained on pre-treatment with NaOH and NaCl that resulted in only 18.58% and 44.48% removal, respectively. The acidic and alkaline groups such as amino, carboxyl, and phosphate were responsible for the biosorption of the Cr(VI), while amino, carboxyl, sulphonyl, carbonyl, and phosphate groups were involved in the biosorption process of heat inactivated fungus. Kinetic studies and surface chemistry indicated that the bioremoval process was divided into two sequential steps of biosorption by electrostatic force by amino, carboxyl, and phosphate groups and then bioreduction via hydroxyl and carbonyl groups on the fungal biomass surface (Ramrakhiani et al. 2011).

In a study by Prigione et al. (2009), a tanning effluent was reported as a source of various fungi such as *Fusarium solani*, *Mucor circinelloides*, *Cunninghamella elegans*, *Rhizomucor pusillus*, and *Rhizopus stolonifer* that exhibited Cr(III) bioremoval. The highest Cr(III) removal percentages were exhibited by *C. elegans* and *R. pusillus* at 30% and 28%, respectively. NaOH pre-treated *C. elegans* biomass exhibited the highest Cr(III) bioremoval of up to 93.49±0.34% after 24 hours of incubation. Chemical treatment enhanced the fungal cell biosorption capacity due to the eventual removal of the cell wall and subsequent increase in surface area and porosity (Prigione et al. 2009). In another study conducted by Singh et al. (2016), *Aspergillus flavus* biomass was used for Cr(VI) removal in the presence of various forms of Fe. It is important to note that pH played a crucial role in the biotransformation process. At pH 4 and pH 8, a biotransformation up to 96.27% and 71.33% was achieved, respectively, using *A. flavus* biomass (Singh et al. 2016).

10.5.3 Yeast Mediated Bioremoval

Saccharomyces cerevisiae is one of the most popular yeast studies for the removal of toxic metals. It can be genetically manipulated easily and is commonly used in fermentation and by pharmaceutical industries. They are produced in abundance as a by-product and disposed as waste. Additionally, they are easy to grow using inexpensive techniques and resources. *S. cerevisiae* has been used in different forms that include living/dead cells, intact/inactivated cells, and as industrial/laboratory waste for the bioremoval of different metal ions which implies its feasibility as a biosorbent. The enhancement of Cr(VI) removal was observed when *S. cerevisiae* biomass was modified with a cationic surfactant, cetyl trimethyl ammonium bromide (CTAB). The biosorption efficiency of CTAB-modified cells was higher than 80% at all the pH values, reaching a maximum of 99.5% at a pH of 5.5. Also, low initial concentrations of CrO_2^{-4} (<30 mg/L) resulted in removal efficiencies above 90% (Bingol et al. 2004).

Goyal et al. (2003) showed that an increase in temperatures from 25–45°C resulted in the enhancement of the heavy metal removal capacity which might be attributed to either the higher affinity of sites for metal or an increase in the binding sites on the relevant cell mass. However, at temperatures higher than 45°C, distortion in the surface chemistry and subsequent reduction in binding sites led to lower metal removal. Ferraz et al. (2004) reported that the ideal contact time for the biosorption of heavy metals by *S. cerevisiae* from brewery effluent was a minimum of 30 minutes. Generally, with an increase in contact time, the metal removal efficiency also increases. However, this can be affected by the presence of interfering ions. The adsorptive capacity of dried *S. cerevisiae* was significantly reduced in the presence of even low concentrations of Fe(III). The presence of supplements such as cysteine, glucose, and ammonium sulfate in the fermentation media enhanced the Cr(VI) uptake. This is due to their contribution in functional groups associated with the surface of the fungal cell (Goyal et al. 2003). The gene Arr4p, a homolog for bacterial arsenic resistance gene (ars), identified in S. cerevisiae encodes an ATPase that facilitates tolerance to Cr^{3+} metal ion (Shen et al. 2003).

Yeast strains, Pichia jadinii M9 and Pichia anomala M10, isolated from textile dye polluted with Cr(VI) were studied for their role in chromate reductase activity in chromium removal. Prior to this, the two strains were able to substantially remove chromate pollutants from tannery effluents (Fernández et al. 2010a, b). A pH of 6.0 and temperature of 60°C was found to be optimum for the reductase activity of P. jadinii, while a pH of 7.0 and temperature of 50°C was found to show maximum reductase activity for P. anomala. Also, an addition of NAD(P)H as an electron donor enhanced the activity of reductase.

The process of immobilization of biosorbents is vastly studied for the bioremediation of toxic metals. In this, the whole cell or any required metabolite is attached to a polymeric surface, such as calcium alginate, allowing advantages over freely suspended cells. They aid in the maintenance of the biomass density, offer easier handling, and the microbe is able to tolerate high amounts of environmental stress such as changes in pH and temperature. Biomass/polymer matrices beads (BPMBs) are one such class of polymers which were used with S. cerevisiae in order to study the biosorption capacity of Cr(VI) from tannery wastes. The deduced optimum pH value was 3.5 with a maximum biosorption capacity at 154 mg/g with an initial Cr(VI) concentration of 200 ppm. At the end of the incubation period, 85% of the Cr(VI) was found to be removed from the tannery wastes. An important benefit of using BPMBs was that they can be recycled and reused, diminishing treatment costs with a minimum decrease in efficiency (Mahmoud and Mohamed 2017).

Gelatin is a natural biopolymer and extensively used for treatment studies. A combination of gelatin and baker's yeast S. cerevisiae (Gel-Yst) was developed as a novel biosorbent with various functional groups for solid phase extraction and the biosorption of Cr(VI). A high adsorption capacity value of 100% was observed at pH 1.0–2.0, while a minimum of 11.94% removal was seen at pH 7.0. Hence, polymer-entrapped biosorbents can be further explored for their potential applications in heavy metal removal (Mahmoud 2015).

10.5.4 Genetically Engineered Microorganisms (GEMs)

Advances in molecular biology have led to the understanding of mechanisms of chromium removal up to genetic level. The optimization of factors such as pH, temperature, concentration of the heavy metal, time of interaction, and nutrients in the culture medium can enhance the metal removal up to a certain level. However, beyond the threshold capacity, genetic engineering can help in enhancing the metal removal capacity by generating transgenic microbes.

Chromium removal improved by 20 fold in the bacteria Pseudomonas putida KT 2441 strain which was genetically modified by inserting a plasmid conferring chromium resistance (Chaturvedi et al. 2015). E. coli was genetically engineered to encode for trehalose which is a biochemical protectant. It was found that the engineered bacteria were able to reduce 1 mM of Cr(VI) to Cr(III), which was double that compared to the wild type of bacteria and could be attributed to the trehalose production that might have protected the genetically modified bacteria from the side effects of chromate. It would be feasible to identify bacteria which can inherently produce high amounts of trehalose, such as Corynebacterium hoagii, and explore their bioremediation capacity (Frederick et al. 2013).

Shewanella sp. are able to reduce chromate, which has been attributed to chromate reductase enzyme encoded by the gene chrBAC (Chourey et al. 2006). The plasmid containing this gene was transferred to E. coli and P. aeruginosa and studied for the induction of chromate reduction. The chrA gene was able to induce chromate reductase activity in both the bacterial species in both low and high copy numbers of the plasmid. It was also found that the chromate was not taken up by the cells but in contrast, was expelled out from the cytoplasm and therefore possessed an efflux mechanism that did not allow the accumulation of the chromate after reduction within the bacterial cells (Aguilar-Barajas et al. 2008).

Not much research has been performed regarding transgenic chromium-reducing microbes due to the constraints regarding environmental and human health. However, based on existing data it can be concluded that this field can be explored in the following ways:

- Over-expression of enzymes that are able to reduce heavy-metal induced stress
- Expression of proteins on the surface and cytoplasm of cells that have the ability to bind heavy metals
- Exploring genetic materials, such as plasmids, conferring chromium resistance and their transformation into suitable microbes

10.6 MECHANISMS OF MICROBIAL CHROMIUM REMOVAL

Microbial removal of chromium consists of various mechanisms, as schematically represented in **Figure 10.2**. These mechanisms help to change the toxicity, mobility, and speciation of chromium that can serve as the basis for the development of technologies for chromium bioremediation. The predominant processes for the bioremoval of chromium by microbes are biosorption, bioaccumulation, and biotransformation/bioreduction (Fernández et al. 2018).

Bioreduction is essentially the conversion of Cr(VI) into Cr(III) in the presence of reductase enzymes. During the transformation of Cr(VI), intermediate oxygen radicals may be formed in addition to toxic chromium species such as Cr(V) and Cr(IV). However, microorganisms possess unique chromate resistant plasmids and an iron efflux system which aid in minimizing additional problems related to chromium bioreduction (Dhal et al. 2013). Generally, the bioreduction mechanism varies from microbe to microbe depending on the nutritional requirements and bio-geochemical activity associated parameters, such as contact time, temperature, pH, initial concentration, and biomass.

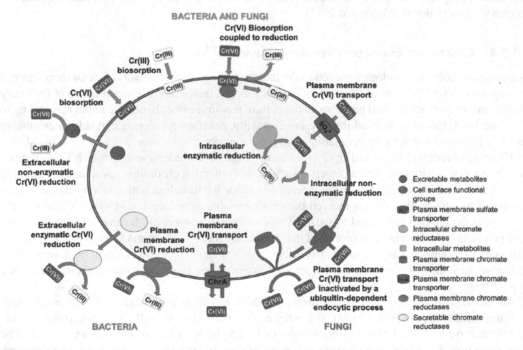

FIGURE 10.2   Schematic summary of microbial interactions with chromium. Reprinted with permission from Gutiérrez-Corona, J.F.; Romo-Rodríguez, P.; Santos-Escobar, F.; Espino-Saldaña, A.E.; and Hernández-Escoto, H., 2016. Microbial interactions with chromium: basic biological processes and applications in environmental biotechnology. World J. Microbiol. Biotechnol. 32, 191. Copyright © 2016 Springer Science+Business Media Dordrecht 2016.

Pattanapipitpaisal et al. (2001) stated that, essentially, microbes display three major processes of bioreduction which are as follows:

i) Aerobic conditions using chromate reductase that use NADH and NAD(P)H as factors
ii) Anaerobic conditions that use Cr(VI) as an electron acceptor
iii) Reduction associated with intra/extracellular compounds such as amino acids, sugars, and glutathione

The best characterized mechanisms identified for chromium bioremoval are discussed in the following section.

10.6.1 Efflux Mechanism

Most of the microorganisms associated with chromium reduction contain plasmids that confer chromate resistance, the most important being *Shewanella* sp. with the *chrBac* genes (Chourey et al. 2006). Evidence for efflux mechanisms was given by Alverez et al. (1999), in which they reported the formation of vesicles containing four-fold more of the toxic CrO_4^{2-} in *Pseudomonas aeruginosa* which had the *ChrA* genes. They also found evidence of the dependence of chromate uptake by cellular vesicles due to NADH oxidation and eventually its removal from the cell by chromate analogue sulfates. Another microbe, *Ochrobactrum tritici* was able to tolerate greater amounts of chromium owing to its transposons. Transposable element *TnOtChr* possesses an array of genes, such as *chrA*, *chrB*, *chrC*, and *chrF*, among which *chrA* and *chrB* were the genes responsible for chromate reductase. However, they are able to bioremediate only Cr(VI) (Branco et al. 2008). The *chr* gene family has been widely explored in several other bacteria, and evidence of their function has been validated. Other genes which have been identified which confer chromate reductase are the *nitR* and *yieF* genes. They encode chromate reductase and were found in the bacteria *Lysinibacillus fusiformis* ZC1 strain (He et al. 2011). In cases where the bacteria lack a potential chromate reductase system, then the efflux mechanism is activated in the cell by the induction of the *chr* operon. This repulses the toxic Cr(VI) from the bacterial cell (Joutey et al. 2015).

10.6.2 Chromate Reduction

10.6.2.1 Aerobic Chromate Reduction

Chromate reduction in a microbial system can occur in both aerobic and anaerobic conditions. In the aerobic mode of reduction, NADH and NADPH reserves of the cell are utilized as electron donors. The reduction of Cr^{6+} occurs in a transient manner: from Cr^{5+} to Cr^{4+} and finally to the non-toxic Cr^{3+}. A one-shuttle system converts Cr^{6+} to Cr^{5+} and a two-shuttle system converts it to Cr^{3+}. Pentavalent Cr is also a known cell toxicant and therefore its accumulation can prove to be hazardous. However, a two-electron transfer catalyzed by ChR reduces it spontaneously and releases ROS. The enzyme YieF, however, is able to convert Cr^{6+} to Cr^{3+} in a four-step electron transfer process; hence the generation of ROS is minimal, making this a better option as compared to ChR (Park et al. 2000, 2003).

10.6.2.2 Anaerobic Chromate Reduction

A number of chromate-resistant, facultative anaerobes have been isolated such as *B. cereus*, *B. subtilis*, *P. aeruginosa*, and *E. coli* along with sulfate-reducing bacteria (SRB). Microorganisms such as *Desulfomicrobium norvegicum*, *Desulfotomaculum reducens*, and *Desulfovibrio vulgaris* are some of the most prominently studied bacterial species for this purpose. They possess unique mechanisms including soluble cytochromes, hydrogenase, and cytochrome catalyzed and electron acceptance for Cr^{6+} reduction, which is found in the SRB group of microbes (Cheung and Gu 2003). Other facultative microbial species such as extremophiles have also shown similar potential. *Deinococcus radiodurans* R1, a radiation-resistant bacterium, and *Pyrobaculum islandicum*, a thermophilic bacterium, are known to reduce Cr(VI) and degrade benzoate. Generally, basic metabolites of SRBs, such as hydrogen sulfide, are capable of Cr(VI). Additionally, in an environment lacking oxygen, Cr(VI) can serve as an electron acceptor for electron donors such as NAD(P)H (Joutey et al. 2015).

10.6.3 CHROMATE UPTAKE

Bioaccumulation includes several processes which basically include the uptake of metal ions by the living cell. In this case, uptake may be in the passive or active form. The passive way of metal uptake does not require energy and usually occurs by entrapment and the biosorption of the ion by cell organelles. On the other hand, the active method of metal uptake requires metabolic energy for transmission across the membrane (Joutey et al. 2015).

10.6.3.1 Biosorption

Functional groups present on the surface of the microbial cell, such as amides and carbonyls, interact with the negatively-charged chromate or dichromate ion and facilitate biosorption. This has been observed in the bacterial species *P. putida*. At low pH levels the chromium uptake was seen to be enhanced as compared to alkaline conditions. This is due to the protonation of the surface functional groups at acidic conditions. Further, upon partial reduction of chromium, the unreduced part is adsorbed by the cell while the non-toxic Cr(III) remains in the solution (Joutey et al. 2015).

10.6.3.2 Bioaccumulation

Cr(III) is permeable to the biological cell. However, it can also form compounds with ligands present in aqueous solutions and can be readily taken up by the cell. Cr(VI) occurs in the form CrO_4^{2-} which is analogous to sulfate ions. Thus, it can be taken up by sulfate transporter channels by competitive selection. Inside the biological cell, Cr(VI) is rapidly converted into Cr(III). With the consideration of a continuous reducing process, bioaccumulation serves as an excellent mechanism for chromium uptake by the microbial cell (Joutey et al. 2015).

10.6.4 INTRACELLULAR AND EXTRACELLULAR REDUCTION

Several inherent components of the microbial cell are able to reduce Cr(VI). NADH, NAD(P)H, and certain flavoproteins are able to reduce Cr(VI) to Cr(III), which are generally present in the protoplasm of the cell. Cytoplasmic fractionation is therefore expected to be able to reduce chromium levels. As mentioned previously, this involves electron reduction steps generating transient chromium species such as Cr(V) and Cr(IV). This process generates ROS which can damage the DNA of a cell. The cytoplasmic fraction of certain *Bacillus* sp. is known to contain reductase activity. Further, certain *Pseudomonas* sp. contain the same in both cytosolic as well as bacterial supernatants (Joutey et al. 2015).

The extracellular reduction of chromate is a comparatively safer mechanism. In this, the cell releases enzymes such as reductases into the solution containing Cr(VI). The proteins are released only in the presence of the chromate ions and reduction takes place in the aqueous solutions. This minimizes the energy required for the uptake of chromium by the cell. Also, chromium species readily interact with the nucleic acid of the cell causing severe DNA damage. An extracellular mode of reduction protects the cell from any such damage. Additionally, extracellular mechanisms make it easier for the separation of reduced chromium, and the microbial cell can be reused and recycled (Joutey et al. 2015).

10.7 CONCLUSION AND FUTURE PERSPECTIVES

Amongst the suggested methods for chromium bioremediation, microbial removal is the most effective and sustainable. The potential of bioremediation by microbes has been studied extensively in various biological systems. These systems rely on and are influenced greatly by several factors such as biomass, viability, and bio-geochemical factors. This low-cost emerging biotechnological tool is extremely versatile as a biosorbent as live, immobilized, as well as dead biomass can be employed for chromium removal. The various parameters like time, temperature, pH, metal concentration, and biomass must be optimized in order to generate a robust bioremediation system. Mixed culture processes with multiple microbes may be developed which will ensure simultaneous bioconversion and removal of chromium. Recent genomic, transcriptomics, and proteomic approaches can be employed to develop genetically engineered microbes with potential chromium removal ability due to the over-expression of mutant reductase enzymes.

Additionally, approaches such as the development of microbial mats and consortiums offer a greater impact by incorporating multiple species and strains of microbial species, each providing individual advantages. This *ex-situ* method, apart from being highly economical, also holds the upper hand for being environmentally friendly. Further, microbes can be immobilized in biopolymeric beads and used for the treatment of chromium-contaminated industrial effluents. The bioremediation technology can therefore focus on three main benchmarks: (i) the effect of abiotic factors on the Cr(VI) reduction capacity and specific resistance mechanism, (ii) the conditions required for the activation of inherent mechanisms of indigenous microbes for bioreduction, and (iii) exploring the functionality of native species residing in contaminated sources. Taking all this into account, large-scale bioremediation applications for chromium removal can be designed to achieve clean and green environments.

ACKNOWLEDGMENTS

Dr. Sougata Ghosh acknowledges the Department of Science and Technology (DST), the Ministry of Science and Technology, the Government of India, and the Jawaharlal Nehru Centre for Advanced Scientific Research, India for funding under Post-doctoral Overseas Fellowship in Nano Science and Technology (Ref. JNC/AO/A.0610.1(4) 2019-2260 dated August 19, 2019).

REFERENCES

Abed, R.M.M., Shanti, M., Muthukrishnan, T., Al-Riyami, Z., Pracejus, B., Moraetis, D., 2020. The role of microbial mats in the removal of hexavalent chromium and associated shifts in their bacterial community composition. *Frontiers in Microbiology* 11, 12.

Aguilar-Barajas, E., Paluscio, E., Cervantes, C., Rensing, C., 2008. Expression of chromate resistance genes from *Shewanella* sp. strain ANA-3 in *Escherichia coli*. *FEMS Microbiology Letters* 285(1), 97–100.

Alvarez, A.H., Moreno-Sanchez, R., Cervantes, C., 1999. Chromate efflux by means of the ChrA chromate resistance protein from *Pseudomonas aeruginosa*. *Journal of Bacteriology* 181, 7398–7400.

Bennett, R.M., Cordero, P.R.F., Bautista, G.S., Dedeles, G.R., 2013. Reduction of Hexavalent chromium using fungi and bacteria isolated from contaminated soil and water samples. *Chemistry and Ecology* 29(4), 320–328.

Bingol, A., Ucun, H., Bayhan, Y.K., Karagunduz, A., Cakici, A., Keskinler, B., 2004. Removal of chromate anions from aqueous stream by a cationic surfactant-modified yeast. *Bioresource Technology* 94(3), 245–249.

Branco, R., Chung, A.P., Johnston, T., Gurel, V., Morais, P., Zhitkovich, A., 2008. The chromate-inducible *chrBACF* operon from the transposable element Tn*OtChr* confers resistance to chromium(VI) and superoxide. *Journal of Bacteriology* 190(21), 6996–7003.

Chaturvedi, A. D., Pal, D., Penta, S., Kumar, A. 2015. Ecotoxic heavy metals transformation by bacteria and fungi in aquatic ecosystem. *World Journal of Microbiology and Biotechnology* 31(10), 1595–1603.

Cheung, K.H., Gu, J.-D., 2003. Reduction of chromate (CrO42–) by an enrichment consortium and an isolate of marine sulphate-reducing bacteria. *Chemosphere* 52(9), 1523–1529.

Cheung, K.H., Gu, J.D. 2007. Mechanism of hexavalent chromium detoxification by microorganisms and bioremediation application potential: A review. *International Biodeterioration and Biodegradation* 59(1), 8–15.

Chourey, K., Thompson, M.R., Morrell-Falvey, J., VerBerkmoes, N.C., Brown, S.D., Shah, M., Zhou, J., Doktycz, M., Hettich, R.L., Thompson, D.K., 2006. Global molecular and morphological effects of 24-hour chromium(VI) exposure on *Shewanella oneidensis* MR-1. *Applied and Environmental Microbiology* 72(9), 6331–6344.

Deng, P., Tan, X., Wu, Y., Bai, Q., Jia, Y., Xiao, H., 2014. Cloning and sequence analysis demonstrate the chromate reduction ability of a novel chromate reductase gene from *Serratia* sp. *Experimental and Therapeutic Medicine* 9(3), 795–800.

Dhal, B., Thatoi, H.N., Das, N.N., Pandey, B.D., 2013. Chemical and microbial remediation of hexavalent chromium from contaminated soil and mining/metallurgical solid waste: A review. *Journal of Hazardous Materials* 250–251, 272–291.

Fernández, P.M., Fariña, J.I., Figueroa, L.I.C., 2010a. The significance of inoculum standardization and cell density on the Cr(VI) removal by environmental yeast isolates. *Water, Air, and Soil Pollution* 212(1–4), 275–279.

Fernández, P.M., Figueroa, L.I.C., Fariña, J.I., 2010b. Critical influence of culture medium and Cr(III) quantification protocols on the interpretation of Cr(VI) bioremediation by environmental fungal isolates. *Water, Air, and Soil Pollution* 206(1–4), 283–293.

Fernández, P.M., Viñarta, S.C., Bernal, A.R., Cruz, E.L., Figueroa, L.I.C., 2018. Bioremediation strategies for chromium removal: Current research, scale-up approach and future perspectives. *Chemosphere* 208, 139–148.

Ferraz, A.I., Tavares, T., Teixeira, J.A., 2004. Cr(III) removal and recovery from *Saccharomyces cerevisiae*. *Chemical Engineering Journal* 105(1–2),11–20.

Frederick, T.M., Taylor, E.A., Willis, J.L., Shultz, M.S., Woodruff, P.J., 2013. Chromate reduction is expedited by bacteria engineered to produce the compatible solute trehalose. *Biotechnology Letters* 35(8), 1291–1296.

Ghosh, S., 2020a. Toxic Metal Removal Using Microbial Nanotechnology. In: Rai, M., Golinska, P. (eds.), *Microbial Nanotechnology*. CRC Press, Boca Raton, FL. eBook ISBN: 9780429276330.

Ghosh, S., Selvakumar, G., Ajilda, A.A.K., Webster, T.J., 2020b. Microbial biosorbents for heavy metal removal. In: Shah, M.P., Couto, S.R., Rudra, V.K. (eds.), *New Trends in Removal of Heavy Metals from Industrial Waste Water*. Elsevier B.V., Amsterdam, Netherlands. (In Press).

Ghosh, S., Sharma, I., Nath, S., Webster, T.J., 2020c. Bioremediation: The natural solution. In: Shah, M.P., Couto, S.R., (eds.), *Microbial Ecology of Waste Water Treatment Plants (WWTPs)*, Elsevier, Amsterdam, Netherlands. (In Press).

Ghosh, S., Joshi, K., Webster, T.J., 2020d. Removal of heavy metals by microbial communities. In: Shah, M.P., Couto, S.R., (eds.), *Microbial Community Structure of Waste Water Treatment Reactors*, Elsevier, Amsterdam, Netherlands. (In Press).

Ghosh, S., Bhattacharya, J., Nitnavare, R., Webster, T.J., 2020e. Heavy metal removal by *Bacillus* for sustainable agriculture. In: Islam, M.T., Rahman, M., Pandey, P., (eds.), *Bacilli and Agrobiotechnology: Plant Stress Tolerance, Bioremediation, and Bioprospecting*, Springer Nature, Cham. (In Press).

Goyal, N., Jain, S.C., Banerjee, U.C., 2003. Comparative studies on the microbial adsorption of heavy metals. *Advances in Environmental Research* 7(2), 311–319.

Guillén-Jiménez, F.d.M., Netzahuatl-Muñoz, A.R., Morales-Barrera, L., Cristiani-Urbina, E., 2009. Hexavalent chromium removal by *Candida* sp. in a concentric draft-tube airlift bioreactor. *Water, Air, and Soil Pollution* 204, 43.

He, M., Li, X., Liu, H., Miller, S.J., Wang, G., Rensing, C., 2011. Characterization and genomic analysis of a highly chromate resistant and reducing bacterial strain *Lysinibacillus fusiformis* ZC1. *Journal of Hazardous Materials* 185(2–3), 682–688.

Hora, A., Shetty, V.K., 2015. Partial purification and characterization of chromate reductase of a novel *Ochrobactrum sp.* strain Cr-B4. *Preparative Biochemistry and Biotechnology* 45(8), 769–784.

Horton, R.N., William A.A., Thompson, V.S., Sheridan, P.P., 2006. Low temperature reduction of hexavalent chromium by a microbial enrichment consortium and a novel strain of *Arthrobacter aurescens*. *BMC Microbiology* 6, 5.

Huang, H., Wu, K., Khan, A., Jiang, Y., Ling, Z., Liu, P., Chen, Y., Tao, X., Li, X., 2016. A novel *Pseudomonas gessardii* strain LZ-E simultaneously degrades naphthalene and reduces hexavalent chromium. *Bioresource Technology* 207, 370–378.

Joutey, N.T., Sayel, H., Bahafid, W., El Ghachtouli, N., 2015. Mechanisms of hexavalent chromium resistance and removal by microorganisms. In: Whitacre, D. (eds.), *Reviews of Environmental Contamination and Toxicology (Continuation of Residue Reviews)*, vol 233, pp 45–69. Springer, Cham.

Ksheminska, H., Fedorovych, D., Babyak, L., Yanovych, D., Kaszycki, P., Koloczek, H., 2005. Chromium (III) and (VI) tolerance and bioaccumulation in yeast: A survey of cellular chromium content in selected strains of representative genera. *Process Biochemistry* 40, 1565–1572.

Kulkarni, S., Yerwa, S., Thakkar, B., Ghosh, S., 2018. Petroleum contaminated soil as rich source of bioemulsifier producing *Enterobacter cloacae*. *World Journal of Pharmaceutical Research* 7(4), 173–182.

Kurniawan, T.A., Gilbert, Y.S.C., Lo, W-H., Babel, S., 2006. Physico-chemical treatment techniques for wastewater laden with heavy metals. *Chemical Engineering Journal* 118(1–2), 83–98.

Liu, T., Li, H., Li, Z., Xiao, X., Chen, L., Deng, L., 2007. Removal of hexavalent chromium by fungal biomass of *Mucor racemosus*: Influencing factors and removal mechanism. *World Journal of Microbiology and Biotechnology* 23(12),1685–93.

Luikham, S., Malve, S., Gawali, P., Ghosh, S., 2018. A novel strategy towards agro-waste mediated dye biosorption for water treatment. *World Journal of Pharmaceutical Research* 7(4), 197–208.

Mahmoud, M.E., 2015. Water treatment of hexavalent chromium by gelatin-impregnated-yeast (Gel-Yst) biosorbent. *Journal of Environmental Management* 147, 264–270.

Mahmoud, M.S., Mohamed, S.A., 2017. calcium alginate as an eco-friendly supporting material for Baker's yeast strain in chromium bioremediation. *HBRC Journal* 13(3), 245–254.

Martín-Domínguez, A., Rivera-Huerta, M.L., Pérez-Castrejón, S., Garrido-Hoyos, S.E., Villegas-Mendoza, I.E., Gelover-Santiago, S.L., Drogui, P., Buelna, G., 2018. Chromium removal from drinking water by redox-assisted coagulation: Chemical versus electrocoagulation. *Separation and Purification Technology* 200, 266–272.

Morales, D.K., Ocampo, W., Zambrano, M.M., 2007. Efficient removal of hexavalent chromium by a tolerant *Streptomyces* sp. affected by the toxic effect of metal exposure. *Journal of Applied Microbiology* 103(6), 2704–2712.

Morales-Barrera, L., Cristiani-Urbina, E., 2006. Removal of hexavalent chromium by *Trichoderma viride* in an airlift bioreactor. *Enzyme and Microbial Technology* 40(1),107–113.

Okeke, B.C., 2008. Bioremoval of hexavalent chromium from water by a salt tolerant bacterium, *Exiguobacterium* sp. GS1. *Journal of Industrial Microbiology & Biotechnology* 35(12), 1571–1579.

Owlad, M., Aroua, M.K., Daud, W.A.W., Baroutian, S., 2009. Removal of hexavalent chromium-contaminated water and wastewater: A review. *Water, Air, and Soil Pollution* 200(1–4), 59–77.

Ozdemir, G., Ozturk, T., Ceyhan, N., Isler, R., Cosar, T., 2003. Heavy Metal biosorption by biomass of *Ochrobactrum anthropi* producing exopolysaccharide in activated sludge. *Bioresource Technology* 90(1), 71–74.

Pakade, V.E., Tavengwa, N.T., Madikizela, L.M., 2019. Recent advances in hexavalent chromium removal from aqueous solutions by adsorptive methods. *RSC Advances* 9(45), 26142–26164.

Park, C.H., Keyhan, M., Wielinga, B., Fendorf, S., Matin, A. 2000. Purification to homogeneity and characterization of a novel *Pseudomonas putida* chromate reductase. *Applied and Environmental Microbiology* 66(5), 1788–1795.

Park, J.K., Lee, J.W., Jung, J.Y., 2003. Cadmium uptake capacity of two strains of Saccharomyces cerevisiae cells. *Enzyme and Microbial Technology* 33(4), 371–378.

Pattanapipitpaisal, P., Brown, N.L., Macaskie, L.E., 2001. Chromate reduction and 16S rRNA identification of bacteria isolated from a Cr(VI)-contaminated site. *Applied Microbiology and Biotechnology* 57(1–2), 257–261.

Prigione, V., Zerlottin, M., Refosco, D., Tigini, V., Anastasi, A., Varese, G.C., 2009. Chromium removal from a real tanning effluent by autochthonous and allochthonous fungi. *Bioresource Technology* 100(11), 2770–2776.

Qu, Y., Zhang, X., Xu, J., Zhang, W., Guo, Y., 2014. Removal of hexavalent chromium from wastewater using magnetotactic bacteria. *Separation and Purification Technology* 136, 10–17.

Ramrakhiani, L., Majumder, R., Khowala, S., 2011. Removal of hexavalent chromium by heat inactivated fungal biomass of *Termitomyces clypeatus*: Surface characterization and mechanism of biosorption. *Chemical Engineering Journal* 171(3), 1060–1068.

Saha, R., Nandi, R., Saha, B., 2011. Sources and toxicity of hexavalent chromium. *Journal of Coordination Chemistry* 64(10), 1782–1806.

Sathvika, T., Balaji, S., Chandra, M., Soni, A., Rajesh, V., Rajesh, N., 2019. A co-operative endeavor by nitrifying bacteria *Nitrosomonas* and zirconium based metal organic framework to remove hexavalent chromium. *Chemical Engineering Journal* 360, 879–889.

Sharma, S.K., Petrusevski, B., Amy, G., 2008. Chromium removal from water : A review. *Journal of Water Supply: Research and Technology-Aqua* 57(8), 541–553.

Shen, J., Hsu, C.M., Kang, B.K., Rosen, B.P., Bhattacharjee, H., 2003. The *Saccharomyces cerevisiae* Arr4p is involved in metal and heat tolerance. *BioMetals* 16(3), 369–378.

Singh, R., Kumar, M., Bishnoi, N.R., 2016. Development of biomaterial for chromium (VI) detoxification using *Aspergillus flavus* system supported with iron. *Ecological Engineering* 91, 31–40.

Srivastava, S., Thakur, I.S., 2006. Evaluation of bioremediation and detoxification potentiality of *Aspergillus niger* for removal of hexavalent chromium in soil microcosm. *Soil Biology and Biochemistry* 38(7), 1904–1911.

Srivastava, S., Thakur, I.S., 2007. Evaluation of biosorption potency of *Acinetobacter* sp. for removal of hexavalent chromium from tannery effluent. *Biodegradation* 18(5), 637–646.

U.S. Environmental Protection Agency (U.S. EPA) (1998) Toxicological review of hexavalent chromium. U.S. EnvironmentalProtection Agency, Washington, DC. https://cfpub.epa.gov/ncea/iris/iris_documents/doc uments/toxreviews/0144tr.pdf

Vendruscolo, F., Ferreira, G.L.R., Filho, N.R.A., 2017. Biosorption of hexavalent chromium by microorganisms. *International Biodeterioration and Biodegradation* 119, 87–95.

Viamajala, S., Peyton, B.M., Apel, W.A., Petersen, J.N., 2002. Chromate reduction in *Shewanella oneidensis* MR-1 is an inducible process associated with anaerobic growth. *Biotechnology Progress* 18(2), 290–295.

Vimala, R., Das, N., 2011. Mechanism of Cd(II) adsorption by macrofungus *Pleurotus platypus*. *Journal of Environmental Sciences* 23(2), 288–293.

Xu, W., Duan, G., Liu, Y., Zeng, G., Li, X., Liang, J., Zhang, W., 2018. Simultaneous removal of hexavalent chromium and o-dichlorobenzene by isolated *Serratia marcescens* ZD-9. *Biodegradation* 29(6), 605–616.

Zahoor, A., Rehman, A., 2009. Isolation of Cr(VI) reducing bacteria from industrial effluents and their potential use in bioremediation of chromium containing wastewater. *Journal of Environmental Sciences* 21(6), 814–820.

Zhao, C., Yang, Q., Chen, W., Teng, B., 2012. Removal of hexavalent chromium in tannery wastewater by *Bacillus cereus*. *Canadian Journal of Microbiology* 58(1), 23–28.

Zheng, Z., Yabo, L., Zhang, X., Liu, P., Ren, J., Wu, G., Zhang, Y., Chen, Y., Li, X., 2015. A *Bacillus subtilis* strain can reduce hexavalent chromium to trivalent and an *nfrA* gene is involved. *International Biodeterioration and Biodegradation* 97, 90–96.

Zhitkovich, A., 2011. Chromium in drinking water: Sources, metabolism, and cancer risks. *Chemical Research in Toxicology* 24(10), 1617–1629.

11 Bioremediation of Cr(VI)-Contaminated Soil using Bacteria

Vandita Anand and Anjana Pandey

CONTENTS

DOI: 10.1201/9781003204442-11

11.1 INTRODUCTION

Population growth and rapid industrialization have led to the overexploitation of available resources and unregulated disposal of industrial wastes in the environment. Any metallic element with relatively high density compared to water and toxic even at low concentrations is termed "heavy metal" (Lenntech, 2004). Industrial effluents are produced by the incorporation of organic and inorganic contaminants, as well as by discharged heavy metals such as chromium, copper, cadmium, lead, and selenium into the environment without appropriate treatment, resulting in a worldwide socio-environmental problem (Wang and Chen, 2006). Heavy metal contamination is a severe problem to the environment because the anthropogenic activities from mining, processing, and the application of these metals have increased enormously during the past few decades and have become a challenge for life on Earth. Hence, their removal/remediation has become all the more necessary. Thus, the removal of metal must be applied effectively and without causing an impact on the environment. The most widely used methods are the conventional physico-chemical processes such as reverse osmosis, the electrochemical process, ion exchange, adsorption on activated carbon, excavation, and solidification/stabilization. These technologies reduce the effects of heavy metals, but they present significant disadvantages such as generating toxic waste sludge, high energy requirements, or incomplete removal (Bahi et al., 2012).

The potential for the adaptation and growth of heavy metal-resistant microorganisms (e.g., bacteria, fungi, and yeasts) led to the hypothesis that biological removal methods would be a sustainable alternative technology with lower impact on the environment. Different microorganisms have been isolated and identified as having the capacity to remove heavy metal contamination by different biological methods (e.g., biosorption, bioaccumulation, and bioreduction). Bioremediation technology has been reported to be more effective for the removal of soluble and particulate forms of metals, especially from dilute solutions. Hence, bioaccumulation and microbe-based technologies can provide an alternative to the conventional techniques of metal removal/recovery. Bioremediation is used to transform toxic heavy metals using microbes or their enzymes to clean up the polluted environment into a less harmful state. The technique is environmentally friendly and cost effective in revitalizing the environment. The bioremediation of heavy metals has limitations. Among these are microbes producing toxic metabolites and heavy metal non-biodegradability.

Chromium is one of the most frequently used metal contaminants and is considered one of the top 20 contaminants on the Superfund priority list of hazardous substances for the past 15 years (Benedetti, 1998). Chromium is commonly distributed in the environment and derives from both natural and anthropogenic sources. A significant number of industrial activities, such as electroplating, chromate processing, leather tanning, and wood preservation, release chromium into the environment. The untreated release of Cr induces significant anthropogenic pollution because of its non-degradable and persistent properties. Untreated industrial effluents are also a critical threat to public health because heavy metals have biomagnification properties and accumulate in the food chain, causing toxicity at a cellular level. While chromium exists in nine valence states ranging from -2 to $+6$, because of its stability in the natural environment, Cr(III) and Cr(VI) are of great environmental significance. Cr(VI) is toxic, carcinogenic, and mutagenic to animals and humans and is associated with decreased plant growth and changes in plant morphology (James, 2002).

Chromium-bearing ores are found in many forms, but the economically extractable form is the mineral chromite. Mineral chromite in its spinel form is inert and insoluble in water. Its total world mine production in 2010 and 2011 was 23,700 and 24,000 $\times 10^3$ metric tons, respectively (gross weight of marketable chromite ore). The total world reserve is estimated at $>480 \times 10^6$ metric tons of shipping-grade chromite ore with $\sim45\%$ Cr_2O_3, and major resources are located ($\times10^6$ tons) in Kazakhstan (220), Southern Africa (200), India (54), and the United States (0.62). They are non-degradable toxic pollutants and thus are persistent and accumulate in the food chain, which with time reach harmful levels in living systems, resulting in several diseases such as irritation or cancer

in the lungs and digestive tract, low growth rates in plants, death of animals, and other health alterations (Modoi et al., 2014).

In both of its prevalent forms, trivalent, and hexavalent chromium, chromium can cause allergic contact dermatitis. It has been estimated that Cr(VI) is 100 times more toxic and 1,000 times
more mutagenic than Cr(III) due to its high solubility, availability, and mobility in the soil, as well
as its ability to penetrate biological membranes. In its hexavalent form, the U.S. Environmental
Protection Agency (USEPA) has classified chromium as a Group "A" human carcinogen and is one
of the main pollutants. Worldwide chromate, which is the most prevalent form of Cr(VI) present in
solid/liquid waste due to human activities, such as electroplating, steel and automobile manufacturing, mining, leather tanning, cement and metal processing, textiles and the production of paint
pigments and dyes, and wood preservation (Kamaludeen et al., 2003). CrO_4^{2-} and $HCrO_4^-$ ions are
the most mobile forms of chromium in soils. They can be taken up by plants and easily leached
into the deeper soil layers, leading to ground and surface water pollution. Cr(VI) toxicity in metal
complexes can easily cross cellular membranes and trigger intracellular reactive oxygen species
(ROS) accumulation altering cell structures. Since several microorganisms possess the capability
to reduce Cr(VI) to relatively less toxic Cr(III), bioremediation gives immense opportunities for the
development of technologies to detoxify Cr(VI)-contaminated soils as an alternative to the existing
physico-chemical technologies.

Considering the socio-environmental impact of Cr, the present chapter highlights the chemistry of
chromium, its toxic effects on human and animal health, and its bioremediation by microorganisms.

11.2   CHROMIUM

11.2.1   Chemistry of Chromium

Chromium (Cr) is an element that belongs to the transition metals, Group 6 (VIb), and an electronic
configuration (Ar) $4d^5s^1$ is presented in the elementary oxidation state of the periodic table. The
chromium element (from the Greek *chrōmos*, "color") and varied chromium compound colorations.
It is naturally existing with atomic number 24 and an atomic mass of 51.996 amu. Chromium is the
seventh most abundant element in the Earth's crust, and is naturally found in rocks, animals, plants,
volcanic dust, and soil at concentrations ranging from 100–300 µg g^{-1}. Chromium is hard, sheen,
odourless, and anti-corrosive. It also has high melting and boiling points: 2,180 K and 2,944 K,
respectively.

In nature, Cr is found in the form of its compounds, and chromite (Fe, Mn) Cr_2O_4 is the most
important chromium ore (McGrath et al., 1990). It is a hard, steel-grey metal that takes a high polish and is used in alloys to increase strength and corrosion resistance. Chromium alloys are used to
manufacture products such as oil tubing, automobile trimming, and cutlery. Chromite is used as a
refractory and as a raw material for chromium chemical processing.

Chromium exists in different oxidation states; the most stable and standard forms are trivalent
[Cr(III)] and hexavalent [Cr(VI)], which have very different chemical properties. Cr(III), in the form
of oxides, hydroxides, or sulfates, exists frequently bound to the organic matter in soil and aquatic
environments. Cr(VI) is typically associated with oxygen as chromate (CrO_2^{-4}) or dichromate
($Cr_2O_2^{-7}$) ions. Cr(VI) is a strong oxidizing agent, and in the presence of organic matter is reduced
to Cr(III). This transformation is more rapid in acidic environments, such as acidic soils (McGrath
et al., 1990). However, high levels of Cr(VI) may block the reducing abilities of the environment,
and thus persist in this form.

Chromium is an essential micronutrient for living organisms, as it is indispensable for the carbohydrate, lipid, and protein metabolism of mammals and yeasts (Mordenti and Piva, 1997). Its
deficiency in the diet causes alteration to lipid and glucose metabolism in animals and humans.
Chromium is included in the complex named glucose tolerance factor (GTF). The biologically active

component of chromium, the glucose tolerance factor, is an important dietary agent that potentiates insulin action and thus functions in regulating the metabolism of carbohydrates.

11.2.2 USE OF CHROMIUM AND ENVIRONMENTAL CONTAMINATION

Chromium and its compounds are useful in everyday life. It is present in effluents originated from the different activities. It represents a serious pollutant of sediments, soil, water, and air which results in significant quantities of Cr(VI) in the environment, and which may create a toxicological hazard to humans, animals, and plants (Saha et al., 2011). Due to its corrosion-resistant consistency and hardness, Cr has tremendous industrial applications. It is used at a large scale in various industries, including metallurgical, electroplating, tanning, wood preservation, manufacturing of stainless steel, production of dye, ink, pulp, and paper (McGrath et al., 1990).

In the total production of chrome ore, 90% is used in the metallurgical industry for steel, alloy, and non-ferrous alloy production; 5% is used in the refractory industry for iron and steel, cement, glass, ceramics, and machinery; and 5% is used in the chemical industry for leather tanning, plating, wood preservation, and pigments (see Figure 11.1). Cr(VI) is widely distributed in sediments and surface waters, and is more soluble, mobile, and bioavailable than Cr(III) and other types of chromium. Chromium is resistant to ordinary corrosive agents at room temperature, responsible for its use in electroplating for protective coating. These steels have a wide range of mechanical properties, besides being corrosion and oxidation resistant. Cast iron contains chromium at 0.5–30%, provides hardness and toughness, as well as resistance to corrosion and wear. Chromium is also used in non-ferrous alloys combined with nickel, iron-nickel, cobalt, aluminum, titanium, and copper (Bielicka et al., 2005). Other applications include leather tanning, metal corrosion inhibition, drilling mud, textile dyes, catalysts, wood, and water treatment. Chromite is used in the refractory industry to make bricks, mortar, and ramming and gunning mixes. Chromite increases their volume stability, strength, and resistance to thermal shock and slag.

Cr(VI), in the forms of anions chromate (CrO_4^{2-}) and bichromate ($HCrO_4^{-}$), is extremely soluble and mobile in the presence of oxidizing conditions. In anaerobic environments, under reduced conditions, C (VI) may rapidly convert to Cr(III) in the presence of reducing agents such as sulfides, ferrous iron, and organic matter, which are some of the organic and inorganic constituents. Again, a bacterially mediated reduction of Cr(VI) in the biogeochemical process of chromium has also been considered.

In the chemical industry, chromium is used as oxidizing agents and in the production of other chromium compounds. About 80–90% of leather is tanned with chromium chemicals (Papp, 2004),

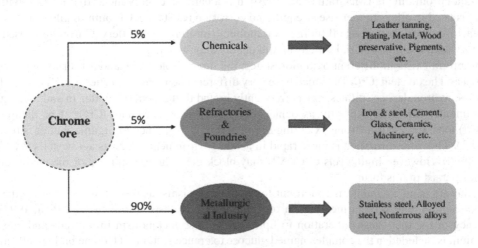

FIGURE 11.1 Percentage use of chromium in different industries.

from which about 40% of the chromium used is discharged as Cr(VI) and Cr(III) into the effluents. The tannery industry is one of the largest industries that uses Cr compounds to produce tan hides (animal skins), and is therefore a major contributor to chromium contamination (Saha et al., 2011). The Cr used in the process is not completely absorbed by the leather, and a large proportion of it runs off into the effluent.

Approximately 40 million tons of waste are produced annually by tanneries around the world that produce Cr, a chief contaminant of heavy metals (Papp, 2004). This waste is mainly disposed of by industries on land and in water bodies without receiving any required treatment. Severe contamination of productive agricultural land and water bodies due to the disposal of tannery waste was observed in Australia (Megharaj et al., 2003), Bangladesh (Park et al., 2000), and India (Thacker and Datta, 2006).

According to USEPA (2018), the permissible concentration for all forms of Cr, inclusive of Cr(VI), is 0.1 mg L^{-1}. The concentration of Cr(VI) and total Cr prescribed under the Indian Standards Specification for drinking water is 0.05 mg L^{-1}. Realizing the seriousness of chromium pollution issues, the Blacksmith Institute has identified some sites as contamination hot spots and recommended them for instant remediation. Over 300 Cr-contaminated sites around the world have been identified through the toxic site identification program. Pure Earth has estimated that about 16 million people are at risk from chromium exposure (Jobby and Desai, 2017).

11.2.3 Toxicity of Chromium

Elevated levels of chromium are always toxic, although the toxicity level is related to the chromium oxidation state. Not only is Cr(VI) highly toxic to all forms of living organisms, but it is mutagenic in bacteria, and carcinogenic in humans and animals (Losi et al., 1994b). Moreover, it is linked to congenital disabilities and reproductive health issues. LD50 (the dose that causes the death of 50% of a defined animal population) for oral toxicity in rats is from 50–100 mg/kg for hexavalent chromium, and 1,900–3, mg/kg for trivalent chromium (Deflora et al., 1990). Cr(VI) toxicity is related to its easy diffusion across the cell membrane in prokaryotic and eukaryotic organisms, and the subsequent Cr(VI) reduction in cells releases free radicals that might directly cause DNA alterations as well as have toxic effects. Cr(VI) exposure is highly toxic as it causes severe ill effects on human and animal health like diarrhea, ulcers, eye and skin irritations, perforation of the eardrum, respiratory tract disorders, kidney dysfunction, and lung carcinoma. It can also accumulate in the placenta, damaging fetal development. Cr(VI) may cause death in animals and humans if ingested in large doses. Considering the hazardous effects of Cr(VI), it has been classified as a priority pollutant and listed as a Class A human carcinogen by the USEPA.

The toxicity of Cr(VI) is attributed to its ability to penetrate cell membranes easily. In both eukaryotic and prokaryotic cells, cell membranes are damaged caused by oxidative stress induced by Cr(VI), with effects such as loss of membrane integrity or inhibition of the electron transport chain. Moreover, Cr(VI) enters the cells of organisms using the sulfate transport system of the membrane (Cervantes et al., 1992). If Cr(VI) has entered the cells, spontaneous reactions occur with ascorbate and glutathione intracellular reductants, producing the short-lived intermediates Cr(V) and Cr(IV), free radicals, and the end product Cr(III) (Xu et al., 2006). Cr(V) in the cytoplasm is oxidized to Cr(VI), and the process creates an ROS, which is quickly mixed with DNA-protein complexes. By contrast, Cr(IV) may bind to cellular materials, modifying their normal physiological functions (Cervantes et al., 2001). Cr(VI) species and hydroxyl radicals are able to cause DNA lesions in vivo. The intermediates produced from the action of Cr(VI) are dangerous to cell organelles, proteins, and nucleic acids.

Browning and Wise (2017) discovered the mechanism of carcinogenesis in lung carcinoma, which is induced by particulate Cr(VI). A stable, heritable structural change in chromosome resulted in the formation of lung carcinoma. Cr(VI) has been found to interrupt the normal pathway of homologous recombination (HR) repair of DNA double-strand breaks by changing the subcellular localization of protein RAD51, an essential protein for the HR repair mechanism. An organism may encounter Cr(VI) from contaminated food or food supplements, water, or air. The mean daily dietary intake of Cr is estimated to be <0.2–0.4, 2.0, and 60 mg from the air, water, and food, respectively (ATSDR). Cr inhalation can cause nasal perforation, thus increasing the risk of respiratory tract diseases. Though most reports are based on the toxic effects of Cr(VI), Cr(III), though less toxic, is also reported to cause damage to lymphocyte DNA.

Cr(VI) environmental degradation changes the composition of soil microbial communities, reducing microbial growth and associated enzymatic activity, resulting in the persistence of organic matter in soils and Cr(VI) accumulation (Zhou et al., 2002). In many studies, it has been challenging to assess the toxicity of chromium to soil microorganisms because the environments examined were often simultaneously polluted with organic pollutants or different heavy metals. In a soil chronically polluted with chromium (about 5,000 mg/kg of soil) by leather tannery activities, the oxygenic phototrophic microorganisms and heterotrophic bacterial communities were affected by chromium (Viti et al., 2001).

Cr(VI) is also associated with a reduction in nutrient uptake and photosynthesis, which results in slow plant growth. It severely affects several physiological, morphological, and biochemical processes, and induces the generation of ROS in plant cells. The toxicity of Cr is indicated in the form of chlorosis and necrosis in plants (Shahid et al., 2017). Additionally, its tendency to bioaccumulate in living tissues as it travels through the food chain creates a concern for its removal from polluted sites. There are also adverse effects to tannery workers of iron metabolism due to the excessive accumulation of Cr in the body.

11.3 TRANSFORMATIONS OF CHROMIUM IN SOIL: MOBILITY AND BIO-AVAILABILITY

Chromium, a steel-grey, lustrous, hard, and brittle metal, occurs in nature in the bound form that constitutes 0.1–0.3 mg kg^{-1} of the Earth's crust in the form of chromite ore. Naturally, Cr is present in trace amounts in all sorts of components of the environment, including air, water, and soil. It is a transition metal and belongs to group VI in the periodic table as the first element of the group. It occurs in various oxidation states, from Cr(II) to Cr(VI). As mentioned above, Cr(III) and Cr(VI) are the most common and highly stable in soil, possess different chemical and physical characteristics. Cr (VI) is relatively immobile because it has a strong affinity for negatively charged ions and colloids in soils, and gives sparingly soluble compounds such as $Cr(OH)_3$. Such products dominate in the pH range of 4–8. Cr(VI) is an anion form under most environmental conditions. At pH values higher than 6.4, it is primarily present as chromate (CrO_4^{-2}), whereas below the pH value of 6.4, it is present principally as bichromate ($HCrO_4^-$) (James, 2002).

Living organisms, ferrous iron, sulphide, and organic matter can reduce hexavalent chromium. Losi et al. (1994b) have demonstrated that organic matter in the soil plays an essential role in the reduction of Cr(VI) to Cr(III) by creating reducing conditions by increasing thr activities of microbial communities, by acting as an electron donator, and by indirectly lowering the oxygen level of the soil (oxygen is depleted through an increase of microbial respiration). Therefore, the presence of an available carbon source to specific bacterial populations is fundamental to alleviate an environment from hazardous forms of chromium.

11.3.1 THE CHROMIUM CYCLES

Cr(VI) is the most oxidized, mobile, reactive, and toxic form of chromium, and it would be the only existing form if all chromium were to be in thermodynamic equilibrium with the atmosphere. Small concentrations can be the result of the oxidation of natural Cr(III), but larger concentrations usually are the result either of pollution with Cr(VI) or the oxidation of Cr(III). In soils, Cr(VI) is more soluble and bio-available than Cr(III). The ionic form of chromium that is absorbed by plants is Cr^{3+}

and Cr^{6+}; Cr^{3+} is absorbed more rapidly than Cr^{6+}. The different oxidation states (stable/unstable) found in the environment (Cr compounds) are 0 [$Cr(CO)_6$]; +1 (unstable); +2 (chromous) (unstable) [$Cr(CH_3COO)_4$, CrO, $CrSO_4$]; +3 (chromic) (stable) [$CrCl_3$, Cr_2O_3, $Cr_2(SO_4)_3$]; +4 (unstable) CrO_2; +5 (unstable) (CrF_5), and +6 (stable) ($K_2Cr_2O_7$, $K_2Cr_2O_4$, CrO_3) (Thacker et al., 2006). The characteristics of Cr(III) forms limit their bioavailability and mobility in waters and soils. The concentrations of soluble Cr(III) in equilibrium with insoluble compounds are $<10^{-9}$M (0.05 parts per billion) in the water at a pH value of 6, to less than 10^{-15}M at a pH value of 8. The reduction of Cr(VI) to Cr(III) is easier compared to the oxidation of Cr(III) to Cr(VI) by oxidizing agents in soils. Soils and sediments in partial equilibrium with atmospheric oxygen contain both oxidized manganese and reduced carbon. Cr(VI) naturally exists in the environment due to the oxidation of Cr(III). But when increased concentrations are released by various industrial effluents, this leads to an excess of Cr(VI) taken up in the biosphere and leads to toxicity. Figure 11.2 depicts the natural Cr cycle in the environment (Dhal et al., 2013).

The oxidation of Cr(III) to Cr(VI) by manganese oxides (see Equation 10.1) and the reduction of Cr(VI) to Cr (III) by soil carbon compounds are both thermodynamically spontaneous reactions (see Equation 10.2). The interesting aspect of the chromium cycle in soil is that oxidation and reduction can occur simultaneously. Some of the Cr(III) added to a sample of an aerobic soil will be oxidized, and some of the Cr(VI) added to the same soil will be reduced.

The oxidation (Apte et al., 2005) of Cr(III) to Cr(VI) by manganese oxides through surface oxidation under neutral pH conditions proceeds as:

$$Cr^{3+} + 1.5\,MnO_2 + H_2O \rightarrow HCrO_4^- + 1.5\,Mn^{2+} + H^+ \tag{10.1}$$

Whereas reduction by organic compounds, e.g., hydroquinone (with the formation of quinine) (James, 2002) proceeds as:

$$C_6H_6O_2 + CrO_4^{2-} + 2H_2O \rightarrow 0.5Cr_2O_3 + 1.5C_6H_4O_2$$
$$+ 2.5H_2O + 2OH^{-\,\text{''}}\ G298^n = -427\ kJ\ mol^{-1} \tag{10.2}$$

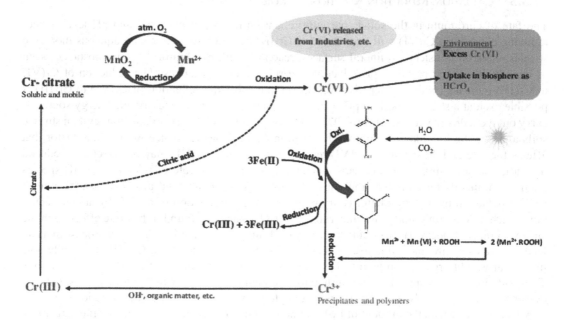

FIGURE 11.2 Chromium cycle.

The potential chromium oxidation score (PCOS) developed by James et al. (1997) in order to design remediation processes by reduction strategies is based on four interacting parameters: solubility and form of Cr (III), reactive soil manganese, soil potential for Cr (VI)-reduction, and soil pH as a modifier of the first three parameters. Such parameters can be quantified and ranked numerically; the sum of their values gives the PCOS. The PCOS ranges from 10 to 40; high scores indicate an elevated probability for Cr(III) oxidation and the persistence of Cr(VI). Therefore, the steps for developing remediation processes of chromium contaminated soils should consider the possibility that certain forms of Cr(III) can be more favorably oxidized to Cr(VI).

11.3.2 Speciation of Cr(VI)

Cr(VI) is dominant under oxidizing conditions, while Cr(III) predominates under reducing conditions. In aqueous solutions, Cr(VI) can form several species, namely $Cr_2O_7^{2-}$, CrO_4^{2-}, H_2CrO_4, and $HCrO_4^-$. This distribution depends on the pH of the solution, the total Cr concentration, the oxidizing and reducing compounds, the redox potential, and the kinetics of the redox reactions. H_2CrO_4 is a strong acid (Sperling et al., 1992), and at pH >1 a deprotonated form of Cr(VI) is seen. Above pH 7, only CO_4^{2-} ions exist in solution throughout the concentration range. In the pH between 1 and 6, $HCrO_4^-$ is the predominant species.

$$H_2CrO_4 \rightarrow H^+ + HCrO_4^-, \quad K1 = 10 - 0.75 \tag{10.3}$$

$$HCrO_4 \rightarrow H^+ + CrO_4^{2-}, \quad K2 = 10 - 6.45 \tag{10.4}$$

$$2HCrO_4^- \rightarrow Cr_2O_7^{2-} + H_2O, \quad K3 = 10 - 2.2 \tag{10.5}$$

As mentioned earlier Cr(VI) compounds are quite soluble and thus mobile in the environment. However, Cr(VI) oxyanions are readily reduced to trivalent forms by electron donors such as organic matter or reduced inorganic species, which are ubiquitous in soil, water, and atmospheric systems.

11.3.3 Oxidation/Reduction Reactions in Soil

The fate of chromium in the soil is partially reliant on the redox potential and pH level. Under reducing conditions, Cr(VI) will be reduced to Cr(III) by redox reactions with aqueous inorganic species, electron transfers at mineral surfaces, reactions with non-humic organic substances such as carbohydrates and proteins, or reductions by soil humic substances. The reduction of Cr(VI) to Cr(III) increases with a lower pH level. In aerobic soils, the reduction of Cr(VI) to Cr(III) is possible even at a slightly alkaline pH if the soil contains an appropriate organic energy source to carry out the redox reaction (Bartlett, 1976). Sub-soils have much lower organic matter than surface soils and may be less likely to inhibit chromate mobility. However, since pH is a key factor that affects the rate and the extent of Cr(VI) reduction in subsoils, Cr(VI) may be effectively reduced in acidic sub-soils simply because acidic conditions enhance the rate of release of Fe(II) species from soil minerals for their reaction with aqueous Cr(VI) species, and also increase the rate of Cr(VI) reduction by organic matter (Eary et al., 1991). The reduction of Cr(VI) by abiotic reductants, such as Vitamin C and nano-materials (UV/TiO$_2$), has been found to be quite effective in the entire pH range. Under certain circumstances, Cr(III) may be oxidized to Cr(VI), a process that can lead to serious environmental consequences (Xu, 2006). The oxidation of Cr(III) chelated by low molecular weight organic compounds is slower than freshly precipitated Cr(III) (less crystalline). The oxidation process appears to be limited to oxidation by oxygen or manganese oxides. Oxygen does not react appreciably with Cr(III), according to Eary and Rai (1991). The oxidation of Cr(III) may be correlated with the amount of hydroquinone reduced manganese in the soil (Bartlett, 1976) and does not occur in dry soils. The drying of the soil alters the manganese surface, decreasing its ability to oxidize Cr(III). The oxidation of Cr(III) to Cr(VI) is also limited by the concentration of water-soluble chromium, pH, initially-available surface area, and ionic strength (Apte, 2005). A large portion of chromium in the soil will not be oxidized to Cr(VI) even in the presence of manganese oxides and favorable pH conditions, due to the unavailability of mobile Cr(III). Cr(VI) is reduced to Cr(III) in reduction reactions, or Cr(III) is oxidized to Cr(VI) in oxidation reactions. The pH, oxygen concentration, and presence and concentration of reducers are important in these processes. Cr(VI) mobile forms that are $HCrO_4^-$ and CrO_4^{2-} can be reduced by different inorganic reducers such as Fe(II) or S^{2-}; this process, called "dechromification," is quite important. In the absence of such a process, atmospheric oxygen could be converted into chromate, which would pose a threat to life on Earth. In the presence of oxidizing agents such as Mn and Pb, the oxidation processes can oxidize Cr(III) to Cr(VI) in the presence of H_2O and free O_2.

11.3.4 Bacterial Interaction with Chromium in Soils

Soil contamination with heavy metals is sometimes irreversible and may repress or eliminate parts of the microbial community, so exposure to metals is usually believed to result in the development of a tolerant/resistant microbial population. The terms "resistance" and "tolerance" are often used interchangeably, but their significance is different. Gadd (1992) defined resistance as "the ability of a microorganism to survive toxic effects of metal exposure through a detoxification mechanism produced in direct response to the metal species concerned," and defined tolerance as "the ability of a microorganism to survive metal toxicity by means of intrinsic properties and or environmental modification of toxicity."

The bacterial chromate resistance is generally linked to plasmids, but it can also be due to chromosomal mutations (Ohta et al., 1971). Chromosomal and plasmid determinants function through different mechanisms. Chromosomal mutation usually affects the sulfate transport system, through which chromate enters the cells of many bacteria. Plasmid-determined resistance results in decreased chromate accumulation in cells without involving sulfate transport systems. The plasmid-coded chromate-resistance has mainly been thought to be based on the efflux of chromate (Cervantes et al., 2001). However, the mechanisms of chromate-resistance have not yet been fully elucidated. The capability of Cr(VI) reduction is suggested as an additional chromosome or plasmid-resistance mechanism and represents a potentially useful detoxification process for several bacteria. Thereby, the bacterial property, which is particularly useful for an effective bioremediation approach, combines high tolerance/resistance with the ability to reduce Cr(VI) to less toxic Cr(III). Microbial Cr(VI) reduction may occur directly through enzymatic activity or indirectly through producing a compound that can reduce Cr(VI) (Fendorf and Li, 1996). The ability of direct Cr(VI) reduction has been found in many bacterial genera, including *Pseudomonas, Micrococcus, Bacillus, Achromobacter, Microbacterium, Arthrobacter*, and *Corynebacterium*. The capability of Cr(VI) reduction is not uncommon in the Cr(VI)-resistant bacteria of soils. Losi et al. (1994b) found that 9 out of 20 Cr(VI)-resistant bacterial strains, isolated from organic-amended and Cr(VI)-acclimated soils, showed the capability to actively reduce Cr(VI) to Cr(III).

Some bacterial species are capable of both anaerobic and aerobic hexavalent chromium reductions, and others are capable of either anaerobic or aerobic hexavalent chromium reductions. The mechanisms through which bacterial strains reduce Cr(VI) to Cr(III) are variable and species-dependent (McLean et al., 2000). Anaerobic bacteria may use chromate as a terminalelectron acceptor, or hydrogenase or cytochrome c3 may reduce chromate in periplasmatic spaces. In aerobic bacteria, Cr(VI) reduction may be carried out by cellular reducing agents (the primary reductant is glutathione) and NADH-dependent chromate reductase (Puzon et al., 2002). It is yet unknown, although some hypotheses have been formulated, whether an enzymatic or non-enzymatic reduction of chromate is dominant in bacterial cells under aerobic conditions, and it also remains unsolved whether the NADH-dependent reductases are specific to chromate. Moreover, it is also unclear whether anaerobic bacterial growth is supported at the expense of chromate as the only electron acceptor.

The mechanisms for Cr(VI) reduction might be a secondary utilization or co-metabolism, as suggested for *Shewanella onoidensis* MR-1. Therefore, under anaerobic conditions, Cr(VI) reduction may be an activity of the reductases that have evolved on other substrates (Kamaludeen et al., 2003). The ability of a bacterial strain to reduce hexavalent chromium, although the mechanism of Cr(VI) reduction may differ from strain to strain, is an attractive feature in order to plan a biological strategy for effective chromate detoxification. However, high concentrations of chromate in the environment can often repress the microbial activity and growth (Megharaj et al., 2003). Therefore, before using a selected microorganism or an indigenous microbial mixed-culture for devising bioremediation strategies for Cr(VI)-contaminated soils, there is a need to understand Cr(VI)-resistance mechanisms in these microorganisms.

Nevertheless, it has recently been demonstrated that Cr(VI) is toxic even at low concentrations (0.015 mM) to *Shewanella oneidensis* MR-1, a good Cr(VI) reducer. It is important to remember that chromate resistance and reduction are not necessarily interrelated. Hexavalent chromium may be reduced by both bacterial Cr(VI)-sensitive and Cr(VI)- resistant strains, and not all Cr(VI)-resistant bacteria reduce hexavalent chromium to trivalent forms.

The reduction of Cr(VI) can also occur indirectly by bacterial activity. For instance, $Fe(II)$ and HS^-, metabolic end products of iron and sulfate-reducing bacteria, can catalyze the reduction of Cr(VI) (Arias and Tebo, 2003). The indirect reduction of chromate using iron-reducing bacteria consists of two reactions. The $Fe(II)$ produced by reducing bacteria is cycled back to $Fe(III)$ by abiotic chromate reductions. At the ecological level, this process represents a significant role, because it permits the continuous regeneration of the $Fe(III)$ terminal electron acceptor in anaerobic conditions.

In sulfate-rich soil environments, when anaerobic conditions are present, such as in flooded compacted soils, the reduction of Cr(VI) by sulfide produced through sulfate-reducing bacteria, which couple the oxidation of organic sources to the reduction of sulfate, is an important mechanism to detoxify the environment from hexavalent chromium (Losi et al., 1994b).

11.4　BIOREMEDIATION OF Cr(VI) BY MICROORGANISMS

Conventional methods for removing metals from polluted environments include chemical precipitation, ion exchange, oxidation/reduction, filtration, membrane use, activated carbon, alum, kaolinite, and ash evaporation and adsorption (Ohta, 1971). However, most of these methods require high energy or large amounts of chemical reagents, with potential secondary pollution production. With regard to the removal of Cr(VI), conventional methods include chemical reduction followed by precipitation, ion exchange, and adsorption on activated carbon, alum, kaolinite, and ashes. The majority of these methods require high energy and large amounts of chemical reagents (Komori et al., 1990). In addition, the safe disposal of toxic sludge, incomplete reduction of Cr(VI), and high cost of reduction of Cr(VI), in particular for removal of relatively low Cr(VI) concentrations, are economically inconvenient.

Bioremediation, which uses a microorganism's metabolic ability to extract radioactive metals, represents an effective method to decontaminate contaminated areas. It is classified as either in-situ or ex-situ, depending on whether the intervention is carried out directly on the polluted site with suitable bacteria or on portions of environmental matrices, such as water, sediment, or soil, after removal and transportation to appropriate treatment facilities (Komori et al., 1990).

Cr(VI)-resistant microorganisms offer a significant opportunity to provide safe, economical, and environmentally-sustainable methods to reduce Cr(VI) to Cr(III) for potential applications for bioremediation. Reducing Cr(VI) to Cr(III) is then a potentially useful method for the recovery of Cr(VI)-contaminated sites. Cr(VI) removal based on microorganisms is now considered a significant alternative method to conventional methods, and the potential application for bioremediation is receiving considerable attention. Considering that Cr(III) insolubility facilitates its precipitation and removal, the biotransformation of Cr(VI) to Cr(III) was considered an alternative process for the treatment of Cr(VI)-contaminated wastes. The microbial reduction of Cr(VI) is cost effective and eco-friendly among the various biotechnological approaches and can offer a viable alternative.

Chromium-resistant microorganisms are responsible for the biological reduction of Cr(VI) to less mobile Cr(III), and the consequent precipitation may constitute an efficient method of the detoxification of polluted sites with Cr(VI) which could potentially be used in bioremediation (Jain et al., 2012). The biosorption and bioaccumulation of chromium have been demonstrated for bioremediation purposes. From this aspect, bacteria have been studied most widely, and the mechanisms of chromium tolerance or resistance of selected microbes are of particular importance in bioremediation technologies. There has been extensive study of the mechanisms of chromium toxicity and detoxification in bacteria, and some promising results have emerged in this field (Poljsak et al., 2010).

The bacterial remediation of Cr is characterized as a fast, economical, non-chemical, and less energy-intensive process that can be accomplished through the use of native, non-hazardous bacterial strains. Bacteria are proven to be most potent in converting Cr(VI) into Cr(III), its less toxic form (Kanmani et al., 2012). All gram-negative and gram-positive bacteria have shown the ability to remove Cr(VI) by biosorption or biotransformation. Microorganisms with the capabilities to tolerate and reduce Cr(VI) may be efficiently applied for Cr(VI) bioremediation (Camargo et al., 2003). The ability to reduce toxic hexavalent Cr has been reported in many species, including PGP rhizobacteria such as *Pseudomonas* sp., *Enterobacter* sp., *Bacillus* sp., *Burkholderia cepacian*, *Providencia* sp., *Ochrobactrum intermedium*, and *Staphylococcus* sp. under both aerobic and anaerobic conditions. *Corynebacterium paurometabolum* shows that both biosorption and biotransformation equally lead to reducing the toxicity of chromium compounds. *Bacillus coagulans* reduced Cr(VI) using a soluble enzyme and malate as an external donor of electrons. *B. circulans*, *B. coagulans*, and *B. megaterium* showed the ability to biosorb chromium at 34.5 mg Cr g^{-1}, 39.8 mg Cr g^{-1}, and 32 mg Cr g^{-1} of dry weight, respectively (Srinath et al., 2002). Moreover, the biosorption ability of living and dead cells of *B. coagulans* and *B. megaterium* were compared, and dead cells were found to be more efficient for chromium absorption. Compared to live cells, the improved output of inactive/dead biomass is due to their resistance to the toxic effects of metal ions, which can lead to cell death during the metal removal phase. Bacterial species belonging to the genus *Lactobacillus* are known for their capability to bind metals, including Cr(VI), and detoxify them. The involvement of chromate reductase enzymes during the conversion of Cr(VI) to Cr(III) by *Pseudomonas putida* was indicated, and it was found to use NADH or NADPH as electron donors (Kapoor et al., 1999).

A method for the bioremediation of metal-contaminated sites, including chromium, is represented by bioaugmentation-assisted phytoextraction, in which bacteria associated with plants that can accumulate metals were analyzed using a proposed bioremediation approach as a bioprocess. Bioaugmentation implementation favoring microbial survival was suggested to improve the relationship between microbial plants and process performance (Lebeau et al., 2008). The biomineralization process is a mechanism by which microorganisms convert the precipitation of aqueous metal ions, including chromium, into amorphous or crystalline. Biomineralization is considered a feasible and cost-effective method for chromium pollution remediation. The biomineralization process is a mechanism by which microorganisms convert the precipitation of aqueous metal ions, including chromium, into amorphous or crystalline. Further work on the bacterial biosorption and biotransformation of hazardous Cr(VI) to Cr(III) has been reported. A few of them have been summarized in Table 11.1.

11.5   BACTERIAL RESISTANCE TO Cr(VI)

Several microorganisms have been identified from contaminated chromate environments as well as uncontaminated habitats that exhibit Cr(VI)-reducing activities and resistance. Microorganisms capable of reducing Cr (VI) are typically called chromium-reducing bacteria (CRB). Among CRB, the gram-positive bacteria are shown to have significant tolerance to Cr(VI) toxicity at relatively high

TABLE 11.1

Remediation of Chromium (VI) Using Bacteria

Sl. no.	Name of Organism	Mechanism of Chromium Removal	Initial Cr(VI) Concentration	% Remediation	Time (in hrs)	Reference
1	B. coagulans	Biotransforma-tion	26 mg L^{-1}	100	72	Philip et al. (1998)
2	B. circulans	Biosorption, Bioaccumulation	50 mg L^{-1}	69	24	Srinath et al. (2002)
3	B. megaterium	Biosorption, Bioaccumulation	50 mg L^{-1}	64	24	Srinath et al. (2002)
4	B. coagulans (live cells)	Biosorption	50 mg L^{-1}	47.6	24	Srinath et al. (2002)
5	B. coagulans (dead cells)	Biosorption	50 mg L^{-1}	79.8 24	24	Srinath et al. (2002)
6	Bravibacterium sp. CrT-12 (live cells)	Reduction	100 mg L^{-1}	100	72	Faisal and Hasnain (2004)
7	Arthrobacter oxydans	Reduction	35 mg L^{-1}	100	10 d	Asatiani et al. (2004)
8	Providencia sp.	Reduction	400 mg L^{-1}	99.3	N/A	Thacker et al. (2006)
9	Pseudomonas stutzeri L1	Biosorption, Bioreduction	100–1,000 mg L^{-1}	97	24	Sathishkumar et al. (2016)
10	Acinetobacter baumannii L2	Biosorption, Bioreduction	1,000 mg L^{-1}	99.58	24	Sathishkumar et al. (2016)

concentrations, whereas gram-negative bacteria are much more sensitive to Cr(VI). Microorganisms found in metal-contaminated environments are naturally resistant to such metals.

It is known that resistance to chromate and reduction are not necessarily interrelated, and not all bacteria resistant to Cr(VI) can reduce Cr(VI) to Cr(III). Thus, both chromium resistance and reduction are found to be independent properties of bacteria. Despite Cr(VI) toxicity, some microorganisms show resistance to this heavy metal, indicating the ability to reduce Cr(VI) to Cr(III), as first reported for *Pseudomonas* spp., and the characterization of Cr(VI) reduction-capable bacteria was successively reported in 1979. Since then, numerous bacteria have indicated their ability to reduce Cr(VI) to Cr(III) as a resistance mechanism to Cr(VI).

The chromosomal resistance function in bacteria makes use of approaches such as specific or unspecific Cr(VI) reduction, free radical detoxification activities, DNA damage repair, and sulfur or iron homeostasis-related processes (Morais et al., 2011). Many microorganisms can endure toxic metal-polluted environments by developing mechanisms to avoid metal toxicity such as metal efflux, adsorption uptake, DNA methylation, and biotransformation of metals either directly by enzymatic reduction to less mobile and hazardous forms or indirectly by making metabolite complexes (such as H_2S). From a bioremediation point of view, the microbial reduction of Cr(VI) to Cr(III) is significant, which can be considered an additional mechanism for chromate resistance (Cervantes et al., 2001). Both aerobic and anaerobic microorganisms with high Cr(VI)-reducing potential have been reported, including *Pseudomonas, Bacillus, Enterobacter, Deinococcus, Shewanella, Agrobacterium, Escherichia, Thermus*, and other species (Ohtake et al., 1987). It has been reported that both chromate resistant and non-resistant strains can reduce chromate, but the continued growth is significantly inhibited at higher concentrations of chromate.

Furthermore, chromate reduction is completed by chromate reductase from different bacterial species generating Cr(III), which may be the target of detoxification because of other mechanisms. The most specific enzymes belong to the large family of NAD(P)H-dependent flavoprotein reductase. Many mechanisms of bacterial chromium resistance have been shown, and these mechanisms have been related to the expression of components of the DNA repair mechanism systems and are related to the processes of iron and sulfur homeostasis.

Bacteria represent different resistance mechanisms to overcome the Cr(VI) toxicity in the environment, which include the reduced uptake of Cr(VI), extracellular Cr(VI) reduction, ROS detoxifying enzyme/intracellular Cr(VI) reduction, DNA repair enzymes, efflux of Cr(VI) from the cell, and ROS scavenging are depicted in Table 11.2 and discussed below.

11.5.1   REDUCED UPTAKE OF Cr(VI)

Some of the successful defensive mechanisms against the lethal effects of Cr(VI) are probably associated with a reduction Cr(VI) absorption, similar to the sulfate uptake process, and with the homeostasis of sulfur or iron. Since chromate ion (CrO_4^{2-}) has a structural resemblances to tetrahedral sulfate ion (SO_4^{2-}), it can easily move through cell membranes through the SO_4^{2-} transportation pathway using non-specific anionic (SO_4^{2-}, PO_4^{3-}) transporters (Wenbo et al., 2000).

Chromate transport is reduced if the sulfate uptake pathway encoded in the chromosome is mutated in bacteria (Ramirez-Diaz et al., 2008). The microorganisms present in a metal-contaminated environment undergo rapid mutation to develop resistance to Cr(VI), leading to the reduced uptake of Cr(VI) by sulphate transport pathways. Susceptible organisms may become insensitive by mutation or by incorporation of the genetic information which encodes the resistance.

11.5.2   EXTRACELLULAR Cr(VI) REDUCTION

Another mechanism of resistance is the extracellular reduction of Cr(VI) to Cr(III) followed by binding to functional groups on the surface of bacterial cells (Ngwenya and Chirwa, 2011). The binding of reduced Cr(III) to the surface of bacterial cells helps to remove it quickly from the

TABLE 11.2

Bacterial Mechanisms for Cr(VI) Resistance

Enzyme/System	Bacterial Species	Function	Reference
	Transport		
ChrA transporter	*P. aeruginosa*	Efflux of cytoplasmic chromate	Alvarez et al. (1999)
Cys operon products	*Shewanella oneidensis*	Sulphate transport	Brown et al. (2006)
TonB receptor, hemin transporter	*Shewanella oneidensis*	Iron transport	Brown et al. (2006)
	Reduction		
chrB, chrC/chrF	*Ochrobactrum tritici* 5bvl1	Resistance to superoxide generating	Morais et al. (2011)
SOD, catalase	*Escherichia coli*	Combat of oxidative stress	Ackerley et al. (2004)
Outer membrane proteins	*Caulobacter crescentus*	General stress response	Hu et al. (2005)
	DNA repair		
RuvB	*Ochrobactrum tritici* 5bvl1	Repair of DNA damage	Morais et al. (2011)
RecG and RuvB DNA helicases	*Pseudomonas aeruginosa*	Repair of DNA damage	Miranda et al. (2005)
SO0368, UvrD, and HrpA helicases	*Shewanella oneidensis*	Repair of DNA damage	Chourey et al. (2006)
	Other mechanisms		
Cys operon products	*Shewanella oneidensis*	Sulphur metabolism	Brown et al. (2006)
Adenylyl sulphate kinase	*Shewanella oneidensis*	Sulphur metabolism	Brown et al. (2006)
Ferritin	*Shewanella oneidensis*	Iron binding	Brown et al. (2006)

polluted area. Peptidoglycan components found to be a potent binder of Cr(III) within the cell walls of bacteria. This has shown that certain bacterial species possess adsorptive properties which facilitate the removal of metal species from aquatic solutions. These properties mostly depend on the distribution on the cell wall surface of bacteria of reactive functional groups such as carboxyl, amine, hydroxyl, phosphate, and sulfhydryl (Parmar et al., 2000). Therefore, when their reduction happens extracellularly, there is no apparent entry of Cr(VI) in the cell.

11.5.3 ROS Detoxifying Enzymes/Intracellular Cr(VI) Reduction

During Cr(VI) reduction to Cr(III) a short-lived, highly reactive intermediate Cr(V) radical is generated, which redox cycles. Therefore, Cr(V) is oxidized back to Cr(VI), giving its electron to dioxygen and generating ROS, which generate oxidative stress in the bacteria. In this process, chromate also induces the bacterial proteins in defense against oxidative stress, which leads to an additional chromate resistance mechanism (Ramirez-Diaz et al., 2008). The oxidative stress caused by ROS is therefore completely nullified by detoxifying enzymes such as glutathione transferase, superoxide dismutase (SOD), catalase, etc. (Ackerley et al., 2004).

11.5.4 DNA Repair Enzymes

The protection of bacterial cells by DNA repair enzymes of damaged DNA caused by Cr(VI) is another defense shield. Cr(VI) reaches the bacterial cell and is readily reduced to Cr(III) through various enzymatic or non-enzymatic activities leading to ROS production, which has deleterious effects on the protein and DNA in the cell. The generated ROS causes DNA damage, such as base modification, single-strand breaks, and double-strand breaks. Such damage to DNA can be repaired through special DNA repair mechanisms such as the SOS response enzymes (RecA, RecG, RuvB) (Hu et al., 2005). For example, in *Pseudomonas aeruginosa*, DNA helicases such as RecG and RuvB, components of the recombinational DNA repair mechanism, have shown to be involved in response to DNA damage caused by chromate (Miranda et al., 2005). Similarly, in *Escherichia coli*, Cr(VI) has long been known to induce the SOS repair system that protects DNA from oxidative damage. Cellular Cr(VI) reduction in the activation mechanism that produces redox-active Cr(V/IV) and stable Cr(III) intermediates that form Cr-DNA adducts, the most abundant type of DNA damage that causes mutations and chromosomal breaks.

11.5.5 EFFLUX OF Cr(VI) FROM CELLS

Another mechanism of resistance found in bacteria is the efflux of chromate ions from the cytoplasm of cells, mediated by transporters encoded by unique plasmid-borne genes. Chromate efflux is considered to be an effective and widespread resistance mechanism that prevents the accumulation of toxic ions within the bacterial cells (Ramirez-Diaz et al., 2008). The best-understood chromate resistance system conferred by *P. aeruginosa* ChrA protein belongs to the chromate ion transporter CHR superfamily. ChrA, is a hydrophobic membrane protein, encoded by *P. aeruginosa* plasmids pUM505 and *Cupriavidus metallidurans* pMOL28 (formerly *Ralstonia metallidurans*), which are involved in chromate resistance by the chromate efflux process (Alvarez et al., 1999). The ChrA protein acts as a chemiosmotic pump transporting chromate from the cytoplasm or periplasm to the outside driven by the proton motive force. CHR proteins from several bacteria have been demonstrated as involved in chromate resistance by chromate efflux mechanisms (Ramirez-Diaz et al., 2008).

11.5.6 ROS SCAVENGING

After entering the cell, electron donors like NAD(P)H or other organic compounds (like glucose) donate electrons to Cr(VI), leading to the formation of relatively unstable toxic intermediate Cr(V). Although the chromate reductases further reduce Cr(V) to Cr(III) by a two-electron transfer (via a "semi-tight" mechanism), sometimes this reaction is not very rapid. As a result, a portion of the intermediate Cr(V) is rapidly reoxidized to Cr(VI), thus producing ROS via a Fenton-like reaction. Molecular oxygen is reduced to O^{2-} radicals during the reduction process, which generates H_2O_2 via dismutation. Hexavalent chromium reacts with H_2O_2 to generate OH radicals. This mechanism is similar to the oxidation of Fe(II) with H_2O_2 in the Fenton reaction as the production of OH from Fe(II) via the Fenton reaction is significantly facilitated by the formation of Fe(II) complexes that have vacant sites for H_2O_2 coordination.

11.6 GENETIC MECHANISM OF Cr(VI) RESISTANCE AND TOLERANCE

The bacterial species able to grow in the toxic conditions prevalent in Cr(VI)-contaminated environments are generally assumed to be tolerant/resistant to chromium (Viti and Giovannetti, 2001). The terms "resistance" and "tolerance" are often used interchangeably, although their significance is different. Resistance is the ability of a microorganism to survive the toxic effects of metal exposure through a detoxification mechanism produced in direct response to the metal species concerned, whereas tolerance is well-defined as the ability of a microorganism to survive metal toxicity through intrinsic properties and environmental modification of toxicity.

High Cr(VI) resistance and its high reduction potential make a bacterial strain effective for remediation. The capability of Cr(VI) reduction is affected by an additional chromosome or plasmid resistance mechanism that represents a potentially useful detoxification process for several bacteria (Cervantes et al., 2001). Chromate-resistance determinants (CRDs) have been identified in bacteria and consist of genes belonging to the chromate ion transport (CHR) superfamily. Usually, CRDs include the chrA gene, which encodes a putative chromate efflux protein driven by the membrane potential. It has been reported that genes for the reduction of Cr(VI) can be either borne with plasmid, as observed in several *Pseudomonas* species, or located on the chromosome DNA, as found in several *Bacilli* and *Enterobacteriaceae* species (Li and Krumholz, 2007).

Several chromosomal genes were found to be responsible for the impartition of Cr(VI) resistance in bacteria. Chromate resistance was conferred by the chrR gene located on the *P. aeruginosa* chromosome (Aguilar-Barajas et al., 2008). In bacteria, the chrA genes can be located on a plasmid or chromosomal DNA or both, and they are generally organized in operons with other chr genes.

Ochrobactrum tritici contains several chromate resistance genes, namely chrB, chrA, chrC, and chrF, in the chromosomal DNA of which chrB acts as a chromium-sensitive regulator. The activity of chrC and chrF genes did not affect chromate resistance. Genetic studies also showed that the ruvB gene of *O. tritici* is related to chromate resistance (Morais et al., 2011). In yet another bacterium, *Ralstonia metallidurans*, the chromate resistance determinant chr2 (comprising genes chrB2, chrA2, and chrF2) was reported on the chromosome (Jhunke et al., 2002).

Apart from the chromosomes, bacterial plasmids also contain genes that are resistant to many toxic metals and metalloids, and encode systems devoted to protecting bacterial cells from the oxidative stress caused by chromate. Encoded membrane transporters are resistance systems related to plasmid genes, which mediate the efflux of chromate ions across the cytoplasmic membrane. In *P. aeruginosa*, the chromate transporter chrA functions as a chemiosmotic pump which extrudes chromate using the proton motive force (Alvarez et al., 1999). Genetic analyses of chromate resistant *P. aeruginosa* and *A. eutrophus* showed that the reduced accumulation of Cr(VI) is plasmid-mediated. Genes for a hydrophobic polypeptide, chrA, have been identified in *P. aeruginosa* and *A. eutrophus* plasmids with chromate resistance. The polypeptide chrA is assumed to be responsible for the translocation of chromate anions outwards from the membrane (Cervantes and Silver, 1992). In another instance, plasmid pMOL28 from *R. metallidurans* was found to encode the chrA chromate efflux pump, in addition to chrC and chrE proteins that seem to be involved in chromate resistance (Jhunke et al., 2002). Plasmid-mediated chromate resistance has also been reported in *Pseudomonas putida*, *Shewanella* sp., and *Streptococcus lactis* (Efstathiou and McKay, 1977).

In some cases, the sequence elements carried on the plasmid DNA are transposable across species. A transposon-based chrBACF operon as the key determinant for high chromate tolerance of *Ochrobactrum tritici* strain 5bvl1 (Branco et al., 2008). The activation of this operon was highly selective and gave resistance mainly through active chromate extrusion. For instance, the plasmid pLHB1 from *P. fluorescens* LB300 carrying the Cr(VI)-reducing genes could be transferred to *E. coli* to produce *E. coli* ATCC 33456 by conjugation. The gene encoding thioredoxin oxidoreductase in *Desulfovibrio desulphuricans* G20, located on an mre operon, was also involved in reducing Cr(VI) (Zou et al., 2013).

The regulation of Cr(VI) reduction in an operon structure was also observed in *Bacillus cereus* SJ1 and *Bacillus thuringiensis*. The Cr(VI) reduction genes were found to be upward regulated by the promoter chrI that regulated the Cr (VI) resistance gene chrA1 and arsenic resistance genes arsR and arsB. In *B. cereus* SJ1 putative chromate transport operon, chrIA1 and two additional chrA genes encoding putative chromate transporters imparted resistance to chromate. ChrA1 and chrI were found to be inducible genes. Azoreductase genes azoR and four nitroreductase genes nitR were also found to be involved in the chromate reduction of *B. cereus* SJ1 species (He et al., 2010).

11.7 BIOLOGICAL Cr(VI) REDUCTION PATHWAYS

Several components of the protoplasm of bacterial cells such as NADH [NAD(P)H in some species], flavoproteins, and other hemeproteins readily reduce Cr(VI) to Cr(III) (Ackerley et al., 2004). The bacterial reduction of Cr(VI) can occur directly through enzymatic activity or indirectly through non-enzymatic pathways by producing compounds such as glutathione, cysteine, etc. that can reduce Cr(VI) (Fendorf and Li, 1996).

11.7.1 Non-Enzymatic Cr(VI) Reduction

In the non-enzymatic process, Cr(VI) reduces to Cr(III) in the presence of different chemical compounds produced in the bacterial metabolic process. For instance, it is well known that Fe(II) and HS⁻, the anaerobic metabolic end products of iron and sulfate-reducing bacteria, can reduce hexavalent chromium. The most potent non-enzymatic chromate reductants could be ascorbic acid, glutathione (GSH), cysteine, hydrogen peroxide (H_2O_2) for microbial cells, and ascorbate for higher organisms (Poljsak et al., 2010). The reduction of Cr(VI) may also occur by chemical reactions associated with compounds present in intra/extracellular compounds such as amino acids, nucleotides, sugars, vitamins, organic acids, or glutathione (Dhal et al., 2013).

11.7.2 Enzymatic Cr(VI) Reduction

The enzymes from different organisms, ranging from bacteria to mammals, take part in the reduction of hexavalent chromium. As far as mammalian sources are concerned, enzymes like aldehyde oxidases, cytochrome p450, and Dt-diaphorase are involved in Cr(VI) reduction. However, microbial enzymes are most important for the reduction of Cr(VI) from an environmental point of view. There are several enzymes involved in Cr(VI) reduction in bacteria. Several oxidoreductases, with otherwise different metabolic functions that have been reported to catalyze Cr(VI) reduction in bacteria, include nitroreductases, iron reductases, quinone reductases, hydrogenases, flavin reductases, as well as NAD(P)H-dependent reductases (Puzon et al., 2002). The type of bacterial chromate reductase depends on the nature of bacteria carrying out the reduction reaction, i.e., aerobic or anaerobic. Bacterial chromate reductases are either localized in the membrane fraction or the cytosolic fractions of the chromate-reducing bacteria (Cheung et al., 2007). The enzymatic reduction of Cr(VI) in bacteria is accomplished in different ways. Soluble reductases can participate either in the extracellular or intracellular reduction of Cr(VI), whereas membrane-bound reductase reduces by extracellular means.

11.7.2.1 Extracellular

In some cases, the Cr(VI)-reducing enzymes, produced by bacterial cells, are exported into the media to reduce Cr(VI). These enzymes are extracellular. A couple of examples are available in support of this observation. The chromate reductases originating in the cytoplasm reduce Cr(VI) extracellularly in *P. putida*. Extracellular enzymes (cytosolic proteins) are soluble chromate reductases such as flavin reductases, nitrate reductases, flavin proteins, and ferrireductases (Cheung and Gu, 2007). The bacteria with membrane-bound reductases can also reduce Cr(VI) to Cr(III) by extracellular processes using electron shuttling compounds coupled to membrane reduction. Most of these enzymes are produced inductively, i.e., they are produced in the presence of Cr(VI) in the solution and are therefore highly regulated (Cheung and Gu, 2007). This regulation is controlled by regulatory elements (e.g., starvation promoter) that allow massive synthesis under the unfavorable in situ environment. The extracellular Cr(VI) reduction pathways in sulfate-reducing bacteria is done through a mass balance in which 90% of the reduced Cr was detected in the supernatant (Smith and Gadd, 2000).

11.7.2.2 Intracellular

In intracellular processes, Cr(VI) is reduced in the cytosol using cytoplasmic soluble reductase enzymes. These enzymes play an intermediate role between associated biological electron donors. The electron donors involved in an intracellular Cr(VI) reduction are NADH and NADPH. Cr(VI) was reduced intracellularly in the cytoplasm by a bacterial enzyme, using NADH as the reductant (Puzon et al., 2002). Many bacteria are also known to participate in the intracellular reduction of Cr(VI). Some examples of gram-negative bacteria that reduce Cr(VI) intracellularly are *P. aeruginosa*, *Pseudomonas* sp. CRB5, *E. coli* ATCC 33456, *Rhodobactersphaeroides*, *Alcaligenes*, *Enterobacter*, and *P. fluorescens* LB300. A gram-positive bacteria *B. subtilis* was reported to carry out the intracellular reduction of Cr(VI) (Ohtake et al., 1987).

11.8   MECHANISMS OF ENZYMATIC Cr(VI) REDUCTION

Microorganisms can remove several metallic and metalloid species from the environment by reducing them to a lower oxidation state. Most of the toxic compounds, especially heavy metals, ideally follow the reduction pathway rather than an oxidative one by native microbes (with a few exceptions) as their reduced forms are comparatively less toxic. In the case of metals, higher oxidation states are always more toxic (10–100 times) than lower oxidation states. The microbes can reduce almost all metallic/metalloid species in higher oxidation states, and the microbial reduction of Cr(VI) to Cr(III) forms the most widely studied example of metal bioremediation. Biologically and ecologically, bacterial Cr^{6+} reduction mechanisms are of significant relevance as they form noxious and mobile chromium derivatives into innocuous and immobile reduced species.

A wide diversity of microorganisms is known to have evolved biochemical pathways through aerobic and anaerobic processes to eradicate toxic compounds. Aerobic bioremediation involves microbial reactions that require oxygen to step forward. The bacteria use a carbon substrate as the electron donor and oxygen as the electron acceptor. The aerobic reduction is a co-metabolic process in which bacteria do not obtain energy or carbon from the degradation of a contaminant, but rather, the contaminant is reduced by a side reaction (EPA, 2006). On the other hand, anaerobic bioremediation involves microbial reactions occurring in the absence of oxygen. This requires many methods, including oxidation, reductive dechlorination, methanogenesis, and the reduction of sulfate and nitrate levels, depending on the contaminant. In anaerobic metabolism, sulfate, nitrate, carbon dioxide, oxidized materials, or organic compounds can substitute oxygen as the electron acceptor. Most of the microorganisms catalyzing redox reactions use the metals or metalloids in anaerobic respiration as terminal electron acceptors. Such microorganisms are both phylogenetically and physiologically diverse, classified as dissimilatory metal-reducing bacteria. Among the different groups of microorganisms, Cr(VI) reduction was investigated in a large number of bacterial species. The mechanisms by which bacterial strains reduce Cr(VI) to Cr(III) are variable and dependent on the species (McLean et al., 2000). Microbial Cr(VI) reduction occurs in two different processes: aerobic conditions and anaerobic conditions. In aerobic conditions, Cr(VI) reduction is found to be co-metabolic (not participating in energy conservation), whereas it is predominantly dissimilatory under anaerobic conditions (Ishibashi et al., 1990).

11.8.1   AEROBIC Cr(VI) REDUCTION

Aerobic Cr(VI) reduction is generally associated with soluble proteins and requires NAD(P)H as an electron donor (see Figure 11.3). These soluble proteins are localized as cytosolic proteins in *E. coli*, *Pseudomonas ambigua*, *P. putida*, and *Bacillus coagulans*. *Bacillus subtilis*, a gram-positive aerobe, reduced hexavalent chromium by using cell-free extracts of the bacterium (Ishibashi et al., 1990). Other aerobes such as *A. eutrophus*, *Paracoccus* sp., *Ochrobactrum* sp., *Shewanella alga* BrY-MT, etc., can produce soluble Cr^{6+} reductases that utilize diverse electron donors and can be located either inside or outside the bacterial cell.

Gram-positive aerobe *Vigribacillus* sp. can reduce Cr(VI) at higher concentrations in a salt medium. As exceptions, *P. maltophilia* O-2 and *Bacillus megaterium* TKW3 were found to utilize membrane-associated reductases for Cr(VI) reduction, despite being aerobes.

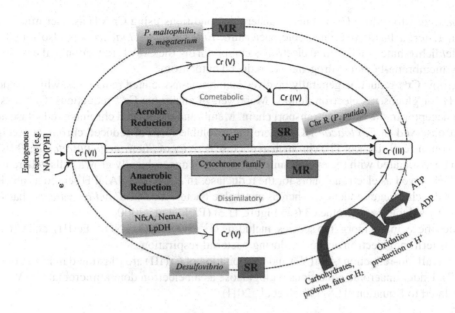

FIGURE 11.3 Mechanisms of enzymatic Cr(VI) reduction under aerobic and anaerobic conditions. (MR: membrane bound chromate reductase; SR: soluble chromate reductase). Dotted lines depict the exceptional reduction cases.

In the presence of oxygen, the enzymes from aerobes responsible for reducing Cr(VI) to Cr(III) require NAD(P)H. Two reaction steps have been suggested to be involved in the reduction reactions: first, Cr(VI) accepts one electron from one molecule of NADH to generate Cr(V) as an intermediate (Equation 11.4), and then Cr(V) accepts two electrons to form Cr(III) (Equation 11.5).

$$Cr^{6+} + e^- \rightarrow Cr^{5+} \tag{11.4}$$

$$Cr^{5+} + 2e^- \rightarrow Cr^{3+} \tag{11.5}$$

In the ability by aerobic bacteria such as *P. aeruginosa* and *P. synxantha* to reduce Cr(VI), it was found that the resistance to Cr(VI) is proportional to the increased efflux of the heavy metal by the cell membrane (Ohtake et al., 1987).

Since reduction mediated by such reductases is an energy-requiring and highly regulated process, these enzymes are produced constitutively. Due to its independence from transport mechanisms for Cr^{6+}/Cr^{3+} intake and expulsion, extracellular Cr^{6+} reduction is advantageous for the bacterial cells as it protects the cell from Cr^{6+}/Cr^{3+}-induced DNA damage.

11.8.2 Anaerobic Cr(VI) Reduction

The bacterial anaerobic reduction is associated with membrane-bound reductases such as flavin reductases, cytochromes, and hydrogenases that can be part of the electron transport system and use chromate as the terminal electron acceptor. Under anaerobic conditions, numerous components of the cell's protoplasm act as electron donors and reduce Cr^{6+}, such as amino acids, nucleotides, carbohydrates, vitamins, organic acids, glutathione; and hydrogen NADH (NADPH in some species), flavoproteins, and hemeproteins serves as terminal electron acceptors. Microbial Cr(VI) reduction was first reported in gram-negative, motile, facultative anaerobe bacterium *Pseudomonas dechromaticans* to reduce Cr(VI) under anaerobic conditions using Cr(VI) as a terminal electron acceptor. Later, a facultative anaerobic bacterium, *Enterobacter cloacae*, was also found to use chromate/dichromate as a terminal electron acceptor during the Cr(VI) reduction in the periplasmic space by membrane-bound hydrogenase or reduced cytochrome.

Anaerobic Cr^{6+} reduction generally involves membrane-associated reductases, which sometimes require H_2 or glucose as electron donors. Moreover, in anaerobic Cr^{6+} reductions, Cr^{6+} acts as an electron acceptor in the electron transport chain. Membrane-associated chromate reductase activity was first observed in *E. cloacae* HO1 where the insoluble form of reduced chromate precipitates was seen on the cell surface. The chromate reductase activity of *Shewanella putrefaciens* MR-1 was found to be associated with the cytoplasmic membrane of anaerobically grown cells, where formate and NADH served as electron donors for the reductase. In *P. putida*, NADPH served as an electron donor for the chromate reductase, whereas anaerobic bacterium *Desulfovibrio vulgaris* has Cr(VI) reduction by soluble cytochrome c3 (see Figure 11.3) (Park et al., 2000).

Besides these, many inorganic species, including O^{2-}, SO_4^{2-}, NO_2^-, NO_3^-, Fe(III), and Cr(VI) can also act as terminal electron acceptors during bacterial respiration.

The overall bio-reduction of Cr(VI) and precipitation of Cr(III) are illustrated in Equations (11.6) and (11.7). Under anaerobic conditions with glucose as an electron donor, microbial Cr (VI) reduction is related to Equation (11.8) (Singh et al., 2011).

$$CrO_4^{2-}\left(aq\right)+8H^+\left(aq\right)+3e^- \rightarrow Cr^{3+}\left(aq\right)+4H_2O \tag{11.6}$$

$$Cr^{3+}\left(aq\right)+4H_2O \rightarrow Cr\left(OH\right)_3\left(s\right)+3H^+\left(aq\right)+H_2O \tag{11.7}$$

$$C_6H_{12}O_6+8CrO_4^{2-}\left(aq\right)+14H_2O8 \rightarrow Cr\left(OH\right)_3\left(s\right)+10OH^-\left(aq\right)+6HCO^-\left(aq\right) \tag{11.8}$$

Sulfate-reducing bacteria (SRB) and iron-reducing bacteria (IRB) are important members of anaerobic microbial communities with economic, environmental, and biotechnological interest. Cr(VI) reduction by biogenic iron (II) and sulfides generated by IRB and SRB are ~100 times faster than chromium-reducing bacteria (CRB). SRB produce H_2S, which serves as a Cr(VI) reductant. Anaerobic bacteria mainly use Cr(VI) as a terminal electron acceptor or reduce Cr(VI) in the periplasmic space by hydrogenase or reduced cytochrome. For example, in SRB like *D. vulgaris* and *Desulfovibrio fructosovorans*, cyt c3 and NieFe dehydrogenase are involved in Cr(VI) reduction (Chardin et al., 2003). Some other anaerobes capable of reducing Cr(VI) to Cr(III) are *Microbacterium* sp. MP30, *Geobacter metallireducens*, *Pantoea agglomerans* SP1., *Agrobacterium radiobacter* EPS-916, etc. Even bacteria tolerant to radiations, like *Deinococcus radiodurans* R1, are quite efficient in reducing Cr(VI) anaerobically (Fredrickson et al., 2000).

Microorganisms are also capable of indirect Cr(VI) reduction by a biotic–abiotic coupling. A wide range of bacteria are known to couple the oxidation of organic compounds and H_2 for the reduction of Fe(III) and SO_4^{2-} to Fe(II) and H_2S, respectively, under oxygen stress conditions. The reduction of Cr(VI) with bacterially-produced H_2S followed by reduced Cr(III) precipitation is an important mechanism in sulfate-rich soil environments when anaerobic conditions prevail. The H_2S produced by sulfate-respiring bacteria in anaerobic systems diffuses out into the medium and reduces Cr(VI) (Kamaludeen et al., 2003).

11.8.3 Both Aerobic and Anaerobic Reductions

Some bacteria have the ability to reduce chromate under both aerobic and anaerobic conditions; for example, *P. fluorescens* LB300 has the ability of reducing Cr(VI) to Cr(III) both aerobically and anaerobically. Under anaerobic conditions, *P. fluorescens* LB300 utilizes acetate as an electron donor for chromate reduction while the microorganisms use a variety of electron donors for chromate reduction under aerobic conditions. Some other chromate-resistant bacteria, including *Achromobacter* sp., *E. coli*, *P. ambigua*, *P. putida*, *E. cloacae*, *Providencia* sp., *Brucella* sp., and *Bacillus* sp. have also shown the Cr(VI) reduction ability under both aerobic and anaerobic conditions although their rate of reduction varied widely (see Figure 11.3) (Ma et al., 2007).

11.9 ANALYSIS OF ONE- AND TWO-ELECTRON Cr(VI) REDUCTION PROCESSES

During the reduction of Cr(VI), the bacterial cells undergo oxidative stress due to simultaneous ROS generation (Cheung et al., 2007). Oxidative stress affects cell viability and reduces the efficiency of Cr(VI). Based on the magnitude of the oxidative stress produced, the enzymes that reduce hexavalent chromium can be classified as either one-electron reducer or two-electron reducers (Table 11.3).

Cr(VI) gets reduced in one-electron reduction processes to form the intermediate highly unstable Cr(V). In one-electron reduction processes, Cr(VI) gets reduced to form the highly unstable Cr(V) intermediate. In a redox cycle, Cr(V) can become oxidized back to Cr(VI), giving molecular oxygen to its electrons, and thus generating a large amount of ROS. The one-electron reducers, being flavin-dependent enzymes, follow the reaction as given below:

$$FMN(Ox.) + e^- + H^+ \rightarrow FMNH^+(Sq.)$$

Flavin mononucleotide (FMN) is a strong, often covalently bound, and is a part of flavoproteins. The oxidized flavin nucleotide [FMN(Ox.)] accepts one semiquinone-producing electron (a stable free radical) to form FMNH(Sq.).

Some of the identified chromate reducing enzymes belonging to one electron reducers are lipoyl dehydrogenase (LpDH) from *Clostridium kluyveri*, cytochrome c, glutathione reductase, and ferridoxin-NAD (Ackerley et al., 2004). These enzymes have the physiological functions of catalyzing energetic or biosynthetic reactions.

On the other hand, two-electron reduction process of Cr(VI) to Cr(III) proceeds without forming a Cr(V) intermediate. As a result, much less ROS is generated during this process than that of a

TABLE 11.3
Involvement of Chromate Reductases in One- and Two-Electron Reduction Mechanisms

	One Electron Reducer	Two Electron Reducers
Mechanism employed for chromate reduction	Reduction of Cr(VI) to Cr(III) occurs via Cr(V) intermediate. A continuous shuttle between Cr(VI) and Cr(V) forms occurs with Cr(V) transferring 1 e⁻ to O_2, generating ROS (superoxide, O_2^{2-}) in a Fenton-like reaction (Barak et al., 2006).	Transfer of 3 e- to Cr(VI) results in its direct reduction to Cr(III). 1 e⁻ is transferred to O_2, forming ROS (superoxide, O_2^{2-}). No Cr(V) intermediate is involved, hence no redox cycle occurs.
Examples	• LpDH • Cytochrome c • Glutathione reductase • Ferridoxin-NADP oxidoreductase • NfsA (*E. coli*)	• ChrR (*P. putida*) • YieF (*E. coli*)
Effect on cells	Highly detrimental to bacterial cells due to more ROS generation. Chromate reducing efficiency of the bacteria declines.	Much less ROS generated than one-electron reducer, hence less toxic to bacterial cells.

one-electron reduction. Two-electron reducers, being NAD(P)-dependent enzymes, are character-ized by the transfer of hydride ion (H⁻, the equivalent of a proton and two electrons) in a reaction as follows:

$$NAD^+ + 2e^- + 2H^+ \rightarrow NADH + H^+$$

$$NADP^+ + 2e^- + 2H^+ \rightarrow NADPH + H^+$$

Some of the chromate-reducing enzymes behaving as two-electron reducers include ChrR from *P. putida*, YieF, and NfsA from *E. coli* (Barak et al., 2006). Broad classifications of enzymes carrying out Cr(VI) reduction are Class I or Class II reductases enzymes based on sequence homologies (see Table 11.3) (Ackerley et al., 2004).

11.9.1 CLASS I CHROMATE REDUCTASE

Two of the Class I enzymes that are most commonly used are ChrR and YieF (Ackerley et al., 2004). The YieF dimer carries this reduction of Cr(VI) in one step without redox cycling so that the generation of ROS is minimal, while the ChrR dimer tends to reduce chromate by a combination of one and two steps of reduction of the electron, producing more ROS than YieF. Two electrons can be transferred from the enzyme either simultaneously ("tight" as with YieF) or non-simultaneously ("semi-tight" as with ChrR).

YieF from *E. coli* is a dimeric flavoprotein and reduces chromate to Cr(III) (Park et al., 2000). The YieF enzyme is unique as it directly reduces Cr(VI) to Cr(III) through a four-electron transfer, in which three electrons are consumed in reducing Cr(VI), and the fourth electron is transferred to oxygen. Since the quantity of ROS generated by YieF in Cr(VI) reduction is minimal, and is regarded as a more effective reductase than ChrR for Cr(VI) reduction. Unlike ChrR, during chromate reduction, YieF did not show semiquinone flavoprotein generation and consumed only about 25% of NADH electrons to molecular oxygen in ROS generation. The ChrR and YieF, homologs were shown to contain the characteristic signature of the NADH-dh2 family of protein, which consists of bacterial and eukaryotic NAD(P)H oxidoreductases. Another chromate reductase is ChrA protein, from *B. cereus* SJ1 and *B. thuringiensis* strains, which belong to the Class I family (He et al., 2010).

N-ethylmaleimide reductase (NemA), from *E. coli*, is a member of the "old yellow enzyme" (Class I) family of flavoproteins and catalyzes the reduction of N-ethylmaleimide (NEM) to N-ethylsuccinimide. NemA can accept a variety of substrates, including chromate, and catalyzes chromate reduction through the addition of one or two electrons from the cofactors NADH or NADPH (Roldan et al., 2008).

11.9.2 CLASS II CHROMATE REDUCTASE

The Class II chromate reductases, which possess nitro-reductase activity, bear no homology to the Class I enzyme but are homologous to chromate reductase purified from *P. ambigua* (Park et al., 2000). Two other members of the Class II family, namely NfsA protein of *E. coli* and the ChfN protein of *B. subtilis*. NfsA, are semi-tight chromate reducers, and reduce nitro-compounds and quinone by an obligatory two-electron transfer (Ackerley et al., 2004). The Class II enzymes reduce quinones and nitro-compounds effectively but vary in their ability to transform chromate. The NfsA protein has received wide attention because of its ability to detoxify nitro-compounds. This protein possesses therapeutic properties and the ability to activate prodrugs used in cancer chemotherapy. Another enzyme of the Class II family ChfN protein of *B. subtilis*, in the electrophoretically pure state, possesses chromate reductase activity (Park et al., 2000).

11.10   POTENTIAL OF CHROMATE REDUCTASE IN ANTICANCER THERAPY

The chromate-reducing enzymes, with some modifications, can play a role in cancer chemotherapy (Barak et al., 2006). The most elaborate instance of such an enzyme is Y6. It is a modified version of the *E. coli* YieF that results due to the technique of error-prone PCR. Y6 can act as a "prodrug," which means it is non-toxic in native form but highly toxic when reduced. These drugs kill by generating DNA adducts and can target both growing and non-growing cells of a tumor. Error-prone PCR was performed using the yieF gene of *E. coli*, and the evolved genes were screened for the improved activity of chromate reductase. Among the enzymes showing superior chromate reductase activity, Y6 showed improved prodrug reduction activity when tested with HeLa cells (Barak et al., 2006). A class of enzymes used as gene-delivered enzyme prodrug therapy is bacterial nitroreductases, such as NfsA and NfsB from *E. coli*. Y6 has been found to kill HeLa cells with 5-fold greater efficiency than NfsA. Thus, such modified chromate reductases can undoubtedly provide new potentialities into cancer chemotherapy (Barak et al., 2006). The efficiency of bacterial bioremediation can be increased in many ways. One of the approaches is to promote the growth of the specific bacterial species that reduces Cr(VI) by providing specific nutrients for it, while outcompeting the growth of other species.

11.11   CONCLUSIONS

Different industrials and anthropogenic activities contaminate the environment. Cr(VI) and Cr(III) may present different behaviors. The high mobility and solubility of Cr(VI) increase the chances for its diffusion through cell membranes which makes it a carcinogen, teratogen, and mutagen. Cr(III) is non-toxic and relatively insoluble in aqueous systems. The application of an eco-friendly, versatile, and low-cost tool is necessary to remove this heavy metal from water, soil, and sediments. This chapter is focused on chromium production, use, cycle and speciation, toxicity, and the bioremediation of chromium(VI).

The microbial reduction of hexavalent chromium to trivalent chromium, which is relatively insoluble and considerably less toxic, is a potentially valid remediation strategy for chromium-contaminated soils. It is cost-effective and environmentally friendly in comparison to physico-chemical treatments. The capability of indigenous bacteria in reducing Cr(VI) to Cr(III) is to be quantified. The optimal conditions are to be defined to improve the ability of specific bacterial strains to play their role under stressful conditions and those in polluted environments.

With molecular engineering, it will be possible to enhance Cr(VI) reduction activities of indigenous bacterial strains that express such activities at high levels under deficient nutrient and stressful environmental conditions. Biosorption and biotransformation have been proven as efficient, eco-friendly, and cost-effective strategies for the remediation of Cr(VI)-contaminated environments. The biomass of bacteria has been implemented and found to be effective for Cr(VI) transformation. The molecular mechanism suggested in this chapter provides an insight into a better understanding of Cr(VI) removal. This can help in developing the existing technologies of chromium remediation to be more efficient.

The enzymes that play a role in reducing Cr(VI) have been reported from different microorganisms and belong to several classes, like quinone reductases, nitro-reductases, NADPH-dependent, etc., which vary in their ability to transform chromate and follow different pathways. Several bacteria have the ability to reduce Cr(VI) through membrane-bound reductases such as flavin reductase, cytochromes, and hydrogenases. These enzymes can be part of the electron transport system and use chromate as the terminal electron acceptor. Bacterial strains of various genera also possess soluble chromate reductase activity in the cytosol. To reduce Cr(VI), the chromate reductases use NAD(P)H as electron donors. Compared to membrane-bound chromate reductases, soluble reductases are ideal for biocatalyst development for bioremediation because they are more appropriate for protein engineering in compliance with the environmental conditions of polluted sites. In rare cases, genes responsible for chromate reductase production have been identified. Cloning of the genes may allow for the enhanced production of large quantities of pure enzymes using a modern molecular approach for efficient chromate reduction. Unlike free enzymes, enzymes can be immobilized onto a carrier and used for a longer period in the bioremediation program. Apart from these, the enzymes that are efficient in reducing chromate can also play a role in cancer chemotherapy and generate interest in exploiting the enzyme for pharmaceutical application (Barak et al., 2006). This brief chapter summarizes the importance of employing indigenous microorganisms as cost-effective and environmentally friendly methods to reduce chromium toxicity.

11.12 FUTURE PERSPECTIVES WITH POTENTIAL APPLICATIONS IN CHROMIUM REMOVAL

The exposure to heavy metals represents a stress condition; the greater capacity of some strains towards metal tolerance may be attributed to their ability to survive under extreme environments since the expression of some genes could be involved in mechanisms battling against both stress factors.

The development of efficient biological processes (accompanied by a global analysis of macro-molecules) offers numerous opportunities to treat environmental heavy metal pollution. Mostly, results related to the central metabolism and energy production and conversion, as well as the effects on different transporters and DNA metabolism and repair, were obtained. Studies applying genetic engineering have been performed with recombinant microorganisms. It is important to continue developing genetic engineering techniques because it represents a promising approach and economic solution for ex situ Cr(VI) remediation.

The application of immobilized microbial cells and enzymes combined with nanotechnology (e.g., carbon nanotubes impregnated into calcium alginate beads) markedly enhanced the stability of the enzyme and the reduction of Cr(VI). Another process related to remediation at ananometre scale is applying metal-reducing bacteria in combination with nano-materials (siderite) that serve as electron donors in the precipitate-containing enzymatic Cr(VI) reduction and immobilization [Cr (III)].

ABBREVIATIONS

ROS   Reactive oxygen species
GTF   Glucose tolerance factor
USEPA   U.S. Environmental Protection Agency
ASTDR   Agency for Toxic Substances and Diseases
PCOS   Potential chromium oxidation score
CRB   Chromium reducing bacteria
SOD   Superoxide dismutase
CRD   Chromate-resistance determinants

REFERENCES

Ackerley, D.F., Gonzalez, C.F., Keyhan, M., Blake, R., Matin, A. (2004). Mechanism of chromate reduction by the Escherichia coli protein, NfsA and the role of different chromate reductases in minimizing oxidative stress during chromate reduction. *Environ. Microbiol.*, 6 (8), 851–860.
Aguilar-Barajas, E., Paluscio, E., Cervantes, C., Rensing, C. (2008). Expression of chromate resistance genes from Shewanella sp. strain ANA-3 in Escherichia coli. *FEMS Microbiol. Lett.*, 285 (1), 97–100.
Alvarez, A.H., Moreno-Sanchez, R., Cervantes, C. (1999). Chromate efflux by means of the ChrA chromate resistance protein from Pseudomonas aeruginosa. *J. Bacteriol.*, 181 (23), 7398–7400.
Apte, A.D., Verma, S., Tare, V., Bose, P. (2005). Oxidation of Cr (III) in tannery sludge to Cr (VI): field observations and theoretical assessment. *J. Hazard. Mater.*, 121 (1–3), 215–222.

Arias, Y.M., Tebo, B.M. (2003). Cr (VI) reduction by sulfidogenic and nonsulfidogenic microbial consortia. *Appl. Environ. Microbiol.*, 69, 1847–1853

Asatiani, Z., Abuladze, N., Lejava, H. (2004). Effect of chromium VI action on Arthrobacter oxydans. *J. Current. Microbiol.*, 49, 321–326.

Bahi, J.S., Radziah, O., Samsuri, A.W., Aminudin, H., Fardin, S. (2012). Bioleaching of heavy metals from mine tailing by Aspergillus fumigatus. *Bioremediat. J.*, 16, 57–65.

Barak, Y., Ackerley, D.F., Dodge, C.J., Banwari, L., Alex, C., Francis, A.J., Matin, A. (2006). Analysis of novel soluble chromate and uranyl reductases and generation of an improved enzyme by directed evolution. *Appl. Environ. Microbiol.*, 72 (11), 7074–7082.

Bartlett, R.J., Kimble, J.M. (1976). Behavior of chromium in soils. II. Hexavalent forms. *J. Environ. Qual.*, 5 (4), 383–386.

Benedetti, A. (1998). Defining soil quality: introduction to round table. In: S. de Bertoldi, F. Pinzari (Eds.), *COST Actions 831, Joint WCs Meeting, Biotechnology of Soil: Monitoring Conservation and Remediation*, 29–33.

Bielicka, A., Bojanowska, I., Wi´sniewski, A. (2005). Two faces of chromium: pollutant and bioelement, *Polish J. Environ. Stud.*, 14 (1), 5–10.

Branco, R., Chung, A.P., Johnston, T., Gurel, V., Morais, P., Zhitkovich, A. (2008). The chromate-inducible chrBACF operon from the transposable element TnOtChr confers resistance to chromium (VI) and superoxide. *J. Bacteriol.*, 190 (21), 6996–7003.

Brown, S.D., Thompson, M.R., VerBerkmoes, N.C., Chourey, K., Shah, M., Zhou, J., Hettich, R.L., Thompson, D.K. (2006). Molecular dynamics of the Shewanella oneidensis response to chromate stress. *Mol. Cell. Proteomics*, 5, 1054–1071.

Browning, C.L., Wise, J.P. (2017). Prolonged exposure to particulate chromate inhibits RAD51 nuclear import mediator proteins. *Toxicol. Appl. Pharmacol*, 331, 101–107.

Camargo, F.A., Bento, F., Okeke, B.C., Frankenbarger, W.T. (2003). Chromate reduction by chromium resistant bacteria isolated from soil contaminated with dichromate. *J. Environ. Qual.*, 32 (4), 1228–1233.

Cervantes, C., Silver, S. (1992). Plasmid chromate resistance and chromate reduction. *Plasmid*, 27, 65–71.

Cervantes, C., Campos-Garcia, J., Devars, S., Gutierrez-Corona, F., Loza-Tavera, H., Torres-Guzman, J.C., Moreno-Sanchez, R. (2001). Interactions of chromium with microorganisms and plants. *FEMS Microbiol. Rev.*, 25, 335–347.

Chardin, B., Giudici-Orticoni, M.T., De Luca, G., Guigliarelli, B., Bruschi, M. (2003). Hydrogenases in sulfate-reducing bacteria function as chromium reductase. *Appl. Microbiol. Biotechnol.*, 63 (3), 315–321.

Cheung, K.H., Gu, J.D. (2007). Mechanism of hexavalent chromium detoxification by microorganisms and bioremediation application potential: a review. *Int. Biodeterior. Biodegrad.*, 59 (1), 8–15.

Chourey, K., Thompson, M.R., Morrell-Falvey, J., VerBerkmoes, N.C., Brown, S.D., Shah, M., Zhou, J., Doktycz, M., Hettich, R.L., Thompson, D.K. (2006). Global molecular and morphological effects of 24-h chromium (VI) exposure on Shewanella oneidensis MR-1. *Appl Environ. Microbiol.*, 72, 6331–6344.

Deflora, S., Bagnasco, M., Serra, D., Zanacchi, P. (1990). Genotoxicity of chromium compounds: a review. *Mutat. Res*, 238, 99–172.

Dhal, B., Thatoi, H.N., Das, N., Pandey, B.D. (2013). Reduction of hexavalent chromium by Bacillus sp. isolated from chromite mine soils and characterization of reduced product. *J. Chem. Technol. Biotechnol.*, 85 (11), 1471–1479.

Eary, L.E., Rai, D. (1991). Chromate reduction by subsurface soils under acidic conditions. *Soil Sci. Soc. Am. J.*, 55, 676–683.

Efstathiou, J.D., McKay, L.L. (1977). Inorganic salts resistance associated with a lactose-fermenting plasmid in Streptococcus lactis. *J. Bacteriol.* 130 (1), 257–265.

Environmental Protection Agency (2006). Engineering issue. In situ and ex situ biodegradation technologies for remediation of contaminated sites. EPA-625-R-06-015, 22.

Faisal, M., Hasnain, S. (2004). Microbial conversion of Cr (VI) into Cr (III) in industrial effluent. *Afr. J. Biotechnol.*, 3, 610–617.

Fendorf, S.E., Li, G.C. (1996). Kinetics of chromate reduction by ferrous iron. *Environ. Sci. Technol.*, 30, 1614–1617.

Fredrickson, J.K., Kostandarithes, H.M., Li, S.W., Plymale, A.E., Daly, M.J. (2000). Reduction of Fe (III), Cr (VI), U (VI), and TC (VII) by Deinococcus radiodurans R1. *Appl. Environ. Microbiol.*, 66 (5), 2006–2011.

Gadd, G.M. (1992). Metals and microorganisms: a problem of definition. *FEMS Microbiol. Lett.*, 100, 197–204.

He, M., Li, X., Guo, L., Miller, S.J., Rensing, C., Wang, G. (2010). Characterization and genomic analysis of chromate resistant and reducing Bacillus cereus Strain SJ1. *BMC Microbiol.*, 10 (221), 1–10.

Hu, P., Brodie, E.L., Suzuki, Y., Mc Adams, H.H., Andersen, G.L. (2005). Whole-genome transcriptional analysis of heavy metal stresses in Caulobacter crescentus. *J. Bacteriol.*, 187 (24), 8437–8449.

Ishibashi, Y., Cervantes, C., Silver, S. (1990). Chromium reduction in Pseudomonas putida. *Appl. Environ. Microbiol.*, 56 (7), 2268–2270.

Jain, P., Amatullah, A., AlamRajib, S., Mahmud, Reza H. (2012). Antibiotic resistance and chromium reduction pattern among Actinomycetes. *Am. J. Biochem. Biotechnol*, 8, 111–117.

James, B.R. (2002). Chemical transformations of chromium in soils: relevance to mobility, bio-availability and remediation. In: *The Chromium File*. International Chromium Development Association, Paris, 1–8.

James, B.R., Petura, J.C., Vitale, R.J., Mussoline, G.R. (1997). Oxidation-reduction chemistry of chromium: relevance to the regulation and remediation of chromate contaminated soils. *J. Soil Contamin.*, 6, 569–580.

Jhunke, S., Peitzsch, N., Hubener, N., Grobe, C., Nies, D.H. (2002). New genes involved in chromate resistance in Ralstonia metallidurans strain CH34. *Arch. Microbiol.*, 179 (1), 15–25.

Jobby, R., Desai, N. (2017). Bioremediation of heavy metals. In: Kumar, P., Gurjar, B.R., Govil, J.N. (Eds.), *Biodegradation and Bioremediation. Environmental Science and Engineering.* Studium Press, New Delhi, 201–220.

Kamaludeen, S.P., Arunkumar, K.R., Avudainayagam, S., Ramasamy, K. (2003). Bioremediation of chromium contaminated environments. *Indian J. Exp. Biol.*, 41 (9), 972–985.

Kanmani, P., Aravind, J., Preston, D. (2012). Remediation of chromium contaminants using bacteria. *Int. J. Environ. Sci. Technol.*, 9, 183–193.

Kapoor, A., Viraraghvan, T., Roy Cullimore, D. (1999). Removal of heavy metals using Aspergillus Niger. *Bioresour. Technol.*, 70, 95–104.

Komori, K., Rivas, A., Toda, K., Ohtake, H. (1990). A method for removal of toxic chromium using dialysis-sac cultures of a chromate-reducing strain of *Enterobacter cloacae. Appl. Microbiol. Biotechnol.*, 33, 91–121.

Lebeau, T., Braud, A., Jézéquel, K. (2008). Performance of bioaugmentation-assisted phytoextraction applied to metal contaminated soils: a review. *Environ. Pollut.*, 153, 497–522.

Lenntech, (2004). *Water Treatment and Air Purification, Water Treatment.* Lenntech, Rotterdamseweg, Netherlands.

Li, X., Krumholz, L.R. (2007). Regulation of arsenate resistance in Desulfovibrio desulfuricans G20 by an arsRBCC operon and an arsC gene. *J. Bacteriol.*, 189, 3705–3711.

Losi, M.E., Amrhein, C., Frankenberger, W.T. (1994b). Environmental biochemistry of chromium. *Rev. Environ. Contam. Toxicol.*, 136, 91–131

Ma, Z., Zhu, W., Long, H., Chai, L., Wang, Q. (2007). Chromate reduction by resting cells of Achromobacter sp. Ch-1 under aerobic conditions. *Process Biochem.*, 42(6), 1028–1032.

McGrath, S.P., Smith, S. (1990). Chromium and nickel. In: Alloway B.J. (ed.), *Heavy Metals in Soils.* Wiley, New York, 125–150.

McLean, J.S., Beveridge, T.J., Phipps, D. (2000). Isolation and characterization of chromium reducing bacterium from a chromated copper arsenate-contaminated site. *Environ. Microbiol.*, 2, 611–619.

Megharaj, M., Avudainayagam, S., Naidu, R. (2003). Toxicity of hexavalent chromium and its reduction by bacteria isolated from soil contaminated with tannery waste. *Curr. Microbiol*, 47, 51–54.

Miranda, A.T., Gonzalez, M.V., Gonzalez, G., Vargas, E., Campos-Garcia, J., Cervantes, C. (2005). Involvement of DNA helicases in chromate resistance by Pseudomonas aeruginosa PAO1. *Mutat. Res.*, 578 (1–2), 202–209.

Modoi, O.C., Roba, C., T€or€ok, Z., Ozunu, A. (2014). Environmental risks due to heavy metal pollution of water resulted from mining wastes in NW Romania. *Environ. Eng. Manag. J.*, 13, 2325–2336.

Morais, P.V., Branco, R., Francisco, R. (2011). Chromium resistance strategies and toxicity: what makes Ochrobactrum tritici 5bvl1 a strain highly resistant. *Biometals*, 24 (3), 401–410.

Mordenti, A., Piva, G. (1997). Chromium in animal nutrition and possible effects on human health. In: Canali S, Tittarelli F, Sequi P (Eds.), *Chromium Environmental Issues.* Franco Angelis.r.l., Milan, 131–151.

Ngwenya, N., Chirwa, E.M.N. (2011). Biological removal of cationic fission products from nuclear wastewater. *Water Sci. Technol.*, 63 (1), 124–128.

Ohta, N., Galsworthy, P.R., Pardee, A.B. (1971). Genetics of sulfate transport by *Salmonella typhimurium. J. Bacteriol.*, 105, 1053–1062.

Ohtake, H., Cervantes, C., Silver, S. (1987). Decreased chromate uptake in *Pseudomonas fluorescens* carrying a chromate resistance plasmid. *J. Bacteriol.*, 169, 3853–3856.

Papp, J.F. (2004). Chromium use by market in the United States. In: Proceedings of the 10th International Ferroalloys Congress, Cape Town, South Africa, 770–778. The South African Institute of Mining and Metallurgy, Marshalltown, South Africa.

Park, C.H., Keyhan, M., Wielinga, B., Fendorf, S., Matin, A. (2000). Purification to homogeneity and characterization of a novel Pseudomonas putida chromate reductase. *Appl. Environ. Microbiol*, 66 (5), 1788–1795.

Parmar, A.N., Oosterbroek, T., Del Sordo, S., Segreto, A., Santangelo, A., Dal Fiume, D., Orlandini, M. (2000). Broad-band BeppoSAX observation of the low-mass X-ray binary X 1822–371. *Astron. Astrophys.*, 356, 175–180.

Philip, L., Iyengar, L. (1998). Chromium (VI) reduction from Bacillus coagulans isolated from contaminated soil. *J. Environ. Eng.*, 124, 165–1170.

Poljsak, B., Pócsi, I., Raspor, P., Pesti, M. (2010). Interference of chromium with biological systems in yeasts and fungi: a review. *J. Basic Microbiol.*, 50, 21–36.

Puzon, G.J., Petersen, J.N., Roberts, A.G., Kramer, D.M., Xun, L. (2002). A bacterial flavin reductase system reduces chromate to a soluble chromium (III)-NAD (+) complex. *BiochemBiophys. Res. Commun.*, 294, 76–81.

Ramirez-Diaz, M.I., Diaz-Perez, C., Vargas, E., Riveros-Rosas, H., Campo-Garcia, J., Cervantes, C. (2008). Mechanisms of bacterial resistance to chromium compounds. *Biometals*, 21 (3), 321–332.

Roldan, M.D., Perez-Reinado, E., Castillo, F., Moreno-Vivian, C. (2008). Reduction of polynitroaromatic compounds: the bacterial nitroreductases. *FEMS Microbiol. Rev.*, 32 (3), 474–500.

Saha, R., Nandi, R., Saha, B. (2011). Sources and toxicity of hexavalent chromium. *J. Coord. Chem.*, 64, 1782–1806.

Sathishkumar, K., Murugan, K., Benelli, G., Higuchi, A., Rajasekar, A. (2016). Bioreduction of hexavalent chromium by Pseudomonas stutzeri L1 and Acinetobacter baumannii L2. *Ann. Microbiol.*, 67, 91–98.

Shahid, M., Shamshad, S., Rafiq, M., Khalid, S., Bibib, I., Khan, N., Niazib, Dumate, C., Rashida, M.I. (2017). Chromium speciation, bioavailability, uptake, toxicity and detoxification in soil-plant system: a review. *Chemosphere*, 178, 513–533.

Singh, R., Misra, V., Singh, R.P. (2011). Synthesis, characterization and role of zerovalent iron nanoparticle in removal of hexavalent chromium from chromium spiked soil. *J. Nanopart. Res.*, 13 (9), 4063–4073.

Smith, W.L., Gadd, G.M. (2000). Reduction and precipitation of chromate by mixed culture sulfate reducing bacterial biofilm. *J. Appl. Microbiol.*, 88 (6), 983–991.

Sperling, M., Xu, S.K., Welz, B. (1992). Determination of chromium (III) and chromium (VI) in water using flow injection on-line preconcentration with selective adsorption on activated alumina and flame atomic absorption spectrometric detection. *Anal. Chem*, 64, 3101–3108.

Srinath, T., Verma, T., Ramteke, P.W., Garg, S.K. (2002). Chromium (VI) biosorption and bioaccumulation by chromate resistant bacteria. *Tannery Technol.*, 48, 427–435.

Thacker, U., Datta, M. (2006). Reduction of toxic chromium and partial localization of chromium reductase activity in bacterial isolate DM1. *World J. Microbiol. Biotechnol.*, 21, 891–899.

USEPA (2018). *IRIS, Toxicological Review of Hexavalent Chromium (2018 External Review Draft)*. U.S. Environmental Protection Agency, Washington, DC.

Viti, C., Giovannetti, L. (2001). The impact of chromium contamination on soil heterotrophic and photosynthetic microorganisms. *Ann. Microbiol.*, 51, 201–213.

Wang, J., Chen, C. (2006). Biosorption of heavy metals by Saccharomyces cerevisiae: a review. *Biotechnol. Adv.* 24, 427–451.

Wenbo, Qi, Reiter, R.J., Tan, D.X., Garcia, J.J., Manchester, L.C., Karbownik, M., Calvo, J.R. (2000). Chromium (III)-induced 8-hydroxydeoxyguanosine in DNA and its reduction by antioxidants: comparative effects of melatonin, ascorbate, and vitamin E. *Environ. Health Perspect.*, 108 (5), 399–402.

Xu, X.R., Li, H.B., Gu, J.D. (2006). Simultaneous decontamination of hexavalent chromium and methyl tert-butyl ether by UV/TiO2 process. *Chemosphere*, 63, 254–260.

Zhou, J., Xia, B., Treves, D.S., Wu, L.Y., Marsh, T.L., O'Neill, R.V., Palumbo, A.V., Tiedje, J.M. (2002). Spatial and resource factors influencing high microbial diversity in soil. *Appl. Environ. Microbiol.*, 68, 326–334.

Zou, L., Liu, P., Li, X. (2013). New advances in molecular mechanism of microbial hexavalent chromium reduction. *Int. J. Biotechnol. Food Sci.*, 1 (3), 46–55.

12 Removal of Refractory Pollutants from Wastewater Treatment Plants

Phytoremediation for the Treatment of Various Types of Pollutants: A Multi-Dimensional Approach

Komal Agrawal and Pradeep Verma

CONTENTS

12.1 INTRODUCTION

The technique of phytoremediation uses plants and associated microbes for the bioremediation of the environment (Salt et al. 1995, 1998; Raskin et al. 1994; Pilon-Smits 2005), i.e., sequestration of both organic and inorganic pollutants. Organic pollutants are mainly man-made and are toxic, carcinogenic, and xenobiotic to living beings. The various sources of these pollutants include oil and solvent spills, the use of explosives and chemical weapons, the unregulated use of pesticides and herbicides, along with chemical and petrochemical waste. On the other hand, the various inorganic pollutants are naturally present in the Earth's crust/atmosphere and can be artificially generated via mining, industrial, and agricultural activities (Nriagu 1979).

Both organic and inorganic pollutants can be remediated using phytoremediation. In case of organic pollutants, phytoremediation of trichloroethylene (TCE an organic solvent) (Moccia et al. 2017), atrazine (herbicide) (Sánchez et al. 2019), trinitrotoluene (TNT and explosives) (Zhang et al. 2019),

DOI: 10.1201/9781003204442-12

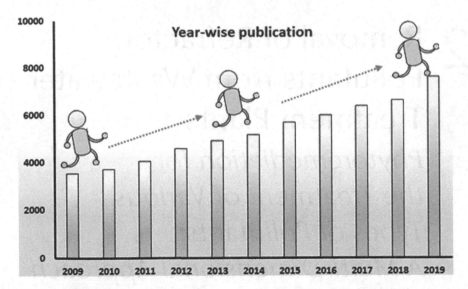

FIGURE 12.1 The average of the year wise increment in the publication of phytoremediation in google scholar and PubMed (https://scholar.google.com/ and https://pubmed.ncbi.nlm.nih.gov/)

oil, gasoline, toluene (petroleum hydrocarbon) (Płociniczak et al. 2017; Cheng et al. 2017), methyl tertiary butyl ether (MTBE and fuel additives) (Hong et al. 2001), and polychlorinated biphenyls (PCBs) (Zeeb et al. 2006) have been reported. Inorganic pollutant phytoremediation has been reported for macronutrients such as nitrate (NO_3^-) (Chen et al. 2015) and phosphate (PO_4^{3-}) (Siswandari et al. 2017), and trace elements such as chromium (Cr) (Revathi et al. 2011), copper (Cu) (Ariyakanon and Winaipanich 2006), iron (Fe) (Yu et al. 2006), manganese (Mn) (Zhang et al. 2020), molybdenum (Mo) (Neunhäuserer et al. 2001), and zinc (Zn) (Wang et al. 2006).

Also, phytoremediation potential has been extended to non-essential elements such as cadmium (Cd) (Ehsan et al. 2014), cobalt (Co) (Mosoarca et al. 2018), fluorine (F) (Banerjee and Roychoudhury 2019; Chen and Xiong 2011), mercury (Hg) (Marrugo-Negrete et al. 2015), selenium (Se) (Pilon-Smits and LeDuc 2009), lead (Pb) (Lim et al. 2004), vanadium (V) (Wang et al. 2018), tungsten (W) (Strigul 2005), radioisotopes uranium (^{238}U) (Stojanović et al. 2012), cesium (^{137}Cs) (Borghei et al. 2011), and strontium (^{90}Sr) (Wang et al. 2017).

Further, the phytoremediation of soil, water, and air that constitutes the solid, liquid, and gaseous phases has too been reported (Burges et al. 2018; Escoto et al. 2019; Pettit et al. 2018). Thus, such a large-scale application of phytoremediation has attracted significant attention globally and the past ten years have seen a significant impact on the research based on phytoremediation (see Figure 12.1). Thus, considering the above aspects the phytoremediation of all three states i.e., soil (soil), liquid (water), and air (gas), will be elaborated along with the potential future application and scope for the betterment of the various Earth elements.

12.2 MECHANISM OF PHYTOREMEDIATION

Phytoremediation is a combination of the Greek and Latin words *phyto* (plant) and *remedium* (to remove/correct the evil) (Cunningham et al. 1996). The technique of phytoremediation is an in-situ pollutant removal technology that presents an efficient replacement to conventional pollutant removal processes that require energy, labor, and cost-intensive processes. In addition, phytoremediation is environmentally friendly and is based on the concept of "using nature to cleanse nature." The plants that are able to thrive in polluted areas without adverse impact on the plants signify that they can effectively phytoremediate pollutants via the integration of agricultural and biotechnological approaches

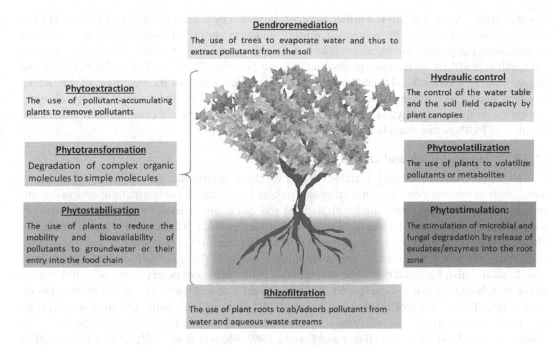

FIGURE 12.2 Schematic representation of the various phytoremediation techniques.

(UNEP undated). Higher plants have been regarded as "*green livers*" as they can metabolize and degrade xenobiotic compounds/pollutants that cause damage to the environment (Schwitzguebel 2000). The method of phytoremediation involves the plantation of plants either naturally or via the construction of wetlands for the remediation/degradation of contaminants/pollutants. Further, plants can then be harvested, processed, disposed and, as they are natural, can be easily decomposed without creating any toxic end products. In literature two types of wetland have been reported i.e., natural and constructed wetland. The prior is static and small in size and the later as it is constructed can be increased in size. However, the efficiency of both systems can be increased with time. The various types of phytoremediation techniques that have been studied include phytoextraction, phytotransformation, phytostimulation, phytovolatilization, rhizofiltration, dendroremediation, phytostabilization, and hydraulic control (see Figure 12.2) (Schwitzguebel 2000).

12.3 PHYTOREMEDIATION OF POLLUTANTS

Phytoremediation has been used effectively for the removal of contaminants from solids (soil), liquids (water), and gases (atmosphere). The various modes of phytoremediation are as follows:

12.3.1 SOIL

The contamination of soil is a major concern globally. As per Xiaomei et al. (2005) and Convard et al. (2005), in China 1/6 of the total agricultural land has been polluted with heavy metals, of which 40% is due to erosion and deforestation. The heavy metal content in farmlands and agricultural soil has exceeded the standard of the environment quality wherein heavy metals such as cadmium (Cd) have been reported to be most commonly present (Environmental Protection Ministry 2015). Soil heavy metal contamination has also impacted European nations where approximately 600,000 ha of brownfield sites are polluted in America; 400,000 in Germany, England, Denmark, Spain, Italy, Netherlands, and Finland; 200,000 in Sweden, France, Hungary, Slovakia, and Austria; and less than or equal to 10,000 in Greece, Poland, Ireland, and Portugal (Perez 2012; De Sousa and

Ghoshal 2012; Orooj et al. 2014). Further, the contamination by persistent organic pollutants (POPs) has also increased due to industrialization, mining, and agricultural activities. POPs are least water-soluble, have the potential to bioaccumulate in food chains, and are not readily biodegradable, having harsh side effects on humans, plants, and animals (Fiedler 2003; Straif et al. 2005; Oliver and Gregory 2015). Conventional treatment techniques were developed for the treatment of this waste, however, due to its adverse effects the focus of researchers has diverted towards biological remediation methods such as phytoremediation. The phytoremediation of soil contaminated with heavy metals and POPs is discussed below:

12.3.1.1 Heavy Metals and Organic Pollutants

The phytoremediation of heavy metals is based on three approaches, i.e., the removal of heavy metal with economic benefit, minimizing the risk of heavy metal contamination, and soil management along with phytoremediation so that the soil can again be used for agricultural purposes with economic benefits (Vangronsveld et al. 2009; Garbisu and Alkorta 2003) (see Table 12.1). Phytoextraction has been employed as a mechanism for the removal of heavy metal from the soil via plant roots where the translocation and accumulation occur in the shoot followed by phytoextraction. It should also be noted that hyperaccumulator plants are more potent in the accumulation of heavy metals as they can accumulate 50–500 times greater concentrations than non-hyperaccumulator plants. The majority of hyperaccumulators (0.2%) identified fall under the angiosperms family: *Asteraceae, Brassicaceae, Caryophyllaceae, Fabaceae, Flacourtaceae, Lamiaceae, Poaceae, Violaceae*, and *Euphorbiaceae* (Baker and Brooks 1989; Memon et al. 2001; McGrath et al. 2001; Garbisu and Alkorta 2003; Halim et al. 2003; Ghosh and Singh 2005; Yang et al. 2005; Prasad

TABLE 12.1
The Phytoremediation of Pollutants From the Soil

S. No	Plants	Pollutants	Removal	References
1	*Jatropha curcas*	Waste lubricating oil (WLO)	• 56.6% and 67.3% loss of waste lubricating oil (WLO) for 2.5% and 1%. • Addition of organic waste (brewery spent grain) increased waste removal to 89.6% and 96.6%.	Agamuthu et al. (2010)
2.	Soybean, green bean, sunflower, Indian mustard, maize, red clover, and ladino clover	Motor oil	• 67% removal with sunflower/mustard, and with the addition of NPK fertilizer the oil was completely removed. • 38% removal with grass/maize, which increased to 67% with fertilizer application.	Dominguez-Rosado and Pichtel (2004)
3.	Barley, cabbage, spinach, sorghum, bean, tomato, and ricinus	Nickel	• Spinach and ricinus appeared to be very effective in the removal of nickel.	Giordani et al. (2005)
4.	Corn	Cadmium and lead	• The plant is an effective accumulator of the pollutant.	Mojiri (2011)
5.	Canola (*Brassica napus* L)	Zinc	• Canola is a zinc hyperaccumulator and high doses do not affect the growth of the plant.	Belouchrani et al. (2016)

2003; Kotrba et al. 2009; Memon and Schroder 2009). The standard that has been successfully used in the reported literature is that of Baker and Brooks (1989). It consists of hyperaccumulator plants that can accumulate more than 100 mg/kg of cadmium (Cd); 1,000 mg/kg of nickel (Ni), copper (Cu), and lead (Pb); and 10,000 mg/kg of zinc (Zn) and manganese (Mn) in the shoots. In addition, hyperaccumulators also consist of those plants that can accumulate more than 1 mg gold (Au) and silver (Ag); 100 mg of selenium (Se) and tantalum (Ta); and 1,000 mg of cobalt (Co), chromium (Cr), uranium (U), and arsenic (As). However, van der Ent et al. (2013) suggested that the criteria for the categorization of a plant as a hyperaccumulator are conservative and should be relaxed. Hyperaccumulators have the potential to survive unfavorable environmental conditions (e.g., droughts), have a high tolerance, are protected against herbivores and pathogens, and survive at high metal concentrations in the soil (Bhargava et al. 2012). It has to be noted that previously the mechanism of hyperaccumulators was not studied in detail; however, with the development of technologies, the study of this mechanism has been possible, paving the way for future research (Verbruggen et al. 2009).

The other mechanism used for the removal of heavy metal via phytoremediation is phytomining. This technique includes growing and harvesting hyperaccumulator plants, followed by the combustion and extraction of the ore. This process/technique has been successfully established for nickel (Ni). The most efficient phytomining has been achieved for gold (Au), thallium (Tl), cobalt (Co), and nickel (Ni), and the least is that for uranium (Ur) (Sheoran et al. 2009). Phytoextraction in combination with phytomining can be used in the future for the production of biofuel. Other phytoremediation methods such as phytostabilization, phytoevaporation, and rhizodegradation have also been employed for the removal of heavy metals from contaminated soil.

The phytoremediation of plants occurs by means of two mechanisms: direct and phytoremediation explanta. The latter relies on the secretion of photosynthetate in the exudates of roots, which supports the growth and diversity of the bacterial and fungal culture in the rhizosphere. The root exudates contain certain compounds that help in the survival of the microbes that may further phytoremediate the pollutant (Shimp et al. 1993; Anderson et al. 1993; Anderson and Coats 1994; Walton et al. 1994; Cunningham and Lee 1995; Salt et al. 1998). It has to be noted that not all plants contain the root exudate and may still be able to degrade the pollutant, which might be due to the presence of polychlorinated biphenyls (PCBs) bacteria harboring in the rhizosphere of the plant. In addition, the microbes in the rhizosphere may also be able to remediate organics, e.g., polynuclear aromatic hydrocarbon (PAHs), by volatilizing or via enhancement in the production of humic substances. A similar process has also been used to remediate trichloroethylene (TCE) and trinitrotoluene (TNT) (Foth 1990; Wolfe et al. 1993; Dec and Bollag 1994; Anderson and Walton 1995; Cunningham and Ow 1996; Alkorta and Garbisu 2001). The other mechanism that includes plants remediating organic pollutants is via enzymes such as laccase (Wang et al. 2004), nitroreductase (Hannink et al. 2001), and peroxidase (González et al. 2006). In a study by Wolfe et al. (1993), plant-derived nitroreductases and laccases in the field efficiently degraded trinitrotoluene, dinitromonoaminotoluene, mononitrodiaminotoluene, and triaminotoluene.

Lastly, in the direct uptake of organic pollutants, remediation depends on the availability of the pollutant and the mechanism of uptake of the pollutant by the plant (Salt et al. 1998). The most common mechanism that has been studied is the uptake of the pollutant (pesticides and herbicides) via the liquid phase of the soil (Kawahigashi et al. 2006; Briggs et al. 1982). Lastly, if phytoremediation of the soil has to be employed then the fate of the parent compound has to be known. Further, the partition of the pollutant varies significantly in the root and shoot of the plant, and the contaminant can be either volatilized, partially, or completely degraded, and may be transformed to non-toxic end products which binds to plants tissues in non-available forms (Field and Thurman 1996; Alkorta and Garbisu 2001). However, much research has to be performed for a better understanding of the system and the mechanism involved in the phytoremediation of organic pollutants from the soil.

12.3.2 WATER

The pollution of water occurs via two sources: pointed and non-pointed. Pointed pollution includes industrial waste such as textile dyes, power-production, runoff, etc. Non-pointed pollution includes pollution by vehicles, which is widespread over a wide geographical area (Wijayawardena et al. 2016). These pollutants cause severe health hazards to humans and animals both on land and in water. Thus, various conventional treatment methodologies were developed for the treatment of water pollutants, such as adsorption, nanotechnology (nanofiltration, nanoparticle, nanofibre, nanocluster, and nanocomposite), electrochemical treatment (precipitation, oxidation, electrodialysis, and membrane electrolysis), biological treatment (biosorption, bacterial and fungal treatment, activated sludge process, biofilter, anaerobic digestion, stabilization ponds, and biofilms), and phytoremediation (Jeevanantham et al. 2019).

12.3.2.1 Heavy Metals and Organic Pollutants

The phytoremediation of water for both heavy metals and organic pollutants can be performed using tropical plants that have been reported for the removal/remediation of pollutants (see Table 12.2). The end products formed are either converted to non-toxic end products or the fate of the remediated pollutant may depend on the type of pollutant being phytoremediated. In the case of water phytoremediation, phytostabilization has been employed wherein the contaminants in groundwater were immobilized in the plants via accumulation/adsorption/precipitation leading to the remediation of the pollutant. The other mechanism includes the secretion of exudates by the plants in the rhizosphere, which have several functions such as pH regulation, the chelation of minerals and nutrients, a chemoattractant for microorganisms, and the detoxification of the pollutants (Girkina et al. 2018; Broeckling et al. 2008; Yan et al. 1996; Dakora and Phillips 2002; Strom et al. 2002;

TABLE 12.2
The Phytoremediation of Pollutants From the Water

S. No	Plants	Pollutants	Removal	References
1.	Water hyacinth (*Eichhornia crassipes*)	Cadmium, lead, copper, zinc, and nickel	• The ratio of translocation in tissues was in the order Cu>Pb>Cd>Ni>Zn. • The highest concentrations in the root tissue in the order Cu>Zn>Ni>Pb>Cd.	Liao and Chang (2004)
2.	*Eleocharis acicularis*	Copper, zinc, arsenic, lead, cadmium	• The concentrations of heavy metals in the shoots were 20,200 mg Cu/kg, 14,200 mg Zn/kg, 1740 mg As/kg, 894 mg Pb/kg, and 239 mg Cd/kg.	Sakakibara et al. (2011)
3.	*Hydrilla* and *Pistia*	Copper	• Th percentage of metal extraction was 69, 43.8, and 33.35% in 2, 5, and 7 ppm concentrations.	Venkateswarlu et al. (2019)
4.	*Myriophyllum aquaticum* (parrot feather) and *Pistia stratiotes* (water lettuce)	Tetracycline (TC) and oxytetracycline (OTC)	• The involvement of root-secreted metabolites in antibiotic modification is suggested.	Gujarathi et al. (2005)
5.	*Ceratophyllum demersum* and *Lemna gibba*	Lead and chromium	• Effective removal of 95% and 84% for lead and chromium was attained, respectively.	Abdallah (2012)

Silva et al. 2004). The other mechanism includes rhizofiltration where the contaminants present in the water are adsorbed or absorbed by the roots of the plants. In this case, precipitation and sequestration are involved to carry out the adsorption and absorption of the metal contaminants into the plant roots (Tangahu et al. 2011; Erakhrumen and Agbontalor 2007) to detoxify the pollutant. In the phytoremediation of water, using microbes and genetic engineering technologies helps improve the phytoremediation potential of the plants and enhance the expression of transport and remediation of the pollutants to various parts of the plants (Dhankher et al. 2012). Phytoremediation also increases the production of biomass via the insertion of accumulator genes in the plant. A genetic engineering approach enables the insertion of aggregator qualities of taller plants into various plants, which further increases the biomass of the plant (Jeevanantham et al. 2019).

12.3.3 AIR

Industrialization is increasing at an uncontrolled rate and has had a significant impact on the environment. Among the various environmental elements, air is an integral part of human survival, and the quality of air that a living being breathes impacts their health either positively or negatively. Nuclear material spills, for example, have serious implications such as altering genetic makeup. However, neither industrialization nor urbanization can be avoided by any nation globally and is continuing at present even at the cost of the ecosystem. Thus, the focus should be on both development as well as sustainability. As a result, various laws and policies have been developed and imposed on various domains to control and regulate pollution and maintain balance. The major air pollutants that are present in the air/atmosphere include oxides of sulfur and nitrogen, carbon monoxide, suspended particulate matter, hydrocarbons, and ozone. Among the various pollutants, 50% is contributed by suspended particulate matter and has been reported to cause respiratory tract diseases in living beings (D'Amato 1999; Freer-Smith 1985). Thus, phytoremediation has been implemented for the removal of these pollutants (Table 12.3).

12.3.3.1 SO_2, NO_2, and Particulate Pollution

SO_2 enters a plant via its stomatal opening and moves into the plant tissues; as a result, plants in sulfate-contaminated areas have high sulfur content. In a study by Manninen and Huttunen (2000), it was observed that in Scots pine and Norway spruce needles, an increase in SO_4^{2-} was observed when exposed to SO_2 and can be due to foliar absorption and an increase in the uptake of sulfur and nitrate from the soil (Reddy and Dubey 2000; Krupa and Legge 1999). Similarly, in a study by Agrawal and Singh (2000), it was observed that near thermal power stations the accumulation potential of compounds rich in sulfur was high in mangoes. It was suggested that the accumulation could be due to the high sensitivity of mango trees and pan-tropical distribution. Mango trees have also been regarded as bio-indicator plants in both tropical and sub-tropical regions. It has also been stated that nitrogen helps in the assimilation of sulfate that has accumulated in plants, and what remains after its utilization is termed as "residual sulfate." This form of sulfate is toxic to plants and might induce chlorosis and an imbalance of nutrients (Keller 1980; Giodano et al. 2005).

In the case of NO_2, it may reach the plant either by foliar deposition or by rainwater/soil deposition. Viskari et al. (2000) studied the effect of NO_2 on spruce seedlings (*Picea abies*), and it was observed that even a short duration of exposure NO_2 might cause severe injury to the ultrastructure at the cellular level along with the needle surface. Further, the penetration of NO_2 via the stomatal opening has also been observed and is governed by multiple factors such as the species of plant, the plant's age, the concentration of NO_2, and various nutritional and environmental factors as well (Okano et al. 1989; Srivastava et al. 1975; Thoene et al. 1991; Singh and Verma 2007). It has been also observed that NO_2 is absorbed via the course of inorganic nitrogen assimilation. The dissolution of NO_2 results in the production of nitrate in the cell sap where the enzymatic machinery of nitrate and nitrite reductase helps in the generation of $NH4^+$, which is then assimilated via a GS-GOGAT pathway to glutamate (Murray and Wellburn 1985; Srivastava and Ormrod 1989; Wellburn 1990;

TABLE 12.3

The Phytoremediation of Pollutants From the Air

S. No	Plants	Pollutants	Removal	References
1	*Sansevieria trifasciata*, *Epipremnum aureum*, and *Dieffenbachia seguine*	Benzene, formaldehyde, styrene, styrene, and phenol	• Plants improved indoor air quality. • Plants also reduced volatile organic compounds.	Sowa et al. (2019)
2.	73 plant species were used	Benzene	• Of the 73 species tested, *Crassula portulacea, Hydrangea macrophylla, Cymbidium Golden Elf., Ficus microcarpa var. fuyuensis, Dendranthema morifolium, Citrus medica var. sarcodactylis, Dieffenbachia amoena cv. Tropic Snow, Spathiphyllum Supreme, Nephrolepis exaltata cv. Bostoniensis,* and *Dracaena deremensis cv. Variegata* have the largest capacity to remove benzene from indoor air.	Liu et al. (2007)
3.	*Syngonium podophyllum*	Volatile organic compounds (VOC) and CO_2	• At a moderate increase in indoor light intensity 61% ±2.2 of 1,000 ppmv of CO_2 from test chambers was removed over a 40-minute period. • The hydroculture growth medium facilitated increased CO_2 removal over potting mix. • Both methods, i.e., conventional potting mix and hydroculture, removed 25 ppmv of VOC from the test chambers within 7 days.	Irga et al. (2013)
4.	28 ornamental species were used	Benzene, toluene, octane, trichloroethylene (TCE), and terpene (α-pinene)	• Of the 28 species tested, *Hemigraphis alternata, Hedera helix, Hoya carnosa,* and *Asparagus densiflorus* had the highest removal efficiencies for all pollutants. • *Tradescantia pallida* displayed superior removal efficiencies for benzene, toluene, TCE, and α-pinene. • *Fittonia argyroneura* effectively removed benzene, toluene, and TCE. • *Ficus benjamina* effectively removed octane and α-pinene. • *Polyscias fruticosa* effectively removed octane.	Yang et al. (2009)
5.	28 plant species representing five general classes (ferns, woody foliage plants, herbaceous foliage plants, Korean native plants, and herbs) were used	Volatile formaldehyde	The most effective species in the individual classes were: • Ferns: *Osmunda japonica, Selaginella tamariscina,* and *Davallia mariesii.* • Woody foliage plants: *Psidium guajava, Rhapis excels,* and *Zamia pumila.* • Herbaceous foliage plants: *Chlorophytum bichetii, Dieffenbachia "Marianne," Tillandsia cyanea,* and *Anthurium andraeanum.* • Korean native plants: *Nandina domestica.* • Herbs: *Lavandula spp., Pelargonium spp.,* and *Rosmarinus officinalis.*	Kim et al. (2010)

Singh and Verma 2007). Further, in the case of particulate matter, the exposed surface of the plant is a natural sink for the settlement of pollutants, and the uptake of the particulates by the plant is categorized based on the relative deposition velocities, the trapping and capturing efficiency via wind tunnels, and relative deposition velocities using micro-meteorological techniques in the field (Singh and Verma 2007). Plants have been used in many countries to combat the pollution caused by particulates. The plants that can tolerate the pollutants have been reported in the literature (Nivane et al. 2001) as having the potential to absorb, adsorb, metabolize, detoxify, and accumulate the pollutant and thus acts as a live filter for the remediation of the atmosphere (Wolverton 2020). In a study by Yang et al. (2005), it was shown that 1,261.4 tons of pollutants were removed from the air due to the urban forest in Beijing. Thus, plants can be effectively used for the remediation of pollutants.

12.4 CURRENT SCENARIO, LIMITATIONS, AND FUTURE PROSPECTS

The concept of phytoremediation has gained the attention of the scientific community and has established itself as an effective technology for environmental bioremediation. As a result, tremendous research has been done in the past decade and enabled a better understanding of the mechanisms involved in the phytoremediation of pollutants. However, gaps exist with respect to its prospective use and would require new research groups with the utilization of advanced technologies to understand the full potential of phytoremediation. A multidisciplinary approach and interaction amongst the experts in various fields can be used for understanding the systems involved in the phytoremediation of pollutants from a molecular level to the ecosystem. This would enable a better understanding of the various mechanisms involved in phytoremediation and would also give a broader sense of understanding from various prospects. Further, phytoremediation research efforts have resulted in the creation of an online database of plant species that can clean up pollutants, e.g., PHYTOPET (useful for the remediation of petroleum hydrocarbons) and PHYTOREM (useful for the remediation of metals and metalloids) (McIntyre 2003). Also, the U.S. Environmental Protection Agency has a website for phytoremediation (http://www.clu-in.org) wherein information is available to both researchers and the public (EPA 1998). Further collaboration between researchers and industries should also be initiated for the future development of bioremediation, e.g., the integration of phytoremediation in landscape architecture will provide an interesting development in the field of phytoremediation (Kirkwood 2003). Also, transgenic plants (e.g., transgenic Indian mustard plants that have an overexpressed enzyme such as sulfate/selenate reduction) too can be used, having enhanced tolerance to pollutants with increased accumulation and biodegradation potential (Pilon-Smits et al. 1999; Zhu 1999; Zhu et al. 1999).

12.5 CONCLUSION

The pollution of the environment comes from both natural and man-made sources and various techniques have been developed to solve the problem, of which phytoremediation has gained interest among research groups globally. The technique of phytoremediation involves the use of plants in remediating the three essential parameters of the ecosystem, i.e., soil, water, and air. Both native and transgenic plants can be used for the phytoremediation of pollutants from the environment. Further research is required using a multidisciplinary approach to further broaden the use of phytoremediation and allow for its practical implementation for a "better and cleaner ecosystem."

REFERENCES

Abdallah, M.A.M., 2012 Phytoremediation of heavy metals from aqueous solutions by two aquatic macrophytes, *Ceratophyllum demersum* and *Lemna gibba L. Environmental Technology* 33:1609–1614

Agamuthu, P., Abioye, O.P. and Aziz, A.A., 2010 Phytoremediation of soil contaminated with used lubricating oil using Jatropha curcas. *Journal of Hazardous Materials* 179:891–894

Agrawal, M. and Singh, J., 2000 Impact of coal power plant emission on the foliar elemental concentrations in plants in a low rainfall tropical region. *Environmental Monitoring and Assessment* 60:261–282

Alkorta, I. and Garbisu, C., 2001 Phytoremediation of organic contaminants in soils. *Bioresource Technology* 79:273–276

Anderson, T.A. and Coats, J.R., 1994 Bioremediation through rhizosphere technology. In ACS symposium series, USA, American Chemical Society

Anderson, T.A. and Walton, B.T., 1995 Comparative fate of [14C] trichloroethylene in the root zone of plants from a former solvent disposal site. *Environmental Toxicology and Chemistry: An International Journal* 14:2041–2047

Anderson, T.A., Guthrie, E.A. and Walton, B.T., 1993 Bioremediation in the rhizosphere. *Environmental Science and Technology* 27:2630–2636

Ariyakanon, N. and Winaipanich, B., 2006 Phytoremediation of copper contaminated soil by Brassica juncea (L.) Czern and Bidens alba (L.) DC. var. radiata. *Journal of Scientific Research Chulalongkorn University* 31:49–56

Baker, A.J.M. and Brooks, R.R., 1989 Terrestrial higher plants which hyperaccumulate metallic elements. A review of their distribution, ecology and phytochemistry. *Biorecovery* 1: 81–126

Banerjee, A. and Roychoudhury, A., 2019 Fluorine: A biohazardous agent for plants and phytoremediation strategies for its removal from the environment. *Biologia Plantarum* 63:104–112

Belouchrani, A.S., Mameri, N., Abdi, N., Grib, H., Lounici, H. and Drouiche, N., 2016 Phytoremediation of soil contaminated with Zn using Canola (*Brassica napus L*). *Ecological Engineering* 95:43–49

Bhargava, A., Carmona, F.F., Bhargava, M. and Srivastava, S., 2012 Approaches for enhanced phytoextraction of heavy metals. *Journal of Environmental Management* 105:103–120

Borghei, M., Arjmandi, R. and Moogouei, R., 2011 Potential of Calendula alata for phytoremediation of stable cesium and lead from solutions. *Environmental monitoring and assessment* 181:63–68

Briggs, G.G., Bromilow, R.H. and Evans, A.A., 1982 Relationships between lipophilicity and root uptake and translocation of non-ionised chemicals by barley. *Pesticide Science* 13:495–504

Broeckling, C.D., Broz, A.K., Bergelson, J., Manter, D.K. and Vivanco, J.M., 2008 Root exudates regulate soil fungal community composition and diversity. *Applied and Environmental Microbiology* 74:738–744

Burges, A., Alkorta, I., Epelde, L. and Garbisu, C., 2018 From phytoremediation of soil contaminants to phytomanagement of ecosystem services in metal contaminated sites. *International Journal of Phytoremediation* 20:384–397

Chen, B., Ma, X., Liu, G., Xu, X., Pan, F., Zhang, J., Tian, S., Feng, Y. and Yang, X., 2015 An endophytic bacterium Acinetobacter calcoaceticus Sasm3-enhanced phytoremediation of nitrate–cadmium compound polluted soil by intercropping Sedum alfredii with oilseed rape. *Environmental Science and Pollution Research* 22:17625–17635

Chen, L. and Xiong, Z., 2011 Phytoremediation in fluoride contaminated water and toxicity of fluoride on plants. *Environmental Science and Technology* 34:60–151

Cheng, L., Wang, Y., Cai, Z., Liu, J., Yu, B. and Zhou, Q., 2017 Phytoremediation of petroleum hydrocarbon-contaminated saline-alkali soil by wild ornamental Iridaceae species. *International Journal of Phytoremediation* 19:300–308

Convard, N., Tomlinson, A. and Welch, C., 2005 *Strategies for Preventing and Mitigating Land-Based Sources of Pollution to Trans-Boundary Water Resources in the Pacific Region* (No. 16). SPREP

Cunningham, S.D. and Lee, C.R., 1995 Phytoremediation: Plant-based remediation of contaminated soils and sediments. *Bioremediation: Science and Applications* 43:145–156

Cunningham, S.D. and Ow, D.W., 1996 Promises and prospects of phytoremediation. *Plant Physiology* 110:715

Cunningham, S.D., Huang, J.W., Chen, J. and Berti, W.R., 1996 August Phytoremediation of contaminated soils: Progress and promise. In *Abstracts of Papers of The American Chemical Society*, The American Chemical Society, Washington, DC, 212

Dakora, F.D. and Phillips, D.A., 2002 Root exudates as mediators of mineral acquisition in low-nutrient environments. In *Food Security in Nutrient-Stressed Environments: Exploiting Plants' Genetic Capabilities*, Springer, Dordrecht, 201–213

D'Amato, G., 1999 Outdoor air pollution in urban areas and allergic respiratory diseases Monaldi archives for chest disease. *Archivio Monaldi per le malattie del torace* 54:470–474

De Sousa, C. and Ghoshal, S., 2012 Redevelopment of brownfield sites In *Metropolitan Sustainability*, Woodhead Publishing, Cambridge, 99–117

Dec, J. and Bollag, J.M., 1994 Use of plant material for the decontamination of water polluted with phenols. *Biotechnology and Bioengineering* 44:1132–1139

Dhankher, O.P., Pilon-Smits, E.A., Meagher, R.B. and Doty, S., 2012 Biotechnological approaches for phytoremediation. In *Plant Biotechnology and Agriculture*, Oxford, Academic Press, 309–328

Dominguez-Rosado, E. and Pichtel, J., 2004 Phytoremediation of soil contaminated with used motor oil: II. *Greenhouse Studies Environmental Engineering Science* 21:169–180

Ehsan, S., Ali, S., Noureen, S., Mahmood, K., Farid, M., Ishaque, W., Shakoor, M.B. and Rizwan, M., 2014 Citric acid assisted phytoremediation of cadmium by *Brassica napus* L. *Ecotoxicology and Environmental Safety* 106:164–172

Environmental Protection Ministry, PR China, 2015 http://www.bbc.com/news/world-asia-china-27076645

EPA, E., 1998 A Citizen's Guide to Phytoremediation. EPA 542-F-98-011 *Technology Innology office*, Office of Solid Waste and Emergency Response, Washington, DC: http://clu-in.org/products/citguide/photo2. htm

Erakhrumen, A.A. and Agbontalor, A., 2007 Phytoremediation: An environmentally sound technology for pollution prevention, control and remediation in developing countries. *Educational Research and Review* 2:151–156

Escoto, D.F., Gayer, M.C., Bianchini, M.C., da Cruz Pereira, G., Roehrs, R. and Denardin, E.L., 2019 Use of Pistia stratiotes for phytoremediation of water resources contaminated by clomazone. *Chemosphere* 227:299–304

Fiedler, H., 2003 Dioxins and Furans (PCDD/PCDF). In *Persistent Organic Pollutants* Springer, Berlin, Heidelberg 123–201

Field, J.A. and Thurman, E.M., 1996 Glutathione conjugation and contaminant transformation *Environmental Science and Technology* 30:1413–1418

Foth, H.D., 1990 Chapter 12: *Plant Soil Macronutrient Relations. Fundamental of Soil Science*, Wiley, New York, 186–209

Freer-Smith, P.H., 1985 The influence of SO2 and NO2 on the growth, development and gas exchange of Betula pendula Roth *New Phytologist* 99:417–430

Garbisu, C. and Alkorta, I., 2003 Basic concepts on heavy metal soil bioremediation. *European Journal of Mineral Processing and Environmental Protection* 3:58–66

Ghosh, M. and Singh, S.P., 2005 A review on phytoremediation of heavy metals and utilization of it's by products. *Asian Journal of Energy and Environment* 6:18

Giordani, C., Cecchi, S. and Zanchi, C., 2005 Phytoremediation of soil polluted by nickel using agricultural crops *Environmental Management* 36:675–681

Giordano, M., Norici, A. and Hell, R., 2005 Sulfur and phytoplankton: Acquisition, metabolism and impact on the environment. *New Phytologist* 166:371–382

Girkin, N.T., Turner, B.L., Ostle, N., Craigon, J. and Sjögersten, S., 2018 Root exudate analogues accelerate CO2 and CH4 production in tropical peat. *Soil Biology and Biochemistry* 117:48–55

González, P.S., Capozucca, C.E., Tigier, H.A., Milrad, S.R. and Agostini, E., 2006 Phytoremediation of phenol from wastewater, by peroxidases of tomato hairy root cultures. *Enzyme and Microbial Technology* 39:647–653

Gujarathi, N.P., Haney, B.J. and Linden, J.C., 2005 Phytoremediation potential of Myriophyllum aquaticum and Pistia stratiotes to modify antibiotic growth promoters, tetracycline, and oxytetracycline, in aqueous wastewater systems. *International Journal of Phytoremediation* 7:99–112

Halim, M., Conte, P. and Piccolo, A., 2003 Potential availability of heavy metals to phytoextraction from contaminated soils induced by exogenous humic substances *Chemosphere* 52:265–275

Hannink, N., Rosser, S.J., French, C.E., Basran, A., Murray, J.A., Nicklin, S. and Bruce, N.C., 2001 Phytodetoxification of TNT by transgenic plants expressing a bacterial nitroreductase *Nature Biotechnology* 19:1168–1172

Hong, M.S., Farmayan, W.F., Dortch, I.J., Chiang, C.Y., McMillan, S.K. and Schnoor, J.L., 2001 Phytoremediation of MTBE from a groundwater plume *Environmental Science and Technology* 35:1231–1239

Irga, P.J., Torpy, F.R. and Burchett, M.D., 2013 Can hydroculture be used to enhance the performance of indoor plants for the removal of air pollutants? *Atmospheric Environment* 77:267–271

Jeevanantham, S., Saravanan, A., Hemavathy, R.V., Kumar, P.S., Yaashikaa, P.R. and Yuvaraj, D., 2019 Removal of toxic pollutants from water environment by phytoremediation: A survey on application and future prospects. *Environmental Technology and Innovation* 13:264–276

Kawahigashi, H., Hirose, S., Ohkawa, H. and Ohkawa, Y., 2006 Phytoremediation of the herbicides atrazine and metolachlor by transgenic rice plants expressing human CYP1A1, CYP2B6, and CYP2C19. *Journal of Agricultural and Food Chemistry* 54:2985–2991

Keller, T., 1980 Der Einfluß bodenbürtiger Sulfationen auf den Schwefelgehalt SO2-begaster Assimilationsorgane von Waldbaumarten, *Angew Botany*, 54, 77–89.

Kim, K.J., Jeong, M.I., Lee, D.W., Song, J.S., Kim, H.D., Yoo, E.H., Jeong, S.J., Han, S.W., Kays, S.J., Lim, Y.W. and Kim, H.H., 2010 Variation in formaldehyde removal efficiency among indoor plant species *HortScience* 45:1489–1495

Kirkwood, N. ed., 2003 *Manufactured Sites: Rethinking the Post-Industrial Landscape* Taylor & Francis, London

Kotrba, P., Najmanova, J., Macek, T., Ruml, T. and Mackova, M., 2009 Genetically modified plants in phytoremediation of heavy metal and metalloid soil and sediment pollution. *Biotechnology Advances* 27:799–810

Krupa, S.V. and Legge, A.H., 1999 Foliar injury symptoms of Saskatoon serviceberry (Amelanchier alnifolia Nutt.) as a biological indicator of ambient sulfur dioxide exposures. *Environmental Pollution* 106:449–454

Liao, S. and Chang, W.L., 2004 Heavy metal phytoremediation by water hyacinth at constructed wetlands in Taiwan. *Photogrammetric Engineering and Remote Sensing* 54:177–185.

Lim, J.M., Salido, A.L. and Butcher, D.J., 2004 Phytoremediation of lead using Indian mustard (Brassica juncea) with EDTA and electrodics. *Microchemical Journal* 76:3–9.

Liu, Y.J., Mu, Y.J., Zhu, Y.G., Ding, H. and Arens, N.C., 2007 which ornamental plant species effectively remove benzene from indoor air? *Atmospheric Environment* 41:650–654

Manninen, S. and Huttunen, S., 2000 Response of needle sulphur and nitrogen concentrations of Scots pine versus Norway spruce to SO2 and NO2. *Environmental Pollution* 107(3):421–436.

Marrugo-Negrete, J., Durango-Hernández, J., Pinedo-Hernández, J., Olivero-Verbel, J. and Díez, S., 2015 Phytoremediation of mercury-contaminated soils by *Jatropha curcas*. *Chemosphere* 127:58–63

Mcgath, S.P. and Zhao, F.J., 2003 Phytoextraction of metals and metalloids from contaminated soils. *Current Opinions in Biotechnology* 14:277–282

McIntyre, T.C., 2003 Databases and protocol for plant and microorganism selection: Hydrocarbons and metals. In: McCutcheon, S.C., Schnoor, J.L. (Eds.), Phytoremediation. *Transformation and Control of Contaminants*. John Wiley & Sons, Hoboken, 887–904.

Memon, A.R. and Schröder, P., 2009 Implications of metal accumulation mechanisms to phytoremediation. *Environmental Science and Pollution Research* 16:162–175

Memon, A.R., Aktoprakligil, D., Özdemir, A. and Vertii, A., 2001 Heavy metal accumulation and detoxification mechanisms in plants. *Turkish Journal of Botany* 25:111–121

Moccia, E., Intiso, A., Cicatelli, A., Proto, A., Guarino, F., Iannece, P., Castiglione, S. and Rossi, F., 2017 Use of Zea mays L. in phytoremediation of trichloroethylene. *Environmental Science and Pollution Research* 24:11053–11060

Mojiri, A., 2011 The potential of corn (Zea mays) for phytoremediation of soil contaminated with cadmium and lead. *Journal of Biological and Environmental Science* 5:17–22

Mosoarca, G., Vancea, C., Popa, S. and Boran, S., 2018 Adsorption, bioaccumulation and kinetics parameters of the phytoremediation of cobalt from wastewater using Elodea canadensis. *Bulletin of Environmental Contamination and Toxicology* 100:733–739

Murray, A.J.S. and Wellburn, A.R., 1985 Differences in nitrogen metabolism between cultivars of tomato and pepper during exposure to glasshouse atmosphere containing oxides of nitrogen. *Environmental Pollution Series A, Ecological and Biological* 39:303–316

Neunhäuserer, C., Berreck, M. and Insam, H., 2001 Remediation of soils contaminated with molybdenum using soil amendments and phytoremediation. *Water, Air, and Soil Pollution* 128:85–96

Ninave, S.Y., Chaudhari, P.R., Gajghate, D.G. and Tarar, J.L., 2001 Foliar biochemical features of plants as indicators of air pollution. *Bulletin of Environmental Contamination and Toxicology* 67:133–140

Nriagu, J.O., 1979 Global inventory of natural and anthropogenic emissions of trace metals to the atmosphere. *Nature* 279:409–411

Okano, K., Machida, T. and Totsuka, T., 1989 Differences in ability of NO2 absorption in various broad-leaved tree species. *Environmental Pollution* 58:1–17

Oliver, M.A. and Gregory, P.J., 2015 Soil, food security and human health: A review. *European Journal of Soil Science* 66:257–276

Orooj Surriya, S.S.S., Waqar, K. and Kazi, A.G., 2014 Phytoremediation of soils: Prospects and challenges. *Soil Remediation and Plants: Prospects and Challenges*, Elsevier, London UK, 1–36

Perez, J., 2012 The soil remediation industry in Europe: The recent past and future perspectives. *Ernst and Young Research Report*

Pettit, T., Irga, P.J. and Torpy, F.R., 2018 Functional green wall development for increasing air pollutant phytoremediation: Substrate development with coconut coir and activated carbon. *Journal of Hazardous Materials* 360:594–603

Pilon-Smits, E., 2005 Phytoremediation. *Annual Review of Plant Biology* 56:15–39

Pilon-Smits, E.A. and LeDuc, D.L., 2009 Phytoremediation of selenium using transgenic plants. *Current Opinion in Biotechnology* 20:207–212

Pilon-Smits, E.A., Hwang, S., Lytle, C.M., Zhu, Y., Tai, J.C., Bravo, R.C., Chen, Y., Leustek, T. and Terry, N., 1999 Overexpression of ATP sulfurylase in Indian mustard leads to increased selenate uptake, reduction, and tolerance. *Plant Physiology* 119:123–132

Płociniczak, T., Fic, E., Pacwa-Płociniczak, M., Pawlik, M. and Piotrowska-Seget, Z., 2017 Improvement of phytoremediation of an aged petroleum hydrocarbon-contaminated soil by Rhodococcus erythropolis CD 106 strain. *International Journal of Phytoremediation* 19:614–620

Prasad, M.N.V., 2003 Phytoremediation of metal-polluted ecosystems: Hype for commercialization. *Russian Journal of Plant Physiology* 50:686–701.

Raskin, I., Kumar, P.N., Dushenkov, S. and Salt, D.E., 1994 Bioconcentration of heavy metals by plants. *Current Opinion in Biotechnology* 5:285–290

Reddy, B.M. and Dubey, P.S., 2000 Scavenging potential of trees to SO2 and NO2 under experimental condition. *International Journal of Ecology and Environmental Sciences* 26:99–106

Revathi, K., Haribabu, T.E. and Sudha, P.N., 2011 Phytoremediation of chromium contaminated soil using sorghum plant. *International Journal of Environmental Sciences* 2:417–428

Sakakibara, M., Ohmori, Y., Ha, N.T.H., Sano, S. and Sera, K., 2011 Phytoremediation of heavy metal-contaminated water and sediment by *Eleocharis acicularis*. *CLEAN–Soil, Air, Water* 39:735–741

Salt, D.E., Blaylock, M., Kumar, N.P., Dushenkov, V., Ensley, B.D., Chet, I. and Raskin, I., 1995 Phytoremediation: A novel strategy for the removal of toxic metals from the environment using plants. *Bio/technology* 13:468–474

Salt, D.E., Smith, R.D. and Raskin, I., 1998 Phytoremediation. *Annual Review of Plant Biology* 49:643–668

Sánchez, V., López-Bellido, J., Rodrigo, M.A. and Rodríguez, L., 2019 Enhancing the removal of atrazine from soils by electrokinetic-assisted phytoremediation using ryegrass (*Lolium perenne L.*). *Chemosphere* 232:204–212

Schwitzguebel, J., 2000, August Potential of phytoremediation, an emerging green technology In ecosystem service and sustainable watershed management in north China. In Proceedings of International Conference, Beijing, PR China, 5

Sheoran, V., Sheoran, A.S. and Poonia, P., 2009 Phytomining: A review. *Minerals Engineering* 22:1007–1019

Shimp, J.F., Tracy, J.C., Davis, L.C., Lee, E., Huang, W., Erickson, L.E. and Schnoor, J.L., 1993 Beneficial effects of plants in the remediation of soil and groundwater contaminated with organic materials. *Critical Reviews in Environmental Science and Technology* 23:41–77

Silva, I.R., Novais, R.F., Jham, G.N., Barros, N.F., Gebrim, F.O., Nunes, F.N., Neves, J.C.L. and Leite, F.P., 2004 Responses of eucalypt species to aluminum: The possible involvement of low molecular weight organic acids in the Al tolerance mechanism. *Tree Physiology* 24:1267–1277

Singh, S.N. and Verma, A., 2007 Phytoremediation of air pollutants: A review. In *Environmental Bioremediation Technologies*, Springer, Berlin, Heidelberg, 293–314

Siswandari, A.M., Hindun, I. and Sukarsono, S., 2017 Phytoremediation of Phosphate content in liquid laundry waste by using Echinodorus paleafolius and Equisetum hyemale used as biology learning resource. *JPBI (Jurnal Pendidikan Biologi Indonesia)* 2:222–230

Sowa, J., Hendiger, J., Maziejuk, M., Sikora, T. and Kamińska, H., 2019 June potted plants as active and passive biofilters improving indoor air quality. *IOP Conference Series: Earth and Environmental Science* 290 (1):012150. IOP Publishing

Srivastava, H.S. and Ormrod, D.P., 1989 Nitrogen dioxide and nitrate nutrition effects on nitrate reductase activity and nitrate content of bean leaves. *Environmental and Experimental Botany* 29:433–438

Srivastava, H.S., Jolliffe, P.A. and Runeckles, V.C., 1975 Inhibition of gas exchange in bean leaves by NO2. *Canadian Journal of Botany* 53:466–474

Stojanović, M.D., Mihajlović, M.L., Milojković, J.V., Lopičić, Z.R., Adamović, M. and Stanković, S., 2012 Efficient phytoremediation of uranium mine tailings by tobacco. *Environmental Chemistry Letters* 10:377–381

Straif, K., Baan, R., Grosse, Y., Secretan, B., El Ghissassi, F. and Cogliano, V., 2005 Carcinogenicity of polycyclic aromatic hydrocarbons *The Lancet Oncology* 6:931–932

Strigul, N., 2005 August Perspectives of phytoremediation of tungsten contaminated sites. In *Abstracts of Papers of the American Chemical Society*, Amer Chemical Soc, Washington, DC, 230, pp. U1738–U1738

Ström, L., Owen, A.G., Godbold, D.L. and Jones, D.L., 2002 Organic acid mediated P mobilization in the rhizosphere and uptake by maize roots. *Soil Biology and Biochemistry* 34:703–710

Tangahu, B.V., Sheikh Abdullah, S.R., Basri, H., Idris, M., Anuar, N. and Mukhlisin, M., 2011 A review on heavy metals (As, Pb, and Hg) uptake by plants through phytoremediation. *International Journal of Chemical Engineering* 2011, 1–31

Thoene, B., Schröder, P., Papen, H., Egger, A. and Rennenberg, H., 1991 Absorption of atmospheric NO2 by spruce (Picea abies L. Karst.) trees: I. NO_2 influx and its correlation with nitrate reduction. *New Phytologist* 117:575–585

United Nations Environment Programme (UNEP), 2019 *Phytoremediation: An Environmentally Sound Technology for Pollution Prevention, Control and Remediation An Introductory Guide to Decision-Makers.* Newsletter and Technical Publications Freshwater Management Series No. 2, United Nations Environment Programme Division of Technology, Industry, and Economics

Van der Ent, A., Baker, A.J., Reeves, R.D., Pollard, A.J. and Schat, H., 2013 Hyperaccumulators of metal and metalloid trace elements: Facts and fiction. *Plant and Soil* 362:319–334

Vangronsveld, J., Herzig, R., Weyens, N., Boulet, J., Adriaensen, K., Ruttens, A., Thewys, T., Vassilev, A., Meers, E., Nehnevajova, E. and van der Lelie, D., 2009 Phytoremediation of contaminated soils and groundwater: Lessons from the field. *Environmental Science and Pollution Research* 16:765–794

Venkateswarlu, V., Venkatrayulu, C.H. and Bai, T.J.L., 2019 Phytoremediation of heavy metal Copper (II) from aqueous environment by using aquatic macrophytes Hydrilla verticillata and Pistia stratiotes. *International Journal of Fish and Aquatic Studies* 7:390–393

Verbruggen, N., Hermans, C. and Schat, H., 2009 Molecular mechanisms of metal hyperaccumulation in plants. *New Phytologist* 181:759–776

Viskari, E.L., Kössi, S. and Holopainen, J.K., 2000 Norway spruce and spruce shoot aphid as indicators of traffic pollution. *Environmental Pollution* 107:305–314

Walton, B.T., Hoylman, A.M., Perez, M.M., Anderson, T.A., Johnson, T.R., Guthrie, E.A. and Christman, R.F., 1994 Rhizosphere microbial communities as a plant defense against toxic substances in soils 563:82–92

Wang, D., Li, H., Wei, Z., Wang, X. and Hu, F., 2006 Effect of earthworms on the phytoremediation of zinc-polluted soil by ryegrass and Indian mustard. *Biology and Fertility of Soils* 43:120–123

Wang, G.D., Li, Q.J., Luo, B. and Chen, X.Y., 2004 Ex planta phytoremediation of trichlorophenol and phenolic allelochemicals via an engineered secretory laccase. *Nature Biotechnology* 22:893–897

Wang, L., Lin, H., Dong, Y. and He, Y., 2018 Effects of cropping patterns of four plants on the phytoremediation of vanadium-containing synthetic wastewater. *Ecological Engineering* 115:27–34

Wang, X., Chen, C. and Wang, J., 2017 Phytoremediation of strontium contaminated soil by Sorghum bicolor (L.) Moench and soil microbial community-level physiological profiles (CLPPs). *Environmental Science and Pollution Research* 24:7668–7678

Wellburn, A.R., 1990 Tansley Review No. 24 Why are atmospheric oxides of nitrogen usually phytotoxic and not alternative fertilizers? *New Phytologist* 115:395–429

Wijayawardena, M.A.A., Megharaj, M. and Naidu, R., 2016 Exposure, toxicity, health impacts, and bioavailability of heavy metal mixtures. *Advances in Agronomy,* Academic Press 138:175–234

Wolfe, N.L., Ou, T.Y. and Carreira, L., 1993 *Biochemical Remediation of TNT Contaminated Soils,* US Army Corps Eng., Washington, DC

Wolverton, B.C., 2020 *How To Grow Fresh Air: 50 Houseplants To Purify Your Home Or Office,* Springer, Orion Spring, UK

Xiaomei, L., Qitang, W. and Banks, M.K., 2005 Effect of simultaneous establishment of Sedum alfredii and Zea mays on heavy metal accumulation in plants. *International Journal of Phytoremediation* 7:43–53

Yan, F., Schubert, S. and Mengel, K., 1996 Soil pH increase due to biological decarboxylation of organic anions. *Soil Biology and Biochemistry* 28:617–624

Yang, D.S., Pennisi, S.V., Son, K.C. and Kays, S.J., 2009 Screening indoor plants for volatile organic pollutant removal efficiency. *HortScience* 44:1377–1381

Yang, X., Feng, Y., He, Z. and Stoffella, P.J., 2005 Molecular mechanisms of heavy metal hyperaccumulation and phytoremediation. *Journal of Trace Elements in Medicine and Biology* 18:339–353

Yu, X.Z., Zhou, P.H. and Yang, Y.M., 2006 The potential for phytoremediation of iron cyanide complex by willows. *Ecotoxicology* 15:461–467

Zeeb, B.A., Amphlett, J.S., Rutter, A. and Reimer, K.J., 2006 Potential for phytoremediation of polychlorinated biphenyl-(PCB)-contaminated soil. *International Journal of Phytoremediation* 8:199–221

Zhang, L., Rylott, E.L., Bruce, N.C. and Strand, S.E., 2019 Genetic modification of western wheatgrass (*Pascopyrum smithii*) for the phytoremediation of RDX and TNT. *Planta* 249:1007–1015

Zhang, M., Chen, Y., Du, L., Wu, Y., Liu, Z. and Han, L., 2020 The potential of Paulownia fortunei seedlings for the phytoremediation of manganese slag amended with spent mushroom compost. *Ecotoxicology and Environmental Safety* 196:110538

Zhu, Y.L., 1999 Overexpression of glutathione synthetase in Brassica juncea enhances cadmium tolerance and accumulation. *Plant Physiology* 119:73–79

Zhu, Y.L., Pilon-Smits, E.A., Tarun, A.S., Weber, S.U., Jouanin, L. and Terry, N., 1999 Cadmium tolerance and accumulation in Indian mustard is enhanced by overexpressing γ-glutamylcysteine synthetase. *Plant Physiology* 121:1169–1178

13 Recent Technological Advances in Tannery Wastewater Treatment

Uttara Mahapatra, Ajay Kumar Manna, and Abhijit Chatterjee

CONTENTS

13.1 INTRODUCTION

Leather is a resistant and durable material that is used in making an extensive array of products like shoes, clothes, furniture, and many others. The leather industry plays a prominent role in the global market as well as in economy of leading leather-producing countries such as China, Brazil, Italy, Russia, India, South Korea, Argentina, and the U.S.A. (Jadhav and Jadhav 2020). The leather processing industry provides a great source of employment, exports, and economic growth.

Leather, the primary raw material for the leather industry, is obtained by the transformation of skins and hides in four stages (as shown in Fig. 13.1): pre-tanning (beamhouse operations), tanning (tanyard operations), post-tanning (wet-finishing operations), and finishing. Figure 13.1 presents a pictorial overview of the processes involved and the resulting pollutants generated in a typical leather industry. Pre-tanning is the simplest technique in which cleaning and air drying operations inhibit bacterial growth on the surface of hide. This process comprises the following steps: soaking, unhairing and liming, deliming and bating, pickling, and degreasing. Pre-tanning is followed by tanning, which gives leather its essential character and stability. Commonly, tanning processes can be of two types: chrome tanning and vegetable tanning. There are different types of tanning agents available in market. Nonetheless, chromium is employed as the most prevalent tanning salt used.

DOI: 10.1201/9781003204442-13

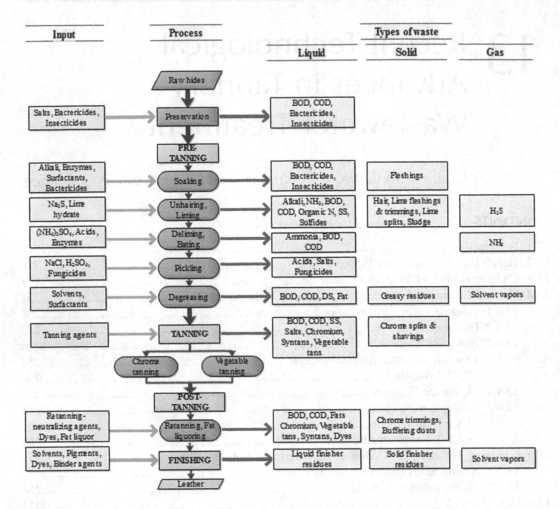

FIGURE 13.1 Flow chart diagram of the general leather processing steps releasing different types of wastes.

Hides tanned with chromium salts are resistant to heat and temperature applied during the tanning process. Also, the penetration capacity of chromium salts inside the skin is higher than that of other tanning substances. However, only a small portion of the chromium salts actually get absorbed into the skin, whereas a huge volume of effluent is generated as a mixture of unabsorbed salts and water during the washing of chrome-tanned leather in a tanning bath (China et al. 2020). The effluent is then sent to a depuration plant where chromium-contaminated sludge is generated. For further quality improvement of tanned skins and hides, post-tanning operations followed by finishing are carried out by the retanning operation, concluded by dying and fatliquoring to obtain smoothness, a uniform color, and a good filling (China et al. 2020). Lastly, hides are polished by using paints and dyes for a softer appearance and uniformity. As evident from this description, leather processing plants generate a huge volume of effluents with various refractory pollutants, including chromium. Accordingly, this chapter summarizes the recent advances in tannery wastewater treatment.

13.2 HEALTH HAZARDS RELATED TO TANNERY WASTES

Chromium is regarded as one of the noxious elements that is always present in tannery waste; this is due to the excessive use of chromium sulfate salts during tanning. Typically, Cr(III) has low solubility and mobility in aquatic systems compared to Cr(VI). The Cr(VI) is more lethal than Cr(III) and slowly gets biomagnified in the food chain of plants and animals, as compared to Cr(III) because of the higher solubility and stability of Cr(VI) in the environment. Nevertheless, the oxidation of Cr(III) to Cr(VI) is thermodynamically feasible in partially anoxic and aerobic environments. Moreover, the presence of MnO_2 enhances the Cr(III)-to-Cr(VI) transformation in the soil environment (Gheju et al. 2016). Recently, studies have been done to minimize the conversion of Cr(VI) into Cr(III) using reducing agents like ascorbic acid (Estrela et al. 2017; Rabelo et al. 2018).

The World Health Organization (WHO) restricted the maximum dose as 0.05 mg L^{-1} of total chromium in drinking water (WHO 2017). The National Institute for Occupational Safety and Health (NIOSH) has recommended that for an 8-hour working period of a worker per day (without exceeding the 40-hours-per-week limit), the exposure of all Cr(VI) compounds should not exceed a concentration of 0.2 µg m^{-3} contaminated air (NIOSH 2013). Usually, chromium is absorbed as dust particles produced during the rubbing, polishing, and powdering of untreated and tanned leathers. The ingestion of high chromium doses can cause lung irritation, affect the upper respiratory tract, obstruct the passageways of the respiratory tract, and cause susceptibility to cancers of the nasal cavity, sinus, or lungs (Rangabhashiyam and Balasubramanian 2018). The general health risks caused by chromium have been linked to higher chances of bronchitis, pharyngitis, asthma, lymph nodes, and polyps in the upper respiratory tract. Additionally, raw hides are a breeding ground for anthrax and can negatively affect humans. Further, when chromium gets absorbed into our skin through unprotected handling, it causes dried, cracked, and scaled skin, leading to allergic dermatitis (Bregnbak et al. 2017; Hedberg et al. 2016).

Other studies have suggested the occurrence of anxiety, short-term memory loss, depression-related behavior (Estrela et al. 2017), liver disorders (Rabelo et al. 2017), reproductive and sexual behavioral disorders, and other neuropsychiatric and cognitive disorders (Rabelo et al. 2017) in male and female Swiss mice when they were subjected to the xenobiotic chemicals present in tannery wastewater (TWW). The toxic impact of TWW has even been reported in plant growth and seed germination (Zaheer et al. 2020).

13.3 CHARACTERIZATION OF TANNERY EFFLUENT

Leather processing industries generate three types of wastes: effluent (liquid), solid waste (solid), and air emission (gas). TWW has been characterized in terms of pH, alkalinity, acidity, hardness, electrical conductivity (EC), total dissolved solids (TDS), biochemical oxygen demand (BOD_5), total solids (TS), chemical oxygen demand (COD), suspended solids (SS), ammonium ions (NH_4^+), phosphate (PO_4^{3-}), nitrate (NO_3^-), chloride (Cl^-), sulfate (SO_4^{2-}), sulfide (S^{2-}), hexavalent chromium [Cr(VI)], oil and grease, and total Kjeldahl nitrogen (TKN). Solid waste obtained during leather processing is contaminated with hair, hide and skin, keratin waste, fat, oil and grease, lime fleshing and split, wet-blue shavings and trimmings, and buffing dust and dry trimmings. Gaseous emissions from tannery industries consist of ammonia (NH_3), hydrogen sulfide (H_2S), volatile organic compounds (VOCs), sulfur dioxide (SO_2), carbon dioxide (CO_2), formic acid (CH_2O_2), chlorine (Cl_2), and ethyl alcohol (C_2H_5OH). The general composition of TWW, described by different authors, is represented in Table 13.1.

13.4 VARIOUS TREATMENT TECHNOLOGIES FOR TANNERY EFFLUENTS

13.4.1 COAGULATION AND FLOCCULATION

Coagulation is the destabilization of a colloidal system in wastewater brought about by the addition of a chemical reagent. The process of promoting destabilized colloidal particles for aggregation to form larger particles (flocs) is known as flocculation. Alum/ aluminum sulfate [$Al_2(SO_4)_3$], ferric chloride ($FeCl_3$), and ferric sulfate [$Fe_2(SO_4)_3$] are some of the commonly-used coagulants.

Chowdhury et al. (2013) treated TWW using a 150 mg L^{-1} $FeCl_3$ solution at a neutral pH which showed reduction percentages for COD, chromaticity, and S^{2-} as 60, 80, and 99, respectively.

TABLE 13.1

Physicochemical Characteristics of Raw Tannery Effluents

							Physico-Chemical Parameters										
pH[a]	Turbidity[b]	BOD	Total COD	Total P	TOC	PO_4^{3-}	Cl^-	SO_4^{2-}	S^{2-}	NO_3^-	NH_4-N	TN	TSS	TDS	Cr(VI)	Total Cr	References
4.08	–	–	–	–	16,022	–	–	–	–	–	–	548	–	–	–	–	(Agustini et al. 2020)
8.04–9.29	–	–	1,497–3,468	–	–	0.6–4.4	1,950–4,950	1,620–3,150	1.6–8	127–366	–	–	–	–	–	–	(Luján-Facundo et al. 2019)
10.66	690	387	3,229	–	–	–	–	–	–	–	–	–	10.5	–	–	6.2	(Pinto et al. 2019)
7.6–9.6	–	–	1,272–3,708	25–29	–	–	–	2,650–3,327	–	–	–	290–367	–	–	0–1.9	0.2–10.1	(Fernandez et al. 2019)
–	2,330–3,107	–	1,702–2,146	–	1,186–1,503	–	–	–	–	–	162–260	892–1,340	–	–	–	–	(Sodhi et al. 2020)
7.5	5,580	2,103	4,291	–	3,861.9	–	–	–	–	–	–	–	–	–	–	–	(Korpe et al. 2019)
11.98–12.54	3,592–3,785	2,540–2,960	1,8800–2,3000	–	3,512–3,868	–	–	–	–	–	–	–	2,5000–3,5200	2,8430–3,2300	–	–	(Saxena et al. 2018)
12.2–12.5	3,260–3,441	1,040–1,100	1,1000–1,2800	–	2,050–2,300	–	–	–	–	–	–	–	1,0860–1,1630	2,2350–2,4870	–	–	(Saxena et al. 2018)
6.5 ± 0.2	54 ± 3	246 ± 16	1,296 ± 84	–	–	–	192 ± 3	–	–	–	–	–	3,200 ± 110	–	–	–	(Borba et al. 2018)
12.04–13	–	–	23,289–40,000	–	–	–	437–1,600	–	576–1,888	–	–	–	–	–	–	–	(Tamersit et al. 2018)
8.19	382	9,333.33	–	33.61	93.32	77.09	–	–	–	23	2.10	110	–	–	–	–	(Rabelo et al. 2017)
7	–	240	320	0.9	–	–	3,650	–	–	–	90	180	–	–	–	–	(dos Santos Moysés et al. 2017)

All the units are in mg L^{-1} except for pH[a]

Turbidity[b] = NTU

Gomes et al. (2016) isolated polyelectrolytes from a cactus plant, *Opuntia ficus indica*, and used this separately with conventional coagulants, $Al_2(SO_4)_3$ and $FeCl_3$, to augment the removal efficiency of physico-chemical parameters present in real TWW. The addition of these polyelectrolytes to typical coagulants such as $Al_2(SO_4)_3$ and $FeCl_3$ increased the COD removal from 77–90% and 91–98%, respectively. Gungor et al. (2016) took an innovative approach and studied the feasibility of replacing commercial alum-based coagulant with wastes generated from the aluminum coating industry for treating TWW. This investigation found that aluminum, regardless of its origin, showed a similar performance in reducing BOD, COD, turbidity, and SS in TWW. The authors claimed that coagulant cost could be reduced by 40% using aluminum etching wastewater as a substitute for alum.

Saxena et al. (2018) conducted an innovative experiment based on conventional coagulation techniques for the pretreatment of tannery effluents. Two different methodologies have been monitored, i.e., with and without adjusting the pH, along with alum dosage variation in the effluent. The first methodology was performed by varying the dosage from 0.125 to 1 g and adjusting the initial pH from 12 to 7 in 100 mL samples. In the second method, the dose was varied from 0.5 to 4 g per 100 mL sample without adjusting the pH. After performing coagulation, the resulting modified effluent was separately subjected to hydrodynamic cavitation (HC) and ultrasonication (US) to compare their removal and energy efficiencies as well. Overall enhancement in the biodegradability index (0.14 to 0.57 for coagulation with HC, and 0.10 to 0.41 for coagulation with US) was observed for both cases. Likewise, the former method proved to be six times more efficient as compared with the latter in terms of energy efficiency.

Pinto et al. (2019) extracted tannin from *Musa* sp. flowers and utilized this as an effective coagulant (avoiding the use of flocculants) for tannery effluent treatment. A degradation of 61.02% Cr(III), 39.43% Cr(VI), 65.4% total chromium, and 78% COD was obtained from this central composite design-based experiment.

13.4.2 ELECTROCOAGULATION

Electrocoagulation is a combined process of electrolysis and chemical coagulation. This process typically uses Fe or Al plates as electrodes. Coagulant formation is generally observed when voltage is applied. The anodes oxidize themselves to produce Al(III) or Fe(II) ions, which form $Al(OH)_3$ or $Fe(OH)_2$, respectively. Electrocoagulation leads to the simultaneous formation of metal hydroxides and coagulation. Due to the highly acidic nature of TWW, electrochemical approaches are more suitable than conventional methods. Consequently, the degree of electricity and energy consumption by electrochemical reactors is assessed through the monitoring of carbon footprint and oxygen consumption.

Deghles and Kurt (2015) applied the electrocoagulation technique for treated TWW by using Al and Fe electrodes separately under various operating conditions (electrolysis time, pH, and current density). The optimization of this process was done by applying the Taguchi orthogonal method, and the operating costs for the removal of color, COD, and total chromium was evaluated. The reported operating costs varied from 0.88–0.94 $ m^{-3} depending on the type of electrodes used and the nature of removal from TWW. In another study, a hybrid electrocoagulation/electrodialysis process applying Al and Fe electrodes separately was utilized for TWW treatment. The investigators observed that the removal of COD, NH_3-N, chromium, and color was 92%, 100%, 100%, and 100%, respectively, after effluent treatment by utilizing Al electrodes. However, the removal rates were 87%, 100%, 100%, and 100%, respectively, for the same pollutants by using Fe electrodes (Deghles and Kurt 2016). In a similar study done by Thirugnanasambandham and Sivakumar (2016), the efficiency of the electrocoagulation reactor was verified for TWW treatment using stainless steel electrodes. This experiment was optimized by using a Box–Behnken design combined with the response surface methodology. The percent removal efficiencies for color, COD, and oil and grease were 92%, 80%, and 95%, respectively. Elabbas et al. (2016) used electrocoagulation cells fitted with two kinds of Al-based electrodes

(Al/Cu/Mg alloy and pure Al) for chromium and COD removal from TWW. This Al alloy was noted as more efficient than Al in pristine conditions for chromium and COD removal.

13.4.3 BIOLOGICAL PROCESS

13.4.3.1 Activated Sludge Process

The activated sludge process (ASP) falls under the category of suspended growth processes. The basic principle behind ASP is to confine naturally occurring microorganisms in treatment tanks. Microbes such as protozoa, bacteria, and other microorganisms degrade organic matter present in the liquid effluent. The limitations of this process are the complex nature of ASP due to many variables, the continuous change of bacterial flora in treatment tanks, and elevated concentrations of lethal chemicals that can kill bacteria, thereupon releasing untreated effluent directly into the environment without prior treatment.

Several studies comparing ASP with different treatment technologies for TWW were reported. Goswami and Mazumder (2016) compared the efficiency of the moving bed bioreactor (MBBR) with ASP by analyzing TWW containing SS of 2,000–2,500 mg L^{-1}. More than 85–90% COD removal in MBBR was reported in this study compared to only 60–70% COD removal in the ASP reactor. In another comparative study by Tammaro et al. (2014), the kinetics of the biological activated carbon process (a combination of AC and microbacteria) was found to be faster than ASP. The soluble chromium removal in TWW was 67% and 46% for biological activated carbon and activated sludge, respectively. The ASP performance was reduced by toxic pollutants contaminating the wastewater. Sirianuntapiboon and Chaochon (2016) studied sequencing batch reactor (SBR) performance in TWW treatment. Under optimized conditions, removal efficiencies for Cr(III), COD, BOD$_5$, TKN, and TN were 95.6 ± 0.3%, 98.0 ± 1%, 99 ± 0.0%, 68.6 ± 0.0%, and 60.0 ± 0.1%, respectively.

Fernandez et al. (2019) carried out four different variants of tannery effluent treatment (one controled and three inoculated experiments) with consortium SFC 500-1 for 3 days. Before carrying out the experiment, the growth of SFC 500-1 was evaluated at different temperatures in TWW along with an additional supplement of 25 mg L^{-1} Cr(VI) and 300 mg L^{-1} phenol. The removal of COD, Cr(VI), and phenol was studied, and the maximum removal obtained was 14.1 to 18.6% for Cr(VI), and 100% for phenol within a treatment period of 3 days at 28°C and 150 rpm. Moreover, a reduction of 54.4% to 60.5% in COD was also observed within the same period.

13.4.3.2 Wetlands

As per the Ramsar Convention 1971, wetlands of international importance are defined as the temporary or permanent and natural or constructed areas comprising of water, swamp, or fen of which the depth is not more than 6 meters. Singh et al. (2016) used phyto-rhizoremediation, followed by further analysis with liquid chromatography-mass spectroscopy (LC-MS) for TWW. The combination of *Spirodela polyrrhiza* (algae) and *Micrococcus luteus, Bacillus pumilus, Bacillus flexus*, and *Virgibacillus sediminis* (chromium-resistant bacterial strains) was utilized for this purpose. The reduction of COD, total Cr, Cr(VI), Pb(II), and Ni(II) were 81.2%, 97.3%, 99.3%, 97.0%, and 95.7%, respectively. The treatment of TWW using bioremediation has been performed by Ashraf et al. (2018). The investigators prepared a constructed wetland by incorporating *Leptochloa fusca* (Kallar grass) and three microbial consortium species, namely *Microbacterium arborescens* HU33, *Enterobacter* sp. HU38, and *Pantoea stewartii* ASI11 synergistically. The authors concluded that *L. fusca* is a potent grass to remediate chromium from TWW. However, augmenting the same with the three endophytic bacteria also helped in the removal of organic and inorganic matter from the effluent and decreases the toxicity.

13.4.4 MEMBRANE FILTRATION

In another study, vegetable tanning liquors were purified from waste streams using UF polymeric membranes (Romero-Dondiz et al. 2015). Here, two different membranes (GR60PP and OT050) were compared using feeds obtained from vegetable tanning and synthetic wastewater liquors. The observed rejection coefficient for the OT050 membrane was less than that of GR60PP. Bhattacharya et al. (2013) treated TWW using ceramic microfiltration (MF) membranes followed by reverse osmosis (RO) for the reuse of TWW. The turbidity value was decreased to 0.025 NTU by this process. The use of oxidation-coagulation-filtration techniques for sulfide removal from TWW in liming processes was studied by Hashem et al. (2016). In this study, H_2O_2 (an oxidizing agent) eliminated H_2S, and $Al_2(SO_4)_3$ acted as a coagulant. The percent removal efficiencies of S^{2-}, TDS, TSS, EC, and salinity were respectively 81%, 96%, 86%, and 87% after TWW treatment with H_2O_2 and coagulant using three-layer filter beds. Bhattacharya et al. (2016) compared the biological effect of raw and treated TWW on *Pila globosa* (snail). Two different membrane processes, i.e., MF alone and MF followed by RO, were applied to get the treated effluent from tannery.

Studies on the recovery of Cr(III) salts from TWW were done by using a synthetic solution containing Cr(III), SO_4^{2-}, and Cl^- (Gando-Ferreira et al. 2015). The Cl^- ions were removed by using an ion exchange membrane (anionic resin Diaion PA360). The resulting chromium sulfate solution was then concentrated in a nanofiltration (NF) membrane unit so that the salt could be reused in the tanning bath. The rejection ratio was found to increase from 70% to 90% on decreasing the (Cl^-/SO_4^{2-}) ratio from 2 to 0.5. Dasgupta et al. (2015) treated TWW through a combined process of NF and coagulation. Effluents generated after this combined process showed the following composition: TS (86 mg L^{-1}), TDS (108 mg L^{-1}), COD (142 mg L^{-1}), Cr(VI) (0.13 mg L^{-1}), BOD (65 mg L^{-1}), and EC (14 mho cm^{-1}).

Innovative work has been carried out to avoid the fouling of the anion exchange membrane by organic molecule present in TWW during its treatment by electrodialysis. The anion exchange membrane is surrounded by an ultrafiltration membrane to restrict fouling. The main idea behind this work was to recover the low (amino acids, HS^-, Cl^-, S^{2-}, OH^-, Na^+, and Ca^{2+}) and high molecular weight (peptide and proteins) compounds present in the effluent as well as to recycle the treated water. No fouling was observed in this experiment. An efficient electrolytic behavior is observed for strong electrolytes such as calcium hydroxide, sodium hydroxide, and a mixture of the same (Tamersit et al. 2018).

An integrated approach of oxidation-nanofiltration was suggested by Pal et al. (2020) to treat wastewater from tannery industries. The original TWW contained 10,500 mg L^{-1} COD, 8.2 mg L^{-1} chromium, 4,620 mg L^{-1} Cl^-, 2,560 mg L^{-1} S^{2-}, and 7,420 mg L^{-1} TDS. After treatment with a graphene oxide nanocomposite at a flux rate of 210–220 LMH at 16 bar, almost 99% COD, 99.5% chromium, and 96% TDS were successfully removed.

13.4.5 Adsorption

13.4.5.1 Biosorbents

13.4.5.1.1 Algal Biosorbent

Recently, da Fontoura et al. (2017a) cultivated *Scenedesmus* sp. in TWW at different dilutions. The results demonstrated an improvized biomass population (0.90 g L^{-1}). The percent reduction for phosphorus, nitrogen (ammoniacal), and COD was found to be 96.78%, 85.63%, and 80.33%, respectively at an occurrence of 88.4% dilution of TWW. Another investigation from the same authors, used defatted microalgal biomass, a waste from microalgal biofuel, to adsorb a leather dye (AB-161) (da Fontoura et al. 2017b). Maximum removal was found to be 75.78 mg g^{-1} (25°C) and 83.2 mg g^{-1} (40°C) in synthetic wastewater under experimental conditions. The adsorbent also reduced the dye concentration (76.65%), TOC (50.78%), and TN (19.80%) in the treatment of real TWW.

13.4.5.1.2 Bacterial Biosorbent

Recently, Marzan et al. (2017) found three classes of bacteria: *Micrococcus* sp., *Gemella* sp., and *Hafnia* sp., capable of detoxifying and reducing Cr, Cd, and Pb contaminations in TWW. *Micrococcus* sp. and *Gemella* sp. showed insensitivity to Cr, Cd, and Pb, whereas *Hafnia* sp. showed sensitivity to Cd. Bharagava and Mishra (2018) have isolated and identified a bacterium named

Cellulosimicrobium sp. (KX710177) from TWW. This bacterium can reduce different Cr(VI) concentrations, i.e., 50 mgL^{-1}, 100 mgL^{-1}, 200 mgL^{-1}, and 300 mgL^{-1}. The percentage removals for these concentrations were found to be 99.33%, 96.98%, 84.62%, and 62.28%, respectively. In an eco-friendly approach, Zhang et al. (2017) have made synergistic use of *Penicillium* sp. to degrade chromium tanned leather and hide powder. Here *Penicillium* sp. performes two main roles. Firstly, these bacteria grow by gaining nutrients from chrome-tanned leather. Secondly, the simultaneous production of protease in this process will degrade the peptide chains of chrome tanned leather.

13.4.5.1.3 Fungal Biosorbent

Sharma and Malaviya (2016) isolated four chromium resistant fungi, namely *Cladosporeumperangustum*, *Penicillium commune*, *Fusarium equiseti*, and *Paecilomyces lilacinus* from tannery effluent contaminated soil for the reduction of pollutants present in TWW. The four fungal species were supported to a nylon mesh with the best provision for these consortiums' immobilization. The removal efficiencies for COD, color, Cr, Total Cr, and Pb were 82.52%, 86.19%, 100%, 99.92%, and 95.9%, respectively. Ejechi and Akpomie (2016) conducted research works to compare between live mycelium (microfungi) and dead mycelia (basidiomycete) in removal of Cr(VI) from TWW. The biosorption capacity of live mycelium was observed to be greater than that of dead mycelia. Further, the recovery of Cr(VI) was around 65.4 to 87.7% by using microfungi.

13.4.5.2 Nanosorbents

Biocomposites developed from different types of biomass and nanoparticles have become more popular in recent times. For example, activated carbon produced from almond husk was impregnated with cobalt ferrite magnetic nanoparticles to increase the adsorptive removal of Cr and Pb(II) ions from TWW (Yahya et al. 2020). Similarly, the adsorption capacity of a commercial filter aid material prepared from diatomite was increased by embedding nanopyroxene into it (Hethnawi et al. 2020). The adsorbent was found to be effective in the removal of Cr(VI) from synthetic and real tannery effluents in both batch and fixed-bed systems. Another composite adsorbent was prepared by using olive stone as support for zero-valent iron and magnetite nanoparticles to remove chromium, organic matter, and total phenols from TWW (Vilardi et al. 2018). The batch process showed removal of 12.66 mg g^{-1} of total chromium [8.37 Cr(III) and 4.29 mg g^{-1} Cr(VI)] within 2 hours at an adsorbent dosage of 4 g L^{-1}. A sustainable bio-nanocomposite was developed for the treatment of TWW by doping FeS and MgS into cellulose nanofibers (CNFs), leading to a high chromium adsorption capacity (142.8 mg g^{-1}) (Sankararamakrishnan et al. 2019). The CNFs prevented the agglomeration of nanoparticles whereas MgS stabilized FeS by suppressing the aerial oxidation of iron. Chromium removal occurs through a number of mechanisms such as electrostatic attraction, ion-exchange followed by the reduction of Cr(VI) to Cr(III), and the immobilization of Cr(III) as chromic oxide and Fe–Cr mixed oxide. Another nano-adsorbent was synthesized using the "green chemistry approach" by incorporating zinc oxide nanoparticles into activated carbon produced from a parthenium weed plant for the treatment of TWW (Kamaraj et al. 2020). The article concluded that the increase in efficiency of Cr(VI) and methylene blue removal was due to the combined reaction of photocatalysis and adsorption. Guan et al. (2017) used a composite nanomaterial made up of four compounds [tetraethylorthosilicate (3-aminopropyl), triethoxysilane, gallic acid, and Fe$_3$O$_4$] as a nanoadsorbent for the removal of Al(III) from TWW by chemisorption through the coordination of Al(III) with gallic acid. The adsorption capacity of nanocomposites was investigated by varying the factors such as coexisting substances, concentration of initial Al(III), dose of adsorbent, and pH. The average Al(III) removal was reported to be 94%. The exhausted nanocomposite was regenerated through chemical desorption. Another recent study focused on the removal of S^{2-} ions from TWW using carbon/alumina nanocomposites synthesized by controlled pyrolysis of aluminum carboxylates/carbon mixture (Balasubramani et al. 2017). The alumina nanoparticles acted as catalysts in the oxidation of the adsorbed sulfide ions within the carbon surface. The adsorption capacity was in the range of 20–40 mg g^{-1} and was found to increase in the presence of surfactants and sodium chloride. A field study using real TWW showed a removal efficiency similar to the synthetic solution used in the simulation.

13.5 EMERGING TECHNOLOGIES: COMBINATORIAL METHODS/ HYBRID TECHNOLOGY

13.5.1 ADVANCED OXIDATION PROCESSES

In advanced oxidation processes (AOPs), a potent oxidizing agent (e.g., ozone, H_2O_2) and/or catalysts (e.g., Fe, Mn, TiO_2) is used in the presence of high-strength radiation (e.g., UV light, gamma rays, and ultrasound waves) for in situ formations of hydroxyl radicals which degrades recalcitrant and refractory organic pollutants into water and carbon dioxide. For TWW treatment, Fenton oxidation, photo-Fenton oxidation, photo-oxidation, photocatalysis, ozonation, and electrochemical methods may be applied. The main benefits are minimal discarded waste, spontaneous reaction rates, easy automation and control, and less labor. However, there is a possibility of transformation of Cr(III) to Cr(VI) in AOPs.

Bletterie et al. (2012) investigated the minimization of foam formation in the transboundary river of Austria which receives effluents from three tanneries. Ozonation followed by the biological treatment of these effluents was conducted on a laboratory scale. A 50% COD reduction and more than 70% of foam removal water were achieved, but neither of these processes could reduce foam formation. Srinivasan et al. (2012) compared the performance of ozonation by taking two (one as secondary treated effluent and the other as primary and secondary treated effluents) treated TWW as feeds with and without using sequential batch reactors (SBR). The percent of COD removal was found to be about 34% when the feed (secondary treated TWW) was treated by ozonation without using SBR; ozonation on the primary treated effluent followed by SBR treatment could remove 64% COD; whereas only 48% COD reduction was recorded after the treatment of the primary treated effluent in SBR alone.

TWW loaded with high organic pollutants was treated in batch mode by electro-oxidation, biooxidation, and by coupling of both processes for COD removal. Electrooxidation at current density 1.5 A dm^{-2} successfully removed 73.1% of COD from the raw effluent. In contrast, the highest COD removal (91.5%) from a diluted effluent was achieved by biological oxidation alone at pH 6.0, in contrast to combined processes that removed 76.6% and 66.2% COD from the diluted and raw effluents, respectively (Kanagasabi et al. 2013).

A hybrid process comprising Fenton and aerobic biological (using *Thiobacillus ferrooxidans*) oxidation was applied on TWW at 30°C and an acidic pH in batch mode for comparison with the individual treatments. The removal rates for BOD$_5$, COD, color, S^{2-}, and total chromium were 72%, 69%, 100%, 5%, and 88% by applying the Fenton process; 80%, 77%, 89%, 52%, and 85% by employing the biological process; and 98%, 93%, 100%, 62%, and 72% by combined treatments, respectively (Mandal et al. 2010). An integrated Anoxic/Oxic (A/O) batch process was compared with Fenton process for removal of organic pollutants from TWW generated in wet-blue fur processing (Wang et al. 2014). The percent of COD removal by A/O and the Fenton process (which was conducted at 3 hours retention time, pH 4.0, dose of H_2O_2 = 14.0 mM, and molar ratio of H_2O_2:Fe(II) = 10.6) were 80% and 55.87%, respectively. Isarain-Chávez et al. (2014) compared TOC removal using hybrid technologies comprising electrocoagulation, electro-oxidation, electro-Fenton, and photoelectron-Fenton. This research was performed by applying different electrodes for each technology in a stirred tank reactor with 250 ml of TWW and current density of 65 or 111 mA cm^{-1}, and 6 W UV-A irradiation of light only in the case of the photoelectro-Fenton method. TOC removal obtained from this study followed the order: electro-oxidation < electrocoagulation < electro-Fenton < photoelectron-Fenton.

Statistical optimization of process variables (pH, reaction time, H_2O_2 dose, and current density) by use of the response surface methodology approach was applied during TWW treatment using Fe electrodes in electro-Fenton and electrocoagulation processes. The percent TSS and COD removal in electro-Fenton (pH 3.31, H_2O_2 dose: 0.14 g L^{-1}, current density: 53.72 mA cm^{-2}, and reaction time: 5.0 minutes) and electrocoagulation (pH 7.0, current density: 50.9 mA cm^{-2}, and reaction time: 40.4 min) were analyzed to be 88% and 87.3%, and 86% and 54.8%, respectively (Varank et al. 2016).

The biological oxidation was carried out by ozone, with and without the catalyst (Al_2O_3/Mn-Cu), during TWW treatment by an integrated multistage bioreactor system. Catalytic ozonation of a heterogeneous nature was investigated to assess COD removal/COD content, using tertiary butanol to identify the reaction mechanism and rate kinetics. The results were reported as 88% COD removal with the catalyst, 70.7% COD removal without the catalyst, the intermediate formation of OH$^\bullet$ during the catalytic reaction, and a 2.3 times faster biological oxidation rate using the catalyst (Huang et al. 2016).

Korpe et al. (2019) investigated an integrated approach of cavitation with AOP on reducing the suspended TOC in TWW. The objective of the experiment was to optimize the H_2O_2 dosage required. The reduction in suspended TOC was found to be 72% and 56.62%, respectively, with cavitation and magnetic stirring using 2 mL of H_2O_2 per liter sample. However, an increase in H_2O_2 dosage from 2 to 6 mL L^{-1} led to an enhancement of suspended TOC removal (87% and 63.78%, respectively, with the cavitation and stirring methods).

A recent investigation on the removal of three dyes (Brown DGI, Bismark Brown R, and Bismark Brown G) present in TWW with the help of three AOPs namely photoelectro-Fenton (PEF), electro-Fenton (EF), and electrochemical oxidation (EO) showed that the reduction in dye discoloration followed as PEF (97%) > EF (92%) > EO (80%). A similar order was also observed for COD: PEF (95%) > EF (76%) > EO (58%) (Medrano-Rodríguez et al. 2020).

In another integrated approach, the removal efficiency of TWW pollutants was tested by applying ozonation coupled with phycoremediation using *Nannochloropsis oculata* (microalgae). While 60% COD was removed through ozonation using 1.5 g of ozone/g of COD, subsequent phycoremediation led to enhanced removal of color (60%), COD (84%), inorganic carbon (90%), TDS (10%), chromium (97%), NH_4^+-N (82%), and PO_4-P (100%) (Saranya and Shanthakumar 2020a). An ozone pretreated tannery effluent was utilized to grow three different microalgae: *Nannochloropsis oculata* (0.67 g L^{-1}), *Chlorella sorokiniana* (1.06 g L^{-1}), and *Chlorella vulgaris* (0.85 g L^{-1}) (Saranya and Shanthakumar 2020b). Among the three species, *C. sorokiniana* showed excellent removal of P (100%), chromium (100%), C (90%), N (90%), and COD (82%) at 27.5°C, 30% (v/v) inoculum, 16 hours of light, and 150 μmol m^{-2} s^{-1} light intensity.

13.5.2 Photocatalytic Oxidation

Photocatalytic oxidation is a proficient and cost-effective AOP which breaks down the recalcitrant pollutant into water, CO_2, and inorganic salts.

Bordes et al. (2015) demonstrated the effective removal of aromatic compounds from TWW by spraying atmospheric plasma on stainless steel coupons. This process led to the generation of different varieties of TiO_2 particles used for methylene blue degradation. A leather industry effluent pretreated with lime was utilized as feed in a sonocatalytic reactor, where TiO_2 acts as a catalyst. This experiment was conducted by variations in time, catalyst load, and effluent concentration to evaluate the catalytic reactor performance. The BOD, COD, and TDS removals were 87.35%, 89.53%, and 92.63%, respectively (Kandasamy et al. 2016). TiO_2 nanoparticles, extracted from the leaf of *Jatropha curcas* L., were applied for chromium and COD removal in secondary TWW by using the photocatalytic method. The chromium and COD removal were assessed to be 76.48% and 82.26%, respectively (Goutam et al. 2017). In another study, the TiO_2 catalyst was prepared from butyl titanate in the presence of surfactants by the sol-gel method. The photocatalytic organic degradation efficiency of TWW under sunlight and UV-light was found to depend on H_2O_2 concentration, the dose of catalyst, and the pH (Zhao et al. 2017a). Tributyltin compounds (a group of recalcitrant organic pollutants present in TWW) were

TABLE 13.2

Application of Different Technologies in the Treatment of Tannery Wastewater

	pH[a]	Turbidity[b]	DO	TOC	TKN	TDS	TSS	COD	BOD	PO₄³⁻	Cl⁻	SO₄²⁻	S²⁻	Cr(VI)	Total Cr	NO₃⁻	NH₄⁺-N	References
																		Physico-Chemical Parameters
Untreated	–	> 999	–	–	–	50,000 ± 5,000	6,000 ± 600	14,000 ± 1,400	–	–	–	–	–	–	–	–	–	(Villalobos-Lara et al. 2021)
Treated (electrocoagulation)	–	5.1 ± 0.51	–	–	–	41,500 ± 4,150	420 ± 42	4,200 ± 420	–	–	–	–	–	–	–	–	–	
Untreated	11.03	–	0.02	2,942.5	–	41,200	–	7,475	–	38	17,000	149	3080	–	–	–	–	(Selvaraj et al. 2020)
Treated (electrochemical oxidation)	8.3	–	2.7	2,531	–	54,100	–	2,000	–	1.2	12,762	9,890	–	–	–	–	–	(Selvaraj et al. 2020)
Untreated (photo-electrochemical oxidation)	11.03	–	0.02	2,942.5	–	41,200	–	7,475	–	38	17,000	149	3080	–	–	–	–	(Selvaraj et al. 2020)
Treated (photo-electrochemical oxidation)	8.43	–	5.5	944	–	49,200	–	480	–	1.32	10,635	9,700	–	–	–	–	–	
Untreated	9.3–12.1	–	–	–	60 ± 20	–	780 ± 350	1,500 ± 400	–	–	–	–	–	–	360 ± 110	–	48 ± 12	(Zhou et al. 2020)
Treated (A/O-MBR)	7.3–8.2	–	–	–	0.8 ± 0.5	–	Nil	90 ± 10	–	–	–	–	–	–	0.3 ± 0.2	–	0.5 ± 0.1	
Untreated	7.6 ± 0.20	–	–	1,097 ± 9.8	310 ± 4.11	24,320 ± 103	–	3,320 ± 35	669 ± 11	20 ± 0.78	9,967 ± 12.4	1,296 ± 4.54	253 ± 12	Nil	5.73 ± 0.05	Nil	114 ± 2.8	(Saranya and Shanthakumar 2020a)
Treated (ozone)	7.9 ± 0.1	–	–	526 ± 7.2	121 ± 2.7	21,170 ± 86	–	1,364 ± 21	882 ± 18	19 ± 0.83	9,652 ± 10	821 ± 3.78	Nil	1.16 ± 0.01	5.73 ± 0.05	39 ± 2.5	286 ± 1.73	
Untreated	7.6	–	–	–	–	26,355	–	3,200	–	9.85	–	–	253	Nil	6.88	–	322	(Saranya and Shanthakumar 2020b)
Treated (ozonation)	7.9	–	–	–	–	23,928	–	1,142	–	9.2	–	–	Nil	2.2	6.88	–	269	(Saranya and Shanthakumar 2020b)
Untreated	7.6	–	–	–	–	26,355	–	3,200	–	9.85	–	–	–	–	6.88	–	322	(Saranya and Shanthakumar 2020b)
Treated (ozonation and phycoremediation)	8	–	–	–	–	23,428	–	810	–	2.5	–	–	–	–	1.8	–	164	
Untreated	–	–	–	–	–	7,420	–	10,500	–	–	4,620	–	2,560	–	8.2	–	–	(Pal et al. 2020)
Treated (Fenton)	8.9	–	–	–	–	1,925	–	6,170	–	–	1,983	–	65	–	8.2	–	–	
Untreated	–	–	–	–	–	7420	–	10,500	–	–	4,620	–	2,560	–	8.2	–	–	(Pal et al. 20200)
Treated (GO-based nanocomposite membrane)	8.7	–	–	–	–	235	–	95.5	–	–	85	–	35.2	–	0.04	–	–	(Pal et al. 20200)

(Continued)

TABLE 13.2 (CONTINUED)
Application of Different Technologies in the Treatment of Tannery Wastewater

	Physico-Chemical Parameters																	References
	pH[a]	Turbidity[b]	DO	TOC	TKN	TDS	TSS	COD	BOD	PO$_4^{3-}$	Cl⁻	SO$_4^{2-}$	S²⁻	Cr(VI)	Total Cr	NO$_3^-$	NH$_4^+$-N	
Untreated	3.6 ± 0.3	–	–	–	–	44.3 ± 0.6	–	5,213 ± 43	3,621 ± 34	–	19,015 ± 0.4	–	–	–	3,190.12 ± 0.7	–	–	(Hashem et al. 2020)
Treated (water hyacinth biochar)	8.3 ± 0.1	–	–	–	–	48.9 ± 0.2	–	387 ± 7.2	239 ± 19.3	–	8,366.6 ± 0.2	–	–	–	27.33 ± 0.4	–	–	
Untreated	–	2,231 ± 110	0.18 ± 0.06	–	–	–	1,208 ± 81.6	1,922 ± 182	–	–	–	–	–	–	–	–	–	(Sodhi et al. 2018)
Treated (MANODOX)	–	12.5 ± 3.35	1.62 ± 0.34	–	–	–	62.7 ± 22.4	99 ± 20.4	–	–	–	–	–	–	–	–	–	
Untreated	–	2,231 ± 110	0.18 ± 0.06	–	–	–	1,208 ± 81.6	1,922 ± 182	–	–	–	–	–	–	–	–	–	(Sodhi et al. 2018)
Treated (p-CAS)	–	56.8 ± 9.5	1.12 ± 0.16	–	–	–	196.8 ± 41.9	518 ± 72.1	–	–	–	–	–	–	–	–	–	
Untreated	11.41	902	–	–	–	24,267	12,518	20,600	–	–	–	–	66.13	6.95	30.11	–	–	(Moradi and Moussavi 2019)
Treated (electrocoagulation)	8.1	< 0.1	–	–	–	9,117	< 2.1	5,800.62	–	–	–	–	51.9	4.49	18.13	–	–	
Untreated	11.41	902	–	–	–	24,267	12,518	20,600	–	–	–	–	66.13	6.95	30.11	–	–	(Moradi and Moussavi 2019)
Treated (EC+UVC/ VUV-treated effluent)	9.11	< 0.1	–	–	–	612	4.09	9,689.02	–	–	–	–	1.14	Nil	Nil	–	–	
Untreated	–	–	–	1,851 ± 46	–	12,928 ± 1549	–	6,066 ± 1335	3,860 ± 612	20 ± 0.3	5,600 ± 1876	2,693 ± 203	–	–	247 ± 5.8	590 ± 22	–	(Ashraf et al. 2018)
Treated (constructed wetlands + *Leptochloa fusca*)	–	–	–	148 ± 11	–	646 ± 6.2	–	728 ± 23	392 ± 5	8.9 ± 5.1	927 ± 23	178 ± 4.5	–	–	3.3 ± 0.05	373 ± 13	–	
Untreated	–	–	–	1851 ± 46	–	12,928 ± 1549	–	6,066 ± 1335	3,860 ± 612	20 ± 0.3	5,600 ± 1876	2,693 ± 203	–	–	247 ± 5.8	590 ± 22	–	(Ashraf et al. 2018)
Treated (constructed wetlands + *Leptochloa fusca* + *Microbacterium arborescens* HU33 + *Pantoea stewartii* ASI11 + *Enterobacter* sp. HU38)	–	–	–	16 ± 0.3	–	130 ± 5	–	152 ± 13	57 ± 3	7.8 ± 0.4	489 ± 2.3	54 ± 1.22	–	–	0.9 ± 0.01	236 ± 16	–	

(Continued)

TABLE 13.2 (CONTINUED)
Application of Different Technologies in the Treatment of Tannery Wastewater

	Physico-Chemical Parameters																References
	pH[a]	Turbidity[b]	DO	TOC	TKN	TDS	TSS	COD	BOD	PO_4^{3-}	Cl^-	SO_4^{2-}	S^{2-}	Cr(VI)	Total Cr	NO_3^-	NH_4^+-N	
Untreated	–	–	–	1,851 ± 46	–	12,928 ± 1549	–	6,066 ± 1335	3,860 ± 612	20 ± 0.3	5,600 ± 1876	2,693 ± 203	–	–	247 ± 5.8	590 ± 22	–	(Ashraf et al. 2018)
Treated (constructed wetlands without vegetation)	–	–	–	7,77 ± 33	–	5,542 ± 27	–	2,880 ± 41	1,729 ± 25	12.5 ± 1.3	2,831 ± 22	1,315 ± 14.1	–	–	142 ± 0.03	492 ± 23	–	
Untreated	4.86	–	–	3,171.5	264.6	–	–	–	–	–	–	–	–	–	–	–	–	(da Fontoura et al. 2017)
Treated (defatted microalgal biomass)	5.32	–	–	1,561	212.2	–	–	–	–	–	–	–	–	–	–	–	–	
Untreated	4.4 ± 0.25	–	–	–	–	–	–	4,800	–	–	36,000	28,480	–	Nil	2,481	–	–	(Selvaraj et al. 2017)
Treated (membrane electrofloatation)	12.47 ± 0.17	–	–	–	–	–	–	2,400	–	–	36,000	28,480	–	Nil	60	–	–	

All units are in mg L^{-1} except for pH[a]
Turbidity[b] = NTU

removed by using the photocatalytic degradation process induced by an electric field in the presence of self-synthesized microspheres of TiO_2 (Zhao et al. 2017b). This investigation compared normal photocatalysis with the photocatalytic process induced by an electric field. The rate constant of the degradation process of the tributyltin compound was nine times more in the latter case than the former.

In another experiment, a heterogeneous catalyst was impregnated onto cobalt oxide nanoporous activated carbon (Co-NPAC) for the removal of organic dye present in TWW through the Fenton reaction. The process was optimized by varying the concentration of H_2O_2, the dose of Co-NPAC, and the temperature. The maximum removal of COD was reported to be 77% (Karthikeyan et al. 2015).

Experimental observations of parameters studied by several researchers, before and after TWW treatment, are shown in Table 13.2.

13.6 CONCLUSION

Simple conventional approaches such as adsorption, membrane filtration, coagulation-flocculation, electrocoagulation, activated sludge, constructed wetlands, and upflow anaerobic sludge blanket can remove different refractory chemicals such as heavy metals and dyes from tannery effluents. Utilizing microbial consortia such as algae, bacteria, and fungi as adsorbents proved to be cost-effective and eco-friendly in treating tannery effluents. Combinatorial methods or hybrid treatment technologies, such as advanced oxidation processes, catalytic approaches, and nanomaterials along with biological treatment, should be applied by tannery industries to curb pollution. These processes are far more efficient, effective, economical, and environmentally friendly for maintaining sustainability. Synergistic technologies will yield cleaner effluents than the application of an individual conventional method. However, the operation of these fabricated processes require additional technical support.

Based on this review, it can be concluded that in the forthcoming years, an improved plan should emerge for the enhanced treatment of tannery industrial effluents. Toxicological studies on tannery waste must continue as this will be helpful to identify the occupational exposure of harmful chemicals on workers in tannery localities. Overall, research is needed for finding non-conventional and cost-effective ways for the treatment of tannery waste.

REFERENCES

Agustini, C.B., M. da Costa, and M. Gutterres. 2020. Tannery wastewater as nutrient supply in production of biogas from solid tannery wastes mixed through anaerobic co-digestion. *Process Safety and Environmental Protection* 135: 38–45. https://doi.org/10.1016/j.psep.2019.11.037.
Ashraf, S., M. Afzal, M. Naveed, M. Shahid, and Z. Ahmad Zahir. 2018. Endophytic bacteria enhance remediation of tannery effluent in constructed wetlands vegetated with *Leptochloa Fusca*. *International Journal of Phytoremediation* 20, no. 2: 121–128. https://www.tandfonline.com/doi/full/10.1080/15226 514.2017.1337072.
Bakers, A.J. 1981. Accumulators and excluders strategies in the response of plants to heavy metals. *Journal of Plant Nutrition* 3(1–4): 643–654.
Balasubramani, U., R. Venkatesh, S. Subramaniam, G. Gopalakrishnan, and V. Sundararajan. 2017. Alumina/ activated carbon nano-composites: Synthesis and application in sulphide ion removal from water. *Journal of Hazardous Materials* 340: 241–252. http://dx.doi.org/10.1016/j.jhazmat.2017.07.006.
Bharagava, R.N., and S. Mishra. 2018. Hexavalent chromium reduction potential of *Cellulosimicrobium* sp. isolated from common effluent treatment plant of tannery industries. *Ecotoxicology and Environmental Safety* 147, no. August 2017: 102–109. http://dx.doi.org/10.1016/j.ecoenv.2017.08.040.
Bhattacharya, P., A. Roy, S. Sarkar, S. Ghosh, S. Majumdar, S. Chakraborty, S. Mandal, A. Mukhopadhyay, and S. Bandyopadhyay. 2013. Combination technology of ceramic microfiltration and reverse osmosis for tannery wastewater recovery. *Water Resources and Industry* 3: 48–62. http://dx.doi.org/10.1016/j.wr i.2013.09.002.
Bhattacharya, P., S. Swarnakar, A. Mukhopadhyay, and S. Ghosh. 2016. Exposure of composite tannery effluent on snail, *Pila Globosa*: A comparative assessment of toxic impacts of the untreated and membrane treated effluents. *Ecotoxicology and Environmental Safety* 126: 45–55.

Bletterie, U., K. Schilling, L. Delgado, and M. Zessner. 2012. Ozonation as a post-treatment step for tannery wastewater to reduce foam formation in a river. *Ozone: Science and Engineering* 34, no. 5: 37–41.

Borba, F.H., L. Pellenz, F. Bueno, J.J. Inticher, L. Braun, F.R. Espinoza-Quiñones, D.E.G. Trigueros, A.R. De Pauli, and A.N. Módenes. 2018. Pollutant removal and biodegradation assessment of tannery effluent treated by conventional Fenton oxidation process. *Journal of Environmental Chemical Engineering* 6, no. 6: 7070–7079. https://doi.org/10.1016/j.jece.2018.11.005.

Bordes, M.C., M. Vicent, R. Moreno, J. García-Montaño, A. Serra, and E. Sánchez. 2015. Application of plasma-sprayed TiO_2 coatings for industrial (tannery) wastewater treatment. *Ceramics International* 41, no. 10: 14468–14474. http://linkinghub.elsevier.com/retrieve/pii/S0272884215013735.

Bregnbak, D., J.P. Thyssen, M.S. Jellesen, C. Zachariae, and J.D. Johansen. 2017. Experimental patch testing with chromium-coated materials. *Contact Dermatitis* 76, no. 6 (June): 333–341.

China, C.R., M.M. Maguta, S.S. Nyandoro, A. Hilonga, S. V. Kanth, and K.N. Njau. 2020. Alternative tanning technologies and their suitability in curbing environmental pollution from the leather industry: A comprehensive review. *Chemosphere* 254: 126804. https://doi.org/10.1016/j.chemosphere.2020.126804.

Chowdhury, M., M.G. Mostafa, T.K. Biswas, and A.K. Saha. 2013. Treatment of leather industrial effluents by filtration and coagulation processes. *Water Resources and Industry* 3: 11–22. http://www.sciencedirect.com/science/article/pii/S2212371713000085.

da Fontoura, J.T., G.S. Rolim, M. Farenzena, and M. Gutterres. 2017a. Influence of light intensity and tannery wastewater concentration on biomass production and nutrient removal by microalgae *Scenedesmus* Sp. *Process Safety and Environmental Protection* 111: 355–362. http://linkinghub.elsevier.com/retrieve/pii/S0957582017302380.

da Fontoura, J.T., G.S. Rolim, B. Mella, M. Farenzena, and M. Gutterres. 2017b. Defatted microalgal biomass as biosorbent for the removal of Acid Blue 161 dye from tannery effluent. *Journal of Environmental Chemical Engineering* 5, no. 5: 5076–5084. http://dx.doi.org/10.1016/j.jece.2017.09.051.

Dasgupta, J., D. Mondal, S. Chakraborty, J. Sikder, S. Curcio, and H.A. Arafat. 2015. Nanofiltration based water reclamation from tannery effluent following coagulation pretreatment. *Ecotoxicology and Environmental Safety* 121: 22–30.

Deghles, A., and U. Kurt. 2015. Treatment of raw tannery wastewater by electrocoagulation technique: Optimization of effective parameters using Taguchi method. *Desalination and Water Treatment* 3994, no. September: 1–12.

Deghles, A., and U. Kurt. 2016. Treatment of raw tannery wastewater by electrocoagulation technique: Optimization of effective parameters using Taguchi method. *Desalination and Water Treatment* 57, no. 32: 14798–14809. https://doi.org/10.1080/19443994.2015.1074622.

dos Santos Moysés, F., K. Bertoldi, G. Lovatel, S. Vaz, K. Ferreira, J. Junqueira, P.B. Bagatini, M.A.S. Rodrigues, L.L. Xavier, and I.R. Siqueira. 2017. Effects of tannery wastewater exposure on adult *Drosophila Melanogaster*. *Environmental Science and Pollution Research* 24, no. 34: 26387–26395.

Ejechi, B.O., and O.O. Akpomie. 2016. Removal of Cr (VI) from Tannery effluent and aqueous solution by sequential treatment with microfungi and basidiomycete-degraded sawdust. *Journal of Environmental Protection* 7, no. May: 771–777.

Elabbas, S., N. Ouazzani, L. Mandi, F. Berrekhis, M. Perdicakis, S. Pontvianne, M.N. Pons, F. Lapicque, and J.P. Leclerc. 2016. Treatment of highly concentrated tannery wastewater using electrocoagulation: Influence of the quality of aluminium used for the electrode. *Journal of Hazardous Materials* 319: 69–77. http://dx.doi.org/10.1016/j.jhazmat.2015.12.067.

Estrela, F.N., L.M. Rabelo, B.G. Vaz, D.R. de Oliveira Costa, I. Pereira, A.S. de Lima Rodrigues, and G. Malafaia. 2017. Short-term social memory deficits in adult female mice exposed to tannery effluent and possible mechanism of action. *Chemosphere* 184: 148–158. http://dx.doi.org/10.1016/j.chemosphere.2017.05.174.

Fernandez, M., C.E. Paisio, R. Perotti, P.P. Pereira, E. Agostini, and P.S. González. 2019. Laboratory and field microcosms as useful experimental systems to study the bioaugmentation treatment of tannery effluents. *Journal of Environmental Management* 234, no. January: 503–511. https://doi.org/10.1016/j.jenvman.2019.01.019.

Gando-Ferreira, L.M., J.C. Marques, and M.J. Quina. 2015. Integration of ion-exchange and nanofiltration processes for recovering Cr(III) salts from synthetic tannery wastewater. *Environmental Technology* 36, no. 18: 2340–2348.

Gheju, M., I. Balcu, and C. Vancea. 2016. An Investigation of Cr(VI) Removal with metallic iron in the co-presence of sand and/or MnO_2. *Journal of Environmental Management* 170, no. April: 145–151.

Gomes, L., E.P. Troiani, G.R.P. Malpass, and J. Nozaki. 2016. *Opuntia Ficus Indica* as a polyelectrolyte source for the treatment of tannery wastewater. *Desalination and Water Treatment* 57, no. 22: 10181–10187. http://www.tandfonline.com/doi/full/10.1080/19443994.2015.1035677.

Goswami, S., and D. Mazumder. 2016. Comparative study between activated sludge process (ASP) and moving bed bioreactor (MBBR) for treating composite chrome tannery wastewater. *Materials Today: Proceedings* 3, no. 10: 3337–3342. http://dx.doi.org/10.1016/j.matpr.2016.10.015.

Goutam, S.P., G. Saxena, V. Singh, A.K. Yadav, R.N. Bharagava, and K.B. Thapa. 2017. Green Synthesis of TiO_2 nanoparticles using leaf extracts of *Jatropha Curcas L.* for photocatalytic degradation of tannery wastewater, 386–396. *Chemical Engineering Journal.* http://linkinghub.elsevier.com/retrieve/pii /S1385894717321423.

Guan, X., S. Yan, Z. Xu, and H. Fan. 2017. Gallic acid-conjugated iron oxide nanocomposite: An efficient, separable, and reusable adsorbent for remediation of Al (III)-Contaminated tannery wastewater. *Journal of Environmental Chemical Engineering* 5, no. 1: 479–487. http://dx.doi.org/10.1016/j.jece.2016.12.010.

Gungor, K., N. Karakaya, Y. Gunes, S. Yatkin, and F. Evrendilek. 2016. Utilizing aluminum etching wastewater for tannery wastewater coagulation: Performance and feasibility. *Desalination and Water Treatment* 57, no. 6: 2413–2421.

Hashem, M.A., M. Hasan, M.A. Momen, S. Payel, and M.S. Nur-A-Tomal. 2020. Water hyacinth biochar for trivalent chromium adsorption from tannery wastewater. *Environmental and Sustainability Indicators* 5, no. January: 100022.

Hashem, M.A., M.S. Nur-A-Tomal, and S.A. Bushra. 2016. Oxidation-coagulation-filtration processes for the reduction of sulfide from the hair burning liming wastewater in tannery. *Journal of Cleaner Production*, 127: 339–342. http://dx.doi.org/10.1016/j.jclepro.2016.03.159.

Hedberg, Y.S., C. Lidén, and M. Lindberg. 2016. Chromium dermatitis in a metal worker due to leather gloves and alkaline coolant. *Acta Dermato-Venereologica* 96, no. 1 (January): 104–105.

Hethnawi, A., W. Khderat, K. Hashlamoun, A. Kanan, and N.N. Nassar. 2020. Enhancing chromium (VI) removal from synthetic and real tannery effluents by using diatomite-embedded nanopyroxene. *Chemosphere* 252: 126523.

Huang, G., F. Pan, G. Fan, and G. Liu. 2016. Application of heterogeneous catalytic ozonation as a tertiary treatment of effluent of biologically treated tannery wastewater. *Journal of Environmental Science and Health. Part A, Toxic/Hazardous Substances & Environmental Engineering* 51, no. 8: 626–33. http:// www.scopus.com/inward/record.url?eid=2-s2.0-84963815269&partnerID=tZOtx3y1.

Isarain-Chávez, E., C. De La Rosa, L.A. Godínez, E. Brillas, and J.M. Peralta-Hernández. 2014. Comparative study of electrochemical water treatment processes for a tannery wastewater effluent. *Journal of Electroanalytical Chemistry* 713: 62–69.

Jadhav, N.C., and A.C. Jadhav. 2020. *Waste and 3R's in Footwear and Leather Sectors.* Springer, Singapore. http://dx.doi.org/10.1007/978-981-15-6296-9_10.

Kamaraj, M., N.R. Srinivasan, G. Assefa, A.T. Adugna, and M. Kebede. 2020. Facile development of sunlit ZnO nanoparticles-activated carbon hybrid from pernicious weed as an operative nano-adsorbent for removal of methylene blue and chromium from aqueous solution: Extended application in tannery industrial wastewater. *Environmental Technology and Innovation* 17: 100540. https://doi.org/10.1016/j. eti.2019.100540.

Kanagasabi, S., Y.L. Kang, M. Manickam, S. Ibrahim, and S. Pichiah. 2013. Intimate coupling of electro and biooxidation of tannery wastewater. *Desalination and Water Treatment* 51, no. 34–36: 6617–6623.

Kandasamy, K., K. Tharmalingam, and S. Velusamy. 2016. Treatment of tannery effluent using sono catalytic reactor. *Journal of Water Process Engineering*: 15: 72–77. http://dx.doi.org/10.1016/j.jwpe.2016.09.001.

Karthikeyan, S., R. Boopathy, and G. Sekaran. 2015. In situ generation of hydroxyl radical by cobalt oxide supported porous carbon enhance removal of refractory organics in tannery dyeing wastewater. *Journal of Colloid and Interface Science* 448: 163–174. http://dx.doi.org/10.1016/j.jcis.2015.01.066.

Korpe, S., B. Bethi, S.H. Sonawane, and K. V. Jayakumar. 2019. Tannery wastewater treatment by cavitation combined with advanced oxidation process (AOP). *Ultrasonics Sonochemistry* 59, no. June 2018: 104723. https://doi.org/10.1016/j.ultsonch.2019.104723.

Luján-Facundo, M.J., J.A. Mendoza-Roca, J.L. Soler-Cabezas, A. Bes-Piá, M.C. Vincent-Vela, and L. Pastor-Alcañiz. 2019. Use of the osmotic membrane bioreactor for the management of tannery wastewater using absorption liquid waste as draw solution. *Process Safety and Environmental Protection* 131: 292–299.

Mandal, T., D. Dasgupta, S. Mandal, and S. Datta. 2010. Treatment of leather industry wastewater by aerobic biological and Fenton oxidation process. *Journal of Hazardous Materials* 180, no. 1–3: 204–211.

Marzan, L.W., M. Hossain, S.A. Mina, Y. Akter, and A.M.M.A. Chowdhury. 2017. Isolation and biochemical characterization of heavy-metal resistant bacteria from tannery effluent in Chittagong City, Bangladesh: Bioremediation viewpoint. *Egyptian Journal of Aquatic Research* 43, no. 1: 65–74. http://dx.doi.org/10. 1016/j.ejar.2016.11.002.

Medrano-Rodríguez, F., A. Picos-Benítez, E. Brillas, E.R. Bandala, T. Pérez, and J.M. Peralta-Hernández. 2020. Electrochemical advanced oxidation discoloration and removal of three brown diazo dyes used in the tannery industry. *Journal of Electroanalytical Chemistry* 873: 114360. http://www.sciencedirect.com/science/article/pii/S1572665720305877.

Moradi, M., and G. Moussavi. 2019. Enhanced treatment of tannery wastewater using the electrocoagulation process combined with UVC/VUV photoreactor: Parametric and mechanistic evaluation. *Chemical Engineering Journal* 358, no. October 2018: 1038–1046. https://doi.org/10.1016/j.cej.2018.10.069.

Occupational exposure to hexavalent chromium. n.d. *Criteria for a Recommended Standard.* National Institute for Occupational Safety and Health: 168.

Pal, M., M. Malhotra, M.K. Mandal, T.K. Paine, and P. Pal. 2020. Recycling of wastewater from tannery industry through membrane-integrated hybrid treatment using a novel graphene oxide nanocomposite. *Journal of Water Process Engineering* 36, no. April: 101324. https://doi.org/10.1016/j.jwpe.2020.101324.

Pinto, M.B., G.R.L. Samanamud, E.P. Baston, A.B. França, L.L.R. Naves, C.C.A. Loures, and F.L. Naves. 2019. Multivariate and multiobjective optimization of tannery industry effluent treatment using *Musa* Sp. flower extract in the coagulation and flocculation process. *Journal of Cleaner Production* 219: 655–666. https://doi.org/10.1016/j.jclepro.2019.02.060.

Rabelo, L.M., F.N. Estrela, B.C. e Silva, B. de O. Mendes, B.G. Vaz, A.S. de L. Rodrigues, and G. Malafaia. 2017a. Protective effect of vitamin C in female Swiss mice dermally-exposed to the tannery effluent. *Chemosphere* 181: 492–499. http://dx.doi.org/10.1016/j.chemosphere.2017.04.130.

Rabelo, L.M., A. Tiago, B. Guimarães, and J.M. De Souza. 2018. Histological liver chances in Swiss mice caused by tannery effluent. *Environmental Science Pollution Research* 25: 1943–1949. https://doi.org/10.1007/s11356-017-0647-1

Rangabhashiyam, S., and P. Balasubramanian. 2018. Adsorption behaviors of hazardous methylene blue and hexavalent chromium on novel materials derived from *Pterospermum Acerifolium* shells. *Journal of Molecular Liquids* 254: 433–445. https://doi.org/10.1016/j.molliq.2018.01.131.

Revees, R.D., A.J.M. Baker, I. Raskin, and B.D. Ensley. 2000. *Phytoremediation of toxic metals using plants to clean up Environment.* John Wiley & Sons, New York, 193–229.

Romero-Dondiz, E.M., J.E. Almazán, V.B. Rajal, and E.F. Castro-Vidaurre. 2015. Removal of vegetable tannins to recover water in the leather industry by ultrafiltration polymeric membranes. *Chemical Engineering Research and Design* 93: 727–735. http://dx.doi.org/10.1016/j.cherd.2014.06.022.

Sankararamakrishnan, N., A. Shankhwar, and D. Chauhan. 2019. Mechanistic insights on immobilization and decontamination of hexavalent chromium onto nanoMgS/FeS doped cellulose nanofibres. *Chemosphere* 228: 390–397. https://doi.org/10.1016/j.chemosphere.2019.04.166.

Saranya, D., and S. Shanthakumar. 2020a. An integrated approach for tannery effluent treatment with ozonation and phycoremediation: A feasibility study. *Environmental Research* 183: 109163.

Saranya, D., and S. Shanthakumar. 2020b. Effect of culture conditions on biomass yield of acclimatized microalgae in ozone pre-treated tannery effluent: A simultaneous exploration of bioremediation and lipid accumulation potential. *Journal of Environmental Management* 273, no. April: 111129. https://doi.org/10.1016/j.jenvman.2020.111129.

Saxena, S., S. Rajoriya, V.K. Saharan, and S. George. 2018. An advanced pretreatment strategy involving hydrodynamic and acoustic cavitation along with alum coagulation for the mineralization and biodegradability enhancement of tannery waste effluent. *UltrasonicsSonochemistry* 44: 299–309. https://doi.org/10.1016/j.ultsonch.2018.02.035.

Selvaraj, H., P. Aravind, H.S. George, and M. Sundaram. 2020. Removal of sulfide and recycling of recovered product from tannery lime wastewater using photoassisted-electrochemical oxidation process. *Journal of Industrial and Engineering Chemistry* 83: 164–172. https://doi.org/10.1016/j.jiec.2019.11.024.

Selvaraj, R., M. Santhanam, V. Selvamani, S. Sundaramoorthy, and M. Sundaram. 2017. A membrane electroflotation process for recovery of recyclable chromium(III) from tannery spent liquor effluent. *Journal of Hazardous Materials* no. 3. http://linkinghub.elsevier.com/retrieve/pii/S0304389417308749.

Sharma, S., and P. Malaviya. 2016. Bioremediation of tannery wastewater by chromium resistant novel fungal consortium. *Ecological Engineering* 91: 419–425. http://dx.doi.org/10.1016/j.ecoleng.2016.03.005.

Singh, A., D. Vyas, and P. Malaviya. 2016. Two-stage phyto-microremediation of tannery effluent by *Spirodela Polyrrhiza (L.) Schleid* and chromium resistant bacteria. *Bioresource Technology* 216: 883–893. http://dx.doi.org/10.1016/j.biortech.2016.06.025.

Sirianuntapiboon, S., and A. Chaochon. 2016. Effect of Cr^{+3} on the efficiency and performance of the sequencing batch reactor (SBR) system for treatment of tannery industrial wastewater. *Desalination and Water Treatment* 57, no. 12: 5579–5591.

Sodhi, V., A. Bansal, and M.K. Jha. 2018. Excess sludge disruption and pollutant removal from tannery efflu-
    ent by upgraded activated sludge system. *Bioresource Technology* 263: 613–624. https://doi.org/10.1016/
    j.biortech.2018.04.118.
Sodhi, V., A. Bansal, and M.K. Jha. 2020. Minimization of excess bio-sludge and pollution load in oxic-
    settling-anaerobic modified activated sludge treatment for tannery wastewater. *Journal of Cleaner
    Production* 243: 118492. https://doi.org/10.1016/j.jclepro.2019.118492.
Srinivasan, S. V., G. PreaSamita Mary, C. Kalyanaraman, P.S. Sureshkumar, K. Sri Balakameswari, R.
    Suthanthararajan, and E. Ravindranath. 2012. Combined advanced oxidation and biological treatment
    of tannery effluent. *Clean Technologies and Environmental Policy* 14, no. 2: 251–256.
Tamersit, S., K.E. Bouhidel, and Z. Zidani. 2018. Investigation of electrodialysis anti-fouling configuration
    for desalting and treating tannery unhairing wastewater: Feasibility of by-products recovery and water
    recycling. *Journal of Environmental Management* 207: 334–340. https://doi.org/10.1016/j.jenvman.2017.
    11.058.
Tammaro, M., A. Salluzzo, R. Perfetto, and A. Lancia. 2014. A comparative evaluation of biological activated
    carbon and activated sludge processes for the treatment of tannery wastewater. *Journal of Environmental
    Chemical Engineering* 2, no. 3: 1445–1455. http://dx.doi.org/10.1016/j.jece.2014.07.004.
Thirugnanasambandham, K., and V. Sivakumar. 2016. Removal of ecotoxicological matters from tannery
    wastewater using electrocoagulation reactor: Modelling and optimization. *Desalination and Water
    Treatment* 57, no. 9: 3871–3880. http://www.tandfonline.com/doi/full/10.1080/19443994.2014.989915.
Varank, G., S. YaziciGuvenc, G. Gurbuz, and G. Onkal Engin. 2016. Statistical optimization of process
    parameters for tannery wastewater treatment by electrocoagulation and electro-Fenton techniques.
    Desalination and Water Treatment 57, no. 53: 25460–25473. https://www.tandfonline.com/doi/full/10.
    1080/19443994.2016.1157042.
Vilardi, G., J.M. Ochando-Pulido, M. Stoller, N. Verdone, and L. Di Palma. 2018. Fenton oxidation and
    chromium recovery from tannery wastewater by means of iron-based coated biomass as heterogeneous
    catalyst in fixed-bed columns. *Chemical Engineering Journal* 351: 1–11. https://doi.org/10.1016/j.cej.
    2018.06.095.
Villalobos-Lara, A.D., F. Álvarez, Z. Gamiño-Arroyo, R. Navarro, J.M. Peralta-Hernández, R. Fuentes, and
    T. Pérez. 2021. Electrocoagulation treatment of industrial tannery wastewater employing a modified
    rotating cylinder electrode reactor. *Chemosphere* 264: 128491. https://doi.org/10.1016/j.chemosphere.20
    20.128491.
Wang, Y., W. Li, A. Irini, and C. Su. 2014. Removal of organic pollutants in tannery wastewater from wet-
    blue fur processing by integrated Anoxic/Oxic (A/O) and Fenton : Process optimization. *Chemical
    Engineering Journal* 252: 22–29. http://dx.doi.org/10.1016/j.cej.2014.04.069.
World Health Organization. 2017. *Guidelines for Drinking-Water Quality*. 4th ed. WHO, Geneva, Switzerland.
Yahya, M.D., K.S. Obayomi, M.B. Abdulkadir, Y.A. Iyaka, and A.G. Olugbenga. 2020. Characterization of
    cobalt ferrite-supported activated carbon for removal of chromium and lead ions from tannery waste-
    water via adsorption equilibrium. *Water Science and Engineering* 13: 202–213. https://doi.org/10.1016
    /j.wse.2020.09.007.
Zaheer, I.E., S. Ali, M.H. Saleem, M. Imran, G.S.H. Alnusairi, B.M. Alharbi, M. Riaz, Z. Abbas, M. Rizwan,
    and M.H. Soliman. 2020. Role of iron–lysine on morpho-physiological traits and combating chromium
    toxicity in rapeseed (*Brassica Napus L.*) plants irrigated with different levels of tannery wastewater.
    Plant Physiology and Biochemistry 155: 70–84. http://www.sciencedirect.com/science/article/pii/S
    0981942820303685.
Zhang, J., Z. Han, B. Teng, and W. Chen. 2017. Biodeterioration process of chromium tanned leather with
    Penicillium Sp. *International Biodeterioration and Biodegradation* 116: 104–111. http://dx.doi.org/10.
    1016/j.ibiod.2016.10.019.
Zhao, W., X.He, Y. Peng, H. Zhang, D. Sun, and X. Wang. 2017a. Preparation of mesoporous TiO_2 with
    enhanced photocatalytic activity towards tannery wastewater degradation. *Water Science and
    Technology* 75, no. 6: 1494–1499. http://wst.iwaponline.com/lookup/doi/10.2166/wst.2017.018.
Zhao, Y., Z. Huang, W. Chang, C. Wei, X. Feng, L. Ma, X. Qi, and Z. Li. 2017b. Microwave-assisted solvo-
    thermal synthesis of hierarchical TiO_2 microspheres for efficient electro-field-assisted-photocatalytic
    removal of tributyltin in tannery wastewater. *Chemosphere* 179: 75–83. http://www.sciencedirect.com/
    science/article/pii/S0045653517304551.
Zhou, L., W. Zhang, Y.G. De Costa, W.Q. Zhuang, and S. Yi. 2020. Assessing inorganic components of cake
    layer in A/O membrane bioreactor for physical-chemical treated tannery effluent. *Chemosphere* 250:
    126220. https://doi.org/10.1016/j.chemosphere.2020.126220.

14 Sustainable Bioremediation Strategies to Manage Environmental Pollutants

Prasenjit Debbarma, Divya Joshi, Damini Maithani,
Hemant Dasila, Deep Chandra Suyal,
Saurabh Kumar and Ravindra Soni

CONTENTS

DOI: 10.1201/9781003204442-14

14.1 INTRODUCTION

The quality of life on Earth is dependent on or influenced by the quality of an environment. However, due to the blooming of industrialization, anthropogenic activities, intensive agricultural practices, and few natural activities, an array of toxic pollutants (e.g., heavy metals, synthetic polymers, hazardous waste, xenobiotics, chemicals, and nano-materials) have been released ubiquitously into the environments by different modes [1, 2]. Therefore, the accumulation of these pollutants disturb the environment and the hazardous effects are exposed/transmitted to living creatures. Moreover, with modernization and development, the level and complexity of contaminants have also been raised to an intolerable level imparting enormous deleterious consequences, such as environmental degradation, global warming, ozone layer depletion, infertile land, and human illness; thus becoming a matter of great global concern [3].

Moreover, according to a recent study by the Organization of Economic Co-Operation and Development (OECD), developing countries have to invest trillions of U.S. dollars for waste management which is not easily accessible [4]. In this context, the need of the hour is the exploitation of cost-effective solutions for environmental clean-up problems. These solutions involve the transformation of complex or simple chemical compounds into non-hazardous forms by biological agents to combat various pollutants and keep the environment healthy and invulnerable. Various micro-organisms and plants belonging to the versatile domains are found as the most attractive options, having a great diversity of members with various potentials in different spheres of life (see Figure 14.1).

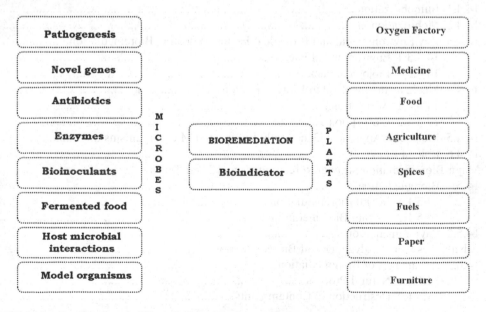

FIGURE 14.1 Exploitation profiling of microorganisms and plants.

Among the various remediation processes available, phytoremediation and bioremediation are well known, significant, widely accepted, and operative since these are natural and cost-effective [5]. Phytoremediation makes use of plants, whereas bioremediation is based on the action of micro-organisms, i.e., fungi, bacteria, algae, and protozoa for the decontamination of pollutants. The bioremediation of pollutants has previously been reported for various microbes and plants [6–8]. Conclusively, bioremediation processes can be considered as the most effective and eco-friendly waste management tools by making use of biological interventions to resolve this global issue. Thus, the goal of this review is to give an overview of the different bioremediation processes that can be used to act against the hyper accumulation of pollutants and to minimize the environmental burden to keep the environment sustainable, dynamic, and shielded.

14.2 THE SOURCES AND GLOBAL STATUS OF DIFFERENT POLLUTANTS

Anything that disrupts the natural functioning of the environment is termed as a pollutant. Pollutants create potential health hazards to living organisms with the formation of different types of chemical agents above their permissible limit in natural conditions [9]. Pollution can arise from both point sources and non-point sources. Broadly, pollutants can be classified as follows:

14.2.1 HEAVY METALS

Heavy metals are toxic elements having relatively high density and are harmful because of the fact that even the vestiges of these elements can be poisonous [10]. Elements like lead (Pb), cadmium (Cd), mercury (Hg), zinc (Zn), arsenic (As), silver (Ag), chromium (Cr), copper (Cu), iron (Fe), and platinum group elements constitute heavy metals.

Though some heavy metals have importance as trace elements in the body, elements like lead, mercury, and cadmium do not benefit human physiology in any known manner [11]. Heavy metals can be introduced into the environment by natural as well as anthropogenic means. Naturally, they form the constituent of the Earth's crust and are also present in rocks in different chemical forms. Mining practices, forest fires, and agricultural wastes form the main anthropogenic cause of heavy metal pollutants [12].

14.2.2 SYNTHETIC POLYMERS

A synthetic polymer is a man-made macromolecule that is made up of thousands of repeating units. Nylon, Teflon, bakelite, LDPE, HDPE, and PVC are some of the common household synthetic polymers being widely used due to their versatile applicability. The sources of plastic accumulation range from direct dumping to blowing from landfill sites and losses in the source site [13]. Thus, properties like its recalcitrant structure, hydrophobic nature, and high molecular weight results in its increasing accumulation in the environment [14].

14.2.3 HERBICIDES/PESTICIDES/WEEDICIDES

The chemicals used to kill unwanted pests from the soil are DDT, chloradane, pyrethroid, carbamate, etc. The main source of such chemical contamination is accumulation and runoff from agricultural fields, which pose serious health risks to farmers as well as consumers. Their residues in food and drinking water raise serious concerns [15, 16]. Being water soluble and heat stable, it is difficult to remove them from the environment. They affect soil, water, and the environment and ultimately lead to negative effects on human health [17].

14.2.4 NANOMATERIALS

The materials in the size range of 1–100 nm are known as nanomaterials. Their physical properties enable their use as nanorods and nanowires (used widely in modern research and the development sector) releasing into the environment. Their mobility is dependent on their aggregation [18]. Further, major sources of their release include mining sites, etc., they may carry contaminants which make them more toxic [19, 20].

14.2.5 HAZARDOUS WASTE AND CHEMICALS

Hazardous waste includes paints, solvents, electronics, and other household or industrial wastes. It is highly toxic, reactive, and corrosive and has become a major concern in terms of environmental health. Sources includes mining activities, electroplating companies, petrochemicals, the metal refining industry, and the fertilizer industry. It is hazardous to human health and can cause side effects such as headaches, nausea, lead poisoning, or even cancer [21]. The Hazardous Wastes Management and Handling Rules, 1989 set the guidelines for the regulation of these waste materials.

These pollutants are non-biodegradable and persist in the environment for a long time. Considering this, it is important to develop reliable and cost-effective methods for their easy identification and monitoring.

14.3 THE CANDIDATE BIOLOGICAL AGENTS WITH BIOREMEDIATION POTENTIAL AGAINST RESPECTIVE POLLUTANTS

The increasing levels of pollutants from different classes are a matter of global concern due to their hazardous effects on the environment and human life. These contaminants are either organic or inorganic chemical compounds that are introduced into our ecosystem from natural as well as anthropogenic sources [22]. The accumulation of different types of contaminants in the environment can be due to agricultural run-offs, industrial discharges, domestic discharges, and improper disposal practices. These lead to a disturbance in the stability of the aquatic ecosystem and negatively affect human health.

Bioremediation of toxic substance is an eco-friendly option that offers a promising possibility to decontaminate the environment by removing hazardous contaminants from the environment using biological activities in a sustainable manner [23]. This technique uses inherent mechanisms of biological agents including microorganisms (e.g., bacteria, fungi, yeast, and algae), plants, and animals to transform hazardous waste material into less hazardous substances. These biological agents either degrade, detoxify, transform, or reduce the pollutant from the environment. Microorganisms are often used in the process of bioremediation of various environmental contaminants due their metabolic versatility and multiple biochemical pathways. For instance, microorganisms that have a good endurance towards heavy metals have been extensively used in the removal of toxic metal ions from the environment [22]. Mechanisms opted by them for the process of decontamination include precipitation, complexation, sequestration by either cell wall components or by intracellular proteins, enzymatic conversion of toxic metal ions into innocuous ones, and blocking metal uptake by altering the biochemical pathways [24]. Plants have also been reported to be effective candidates for the decontamination of polluted sites using different mechanisms. Transgenic plants and microorganisms are also effective in the process of detoxification and degradation of pollutants, but their use is subject to stringent biosafety measures and their performance is highly variable with respect to environmental conditions. Further, the risk of horizontal gene transfer to indigenous microbial populations is always associated with them [22]. The groups of organisms with bioremediation potential are discussed below.

14.3.1 BACTERIA

Microorganisms are utilized in the process of bioremediation because of their ability to degrade environmental pollutants due to their metabolism via biochemical pathways. Various bacterial genera have been reported in the decontamination of polluted terrestrial and aquatic ecosystems using their inherent biological mechanisms because they are able to grow under controlled conditions and can withstand intense environmental conditions. There are several reports that indicate their ability to remove both organic pollutants and heavy metals from the environment. For instance, *Enterobacter cloacae* strain PB2 has been reported to remove 2,4,6-trinitrotoluene (TNT) from seawater. It uses this compound as a nitrogen source using the enzyme pentaerythrotol tetranitrate reductase [25]. Biosurfactant-producing bacterial strains have been reported to degrade and remove a range of hydrocarbons from contaminated sites. *Brevibacterium* sp. PDM-3 was reported to degrade a range of polyaromatic hydrocarbons such as anthracene and fluorine. It could degrade phenanthrene by 93.92% [26]. Similarly, biosurfactant from bacteria *Lactobacillus pentosus* was able to reduce octane from the medium [27]. *Rhodococcus* genus is able to degrade alkanes up to a chain length of 36 carbons and is a potential candidate for the removal of heavy alkanes from fuels [28]. Various bacterial strains have been reported to degrade different classes of pesticides, which are one of the major classes of organic pollutants in nature. These include *Pseudomonas* sp., *Micrococcus luteus*, *Arthrobacter globiformis*, *Alcaligenes eutrophus*, *Streptomyces capoamus*, *Streptomyces galbus*, *Bacillus* sp., *Pseudomonas* sp., *Exiguobacterium aurantiacum*, *Pandoraea* sp., *Pseudomonas pseudoalcaligenes*, *Agrobacterium tumefaciens*, *Alcaligenes xylosoxidans*, *Streptomyces bikiniensis*, *Dehalospirilum multivorans*, *Escherichia coli*, *Pseudomonas putida*, *Pseudaminobacter* sp., *Providencia rustigianii*, *Ralstonia basilensis*, *Rhodococcus* sp., *Erythropolis* sp., *Rhodococccus erythropolis*, *Acinetobacter radioresistens*, *Arthrobacter* sp., *Flavobacterium* sp., *Pseudomonas putida*, and *Sphingomonas paucimobilis* [29]. Their versatility of pesticide degradation is due to the bioenzymes (hydrolase, dehydrogenase, ehydro-chlorinase, etc.) produced by them for the breakdown of complex pesticides.

Microorganisms are also capable of heavy metal removal by either using them as electron acceptors or remediating them from the environment via bioabsorption or bioaccumulation. *Desulfovibrio desulfuricans* converts soluble heavy metals such as cadmium and zinc to insoluble metal sulphides, thus reducing their bioavailability in nature. Microbial cell surfaces contain various negatively charged functional groups that offer active binding sites for positively charged heavy metal ions; thus, microbial biomass is a useful biosorbent for heavy metals [22]. Bacterial strains reported in the bioremediation of heavy metals include *Pseudomonas aeruginosa*, *Enterobacter* A47, *Sphingomonas paucimobilis*, *Azotobactervinelandii*, *Pasteurella multocida*, *Pseudomonas oleovorans*, *Xanthomonas campestris*, *Bacillus cereus*, *Kocuria flava*, *Sporosarcina ginsengisoli*, *Pseudomonas veronii*, *Pseudomonas putida*, *Enterobacter cloacae*, and *Bacillus subtilis* [30, 31]. Through the process of co-metabolism, microorganisms are able to degrade to harmless end products, thereby remedying or eliminating hazardous substances found in polluted environments. Bacterial consortiums have been reported to be more efficient in the removal of heavy metals than single strains. For instance, a bacterial consortium consisting of *Viridibacillus arenosi* B-21, *Sporosarcina soli* B-22, *Enterobacter cloacae* KJ-46, and *E. cloacae* KJ-47 was reported to remediate heavy metal contamination from the soil with greater efficiency in comparison to using a single strain [32]. Some genetically modified strains such as *Escherichia coli* strain M109, *Deinococcus geothemalis*, *Cupriavidus metallidurans*, and *Pseudomonas* spp. have been engineered for the effective decontamination of heavy metal polluted sites [22]. Many bacterial strains have been reported in radioactive waste management such as *Rhodococcus* sp., *Deinococcus radiodurans*, *Nocardia* sp., *Pseudomonas putida*, *Shewanella putrefaciens* CN32, *Desulfosporosinus* spp., *Anaeromyxobacter*, *Geobacter* species, *D. desulfuricans* G20, *Geobacter sulfurreducens*, and *Arthrobacter ilicis* [33].

14.3.2 FUNGI

Bioremediation carried out by fungi is often termed as mycoremediation. These fungi have been reported as suitable candidates in the removal of organic pollutants, metals, radionucleotides, etc., by either secreting chemicals and enzymes for modification or reducing their bioavailability [34]. Fungi are able to withstand and detoxify metal ions by active accumulation, intracellular and extracellular precipitation, and valence transformation, hence they are potential biocatalysts for the bioremediation of heavy metals. Fungi produce a variety of hydrolytic enzymes such as amylase, lipase, nuclease, protease, etc., which help to degrade organic pollutants in the environment. These enzymes are less specific and thus broaden the range of compounds they can metabolize. Fungi have been reported to degrade a variety of waste materials including wood, plastic waste, textile, paper, leather, poly aromatic hydrocarbons, pesticides, and dyes. For instance, *Penicillium simplicissimum* YK, *Mucor roxii*, white-rot fungi, and *Aspergillus flavus* have been reported to change the mechanical properties and weight of polyethylene bags, indicating their polyethylene degrading potential [35]. *Resinicum bicolor* was able to detoxify ground waste tire rubber [36]. *Aspergillus niger, Armilleria gemina*, and *Pholiota adipose* are able to degrade organic compounds and have been used for solid waste management [34]. One of the major benefits of fungi is the degradation of lignocellulolytic waste (e.g., wood, litter, and agriculture waste), which consists of three types of polymers: cellulose, hemicelluloses, and lignin [37]. Lignin is the most complex and recalcitrant component out of the three. A group of fungi known as white-rot fungi possess lignin degrading abilities. They are also capable of degrading other pollutants like hydrocarbons, dyes, synthetic polymers, etc. Brown-rot fungi, on the other hand, are able to degrade cellulosic material and merely cause modification in the structure of lignin. Fungal genera like *Trichoderma, Aspergillus, Fusarium, Penicillium*, etc., produce cellulase enzymes and thus are capable of degrading cellulolytic waste [38]. *Strobilurus ohshimae, Phanerochaete chrysosporium, Trametes versicolor, Bjerkandera adusta, Clonostachys rosea, Fusarium merismoides, Pycnoporus cinnabarinus, Postia placenta, Schizophyllum commune, Coprinopsis cinerea, Serpula lacrymans, Pleurotus ostreatus, Agaricus bisporus*, and *Ceriporiopsis subvermispora* are reported to be suitable for the biodegradation of lignocellulosic waste [34, 37]. Effluents from the dairy industry are rich in organic matter and thus are capable of increasing organic waste load into the environment. Fungi such as *Geotrichum candidum* and *Candida bombicola* have been reported to be used in dairy waste management [39]. Fungi such as *Aspergillus versicolor, Aspergillus fumigatus, Gloeophyllum sepiarium*, and *Rhizopus oryzae* have been reported to decontaminate heavy metal polluted sites [22]. *Phanerochaete chrysosporium* is able to degrade a wide range of harmful chemicals including benzene, xylene, organochlorines, heterocyclic explosives, and a large number of polyaromatic hydrocarbons [40]. *Coriolus versicolor, Inonotus hispidus*, and *Phlebia tremellosa* have been studied for the degradation of dyes [41]. *Aspergillus niger* and *Mycobacterium chlorophenolicum* have been reported to be effective in the removal of pentachlorophenol (PCP) using the mechanism of biosorption [22].

14.3.3 ALGAE

Recently, algae (micro and macro algae) have been gaining increasing attention as potential candidates for the process of bioremediation in a clean, green, and cost-effective manner. They have proven useful in the removal of xenobiotics and heavy metals due to their effective capability in hyperaccumulating toxic compounds. These can accumulate toxic metal ions and this process occurs in two steps. Firstly, the physical adsorption of toxic metal ions occurs at the cell surface, which is then followed by a second step called chemisorption. The dead cells of algae are also used for bioremediation purposes, which sometimes perform better than live cells [42]. Thus, algae can be used for the process of decontamination in a cost-effective manner by using either the products derived from algae or live, dead, or immobilized forms of algae itself. *Thalassiosira rotula, Cricosphaere elongate, Selenastrum* sp., *Ankiistrodesmus* sp., *Fragilaria crotonensis, Dunaliella tertiolecta, Scenedesmus obliquus, Chlamydomonas reinhardtii, Chlorella vulgaris, Spirogyra* spp., *Cladophora* spp., *Spirullina* spp., *Hydrodictylon, Oedogonium*, and *Rhizoclonium* spp. are some algal species reported for the biosorption and bioremediation of heavy metals like uranium, technitium, molybdenum, lead, cadmium, copper, nickel, and chromium [22, 42]. Exudates from algae in aquatic systems help in the sequestration of the dissolved metal ions and thus reduce the free metal ion concentrations in natural waters. Cyanophytes, for example, release polyhydroxamatesiderophores that are strong metal ion chelators. Brown algae such as *Sargassum* and *Ascophyllum* are considered effective for the purpose of biosorption because of their biomass [42], and they have been used for the treatment of colored wastewater, which has gained attention in recent years [28]. A red alga *Portieria hornemannii* has been reported to remove TNT from seawater [43]. In another study, three species of marine algae, viz. *Acrosiphoniacoalita* (green alga), *Porphyrayezoensis*, and *Portieriahornemannii* (red algae), were found to convert the aforementioned explosive compound into 2-amino-4,6- dinitrotoluene and 4-amino-2,6-dintrotoluene following the same metabolic route [44]. Other species of algae including *Chlamydomonas reinhardtii*, *Scenedesmus quadricauda*, *Monoraphidium braunii*, and *Chlorococcum* sp. have been reported to remediate several fungicides and herbicides from the environment [28]. *Chlorella sorokiniana* was able to degrade significant amounts of phenanthrene in silicone oil when used in combination with *Pseudomonas migulae* [45]. Algal-based products are available commercially, such as AlgaSORB™, which is an immobilized mass of *Chlorella vulgaris* in silica or polyacrylamide gel. It is an efficient system used for the removal of heavy metals from concentrations of 100 mg/l using an ion exchange mechanism. Similarly, Biofix, another algal-based product, is granular in nature containing algal biomass for the purpose of biosorption encapsulated in polypropylene beads. The biomass is blended with xanthan gum and guar gum [42].

14.3.4 PLANTS

Using plants for the process of detoxification or the removal of contaminants is termed as "phytoremediation." Plants contain many enzymes, such as nitroreductase, dehalogenase, and laccase, that allow them to be used as suitable candidates for the purpose of environmental clean-up [28]. Phytoremediation is used in the removal of several contaminants including hydrocarbons, poisonous solvents, explosives, heavy metals, toxic and persistent pesticides, and radionuclides. These are also used in wastewater treatment. *Helianthus annuus*, *Phragmites karka*, *Datura innoxia*, *Brassica juncea*, *Alternanthera sessilis*, and *Zea mays* have been used for the treatment of effluents contaminated with metals, industrial discharge, and sewage sludge [46]. Plants mediate detoxification through various mechanisms such as phytoextraction/phytoacumulation, phytofiltrtaion, phytostimulation, and phytostabilization. Plant genera from the families Brassicaceae, Caryophyllaceae, Fabaceae, Euphorbiaceae, Lamiaceae, Asteraceae, Cyperaceae, and Poaceae are well known for having phytoremediation potential. In a previous study, three aquatic plant species, namely *Leman minor*, *Elodea Canadensis*, and *Cabomba aquatic*, were reported to assimilate copper sulphate, flazasulfuron, and dimethomorph [47]. *Thlaspi caerulescens* (Brassicaceae) has been reported to accumulate Zn, Pb, and Cd up to 3%, 0.5%, and 0.1%, respectively. Trees such as *Agrostis capillaris*, *Salix* (willow), *Betula* (birch), *Populus* (poplar), *Alnus* (alder), and *Acer* (sycamore) also possess bioremediation capabilities and are grown in heavy metal contaminated areas [48]. Other plant species reported in the removal of heavy metals such as Ni, Pb, Zn, Cd, Hg, As, and Cr include *Berkheya coddii*, *Helianthus annuus*, *Alyssum bertolonii*, *Alyssum murale*, *Arabidopsis halleri*, *Minuartia verna*, *Sedum alfredii*, *Euphorbia cheiradenia*, *Astragalus racemosus*, *Medicago sativa*, *Spartina argentinensis*, *Pterisvittata*, and *Viola boashanensis* [22, 49–51]. Various transgenic plants with high extraction abilities and higher biomass such as *Lycopersicon esculentum*, *Arabidopsis thaliana*, *Nicotiana tabaccum*, *Brassica oleraceae*, and *Brassica juncea* have been used for the process of bioremediation [22].

14.3.5 ANIMALS IN BIOREMEDIATION

Animals are not as prominent as candidates in the process of bioremediation as what plants and microorganisms are due to their limited metabolic capabilities. However, there are a few candidates such as enchytraeids, mites, and earthworms that are able to use organic waste using their metabolism and help in increasing the number of microorganisms capable of the degradation of contaminants [52]. They contribute to the process of bioremediation in two ways: either by taking part in the process as well as increasing the microbial activity in the soil, or by contributing in an indirect manner by being useful indicators of contaminants and used for risk assessments of the toxicity of pollutants. Organic waste (e.g., vegetable waste, post-harvest residues, and cane straw) disposal and management is a serious problem in developing countries like India. The treatment of such waste using biological agents offers a safe and eco-friendly option [53]. For instance vermicomposting utilizes earthworms for transforming organic solid waste into compost rich in nitrogen, phosphorous, and other valuable nutrients [54]. Earthworms are suitable candidates for the process of bioremediation of organic waste using their physical, chemical, and biological activities. They decompose the dead organic matter and release nutrients that are then available to plants. In addition, earthworms have an additive positive effect on beneficial bacteria and fungi in the soil. They are also helpful in the dispersal of other microorganisms with degradation potential [52].

14.4   BIOREMEDIATION: POTENTIAL AND SUSTAINABLE PROCESS FOR ENVIRONMENTAL CLEANUP

Bioremediation is an eco-friendly alternative to harmful conventional practices. It employs many different microbes (parallel or series) to alter/detoxify contaminants, which can be achieved via the involvement of microorganisms (e.g., bacteria and fungi), plants (phytoremediation), or microbial enzymes that transform toxic contaminants into CO_2, H_2O, inorganic salts, and other less harmful metabolites [22]. In recent times, the microbial based approaches for the degradation and detoxification of pollutants has been rapidly gaining popularity and is preferred over conventional techniques for the cleanup or restoration of contaminated sites back to their natural sustainable form [55].

14.4.1   IMMOBILIZATION

Bioremediation processes employ various immobilization methods. Immobilization involves limiting the mobility of the microbial cells/enzymes and simultaneously preserving their viability and catalytic activity. This method adds several beneficial properties to bioremediation, such as efficient pollutant degradation, the use of biocatalysts, low costs, modified techniques to ensure a stable environment for cells/enzymes, the development of resistance to forces present in bioreactors, the increased survival of biocatalysts during storage, and enhanced tolerance to high contaminant concentrations. There are five techniques of immobilization, namely adsorption, binding on a surface, flocculation, entrapment, and encapsulation [56].

14.4.2   MOBILIZATION

The bioremediation process is aided by the conversion of an element from its insoluble/stationary form into its soluble/mobile form. Mobilization can sometime have deleterious impacts when toxic metal ions are redistributed and released. Microorganisms carry out metal mobilization by redox reactions and hence impact bioremediation processes [57]. Heavy metals like Fe, As, Cr, and Hg undergo oxidation and reduction cycles. Microorganisms have the ability to adapt and resist heavy metals in contaminated areas, thus making them perfect candidates to alleviate the problem and detoxify the surroundings. Substances like extra-cellular polymeric substances available on the biomass cell wall can attach to heavy metals by mechanisms like proton exchange or the micro-precipitation of metals.

14.4.3   KEY ENVIRONMENTAL AND BIOLOGICAL FACTORS AFFECTING BIOREMEDIATION

The process of bioremediation depends on biological agents to reduce pollutants in the environment. In view of this, based on the site of application, it is categorized as ex situ or in situ. Generally, the nature of the pollutant, the degree of pollution, the type of environment, location, cost, and environmental factors are the key factors when choosing any bioremediation technique [58]. The result of each degradation process also depends on the microbial (e.g., biomass concentration, diversity of microbes, and enzyme activities), the substrate (e.g., physico-chemical characteristics, chemical structure, and concentration), and several environmental factors (e.g., pH, temperature, moisture content, availability of electron acceptors and carbon, and energy sources). These parameters govern the acclimation period of any microbes to the respective substrate.

14.4.3.1 Environmental Factors

The growth and activity of microorganisms present in contaminated soil must be stimulated for optimum efficiency of bioremediation. This usually involves enrichment of the indigenous microorganisms by the addition of oxygen and nutrition. This will allow the specialized microbes to create biomolecules to break down the pollutants. Microbial growth and activity are affected by factors like pH, temperature, and moisture. Microbial populations inhabit a wide range of habitats in nature but most of them grow best over a narrow range. These need to be maintained to achieve the desired biomass and simultaneously their product [59]. For instance, for soils rich in acid, it is possible to rinse the pH by the addition of lime [60]. Temperature affects biochemical reactions and above a certain temperature either the enzyme activity comes to a halt or the cells die. Water is essential for all living organisms and is needed to achieve optimal moisture levels for the cells to function normally. The amount of oxygen will generally determine whether a system is aerobic or anaerobic. This is important as hydrocarbons are readily degraded under aerobic conditions while chlorinate compounds require anaerobic conditions. To ensure aeration in soil it is possible to till or sparge air. Hydrogen peroxide or magnesium peroxide have also been reported in some situations to achieve the same results. The structure of soil also controls the delivery of air, water, and nutrients into it and thus materials such as gypsum or organic matter are used to improve soil structure [61].

14.4.3.2 Biological Factors

14.4.3.2.1 Source of Energy

One of the primary factors affecting bacterial activity in soil is the availability of reduced organic materials to serve as energy sources. The average oxidation state of carbon in the material determines whether a contaminant will serve as an effective energy source for an aerobic heterotrophic organism. A higher oxidation state relates to lower energy yields, providing less energetic incentive for microorganism-based degradation processes. These parameters have an important role in the acclimation period of microbes towards its substrate [62]. As discussed earlier, the molecular structure and concentration of a pollutant are pivotal in determining the feasibility of bioremediation as well as the type of microbial transformation that will take place.

14.4.3.2.2 Bioavailability

The pollutant removal efficiency of the microbes depends upon the rate of contaminant uptake and its metabolism inside the cell. However, these processes are not necessarily be dependent on each other. This is common in most contaminated soils and sediments [58]. The bioavailability of a pollutant is regulated by various processes such as sorption and desorption, diffusion, and dissolution [63]. A low bioavailability of contaminants in soil is due to a slow mass transfer to the microbes responsible for degradation. For instance, if the rate of mass transfer is zero, contaminants become unavailable. These problems related to bioavailability can be controlled by using surfactants which results in the increased contaminant degradation [64].

14.4.3.2.3 Bioactivity and Biochemistry

Bioactivity is the operating state of microbiological processes. Improving bioactivity means modifying the system conditions to achieve optimized biodegradation of the contaminant. For example, if the use of bioremediation requires the adjustment of certain parameters, optimization becomes a key element and the respective bioremediation system has a substantial advantage over one that does not require adjustment. Under natural conditions, the ability of organisms to transform pollutants to simpler and complex molecules shows a wide range of diversity [8]. With the limited capacity to control any biochemical mechanism, the feasibility or non-feasibility of the reaction are evaluated in terms of whether the parent compounds are removed, and whether the bioremediation process resulted in more toxic by-products. Sometimes the elements in the parent compound are evaluated for conversion to detectable metabolites.

14.4.4 Mechanisms Involved in Bioremediation

14.4.4.1 Adsorption

The process of immobilizing microbial cells and enzymes is known as adsorption. It takes place via the physical interaction of these entities with the surface of the carriers. Adsorption is commonly used in bioremediation processes as it is quick, eco-friendly, and cost-effective. This process includes interaction on the carrier surface by the formation of weak bonds [65]. For this reason there is the possibility of cells leaking from the carrier into surroundings. Thus, carriers should have a high porosity to ensure a larger contact area of the immobilized material and the carrier.

14.4.4.2 Biosorption

Bioremediation can be classified into two categories: biosorption and bioaccumulation. Biosorption is a passive, reversible adsorption mechanism that occurs rapidly. The metals are retained by the presence of physicochemical interaction like ion exchange, adsorption, precipitation, and crystallization. This interaction occurs between the metal and the functional groups available on the cell surface. Biosorption is affected by pH, ionic strength, biomass concentration, temperature, particle size, and the presence of other ions in the solution [66]. Both living and dead biomass can be used for this process as it is independent of cell metabolism.

14.4.5 Molecular Approach: Genetically Engineered Microorganisms

Several naturally existing microbes are regularly employed for bioremediation. They can be used individually for their specific properties or in consortia according to the environmental conditions and nature of the contaminants. This provides an idea about the relation of metabolism in biodegradation. But sometimes the existing microbes are genetically altered or modified to make optimum use of their remediation properties. For this, various techniques are used to produce genetically engineered microorganisms (GEMs) with enhanced or improved bioremediation efficacy [67]. For instance, engineering approaches such as rational designing and directed evolution have been successfully used to genetically modify microorganisms and their enzymes, which aids in the degradation of recalcitrant pollutants like polycyclic aromatic hydrocarbons, polychlorinated biphenyls, and harmful pesticides [68]. In recent times, developments in recombinant DNA technologies have also been used to achieve safe and efficient detoxification of contaminated sites (see Figure 14.2). One such example is the development of "suicidal-GEMs" (S-GEMs) [69]. This concept stems from the fact that during the course of evolution, the indigenous microorganisms of contaminated sites have gained peculiar abilities to degrade complex chemical compounds due to constant physiochemical pressure in those environments. Thus, the potential of this group of organisms can be employed by the identification of specific biomolecules responsible for the process and engineering catabolic gene pools. Indigenous populations are always preferred over exotic strains for the construction of recombinant microorganisms due to their ability to interact with and withstand the total population as well as successfully tolerate the complex a and stressful conditions of the particular site [70]. There are very few reports where GEMs have been applied and proven to be more efficient than natural microbes in the elimination of recalcitrant compounds under natural conditions [58]. However, efforts have been made to expand the range of contaminants that can be removed or degraded using recombinant technology. The successful application of genetically engineered microbes for bioremediation depends on the successful establishment of the engineered microorganisms in the

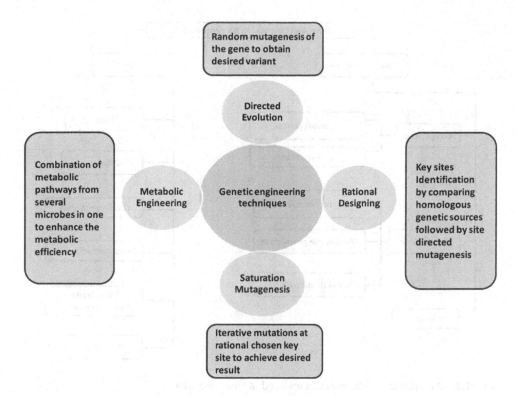

FIGURE 14.2 Methods to develop genetically modified microorganisms for bioremediation.

environmental conditions and their successful removal mechanisms from the site of action after the task has been completed [71].

14.4.6 ROLE OF ENZYMES

Enzyme-based bioremediation refers to the use of microbial or plant enzymes in microorganisms or plants to reduce harmful, undesirable, and recalcitrant environmental pollutants. Bioremediation can be effective under favorable environmental conditions for microbial growth and activity. It often involves the manipulation of environmental parameters to promote microbial growth and for degradation to occur at a faster rate. A number of microbial enzymes have been isolated, characterized, and well documented from different natural sources and have been reported to have tremendous potential for bioremediation. However, low production is one of the limiting factors in their use [68]. The small size of enzymes offer a high surface-to-mass ratio which enables them to make contact with contaminants easily and efficiently, and to promote quicker mobility and more targeted rapid degradation to a neutral or a less harmful state [72]. Enzymes such as oxygenases, laccases, peroxidases, haloalkane dehalogenases, carboxylesterases, phosphotriesterases, lipases, and cellulases have been explored in depth in the degradation pathways of various pollutants (see Figure 14.3) [73]. The enzymes used in the bioremediation of harmful chemical compounds along with their activity are summarized in Table 14.1.

Over the past few years, researchers have focused on the concept of using artificial enzymes to imitate the structures and functions of natural enzymes. Nanozymes or artificial enzymes are nanomaterials ranging in size from 0–100 nm and exhibit enzyme-like characteristics. In recent times, they have gained the attention and interest of researchers. They connect nanomaterials such as clusters, crystals, nanorods, nanobelts, and nanowires with biological systems. They exhibit excellent robustness, stability, a wide range of pollutant degrading abilities, low-cost production, and ease in

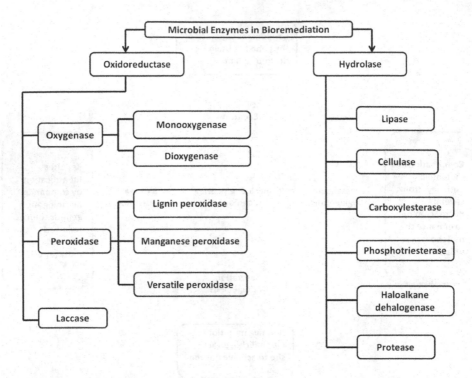

FIGURE 14.3   An overview of the enzymes involved in bioremediation.

scaling-up [74]. These give an added economic advantage over natural enzymes. Some examples are metal oxide-based nanozymes (chemically and biologically inert), metal-based nanozymes [e.g., gold nanoparticles (Au NPs) and platinum nanoparticles (Pt NPs)], and carbon-based nanozymes (e.g., fullerene, carbon nanotubes, graphene oxide, and carbon dots) [75].

To sum up, various physiochemical and biological methods are being widely studied and employed for the bioremediation of environmental contaminants. Considering all the strategies, from microbial cell-based to enzyme-mediated to nanozymes, there are several technical and/or economical challenges posed which must be addressed to achieve optimum results. These obstacles should be rectified before introducing any of these methods for widespread societal acceptance and use.

14.5 MAJOR BIOREMEDIATION STRATEGIES/TECHNIQUES AND THEIR TYPES

14.5.1 BIOREMEDIATION

The process of reducing (i.e., degrade, detoxify, transform, or mineralize) pollutants via biological mechanisms to an innocuous state is known as bioremediation. Pollutant removal primarily depends upon the nature of the pollutant which may include chlorinated compounds, agrochemicals, dyes, greenhouse gases, heavy metals, plastic, sewage, and nuclear waste. Depending upon the site of action and techniques used in the bioremediation process, bioremediation can be categorized as either in situ or ex situ.

14.5.1.1 In-situ Bioremediation

In-situ bioremediation processes include biological methods to treat contaminated soil without excavating the soil. The efficiency of in-situ bioremediation highly depends upon the microbial interaction with the contaminants because, in most cases, contamination is not limited to topsoil; it appears that deeper soil layers also get contaminated due to leaching. Since topsoil is easily accessible it is

TABLE 14.1
Enzymes Used in the Bioremediation of Different Harmful Chemical Compounds

Enzymes	Source Organism	Contaminant/Substrate	Mechanism	Application
Monooxygenase	*Arthrobacter*	Alkane, steroids, fatty acid, and aromatic compounds	Incorporation of oxygen atom to substrate and utilize substrate as reducing agent. Desulfurization, dehalogenation, denitrification, ammonification, and hydroxylation of substrate.	Protein engineering, bioremediation, etc.
Dioxygenase	*Mycobacterium*	Aromatic compounds	Introduction of two oxygen atoms to the substrate results in intradiol cleaving and extradiol cleaving with the formation of aliphatic product.	Pharmaceutical industry, bioremediation, etc.
Laccase	*Trametes versicolor*	Ortho and paradiphenols, aminophenols, polyphenols, polyamines, lignins, and aryldiamines	Oxidation, decarboxylation, and demethylation of substrate.	Food industry, textile industry, nanotechnology, synthetic chemistry, bioremediation, cosmetics, etc.
Lignin peroxidase	*Phanerochaete chrysosporium*	Halogenated phenolic compounds, polycyclic aromatic compounds, and other aromatic compounds	Oxidation of substrate in the presence of co-substrate H_2O_2 and mediator like veratryl alcohol.	Food industry, paper and pulp industry, pharmaceutical industry, bioremediation, etc.
Manganese peroxidase	*Bacillus pumilus*	Lignin and other phenolic compounds	In the presence of Mn^{2+} and H_2O_2 the co-substrate catalyzes oxidation of Mn^{2+} to Mn^{3+} which results in an Mn^{3+} chelateoxalate, which in turn oxidizes the phenolic substrates.	Food industry, paper and pulp industry, textile industry, bioremediation, etc.
Versatile peroxidase	*Pleurotus ostreatus*	Methoxybenzenes and phenolic aromatic	Catalyzes the electron transfer from an oxidizable substrate, with the formation and reduction of compound I and compound II intermediates.	Industrial biocatalyst, bioremediation, etc.
Lipase	*Candida antarctica, Rhizomucor miehei*	Organic pollutants such as oil spills	The hydrolysis of triacylglycerols to glycerols and free-fatty acids.	Control of oil spills, detergent production, personal care products, etc.

(Continued)

TABLE 14.1 (CONTINUED)
Enzymes Used in the Bioremediation of Different Harmful Chemical Compounds

Enzymes	Source Organism	Contaminant/Substrate	Mechanism	Application
Cellulase	*Cellulomonas, Cellvibrio*	Cellulosic substance	Hydrolyzes the substrate to simple carbohydrates.	Textile manufacturing, detergent production, paper and pulp industry, bioremediation, etc.
Protease	*Bacillus* sp.	Proteins	Enzymes that hydrolyze peptide bonds in aqueous environments.	Leather, laundry, biocatalyst, bioremediation, etc.
Carboxylesterases	*Pseudomonas aeruginosa* PA1	Malathion and parathion	Hydrolysis of carboxyl ester bonds present in synthetic pesticides.	Agriculture, environment monitoring, bioremediation, etc.
Phosphotriesterases	*Brevundimonas diminuta*	Organophosphorus compounds	Hydrolysis of phosphotriesters.	Bioremediation and detoxification, enzyme therapy, biosensors, etc.
Haloalkane dehalogenases	*Xanthobacter autotrophicus* GJ10	Halogenated aliphatic compounds	Biodegradation of halogenated aliphatic compounds.	Industrial biocatalysts in synthesis of chiral compounds, bioremediation, etc.

easily bioremediated, but there are major challenges associated with the bioremediation of deeper soil.

There are two major pathways of bioremediation that constitute in-situ bioremediation: unsaturated zone treatment and saturated zone treatment.

14.5.1.1.1 Unsaturated Zone Treatment Method

Unsaturated zones comprise the soil zones that are above ground level, also termed as vadose zones. Thus, the bioremediation of vadose regions is highly reliant on aerobic biodegradation; however, the majority of the contaminants are present below ground water level and so a natural attenuation process (without human intervention) is used to degrade it. There are some important processes that promote unsaturated zone biodegradation. These are discussed below:

a) **Natural attenuation.** The process of biodegradation without human intervention is called natural attenuation. In this method, a chemical analysis is undertaken of degraded contaminants which are present in the subsurface and surface of soil. The rate of biodegradation is dependent on the amount of oxygen consumed and carbon dioxide released [76]. This method is also very useful in estimating low temperature bioremediation. Research groups have reported high contaminant levels due to petroleum hydrocarbon spills in the arctic region by measuring the rates of oxygen and carbon dioxide [77].

b) **Bioventing.** This process involves enhancing the aerobic biodegradation rate in unsaturated zones via pumping air or oxygen into the subsurface region of soil by gas injection wells that are already planted into the soil. This method is highly useful in degrading volatile compounds like BTEX and chlorinated solvents from soil in the gas phase. The bioventing process is often followed by respiration rate analysis, as seen in nano-aromatic hydrocarbon biodegradation in Austria [78].

c) **Nutrients infiltration.** This process involves the addition of nutrients that drastically increased the rate of aerobic biodegradation in contaminants sites [79].

14.5.1.1.2 Saturated Zone Treatment Method

Saturated zones are the regions below ground water level. Saturated zone contamination occurs either by leakage of contaminants from unsaturated regions or by direct leaching, for example water soluble compounds and oil hydrocarbon, which are even lighter than water soluble compounds, can directly percolate through the soil. Oxygen levels, however, are low when compared to unsaturated zones. Two very important methods are used in saturated zone treatment: enhanced aerobic natural attenuation and enhanced an-aerobic natural attenuation.

a) **Enhanced aerobic natural attenuation.** This method directly influences the rate of biodegradation in saturated zones by adding oxygen with the help of oxygen sparging [80]. The most useful application of injecting oxygen is seen in case of methyl tetra butyl ether (MTBE) bioremediation with or without adding bioinoculants. In a pilot study, it was observed that, with the addition of oxygen, there is a sharp increase in the degradation of MTBE pollutants from the soil [81]. Other target compounds include tri chloro ethane (TCE), polycarboxylate ether (PCE), and vinyl chloride (VC).

b) **Enhanced an-aerobic natural attenuation.** Since there is a small amount of oxygen at the subsurface regions, biodegradation is challenging. But in this method, the addition of inorganic electron acceptors like sulfate or nitrate stimulates the biodegradation ability of microorganisms. The addition of nitrate as an electron acceptor in contaminated sites has been reported. Studies suggest that the addition of nitrate results in an increase in the inorganic carbon content by estimating the number of stable radioisotopes generated by the degradation of petroleum hydrocarbon, whereas nitrate recycled to nitrogen via the nitrogen cycle [82].

14.5.1.1.3   Genetically Engineered Microorganisms

As we know, microorganisms are able to adapt to various stress conditions, e.g., salinity, osmotic, and heat and cold stresses. Various genes have been targeted which can be later amplified and expressed under suitable hosts for the biodegradation of contaminants. The only microorganism used in field conditions, which was genetically modified by introducing lux genes via transposon, was *Pseudomonas fluorescens* HK44 [83]. However, the use of genetically modified microorganisms in in-vitro bioremediation is not entirely successful due the low survival rate and high demand of energy input due to the presence of extragenetic material.

14.5.1.1.4   Factors Affecting In-Situ Degradation

a. **Low temperatures:** Low temperatures provide the suitable environment for microbial bioremediation in soil, as can be seen in most of the countries in Europe and North America.
b. **Anaerobic conditions:** Bioremediation in anaerobic conditions is low when compared to aerobic conditions, which results in the accumulation of xenobiotic compounds.
c. **Nutrient availability:** The availability of nutrients in contaminated soil regions drastically increases the metabolic potential of microorganisms, which results in high bioremediation rates.

14.5.1.2   Ex-Situ Bioremediation

Ex-situ bioremediation involves the treatment of excavating soil. Ex-situ technology provides a number of advantages, one which is better control in the bioremediation process because the treatments are under stringent control. Some of the major processes in ex-situ bioremediation include slurry bioreactor and solid phase bioreactor.

i. **Slurry bioreactor:** Slurry bioreactor is a highly engineered and efficient type of bioreactor. Bioremediation using slurry bioreactors can be efficiently regulated by controlling critical operating parameters. Slurry bioreactors require the regular availability of microbial inoculums that are contaminant specific. In slurry bioreactors contaminated soil or wastewater is mixed with water to provide a slurry of predetermined consistency with aqueous suspensions of generally 10–30% w/v (weight/volume). The system can be run under anoxic, aerobic, and anaerobic conditions with different feeding modes, i.e., batch, continuous, and semi-continuous. However, batch culture is commonly used in the treatment of soil or sediments in slurry bioreactors. The modified form of batch bioreactors is sequenced batch bioreactors (SBRs) with a work cycle consisting of a reaction phase and feed phase. Part of the slurry left in the bioreactor is used for the next operation. Slurry bioreactor operations require screening of the crushing of coarser fractions (0.85–4 mm) with fine fractions of <0.85 mm being fed to the bioreactor for treatment. This modified form of bioreactor is usually used for pollutants that are rich in fine particles. Slurry bioreactors have been extensively used in the removal of poly hydroxyl alkane (PAHs) [84]. The substrate load in bioreactors is 0.084 (gm chrysene/kg soil d) in a single metabolic operation with anaerobic sludge or microflora alone. Data envelopment analysis is used monitor chrysene biodegradation, with the results indicating that best method in anoxic conditions. For the successful bioremediation of PAH, aerobic processes are used, but there is also evidence of removal of PAH under anaerobic conditions by placing sulfates and nitrates as terminal electron acceptors in slurry bioreactors [85].
ii. **Solid phase bioreactor:** In ex-situ solid phase bioreactors, contaminated soil is excavated and placed into piles along with bacterial inoculum. Bacterial growth stimulation is promoted through network off pipes that are randomly distributed through piles [86]. Ventilation is provided through these pipes for microbial respiration. Solid phase bioreactors require large amounts of inbuilt space and clean-ups as compared to slurry phase

processes. Solid phase bioremediation involves three major processes that includes composting, soil bio-piles, and landfarming.

a) **Composting.** In compositing, contaminated soil is combined with non-hazardous organic matter such as agricultural wastes or manure. The presence of high organic material in the soil supports the development of microbial populations. High temperature during the bioremediation process is a characteristic feature of composting process [87]. Bulking agents such as hay or straw are also mixed with contaminated soil to provide water and air to the microorganisms. Some of the most common composting processes include mechanically agitated composting, static pile composting, and window composting.

b) **Bio-piles.** Bio-piles are a hybrid of composting and land farming. Bio-piles provide favorable environmental conditions for both anaerobic and aerobic microorganisms. This technique, also called bio-cells, promotes the bioremediation of soil contaminated with petroleum hydrocarbon. Bio-pile techniques also involve the addition of contaminated soil into piles, and microbial growth is stimulated either aerobically or by adding nutrients, moisture, or minerals. Generally, the typical height of bio-piles is about 3–10 feet. Perforated piping is placed within the piles through which air is forcefully entered, and bulking agents such as straw are also added to improve aeration which results in high microbial growth [88].

c) **Landfarming.** This is one of the simplest techniques in solid phase bioreactors which promotes soil excavation spread over a prepared bed to stimulate pollutant degradation. The basic aim is to promote indigenous microorganisms to stimulate biodegradation aerobically. Around 10–35 cm of superficial soil is treated. Landfarming has the potential to reduce monitoring and maintenance costs and also provides an alternative method for disposal [87].

14.5.2 PHYTOREMEDIATION

The term phytoremediation has been used since 1991. The word derives from the Greek *phyto*, meaning "plant," and the Latin *remedium,* meaning "to remedy." General information about phytoremediation comes from a variety of research work covering oil spills, wetlands, and agriculture plants with an accumulation of heavy metals. Different definitions of phytoremediation have been given by researchers but the general definition is that it is an emerging technology that uses selected plants to clean up the contaminated environment to improve environmental quality. In recent years, phytoremediation technology has been receiving worldwide attention due to its innovative technique, in addition to the fact that it is eco-friendly and highly cost effective which makes it a great alternative to already-established treatments [89]. There are numerous mechanisms that are adopted by plants to remove heavy metal contaminants but there are six major mechanisms that are opted by plants for the bioremediation of heavy metals: phytoextraction (accumulation of metal in shoots), phytovolatilization (converts metal into volatile form), phytofiltration (sequestration of metals from water), phytostabilization (limits mobility and availability of metals in soil), phytodegradation (degradation of metal within tissue), and rhizosphere degradation (breakdown of heavy metals present in microorganisms). Some of the plants that are used to bioremediate heavy metal from contaminated sites are listed in Table 14.2.

14.6 ADVANTAGES AND DISADVANTAGES OF BIOREMEDIATION

14.6.1 ADVANTAGES OF BIOREMEDIATION

Bioremediation is the most proficient techniques to clean environmental pollutants. However, similar to any other process, it has its pros and cons. The biggest advantage of bioremediation is that it

TABLE 14.2

Different Plants Used in Phytoremediation to Bioremediate Heavy Metals

Plants Used	Source of Contamination	Heavy Metal Contamination	References
Brassica juncea L. (Indian mustard)	Soil and water	Cd, Cu, Zn, Pb	[90]
Allium schoenoprasum L. (Chive)	Soil	Ni, Co, Cd	[91]
Lantana camara L. (lantana)	Soil	Pb	[92]
Medicago sativa L. (alfalfa)	Soil	Cd	[93]
Oryza sativa L. (rice)	Soil	Cu, Cd	[94]
Pisum sativum L. (pea)	Soil	Pb, Cu, Zn, Fe, Cd	[95]

is environmentally friendly and can be considered as part of the green movement. Other advantages of bioremediation are summarized below:

14.6.1.1 Natural Process

Bioremediation involves living organisms to clean to the environment and therefore is a purely natural process. It involves the use of plants (phytoremediation), fungi (mycoremediation), microorganisms (microbial remediation), amongst others [55]. Being eco-friendly in nature, it is well accepted by administrations as well as the public.

14.6.1.2 Destruction of Contaminants

Bioremediation is known to completely degrade several contaminants. Moreover, it is able to transform a wide range of hazardous materials into less harmful forms. Bioremediation is widely used in wastewater treatment plants where it has been proved very efficient [96].

14.6.1.3 In-situ Treatment

Bioremediation can be used in situ for the removal of the contaminants. It does not cause any side effects to the natural flora and fauna. Moreover, it reduces the overall cost of environmental cleaning [8].

14.6.1.4 Cost-Effective Process

Bioremediation is considered less expensive in comparison to its counterparts. Due to its ability to work in situ, it reduces the cost of excavation, transportation, storage, treatment, etc. [8, 97].

14.6.2 DISADVANTAGES OF BIOREMEDIATION

Despite having an eco-friendly nature, bioremediation is limited mostly to biodegradable compounds. Moreover, some reports have reported having more persistent and/or toxic by-products of bioremediation [98]. In addition to these, the following limitations can be considered:

14.6.2.1 Requires Specificity

Bioremediation requires specific metabolically active organisms and optimum growth conditions. In the absence of these, it cannot be used successfully. On the other hand, chemical-based bioremediation technologies do not rely on these aspects.

14.6.2.2 Requires Modern technologies

Recent technological advancements and innovations are required to conduct and/or monitor the bioremediation process. Nowadays genetic engineering is being explored to improve the bioremediation efficiency of microorganisms [67]. It involves a great deal of monetary and technological support.

14.6.2.3   Time-Consuming Process

Bioremediation is considered a time-consuming process as it depends on the metabolic potential of the organisms. Therefore, it takes a longer time to complete the process [8]. However, recent studies are focusing on improving biodegradation efficiency through metabolic engineering, genetic engineering, and omics technologies [3].

14.7   CONCLUSION

Bioremediation is a highly accepted technology for environmental cleanup due to its eco-friendly, non-hazardous, safe, and cost-effective nature. However, further research and innovations are required to further enhance its efficiency and potential.

REFERENCES

1. Arora, N.K., Bioremediation: a green approach for restoration of polluted ecosystems. *Environmental Sustainability*, 2018. **1**(4): p. 305–307.
2. Bilal, M. and H.M.N. Iqbal, Microbial bioremediation as a robust process to mitigate pollutants of environmental concern. *Case Studies in Chemical and Environmental Engineering*, 2020. **2**: p. 100011.
3. Debbarma, P., et al., Comparative in situ biodegradation studies of polyhydroxybutyrate film composites. *3 Biotech*, 2017. **7**(3): p. 178.
4. Ferronato, N. and V. Torretta, Waste mismanagement in developing countries: a review of global issues. *International Journal of Environmental Research and Public Health*, 2019. **16**(6): p. 1060.
5. Pant, D., A. Giri, and V. Dhiman, Bioremediation techniques for E-waste management. In *Waste Bioremediation*, S.J. Varjani, et al., Editors. 2018, Springer: Singapore. p. 105–125.
6. Tahri Joutey, N., et al., Biodegradation: Involved Microorganisms and Genetically Engineered Microorganisms, 2013. Biodegradation - Life of Science, Rolando Chamy and Francisca Rosenkranz, IntechOpen, DOI: 10.5772/56194.
7. Meril, D., et al., Bioremediation: an eco-friendly tool for effluent treatment: a review. *International Journal of Applied Research*, 2015. **1**: p. 530–537.
8. Azubuike, C.C., C.B. Chikere, and G.C. Okpokwasili, Bioremediation techniques: classification based on site of application: principles, advantages, limitations and prospects. *World Journal of Microbiology and Biotechnology*, 2016. **32**(11): p. 180.
9. Ali, H., E. Khan, and I. Ilahi, Environmental chemistry and ecotoxicology of hazardous heavy metals: environmental persistence, toxicity, and bioaccumulation. *Journal of Chemistry*, 2019. **2019**: p. 6730305.
10. Singh, R., et al., Heavy metals and living systems: an overview. *Indian Journal of Pharmacology*, 2011. **43**(3): p. 246–253.
11. Tchounwou, P.B., et al., Heavy metal toxicity and the environment. *Experientia supplementum*, 2012. **101**: p. 133–164.
12. Naveedullah, et al., Risk assessment of heavy metals pollution in agricultural soils of siling reservoir watershed in Zhejiang Province, China. *BioMed Research International*, 2013. **2013**: p. 590306.
13. Barnes, D.K.A., et al., Accumulation and fragmentation of plastic debris in global environments. *Philosophical Transactions of the Royal Society of London. Series B, Biological Sciences*, 2009. **364**(1526): p. 1985–1998.
14. Thompson, R.C., et al., Plastics, the environment and human health: current consensus and future trends. *Philosophical Transactions of the Royal Society of London. Series B, Biological Sciences*, 2009. **364**(1526): p. 2153–2166.
15. Hamilton, D. and S. Crossley, *Pesticide Residue in Food and Drinking Water: Human Exposure and Risks*, 2004. John wiley & Sons Ltd, USA.
16. Aktar, M.W., D. Sengupta, and A. Chowdhury, Impact of pesticides use in agriculture: their benefits and hazards. *Interdisciplinary Toxicology*, 2009. **2**(1): p. 1–12.
17. Nicolopoulou-Stamati, P., et al., Chemical pesticides and human health: the urgent need for a new concept in agriculture. Frontiers in *Public Health*, 2016. **4**: p. 148.
18. Karn, B., T. Kuiken, and M. Otto, Nanotechnology and in situ remediation: a review of the benefits and potential risks. *Environmental Health Perspectives*, 2009. **117**(12): p. 1813–1831.

19. Wiesner, M.R., et al., Assessing the risks of manufactured nanomaterials. *Environmental Science and Technology*, 2006. **40**(14): p. 4336–45.

20. Stander, L. and L. Theodore, Environmental implications of nanotechnology: an update. *International Journal of Environmental Research and Public Health*, 2011. **8**(2): p. 470–479.

21. Dutta, S., V.P. Upadhyay, and U. Sridharan, Environmental management of industrial hazardous wastes in India. *Journal of Environmental Science and Engineering*, 2006. **48**: p. 143–50.

22. Ojuederie, O.B. and O.O. Babalola, Microbial and plant-assisted bioremediation of heavy metal polluted environments: a review. *International Journal of Environmental Research and Public Health*, 2017. **14**(12): p. 1504.

23. Yadav, K.K., et al., Bioremediation of heavy metals from contaminated sites using potential species: a review. *Indian Journal of Environmental Protection*, 2017. **37**(1): p. 65.

24. Jan, A.T., et al., Prospects for exploiting bacteria for bioremediation of metal pollution. Critical Reviews in Environmental Science and Technology, 2014. **44**(5): p. 519–560.

25. Dhankher, O.P., et al., Biotechnological approaches for phytoremediation. In *Plant Biotechnology and Agriculture*. 2012, Elsevier: Amsterdam. p. 309–328.

26. Reddy, M.S., et al., Biodegradation of phenanthrene with biosurfactant production by a new strain of Brevibacillus sp. Bioresource *Technology*, 2010. **101**(20): p. 7980–7983.

27. Makkar, R.S. and K.J. Rockne, Comparison of synthetic surfactants and biosurfactants in enhancing biodegradation of polycyclic aromatic hydrocarbons. Environmental Toxicology and Chemistry: An International Journal, 2003. **22**(10): p. 2280–2292.

28. Chekroun, K.B., E. Sánchez, and M. Baghour, The role of algae in bioremediation of organic pollutants. *International Research Journal of Public and Environmental Health*, 2014. **12**: p. 19–23.

29. Umadevi, S., P. Ayyasamy, and S. Rajakumar, Biological perspective and role of bacteria in pesticide degradation. In *Bioremediation and Sustainable Technologies for Cleaner Environment*. 2017, Springer: Singapore. p. 3–12.

30. Gupta, P. and B. Diwan, Bacterial exopolysaccharide mediated heavy metal removal: a review on biosynthesis, mechanism and remediation strategies. *Biotechnology Reports*, 2017. **13**: p. 58–71.

31. Coelho, L.M., et al., Bioremediation of Polluted Waters Using Microorganisms. Advances in Bioremediation of Wastewater and Polluted Soil, Naofumi Shiomi, IntechOpen, 2015. DOI: 10.5772/60770.

32. Kang, J.W., Removing environmental organic pollutants with bioremediation and phytoremediation. Biotechnol Lett, 2014. **36**(6): p. 1129–1139.

33. Lloyd, J.R. and L.E. Macaskie, Bioremediation of radionuclide-containing wastewaters. In *Environmental Microbe-Metal Interactions*. 2000, John Wiley & Sons, USA. p. 277–327.

34. Singh, H., *Mycoremediation: Fungal Bioremediation*. 2006, Wiley: Hoboken, NJ.

35. Yamada-Onodera, K., et al., Degradation of polyethylene by a fungus, Penicillium simplicissimum YK. Polym Degrad Stab, 2001. **72**(2): p. 323–327.

36. Bredberg, K., et al., Microbial detoxification of waste rubber material by wood-rotting fungi. Bioresour Technol, 2002. **83**(3): p. 221–224.

37. Sánchez, C., Lignocellulosic residues: biodegradation and bioconversion by fungi. Biotechnol Adv, 2009. **27**(2): p. 185–194.

38. Kadarmoidheen, M., P. Saranraj, and D. Stella, Effect of cellulolytic fungi on the degradation of cellulosic agricultural wastes. Int J App Microbiol Sci, 2012. **1**(2): p. 13–23.

39. Punnagaiarasi, A., et al., Bioremediation: a ecosafe approach for dairy effluent treatment. In *Bioremediation and Sustainable Technologies for Cleaner Environment*. 2017, Springer: Singapore. p. 45–50.

40. Kües, U., Fungal enzymes for environmental management. Curr Opinion Biotechnol, 2015. **33**: p. 268–278.

41. Singh, R.K., et al., Fungi as potential candidates for bioremediation. In *Abatement of Environmental Pollutants*. 2020, Elsevier: Amsterdam. p. 177–191.

42. Kaur, I. and A. Bhatnagar, Algae-dependent bioremediation of hazardous wastes. *Progress in Industrial Microbiology*, 2002. **36**: p. 457–516.

43. Cruz-Uribe, O. and G.L. Rorrer, Uptake and biotransformation of 2, 4, 6-trinitrotoluene (TNT) by microplantlet suspension culture of the marine red macroalga *Portieria hornemannii*. *Biotechnology and Bioengineering*, 2006. **93**(3): p. 401–412.

44. Cruz-Uribe, O., D.P. Cheney, and G.L. Rorrer, Comparison of TNT removal from seawater by three marine macroalgae. Chemosphere, 2007. **67**(8): p. 1469–1476.

45. Munoz, R., et al., Phenanthrene biodegradation by an algal-bacterial consortium in two-phase partitioning bioreactors. Appl Microbiol Biotechnol, 2003. **61**(3): p. 261–267.
46. Arivoli, A., T. Sathiamoorthi, and M. Satheeshkumar, Treatment of textile effluent by phytoremediation with the aquatic plants: *Alternanthera sessilis*. In *Bioremediation and Sustainable Technologies for Cleaner Environment*. 2017, Springer: Singapore. p. 185–197.
47. Olette, R., et al., Toxicity and removal of pesticides by selected aquatic plants. Chemosphere, 2008. **70**(8): p. 1414–1421.
48. Pulford, I. and C. Watson, Phytoremediation of heavy metal-contaminated land by trees: a review. Environment Int, 2003. **29**(4): p. 529–540.
49. Chibuike, G.U. and S.C. Obiora, Heavy metal polluted soils: effect on plants and bioremediation methods. Appl Environmen Soil Sci, 2014. 2014:752708.
50. Angelova, V.R., et al., Potential of sunflower (Helianthus annuus L.) for phytoremediation of soils contaminated with heavy metals. Int J Environmen Ecol Eng, 2016. **10**(9): p. 1–8.
51. Zhang, Z., et al., Higher accumulation capacity of cadmium than zinc by Arabidopsis halleri ssp. Germmifera in the field using different sowing strategies. Plant Soil, 2017. **418**(1–2): p. 165–176.
52. Prakash, S., et al., The role of decomposer animals in bioremediation of soils. In *Bioremediation and Sustainable Technologies for Cleaner Environment*. 2017, Springer: Singapore. p. 57–64.
53. Girija, J., et al., Stabilization of market vegetable waste through the process of vermicomposting by *Eisenia foetida*. In *Bioremediation and Sustainable Technologies for Cleaner Environment*. 2017, Springer: Singapore. p. 35–43.
54. Mall, A., A. Dubey, and S.J.A.N. Prasad, Vermicompost: an inevitable tool of organic farming for sustainable agriculture. Agrobios News-letter, 2005. **3**(8): p. 10–12.
55. Kumar, V., S. Shahi, and S. Singh, Bioremediation: an eco-sustainable approach for restoration of contaminated sites. In *Microbial Bioprospecting for Sustainable Development*. 2018, Springer: Singapore. p. 115–136.
56. Dzionek, A., D. Wojcieszyńska, and U. Guzik, Natural carriers in bioremediation: a review. *Electronic Journal of Biotechnology*, 2016. **23**: p. 28–36.
57. Kapahi, M. and S. Sachdeva, Bioremediation options for heavy metal pollution. J Health Pollut, 2019. **9**(24): p. 191203.
58. Abatenh, E., et al., The role of microorganisms in bioremediation: a review. *Open Journal of Environmental Biology*, 2017. **1**: p. 038–040.
59. Moxley, E., et al., Influence of abiotic factors temperature and water content on bacterial 2-chlorophenol biodegradation in soils. Front Environ Sci, 2019. **7**: p. 41.
60. Goulding, K.W., Soil acidification and the importance of liming agricultural soils with particular reference to the United Kingdom. *Soil Use and Management*, 2016. **32**(3): p. 390–399.
61. Alkorta, I., L. Epelde, and C. Garbisu, Environmental parameters altered by climate change affect the activity of soil microorganisms involved in bioremediation. FEMS Microbiol Lett, 2017. **364**(19): p. fnx200.
62. Rashid, M.I., et al., Bacteria and fungi can contribute to nutrients bioavailability and aggregate formation in degraded soils. Microbiol Res, 2016. **183**: p. 26–41.
63. Maletić, S., B. Dalmacija, and S. Rončević, Petroleum hydrocarbon biodegradability in soil: implications for bioremediation. IntechOpen, 2013. 2013: p. 43–51.
64. Juwarkar, A.A., et al., A comprehensive overview of elements in bioremediation. Rev Environ Sci Biotechnol, 2010. **9**(3): p. 215–288.
65. Tarekegn, M.M., et al., Microbes used as a tool for bioremediation of heavy metal from the environment. Food Sci Technol, 2020. **6**(1): p. 1783174.
66. Crini, G., et al., Adsorption-oriented processes using conventional and non-conventional adsorbents for wastewater treatment. In *Green Adsorbents for Pollutant Removal*. 2018, Springer: Singapore. p. 23–71.
67. Janssen, D.B. and G. Stucki, Perspectives of genetically engineered microbes for groundwater bioremediation. Environ Sci Processes Imp, 2020. **22**(3): p. 487–499.
68. Sharma, J.K., et al., Advances and perspective in bioremediation of polychlorinated biphenyl-contaminated soils. Environ Sci Pollut Res, 2018. **25**(17): p. 16355–16375.
69. Pandey, J. and A. Chauhan, and R.K. Jain, Integrative approaches for assessing the ecological sustainability of in situ bioremediation. FEMS Microbiology Reviews, 2009. **33**(2): p. 324–375.
70. Megharaj, M., et al., Bioremediation approaches for organic pollutants: a critical perspective. Env Int, 2011. **37**(8): p. 1362–1375.
71. Kumar, S., et al., Genetically modified microorganisms (GMOs) for bioremediation. In *Biotechnology for Environmental Management and Resource Recovery*. 2013, Springer: Singapore. p. 191–218.

72. Rao, M., et al., Role of enzymes in the remediation of polluted environments. J Soil Sci Plant Nutr, 2010. **10**(3): p. 333–353.
73. Dangi, A.K., et al., Bioremediation through microbes: systems biology and metabolic engineering approach. Crit Rev Biotechnol, 2019. **39**(1): p. 79–98.
74. Wu, J., et al., Nanomaterials with enzyme-like characteristics (nanozymes): next-generation artificial enzymes (II). Chem Society Rev, 2019. **48**(4): p. 1004–1076.
75. Meng, Y., et al., Applications of nanozymes in the environment. Env Sci Nano, 2020. **7**(5): p. 1305–1318.
76. Margesin, R. and F. Schinner, *Manual for Soil Analysis-Monitoring and Assessing Soil Bioremediation.* Vol. 5. 2005: Springer: Singapore.
77. Rike, A.G., et al., In situ biodegradation of petroleum hydrocarbons in frozen arctic soils. Cold Reg Sci Tech, 2003. **37**(2): p. 97–120.
78. Aichberger, H., et al., Potential of preliminary test methods to predict biodegradation performance of petroleum hydrocarbons in soil. Biodegradation, 2005. **16**(2): p. 115–125.
79. Delille, D., B. Delille, and E. Pelletier, Effectiveness of bioremediation of crude oil contaminated subantarctic intertidal sediment: the microbial response. Microb Ecol, 2002. **44**(2): p. 118–126.
80. Knapp, R.B. and B.D. Faison, A bioengineering system for in situ bioremediation of contaminated groundwater. J Ind Microbiol Biotechnol, 1997. **18**(2–3): p. 189–197.
81. Salanitro, J.P., et al., Field-scale demonstration of enhanced MTBE bioremediation through aquifer bioaugmentation and oxygenation. Environ Sci Technol, 2000. **34**(19): p. 4152–4162.
82. Hunkeler, D., et al., Engineered in situ bioremediation of a petroleum hydrocarbon-contaminated aquifer: assessment of mineralization based on alkalinity, inorganic carbon and stable carbon isotope balances. J cont Hydro, 1999. **37**(3–4): p. 201–223.
83. Sayler, G.S. and S. Ripp, Field applications of genetically engineered microorganisms for bioremediation processes. Curr Opin Biotechnol, 2000. **11**(3): p. 286–289.
84. Mohan, S.V., et al., Ex situ bioremediation of pyrene contaminated soil in bio-slurry phase reactor operated in periodic discontinuous batch mode: influence of bioaugmentation. Int Biotet Biodegrad, 2008. **62**(2): p. 162–169.
85. Sayara, T., et al., Anaerobic bioremediation of PAH-contaminated soil: assessment of the degradation of contaminants and biogas production under thermophilic and mesophilic conditions. *Environmental Engineering and Management Journal*, 2015. **14**: p. 153–165.
86. Robles-González, I.V., F. Fava, and H.M. Poggi-Varaldo, A review on slurry bioreactors for bioremediation of soils and sediments. *Microbial Cell Factories*, 2008. **7**: p. 5–5.
87. Kumar, A., et al., Review on bioremediation of polluted environment: a management tool. Int J Environ Sci, 2011. **1**(6): p. 1079–1093.
88. Pavel, L.V. and M. Gavrilescu, Overview of ex situ decontamination techniques for soil cleanup. 2008. Environmental Engineering and Management Journal, 2008. **7**(6): p. 35993214.
89. Ali, H., E. Khan, and M.A.J.C. Sajad, Phytoremediation of heavy metals: concepts and applications. *Chemosphere*. 2013. **91**(7): p. 869–881.
90. Belimov, A., et al., Cadmium-tolerant plant growth-promoting bacteria associated with the roots of Indian mustard (Brassica juncea L. Czern.). Soil Biology and Biochemistry. 2005. **37**(2): p. 241–250.
91. Golan-Goldhirsh, A., Plant tolerance to heavy metals, a risk for food toxicity or a means for food fortification with essential metals: the Allium schoenoprasum model. In *Soil and Water Pollution Monitoring, Protection and Remediation*. 2006, Springer: Singapore. p. 479–486.
92. Alaribe, F. and P. Agamuthu, Assessment of phytoremediation potentials of Lantana camara in Pb impacted soil with organic waste additives. Ecological Engineering. 2015. **83**: p. 513–520.
93. Ghnaya, T., et al., Nodulation by Sinorhizobium meliloti originated from a mining soil alleviates Cd toxicity and increases Cd-phytoextraction in Medicago sativa L. Frontiers in *Plant Science*. 2015. **6**: p. 863.
94. Ping, L., et al., Effects of several amendments on rice growth and uptake of copper and cadmium from a contaminated soil. *Journal of Environmental Sciences*. 2008. **20**(4): p. 449–455.
95. Garg, N., et al., Metal uptake, oxidative metabolism, and mycorrhization in pigeonpeaand pea under arsenic and cadmium stress. *Turkish Journal of Agriculture and Forestry*. 2015. **39**(2): p. 234–250.
96. Ariste, A.F. and H. Cabana, Challenges in applying cross-linked laccase aggregates in bioremediation of emerging contaminants from municipal wastewater. In *Laccases in Bioremediation and Waste Valorisation*. 2020, Springer: Singapore. p. 147–171.
97. Alegbeleye, O.O., B.O. Opeolu, and V.A.J.E.m. Jackson, Polycyclic aromatic hydrocarbons: a critical review of environmental occurrence and bioremediation. *Environmental Management*. 2017. **60**(4): p. 758–783.
98. Raman, J.K. and E. Gnansounou, A review on bioremediation potential of vetiver grass. In *Waste Bioremediation*. 2018, Springer: Singapore. p. 127–140.

15 Viability of Membrane Separation Techniques in the Detoxification of Recalcitrant Pollutants

Joydeep Das and Soma Nag

CONTENTS

15.1 INTRODUCTION

Water is an essential element to any living being, and life cannot be imagined in the absence of water. With the rapid increase in population, gallons of wastewater are released in freshwater bodies each day from various industries, agricultural paddy fields, oil refineries, and domestic households. However, very few attempts have been made to replenish freshwater bodies, which would meet the water requirements of an ever-increasing population. This neglig ence of managerial authorities has led to a shortage of usable water and has opened up unfair competition and uneven distribution among the various industries and the agricultural sector is not excluded. Proof of this scenario is observed around the world, particularly within Africa and its adjacent Middle Eastern countries. The details are brutal: approximately 2 billion people are residing with no safe water for consumption at home and, apart from this, almost 4 billion individuals encounter severe drinking water scarcity for more than one month in each calendar year (Pavon, 2020; United Nations, 2019).

The formation of wastewater can not be avoided as it a fundamental component in every processing unit, including the manufacturing industry and the power sector. The formation of domestic wastewater cannot be avoided either. In any refinery, each barrel of crude oil to be refined costs around ten barrels of wastewater (Pendashteh et al., 2011). Wastewater is nothing but clean water contaminated with toxic elements and impure substances. The implementation of a systematized water treatment technology appears to be the most palpable means of combating water scarcity. In this vein, a number of attempts have been made over the years to counter the stumbling block, such as traditional filtration techniques, coagulation, and flocculation alongside biological treatments. In

DOI: 10.1201/9781003204442-15

addition, attempts have been made at improving currently existing solutions to meet the standard discharge limit of the released effluents.

Among the wastewater treatment solutions available, which has seen a significant increase over the course of time, is membrane-based water treatment technology. In the past few years, membrane technology has made notable headway due to its beneficiation ability in water and wastewater treatment with a substantial reduction in equipment size, energy consumption, and low capital investment. Membrane processing provides numerous options in the handling of wastewater (Quist-Jensen et al., 2015). In recent times, membrane technology has gained considerable attention in the treatment of groundwater as an alternative to many age-old methods (Bódalo-Santoyo et al., 2003; Luo et al., 2012; Ochando-Pulido et al., 2012). The technology is well equipped for the reclamation of water from various sources, especially from agricultural-based wastewater (Stoller, 2011; Stoller and Bravi, 2010; Stoller, 2009). Membrane engineering has the ability to overthrow the high chemical consumption, which is common in most of the separation processes, and can provide an eco-friendly, sustainable solution to a sheer portion of population. In recent times, especially for wastewater treatment processes, membrane technology has proved to be a much more positive course of action in comparison to traditional approaches.

Although membrane engineering is not a brand new concept, the altering complexity and nature of wastewater tend to make space for even more enhancements in regard to effectiveness, space needs, power, quality of permeate, along with specialized abilities requirements. Once again, the constant evolution of membrane modules and membrane elements has become a major challenge for membrane processes to improve the significant drop in membrane fouling. The potential of amalgamating two or maybe more membrane systems with one another, or perhaps along with other types of methodologies such as adsorption, coagulation, or flocculation in a hybrid manner, is continually being investigated, established, and used in numerous water remediation operations (Hankins and Singh, 2016; Stoquart et al., 2012). As of late, the requirements for sustainable resources have quickened the advancement of membrane technology-based research in the scholarly world, enterprises, and laboratories subsidized by the government. The application of membrane technological innovation in wastewater treatment is reviewed in this chapter. It also considers the pros and cons of this process.

15.2 CLASSIFICATION OF MEMBRANE PROCESSES

A basic understanding of membrane processes is significant for the improvement of membrane technology in water remediation. The word membrane is coined from the Latin word *membrane*, meaning "a skin, membrane, or parchment". Its literal meaning in the realm of purification is the obstruction to segregate particles from the same or different phases and allow the selective passage of specific components. Membranes may be categorized by their composition, design, structure, preparation techniques, and processes.

In this chapter, we will emphasize on the classification of membrane processes in the light of wastewater treatment. In the domain of water treatment, there are numerous ways to categorize the membrane process, which is shown diagrammatically in Figure 15.1. Pressure-driven membrane procedures, especially high pressure- and low pressure-driven processes, are undoubtedly the most commonly used processes using membranes in the field of water purification, starting from effluents to decontaminated water.

Microfiltration (MF), ultrafiltration (UF), nanofiltration (NF), and reverse osmosis (RO) are the four primary types of pressure-driven membrane filtrations distinguished according to pore measurements (see Table 15.1). The details about these filtration techniques along with the nature of particles to be separated is illustrated in Figure 15.2.

Membrane pore sizes range from 0.1–10 µm in the microfiltration membrane. They are used primarily for the removal of large living or non-living objects from large feedstock solutions. In the

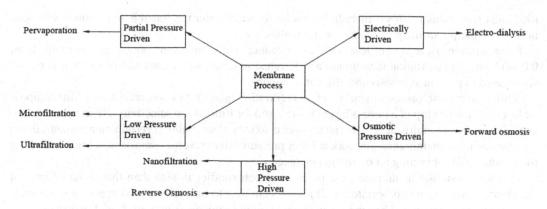

FIGURE 15.1 A chart of various membrane processes.

TABLE 15.1

Overview of Acknowledged Membrane Technologies (Banerjee et al., 2018; Basile et al., 2015)

Method	Microfiltration	Ultrafiltration	Nanofiltration	Reverse Osmosis
Pressure (in bar)	0.1–3	0.5–10	2–40	7–75
Types particles separated	Suspended solids, microorganisms	Colloidal particles	Nano-sized pollutants, glucose	Different salts
Mechanism	Molecular sieve	Molecular sieve	Solution diffusion	Solution diffusion
Membrane category	Symmetric polymer, ceramic membranes	Asymmetric polymer composite, ceramic membrane	Asymmetric polymer, Thin-film composite membrane	Thin-film composite membrane

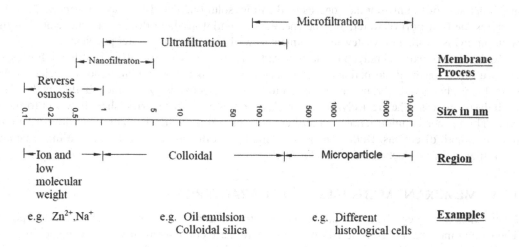

FIGURE 15.2 Pressure-driven membrane separation processes along with specific particle sizes.

food and drink industry, this is incredibly popular for wastewater treatment before being discharged into the municipal drainage system or water bodies.

Ultrafiltration is a microfiltration-like procedure with narrower pore sizes varying from 0.01–0.1 μm. Ultrafiltration membranes are designed to eradicate viruses and proteins. It is extensively used in protein synthesis and effluent treatment.

Similar to reverse osmosis membranes, nanofiltration membranes comprise a thin-film composite layer (<1 μm) on top of a porous layer (50–150 μm) for limited ion selectivity. NF membranes are capable of refusing multivalent salts and non-polar solutes, thus enabling certain monovalent salts to pass through. NF membranes can work at lower pressures than reverse osmosis membranes, making them suitable for obtaining an optimum outcome.

In reverse osmosis membranes, the pores are much smaller in size than that of nanofiltration membranes, making it appropriate for all monovalent ions to be excluded and enabling water molecules to move through it. They may even extract microorganisms from bulk feed. Reverse osmosis filtration is mostly used in softening brackish water and effluent disposal. It is necessary to remember that the average yield is comparatively smaller than that of MF and UF membranes as the operating pressure for RO and NF is far greater than the pressure utilized by MF and UF ("Synder filtration," 2020).

Out of all these membrane processes, reverse osmosis is well regarded for its proficiency to eliminate tiny particles like microorganisms, metal ions, etc., with over 99% productivity (Muro et al., 2012). Different treatment operations have incorporated several combinations of the above-mentioned pressure-driven membrane processes. In many instances, they end up serving as preliminary treatments to many other treatment systems. Nataraj et al. (2006) developed a fused system of NF and RO in a study to process a distillery effluent with an average removal percentage of over 95% of toxic substances including total dissolved solid (TDS) and chemical oxygen demand (COD). In another experiment, a pilot-scale plant merged UF and RO to tackle wastewater from the dye industry. The permeate did not meet discharge standards after UF, but the RO permeate was able to reach the discharge criteria and was also fit for reuse. Contaminants, including urea, sodium alginate, reactive dye, and oxidants were effectively extracted (Šostar-Turk et al., 2005).

In forward osmosis (FO), a semipermeable membrane detaches two solutions, and the concentration difference of the solutions, i.e., osmotic pressure, is the chief motivating force in this process (Cath et al., 2006; Shaffer et al., 2015). Between the two differential concentrated solutions on either side of the membrane, the less concentrated solution is known as the feed solution, whereas the draw solution is more concentrated. The active side of the membrane is confronted by the feed solution, and other side of the membrane consists of the draw solution. The difference in osmotic pressure operates the transport of water, and the chemical potential gradient decides the movement of the solution. FO is applied in wastewater treatment as well as in drinking water purification.

To separate or extract charged particles from feed solutions, electrodialysis is utilized. It is especially utilized for the desalination of water as electrodialysis is nothing but removal of electrolytic salts. Ion exchange membranes are arranged to form an electrodialyzer, which facilitates the electrodialysis process. The electrolyzer can be classified into two categories: sheet flow and tortuous type. The former is unidirectional in a direct current and the latter is controlled with direct current from both directions. Both systems have highly specific characteristics and are implemented according to the product criteria (Tanaka, 2015).

15.3   MEMBRANE MATERIALS AND CHARACTERISTICS

Membranes are categorized as isotropic or anisotropic, especially for characterization purposes. The membranes that are considered isotropic in nature are compositionally and mechanically standardized throughout the cross-section. Anisotropic membranes are inhomogeneous throughout the membrane sample, usually consisting of surfaces differing in shape and chemical structure. Again, isotropic membranes are subdivided into separate subcategories. Isotropic membranes, for example, can be microporous. Microporous membranes are mostly constructed of unbending polymer composites with large porosity and interconnected with the pores (Baker, 2012). Phase inversion membranes are the most popular type of microporous membrane. They are formed by casting polymer film on a solvent by plunging the casted film in a specific solvent which acts as a non-solvent for the polymer. The majority of polymers employed for such practices are hydrophobic, with rendering water being the most popular non-solvent (Smith, 1910). The polymer begins to form a membrane by precipitation when coming into contact with the aqueous medium. There is another type of microporous membrane which is known as a track-etched membrane. This special type of membrane is produced when charged polymer film strikes the parent chain and broken molecules are left behind. It leads to the formation of a membrane after dissolving this mangled portion of molecules into an etching solution through which the film is passed. The shape of the membranes produced through this process is almost cylindrical. Expanded film membranes are another type of microporous membrane which are barely used. In this type of membrane, pores or voids are generated by extrusion and elongation, respectively.

There are two major forms of anisotropic membranes: membranes for phase separation and thin-film membrane composites. The anisotropic phase separation membranes are also known as Loeb–Sourirajan membranes, named after the individuals who introduced them. The structural homogeneity is not found in this type of membrane. This membrane is produced by employing the phase inversion method, which yields variant dimensions of pore and porosity throughout the thickness of the membrane. The thin-film membrane is another type of anisotropic membrane which is heterogeneous in composition as well as structurally. This thin film comprises significantly porous materials in which the outside of the surface is coated with some other polymer. They can be produced by different methods, namely plasma polymerization, interfacial polymerization, surface treatment, or solvent coating – the explanation of the aforesaid membranes or of flat sheet configuration. Nevertheless, the hollow configuration is also there for isotropic or anisotropic membranes, which is produced more than in the flat sheet because of a high surface-to-volume ratio. For this reason, anisotropic membranes are practical for use in real-world scenarios (El-Ghaffar and Tieama, 2017).

A large portion of the membrane is produced from synthetic polymers derived from organic compounds. Generally, MF and UF membranes are made out of identical materials, but the preparation procedure is different for each of them as the process has to yield a product of variant pore sizes ("American chemical society," 1986). The polymeric materials mostly used for the preparation of MF and UF are poly vinylidene fluoride (PVF), poly acrylonitrile (PAN), and poly vinyl chloride (PVC). Poly tetrafluoroethylene (PTFE), cellulose acetate, nitrate blends, and nylons can be used as raw materials for MF manufacturing. Cellulose acetate and aromatic polyamide-coated polysulfone are regular materials utilized for RO membrane production. Similarly, cellulose acetate mixtures or polyamides can also be used to construct NF membranes (Nunes and Peinemann, 2006). Inorganic substances like ceramic or metals can also be used as raw material for membrane construction because of their unique features such as thermal stability, chemical resistivity, and ability to form highly porous structures. Among the metals, stainless steel is the most utilized raw material for membrane as it enables the production of uniformly porous textures and can withstand high temperatures. This type of membrane is generally used in gas separation and drinking water filtration processes ("Product Information: Pall Corporation Inorganic membranes," 2004).

15.4 APPLICATION IN WASTEWATER TREATMENT

The disposal of large volumes of heavy metals and pigments into water bodies is a matter of serious concern because it gives rise to dangerous threats to human health. Heavy metals are one of the most hazardous components and are mostly found in industrial effluents. The most carcinogenic heavy metals like Cr(VI), Cd(II), As(III), Pb(II), and some others induce significant contamination and health concerns. Due to their sharp sieving abilities and exposed charged surfaces, nanofiltration and reverse osmosis can be used to effectively eliminate these metal ions (Nedzarek et al., 2015). Due to high energy consumption and poor efficiency, attempts were made to adjust ultrafiltration and microfiltration membrane processes to increase membrane selectivity against toxic metal ions. As an example, a study has been performed to improve the efficiency of ultrafiltration towards the removal of Cr^{6+} ions by coupling it with TiO_2 nanoparticles (Gebru and Das, 2018). The nanoparticles created more functioning spots and increased the residence period of the metal ions on the membrane surface. Consequently, the high adsorption capacity of the metal ions on the suitable membrane surface was observed.

FO is another in-demand method used for heavy metal removal, which uses less energy to get the work done when compared with the other processes. Liu et al. (2013) used an efficient technique called the layer-by-layer method to develop a polymer network of FO membranes. The resulting FO membranes have been checked for their efficiency to remove heavy metals, i.e., Pb^{2+}, Cd^{2+}, Cu^{2+}, Zn^{2+}, Ni^{2+}, from aqueous solutions. The high elimination rate by the FO membrane resulted in a high optimal number of bilayers that were formed with adequate thickness and compactness to prevent the passage of heavy metal ions without sacrificing the flow. A maximum elimination rate of 99.3% was achieved. The electrostatic repulsive force between the metal ions and the surface developed a barrier for the rejection of incoming heavy metal ions (Liu and Cheng, 2013).

Subhramaniam et al. (2018) have introduced a photocatalytic polyvilylidene fluoride (PVDC) hollow fiber membrane integrated with different loadings of tubular-shaped titanium dioxide nanotubes (TNTs) where the function of the TNT is to enhance the hydrophilicity of the membrane (Subramaniam et al., 2018). The resultant nanocomposite was used to decolorize the effluent originating from the palm oil industry, which looked brown in color and showed a maximum filtration capacity with a color removal efficiency of 59%.

15.5   MEMBRANE FOULING AND PRETREATMENT

Membrane fouling happens because of the accumulation of various organic compounds, suspended solids, bacteria, etc., on the membrane surface as well as in the pores, resulting in declined permeate flux (Speth et al., 1998). Fouling can be of two types based on the mechanism, namely reversible and irreversible. In reversible fouling, residues accumulate into the pores of the membrane. In the case of irreversible fouling, foulants posited on the film surface affects the movement of the permeate through the membrane and forms a layer of cake on the surface (Guo et al., 2012). The fouling influences the efficiency of the membrane as permeate circulation is significantly hampered. For the transit of permeates through the membrane, higher than usual pressure is required. Higher pressures are required for increases in fouling (Kucera, 2010). Membrane fouling has negative consequences for the overall efficiency of the membrane as it causes a reduced flow rate through the membrane, a high energy usage, a slower filtration speed, etc. (Le-Clech et al., 2006). Based on the foulant, there are multiple forms of fouling: colloidal fouling, chemical fouling, and organic and inorganic fouling (Amy, 2008). It must be borne in mind that membrane fouling is largely driven by the pH of the solution, the acidic or basic strength, the concentration of the feed, the membrane morphology, and the operating conditions. All these parameters interact in some way to promote membrane fouling (Li and Elimelech, 2004; Zhao et al., 2000).

The pretreatment of feed wastewater is a much-needed step before membrane separation procedures are applied. The pretreatments not just done to minimize fouling of the membrane, it also brings down the process's energy consumption (Huang et al., 2009). Different approaches are applied for the pretreatment of effluent or feed, including coagulation, softening, and adsorption (Tong et al., 2019).

Prefiltration is another method of pretreatment to the feed and is accomplished by a low-pressure membrane, filter clothes, or using packed bed filters (Huang et al., 2009). Dissolved air flotation and biological processes are extensively-used pretreatment techniques (Huang et al., 2009; Kim et al., 2015;

Riley et al., 2018, 2016; Shi et al., 2017; Tong et al., 2019; Zhao et al., 2000). Whichever pretreatment process is to be implemented will depends upon the characteristics of the feed.

15.6 ADVANTAGES AND DISADVANTAGES

The advantages and disadvantages of membrane separation techniques are listed below.

15.6.1 ADVANTAGES

a. The appropriate membrane can produce better quality, potable water.
b. Less amounts of additives are used during the purification process.
c. The energy required during the process and maintenance are comparatively less.
d. The process facilitates the development and construction of a simpler device.
e. It has the potential of generating safe and uniform quality water.
f. It has the capacity to extract a significant variety of contaminants.
g. The issue of variant raw water quality is supplemented by membrane operation.
h. The removal efficiency of the membrane is independent of flux and pressure for particle dimensions larger than membrane pores.
i. The deposition of large particles on the surface of the membrane filter can resist the flow of smaller particles through it.

15.6.2 DISADVANTAGES

a. Disinfection of water by membrane leads to the growth of microorganisms, which cannot be ignored. At the same time, this issue can be resolved by increasing the chlorine dosage (Madsen, 1987).
b. The ability to hold tiny species like a virus or bacteria relies on the operating conditions and nature of the solutions, which are highly unpredictable.
c. Liquid that flows through the pores is easily hindered owing to fouling, which is shown in Figure 15.3.

15.7 CONCLUSION AND FUTURE PROSPECTS

Pretty much across the board, it can be stated that membrane processing is a very effective option for the handling of wastewater. However, noteworthy developmental steps are required for shifting membrane technology from laboratory to feasible industrial scale. These recommendations are listed below.

1. The development of a membrane module that can provide a high surface area that should be efficiently functional.
2. The membrane module should be enabled to separate identical particles with specific selectivity.
3. Morphological simulations to control the microscopic transport of the permeate.
4. The development of membranes by economical methods using affordable and readily available materials.

The scope of membrane technology in wastewater treatment is infinite. This chapter sought to outline all the aspects of membrane technology broadly, citing instances of their usages. Expectantly, the information contained in this chapter will provide further insight into the application of membrane-based wastewater management.

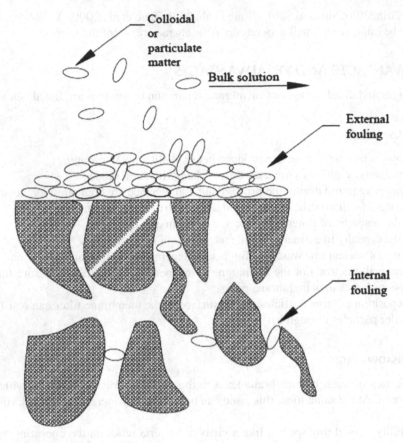

FIGURE 15.3 Fouling mechanism on the membrane surface.

REFERENCES

American Chemical Society, 1986. *Colloids and Surfaces*. American Chemical Society, Washington, DC. https://doi.org/10.1016/0166-6622(86)80276-1

Amy, G., 2008. Fundamental understanding of organic matter fouling of membranes. *Desalination*, 231(1–3), 44–51. https://doi.org/10.1016/j.desal.2007.11.037

Baker, R.W., 2012. *Membrane Technology and Applications, Membrane Technology and Applications*. Wiley, Hoboken, NJ. https://doi.org/10.1002/9781118359686

Banerjee, P., Das, R., Das, P., Mukhopadhyay, A., 2018. Membrane technology. *Carbon Nanostructures*, Springer International Publishing, 127–150. https://doi.org/10.1007/978-3-319-95603-9_6

Basile, A., Cassano, A., Rastogi, N.K., 2015. *Advances in Membrane Technologies for Water Treatment: Materials, Processes and Applications*. Elsevier, Amsterdam. https://doi.org/10.1016/C2013-0-16469-0

Bódalo-Santoyo, A., Gómez-Carrasco, J.L., Gómez-Gómez, E., Máximo-Martín, F., Hidalgo-Montesinos, A.M., 2003. Application of reverse osmosis to reduce pollutants present in industrial wastewater. *Desalination*. https://doi.org/10.1016/S0011-9164(03)00287-X

Cath, T.Y., Childress, A.E., Elimelech, M., 2006. Forward osmosis: Principles, applications, and recent developments. *J. Membr. Sci.* https://doi.org/10.1016/j.memsci.2006.05.048

El-Ghaffar, M.A.A., Tieama, H.A., 2017. *A Review of Membranes Classifications, Configurations, Surface Modifications, Characteristics and Its Applications in Water Purification*. Http://Www.Sciencepublishinggroup.Com 2, 57. https://doi.org/10.11648/J.CBE.20170202.11

Gebru, K.A., Das, C., 2018. Removal of chromium (VI) ions from aqueous solutions using amine-impregnated TiO2 nanoparticles modified cellulose acetate membranes. *Chemosphere*. https://doi.org/10.1016/j.chemosphere.2017.10.107

Guo, W., Ngo, H.H., Li, J., 2012. A mini-review on membrane fouling. *Bioresour. Technol.* https://doi.org/10.1016/j.biortech.2012.04.089

Hankins, N.P., Singh, R., 2016. *Emerging Membrane Technology for Sustainable Water Treatment*, Elsevier Science; 1st edition (March 31, 2016). https://doi.org/10.1016/C2012-0-07949-5

Huang, H., Schwab, K., Jacangelo, J.G., 2009. Pretreatment for low pressure membranes in water treatment: A review. *Environ. Sci. Technol.* https://doi.org/10.1021/es802473r

Kim, Y., Choi, D., Cui, M., Lee, J., Kim, B., Park, K., Jung, H., Lee, B., 2015. Dissolved air flotation separation for pretreatment of membrane bioreactor in domestic wastewater treatment. *J. Water Supply Res. Technol.: AQUA.* https://doi.org/10.2166/aqua.2014.003

Kucera, J., 2010. *Reverse Osmosis, Industrial Applications and Processes.* Wiley, Hoboken, NJ.

Le-Clech, P., Chen, V., Fane, T.A.G., 2006. Fouling in membrane bioreactors used in wastewater treatment. *J. Membr. Sci.* https://doi.org/10.1016/j.memsci.2006.08.019

Li, Q., Elimelech, M., 2004. Organic fouling and chemical cleaning of nanofiltration membranes: Measurements and mechanisms. *Environ. Sci. Technol.* https://doi.org/10.1021/es0354162

Liu, C., Cheng, H.M., 2013. Carbon nanotubes: Controlled growth and application. *Mater. Today.* https://doi.org/10.1016/j.mattod.2013.01.019

Luo, J., Ding, L., Wan, Y., Jaffrin, M.Y., 2012. Threshold flux for shear-enhanced nanofiltration: Experimental observation in dairy wastewater treatment. *J. Membr. Sci.* https://doi.org/10.1016/j.memsci.2012.03.065

Madsen, R.F., 1987. Membrane technology as a tool to prevent dangers to human health by water-reuse. *Desalination.* https://doi.org/10.1016/0011-9164(87)90257-8

Muro, C., Riera, F., Carmen Diaz, M. del, 2012. Membrane separation process in wastewater treatment of food industry. In: *Food Industrial Processes: Methods and Equipment.* https://doi.org/10.5772/31116

Nataraj, S.K., Hosamani, K.M., Aminabhavi, T.M., 2006. Distillery wastewater treatment by the membrane-based nanofiltration and reverse osmosis processes. *Water Res.* https://doi.org/10.1016/j.watres.2006.04.022

Nedzarek, A., Drost, A., Harasimiuk, F.B., Tórz, A., 2015. The influence of pH and BSA on the retention of selected heavy metals in the nanofiltration process using ceramic membrane. *Desalination.* https://doi.org/10.1016/j.desal.2015.04.019

Nunes, S.P., Peinemann, K.V., 2006. *Membrane Technology: In the Chemical Industry*, Wiley-VCH Verlag GmbH & Co. KGaA. https://doi.org/10.1002/3527608788

Ochando-Pulido, J.M., Rodriguez-Vives, S., Martinez-Ferez, A., 2012. The effect of permeate recirculation on the depuration of pretreated olive mill wastewater through reverse osmosis membranes. *Desalination.* https://doi.org/10.1016/j.desal.2011.10.041

Pavon, C., 2020. *Water Scarce Countries, Present and Future.* World data lab. Lindengasse. Vienna, Austria.

Pendashteh, A.R., Fakhru'l-Razi, A., Madaeni, S.S., Abdullah, L.C., Abidin, Z.Z., Biak, D.R.A., 2011. Membrane foulants characterization in a membrane bioreactor (MBR) treating hypersaline oily wastewater. *Chem. Eng. J.* https://doi.org/10.1016/j.cej.2010.12.053

Product Information: Pall Corporation Inorganic Membranes, 2004. [WWW Document], URL http://www.pall.com/AccuSep.asp.

Quist-Jensen, C.A., Macedonio, F., Drioli, E., 2015. Membrane technology for water production in agriculture: Desalination and wastewater reuse. *Desalination.* https://doi.org/10.1016/j.desal.2015.03.001

Riley, S.M., Ahoor, D.C., Regnery, J., Cath, T.Y., 2018. Tracking oil and gas wastewater-derived organic matter in a hybrid biofilter membrane treatment system: A multi-analytical approach. *Sci. Total Environ.* https://doi.org/10.1016/j.scitotenv.2017.09.031

Riley, S.M., Oliveira, J.M.S., Regnery, J., Cath, T.Y., 2016. Hybrid membrane bio-systems for sustainable treatment of oil and gas produced water and fracturing flowback water. *Sep. Purif. Technol.* https://doi.org/10.1016/j.seppur.2016.07.008

Shaffer, D.L., Werber, J.R., Jaramillo, H., Lin, S., Elimelech, M., 2015. Forward osmosis: Where are we now? *Desalination.* https://doi.org/10.1016/j.desal.2014.10.031

Shi, Y., Ma, J., Wu, D., Pan, Q., 2017. Dissolved air flotation in combination with ultrafiltration membrane modules in surface water treatment. *J. Harbin Inst. Technol. (New Ser.* https://doi.org/10.11916/j.issn.1005-9113.15307

Smith, A., 1910. American chemical society. *Ind. Eng. Chem.* https://doi.org/10.1021/ie50023a025

Šostar-Turk, S., Simonič, M., Petrinić, I., 2005. Wastewater treatment after reactive printing. *Dyes Pigment.* https://doi.org/10.1016/j.dyepig.2004.04.001

Speth, T.F., Summers, R.S., Gusses, A.M., 1998. Nanofiltration foulants from a treated surface water. *Environ. Sci. Technol.* https://doi.org/10.1021/es9800434

Stoller, M., 2009. On the effect of flocculation as pretreatment process and particle size distribution for membrane fouling reduction. *Desalination*. https://doi.org/10.1016/j.desal.2007.12.042

Stoller, M., 2011. Effective fouling inhibition by critical flux based optimization methods on a NF membrane module for olive mill wastewater treatment. *Chem. Eng. J.* https://doi.org/10.1016/j.cej.2011.01.098

Stoller, M., Bravi, M., 2010. Critical flux analyses on differently pretreated olive vegetation waste water streams: Some case studies. *Desalination*. https://doi.org/10.1016/j.desal.2009.09.027

Stoquart, C., Servais, P., Bérubé, P.R., Barbeau, B., 2012. Hybrid membrane processes using activated carbon treatment for drinking water: A review. *J. Membr. Sci.* https://doi.org/10.1016/j.memsci.2012.04.012

Subramaniam, M.N., Goh, P.S., Lau, W.J., Ng, B.C., Ismail, A.F., 2018. AT-POME colour removal through photocatalytic submerged filtration using antifouling PVDF-TNT nanocomposite membrane. *Sep. Purif. Technol.* https://doi.org/10.1016/j.seppur.2017.09.042

Synder Filtration, 2020. [WWW Document]. URL https://synderfiltration.com/learning-center/articles/introduction-to-membranes/pressure-driven-membrane-filtration-processes/

Tanaka, Y., 2015. *Electrodialysis, Progress in Filtration and Separation*. Elsevier Ltd. https://doi.org/10.1016/B978-0-12-384746-1.00006-9

Tong, T., Carlson, K.H., Robbins, C.A., Zhang, Z., Du, X., 2019. Membrane-based treatment of shale oil and gas wastewater: The current state of knowledge. *Front. Environ. Sci. Eng.* https://doi.org/10.1007/s11783-019-1147-y

United Nations, 2019. *UN Water. World Water Day-Factsheet*. United Nations: New York.

Zhao, Y.J., Wu, K.F., Wang, Z.J., Zhao, L., Li, S.S., 2000. Fouling and cleaning of membrane: A literature review. *J. Environ. Sci.* English Ed. 12. 241–251.

16 Treatment of Emerging Water Pollutants by Fenton and Modified Fenton Processes

Sachin Rameshrao Geed and Damodhar Ghime

CONTENTS

16.1 INTRODUCTION

Nowadays, continued research efforts attempt to create more capable processes to treat emerging water pollutants. Emerging water pollutants include phenolic compounds, pesticides, textile dyes such as azo dyes, and pharmaceuticals (Geed et al. 2017; Mailler et al. 2016). These pollutants show several health effects on human beings, such as neurological, carcinogenic, loss of memory, loss of coordination, and many more. These emerging pollutants arise from the chemical and petrochemical industries and contaminate mostly soil and water through the disposal of effluents into the environment. Hence, their treatment is a high priority. Mailler et al. (2016) reported several physicochemical and biological techniques for treating these emerging pollutants; each methods presents its own merits and shortcomings. This chapter mainly focuses on more efficient treatment processes for removing these emerging contaminants in effluents. The pollutants need to be mineralized completely with advanced oxidation processes that are capable of efficiently degrading pollutants (O'Dowd and Pillai 2020). In advanced oxidation processes, hydroxyl (HO•) radicals are produced, which can oxidize contaminants. Over the past few decades, advanced oxidation processes (AOPs) have received great interest in decreasing the concentrations of a wide range of wastewater pollutants. AOPs are generally categorized as Fenton oxidation, heterogeneous Fenton, UV-Fenton, and electro-Fenton processes.

16.1.1 FENTON AND MODIFIED FENTON PROCESSES

In the year 1894, the Fenton process was discovered by research chemist Henry John Horstman Fenton. He invented Fenton's reagent, which was prepared by using a strong oxidant, i.e., hydrogen

DOI: 10.1201/9781003204442-16

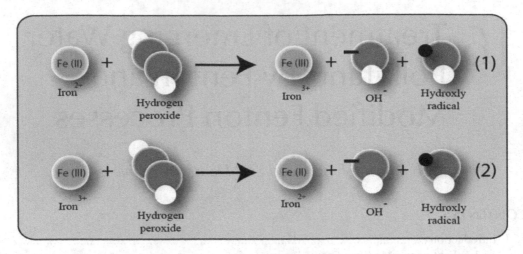

FIGURE 16.1 Schematic representation of the generation of hydroxyl (•OH) radicals from Fenton's reagent. (Source- O'Dowd and Pillai 2020 @Elsevier.)

peroxide (H_2O_2), and ferrous salt (Fe^{2+}). For the first time, Fenton in 1894 successfully oxidized tartaric acid to dihydroxy maleic acid using Fenton's reagent. After that, several studies have reported on the basic Fenton process for the oxidation of emerging pollutants in soil and wastewater, mainly industrial effluents. In the Fenton process, the primary oxidizing agents are the hydroxyl radicals (•OH) (O'Dowd and Pillai 2020). The production of hydroxyl radicals is schematically explained by O'Dowd and Pillai (2020), as shown in Figure 16.1. The H_2O_2 (hydrogen peroxide) is decomposed by Fe^{2+} (ferrous ions). The •OH radicals are powerful oxidizing agents and efficient remove the stubborn organic contaminants in effluents. The hydroxyl radicals are unselective and extract a large variety of emerging contaminants present in wastewater. Still, the Fenton process has some limitations, namely lower process efficiency, and slower reaction kinetics and reaction rates. Therefore, many researchers have been working on modifying the Fenton processes. The modified Fenton processes consist of the methods, including heterogeneous Fenton, photo-Fenton, and electro-Fenton. Certain modifications have been made in these processes to enhance the degradation rate, reaction rate, and removal efficiency and kinetics.

16.1.2 LIMITATIONS OF BASIC FENTON PROCESS

The basic Fenton oxidation possesses few limitations that hamper Fenton oxidation in the mineralization of industrial effluents. Bello et al. (2019) have observed that the Fenton oxidation process's efficiency was affected by the treating solution's pH. The acidic pH around 3.0 is favourable for better removal of contaminants from wastewater. Usman et al. (2016) investigated the influence of solution pH and the complexity of other operational parameters on the Fenton oxidation reaction, limiting the process's industrial applications. A higher value of pH affects the Fenton mechanism, which can cause the precipitation of iron oxides and further generate excessive sludge (see Table 16.1). Therefore, iron sludge generation is the second major limitation of the Fenton oxidation process (Pouran et al. 2014; Clarizia et al. 2017; Ma et al. 2017). The creation of iron sludge causes hazards of minor contamination, whereby additional treatments for sludge removal and disposal are required. Neyens and Baeyens (2003) stated that sludge management expenditure could make up about 35–50% of the wastewater management's full functioning cost. The additional limitations of the Fenton oxidation process are the insecurity of the Fenton reaction reagent, sponging reactions, embracing more chemical expenditure, the impenetrability in optimizing the reagent concentration, and the loss of oxidants to neutralize the emerging water pollutant (Asghar et al. 2015;

TABLE 16.1

Major Limitations of the Fenton Oxidation Process (Source- Bello et al. 2019 @Elsevier)

Limitations of Basic Fenton Process	Reasons	Probable Modifications	References
The requirement of acidic Ph	• The production of $^{\bullet}$OH requires acidic pH (2.5–3.5) • Fe^{2+} in soluble form as acidic pH is required • The complex formation and excess sludge generation as above pH 4 value	• Use of heterogeneous catalysts (e.g., iron oxide, transition metal, and composites) • Use of chelating agents (e.g., EDTA and oxalate)	(Pouran et al. 2014; Usman et al. 2016; Clarizia et al. 2017; Ma et al. 2017).
Excessive sludge generation	• Precipitation of iron oxide • Neutralization of the Fenton reaction	• Fluidized bed Fenton process • Heterogeneous Fenton • Electro-Fenton	(Garcia-Segura et al. 2016; Bello et al. 2017; Garcia-Segura et al. 2016; Guo et al. 2017; Lima et al. 2017).
Chemical consumption/cost of chemicals (H_2O_2)	• The large amount of H_2O_2 may be necessary to mineralize pollutants completely • Scavenging of H_2O_2 anions/cations present	• Use of electro-Fenton/bioelectro-Fenton and other technologies with in- situ generations of H_2O_2 • Other alternative oxidants	(Asghar et al. 2015; Li et al. 2016; Wang et al. 2016, 2017).

Li et al. 2016). There is a keen research interest in using the Fenton process where possible modifications have emerged as better solutions, as highlighted in Table 16.1.

16.1.3 Modifications Carried Out for the Generation of Hydroxyl Radicals

Modification to the basic Fenton process were carried out to overcome the limitations of the process. Changes are carried out for the generation of hydroxyl radicals ($^{\bullet}$OH), and methods consist of the photo-Fenton, heterogeneous Fenton processes, and electro-Fenton processes. O'Dowd and Pillai (2020) enhanced the photo-Fenton process using a UV light to improve radical hydroxyl production. Figure 16.2 offers a schematic representation of the generation of $^{\bullet}$OH radicals in the presence of sunlight. In a heterogeneous Fenton process, the dynamic site was designed to generate iron oxides, ferrous iron, or other metal-supported catalytic materials. This transition metal incorporation in the catalysts plays the role of active sites to enhance process efficiency. The ferrous (Fe^{2+}) ions can be toothless inside the metal catalyst. Other advantages of the heterogeneous Fenton process include reducing the generation of catalysts, ferric oxide sludge, and pH operating parameters.

A further modification was applying the Fenton process's electrochemical concept to enhance the generation of hydroxyl ($^{\bullet}$OH) radicals. Two different approaches were used in the electrochemical Fenton process: the first was the electro-generation of hydrogen peroxide on the cathode and the second includes the anodic Fenton oxidation process, which involves using an iron anode for Fe^{2+} electro-generation. Simultaneously, H_2O_2 might have been produced at the cathode or sometimes added externally to carry out the Fenton process.

16.2 MECHANISMS INVOLVED IN THE AOPs

The hydroxyl ($^{\bullet}$OH) radicals are produced by the breakdown of oxidant hydrogen peroxide (H_2O_2) by ferrous iron (Fe^{2+}), according to Equation (16.1) (Alalm et al. 2015). The hydroxyl radicals

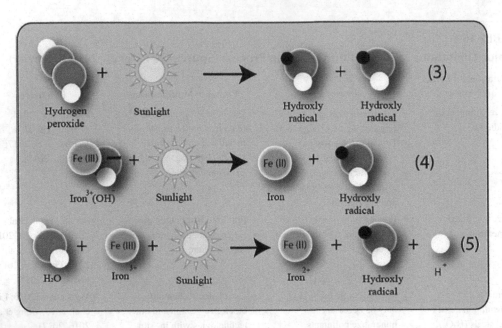

FIGURE 16.2 The modified photo-Fenton reaction for the generation of •OH radicals with UV light. (Source- O'Dowd and Pillai 2020 @Elsevier.)

oxidize the targeted organic pollutant in wastewater as per Equation (16.2). Fenton oxidation's primary phase is the rapid breakdown of H_2O_2 with Fe^{2+}, which produces a considerable quantity of hydroxyl radicals and converted Fe^{2+} ions into Fe^{3+} ions, as shown in Equation (16.3). Munoz et al. (2015) have reported the Fenton oxidation, which was generally discussed at the start of the process, propagating the Fenton reactions and termination of Fenton reactions.

$$H_2O_2 + Fe^{2+} \rightarrow Fe^{3+} + \text{•OH} + OH^- \tag{16.1}$$

$$\text{•OH} + \text{Organic matter} \rightarrow \text{oxidised products} \tag{16.2}$$

$$\text{•OH} + Fe^{2+} \rightarrow Fe^{3+} + OH^- \tag{16.3}$$

The catalytic Fenton process involves the use of a solid catalyst in a different medium. Iron is generally used in the solid phase, later placed into water as a liquid medium. The support structure in this example was the suspends, i.e., the iron catalyst. Heterogeneous Fenton oxidation involves a complex mechanism during the breakdown of emerging water pollutants. These processes consist of chemical reactions involving hydroxyl (•OH) radicals and targeted organic pollutants; another method is the physical route, which occurs at the catalyst's surface. Both processes may coexist during heterogeneous catalysis. The H_2O_2 reacts with iron at the interface of the catalyst and generates hydroxyl radicals in the solution. The emerging pollutant from wastewater and hydroxyl radicals might have adhered to the catalyst through chemisorption. The photo Fenton process, $Fe^{2+}/H_2O_2/$ UV–vis in this Fenton reaction, uses the light ranges "near-UV to the range of visible region" up to 600 nm wavelength to improve the radical hydroxyl production, as shown in Figure 16.2 and towards speedily diminishing Fe^{3+} back to Fe^{2+}. This conversion (Fe^{3+} back to Fe^{2+}) results in no additional amount of iron salt being needed for the Fenton process, seeing that the iron is continuously reduced. The decrease of Fe^{3+} ions produces another hydroxyl •OH molecule, which shows oxidant, H_2O_2 is wholly converted into •OH radicals.

The electro-Fenton process is designed so that there is a continuous electro-generation of H_2O_2 by the cathode. In the primary process, the air/O_2 is continuously supplied to the cathode. At the cathode, the oxidant, hydrogen peroxide (H_2O_2), is generated by the reduction of O_2 in an acidic medium (around pH 2–3), as given in Equation (16.4) (Pliego et al. 2015). By adding Fe^{2+} salt in the reaction, the catalytic decomposition of H_2O_2 produces hydroxyl ($^\bullet OH$) radicals. In the Fenton reaction, the continuous production of Fe^{2+} at the cathode surface is shown in Equation (16.5) (Pliego et al. 2015). The electro-Fenton process is the peroxi-coagulation process in which the water pollutant is treated using a combined oxidation and coagulation induced by the presence of $Fe(OH)_3$.

$$O_2 + 2H^+ + 2e- \rightarrow H_2O_2 \tag{16.4}$$

$$Fe^{3+} + e- \rightarrow Fe^{2+} \tag{16.5}$$

Fe ions and other soluble catalysts like copper ion or manganese ion have been used to treat a variety of organic pollutants under different experimental operating conditions in the homogeneous Fenton process. The application of the homogeneous Fenton processes is to develop the combining hybrid strategy or integrated operations with other technologies to improve the process's removal efficiency. Many studies have been undertaken in the Fenton processes performed with an iron catalyst to treat emerging water pollutants. The Fenton reaction's iron catalyst is to produce the strong oxidant $^\bullet OH$ in lab-scale experiment studies. The operating conditions are comparative to a Fenton with H_2O_2/Fe^{2+} and Fenton-like process using H_2O_2/Fe^{3+}; the former process becomes faster reaction was much slower than Fenton's reaction. The Heterogeneous Fenton processes are used for the treatment of a variety of pollutants. Some studies have been reported on homogeneous Fenton processes where iron catalysts are replaced with other catalysts such as copper ion or manganese ions. Copper ion or manganese ion have been utilized as alternatives or in concomitance with iron ion as a Fenton-like treatment catalyst. Few studies have been reported the comparative result of the copper catalyst-driven Fenton process, and conventional iron catalysts used in the Fenton process show catalytic responses. Liu et al. (2017) studied phenol degradation performance. In a stirred tank reactor, 0.1 mM of phenol with 10 mM H_2O_2 can be mineralized completely at pH 3.0 for 60 minutes using 0.1 mM Fe^{3+}; at a higher pH, phenol was not degraded in higher efficiency because of the decrease in the rate of reaction by the precipitation of $Fe(OH)_3$. The total phenol removal was extended up to pH 5 in the presence of 0.1 mM Fe^{2+} when Fenton's reaction was still able to produce enough $^\bullet OH$ radicals, before $Fe(OH)n$ precipitation.

In contrast, the pollutant was partially removed between pH 5 and 8 with 0.1 mM Cu^{2+}, reaching 73% decay at pH 6. The slower kinetics can be related to the limited transformation of Cu^{2+} into Cu^+ with H_2O_2 via a Fenton-like reaction followed by the fast regeneration of Cu^{2+} ion from $Cu+$ with the strong generation of oxidant $^\bullet OH$ (Lee et al. 2018). As the pH range increases, the solubility of copper is also increased. The more comprehensive pH range is used in Fenton processes for the treatment of emerging pollutants. The higher concentrations of catalysts used in the Cu-driven Fenton-like process presupposes the further removal of toxic Cu^{2+} from the treated effluent. The process capital cost increases when considering the continuous operation to treat a wide range of pollutants. The limitations of homogeneous Fenton processes can be overcome by using heterogeneous Fenton oxidation processes. The heterogeneous Fenton oxidation does not require a novel set of equipment (like UV lamp or electrochemical setup). Heterogeneous Fenton oxidation can be performed better at a neutral pH because it relies on solid-based Iron catalysts to generate $^\bullet OH$ radicals instead of soluble Fe ions. Thus, its use can be advantageous as compared to the homogeneous process. Several researchers have studied the heterogeneous Fenton oxidation to treat emerging pollutants using a variety of catalysts. Pham et al. (2018) developed a composite catalyst made up of copper and magnetite, which maintained catalytic capability at a higher pH range. The iron–manganese-based binary oxide has been loaded onto zeolite to catalyze Fenton oxidation, which removed emerging organic pollutants from landfill leachates (Sruthi et al. 2018). Several studies are available on applying a variety of iron-based catalysts as Fe minerals, e.g., iron shavings and nano zero-valent iron (nZVI) (Martins et al. 2012; Ertugay et al. 2017; Wang et al. 2019). The homogeneous Fenton oxidation process produced a considerable quantity of ferric sludge. These drawbacks can be overcome by using solid catalysts.

The use of solid Fe minerals is a worthy process due to its ability to be reused for repeated applications. In AOPs, the potential for hydroxyl radical (•OH) production has great importance for the treatment of a wide range of organic contaminants in a non-selective manner. In recent years, integrated Fenton processes have been used for the treatment of emerging pollutants. The sono-Fenton process was used for the treatment of water and wastewater. The ultrasonic waves and Fenton have been used separately for the treatment of emerging wastewater pollutants. Rahmani et al. (2019) have reported the use of the sono-Fenton process to treat poultry slaughterhouse wastewater in the stabilization and reduction of sludge volume. In addition, the stabilization and improvement of the dewatering of excess activated sludge in poultry slaughterhouse wastewater was achieved using a combined Sono-Fenton process. The stabilization takes place at a short retention time with high efficiency. Ke et al. (2018) reported an integrated ultrasound Fenton process for the treatment of PAHs.

The influence of organic matter on the degradation of polycyclic aromatic hydrocarbons in textile dyeing sludge by ultrasound Fenton process has been studied. Sludge organic matter was characterized, and the oxidation efficacies of PAHs at different intensities were reported. The obtained results showed that 75.52–84.40% of PAHs and 16.32–31.13% of sludge organic matter had degraded after ultrasound-Fenton treatment, confirming the competitive relationship between both of them for degradation. Within this context, due to its high efficiency and low selectivity in the degradation of compounds in solution, the electrochemical advanced oxidation processes (EAOPs) based on the production of hydroxyl radicals (•OH) become an exciting alternative for studies. These processes have as a central mechanism the Fenton reaction that can be assisted by photo- and electro-oxidation, increasing the radical's production rate, and consequently, the efficiency of pollutant degradation. Heterogeneous Fenton and integrated Fenton processes are promising alternative advanced oxidation technologies for the treatment of emerging pollutants. The integrated processes have great unique therapy in recent years. The removal efficiency of integrated biological processeses can be improved because of the mineralization of emerging contaminants and the possible improvement in the biodegradability of wastewater. The biological processes could be coupled with integrated Fenton processes in series. The Fenton process produces toxic by-products; this toxic intermediate is further mineralized with natural methods. The bioreactor is associated with the Fenton process for the mineralization of toxicants. In the first step, a high concentration of emerging pollutants is reduced, and then a bioreactor is a combined for the further treatment of emerging contaminants. The toxicity of emerging pollutants affects the microbe's growth and that is why toxic pollutants are firstly treated in an integrated process, then in the bioreactor. The process has been modified to overcome the limitations of homogeneous Fenton by iron ions or oxides immobilized onto solid supports as catalysts, which can be used in a broad working pH range without sludge disposal problems, more suitable for engineering application (Li et al. 2014).

Kotsou et al. (2004) have studied the integrated aerobic biological treatment and chemical oxidation with Fenton's reagent to process green table olive wastewater. Table olive processing wastewater (TOPW) is unsuitable for disposal at municipal or industrial wastewater treatment plants because of the high organic loads. Aerobic biological treatment using an *Aspergillus Niger* strain in a bubble column bioreactor combined with chemical oxidation was studied to manage TOPW to a quality that corresponds to the input standards for wastewater treatment plants (Chemical Oxygen Demand (COD) <1,200 ppm, Biochemical oxygen demand (BOD) <500 ppm). After two days of biological treatment, COD was reduced by 70%, while the total and straightforward phenolic compounds were decreased by 41% and 85%, respectively. In the chemical treatment step, the effect of different H_2O_2 concentrations on COD and phenol reduction patterns was studied. Changotra et al. (2019) reported real pharmaceutical wastewater treatment using a combined approach of Fenton applications and aerobic biological treatment. The pharmaceutical wastewater of different organic loads, i.e., high strength wastewater (HSW) and low strength wastewater (LSW) were collected from the bulk pharmaceutical industry and subjected to various applications of Fenton's treatment followed by subsequent biological treatments. For both the HSW and LSW, applications of Fenton, such as dark-Fenton (DF), solar-driven photo-Fenton (PF), and electro-Fenton (EF), were utilized as pre-treatment technologies to improve the biodegradability and reduce the organic load of wastewater by combined oxidation and coagulation.

16.3 APPLICATION OF FENTON AND MODIFIED FENTON OXIDATION PROCESSES

Fenton and modified Fenton processes are used to treat a full range of pollutants found in industrial effluents. In the textile industry, the occurrence of dark dyes in the released effluent is a major problem. Most of the dyes are easily degraded using Fenton and modified Fenton processes. Fenton processes are applied to β-blockers, anti-inflammatories, analgesics, antibiotics, and diuretics. Waste from the pharmaceutical industry have been treated using Fenton and modified Fenton processes for the last few years. The presence of pesticides and poly-aromatic hydrocarbon in industrial wastewater is an emerging water pollution problem that can cause several health effects and contaminate the air, soil, and water. Most of the organic pollutants affect the aquatic system as well as human life. Hence, its treatment before final disposal is necessary using low-cost, highly-efficient technologies like Fenton and modified Fenton processes. The Fenton and Fenton-like processes are considered separately. These processes have the potential for inherent concurrence and conversion of the targeted water contaminants. The comparative assessment of both processes is significant to clarify the roles of each process. For the Fenton-like process, the intermediates produced by the degradation of the phenolic compounds may promote the conversion of Fe(III) to Fe(II) in addition to the uni-molecular decomposition of Fe(III)-hydroperoxyl complexes (Jiang et al. 2010). The Fenton-like reactions are highly sensitive to pH and more dependent on the iron dosage because of the demand for Fe(III) to Fe(II). The Fenton-like reaction is a prototype for the developed Fenton process. The advanced oxidation processes are typically operated with less energy than the direct oxidation of water contaminants. These processes are generally performed at ambient temperatures and pressures, which involve hydroxyl radicals to affect the water's purification. These processes are classified based on the reactive phase (homogeneous and heterogeneous) and methods of the hydroxyl radical's generation (chemical, electro-chemical, sonochemical, and photochemical) (Babuponnusami and Muthukumar 2013). The Fenton-like processes, such as photo-electro-Fenton and sono-electro-Fenton, are the combined with conventional advanced oxidation processes. Simultaneously, the non-conventional AOPs are ionized by radiation, microwaves, and pulsed plasma techniques. Solar-irradiated processes have also been studied to decrease the costs associated with artificial light sources. Many water contaminants are considered as harmful, even when they are present at low concentrations. Biological processes are inefficient for removing these toxic pollutants in water due to their recalcitrant nature. Therefore, oxidation processes are applied for the effective degradation of water pollutants. Natural oxidation processes are also ineffective for complete removal because they demand specific operating conditions to oxidize the organic pollutants and therefore increase the operational costs. The Fenton and Fenton-like processes have been successfully applied to remove certain chlorinated compounds, like chloral hydrate or 1,1,1-trichloroethane, and for the treatment of real wastewater (Prousek et al. 2007). Ferrous salts such as ferrous sulfate and Mohr's salt $[(NH_4)_2Fe(SO_4)_2.6H_2O]$ could also be used as sources of ferrous ion sources. Both of the advanced oxidation processes can be operated with Mohr's salt at a pH value ranging from 3 to 7. Sometimes, the Fenton-like reaction uses zero-valent iron (Fe^0) instead of a ferrous salt. The reactive products of the Fenton process are the hydroxyl radicals or ferryl. Thus, the chemistry involved in Fenton's reagent usage is, above all, radical chemistry. The organic and inorganic water pollutants present in the industrial effluent greatly affect the Fenton and Fenton-like reaction propagation step. These pollutants react with the iron species and influence the reaction mechanism and the kinetics of the Fenton process. The organic water pollutants such as dyes, tenzides, and detergents react with the Fenton's reagent to remove water toxicity and COD. Each step of the Fenton process could be influenced by the components involved in the process. The process modification can principally develop some economic parameters of the process. The photo-Fenton and other Fenton-like processes were reported for the degradation and mineralization of the antineoplastic drug 5-fluorouracil. Researchers investigated the effects of important experimental parameters such as 5-fluorouracil concentration, ferric ions, and oxidant concentrations using solar irradiation on the treatment efficiency. They obtained lower toxicity in the solution at the end of the process (Koltsakidou et al. 2016). The photo-Fenton and other Fenton-like processes for removing 5-fluorouracil avoid the generation of highly toxic transformation products during treatment. Rodríguez-Narvaez et al. (2018) compared the Fenton and other Fenton-like reactions for the degradation of L-proline under different experimental conditions with and without using UV irradiation. The heterogeneous Fenton-like process was found very effective for the degradation of L-proline with no risk of increased toxicity with the leachate of cobalt. The amount of leached cobalt was negligible at a neutral and acidic pH value. In recent publications, the Fenton-like process has been developed in a heterogeneous phase where cobalt is supported in a solid matrix that can carry out the reaction without the solution's cobalt. The Fenton-like process, either in the homogeneous or heterogeneous phase, was found as a viable alternative for removing water pollutants because it is capable of high removal efficiencies in a short treatment time and using lower reagent concentrations. The Fenton and Fenton-like reactions were assessed for the first time to clarify their roles for the oxidation of phenol under varying concentrations of the oxidant H_2O_2 (Jiang et al. 2010). In this report, the Fenton-like reaction has proceeded concurrently with a classic Fenton process. The concurrent Fenton reaction played a significant role in the degradation of phenol and phenolic compounds in effluents. From the obtained results, it was found that the Fenton process was more efficient compared to the Fenton-like process at lower levels of iron concentration. In the Fenton-like reactions, great care should be taken to prepare ferric solutions to avoid ferric hydroxide's precipitation. An appropriate amount of ferric salt was diluted in an appropriate volume of ultrapure water to obtain the desired composition of Fe^{3+} ions in the solution. Jiang et al. (2010) used freshly prepared solutions of the ferrous and ferric salts daily to get accuracy in the results. All the experimental runs must be performed rapidly to avoid generating the polymeric Fe(III) species.

16.4 INTEGRATING AOPs WITH BIOLOGICAL AND OTHER CHEMICAL PROCESSES

Fenton and modified Fenton processes are widely used to treat industrial wastewater; still, these methods have some limitations associated with basic Fenton oxidation. To minimize the Fenton process's defects, there are several modifications and integration of AOPs with Fenton and other chemical and biological processes. The combination of Fenton and modified Fenton processes enhances the percentage removal efficiency of the toxic pollutants. The application of a fluidized bed reactor system coupled with the Fenton process can reduce excessive sludge production and recover the process's performance. The heterogeneous Fenton process is extensively used for addressing the limitations of basic Fenton oxidation. The main advantage of material science is to develop a variety of materials that help to advance the different elements, i.e., a catalyst that can be used in a heterogeneous Fenton process. Still, the mass transfer drawback problems are a significant issue confronted with the use of catalysts in the heterogeneous Fenton oxidation process. Various materials have been prepared to replace catalysts used in traditional heterogeneous Fenton oxidation, which is mainly based on iron-oxides. However, the cost of these materials is higher than iron oxides. Hence, the cost factor should be considered to select particular material as a heterogeneous catalyst for the Fenton oxidation process. Another significant deliberation is the stability of the used catalyst in the heterogeneous Fenton process. The catalyst retains its strength due to a lower or negligible amount of iron leaching during treatment. However, many studies have reported the low-level leaching of catalysts, and few materials may be toxic. The hybrid processes are not economically viable for the degradation of large quantums of effluent released by the industries (Babuponnusami and Muthukumar 2013). Therefore, it is advisable to use hybrid methods as pre-treatment to reduce the toxicity to a certain level beyond which the biological processes can be applied. The combined processes are assumed to reduce the reactor size and operating cost of the process too. The oxidants which might have been added during the pre-treatment phase should be completely removed before subjecting to the biological process as they can act as an inhibitor to the microbes. The integration of advanced oxidation processes with the biological processes strongly depends on the characteristics of the effluent released from the industries, concentrations, and the processes' desired efficacies. The coupling of the AOPs with the biological process reactor could reduce industrial effluents (Tabrizi and Mehrvar 2004). As AOPs are expensive treatment methods, optimizing the process's total cost is a more challenging task for environmental researchers. Thus, the appropriate design should consider the ability of this coupling to remove water contaminants and try to obtain the desired results in a cost-effective manner. Some of the parameters, such as the kinetic rates, residence time, capital, and operating costs, play a significant role in considering the integration approach's total cost. The advanced oxidation treatment techniques could also be coupled with wet scrubbers at mild conditions to remove odorous sulfur compounds (Vega et al. 2014). The multicomponent aqueous solution of ethyl mercaptan, dimethyl sulfide, and dimethyl disulfide was chosen for this treatment. The different AOPs, including UV, assisted H_2O_2, Fenton, photo-Fenton, and ozonation, were tested under mild conditions, and the degradation efficiency obtained in each process was accessed. The obtained results showed that the ozonation and the photo-Fenton processes oxidized about 95% of the sulfur compounds from the aqueous medium, and the mineralization percentage reached 75% in 10 minutes. But from the economic analysis, it was found that the Fenton process is the most economical option to be integrated with a wet scrubber for the removal of volatile sulfur compounds. The Fenton catalytic process was possibly integrated with the sonolysis process for the mineralization and volume reduction of activated sludge (Rahmani et al. 2019). The efficiencies of both processes were also investigated separately. The COD removal efficiency was increased to 77% because of the synergistic effect of ultrasonic waves, Fenton's reagent, and more hydroxyl radicals. The efficacy of the process was increased significantly due to the combination of Fenton catalytic and the sonolysis process. It was only possible due to the production of more oxidizing agents and improvement in mass transfer. In the modified Fenton process, chemical Fenton and biochemical Fenton processes have possible advanced approaches for the in situ generation of hydroxyl radicals. The electro-Fenton and bioelectro-Fenton processes produce the Fenton's reagent during treatment in the solution, which reduces the cost related to their external addition. The effective degradation of emerging pollutants is possible with these processes. Sometimes, the quantity of H_2O_2 produced in the solution is low during electrochemical treatment and might not be useful for wastewater treatment. Hence, there is a requirement to enhance the production of insufficient amounts of H_2O_2, possibly achieved from end-to-end modification in cathodic material. The researchers produced useful electrode materials with outstanding properties of three-dimensional and two-dimensional materials with higher surface areas. Another development in the Fenton process is the solar-driven procedure, which is a possible solution to the higher energy consumption associated with the electro-Fenton process. The bioelectro-Fenton process also helps to address this energy challenge. In the bioelectro-Fenton process, microbes have been utilized, which are associated with electricity production. The bioelectro-Fenton process also faces some problems, such as microbial species' slow activity, high internal resistance, membrane fouling, and reactor application. Future research is needed to identify the uses of the bioelectro-Fenton process.

16.5 CONCLUSION

The Fenton process has several disadvantages: it has high chemical inputs and sludge generation, and it is challenging to treat emerging pollutants using a continuous process. These shortcomings warrant using modified Fenton processes. Research is underway to develop modified Fenton processes. Many research articles on the modified Fenton processes have been published to overcome the Fenton process's limitations. The modified Fenton processes include the heterogeneous Fenton, electro-Fenton, and possible integrated Fenton processes to produce hydroxyl ions. This chapter provides a complete overview on the fundamentals, mechanisms, and significant Fenton applications and modified Fenton processes such as photo-Fenton, heterogeneous Fenton, and electrochemical Fenton. Furthermore, the concept of integrating AOPs with modified processes is also discussed for the 100% removal of emerging pollutants from industrial effluents.

16.6 FUTURE PROSPECT

Presently, several research articles have reported on basic Fenton and modified Fenton processes; still, there is room to optimize the process parameters and materials used in Fenton processes. There is scope within the field of material science for the utilization of catalysts that can enhance the method's efficiency and develop cost-effective alternatives. Another investigative opportunity presents to work on using the chemicals and regeneration of the Fenton reagent itself in the processes. The solution pH has is critical to the processes, so related research can be further extended to cover pH dependency. Since hybrid processes have a better performance, there is a strong need to combine physicochemical processes or biological processes with the Fenton and modified Fenton processes.

GLOSSARY

Emerging water pollutants: These are the harmful chemicals that are not commonly monitored in the environment but have the potential to enter the atmosphere and cause adverse effects on human health.

Fenton process: Effective advanced oxidation process for the remediation of emerging water pollutants in industrial effluents.

Fenton's reagent: A mixture of hydrogen peroxide (H_2O_2) and ferrous iron (II) sulfate.

Modified Fenton processes: Processes that can enhance the mineralization, degradation, and decolorization efficiency for the removal of emerging water pollutants.

Photo-Fenton: The process uses hydrogen peroxide as an oxidant and UV light with ferrous ions, hence the so-called photo-Fenton method, and generates more hydroxyl radicals compared to the conventional Fenton process or photolysis, promoting the removal of emerging water pollutants.

Electro-Fenton: A type of electrochemical advanced oxidation process, which is generally based on in situ generations of the hydroxyl radicals in an electro-catalytical way.

Heterogeneous catalysis: Catalysis in which the solid catalyst occupies a different phase from the reactants or products.

Oxidant: A substance that can degrade other materials.

Hydrogen peroxide: A member of the reactive oxygen species.

Ferrous salt: Supplement used to prevent iron deficiency.

Sludge: Semi-solid slurry that can be produced in different industrial processes, from wastewater treatment.

Hydroxyl radicals: Highly-reactive and short-lived radicals.

UV light: A form of electromagnetic radiation with a wavelength from 10 nm to 400 nm.

REFERENCES

Alalm, M. G., Tawfik, A., and Ookawara, S. 2015. Degradation of four pharmaceuticals by solar photo-Fenton process: Kinetics and costs estimation. *Journal of Environmental Chemical Engineering* 3:46–51.

Asghar, A., Raman, A. A. A., and Daud, W. M. A. W. 2015. Advanced oxidation processes for in-situ production of hydrogen peroxide/hydroxyl radicals for textile wastewater treatment: A review. *Journal of Cleaner Production* 87:826–838.

Babuponnusami, A., and Muthukumar, K. 2013. A review on Fenton and improvements to the Fenton process for wastewater treatment. *Journal of Environmental Chemical Engineering* 2:557–572.

Bello, M. M., Raman, A. A. A., and Asghar, A. 2019. A review of approaches for addressing the limitations of Fenton oxidation for recalcitrant wastewater treatment. *Process Safety and Environmental Protection* 126:119–140.

Bello, M. M., Raman, A. A. A., and Purushothaman, M. 2017. Applications of fluidized bed reactors in wastewater treatment– A review of the major design and operational parameters. *Journal of Cleaner Production* 141:1492–1514.

Changotra, R., Rajput, H., and Dhir, A. 2019. Treatment of real pharmaceutical wastewater using combined approach of Fenton applications and aerobic biological treatment. *Journal of Photochemistry and Photobiology, part A: Chemistry* 376:175–184.

Clarizia, L., Russo, D., Di Somma, I., and Marotta, R. 2017. Homogeneous photo-Fenton processes at near-neutral pH: A review. *Applied Catalysis B: Environmental* 209:358–371.

Ertugay, N., Kocakaplan, N., and Malkoc, E. 2017. Investigation of pH effect by Fenton like oxidation with ZVI in treatment of the landfill leachate. *International Journal of Mining, Reclamation and Environment* 31:404–411.

Fenton, H. J. H. 1894. Oxidation of tartaric acid in the presence of iron. *Journal of the Chemical Society, Transactions* 65:899–910.

Garcia-Segura, S., Bellotindos, L. M., and Huang, Y. H. 2016. Fluidized-bed Fenton process as alternative wastewater treatment technology– A review. *Journal of the Taiwan Institute of Chemical Engineers* 67:211–225.

Geed, S. R., Kureel, M. K., and Giri, B. S. 2017. Performance evaluation of Malathion biodegradation in batch and continuous packed bed bioreactor (PBBR). *Bioresource Technology* 227:56–65.

Guo, S., Yuan, N., and Zhang, G. 2017. Graphene modified iron sludge derived from a homogeneous Fenton process as an efficient heterogeneous Fenton catalyst for the degradation of organic pollutants. *Microporous and Mesoporous Materials* 238:62–68.

Jiang, C., Pang, S., and Ouyang, F. 2010. A new insight into Fenton and Fenton-like processes for water treatment. *Journal of Hazardous Materials* 174:813–817.

Ke, Y., Ning, X. A., and Liang, J. 2018. Sludge treatment by integrated ultrasound-Fenton process: Characterization of sludge organic matter and its impact on PAHs removal. *Journal of Hazardous Materials* 343:191–199.

Koltsakidou, A., Antonopoulou, M., and Sykiotou, M. 2016. Photo-Fenton and Fenton-like processes for the treatment of the antineoplastic drug 5-fluorouracil under simulated solar radiation. *Environmental Science and Pollution Research* 24:4791–4800.

Kotsou, M., Kyriacou, A., and Lasaridi, K. 2004. Integrated aerobic biological treatment and chemical oxidation with Fenton's reagent for the processing of green table olive wastewater. *Process Biochemistry* 39:1653–1660.

Lee, H., Seong, J., and Lee, K. M. 2018. Chloride-enhanced oxidation of organic contaminants by Cu (II)-catalysed Fenton-like reaction at neutral pH. *Journal of Hazardous Materials* 344:1174–1180.

Li, N., An, J., and Zhou, L. 2016. A novel carbon black graphite hybrid air-cathode for efficient hydrogen peroxide production in bio-electrochemical systems. *Journal of Power Sources* 306:495–502.

Li, W. G., Wang, Y., and Angelidaki, I. 2014. Effect of pH and H_2O_2 dosage on catechol 568 oxidation in nano-Fe_3O_4 catalysing UV-Fenton and identification of reactive 569 oxygen species. *Chemical Engineering Journal* 244:1–8.

Lima, M. J., Silva, C. G., and Silva, A. M. 2017. Homogeneous and heterogeneous photo-Fenton degradation of antibiotics using an innovative static mixer photoreactor. *Chemical Engineering Journal* 310:342–351.

Liu X., Yin H., and Lin A. 2017. Effective removal of phenol by using activated carbon-supported iron prepared under microwave irradiation as a reusable heterogeneous Fenton-like catalyst. *Journal of Environmental Chemical Engineering* 5:870–876.

Ma, J., Yang, Q., and Wen, Y. 2017. Fe-g-C_3N_4/graphitized mesoporous carbon composite as an effective Fenton-like catalyst in a wide pH range. *Applied Catalysis B: Environmental* 201:232–240.

Mailler, R., Gasperi, J., and Coquet, Y. 2016. Removal of a wide range of emerging pollutants from wastewater treatment plant discharges by micro-grain activated carbon in fluidized bed as tertiary treatment at large pilot scale. *Science of the Total Environment* 542:983–996.

Martins, R. C., Lopes, D. V., and Quinta-Ferreira, R. M. 2012. Treatment improvement of urban landfill leachates by Fenton-like process using ZVI. *Chemical Engineering Journal* 192:219–225.

Munoz, M., Casas, J. A. and Rodriguez, J. J. 2015. Preparation of magnetite-based catalysts and their application in heterogeneous Fenton oxidation: A review. *Applied Catalysis B: Environmental* 176:249–265.

Neyens, E., and Baeyens, J. 2003. A review of classic Fenton's peroxidation as an advanced oxidation technique. *Journal of Hazardous Materials* 98:33–50.

O'Dowd, K., and Pillai, S. C. 2020. Photo-Fenton disinfection at near-neutral pH: Process, parameter optimization, and recent advances. *Journal of Environmental Chemical Engineering* 8:104063.

Pham, V. L., Kim, D. G., and Ko, S. O. 2018. Cu@Fe$_3$O$_4$ core-shell nanoparticle-catalysed oxidative degradation of the antibiotic oxytetracycline in pre-treated landfill leachate. *Chemosphere* 191:639–650.

Pliego, G., Zazo, J. A., and Garcia-Munoz, P. 2015. Trends in the intensification of the Fenton process for wastewater treatment- An overview. *Critical Reviews in Environmental Science and Technology* 45:2611–2692.

Pouran, S. R., Raman, A. A. A., and Daud, W. M. A. W. 2014. Review on the application of modified iron oxides as heterogeneous catalysts in Fenton reactions. *Journal of Cleaner Production* 64:24–35.

Prousek, J., Palackova, E., and Priesolova, S. 2007. Fenton and Fenton-Like AOPs for wastewater treatment: From laboratory-to-plant-scale application. *Separation Science and Technology* 42:1505–1520.

Rahmani, A. R., Mousavi-Tashar, A., and Masoumi, Z. 2019. Integrated advanced oxidation process, sono-Fenton treatment, for mineralization and volume reduction of activated sludge. *Ecotoxicology and Environmental Safety* 168:120–126.

Rodríguez-Narvaez, O. M., Perez, L. S., Yee, N. G. 2018. Comparison between Fenton and Fenton-like reactions for L-proline degradation. *International Journal of Environmental Science and Technology* 16:1515–1526.

Sruthi, T., Gandhimathi, R., and Ramesh, S. T. 2018. Stabilized landfill leachate treatment using heterogeneous Fenton and electro-Fenton processes. *Chemosphere* 210:38–43.

Tabrizi, G. B., and Mehrvar, M. 2004. Integration of advanced oxidation technologies and biological processes: Recent developments, trends, and advances. *Journal of Environmental Science and Health. Part A: Toxic/Hazardous Substances & Environmental Engineering* 39:3029–3081.

Usman, M., Hanna, K. and Haderlein, S. 2016. Fenton oxidation to remediate PAHs in contaminated soils: A critical review of major limitations and counter-strategies. *Science of the Total Environment* 569:179–190.

Vega, E., Martin, M. J., and Gonzalez-Olmos, R. 2014. Integration of advanced oxidation processes at mild conditions in wet scrubbers for odorous sulphur compounds treatment. *Chemosphere* 109:113–119.

Wang, N., Zheng, T., and Zhang, G. 2016. A review of Fenton-like processes for organic wastewater treatment. *Journal of Environmental Chemical Engineering* 4:762–787.

Wang, Y., Feng, C., Li, Y. 2017. Enhancement of emerging contaminants removal using Fenton reaction driven by H$_2$O$_2$-producing microbial fuel cells. *Chemical Engineering Journal* 307:679–686.

Wang, Z., Li, J., and Tan, W. 2019. Removal of COD from landfill leachate by advanced Fenton process combined with electrolysis. *Separation and Purification Technology* 208:3–11.

17 Advanced Technologies for the Removal of Refractory Contaminants from Pulp and Paper Mill Wastewater

Sumedha Mohan and Praveen Dahiya

CONTENTS

17.1 INTRODUCTION

The pulp and paper (P&P) industry is one of the major sources of wastewater generation that has a significant impact on the environment. This industry is the largest buyer and manufacturer of bio-based products, and is considered to play a vital role in the economic growth of countries in Northern America, including the U.S. (Sinclair 1990). The P&P industry produces large amounts of pulp and paper using electricity, dry barks, chemicals like ethanol, biofuels, carbon fibers, etc. In Canada, roughly half of the waste generated from the P&P industry is discharged into Canadian waters. As per Central Pulp and Paper Research Institute, at present there are more than 850 pulp and paper mills present in India with a total installed capacity of 25 MMT. Out of these, approximately 600 mills are currently operational, including around 250 small-scale (<50 tpd), 150 medium-scale (50–100 tpd), and 200 large-scale (100–1,200 tpd and above) mills.

The Indian P&P industry utilizes diverse raw materials including more than 15 species of wood, and non-woods like bagasse, rice and wheat straw, jute, grasses and reeds, and waste papers (white, brown, and mixed type). During the manufacturing process, huge amounts of water is consumed which is further discarded as wastewater. Thus, the P&P industry is categorized as one of the 20 most polluting industries by the Ministry of Environment and Forests, Government of India (Singh et al. 2016). The P&P industry in India utilizes around 100–250 m^3 of freshwater and generates 75–225 m^3 of wastewater per ton of paper produced (Ansari 2004). Based on the raw materials used, the industry is categorized into three sectors: a) wood and bamboo-based mills generating 3.19 million tons b) agro residue-based mills such as bagasse and straw from wheat, rice, or grasses generating 2.2 million tons, and c) wastepaper-based fiber mills generating 4.72 million tons. which

DOI: 10.1201/9781003204442-17

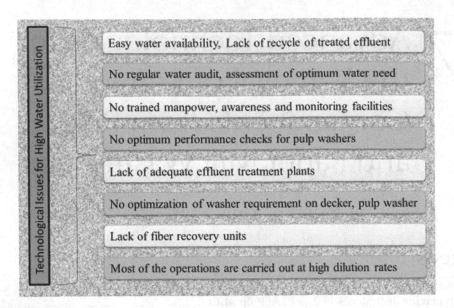

FIGURE 17.1 The reasons for high water utilization in P&P industries.

contributes 47% of total production. There are a total of 538 of recycled paper mills currently in operation. Out of the three sectors, agro residue-based mills raise major environment concern as they produce large volumes of effluents with high loads of suspended solids, biochemical oxygen demand (BOD), and chemical oxygen demand (COD). Due to various technological issues, large quantities of water are consumed in the P&P industries. Some of the major reasons for this high water consumption are represented in Figure 17.1.

P&P industries using virgin raw materials are mainly responsible for producing huge amounts of wastewater. This wastewater, if discharged into water bodies after incomplete treatment, will result in adverse impacts on aquatic organisms, such as scum formation, color development, formation of slime by *Sphaerotilus* sp. microbes, temperature variations, and loss of aesthetic value of the water system (Pokhrel and Viraraghavan 2004). Depending on the choice of raw material selected followed by the production strategy adopted, the wastewater generated will contain a high COD with approximately 700 diverse compounds (organic and inorganic) present and a limited biodegradability. The effluents from wastewater will result in a brown or black color in the receiving water bodies and include major polluting compounds such as lignin and chlorinated organic compounds. The color developed will decrease the dissolved oxygen level and photosynthetic rate in aquatic plants which will ultimately destroy and decay the plants (Singh et al. 2016). The effluents from the P&P industry also have adverse effect on the soil as it contains highly toxic chlorinated compounds which accumulate in the soil and affect plant growth. Moreover, in some developing countries farmers are irrigating their crops with P&P industry wastewater containing several toxic and carcinogenic compounds. Thus, proper treatment processes are required before discharging the effluent into the environment.

Various treatment methodologies such as physico-chemical, biological, and combination methodologies are available for P&P industry wastewater treatment. These include ultrafiltration, precipitation, electrochemical treatment, coagulation, adsorption, aerobic and anaerobic treatment, and fungal treatment. The physico-chemical and combinational treatments are found to be energy intensive and costly and result in the production of secondary sludge (Singhal and Thakur 2009a, b). On the contrary, biological treatments are found to be effective in reducing COD, BOD, color, organic load, and chlorolignins from the effluents (Nagarthnamma et al. 1999). Using microorganisms, wastewater can be treated via the production of various enzymes, such as laccase, lignin peroxidase, and manganese peroxidase. The most commonly utilized biological treatment methodology is the activated sludge process. This process is highly efficient in operating at high organic loads and possesses significant efficiency in the removal of both BOD and COD. Despite all the treatment methodologies adopted, the P&P industry is still considered to be a major polluting industry across the globe. This chapter focuses on the production techniques and various refractory contaminants generated from pulp and paper mill wastewater. Furthermore, various treatment technologies such as physico-chemical, biological, and other emerging treatments for the removal of pollutants from wastewater are also discussed.

17.2 PRODUCTION TECHNIQUES IN PAPER AND PULP MILLS

The P&P industry utilizes wood/virgin fiber as the most abundant raw material, followed by recovered fibers. The production process of pulp includes chemical, mechanical, and combinational techniques in the case of wood, whereas the fiber recovery process is used to get recovered pulp. Using wood as raw material, the production process of making paper/paperboard includes the following stages: a) pulp making, b) pulp processing, and c) paper making. Pulp making starts with debarking which includes the removal of soil, dust, and bark from the wood which is further converted into wood pieces or wood chips. The pieces of wood are treated at higher temperatures or via chemical techniques which separates cellulose from lignin and hemicellulose for making paper. Various organic compounds are removed during the processing of wood by applying wet processes (Vepsäläinen et al. 2011). Mechanical pulping results in higher yields (90–95%) but low product quality when compared to chemical processing (Pokhrel and Viraraghavan 2004). Thus, some modifications in mechanical pulping are done to improve the process which includes thermo mechanical pulping, chemo mechanical pulping, and chemical thermo mechanical pulping techniques. Various studies are carried out to optimize the chemical thermo-mechanical pulping process for fiber production via the biological (Lei et al. 2012) and chemical treatment of wood chips (Pan et al. 2013). Bleaching via elemental chlorine-free and total chlorine-free techniques is done to increase the brightness of paper which includes the use of various bleaching agents like chlorine, oxygen, ozone, hydrogen peroxide, and chlorine dioxide. To remove the bleaching agents used, the produced pulp is then washed with alkali caustic soda. This processed pulp is then added with dyes, resins, sizing agents, paper fillers, etc., to form the paper, followed by the dewatering and drying of sheets by applying heat or air (Santos and Almada-Lobo 2012). The paper production process is an energy-intensive process as it consumes huge amounts of electricity (47.2%) and thermal energy (94%), as reported by Chen et al. (2012) when examining the process of pulping in a P&P mill in Taiwan.

These days, the P&P industry is more focused on the use of recovered fibers as raw materials due to environmental pollution issues. Recycling of recovered fibers can be achieved through processes like pulping, screening, and deinking. Pulping involves the conversion of wastepaper into recovered fibers, deinking involves the removal of ink from the fibers, and screening includes the segregation of fibers based on density. The most important stage in recovered fiber recycling is the deinking step which not only involves the removal of ink particles from cellulose fibers but also includes the removal of ink particles from pulp slurry by adopting washing or flotation techniques (Zhenying et al. 2009). Zhang et al. (2008) reported various toxic chemicals and surfactants which are used as deinking agents to increase the pulp brightness which is difficult to remove. Therefore, enzyme-based deinking using enzymes like cellulase, pectinase, hemicellulose, and xylanase is found to be a better solution which helps in decreasing the overall cost of wastewater treatment. Compared to the chemical deinking process, enzyme-based deinking is eco-friendly as it decreases the COD and BOD values of effluents produced in the case of only partial (50%) replacement of deinking chemicals with the enzyme-based deinking process (Singh et al. 2012).

17.3   REFRACTORY CONTAMINANTS IN PAPER AND PULP WASTEWATER

There are various stages in the manufacturing of paper which contribute to the level of pollution. These include the preparation of wood, pulping, bleaching, and deinking. Other than the pollution load produced, each stage also requires huge amounts of raw materials, water, energy, and chemicals. It was reported that roughly 42% of wood is reclaimed in the form of bleached pulp, 5.25% enters wastewater in a soluble form, 4.2% remains as solid waste, and 2.3% enters wastewater as suspended solids. Various chemical compounds are present in wastewater such as lignin, cellulose, hemicellulose, and wood extractives which are difficult to treat biologically. The diverse compounds from lignin degradation products include mercaptides, enol ethers, quinone derivatives, acetic and formic acids, methanol, furfural, manomeric and chlorinated phenols, and acetaldehyde. More than 300 organochlorine compounds are reported from pulp and paper effluents and almost half of organic matter is refractory in nature. The refractory compounds are organic compounds which can resist biodegradation processes and can survive both the aerobic and anaerobic biological wastewater treatment process (Ortega-Clemente et al. 2009). Thus, the advanced oxidation process is employed for pre- or post-treatment of such organic compounds in which the biodegradation is enhanced during the pre-treatment phase and the removal of the remaining organic content occurs during the post-treatment phase.

Lignin is a polyphenolic polymer which is one of the most difficult to degrade of the biomass components. Lignin and its derivatives are hazardous and refractory in nature which imparts a dark brown color (called black liquor) and high BOD and COD to the wastewater generated during the pulp making process. The lignin and its derivatives will further react with chlorine during pulp bleaching and lead to the formation of various toxic and refractory compounds, including benzaldehydes, catechols, chlorinated lignosulfonic and resin acids, chlorinated phenols, syringo-vanillins, and guaiacols. Some of the most highly toxic and persistent chlorinated compounds present include chlorophenols like tri- and penta-chlorophenols, guaicols, catechols, anisoles, and verathroles. Dioxins and furans are examples of highly dangerous chlorinated compounds available in P&P mill wastewater. These chlorinated compounds are generated in large quantities during the chlorine bleaching stage of kraft pulp in the P&P industry. In order to decrease the generation of these toxic refractory compounds, various measures are adopted during the bleaching stage, including the utilization of oxygen bleaching and the use of chlorine dioxide bleach (elemental chlorine free). The other strategies available include the extended delignification process which can remove more lignin, modified cooking, enhanced delignification efficiency, spill collection systems, and highly efficient washing and reusing condensate facilities. Some researchers also reported the setting up of external treatment plants with modified designs for further reducing the levels of toxic and refractory chlorinated organic contaminants that are released into the water bodies.

The non-chlorinated compounds generated from the P&P industry include fatty acids, tannins, resin acids, diterpene alcohol, and steroids. The resin acids are diterpenoids which are released in higher amounts during the pulping process mainly by mechanical pulp production including dehydroabietic acids, abietic, palustic, levopimaric, and neoabietic. Due to the stable structure of resin acids, they cannot be degraded by the chemical degradation process and thus will contribute to the toxic levels of effluents ending up in the water bodies near P&P mills. Diethylenetriamine penta-acetic acid (DTPA) and ethylenediamine tetra-acetic acid (EDTA) are chelating agents which are mostly used for the ozone and peroxide bleaching of pulp. These are large organic molecules that can bind to metals. The chelating agents are discharged from P&P mills after the bleaching process and are found to be either completely resistant to degradation or undergo a slow degradation process. It was observed that bleach liquor with chelating agents when discharged may avoid the sedimentation of metals and thus result in the spreading of metals over a larger surface area.

The pollutants from the P&P industry influence aquatic organisms, animals, plants, and ultimately human beings. Various reports are available on the harmful effects of P&P effluents on fishes, including damage to the liver, changes in reproduction, delay in sexual maturity, reduced size of gonads, a decrease in juvenile numbers, skin problems, and reduced growth rates. Toxic chlorinated lignin and phenolic compounds contribute to various health issues reported in aquatic organisms such as thyroid dysfunction, reduced fertility, reduced growth, and feminization or masculinization of biota. Chlorinated phenols and catechols are also known to lead to mammalian cell mutagenesis. Refractory compounds such as dioxins can bioaccumulate in the case of fishes, and chlorinated hydrocarbons have been shown to possess significantly high bioaccumulation and biomagnification rates even at higher tropic levels. P&P mill effluents not only affect aquatic organisms but also disturb soil quality and influence crops and livestock. P&P mill wastewater contains high concentrations of elements (e.g., S, Mg, and NaCl) which will decrease crop yields due to nutrient imbalances, changes in the soil structure, and enhanced salinity conditions. Toxic refractory compounds from the P&P industry may accumulate in the soil resulting in enhanced soil pH, changes in the soil color and microbial activities, decreased levels of oxygen, and disturbances to the nutrient availability in the soil which will influence the growth and productivity of various crops. Higher concentrations of toxic elements may affect livestock and human beings.

17.4 TREATMENT METHODOLOGIES OF PAPER AND PULP WASTEWATER

17.4.1 Physico-Chemical Treatments

Physico-chemical procedures are utilized to eradicate suspended solids, poisonous compounds, colloidal particles, colors, and floating matters from wastewater. Such procedures consist of ozonation, screening, electrolysis, coagulation, flocculation, sedimentation, and ultrafiltration. Physico-chemical processes are commonly used to treat wastewater at the preliminary, primary, and tertiary stages. The amount of impurities found in wastewaters and the ideal efficiencies of eradication of these impurities are significant factors in selecting the kind of physico-chemical process for its treatment. The existence of lignin and its derivatives add to strong color in wastewater coming from the P&P industries. These wastewaters simultaneously consist of high levels of floating matter and suspended solids. As a result, using a primary treatment method like sedimentation becomes vital for the overall treatment process. Thompson et al. (2001) found that 80% of the suspended matter from wastewater can successfully be removed by the process of sedimentation. Bhattacharjee et al. (2007) accomplished 87% and 60% total solid removal for the treatment of kraft black liquor by combining sedimentation with ultrafiltration and adsorption, respectively. The P&P industry does not commonly use the technique of ultrafiltration, despite the fact that it is a far superior and efficient method to remove waste. This can be attributed to the high expenses related to the ultrafiltration method. Various methods were compared like hydrogen peroxide oxidation, ozonation, Fenton's oxidation, and combined ozonation for extracting color and COD from the wastewater produced by P&P industries. The outcomes of this study demonstrated that simple ozonation along with ozonation with hydrogen peroxide were able to remove the color. On the other hand, the efficiency for COD removal was not satisfactory. At the same time, the Fenton's oxidation process was the preferred method for efficiency in removing the COD and color. This procedure has a lower expense when compared to simple ozonation and the combination of hydrogen peroxide oxidation with ozonation. De los Santos et al. (2009) utilized chemical precipitation by sulfuric acid which is then followed by ozonation for treating the wastewater formed by paper-making industries with a high concentration of large molecular weight contaminants, which was successfully able to remove 96% and 60–70% of BOD and color, respectively. Albeit not many varieties of tertiary treatment processer are employed in these factories because of the high expenses associated with them. Coagulation and flocculation are used when it comes to further polishing of these effluents and toxins. It was analyzed that coagulation methods are effective in expelling the COD and suspended solids from wastewater formed by the P&P industries. The removal rate for COD in the study ranged between 20% and 96%, based on the characteristics of the wastewater and the treatment method conditions. The rate for the removal of color was about 80% which resulted in treated wastewater being safe to be used again. The treated papermaking wastewater utilized electro-flotation and coagulation to expel suspended solids, COD, and color, but despite the elimination of suspended solids and color, the outcome of BOD and COD removal rates was not totally up to the mark under all tested circumstances. Hence, a combination of the biological process along with these processes must be used for satisfactory results.

There are quite a few existing treatment methods for P&P industrial waste discharge, consisting of adsorption treatment processes, flocculation–coagulation, ozonation, anaerobic, aerobic, photocatalysis, and electrochemical (Amor et al. 2019; Kamali and Khodaparast 2015). The amount features and type of wastewater are important to find the most efficient method for treatment. Removal rates of 90% BOD, 70% COD, and 40–60% absorbable organic halides were achieved with the activated sludge process, but the pulp and paper factory effluents being treated at activated sludge plants make the plant prone to continuous operational issues like blocking and bulking. Peerbhoi et al. (2003) explored the black liquor's anaerobic treatability by an up-flow anaerobic sludge blanket reactor and reported that it is not feasible for treating black liquor anaerobically because the pollutants present were not promptly biodegradable. Outstanding polishing results of effluents from the paper industry treated biologically were reported with regard to color by treating them with activated carbon. Coagulation when coupled with wet oxidation was successful in removing 51% of COD, 83% of color, and 75% of lignin from the wastewater of P&P industries (Büyükkamaci and Koken 2010). Afzal et al. (2008) analyzed the biological and natural treatment of pulp and board wastes by fed batch reactor (FBR), which is subsequently sent for coagulation followed by sand filtration. The combined treatment method reduced COD by 93% and BOD by 96.5%, respectively, and the waste particles could then be released into the water bodies without posing any danger. The coagulants which are most frequently incorporated in treating industrial wastewater are magnesium sulfate ($MgSO_4$), calcium hydroxide Ca $(OH)_2$, polyelectrolytes, aluminium chloride, alum, ferrous sulfate ($FeSO_4$), and lime. The decision to choose a particular coagulation reagent is mainly dependent on its efficiency, followed by the cost of the reagent compared to other available reagents.

In large P&P producing factories, pulp is mainly created by a method called kraft pulping which uses wood-based, raw, unprocessed material as input. In this procedure, the highly alkaline medium is used to digest the raw wood in a very hot environment (150–190°C) to make pulp free of lignin and then is accordingly used in the process of making paper. Also, a significant percentage of hemicellulose is extracted during the digestion of raw material in the process. The procedure leads to the dissolution of lignin and other organic materials in a liquid-based solution commonly called black liquor. This kraft black liquor has a pH value which is considered very high. This liquid is also characterized by low biodegradability. The processing of diluted black liquor by physicochemical processes (e.g., coagulation–flocculation and acid precipitation) leads to the production of concentrated lignin and treated wastewater that is characterized by a high recycling potential. Physico-chemical methods have shown great prospects for the processing of waste streams from various industries (Kamali and Khodaparast 2015). Procedures like coagulation–flocculation and wet oxidation–peroxidation have been used for the pre-treatment of pulp and paper mill waste effluents. The process of oxidation removes the organic impurities present in wastewater and is not reasonable in recovery of lignin when compared to other methods like precipitation/coagulation–flocculation processes. Aluminum and iron salts are commonly used as coagulants in the processes of wastewater treatment. Garg et al. (2010) demonstrated a color and COD reduction of 90% and 63%, respectively, from the diluted black liquor while the coagulant was alum. The use of iron-based coagulants (e.g., ferrous sulfate and ferric chloride) and aluminium-based coagulants (e.g., aluminium chloride, polyaluminium chloride, and alum) alone and when mixed with polymer flocculants (e.g., anionic and cationic polyacrylamide) for the treatment of pulping effluents has been broadly acknowledged. Maximum color and COD removals of 95% and 76%, respectively, were discovered when chlorides of Al and Fe were utilized with polyacrylamide at a pH value under 3. The settling of the sludge was additionally improved drastically by the use of flocculants (Lal and Garg 2017). Another fascinating area of utilizing coagulants before effluents from the P&P industry were treated was explored for its effect on lowering the requirements for aeration and, consequently, increasing the efficiency of the procedure (Elnakar and Buchanan 2019). Sandberg et al. (2018) discovered that using low quantities of ferric iron ($5g/m^3$) reduced the surface movement of surfactants generally found in pulping wastewaters and drastically improved the rate of oxygen transfer, thus leading to the improved biodegradability of the wastes.

17.4.2 BIOLOGICAL TREATMENTS

The many kinds of refractory organic waste pollutants contained in P&P mill waste effluents are damaging to nature and deeply affect abiotic and biotic elements. Some of the major drawbacks of using traditional physico-chemical procedures for waste treatment are large monetary expenditures, high running costs, and increased sludge production. The biotechnological techniques based on biological treatment are used in present treatment systems, in which a large variety of microorganisms consisting of actinomycetes, fungi, enzymes, and bacteria have been implicated. In a recent study it was observed that ligninolytic fungi (white rot and brown rot), because of their powerful lignin-degrading enzyme systems, are effective for cleaning up refractory pollutants from the P&P waste waters.

The majority of industrial plants treating wastewater use anaerobic and/or aerobic biological methods to extract organic impurities which are generally found in wastewater. The majority of P&P-based industries use aerobic methods owing to their comparatively low operational costs and also their relatively low capital and ease of operations. In the P&P industry, aerated lagoons and activated sludge (AS) methods are mostly employed when it comes to the use of anaerobic technology (Pokhrel and Viraraghavan 2004). Despite the fact that the anaerobic processes are not commonly used in the P&P industry, various factories have utilized several anaerobic technologies on account of their low production of sludge, production of sustainable energy (biogas), facilitation in the further degradation of harmful contents, and smaller area requirements. Both anaerobic and aerobic processes have their respective disadvantages which consist of sensitivity of anaerobic bacteria to poisonous materials and the production of large amounts of sludge in aerobic processes. Anaerobic digestion (AD) is the process which consists of the degradation of organic compounds biologically into various final products like hydrogen (5–10%), carbon dioxide (25–50%), nitrogen (1–2%), and methane (50–75%) by a microbial consortium without the presence of air (Siles et al. 2010). This has been broadly utilized for the primary level or secondary level treatment of various factory residues. The reason behind this is the advantages provided by anaerobic digestion AD over the standard biological paper and pulp residue treatment like the notable reduction of the wastes produced and the amount of biogas produced, primarily consisting of methane. However, lignocellulosic materials and other refractory pollutants present in paper and pulp residue have the capability to reduce the rate of hydrolysis and stability of the method, and generate methane. Therefore, so as to take advantage of the benefits offered by various treatment methods, integrated treatment methods have been utilized to treat the wastewaters produced from P&P industries, comprising combined biological procedures operating under a variety of environmental conditions (anaerobic and aerobic), or biological processes and physico-chemical processes.

In recent years, numerous fungal strains have been researched and experimented on to find out more about their abilities in degrading a large amount of structurally diverse toxins. White-rot fungi create a large number of extracellular enzymes like phenol oxidase, lignin peroxidase, manganese independent peroxidase, manganese dependent peroxidase, and laccase which decompose the steady natural compounds like cellulose, lignin, and hemicellulose. A recent experiment took ligninolytic enzymes out of filamentous fungi, as *Aspergillus fumigatus* and *Aspergillus flavus* demonstrated effective removal of color from P&P mill effluents. Extensive evaluation of *Aspergillus foetidus* was done to discover how useful it is in the process of removing lignin, COD, and color from wastewater, and this research displayed positive outcomes. About 90–95% of the overall color was successfully eliminated and the concurrent decline in lignin and color demonstrated a strong relation between the lignolytic processes and decolorization. The waste produced at the pulping stage (black liquor) consists of a number of compounds such as dissolved lignin and its degradation products, resin acid, phenols, fatty acids, hemicelluloses, and tannins. Many researchers have proposed that oleaginous yeasts *Yarrowia lipolytica*, *Candida cylindracea*, *Trichosporon cutaneum*, *Saccharomyces* sp., *Candida rugosa*, and *Candida tropicalis* have been utilized in the elimination of phenol-based compounds and other poisonous substances present in P&P mill wastewaters (Pokhrel and Viraraghavan 2004). Some studies also demonstrate the role of oleaginous yeast *Rhodosporidium kratochvilovae* HIMPA1 in an integrated method to deal with P&P mill wastes for the elimination of poisonous organic substances and the concurring creation of biodiesel from the gathered intracellular lipids. This has provided gentle solutions for the appropriate and legitimate disposal of wastewaters in an eco-friendly way, simultaneously with a useful by-product that is renewable and environmentally friendly. In a novel approach of treating our biological wastewater, the usage of enzyme technology along with biotechnology is becoming extensively common in a variety mills and factories. Xylanase is an example of such an enzyme which has gathered recent fame in this industry over the past few years. These are the protein enzymes which are hydrolytic in nature and which randomly split the β-1,4 strength of the complex wall of the plant cell polysaccharide xylan, which is the most commonly found polysaccharide found in several agro-industrial, agricultural, and wood wastes. Generally, xylanases are generated by *Aspergillus*, *Trichoderma*, *Penicillium*, *Talaromyces* sp., and *Aureobasidium*. The formation of xylanolytic and xylanases enzymes by *Penicillium* has been deeply and studied and explored. Moreover, xylanases free of cellulase are significant in pulp biobleaching as a replacement for the utilization of harmful compounds (which are chlorinated) because of the dangers and threat posed to our environment and the ecosystem which leads to the spreading of many harmful diseases by the release of absorbable organic halogens (Sridevi et al. 2019).

In a study conducted by Sonkara et al. (2019), a strain of novel bacteria separated from P&P industry effluents showed maximum decolorization and degradation when it was mixed with phosphorus and nitrogen in an ordered batch treatment. The treatment was done under an optimized environment and it was found that the greatest decrease in BOD, color, COD, and TOC was 93.33%, 73.01%, 89.50%, and 82.22%, respectively. Also, a decrease in the total lignin and phenol was reported at 64.10% and 88.5%, respectively. A notable degradation of the derivatives of chlorolignin and phenolic was also observed. The outcomes highlighted the decolorization and degradation scope of *Bacillus* sp. IITRDVM-5 and the various applications which can be used in pulp and paper waste treatment in the near future (Sonkara et al. 2019). Different species of bacteria were extensively studied for their decolorization and degradation capabilities in paper and pulp effluents (Hossain and Ismail 2015). Superior microbes like *Micrococcus luteus* and *Bacillus subtilis* were discovered to be capable of lowering lignin content up to 97%, COD up to 94.7%, and BOD up to 87.2% in P&P factory effluents. Under aerobic conditions, the cultures of *Pseudomonas aeruginosa* have the capability to reduce kraft mill effluent color by somewhere between 26% and 54% or even more. Intense studies were done on two strains of bacteria, *Acienetobacter calcoaceticus* and *Pseudomonas putida*, for their degradation abilities of black liquor from a kraft P&P industry in a continuous reactor. The strains of bacteria successfully removed 70–80% of lignin and around 70–80% of COD, while the efficiency of color removal was found to be around 80%. *Ancylobacter Methylobacterium* and *Pseudomonas* strains were studied to evaluate their corresponding abilities of organochlorine removal from paper mill effluents and bleached kraft pulp. They were also experimented for growth on alcohols and chlorinated acetic acids for absorbable organic halogen reduction or, in short, AOX reduction in the batch treatment of sterile bleached kraft mill effluents from a variety of sources. The outcomes of this study showed that *Methylobacterium* exhibits a meagre range of substrates, yet was capable of removing notable quantities of absorbable organic halogens/AOX from both softwood and hardwood effluents. On the other hand, *Ancylobacter* displayed the widest substrate range but was only able to effect large absorbable organic halogens/AOX reductions in softwood effluents. So, in comparison, *Pseudomonas* revealed a finite substrate range and below par range of negligible reduction in levels of AOX from hardwood and softwood effluents. The separated bacteria like *Enterobacter* spp., *Citrobacter* spp., and *Pseudomonas putida* can decolorize effluents up till 97%, and can also lower the phenolics, BOD, sulphide, and COD content by 96.92%, 96.63%, 96.67%, and 96.80%, respectively, and also can remediate heavy metals by 82–99.80% (Tiku et al. 2010).

Effective *in situ* phytoremediation (use of plants) techniques have also been applied for reducing heavy metal loads from paper mill effluents (Kumar et al. 2014). Even though it seems to be optimistic, some problems exist in relation to this method. Phytoremediation requires long-term maintenance and it may be effective only seasonally. Moreover, the remediation efficiency is very low. As an alternative, phycoremediation employing microalgae and macroalgae for the removal of nutrients from wastewater is gaining much attention (Bansal et al. 2018). Studies on the cultivation of microalgae such as *P. nurekis* and *C. reinhardtii* in P&P mills for nutrient removal are broadly recognized. Similarly, the scope of *Planktochlorella* sp. and *C. reinhardtii* for treating P&P industry wastewater, and the subsequent biomass production and lipids with a bio-energy perspective, is gaining much attention (Sasi et al. 2020). Some reports mention that some algae like *Microcystis* spp. are capable of decolorizing diluted bleach kraft factory effluents in a sustainable manner (Iyovo et al. 2010). Studies also raise the fact that mixed and pure algae cultures extract up to 70% of color in under 60 days of being incubated. The decolorization by algae is a result of the metabolic transformation or adsorption mechanism of colored molecules to molecules that are colorless in nature with finite degradation or assimilation of molecular substances (Chandra and Singh 2012). Diverse algal cultures consisting of *Microcystis, Chlamydomonas, Chlorella*, and others were used in removing the absorbable organic halogens and color. A report found that there was an approximate reduction of 70% AOX and 80% color in 30 days in a continuously illuminated environment (Hossain and Ismail 2015).

17.5 OTHER EMERGING TREATMENTS

In addition to various physico-chemical and biological methods, other innovative methods like Fenton's reagent, UV treatment, photo-Fenton procedures, and electrocoagulation are now being incorporated in advanced methods of oxidation to lessen these toxic materials from paper and pulp wastes. Advanced oxidation processes (AOPs) are one of the highly useful and appropriate methods to eliminate the presence of refractory organic material. The idea behind AOP is the creation of hydroxyl free radicals (HO·) which are highly reactive chemical oxidants. This effective advanced treatment process is referred to as Fenton. When hydrogen peroxide (H_2O_2) is reacted with aqueous ferrous ions it leads to the creation of hydroxyl radicals (·OH) which are capable of removing refractory and poisonous organic pollutants found in wastewater (Deng and Zhao 2015). The strong oxidants have the ability to degrade recalcitrant organic toxins and also expel inorganic toxins from wastewater; hence, advanced oxidations processes are used to treat wastewater of various kinds. Thus, the organic compound is a total breakdown by the reaction of HO·/hydroxyl radicals which leads to the value of the pollutants being reduced drastically by AOP from about 100 ppm to a value at almost below 5 ppb. Hydroxyl free radicals have certain characteristics like a nonselective nature, and thus reacts quickly with a number of chemicals and possesses the ability to treat various pollutants at the same time owing to its non-selective oxidation and very high rate of reaction. The driving force behind AOPs is the presence of HO·/ hydroxyl free radicals which are very reactive in nature. From the various AOPs, the Fenton process is determined to be successful in treating unmanageable organic compounds. The AOP is a good method for the treatment of toxins which are generally rich in organic wastes because of their low oxidation and high reactivity rates (Srivastav et al. 2019). Some of the approved AOPs in use are TiO_2 photocatalysis (Affam and Chaudhuri 2013), advanced ozonolysis (Parsa and Negahdar 2012), electrochemical oxidation (Pipi et al. 2014), and Fenton's reaction (Romero et al. 2016). Amongst all the emerging AOPs, the process of Fenton's reaction is found to be amongst the best methods when it comes to the treatment of acidic wastewater that have

POPs. Many studies and tests have been done that show total toxin abatement and large values of mineralization in less reaction times. One drawback of the method is the formation of large quantities of sludge, which would need to be processed and even disposed properly after treatment. Hence, an improved method of fluidized-bed Fenton was incorporated to overcome this bottleneck in which carriers are used to lower the sludge production rate by crystallizing the target toxins onto the surface of the carrier (Segura et al. 2016). In further advancement over the years, nanotechnology has been coupled with the FBF process. Nano-Fe_2O_3 and FBF and photocatalytic nano-TiO_2 have been experimented on pulp and paper mill pollutants, depending on the properties of the pollutants, Anotai et al. (2018) discovered that 70–82% of color and 50–63% of COD can be extracted by this FEF method. At feasible retention times, 53–81% of total iron extraction was recorded. Hybrid processes to Fenton have also been performed over a couple of years. Grötzner et al. (2018) compared the efficiency of combined Fenton in sequence with the coagulation–flocculation–sedimentation (CFS) process to clean chemical thermal mechanical pulping (CTMP) pollutants. Considering only coagulation–flocculation–sedimentation, the process was efficient in removing about 36% lignin, 20% COD, and 78% total organic carbon (TOC) contents, and the Fenton treatment itself successfully reached removal rates of 63% for lignin contents, 52% for COD, and 79% for TOC. But, on merging Fenton and CFS, removal rates of 61% for lignin, 76% for COD, and 95% for TOC contents were achieved. In the same way, the Fenton process and membrane–filtration–coagulation were applied exclusively prior to ultrafiltration (UF) treatment for treating the secondary level pollutants from a recycled paper factory. Outcomes of this suggested that UF with Fenton pre-treatment had the best outputs, and a COD decline of 91.81% was achieved (Xu et al. 2019). In addition, membrane fouling and pore blocking were mitigated with this method (Elnakar and Buchanan 2019).

17.6 CONCLUSION

Despite modifications and recent advances in technologies for P&P industry wastewater treatment, various challenges are still faced by this industry. For sustainable development, all these issues need urgent attention to solving the difficulties faced. The challenges include: a) generation of large quantities of wastewater polluting the environment, b) improper waste disposal and the use of traditional methods of treatment at the factories, c) higher levels of COD, BOD, and TDS leading to toxicity in the aquatic environment, d) the policymakers from government agencies are not well connected with the industries, and e) the improper dumping of sludge and degradation of lignin polluting the water bodies and soil. The P&P industry utilizes various plant bioresources for agricultural waste residues as raw materials including eucalyptus, bamboo, bagasse, jute, wheat and rice straw, and grasses. The requirement of raw materials varies from mill to mill. The processes of pulp washing, bleaching, and cleaning will result in the release of diverse effluents in wastewater. Various chemical compounds present in effluents include lignin, cellulose, and hemicellulose. Effluents also include wood extractives like chlorinated and manomeric phenols, mercaptides, acetic and formic acid, quinone derivatives, methanol, methyl glyoxal, and enol ethers. Approximately 300 diverse organochlorine contaminants have been identified in wastewater from the P&P industry, bur hundreds more remain unidentified. P&P industry wastewater contains high levels of BOD, COD, TDS, and various refractory contaminants. Lignin and tannin will impart color to the effluent which will have a toxic impact on aquatic organisms. Such colored wastewater is effectively treated using fungal strains, chemical oxidation, ozonation, or coagulation techniques. Pulp bleaching will result in the release of chlorinated phenolic compounds which are highly toxic. These phenolic compounds are removed by membrane filtration, ozonation, and adsorption techniques.

In order to overcome the improper treatment of P&P industry effluents, various physical, chemical, and biological treatment methodologies are adopted. Physico-chemical treatments showed the capacity to remove suspended and floating solids and refractory contaminants. Precipitation, coagulation, and electrocoagulation methods were able to efficiently eliminate lignin and organic matter, and reduce turbidity and COD. Combinational technologies prove quite efficient in the treatment of wastewater from P&P industries. They include a combination of techniques such as electrochemical and membrane technologies that can eliminate COD, BOD, TSS, and sludge levels. The biologically resistant contaminants can be degraded by the use of various membrane filtrations as well as physico-chemical treatment methods like oxidation and adsorption. Oxidation is a comparatively costly process which can be used for wastewater treatment and can remove color, COD, BOD, and various other organic contaminants present. An important challenge is to reduce the amount of sludge generated which is achieved by the use of membrane bioreactors, especially the upflow anaerobic sludge blanket reactor. Despite the many treatment techniques available, biological treatment methodologies are most commonly used as they also produce useful by-products which are in demand. Combinational treatment (biological and physico-chemical) processes can provide a long-term solution for dealing with problems faced by P&P industry wastewater treatment so as to develop mills with zero discharge.

REFERENCES

Affam AC, Chaudhuri M 2013 Degradation of pesticides chlorpyrifos, cypermethrin and chlorothalonil in aqueous solution by TiO_2 photocatalysis. *J Environ Manag* 130:160–165.

Afzal M, Shabir G, Hussain I, Khalid ZM 2008 Paper and board mill effluent treatment with the combined biological–coagulation–filtration pilot scale reactor. *Bioresour Technol* 99:7383–7387.

Amor C, Marchão L, Lucas MS, Peres JA 2019 Application of advanced oxidation processes for the treatment of recalcitrant agro-industrial wastewater: A review. *Water* 11:205.

Anotai J, Wasukran N, Boonrattanakij N 2018 Heterogeneous fluidized bed fenton process: Factors affecting iron removal and tertiary treatment application. *Chemosphere* 229(2):40–49.

Ansari PM 2004 Water conservation in pulp and paper, distillery. In: Indo-EU Workshop on Promoting Efficient Water Use in Agro Based Industries. New Delhi. January 15–16.

Bansal A, Shinde O, Sarkar S 2018 Industrial wastewater treatment using phycoremediation technologies and co-production of value-added products. *J Bioremediat Biodegrad* 9(1):1–10.

Bhattacharjee S, Datta S, Bhattacharjee C 2007 Improvement of wastewater quality parameters by sedimentation followed by tertiary treatments. *Desalination* 212:92–102.

Büyükkamaci N, Koken E 2010 Economic evaluation of alternative wastewater treatment plant options for pulp and paper industry. *Sci Total Environ* 408:6070–6078.

Chandra R, Singh R 2012 Decolourisation and detoxification of rayon grade pulp paper mill effluent by mixed bacterial culture isolated from pulp paper mill effluent polluted site. *Biochem Eng J* 61:49–58.

Chen HW, Hsu CH, Hong GB 2012 The case study of energy flow analysis and strategy in pulp and paper industry. *Energy Policy* 43:448–455.

De los SRW, Poznyak T, Chairez I, Cordova RI 2009 Remediation of lignin and its derivatives from pulp and paper industry wastewater by the combination of chemical precipitation and ozonation. *J Hazard Mater* 169:428–434.

Deng Y, Zhao R 2015 Advanced oxidation processes in wastewater treatment. *Curr Pollut Rep* 1(3):167–176.

Elnakar H, Buchanan ID 2019 Pulp and paper mill effluent management. *Water Environ Res* 91:1069–1071.

Garg A, Mishra IM, Chand S 2010 Effectiveness of coagulation and acid precipitation processes for the pretreatment of diluted black liquor. *J Hazard Mater* 180(1–3):158–164.

Grotzner M, Melchiors E, Schroeder LH, dos Santos AR, Moscon KG, de Andrade MA, Xavier CR 2018 Pulp and paper mill effluent treated by combining coagulation-flocculation-sedimentation and fenton processes. *Water Air Soil Pollut* 229(11):1–7.

Hossain K, Ismail N 2015 Bioremediation and detoxification of pulp and paper mill effluent: A review. *Res J Environ Toxicol* 9(3):113–134.

Iyovo GD, Du G, Chen J 2010 Sustainable bioenergy bioprocessing: Biomethane production, digestate as biofertilizer and as supplemental feed in algae cultivation to promote algae biofuel commercialization. *J Microb Biochem Technol* 2:100–106.

Kamali M, Khodaparast Z 2015 Review on recent developments on pulp and paper mill wastewater treatment. *Ecotoxicol Environ Saf* 114:326–342.

Kumar BA, Jothiramalingam S, Thiyagarajan SK, Hidhayathullakhan T, Nalini R 2014 Phytoremediation of heavy metals from paper mill effluent soil using *Croton sparsiflorus*. *Int Lett Chem Phys Astro* 36:1–9.

Lal K, Garg A 2017 Physico-chemical treatment of pulping effluent: Characterization of flocs and sludge generated after treatment. *Sep Sci Technol* 52(9):1583–1593.

Lei X, Zhao Y, Li K, et al. 2012 Improved surface properties of CTMP fibers with enzymatic pretreatment of wood chips prior to refining. *Cellulose* 19:2205–2215.

Nagarthnamma R, Bajpai P, Bajpai PK 1999 Studies on decolourization, degradation and detoxification of chlorinated lignin compounds in kraft bleaching effluents by *Ceriporiopsis subvermispora*. *Process Biochem* 34:939–948

Omid AO, Yerushalmi L, Haghighat F 2015 Wastewater treatment in the pulp-and-paper industry: A review of treatment processes and the associated greenhouse gas emission. *J Environ Manage* 158:146–157.

Ortega-Clemente A, Caffarel-Méndez S, Ponce-Noyola MT, Barrera-Córtes J, Poggi-Varaldo HM 2009 Fungal post-treatment of pulp mill effluents for the removal of recalcitrant pollutants. *Bioresour Technol* 100(6):1885–1894.

Pan Y, Xiao H, Zhao Y, Wang Z 2013 CTMP-based cellulose fibers modified with core–shell latex for reinforcing biocomposites. *Carbohydr Polym* 95:428–433.

Parsa JB, Negahdar SH 2012 Treatment of wastewater containing Acid Blue 92 dye by advanced ozone-based oxidation methods. *Sep Purif Technol* 98:315–320.

Peerbhoi ZM, Mehrotra I, Shrivastava AK 2003 Treatability studies of black liquor by upflow anaerobic sludge blanket reactor. *J Environ Eng Sci* 2:307–313.

Pipi ARF, de Andrade AR, Brillas E, Sirés I 2014 Total removal of alachlor from water by electrochemical processes. *Sep Purif Technol* 132:674–683.

Pokhrel D, Viraraghavan T 2004 Treatment of pulp and paper mill wastewater: A review. *Sci Total Environ* 33:37–58.

Romero V, Acevedo P, Marco J, Esplugas S 2016 Enhancement of Fenton and photo-Fenton processes at initial circumneutral pH for the degradation of the β-blocker metoprolol. *Water Res* 88:449–457.

Sandberg M, Venkatesh G, Granstrom K 2018 Experimental study and analysis of the functional and life-cycle global warming effect of low-dose chemical pre-treatment of effluent from pulp and paper mills. *J Cleaner Prod* 174:701–709.

Santos MO, Almada-Lobo B 2012 Integrated pulp and paper mill planning and scheduling. *Comput Ind Eng* 63:1–12.

Sasi PKS, Viswanathan A, Mechery J, Thomas DM, Jacob JP, Paulose SV 2020 Phycoremediation of paper and pulp mill effluent using *Planktochlorella nurekis* and *Chlamydomonas reinhardtii*: A comparative study. *J Environ Treat Tech* 8(2):809–817.

Segura SG, Bellotindos LM, Huang YH, Brillas E, Lu MC 2016 Fluidized-bed Fenton process as alternative wastewater treatment technology: A review. *J Taiwan Inst Chem Eng* 67:211–225.

Siles JA, Brekelmans J, Martín MA, Chica AF, Martín A 2010 Impact of ammonia and sulphate concentration on thermophilic anaerobic digestion. *Bioresour Technol* 101:9040–9048.

Sinclair WF 1990 *Controlling Pollution from Canadian Pulp and Paper Manufactures: A Federal Perspective*. Canadian Government Publishing Centre, Ottawa.

Singh A, Yadav RD, Kaur A, Mahajan R 2012 An ecofriendly cost effective enzymatic methodology for deinking of school waste paper. *Biores Tech* 120:322–327.

Singh C, Chowdhary P, Singh JS, Chandra R 2016a Pulp and paper mill wastewater and coliform as health hazards: A review. *Microbiol Res Int* 4(3):28–39.

Singh P, Jain P, Verma R, Jagadish RS 2016b Characterization of lignin peroxidase from *Paecilomyces species* for decolourisation of pulp and paper mill effluent. *J Sci Ind Res* 75:500–505.

Singhal A, Thakur IS 2009a Decolourization and detoxification of pulp and paper mill effluent by *Cryptococcus sp. Biochem Eng J* 46(1):21–27.

Singhal A, Thakur IS 2009b Decolourization and detoxification of pulp and paper mill effluent by *Emericella nidulans*. *Biochem Eng J* 171:619–625.

Sonkara M, Kumar M, Dutta D, Kumar V 2019 Treatment of pulp and paper mill effluent by a novel bacterium Bacillus sp. IITRDVM-5 through a sequential batch process. *Biocatal Agric Biotechnol* 20:1–10.

Sridevi A, Narasimha G, Devi PS 2019 Production of xylanase by Penicillium sp. and its biobleaching efficiency in paper and pulp industry. *Int J Pharm Sci Res* 10(3):1307–1311.

Srivastav M, Gupta M, Agrahari SK, Detwal P 2019 Removal of refractory organic compounds from wastewater by various advanced oxidation process: A Review. *Curr Environ Eng* 6:8–16.

Thompson G, Swain J, Kay M, Forster CF 2001 The treatment of pulp and paper mill effluent: A review. *Bioresour Technol* 77:275–286.

Tiku DK, Kumar A, Chaturvedi R, Makhijani SD, Manoharan A, Kumar R 2010 Holistic bioremediation of pulp mill effluents using autochthonous bacteria. *Int Biodeterior Biodegrad* 64:173–183.

Vepsäläinen M, Kivisaari H, Pulliainen M, Oikari A, Sillanpää M 2011 Removal of toxic pollutants from pulp mill effluents by electrocoagulation. *Sep Purif Technol* 81:141–150.

Xu Y, Li Y, Hou Y 2019 Reducing ultrafiltration membrane fouling during recycled paper mill wastewater treatment using pre-treatment technologies: A comparison between coagulation and fenton. *J Chem Technol Biotechnol* 94(3):804–811.

Zhang X, Renaud S, Paice M 2008 Cellulase deinking of fresh and aged recycled newsprint/magazines (ONP/ OMG). *Enzyme Microb Technol* 43(2):103–108.

Zhenying S, Shijin D, Xuejun C, Yan G, Junfeng L, Hongyan W, Zhang SX 2009 Combined de-inking technology applied on laser printed paper. *Chem Eng Process: Process Intensif* 48(2):587–591.

18 Challenges Pertinent to Phytoremediation
A Future Prospect

Nitika Gaurav and Ananya Das

CONTENTS

18.1 INTRODUCTION

The explosive growth of the human population has maximized the need for technologies, and has fueled industries being set up in far-flung areas of the world. The era of modernization and industrialization has transformed countries from developing to developed but has conversely affected the health of man and the ecosystem. In addition, natural catastrophes, such as soil erosion, wildfires, landslides flooding drought, volcanic eruptions etc. The concentration of pollutants may vary from lethal to mild, however, exposure to even minute concentrations could be deleterious to agriculture, human life, and the overall biosphere. The effluxes of natural and anthropogenic activities constitute both organic and inorganic pollutants. Organic pollutants are usually degraded in the ecosystem by microbial action, while inorganic elements remain in undegraded or partially degraded form. These non-degradable inorganic pollutants, particularly heavy metals, are enriched through the food chain and can persist in the soil matrices and water bodies without any intervention. Some significant heavy metals present as pollutants are mercury (Hg), lead (Pb), cadmium (Cd), arsenic (As), nickel (Ni), zinc (Zn), and copper (Cu). Heavy metals are required in small quantities for plant growth, however, excess concentrations can cause cell damage and disturb the physiological and biochemical activities. The consumption of plants with high metal concentrations by humans and animals can lead to different morbid conditions that include cancers, and physiological and neurological disorders. About 16% of the total land area of Europe, >19.4% of agricultural land of China, and

DOI: 10.1201/9781003204442-18

600,000 ha of brown fields of America are affected by heavy metal contamination. Their overabundance in the biosphere is a potential threat to the health of humans and the ecosystem, and thus requires immediate action to clean up the polluted sites (Aguilar et al., 2008).

Several technology-based conventional remediation methods such as soil washing and flushing, chemical and vapor extraction, and excavation have been used to rehabilitate contaminated soil. However, the high maintenance cost and low efficiency (low limit of cleaning process turnover) de-escalates the purpose and eventually results in abandoned contaminated sites. Also, the requirement of soil storage in these processes enhances the risk of secondary site contamination. These are some notable factors that together complicate the use of conventional remediation processes and brought attention to an alternative method of using green technology, i.e., bioremediation. This technique involves the strategy of exploring the natural propensity of micro-organisms and plants to rectify contaminants from the ground. The use of natural processes not only guarantees reduced costs but is also efficient.

The involvement of micro-organisms (bacteria and fungi) in the clean-up process proved to be a good alternative, but the process has mixed outcomes. The presence of optimal environments, and the availability of essential sources and stimulators are some of the factors that determine the success.

Plants, being autotrophic organisms, overcome these limitations by using external sources of energy and food, i.e., sunlight and carbon dioxide. The ability of plants to uptake water and other essential components from the ground is exploited in the bioremediation process to remove contaminants from the soil and water bodies. During the uptake of water and nutrients, plants also take up other elements that include the heavy metals present in the soil. The uptake and accumulation of heavy metals in the shoot system is used as a defense mechanism by some plant species. These factors that highlight the advantage of using plants for soil rehabilitation purposes gave rise to the concept of phytoremediation. Some plants has natural propensity of heavy metal accumulation, Brassica and Eicchornia for instance. However, some plant species are also modifies genetically, to integrate bioaccumulation property or to catalyse the process of heavy metal absorption by the4 plants. I addition to the removal of contaminants from soil and water bodies, plants are also capable of metabolising, volatizing and transforming heavy metals in less toxic forms. The heavy metal absorption by the upper parts of some plant species is also utilized for metal extraction process. The high biomass yield of certain plant species is another key feature that promotes the employment of the plant based remediation technique. The phytoremediation process is sub-categorized into different forms, each involvings different strategies to extract heavy metal pollutants: phyto-stabilization, phytoextraction, phytovolatization, phytomining, and phytofiltration/ rizhofiltation. Phytoextraction and Phytomining are mostly used for metal recovery, while phytovolatization is the strategy exerted releases contaminants in volatile form. Phytostabilization utilizes the relationship with microbial communities to remediate contaminated sites, whereas phytofiltration absorbs or precipitates and prevents the migration of the contaminants. Recently, a method of transformation and genetic manipulation has been employed to ameliorate the performance of phytoremediation. The genes from different plants and animals that possess a unique mechanisms of metal tolerance are used to engineer plants that are fast growing and can survive in different climatic conditions.

Though the phytoremediation process is an eco- and budget-friendly technique, it is also a challenging task owing to its slow speed, technical limitation, and potential environmental risks. This chapter mainly discuss the challenges associated with the application of phytoremediation technology and its implications (see Table 18.1).

18.2 AN IN-DEPTH OVERVIEW OF PHYTOREMEDIATION

Plants are the basis of phytoremediation technology, but not every plant has the ability to survive in metal-contaminated soil, therefore selecting the correct plant species is of prime importance for a fruitful phytoremediation process. The plant species with a natural propensity for heavy metal accumulation are often used for remediation purposes, and these plants are termed

TABLE 18.1
Potential Plants with Metal Uptake Capacity

GM plants	Genes	Metal Tolerance
Nicotiana tabacum	MT2 gene (human)	Cadmium (Cd)
(tobacco)	MT1 gene (mouse)	Cadmium (Cd)
Nicotiana glauca	CUP-1 gene (yeast)	Copper (Cu)
	Cysteine synthetase (rice)	Cadmium (Cd)
	CAX-2 (*Arabidopsis*)	Cadmium (Cd), Calcium (Ca), Manganese (Mn)
	At MHX (*Arabidopsis*)	Magnesium (Mg) and Zinc (Zn)
	Nt CBP4 (tobacco)	Nickel (Ni) and Lead (Pb)
	FRE-1 and FRE-2 (yeast)	Iron (Fe)
	Ferritin (soyabean)	Iron (Fe)
	Phytochelatin synthetase (wheat)	Lead (Pb)
Arabidopsis	Mercury ion reductase gene	Mercury (Hg)
Arabidopsis thaliana	Glutathione-s- transferase (tobacco)	Aluminum (Al), Copper (Cu), Sodium (Na)
	Citrate synthetase (bacteria)	Aluminum (Al)
	Zn transporters ZAT (*Arabidopsis*)	Zinc (Zn)
	Znt A-heavy metal transporters (*E.coli*)	Cadmium (Cd) and Lead (Pb)
	YCF1 (yeast)	Cadmium (Cd) and Lead (Pb)
	Se-cys lyase (mouse)	Selenium (Se)
	Selenocysteine methyltransferase	Selenium (Se)
Populus sp.	Mercuric reductase (bacteria)	Mercury (Hg)
	γ-glutamyl-cysteine synthetase (bacteria)	Cadmium (Cd)
Oryza sp.	Nicotinamine amino transferase-NAAT (barley)	Iron (Fe)
	Ferritin (soyabean)	Iron (Fe)
Brassicaceae	γ-Glutamylcysteine synthetase (*E.coli*)	Cadmium (Cd)
(Indian mustard)	Glutathione synthetase (rice)	Cadmium (Cd)
	Arsenate reductase γ- glutamylcysteine (bacteria)	Arsenic (As)
	ATP sulfurylase CAPS	Selenium (Se)
	Cystathione-gamma synthase (CGS)	Selenium (Se)
	Glutathione reductase	Cadmium (Cd)

as **hyperaccumulators**. Hyperaccumulator plant species can uptake, translocate, and accumulate heavy metals in the root–shoot system of the plant without displaying any phytotoxic symptoms. This phenotype is advantageous to plants in a way to protect them from micro-organisms (pathogens) and herbivores. The content limits of the metals in the dry biomass of the hyperaccumulators vary with different metal elements; thus the metal bioconcentration factors of these species re greater than one.

The other plant group to which the majority of species belong is **excluders**, where the roots of plants absorb metals from contaminated sites. These species of plants prevent the toxic effect induced by heavy metals, restricting the movement of contaminants to above-ground parts of the plants. They limit root-to-shoot translocation either by using the efflux mechanism or by restricting the migration of pollutants to the root system. The majority of plant species are classified as excluders.

Baker (1981) suggested the use of both hyperaccumulators and excluders on the basis of the different strategies adopted by these plant species to survive in polluted areas. The other types of plant species that are sensitive to the presence of metal contamination in soil are known as **indicators**. For remediation purposes, indigenous hyperaccumulators are often preferred over exotic species as they can easily acclimatize in that particular environment. However, the introduction of an exotic species is sometimes required to remove specific metal pollutants.

Plant selection is an important attribute influencing the outcome of phytoremediation. Thus, an ideal hyperaccumulator should have the following characteristics to overcome the limitations of phytoremediation technology (Suman et al., 2018):

- Tolerant to other pollutants present at that particular site
- Tolerant to different adverse climatic situations and environments
- Capable of uptaking, translocating, and accumulating high concentrations of metals
- Have a deep-root system
- Fast growing and produce high biomass yields
- Resistant to disease and pests

The hyperaccumulator species that remain restricted to metalliferous sites are **obligate** hyperaccumulators, while **facultative** hyperaccumulators can be found at both metalliferous sites and non-metalliferous sites.

About 500 plant species are known to have hyperaccumulation properties (as of 2015), where the majority of the plant species belong to the families *Brassicaceae, Euphorbiaceae, Fabaceae, Proteaceae, Scrophulariaceae, Myrtaceae, Caryophyaceae, Rubiaceae, Tiliaceae,* among others.

In general, there are three different types of hyperaccumulators used for phytoremediation purposes: (i) natural hyperaccumulators, (ii) fast-growing plant species with a high biomass production, and (iii) genetically engineered plants. These are discussed below.

- **Natural hyperaccumulators:** These plant species have excellent capacity for heavy metal uptake and accumulation. They are mainly indigenous species and are often specific to the type of metals. However, these species are slow growing and have a low biomass production rate. Due to their autochthonous nature, these species are difficult to grow in non-native areas.
- **Non-hyperaccumulators with high biomass production:** These plant species are fast growing, easy to harvest, and have high biomass yields. They have deep root systems, making them efficient for water uptake and transpiration. These plant species are advantageous as they produce high-value biomass, e.g., biogas and biofuel production. These kind of woody non-edible plants can also prevent soil erosion and thereby transferring the contaminants from one region to other. Also, these kind of plants can be later used for uptake of the contaminants (not accumulated in a large content), but useful and stands as a saviour in many cases.
- **Genetically modified plants:** Naturally occurring hyperaccumulators are efficient in accumulating heavy metals; however, to overcome the challenges of phytoremediation technology, these plants need to have additional features that make them suitable candidates. Thus, to ameliorate the performance and suitability of plants, the strategy of engineering plants is employed either by modifying endogenous genes or by expressing genes of other species with metal tolerance features. The plants used for genetic engineering purposes are usually fast growing with deep root system and are tolerant to different adverse conditions.

The bioconcentration factor is another attribute of plants that is required when evaluating plant species for use in phytoremediation. Bioconcentration defines the ratio of metal present in soil and plants: the ratio >1 shows a higher occurrence of metals in the plant than the soil. It is associated with the ability of a plant to uptake and accumulate metals from the soil against a concentration gradient. The other important factor that determines the detoxification capacity of the plant is the translocation factor (TF). This factor describes the efficient root-to-shoot translocation of heavy metals, with high tolerance limits. The bioconcentration factor (BCF) and TF varies for different types of metals.

The steps involved in phytoremediation include uptake, translocation, sequestration into the vacuole, and metabolization.

18.2.1 Uptake Stage

The uptake of pollutants by plants depends to a great extent on the bioavailable concentrations of pollutants. Bioavailable concentration is referred to as the concentration of certain substances (in this case pollutants) which can be taken up by plants, microbes, genetically engineered organisms, etc. There are many factors that influence the uptake of pollutants from the soil by plants, some of which include the organic matter content, soil, plant age, competitive pollutants, and root exudates (refer to Figure 18.1) (McKeehan, 2000; Prasad, 2011). These factors directly or indirectly dominate the uptake process either by hindering or catalyzing metal leaching in the soil or ascending/descending in the uptake mechanism of the plants (Kumar et al., 2017).

18.2.2 Translocation Stage

Translocation traditionally mean the movement of one particle from one place to another. This movement can be gradient dependent, so the establishment of green plants at contaminated sites involves a few techniques before the actual translocation starts; these are phytostabilization, phytoimmobilization, and phytoextraction, all of which are involved in translocation. The concept of using green plants of optimum age for green remediation measures for the uptake and degradation of harmful pollutants from the soil is an important consideration.

18.2.2.1 Phytostabilization

Phytostabilization is a phenomenon which works with the decrease in bioavailable concentrations of mobile contaminants in the soil (Pilon & Pilon 2002). With passing time, the ability of the plants to accumulate contaminants on the factors relative to accumulation are called phytodeposition and phytosequestration. The multiple mechanism involved in the fixation of the contaminants such as metals are absorbed or accumulated by the roots. Assimilation at the root surface, growth and transpiration, physical stabilisation of the soil, organic binding of the root zones, due to the chemicals produced in the zones of roots are parts of the process of phytostabilizations.(Lister 2020). So, unlike other fast technologies, phytoremediation and phytostabilization are not suitable for the remediation of sites that are contaminated, but are rather used for a reduction in the percentage of

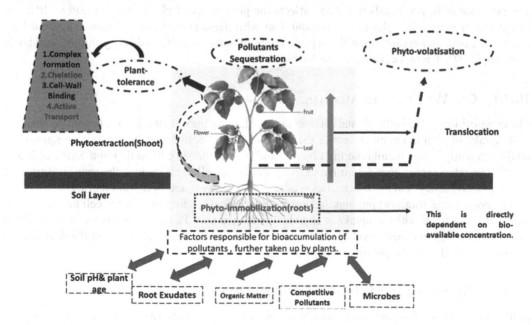

FIGURE 18.1 Factors responsible for the uptake of pollutants from the soil by plant organisms.

the pollutants of that area, which is achieved using different techniques such as sequestration of metal pollutants. Fixation can be enhanced by various processes (organic or inorganic) and additional substances from the soil act as catalysts in this process. The most effective plants that act as agents to hyperaccumulation with the aim of phytostabilization are species that accumulate heavy metals in the ground or below parts which restricts the flow of contaminants in the food chain. In recent times, commercially available plants and grasses are used for phytostabilization. (Panagos et al., 2013). A combination of grasses and trees used for rehabilitation purposes shows potential for phytoremediation. Grasses with branching systems help the surface water run-off and stop wind erosion. More roots in the system support soil formation, and later provides scaling for organic soil profile enrichment. The transpiration process effectively does not help in the loss of metals through ground surface phenomenon.

18.2.2.2 Phytoimmobilization

Plants prevent the transport of dissolved contaminants in the soil. In this process they are able to decrease the bioavailability and movement of metals and other metal-like materials in the root zone, thereby preventing uptake by the plants. There are a few plant-based mechanisms which are described such as the process of adsorption or absorption for the plant-assisted formation of insoluble compounds (Movahed & Maeiyat, 2009). There are certain species of microorganisms that participates in the fixation (immobilization), adsorption and uptake or synthesis of less mobile compounds.

The process of phyto immobilisation is very important , it helps in blocking of the complex metal compound and thus reducing the uptake by the plants. This in turn help in non transfer of the metals to the shoot sections (as their deposition occurs in the roots sections, initially). This ways harmful contamination in the case of edible plants are avoided.

18.2.2.3 Phytoextraction

Phytoextraction involves the use of plants for heavy metal removal from the contaminated matrix (soil and water) through uptake into the harvestable parts of the plant.

18.3 MECHANISMS INVOLVED IN POLLUTANT DETOXIFICATION BY PLANTS

The presence of heavy metals in the soil affects the physiology of plants and can lead to different symptoms in plants. In order to survive and deal with these contaminants, plants employ different mechanisms depending upon the type of contaminants (organic and inorganic) (Coutinho & Barbosa, 2007). These mechanisms are discussed below.

18.3.1 CELL WALL AND CELL MEMBRANE COMPOSITION

The composition of the cell wall and cell membrane plays an important role in determining the possible uptake of contaminants present in the soil. For instance, hydrophobic organic contaminants binds efficiently to the membrane lipid bilayer and hemicellulose cell wall (Pilon-Smits & Pilon, 2002). The other factors that determine the root uptake of pollutants include the transpiration rate, the uptake efficiency, and the concentration of contaminants present in the soil and water.

The presence of transport proteins and intracellular metal binding sites on the cell membranes of plant cells also helps with the uptake of chemicals from the soil. These chemicals are often required by the plant for growth and development. Sometimes, these chemicals also help in the degradation of toxic metals during the phytoremediation process (Yang et al., 2017).

18.3.2 RHIZOSPHERE AND ENZYMATIC TRANSFORMATION

Some plants also exploit their interaction with soil micro-organisms to reduce the negative effects of chemicals present in the soil. Rhizobial strains, particularly those strains that have both metal tolerance and plant growth activities, are useful for phytoremediation. Several rhizobial strains like *Ensifer meliloti, Azorhizobium, Rhizobium, Bradyrhizobium*, and *Mesorhizobium* have demonstrated greater tolerance to different metals. These microbial communities degrade the chemicals in less toxic compounds and thereby prevent the plant from their negative effects. Polychlorinated biphenyls, petroleum, and polycyclic aromatic hydrocarbons are some of the major pollutants degraded by rhizobia (Brígido & Glick, 2015).

The enzymatic transformation of toxic chemicals, particularly organic pollutants, is another strategy of phytoremediation and is considered as phytotransformation. In this process, plants employ three different steps to transform organic pollutants into less toxic metabolites: (1) modification by hydrolysis, oxidation, and reduction; (2) conjugation with sugars and amino acids; and (3) sequestration of these conjugates deposited in plant vacuoles or in the cell wall matrix. For instance, P450 monooxygenases, carboxylesterases, and glutathione S-transferases are some of the plant enzymes that are involved in the enzymatic transformation of toxic chemicals (Cherian & Oliveira, 2005).

18.3.3 COMPLEXATION AND COMPARTMENTALIZATION

The complexation process is mainly associated with heavy metal contaminants and can be an intracellular or extracellular process (Yang et al., 2005). The binding of metal ions present in soil with chelators such as organic acids, phenolics, siderophores increases the bioavailability. The intracellular complexation process involves the chelation of metals in cytosol using peptide ligands such as phytochelatins and metallothioneins. In the extracellular complexation process, the excreted metal chelators such as organic acids and siderophores help in increasing the bioavailability of metal cations by lowering the pH around the root. However, the synthesis of these metal chelators requires a high input of amino acids and other growth-limiting components, and thus their increased synthesis also affects plants growth and sometimes limits their use as phytoremediators. The compartmentalization process involves the vacuoles of plant cells which act as a store house for metals. The ligands (chelators)–metal complexes pumping to the vacuole of plant cells is another method of metal tolerance in phytoremediators.

18.4 ADVANTAGES AND DISADVANTAGES

The phytoremediation process employing plants for remediation purposes is not just an eco-friendly method but is also inexpensive and effective compared to other conventional and technology-based methods that are expensive and not as efficient. In addition, phytoremediation is also safe to use as the method is performed at the primary sites and thereby reduces the chances of secondary contamination. The production of high biomass, the transformation of pollutants into less toxic forms, the minimal site disruption, the prevention of soil removal, the enhanced soil revegetation, and the ability to be used in both terrestrial and aquatic environments are some of the other major advantages of the phytoremediation method. However, the process also has some major limitations that restricts its usage in polluted areas. The process of remediation using plants is very slow, seasonally effective, has low bio productivity, and contains shallow root systems. The method is unsatisfactory for certain chemicals, such as in the case of CCL4 and trichloroethylene, where the pollutant's reduction is very low and the eventual maximum phytotoxic concentration makes plants ineffective. Also, the effectivity of the phytoremediation process depends on the type of contaminated site, as the method is not very effective for places with high concentrations of toxic compounds (Gratão et al., 2005). In addition, the implementation of the phytoremediation process also requires expert knowledge of field and plant species for particular sites.

18.5 PROBLEMS AND CHALLENGES FACED

The utilization of plants for remediation technology has several economic benefits. However, several concerns with respect to environmental and human health arise with the installation and future implications of the use of transgenic plants. The potential risks associated with the application of transgenic plant species are:

- Transgenic plants are more efficient than wild species as the transgenes incorporated in these plants give them the ability to survive in stressed environments. This increases the risk of invasion of genetically modified plants by competitively inhibiting the native species. It can also cause the extinction of native species.
- Transgenic plants san affect the soil embedded microbial system, food chain, and ecosystem functioning.
- The gene flow could also result in altered biodiversity and increased weed problems that are resistant to herbicides.
- Technical and financial implications have made remediation technology a difficult task.

Some basic enrichment factors that can be used for the betterment of the process include:

1. **Fertilizers and its usage:** Fertilization helps to improve the capacity of the plants and enriches its metal collecting capacities (Kandpal & Deep, 2018). The whole yellow marked sentence needs to be changed. The additional treatments of the soil containing the metals with nitrogen, phosphorous, potassium, manure and carbon dioxide gas and limited amount of fertilizers needs to be done to change the basic formulation of the heavy metals and soil complex , and to increase their percentage of essential accumulation. These additional treatments increases the power of the plants to absorb the metals from the soil.
2. **Practices of another agronomic type:** Simultaneous co-cropping can increase the potentiality of phytoremediation. This can increase the remediation potential of the zone, and thus upgradation of the efficiency is done (Kandpal & Deep, 2018). Many studies with experimental procedures have shown that interplanting or intercropping diminishes weed production and also increases pest control along with increasing hyperaccumulators essentiality to upgrade the remediation capacity.
3. **Preparation of soil by earthworms:** Earthworms are natural fertilizers of soil and crops. They can be used to increase the enrichment of the soil by increasing the physical and chemical properties of the soil (Kandpal & Deep, 2018). Thus, shuffling the layers of soil will increase the surface area and helps in the phytoremediation process.
4. **Chemical/compound application, i.e., chelating agents:** Chelating material are used to upgrade the bioavailability of metals in soil. Other man-made synthetic remediating compounds include antacids, chelating agents, plant hormones, and plant supplements. Chelating operators like Nitrilotriactetic acid (NTA) and Ethylene diamine tetraacetic acid (EDTA) here have been used to enhance the dissolvability of metal in contaminated soils (Kandpal & Deep, 2018).
5. **Use of base regulators:** Variations in pH levels impact the accumulation of heavy metals in soil. Some pH controllers used, such as sulphur and calcium hydroxide, enhance the accumulation potential of plants. The addition of citrus extract with a low atomic weight as a natural acid along with hummus can help in increasing the metal concentration absorption of heavy metals from the soil by plants (Kandpal & Deep, 2018).
6. **Other strengthening measures:** The use of essential metal transporters with the characteristics of hyperaccumulators can be effectively used in the process of the extraction of metals. The transgenic development of plants is also used in phytoremediation techniques (a more strengthened method for remediation). Biosurfactants, common acids, and amino acids increase the stability of metal hyperaccumulation in plant development (Kandpal & Deep 2018).

18.6 CONCLUSION

Phytoremediation is a promising green measure in the remediation of heavy metals and other foreign particles from contaminated soil (Zayed & Terry, 2003). However, there are a few disadvantages too. There are many new plants which have been identified for use to add to the social and financial advancement of the local population. Comprehensive information concerning the metal uptake by plants and biomass transfer is still unclear and requires further research.

REFERENCES

Aguilar J.R.P., Cabriales J.J.P., & Vega M.M. (2008). Identification and characterization of sulfur-oxidizing bacteria in an artificial wetland that treats wastewater from a tannery, *International Journal of Phytoremediation*, 10(5), 359–370.

Brígido, C., & Glick, B. R. (2015). Phytoremediation using rhizobia. In: A. A. Ansari, S. S. Gill, R. Gill, G. R. Lanza, & L. Newman (Eds.), *Phytoremediation: Management of Environmental Contaminants*, Volume 2 (pp. 95–114). Springer. https://doi.org/10.1007/978-3-319-10969-5_9

Cherian, S., & Oliveira, M. M. (2005). Transgenic plants in phytoremediation: Recent advances and new possibilities. *Environmental Science & Technology*, 39(24), 9377–9390. https://doi.org/10.1021/es0511341

Coutinho, H. D., & Barbosa, A. R. (2007). Fitorremediação: Considerações Gerais e Características de Utilização. *Silva Lusitana*, 15(1), 103–117. http://www.scielo.mec.pt/scielo.php?script=sci_abstract&pid=S0870-63522007000100008&lng=pt&nrm=iso&tlng=pt

Gratão, P. L., Prasad, M. N. V., Cardoso, P. F., Lea, P. J., & Azevedo, R. A. (2005). Phytoremediation: Green technology for the clean up of toxic metals in the environment. *Brazilian Journal of Plant Physiology*, 17(1), 53–64. https://doi.org/10.1590/S1677-04202005000100005

Kandpal, S., & Deep, S. (2018). Review of various amendments for phytoremediation of tannery waste. *Research Journal of Chemistry and Environment*, 22, 88–92.

Kumar, A., Maiti, S. K., Tripti Prasad, M. N. V., & Singh, R. S. (2017). Grasses and legumes facilitate phytoremediation of metalliferous soils in the vicinity of an abandoned chromite–asbestos mine. *Journal of Soils and Sediments*, 17(5), 1358–1368. https://doi.org/10.1007/s11368-015-1323-z.

Lister, K.H., 2000. Evaluation of remediation alternatives. In: Surammpalli, R.Y. (Ed.), Proceedings of the ASCE National Conference on Environmental and Pipeline Engineering, July 23–26, 2000, Kansas City, MO, pp. 259–268.

McKeehan, P., (2000). The financial, legislative and social aspects of the redevelopment of contaminated commercial and industrial properties.

Movahed, N., & Maeiyat, M.M., (2009). Phytoremediation and sustainable urban design methods (Low carbon cities through phytoremediation). In: 45th ISOCARP Congress. http://www.isocarp.net. Consulted on September, 2009.

Panagos, P., Van Liedekerke, M., Yigini, Y., & Montanarella, L., 2013. Contaminated sites in Europe: Review of the current situation based on data collected through a European network. *Journal of Environmental and Public Health* 2013, 158764, 11 pages. Available from: https://doi.org/10.1155/2013/ 158764.

Pilon-Smits, E., & Pilon, M. (2002). Phytoremediation of metals using transgenic plants. *Critical Reviews in Plant Sciences*, 21(5), 439–456. https://doi.org/10.1080/0735-260291044313.

Prasad, M.N.V., 2011. *A State-of-the-Art Report on Bioremediation, Its Applications to Contaminated Sites in India*. Ministry of Environment & Forests, Government of India.

Sharma, H., (2011). Metal Hyperaccumulation in Plants: A Review Focusing on Phytoremediation Technology. 4(2), 118–138. https://doi.org/10.3923/jest.2011.118.138.

Suman, J., Uhlik, O., Viktorova, J., & Macek, T. (2018). Phytoextraction of heavy metals: A promising tool for clean-up of polluted environment? *Frontiers in Plant Science*, 9, 1476. https://doi.org/10.3389/fpls .2018.01476

Yang, X., Feng, Y., He, Z., & Stoffella, P. J. (2005). Molecular mechanisms of heavy metal hyperaccumulation and phytoremediation. *Journal of Trace Elements in Medicine and Biology*, 18(4), 339–353. https://doi. org/10.1016/j.jtemb.2005.02.007

Zayed, A. M., & Terry, N. (2003). Chromium in the environment: Factors affecting biological remediation. *Plant and Soil*, 249(1), 139–156. https://doi.org/10.1023/A:1022504826342

19 Membrane-Based Technologies for the Removal of Toxic Pollutants

Sanchita Patwardan, Rajesh W. Raut,
Nilesh S. Wagh, and Jaya Lakkakula

CONTENTS

19.1 INTRODUCTION

Water, which covers 71% of Earth's total surface, is one of the most important source of life. However, only 0.03% of water accounts for the freshwater, which includes, freshwater lakes, rivers, and shallow groundwater that is directly accessible by humans. With the rapid increase in the human population and the increased demand of water for domestic, agricultural, industrial, and energy use, water resources are under threat. To cater to the needs of an ever-increasing population and the rapid development in developing countries, many organic and inorganic pollutants are released into the water bodies by various industries, such as paper, pharmaceutical, pesticide, metal plating, tanneries, and mining. Although the effluents released from these industries are treated at wastewater treatment plants, many of the pollutants escape the treatment and reach our water bodies, resulting in disturbances to the ecology of the aquatic environment.

The nature of pollutants in the water may be of a biological, chemical, or physical nature. Chemical pollutants include organic and inorganic chemicals; organic pollutants include fertilizers, pesticides, phenols, hydrocarbons, biphenyls, pharmaceuticals, oils, and greases. (Musteret and Teodosiu 2007). These organic pollutants are poorly biodegradable and exhibit a low value for the ratio of biological oxygen demand (BOD) to chemical oxygen demand (COD) and are referred to as refractory. Refractory pollutants are resistant to or difficult to biodegrade and become toxic, inhibiting the growth of bacteria that are involved in the degradation process. Such pollutants pose a problem in wastewater treatment systems and need to be removed using filtration.

Presently, many technologies such as chemical precipitation, adsorption, ion exchange, membrane filtration, electrochemistry, phytoremediation, and others are used for the removal of organic pollutants (Rivas et al. 2016). Among these, membrane-based technology has the upper hand because of its many advantages, namely its simple operation, the ability to be operated at room temperature, its high efficiency, and its low energy consumption and cost. The International Union of

Pure and Applied Chemistry (IUPAC) defines a membrane as "a structure with lateral dimensions much greater than its thickness, through which mass transfer may occur under a variety of driving forces" (Rivas et al. 2016). Membranes can be categorized into two groups: natural or biological and artificial or synthetic (Velizarov et al. 2004). For the removal of refractory organic pollutants artificial membranes are widely used. Artificial membranes come in two forms: 2D planar sheets and cylindrical (Allabashi et al. 2007). The dimensions of the pores also play an important role in the filtration of refractory toxic pollutants, for instance the microporous membranes with a pore size in the range of 0.1–5 μm have limitations in their application in filtering pollutants. On the other hand, nanoporous membranes having the size range of 1 to few nanometers can filter most pollutants, such as metal salts or ions, organic molecules, and microbes from wastewater. However, with nanoporous membranes, and even with polymeric or ceramic-based membranes, there is always a possibility of membrane fouling. Membrane fouling occurs due to colloids, chemicals, and microbes that have been rejected by the membranes and remain attached to the membranes. These therefore need considerable attention as it always results in an increase in energy demand making the technology more costly (Isawi 2019). Surface modification is one way to reduce the fouling of membranes. In this chapter we will discuss the membranes made of ceramic-based material, polymers, nanomaterial, and carbon-based material along with their ability to remove refractory pollutants.

19.2 MEMBRANE-BASED TECHNOLOGIES

In the past few decades, tremendous efforts have been made by scientists to develop an optimum wastewater treatment to eliminate refractory pollutants from drinking water in order to avoid problems such as bioaccumulation. The use of membrane technologies for the decontamination of water bodies has emerged as a superior alternative to other conventional processes of water purification. This chapter provides a comprehensive review of refractory pollutant removal efficiency by membranes made from different materials (i.e., ceramic, carbon, nanoparticles, and different polymers). It also highlights the various properties, such as stability and durability, of the membranes used for water purification processes.

19.2.1 Ceramic-Based Membranes

Al_2O_3/$CoFe_2O_4$ catalytic ceramic membranes, feasible for the removal of organic micro pollutants, were designed using a low temperature calcination process. The α-Al_2O_3 ceramic membranes were decorated with $CoFe_2O_4$ nano-catalysts by dipping the membrane in a $CoFe_2O_4$ catalyst solution made from 10 mmol $Co(NO_3)_3 \cdot 6H_2O$ and 20 mmol $Fe(NO_3)_3 \cdot 9H_2O$. (The membrane was kept in the solution for 15 seconds and removed vertically at a speed of 3 cm·s^{-1} at increasing temperature followed by cooling at room temperature.) Scanning electron microscopy (SEM) and energy dispersive X-ray (EDX) were used as analyzing techniques for determining the morphology and micro-regional concentration of the membrane. This membrane has average an pore size of 267.5 nm with a thickness of approximately 420 ± 40 mm. A sandwiched α-Al_2O_3/$CoFe_2O_4$ catalytic ceramic membrane, a gravity driven process, showed a remarkable ability for removing persistent organic pollutants, and especially displayed great potential for high removal rates of methylene blue (MB) (~98%) and ibuprofen (~99.5%). The absorbency of the membrane was measured using quartz crystal microbalance along with an Au-coated resonator. Due to several promising characteristics, such as anti-fouling properties, reusability, zero energy consumption, and long-term durability, this membrane can be used on a large scale (Wang et al. 2020a).

A CuO modified ceramic hollow fiber membrane was developed using impregnation and calcination methods. The impregnating solution was made by suspending 1 g of polyethylene glycol (PEG) and 5 g of $Cu(NO_3)_2 \cdot 3H_2O$ into 25 mL of ethanol. The pristine membrane was then rinsed in the impregnating solution for 2 minutes followed by its withdrawal with a speed of ~0.5 mm/s. The membrane was calcinated at a temperature of 500°C for 3 hours and CuO was successfully loaded on the calcinated membrane. CuO was fabricated on the membrane with different concentrations, where MCuO-22 (22 g/m^2) exhibits optimal performance in relation to all the membranes by eliminating MB, bisphenol A (BPA), phenol, 4-chlorophenol (4-CP), and sulfamerazine (SMZ) with removal rates of 93.2%, 94.5%, 91.8%, 95.3%, and 91.6%, respectively. The overall performance and absorption capacity of the membrane was measured using an ultra-high performance liquid chromatography (C18 column) and a photodiode array detection (PDA) detector. Field emission scanning electron microscopy was used to determine the morphology of the membrane. The membrane has a pore size of 0.22 μm with an effective membrane surface area of ~13.8 cm^2. Due to its high efficiency, superior antifouling properties, excellent stability, and reusability, this membrane can be used on a large scale (Wang et al. 2020b).

Three ceramic membranes made from α-Al$_2$O$_3$ and decorated with TiO$_2$ were used for microfiltration and ultrafiltration. Membrane fouling was found to be a disadvantage of this experiment; however, the membrane managed to efficiently reduce 85% of the COD. The membrane also showed high retention capability for other chemical contaminants like nitrogen, phosphorus, ferric chloride, and calcium hydroxide with an effective filter area of 94.2 cm^2. All three membranes were chemically and mechanically stable. However, a significant decrease in the permeability of the membrane was observed due to the formation of cake layer (Waeger et al. 2010).

Ceramic membranes impregnated with cross-linked silylated dendritic and cyclodextrin polymers were manufactured in order to eliminate organic pollutants from wastewater. The basic ceramic sheets created from Al$_2$O$_3$, SiC, and TiO$_2$ were soaked in the form of tubes into a range of triethoxysilylateddendritic concentrations and cyclodextrin compound solutions for 6 hours at a temperature of 60°C in the presence of argon. After 6 hours, those multichannel tubes were removed from the solution and dried for 50 hours under open air. This aeration leads to the chemical breaking of ethoxysilyl polymers into silanols monomers, subsequently followed by condensation into siloxane networks followed by bond formation with the hydroxyl group of the ceramic substrate to produce dendrimeric (DAB-Si), hyper-branched (PEI-Si and PG-Si), and cyclodextrin (CD-Si) derivatives. Ultimately, these membranes are washed with ACN and dried in a vacuum. All the three membranes were highly efficient for the removal of organic contaminants like polycyclic aromatic hydrocarbons (up to 99%), monocyclic aromatic hydrocarbons (up to 93%), trihalogenmethanes (up to 81%), pesticides (up to 43%), and methyl-tert-butyl ether (up to 46%). The filter membrane can be reused by washing the filter and drying it under vacuum. Due to their high stability and regenerative capacity, membranes can be employed for the purification of water on large scales (Allabashi et al. 2007).

An eco-friendly, mullite-carbon nanotubes (CNT) composite membrane with a porous network surface has been constructed for the effective removal of bacteria from contaminated water. The mullite ceramic substrate was produced from industrial coal fly ash and bauxite minerals. Ni catalysts were obtained by dissolving 21 g of Ni(NO$_3$)$_2$.6H$_2$O in 50 mL of deionized water. Ni calcinated mullite ceramic substrates were obtained by drying mullite ceramic substrates along with Ni precursors for 2 hours with a constant elevating temperature up to 500°C till the ratio of NiO to ceramic substrate becomes 0.1:1 g. The chemical vapor deposition (CVD) method was used to prepare a mullite-CNT composite membrane. The NiO-coated mullite ceramic substrate was allowed to cool for 1 hour at 500°C along with a mixture of H$_2$–N$_2$ gas. The reactor was then heated up to an extreme temperature of 650°C with methane gas for 2 hours. The resultant sample was allowed to cool at room temperature under the N$_2$ gas at a flow rate of 20 mL.min^{-1} to obtain the final mullite-CNT composite membrane with an average pore size of 1.02 μm. The fluorescence-based Bacterial Viability Kit along with flow cytometry were used to determine the percentage of the two inactivated bacteria used. A 100% filtration of the two model bacteria, E. coli and S. aureus, was observed. It is a highly-stable, self-supporting, low-cost membrane with low fouling properties and is feasible for large scale use (Li Zhu et al. 2019).

A ceramic-based electrosorption membrane was employed for the removal of metal ions from industrial wastewater using a desalination process. The basic support of the ceramic electrosorption membrane, i.e., α-alumina porous substrate, was carbonized under liquid petroleum gas for 30 minutes and coated with gold to attain appropriate conductivity. Ion adsorption properties were incorporated by submerging the resultant membrane in 20% ZrO_2 solution for 24 hours followed by its desiccation under a furnace for 3 hours at 200°C. Lastly, the membrane was processed with 15% H_3PO_4 followed by air drying and heating the membrane under a furnace for 3 hours at 200°C. This modified ceramic membrane was successful in the significant adsorption of metal ions including sulfate ions (>70%) and calcium ions (>93%), with the same rates being achieved over several cycles. Total ion adsorption was measured using ion chromatography in combination with a Hamilton PRP-X100 anion exchange column for sulfate ions, whereas PU9 100 atomic absorption spectrophotometry was used in the case of calcium ions. Unsatisfactory results in low-grade chemical and thermal stability, and elevated expenses due to higher energy consumption restrict its use only on a laboratory scale (Bladergroen and Linkov 2001).

A novel ceramic microfiltration membrane fabricated with fly ash [along with glycerol and polyvinyl alcohol (PVA)] was used for the successful refining of oil-in-water (O/W) emulsions. A mixture of fly ash, PVA, and glycerol was exposed to high pressures of 20 MPa, leading to the formation of sheet supports having a thickness of 2–3 mm. Mullite fiber-doped fly ash support (FMS) was created by adding an amount of mullite fibers to the fly ash mixture. The FMS was dried at room temperature and was in turn used as a support mechanism of the MF membrane made from α-alumina. The standardized coating suspension of alumina was sprayed on the FMS for 5 seconds for up to 8 spraying cycles and was then air dried at room temperature for 12 hours further, desiccating it at 110°C for the same time span. The resultant dried membrane was sintered by a green specimen at 1,050°C for 2 hours. The crack-free membrane with an average pore size of ~100 nm revealed greater potential by removing >99% of total organic carbon (TOC) from the oil-in-water emulsion. Permeability of the membrane was maintained at 165 $Lm^{-2}h^{-1}bar^{-1}$ with a high performance. The use of low-cost materials and chemicals in combination with superior heat resistance, stability, strength, and flexibility make it a suitable option for commercial water treatment (Zou et al. 2019).

A tubular ceramic microfiltration membrane was designed using naturally-occurring pozzolan. 78 wt.% of pozzolan, 10 wt.% of clay as a plasticizer, 6 wt.% of starch as a porosity agent, 3 wt.% of Amijel as a binder, and 3 wt.% of Methocel as a plasticizer were used as raw materials for the production of an external tubular support. All these materials were mixed for 15 minutes using an electric mixer. The resultant mixture was mixed with 22 wt.% of distilled water which led to the formation of a paste. This paste was molded into a tubular shape with the help of a ram extruder and then dried under a furnace. A heterogeneous mixture was made by combining pozollan powder (10 wt.%) and distilled water (90 wt.%) containing PVA (3 wt.%) and acetone (1 wt.%) by using a magnetic stirrer for 30 minutes at a speed of 250 tr/min. The mixture was allowed to rest for 4 hours. The supernatant network formed was cautiously removed and coated on the inner side of the tubular support using a cross-flow filtration technique. SEM was used to examine the morphological characteristics of the membrane. The modified ceramic membrane has a pore diameter of 1.12 µm and thickness of 5.52 µm. The membrane was crack free with high permeability of 1,444.7 L/h·m²·bar. Ceramic microfiltration membranes can act as a suitable alternative to other conventional methods for seawater treatment. It was able to attain 77.8% of COD rejection as well as a high rejection of turbidity (98.25%). Its low energy consumption, low cost, and environment-friendly properties make it affordable for its large-scale use in water desalination (Achiou et al. 2017).

An ultrafiltration clay-alumina ceramic membrane layered with a titania film was introduced for the removal of dyes from raw textile wastewater, as shown in **Figure 19.1**. The tubular clay-alumina support was made by grinding dry kaolinite clay (25%), alumina (75%), and Methocel powder (6%) together. The paste formed was then shaped into a tubular structure using a single-screw extruder and dried at room temperature followed by sintering up to a temperature of 1,350°C. The active TiO_2 membrane was manufactured via slip casting. The slurry was obtained by adding 4 g of titania powder in an aqueous solution which comprises PVA (12 wt.%) and DolapixCE64 (0.2 wt.%). The resultant TiO_2 layer was casted on the support by capillary suction with an exposure time of

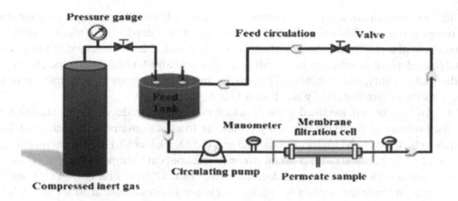

FIGURE 19.1 Ultrafiltration setup for the removal of organic dyes. Reprinted with permission from Oun, Abdallah, Nouha Tahri, Samia Mahouche-Chergui, Benjamin Carbonnier, Swachchha Majumdar, Sandeep Sarkar, Ganesh C. Sahoo, and Raja Ben Amar. 2017. "Tubular Ultrafiltration Ceramic Membrane Based on Titania Nanoparticles Immobilized on Macroporous Clay-Alumina Support: Elaboration, Characterization and Application to Dye Removal." Separation and Purification Technology 188: 126–33. doi:10.1016/j. seppur.2017.07.005.

10 minutes and dried at room temperature for 24 hours. Lastly, the membrane was calcinated at 800°C for 3 hours. Field Emission Scanning Electron Microscopy (FESEM) together with EDX were used to determine the structure of the membrane. The membrane had a thickness of 4.2 μm with a high permeability rate of 117 L.m^{-2} h^{-1} bar^{-1}. The membrane successfully retained 99% of alizarin red dye with an average pore diameter of 0.75 μm. As a result, an inexpensive, crack-free, chemically- and thermally-stable membrane was innovated with improved sustainability for textile wastewater treatment (Oun et al. 2017).

A ceramic membrane customized with natural clay and Moroccan phosphate was designed for wastewater treatment in order to remove a diversity of pollutants as well as for desalination. Red clay (RC) and natural phosphate (as a pore-forming agent), the raw materials for preparation of the membrane, were compressed under a pressure of 150 MPa which shaped it into a flat rectangular sheet. Further, the membrane was treated with sintering at 4 different temperatures using an oven. Primarily, the membrane was desiccated for 1 hour at 250°C followed by increasing exposure to temperatures up to 450°C for the thermal degradation of organic substances. Later, the breakdown of mineral substances present was carried out at 750°C for 1 hour. Finally, the membrane was crystallized at 1,100°C for 2 hours. SEM was used to evaluate the fine structure of the membrane. The permeability of the membrane was found to be 928 L/(h·m^2·bar) with an average pore size of about 2.5 μm. The membrane was confirmed to have promising advantages like durability, reusability, and environmental application due to its low cost. The membrane demonstrated good microfiltration performance by eliminating turbidity from the beam house effluent (99.80%), raw seawater (99.62%), and synthetic salt water (99.86%). The membrane can be considered as an excellent future candidate in wastewater treatment on an industrial scale (Mouiya et al. 2018).

A pre-ceramic reactive self-sacrificed process was used to develop a mesoporous polymer-based ceramic silicon oxycarbide (SiOC) membrane. Polydimethyl siloxane (PDMS) and 0.2 wt.% of Karstedt's catalyst were mixed with D$_4$V (i.e., 2,4,6,8-tetramethyl-2,4,6,8-tetravinylcyclotetra siloxane) using a magnetic stirrer. An equal amount of D$_4$H (2,4,6,8-tetramethylcyclotetrasiloxane) was added to the solution, with continued stirring for 12 hours. The gelatinous sample obtained was degasified in a Teflon mold followed by warming at 80°C and 150°C for 4 hours each. The sample was then thermally decomposed at 1,000°C for a couple of hours under nitrogen gas. The resultant nano-porous SiOC ceramic sheet was washed with ethanol as well as deionized water and air dried before using it for further experimentation. The effective area of the membrane was calculated to be

2.54 × 10^{-4} m^2 with an average pore diameter of 0.63 μm. A UV-1800 spectrophotometer evaluated higher removal rates of rhodamine B (RhB) (95%), whereas a conductivity meter confirmed high desalination by the membrane. The resultant membrane was able to overcome the failures of the other polymer-derived membranes by revealing its excellent anti-fouling properties, chemical resistance, thermal stability, and durability. The scalable method may possibly be helpful in restricting the deteriorating environments of water bodies (Zhang et al. 2020).

A TiO$_2$-GO decorated modified ceramic membrane was introduced as an effective way to resolve the problem of purifying refractory pollutants from wastewater. The coating of TiO$_2$-GO was acquired by mixing (NH$_4$)$_2$TiF$_6$ (0.1 mol/L) and H$_3$BO$_3$ (0.3 mol/L) to a GO homogenous solution (1.6 wt.% of GO content) with constant stirring for 2 hours at a temperature of 60°C. A ceramic membrane coated with ZrO$_2$ sheets was decorated with TiO$_2$-GO nanosheets via a vacuum method. The resultant membrane was washed with deionized water and desiccated at 50°C. Shortly after, the membrane was processed at a high temperature under a furnace for a couple of hours to make the membrane thermally stable and compact. The microstructure of the membrane was evaluated using an FESEM. The membrane showed excellent removal efficiency by removing 74.5% of humic acid (HA), 33.8% of tannic acid (TA), and 67.3% of haloacetic acids formation potential (HAAs-FP). The membrane was also successful for the removal of certain negatively-charged pharmaceuticals including DCF, NAP, IBU, and CBZ. The modified ceramic membrane was reported with enhanced anti-fouling abilities and reusability with remarkable stability and anti-bacterial effects, which makes it a candidate for industrial and municipal wastewater treatment for large-scale application (Chen Li et al. 2019b).

Sintering and phase inversion methods were collectively used to design the hydroxyapatite (HAp)-based bio-ceramic hollow fiber membranes for the decontamination of industrial textile wastewater. A hydroxyapatite powder was prepared from waste cow bone (WCB). Primarily, 10 kg of WCB was manually fragmented and boiled at 100°C to disengage the attached tissues. The bone was dried and crushed into a fine grain followed by calcination at 800°C for 3 hours. The WCB grains were finally strained to obtain 8.3 kg of HAp powder (36 μm). A heterogeneous mixture was prepared by mixing 60% of the HAp powder and 1 w.% of Arlacel to the N-methyl-2-pyroolidone (NMP) solvent and was mixed thoroughly using ZrO$_2$ milling balls. Shortly after, Polyethersulfone (PESf) was added and the mixture was milled for 48 hours. The resultant suspension was exposed to the vacuum process with constant stirring for degasification. The suspension was then transmitted to stainless steel syringes for extrusion under constant suspension and bore fluid flow rates of 10 mL/min. The membrane precursors obtained by the above process were then sintered for 3 hours over a temperature range of 900–1,300°C. SEM was used to examine the microstructure of the modified ceramic membrane. Mercury intrusion porosimetry (MIP) determined the average pore size of membrane to be 0.013 μm. The filtration attributes of the membrane including conductivity, COD, and turbidity were measured using portable conductivity meter EC300, Hach spectrophotometer, and Hach turbidimeter, respectively. The membrane demonstrated an outstanding adsorption capacity by eliminating 100% of hazardous heavy metals, 99.9% of color pigments, 99.4% of turbidity, and 80.1% of COD. Research is underway to improve its anti-fouling ability and mechanical and chemical strength to make the membrane viable for commercial use for water purification (Hubadillah et al. 2020).

19.2.2 POLYMER-BASED MEMBRANES

The polyacrylonitrile/polyvinylidene fluoride (PAN/PVDF) chelating membrane is a device used for the instantaneous elimination of organic pollutants and metal ions from industrial wastewater. Primarily, PAN and PVDF powders (96.5:3.5) were dehydrated for 24 hours at a temperature of 60°C. These powders were suspended in N,N-dimethylacetamide (DMAc) followed by degasification. An L-S phase exchange process was used to prepare the basic membrane and the resultant membrane was rinsed with deionized water. The dispersion solution, made by mixing hydroxylamine hydrochloride and sodium carbonate, was used to customize the PAN/PVDF membrane. The amidoximation reaction results in a partial conversion of nitrile groups into amidoxime groups. The modified PAN/PVDF membrane was cleaned using deionized water. SEM was used to depict the morphology of the membranes. Energy dispersive spectroscopy (EDS) was used to calculate the concentration of the various elements on the chelating membrane. Increments of 7.16% of carbon, 1.12% of nitrogen, and 6.04% of oxygen concentrations were observed in the modified chelating membrane, which clearly indicates that the amidoxime group was effectively inserted into the membrane. The finger-like pore-structured membrane exhibited admirable retention efficiency by eradicating Cu^{2+} (99.77%), Pb^{2+} (95.25%), Cd^{2+} (79.57%), lysozyme (94.36%), and BSA (97.67%) from the mixture of industrial wastewater containing these 3 metals and 2 proteins. The key principle of this lab experiment was to understand the influence of the reaction between metal ions and proteins on pollutant removal rates of PAN/PVDF membranes (Xu et al. 2020).

The integration of the Fenton oxidation process and membrane photo-bioreactor (MPBR) was a highly efficient method for the removal of refractory pollutants from wastewater samples along with a high desalination capability. A graphene oxide based-polyvinylidene fluoride (PVDF) membrane was used in the MPBR. The commercially-available PVDF membrane was pre-oxidized with potassium hydroxide and potassium permanganate. The defluorinated membrane was then rinsed with sodium bisulfite followed by soaking it in a graphene oxide suspension to get the resultant graphene oxide polymer-based membrane. The refractory pollutant sample was first subjected to the Fenton oxidation process for the photocatalytic degradation of the pollutants where Fe–Ti biometallic oxides were used as catalysts. The pollutant sample was then passed through the MPBR with a total filtration area of 0.2 m^2. The highly-stable modified GO/PVDF membrane revealed remarkable features like improved permeability, increased hydrophilic nature, and higher water flux. It was also reusable and easily cleanable. The membrane successfully eliminated ammonium up to 98%. A significant amount of desalination was observed with great reduction in COD up to 95%. The use of this cost-effective technique on a commercial scale is a good alternative for other treatments employed for the seafood processing industries (Chang Li et al. 2019a).

A cross-linked carbon nanotube-based biocatalytic membrane was designed by filtration-coating to facilitate the removal of micropollutants from contaminated water bodies. In an ice bath, 50 mg of CNTs was dissolved in 50 mL of purified water and after 20 minutes the homogenous solution obtained was passed through a PVDF membrane to obtain a CNTs coating. The heterogeneous solution, containing 1 mL of polyvinyl alcohol PVA, 0.5 mL of glutaraldehyde (GLU), and 0.5 mL of hydrochloric acid (HCl), was sieved through the membrane and dried in a hot air oven at a temperature of 110°C for a few minutes. The resultant CNTs–PVDF membrane was rinsed with deionized water to detach the unprocessed chemical material. A reversible laccase immobilization was executed on the CNTs–PVDF membrane. The membrane was primarily exposed to a 10 mL phosphate buffer solution containing 1 mg/mL laccase. The laccase was adsorbed onto the membrane at 4°C and, subsequently, the unreacted laccase was removed by washing the membrane with deinozied water. The laccase was covalently bonded to the CNTs–PVDF membrane by straining 5 mL of 30 mg/mL N-Hydroxysuccinimide (NHS) and 3 mL of 10 mg/mL N-(3-Dimethylaminopropyl)-N'-ethyl carbodiimide hydrochloride (EDC) through the membrane. The membrane was washed with deionized water and was soaked in laccase solution for 48 hours and finally rinsed with deionized water to obtain a cross-linked carbon nanotube biocatalytic membrane. The surface morphology of the membrane was determined using FESEM. The effective membrane area was observed to be 8.4 cm^2 and the average pore size was 20 nm. The membrane efficiently removed the micropollutants including bisphenol-A (BPA), carbamazepine (CBZ), diclofenac (DCF), clofibric acid (CA), and ibuprofen (IBF) with an average removal rate of 95%. The membrane exhibits superior stability and can be reused for 4 cycles without a reduction in the removal rate. The membrane can be easily regenerated by reacting it with a sodium dodecyl sulfate (SDS) solution. The membrane's physical and chemical stability with durability and anti-fouling properties indicates its potential for commercial application (Ji et al. 2016).

A polyelectrolytic modified poly-amide microfiltration membrane with a homogeneous varnish of poly-diallyldimethylammoniumchloride (PDAC) and poly-sodium styrene sulfonate (PSS) multilayered membrane was customized for the effective removal of obidoxime (OBD) from wastewater produced by pharmaceutical industries. Atomic force microscopy and SEM were used to evaluate the structural and functional characteristics of the customized membrane. The membrane, with an average pore size was 0.45 μm and a mean permeation flux of 6.19 m^3/m^2, exhibited great stability and flexibility. Poly-ionic interaction between nine-layered polyelectrolytic membranes rejected 99.26% of obidoxime (OBD) without a reduction in the removal rate in varied surrounding conditions. The hydrophilic nature and higher rejection potential of the membrane makes it feasible for commercial-scale application (Nikhil Chandra and Mothi Krishna Mohan 2020).

A polyether sulfone (PES) ultrafiltration membrane made via phase inversion was customized with a 5% weight fraction of nano-sized alumina (Al_2O_3). The 0.05 Al_2O_3/PES (w/w) with an average particle size of 48 nm was tailored on the plain PES membrane by dissolving the alumina in a N-methyl pyrrolidone (NMP) solution. The membrane was exposed to ultra-sonication at a temperature of 60^0C for 3 days. Further, the membrane was soaked and sonicated in an 18% PES solution to obtain a thin uniform coating for a period of 7 days followed by desiccation. The membrane was water co-angulated for 2 minutes and transferred to a water bath for the detachment of loosely bound substances from the membrane. The anti-fouling properties of the membrane were examined using experimental observations via Sigma software. The average fouling rate was calculated to be between 2.5 and 3 h (dy/dt). The morphological characteristics of the membrane were evaluated using SEM. The membrane has an average pore size of 0.1–1 μm and offers total resistance of 1.8 × $10^7 m^{-1}$. Due to its high mechanical and chemical stability, the membrane was able to eliminate effluents like starch, casein, $(NH_4)_2SO_4$, and KH_2PO_4 from activated sludge (Maximous et al. 2010).

A polytetrafluoroethylene (PTFE) membrane mounted on a polypropylene layer with an average pore diameter of 0.2 mm and membrane thickness of 165 mm was invented to carry out a sweeping gas membrane distillation process (SGMD). This experiment focused on the purification of water samples housing volatile organic compounds including acetone and ethanol. The membrane (43 cm^2), along with filter paper and a polypropylene web, was placed in the middle of a rectangular vessel dividing it into 2 compartments. Nitrogen acted as a stripping agent in this process. Liquid and gas used enter the respective chamber perpendicular to the membrane. Pseudo-binary diffusion led to the mass transfer in the gas phase. The mass transfer in-between these two phases was completely dependent on a fluid-dynamic environment, i.e., the surrounding conditions. The membrane shows superior separation efficiency by removing volatile compounds from the aqueous sample in the chamber. It is also applicable for the generation of ultrapure water from salt solutions and purification in a zeotropic mixture. The hydrophobic nature of the membrane makes it reusable and mechanically stable over a long period of time (Boi et al. 2005).

Seven novel hydrophobic and hydrophilic membranes made from polyvinylidene fluoride (PVDF) and poly-acrylic matter were tested for their filtration efficiencies for primary effluents. The membranes with mean pore sizes of 0.5–4 μm were employed for the removal of certain strains of bacteria including fecal-coli from bacteria like *Escherichia coli* and *Enterococci*. A significant reduction in COD and BOD was observed during the filtration process using different polymer-based membranes. Due to the higher water flux rate, less driving pressure is required, which ultimately reduces the cost of the filtration process. The membrane is stable, highly permeable, durable, and reusable. However, fouling problems are a drawback which may reduce its feasibility for large-scale application. The results illustrated that the best antimicrobial membranes were the hydrophobic polyvinylidene fluoride (PVDF) membrane with a pore size of 0.3 μm and the hydrophilic poly-acrylic membrane with a pore size of 0.8 μm. The hydrophilic membrane with a pore size of 0.2 μm was recommended for industrial-scale application due to its high microbial retention rate and comparatively lower fouling capacity (Modise et al. 2006).

Reverse osmosis was carried out using a thin-film polyamide composite membrane for the effective removal of excess levels of lethal fluoride from underground drinking water. Around

2.5–10 mg/L of sodium fluoride was dispersed in the water and used as an effluent. The fluoride solution was primarily passed through a cartilage filter of 20 μm for its pre-treatment and then filtered using a polyamide-based membrane. The polyamide-based membrane showed optimum results at a temperature of ~30°C, a pressure of 2 bar, a pH of 7.0, and a feed flow velocity of 250 mL/min. The results of this experiment confirm that the membrane can efficiently eliminate fluoride by up to 95% from drinking water to make it potable. The membrane possesses a very high mechanical strength and durability. Due to its high deflouridation capacity and comparatively low cost, it is used on a wide scale in municipal water treatments and at an industrial scale (Arora et al. 2004).

A dynamic membrane bioreactor (DMBR) was manufactured consisting of 3 cake layers, i.e., anaerobic zone (5 L), anoxic zone (5 L), and aerobic zone (10 L), of 20 L effective volume. Extra cellular polymeric substances (EPS), mostly protein, were present in all three layers in different amounts. Water purification using a dynamic membrane occurred in 3 phases. The first phase, i.e., pre-coating, took place in the aerobic zone and lasted for 20 minutes. During this phase, a dynamic membrane is formed from activated sludge and sewage is recycled. During the filtration phase, a total critical flux of 60–120 L/m²h was applied under high pressure up to 40 kPa using a pressure gauge. Backwash is executed from the bottom passage with the help of an air compressor. The flat-film dynamic membrane unit utilized a stainless-steel web with a corresponding aperture of 38 mm that acted as a support layer, and had an effectual filtration region of 0.042 m². The dynamic membrane exhibited a high rejection efficiency by eliminating pollutants like ammonia (NH$_3$-H), total nitrogen (TN), total phosphorous (TP), and some volatile suspended solids. COD was also reduced significantly by the dynamic membrane under anaerobic conditions. The membrane also possesses excellent chemical and mechanical strength and flexibility. DMBR forms part of the latest research in water purification processes and has potential for large-scale application on an industrial scale for water decontamination (Chu et al. 2014).

A polyester thin-film composite (TFC) membrane was introduced to reduce the concentration of organic pollutants like humic acid from wastewater. The ultimate membrane was supported using a polymer polyether sulfone (PES). A mixture containing 18wt.% of polyether sulfone (PES) and 15wt.% of polyvinylpyrrolidone (PVP) was added to the N-methyl-2-pyrrolidone (NMP) dissolvent and heated with constant stirring. The solution was brought to room temperature using a water bath. The yellowish viscous solution obtained was poured onto a glass plate and shortly after was drenched into a water bath to detach the membrane support from the glass plate. The membrane support was white-opaque due to polyether sulfone precipitation. The resultant membrane was incised into small circular forms and soaked in a Triethanolamine (TEOA) solution for half an hour. The TEAO solution was prepared by dissolving 6 g of TEAO in 100 mL of sodium hydroxide solution. Further, interfacial polymerization was carried out by placing the TEOA-encrusted membrane in a trimesoyl chloride (TMC) solution. The resultant membrane was desiccated for 12 hours before using it for nanofiltration. The membrane demonstrated an excellent pollutant removal capacity by eliminating ~75% of humic acid from the given effluent. Nanofiltration by polyester TFC membrane is a cost-effective and environmentally-friendly process with a high rejection rate of pollutants. However, fouling problems hinder its large-scale application at industrial levels (Jalanni et al. 2012).

Membrane processes like ultrafiltration and reverse osmosis were used for the removal of refractory pollutants from palm oil mill effluent (POME). Firstly, the POME was coagulated using customized alum and was neutralized up to a pH of approximately 5.5–6.5 using potassium hydroxide. During the flocculation process, an ionic polymer was added to the coagulated POME and was constantly stirred for 30 minutes. The processed POME was filtered using a PVDF ultrafiltration membrane with an effectual filtration area of 0.9 m². The pressure was maintained at 2 bar using a pressure gauge and the POME run rate was retained at 0.1 m/s. The POME from the ultrafiltration unit was further purified using a thin-film composite (TFC) reverse osmosis membrane. The pressure was raised up to 45 bar at a constant temperature of 25°C. The use of membrane processes successfully improved the quality of the water by reducing turbidity, color, and odor, which was measured using a turbid meter, spectrophotometer, and threshold odor test, respectively.

A significant reduction in COD up to 99% was calculated using a colorimeter at a wavelength of 600 nm. The membrane was able to achieve 78% of water recovery by removing ammonia (99%), minerals, oil, and other heavy metals. The membrane was easy to wash and reuse. It showed excellent anti-fouling properties and chemical stability. The membrane can be used on a commercial scale as it is a relatively inexpensive and easy process (Ahmad et al. 2006).

Polyvinylidene fluoride (PVDF)/chitosan thin-film composite membranes were designed for the removal of dye from polluted water bodies. A homogenous mixture of 0.1% chitosan and 0.1% aq. acetic acid was agitated for 8 hours at a velocity of 400 rpm. Cloisite 15A and 30B were suspended in the solution to obtain varied grades (0.5, 1.2 wt.%) of nano clay in the resulting solution. The PVDF microfiltration membrane was customized with a thin nano-composite layer. The PVDF support membrane (0.22 μm average pore size) was incised into circular disks and placed on a glass plate. It was further washed with deionized water and desiccated for a day. The membrane was made hydrophobic by rinsing it with an NaOH solution. A thin-film composite (TFC) membrane was fabricated using a chitosan solution and rinsed with an NaOH solution. The membranes were thoroughly washed with distilled water and dried for few hours before coating on the PVDF support membrane. The chemical properties of the membrane were determined using Fourier transform infrared spectroscopy (FTIR). It was observed that chitosan was superficially attached to the clay sheets. SEM was used to analyze the physical features of the membrane. The membranes showcased superior potential in water purification by eliminating a higher percentage of dye, including methylene blue (closite 15A membrane) and acid orange 7 (cloisite 30B) from textile industrial effluents. The membrane comprises excellent mechanical stability with higher adsorption rates, despite its fouling issues (Daraei et al. 2013).

The complexation-ultrafiltration process was created for the efficient removal of heavy metals from contaminated water. A heavy metal ion solution (1,000 mg/l), including Cu(II), Ni(II), Cr(III), and carboxymethyl cellulose (CMC), was used as a precursor for the ultrafiltration process. The solution was subjected to the ultrafiltration process using a commercially available polyether sulfone (PES) ultrafiltration membrane with an effective filtration area of 0.26 m^2. After a cycle of 10 minutes, the permeate and retenate were successfully segregated at a constant pressure of 1 bar at pH 7. The metal ion concentration was sequenced as Cu(II) > Cr(III) >> Ni(II) on the basis of observations for their interaction and binding with CMC. Subsequently, there was an optimal retention of metal ions from the given solution. The PES ultrafiltration membrane showed high separation efficiency by removing 97.6% of Cu(II), 99.1% of Ni(II), and 99.5% of Cr(III) ions. The membrane was easy to wash and can be easily regenerated by rinsing the used membrane in a solution consisting of $Na_2S_2O_6$, NaOH, and citric acid. The membrane can be possibly used on a large scale as the process possesses remarkable features including low cost, lower energy consumption, and reusability (Barakat 2008).

The geopolymer-based inorganic membrane provides a promising way to eliminate heavy metal ions from industrial wastewaters. The steps involved in the production of a geopolymer-based membrane and the filtration process using this polymer-based membrane is demonstrated in **Figure 19.2**. The modified glass solution contained SiO_2/Al_2O_3 and Na_2O/Al_2O_3 in the molar ratios of 2.96 and 0.8, respectively. Deionized water along with metakaolin was mixed with the glass solution with constant stirring at a speed of 2,000 r/min. The mixture was held in reserve at room temperature followed by the molding process. The mixture obtained was molded in an oven at 60°C for a day where the geopolymerization reaction takes place. The surface morphology of the membrane was analyzed using SEM along with the Brunauer–Emmett–Teller (BET) test. The porosity of the geopolymer-based membrane was found to be 62.64% with a membrane thickness of 7 mm and an average diameter of 40 mm. The membrane demonstrated superior heavy metal removal efficiency by eliminating nickel (Ni^{2+}) (<95%) and other heavy metal ions. The membrane is rigid, self-sustaining, and environmentally friendly. The membrane shows great potential for its use on a large industrial scale as the membrane is easy to handle, durable, and cost effective (Ge et al. 2015).

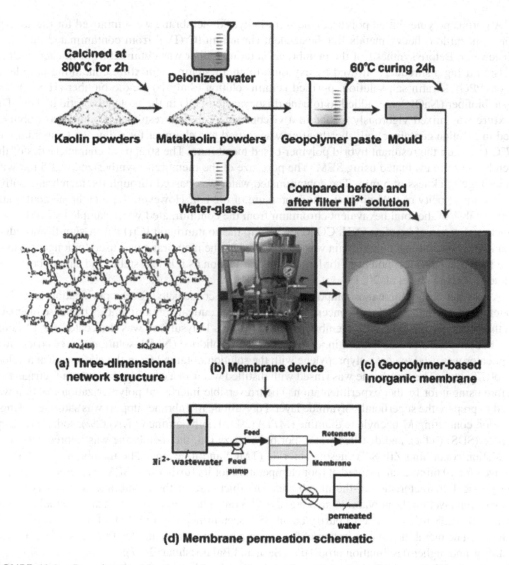

FIGURE 19.2 Steps involved in the production of a geopolymer-based membrane and filtration process. (Reprinted with permission from Ge, Yuanyuan, Yuan Yuan, Kaituo Wang, Yan He, and Xuemin Cui. 2015. "Preparation of Geopolymer-Based Inorganic Membrane for Removing Ni²⁺ from Wastewater." Journal of Hazardous Materials 299. Elsevier B.V.: 711–18. doi:10.1016/j.jhazmat.2015.08.006.)

The vacuum membrane distillation (VMD) process was used for extraction of ammonia from contaminated water. Vacuum membrane distillation was carried out by using a polymer-based polytetrafluoroethylene membrane (PTFE). The membrane was sited at the center of the two chambers with the support of a stainless-steel frame. Environmental conditions like pH, temperature, and pressure were controlled using a pH meter, heating bath, and vacuum pump, respectively, for the efficient removal of ammonia from wastewater. The ammonia solution was filtered using a gear pump at a velocity of 0.84 m/s with an initial concentration of 0.65 M. The membrane had a pore size 0.23 μm with a diffusion rate of 916.2 m⁻¹ and successfully eliminated <90% of ammonia from the given aqueous ammonia sample. The membrane possesses properties like flexibility, reusability, and durability. This membrane technique is relatively cost effective and efficient for purification processes. The vacuum membrane distillation technique is presently used in most of the industries where ammonia is produced on a large scale as a contaminant (EL-Bourawi et al. 2007).

A porous polymer-based polyurethane–keratin hybrid membrane was employed for the adsorption of hazardous heavy metals like hexavalent chromium [Cr(IV)] from contaminated industrial wastewater. Before synthesis of the membrane, keratin biofiber was obtained from chicken quill by incision using a shredder followed by exposure to an air steam. A mixture containing propilenglycol (PPG), keratin salt solution, purified keratin solution (dialysis), acidic biofiber (H_2SO_4), and basic biofiber (NaOH) was added to toluene diisocyanate (TDI) in the molecular ratio of 11:6. The mixture was mixed vigorously to obtain a viscous solution. The resultant mixture was polymerized in a Teflon container and, shortly after, was shifted to a chamber for a day at a temperature of 27°C to obtain the resultant hybrid polymer-based membrane. The structural characteristics of the membrane were evaluated using SEM. The pore size of the membrane synthesized was 5 nm with an average thickness of 5 mm. The contaminated water was passed through the membrane with a maximum velocity of 0.06 cm/s at a high pressure of 6.2 bar. However, the membrane could only remove 38% of the total hexavalent chromium from the contaminated water sample by linking the functional group of keratins (NH, CO, C–S,S–S) to the contaminant [Cr(IV)]. A significant reduction in the removal rate of ammonia was observed when the membrane was used for longer periods. The membrane has less potential for large-scale application as it is less stable and cannot be reused (Saucedo-Rivalcoba et al. 2011).

A polymer-based thin nano-composite membrane was constructed for the purification process of water samples containing large concentrations of pharmaceutical contaminants. The phase inversion method was used to make the membrane support (PSF). Polysulfone was desiccated for 12 hours in a hot air oven and then soaked in an N-Methyl-2-Pyrrolidone (NMP) solution. An electrical film applicator was used to coat polypropylene with the solution obtained from the above step at a velocity of 77 mm/s. The membrane was rinsed with distilled water in a coagulation bath and refrigerated before using it for further experimentation. The irreversible interfacial polymerization reaction was used to prepare the superficial polyamide layer. Polysulfone membrane support was saturated using a solution contacting M-phenylene diamine (MPD) (2%), Triethylamine (TEA) (2%), sodium dodecyl sulfate (SDS) (0.1%), and distilled water. For polymerization, the membrane was impregnated with a solution containing ZIF-8/Trimesoyl chloride (TMC) and n-hexane. The membrane was dried in an oven for 10 minutes and stored at room temperature for its further use. SEM was used to examine the physical characteristics of the membrane. The thickness of the membrane was observed as a 0.4–0.5 µm layer on the polyamide membrane of 2–3 µm. The membrane, with an effectual filtration area of 0.00138 m^2, was able to purify about 55% acetaminophen and 91% of salt from the given solution. The membrane shows potential for use on a commercial scale due to its cost effectiveness, stability, and higher desalination properties (Basu and Balakrishnan 2017).

Sulfonated homo-polyimides and co- polyimide ultrafiltration membranes were designed in order to remove heavy metal ions and selective proteins from wastewater. 0.50wt.% of sPI4homopolyimides and 0.75wt.% of sPI5 copolyimides were mixed with dimethyl sulfoxide (DMSO). Another solution was prepared by dissolving polysulfone (PSf) and polyvinylpyrrolidone (PVP) in an N-Methyl-2-pyrrolidone (NMP) solvent. The two solutions were blended with constant stirring until a uniform solution was formed. The air bubbles in the solution were removed using an ultrasonic bath. The phase inversion technique led to the formation of a flat preliminary membrane. The resultant membrane was exposed to 10% glycerol followed by deionized water to remove excess solutes on the membrane. Atomic force microscopy (AFM) and SEM were used to evaluate the structural characteristics of the membrane. The membrane, with an effective filtration area of 0.25 cm^2, possesses a high permeability of 72.1 L m^{-2} h^{-1} bar^{-1}. The membrane successfully removed heavy metal ions like Pb^{2+} and Cd^{2+} with a high rejection rate of 98% and 92%, respectively. The membrane also eliminated >98% of bovine serum albumin (BSA) and >86% pepsin. The membrane can be used on an industrial scale due to its excellent stability and anti-fouling capacity (Jafar Mazumder et al. 2020).

The distillation-precipitation polymerization method was used to synthesize a polyetherimide (PEI)/customized bentonite nanocomposite membrane. Specifically, 3 g of bentonite clay was assorted in a solution containing 9 mL of ethyl alcohol, 7.5 mL of NH_3, and 90 mL of distilled water. 0.6 mL of Methacryloxypropyltrimethoxysilan (MPS) was added to the bentonite solution and stirred constantly using a magnetic stirrer. Briefly, the mixture was centrifugated and the pellets formed were desiccated in an oven. A mixture of modified bentonite pellets along with styrene (St), Sodium-p- styrene sulfonate (SS), and 2, 2′- Azobisisobutyronitrile (AIBN) was added to the acetonitrile solvent. The sample was again centrifugated and ultimately rinsed in hydrochloric acid (0.1 M). Eventually, grafting sodium4-styrenesulfonate on the bentonite takes place after desiccation process. The customized bentonite was dissolved in an N-methyl pyrrolidone (NMP) solvent for ultra-sonication. The uniform polymeric solution was prepared by adding polyetherimide (PEI) and polyvinylpyrrolidone (PVP) to the above solution and mixed by ultra-sonication. The resultant mixture was then molded into a membrane using a glass plate. The membrane was then set aside in a coagulation bath for day for the absolute casting process. The morphological features of the membrane were studied using SEM. The water uptake capacity of the membrane was relatively high (71.3%) with 62.7% of porosity. The membrane possesses an anti-fouling nature with a fouling percentage of approximately 15%, which is relatively low. The hybrid polymer-based membrane illustrated excellent removal rates by rejecting 87.6% of humic acid along with heavy metals including Cd^{2+} (74.6%) and Pb^{2+} (80.5%). The use of this membrane on a commercial scale would present a greener option for assisting areas in environmental distress (Hebbar et al. 2018).

19.2.3 Nano-Based Membranes

A multifunctional Ag-TiO_2/PVDF-HFP nanocomposite photocatalytic membrane was introduced for the removal of pollutants like norfloxacin and microbes from industrial wastewater. A powder containing TiO_2-P25 (200 micrograms) nanoparticles was added to deionized water and ultrasonicated for an hour. Shortly after, 1,600 micro liters of dilver nitrate solution was added, and a uniform solution was obtained by mixing it vigorously for 10 minutes using a magnetic stirrer. The pH was adjusted to 10 by the NaOH solution. The resultant solution was centrifugated at a speed of 6,000 rpm twice and the pellet obtained was dried for 12 hours in a hot air oven at a temperature of 80°C. The hard pellet was then grinded into powder form using a pestle and mortar. Powder-formed Ag-$TiO_2$was was added to N,N-dimethylformamide (DMF) and ultrasonicated for a couple of hours. The polymer was added to the solution and constantly stirred to get a homogenous solution. The solution was poured into a glass plate and kept at room temperature for 3 days for the evaporation of excess solvent from the solution. The electrospunAg-TiO_2/PVDF-HFP membrane was obtained by putting the above prepared solution into a plastic syringe and exposing it to an electrospinning voltage of 14 kV. The electrospun fibers obtained were accumulated on an aluminum dish. The microstructure of the Ag-TiO_2/PVDF-HFP membrane was evaluated using SEM. The membrane had an average fiber diameter of 0.36 ± 0.05 μm with a pore size of 1–4 μm. The membrane was crack free and flexible with anti-fouling potential. The membrane illustrated excellent degradation (photocatalytic) efficiency of up to 80.7% by degrading refractory pollutants including Methylene Blue, Tartrazine, Tetracycline, Norfloxacin, and other antibiotics. The membrane was also able to eliminate bacteria like *E.coli* and *S. epidermidis*, making it more potable. The membrane showed superior properties including reusability and mechanical stability, as the performance of the membrane was unaffected even after 3 uninterrupted experiments. The Ag-TiO_2/PVDF-HFP membrane can certainly be used for the purification of wastewater contaminated with microbes and antibiotics (Salazar et al. 2020).

A poly ceramic TiO_2 nanofibrous membrane was functionalized for the effective removal of micropollutants from a given sample of contaminated water. The basic solution was made by mixing tetraethoxysilane (TEOS) and ethyl alcohol at a ratio of 1:2. Further, a mixture of hydrochloric acid and water (2:0.01) was added to the solution and the solution was constantly stirred to obtain appropriate viscosity. A solution of polyamide for the electrospinning process was made by adding polyamide in formic acid and acetic acid solutions in equal concentrations. The process of electrospinning was carried out at a voltage of ~25 kV at a temperature slightly lower than room temperature. The electrospinning of SiO_2 was carried out under the same atmospheric conditions but at a higher TCD. This resulted in the formation of a nanofibrous membrane. The TiO_2 nanofibrous solution was prepared by mixing TiO_2 (35wt.%) into a prepared polyamide solution and then ultrasonicated to obtain a uniform solution by adding some polyamide crystals. The electrospinning process assures the cross-linking of tetraethoxysilane, leading to the formation of silicon oxide sheets. The nanofibrous membrane was dip-coated with the TiO_2 nano-solution for 10 minutes, followed by desiccating it for a day at room temperature. With its photocatalytic activity, the membrane effectively eliminated ~99% of isoproturon and methylene blue, along with other micropollutants. The membrane was crack free and mechanically stable. It has a high potential for industrial-scale application due to its high rejection rate of micropollutants (Geltmeyer et al. 2017).

The pre-filtration process of particulate removal from primary and secondary effluents was executed using nano-composite polysulfone (PSU) membranes. A polymeric solution for electrospinning was prepared by mixing 20wt.% polysulfone (PSU) in an N,N-dimethylformamide solvent. The electrospinning process was carried out at a voltage of 12 kV with a regular supply of the PSU solution at a speed of 4 mL/h. The fibrous sheets formed were placed in an aluminum dish and thermally treated for 3 hours at a high temperature of 188°C for proper linking. The membrane with an efficient filter width of 25 mm was mechanically strong and showed excellent anti-fouling properties with pore size of <3 µm. The modified nano-composite ultrafiltration membrane with pore size of 4.6 µm eliminated nearly 99% of particulate contaminants. The membrane can possibly be used as a pre-filter due to its high removal rate of particulate contaminants and its durability (Gopal et al. 2007).

The phase inversion method was used to prepare nano-composite polyvinylidene fluoride (PVDF) membranes for the extraction of certain organic contaminants including dyes from organic wastewater. Vermiculite nanoparticles were graded at a particle size of <5 µm. 0.1 g of these nanoparticles was suspended in 20 mL potassium persulfate in a conical flask and then the flask was exposed to a heat of up to 60°C to attain a uniform mixture. The resultant mixture was centrifuged at a speed of 6,000 rpm for a couple of hours. The supernatant was discarded, and the pellet was washed with distilled water and dried in a hot air oven at a temperature of 50°C. The primary polyvinylidene fluoride (PVDF) membrane was manufactured by forming a homogenous solution containing powdered-form PVDF (12wt.%) in N,N-dimethylformamide (DMF) (88%). Four different membranes containing a diversity of nanoparticles such as Al_2O_3, SiO_2, and CuO along with vermiculite nanoparticles were synthesized. The powdered-form Al_2O_3, SiO_2, CuO, and vermiculite nanoparticles were blended with N,N-dimethylformamide (DMF) using ultrasonication. Shortly after, the PVDF powder was dispersed into the above solutions with constant stirring and was rested for 12 hours for polymerization to occur. The resultant solution was applied to a glass plate with a uniform thickness of 200 µm. The solution was dried and degasified for a minute and instantly transferred into a water bath. The membranes were removed after a couple of minutes and washed thoroughly in deionized water to remove excess solvent. The PVA coating sheet was prepared by dispersing PVA in deionized water via a magnetic stirrer, followed by the addition of a glutaraldehyde solution with constant stirring. The prepared nano-composite membranes were soaked in the PVA solution without any disturbance for 30 minutes at room temperature. The membrane was withdrawn from the solution and degasified under a hot air oven for the appropriate attachment of the PVA coating sheet to the nano-based membrane. The vermiculite nano-composite PVDF membrane, with a maximum water flux of 628.7 $L/m^2.h$ successfully removed 100% of harmful dyes malachite green (MG), methylene blue (MB), Congo red (CR), and safranin O (SO) from organic contaminated water. The membrane effectively rejected 94.56% of humic acid from the given sample. The membrane is a promising candidate for organic pollutants removal on a commercial scale due to its high pollutant rejection rate, anti-fouling properties, and mechanical stability (Isawi 2019).

Phase inversion via the immersion precipitation technique was used to design a magnetic graphene polymer-based nanofiltration membrane for the effective rejection of organic pollutants such as dyes and other heavy metals. Graphene oxide (GO) was synthesized by mixing 1 g of graphite powder and 6 g of potassium permanganate in 120 mL of sulphuric acid with constant stirring at 27^0C. 30wt.% of hydrogen peroxide (3 mL) was added to the solution. The resultant solution was filtered using filter paper and the residue obtained was washed with hydrochloric acid followed by deionized water to obtain a powder form GO. Another solution was prepared by blending graphene oxide with dimethylformamide and N,N'-dicyclohexylcarbodiimide (DCC). Shortly after, metformin was dispersed into the solution with vigorous stirring for a day at a temperature of 27°C. The resultant solution was then added to a mixture containing 1.86 g of $FeCl_3 \cdot 6H_2O$ and 0.6 g of $FeCl_2 \cdot 4H_2O$ under the influence of nitrogen gas. For precipitation to occur, ammonia solution was added to the mixture and, subsequently, a magnetic blackish colored precipitate was attained. The precipitate was thoroughly washed with deionized water and ethyl alcohol followed by desiccating it under a vacuum. The hybrid magnetic graphene-based composite (MMGO) was dispersed in DMAc and ultrasonicated for 30 minutes. For polymerization, appropriate amounts of polyether sulfone (PES) and polyvinyl pyrrolidone (PVP) were constantly stirred at a high speed for 24 hours to obtain a uniform homogenous solution. The resultant solution was cast onto a glass dish and rinsed in deionized water followed by drying it at room temperature. The modified nanofiltration membrane with an effective filtration area of 12.56 cm^2 efficiently eliminated 92% of copper ions without any change in its performance for several cycles. The hydrophilic membrane exhibited greater dye removal potential by rejecting 99% of organic dyes from contaminated water. The membrane is a good candidate for water remediation due to its remarkable features like its anti-fouling potential, surface roughness, and chemical and mechanical stability (Abdi et al. 2018).

An Ag–SiO_2 nanocomposite polyether sulfone (PES) membrane was employed for the removal of organic pollutants, mostly bacteria. A solution for the preparation of Ag-SiO_2 nanocomposite was prepared by adding tetraethoxysilane (TEOS) in a mixture containing ethyl alcohol, water, and ammonia. The mixture was constantly agitated for few hours and dried at a temperature slightly lower than the boiling point of water. A mixture of silver nitrate and ammonia solvent (molar ratio 0.083) was added to the SiO_2 sludge with steady agitation. The resultant Ag–SiO_2 nanocomposite was then washed with ethyl alcohol and desiccated at 27°C. An appropriate amount of Ag–SiO_2 was dispersed in a dimethylacetamide (DMAC) solvent using ultrasonication. Briefly, 18% of polyether sulfone (PES) was mixed with constant agitation to obtain a uniform mixture. The resultant mixture was dried at 60°C in a hot air oven. The non-woven polyester filament was fixed with a glass dish followed by spreading the polymeric mixture with a thickness of 200 μm on the polyester filament. The resultant membrane was then water coagulated at room temperature. Ultimately, the membrane was washed with ultra pure (UP) water for the removal of excess solvent from the solution. The membrane's effective filtration area of 13.4 cm^2 efficiently eliminated *Escherichia coli* and *Pseudomonas* sp., illustrating its notable anti-bacterial properties. The membrane was constructed with the intension of inducing noteworthy features into the membrane such as anti-fouling and anti-biofilm effects and hydrophilicity. The membrane is currently used for the treatment of ground water in municipal corporations (Huang et al. 2014).

An organic polyamide (PA) nanofiltration membrane was utilized for organic water treatment and reducing COD up to standard value. The contaminated solution from the feed chamber was driven into a membrane canal. It provided an effective filtration area of 119.6 cm^2. The membrane with a permeability of $2.42 \times 10^{-11} \text{ m}^3/\text{m}^2\text{sPa}$ successfully diminished the COD up to 94% at a pressure of 276 kPa. The membrane showed an admirable rejection rate for organic dyes by removing 94% Cibacron Black B and 92% Cibacron Red RB from the polluted solution. The process of water remediation using a nanofiltration membrane is a promising approach for large-scale industries where dyes are produced as major contaminants (Chakraborty et al. 2003).

A titania-based photocatalytic membrane was suggested for the purification of water samples compromising organic pollutants, mainly dyes and certain phenolic compounds. Basic TiO_2 (17.5% wt.) hydrosols were taken as a precursor for the membrane formation. A mixture was prepared by mixing basic hydrosols with an ammonia solution at a ratio of 1:9. Hydroxyethyl cellulose (HEC) was then added to the above diluted titania solution. The solution was cast on an alumina support tube of 200 mm × 7 mm for 30 seconds. The modified membrane was then thermally treated at 250°C and 450°C for 24 hours each for the proper binding of the casted solution on the alumina support. The structural morphology of the membrane was studied via SEM. The average thickness of the membrane was evaluated to be 3 μm with 68% porosity and high-water permeability (150 Lh^{-1}m^{-2} bar^{-1}). The membrane was mechanically and chemically established with a reversible fouling effect. Due to its remarkable photocatalytic degradation properties, the membrane demonstrated great potential for the elimination of phenolic contaminants and organic dyes from industrial wastewater residues (Djafer et al. 2010).

A multifunctional Ag-casted TiO$_2$ (Ag–TiO$_2$) photocatalytic nano-based membrane was manufactured for the photocatalytic degradation of bacteria, mainly E.coli and organic dyes like Rhodamine B (RhB). An Ag–TiO$_2$ coating thin film was prepared primarily by mixing Pluronic P-123 into a 2-propanol (IPA) solvent. Another solution was prepared by dispersing AgNO$_3$ in a 2-propanol (IPA) solvent. The AgNO$_3$ solution was mixed with a P-123 solution with constant agitation to attain a uniform solution. Ultimately, the solution containing titanium tetraisopropoxide (TTIP) and water (1:4) was added to the above homogenous solution. The alumina support was layered with polyvinyl alcohol (PVA) and dried at room temperature. The Ag–TiO$_2$ solution was cast on the alumina support using a fabricated dip-coater. The resultant membrane was then desiccated in a hot air oven at 105°C followed by calcination. The modified, cost-effective, nano-composite membrane extracted a large degree of Rhodamine B and successfully removed 7-logs of E.coli bacteria in the photocatalytic membrane bioreactor (PMR). This biocompatible membrane, with a high photocatalytic potential and excellent adsorption capacity, makes it a probable candidate for large-scale application (Goei and Lim 2014).

A PVDF cellulose nano-composite membrane was synthesized for the effective adsorption of crystal violet dye and some nano-particulate compounds like Fe$_2$O$_3$ from toxic industrial wastewaters. The primary polyvinylidene fluoride membrane was prepared by electrospinning. The solution for electrospinning was prepared by dispersing polyvinylidene fluoride pellets into a mixture comprising dimethylformamide and acetone (3:2). The electrospinning method was carried out at the room temperature at a voltage of 15k V. The fibrous membrane obtained after the process was desiccated at a temperature of 150°C. The cellulose nanofibers were dispersed into Meldrum's acid in equal amounts, followed by heating it at a temperature of 110°C with simultaneous agitation using a magnetic stirrer. The customized cellulose nanofibers were rinsed with deionized water and acetone for the removal of excess acid. The tailored CNFs were dehydrated in an oven for an hour. The basic polyvinylidene fluoride membrane obtained from the electrospinning process was placed on a glass filtration unit. Shortly after, the membrane was decorated with CNFs customized with Meldrum acid under very high pressure (200 mm/Hg) at a neutral pH. The resultant membrane was ultimately desiccated at a temperature equal to the boiling point of water for 15 minutes. The membrane effectively eliminated over 99% of CV dye and Fe$_2$O$_3$ from the given contaminated water sample over 4 cycles without any appreciable decrease in its filtration performance. The use of this cost-effective, nanofibrous membrane on a large scale would be an effective step towards the application of green technology in water purification treatment (Gopakumar et al. 2017).

The interfacial polymerization (IP) method was used to prepare a thin-film nano-composite membrane for water desalination and decontamination, as shown in **Figure 19.3**. Firstly, a mixture was prepared by mixing sodium hydroxide in distilled water in a molar ratio of ~1:120. Cetyltrimethylammonium bromide (CTAB) (1 g) was then mixed with a sodium hydroxide solution with constant stirring. A white-colored, paste-like mixture was obtained by adding 5 mL of tetraethyl orthosilicate (TEOS) steadily. The mixture was then centrifuged at a speed of 10,000 rpm. The pellet obtained was rinsed in distilled water followed by calcination to obtain MCM-41 nanoparticles. The silica nanoparticles were synthesized by hydrolyzing tetraethyl orthosilicate (TEOS) (22.4 mL) in a solution comprising ethanol (190 mL) and ammonia (14 g). The resultant

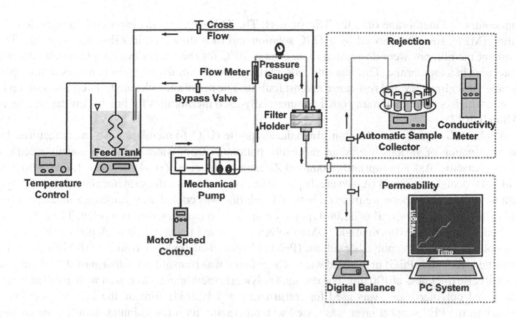

FIGURE 19.3 Diagram depicting the nanofiltration process for wastewater treatment. Reprinted with permission from Yin, Jun, Eun Sik Kim, John Yang, and Baolin Deng. 2012. "Fabrication of a Novel Thin-Film Nanocomposite (TFN) Membrane Containing MCM-41 Silica Nanoparticles (NPs) for Water Purification." Journal of Membrane Science 423–424. Elsevier: 238–46. doi:10.1016/j.memsci.2012.08.020.

mixture was centrifugated at 10,000 rpm and the pellet was rinsed with distilled water to attain the silica nanoparticles. A support film was prepared by dissolving polysulfone pellets (15wt.%) into N,N-dimethylformamide (DMF) at 50°C. The resultant mixture was poured onto a glass plate up to a thickness of 100 μm. The solidified membrane on the glass plate was detached by immersing the glass plate in distilled water at room temperature. The prepared membrane support was rinsed in an m-phenyl diamine (MPD) solvent for a couple of minutes and then transferred into a TMC-hexane solution to be ultrasonicated for an hour. This led to the development of a polyamide layer on the PSU support membrane. The membrane was ultimately rinsed with a hexane solution and desiccated in a hot air oven. The structural and molecular characteristics of the membrane were evaluated using different techniques, including SEM, TEM, and AFM. The membrane demonstrated high water flux up to 46.671.1 L/m^2 with a membrane thickness between 300 nm and 500 nm. The anti-fouling membrane, with an effective filtration area of 100 μm^2, maintained a high rejection rate by removing 99.3% of sodium chloride and 98.5% of sodium sulfate from the given water sample. The membrane can be recommended for industrial application due to its high potential for desalination (Yin et al. 2012).

A reverse osmosis membrane impregnated with ZIF-8 NPs was employed for the desalination process of brackish water. Two solutions were prepared by dispersing Zn(NO$_3$)$_2$.6H$_2$O (3 g) and 2-methylimidazole (6.6 g) in equal volumes of methyl alcohol. The resultant solutions were mixed and then vigorously stirred for an hour. The homogenous solution obtained was then centrifugated and the resultant ZIF-8 nanoparticles produced in the form of pellets were thoroughly rinsed with methyl alcohol and chloroform. Dehydrated polysulfone (PSF) crystals were dispersed into a dimethyl acetamide (DMAc) solvent using a magnetic stirrer. The solution was then spread onto a glass dish with a uniform thickness of 150 μm. The resultant PSF support membrane sheet on the glass dish was then transferred to a water bath for 24 hours before casting ZIF-8 nanocrystals onto it. The resultant PSF membranous sheet was exposed to an m,phenylenediamine (MPD) solvent followed by a trimesoyl-chloride (TMC) solution at room temperature to synthesize the TMC PA membrane onto the PSF support. The same procedure was followed for synthesizing the thin-film nanocomposite membrane onto the TSF support. The TSF support was exposed to a phenylenedi-amine (MPD) solvent followed by a TMC solution and then dissolved in ZIF-8 nanocrystals. The resultant membranes were desiccated in an oven at 70°C for the proper bonding between different molecules in a membrane. The superior hydrophilic properties of the membrane makes it more per-meable, providing elevated resistance against fouling. The membrane elevated salt elimination up to 99.4%, which is more than many other commercially-used membranes for brackish water treatment (Aljundi 2017).

A robust Zr-organo-metallic thin-film nanocomposite (UiO-6) membrane was manufactured for the elimination of various hazardous refractory pollutants from water bodies, primarily selenium (Se) and arsenic (As). Appropriate amounts of Zirconium chloride (ZrCl4), benzene-1,4-dicarboxylic acid (BDC), and acetic acid (AA) were dispersed and reacted with a dimethylformamide (DMF) sol-vent at a constant temperature in an oil bath. The solution was cooled up to atmospheric temperature and centrifugated at a speed of 12,000 rpm to obtain UiO-6 nanocrystals as a pellet. The pellet was washed with a dimethylformamide (DMF) solvent and dried under vacuum. A polymeric mixture was obtained by mixing polyether sulfone (PES) polymer, N-methyl-2-pyrrolidone (NMP), and poly-ethylene glycol 400 (PEG) in distilled water. The solution was transmitted onto a glass dish maintain-ing a precise thickness of 100 μm. The resultant polymeric membrane was rinsed with distilled water. Interfacial polymerization was used for constructing a polyamide film on the PES support layer. Primarily, the PES support layer was rinsed with piperazine for a few minutes, followed by casting 1,3,5-benzenetricarbonyltrichloride (TMC) solvent on the membrane which led to the formation of the thin-film composite membrane. The thin-film nano-composite membrane was produced by wash-ing a PES support with piperazine. Shortly after, the UiO-6 nano-pellets were dissolved into a hexane solvent using ultrasonication. The resultant solution was then cast onto the PES support to form a TFN membrane. The impregnation of UiO-6 nanoparticles on the membrane significantly increased the anti-fouling resistance of the membrane. The membrane demonstrated excellent water flux of 11.5 LMH/bar with greater hydrophilicity and permeability. The membrane successfully adsorbed 96.5% of SeO_3^{2-}, 97.4% of SeO_4^{2-}, and 98.6% of $HAsO_4^{2-}$ over a number of cycles without any deteriora-tion in the filtration performance. The use of UiO-6 incorporated nanofiltration membranes can be a promising approach in future water purification processes (He et al. 2017).

A multifunctional silica-titania nano-tubular membrane was prepared for the successful elimi-nation of sodium dodecylbenzene sulfonate (SDBS) via photocatalytic degradation. A solution was prepared by mixing $Ti(OC_4H_9)_4$ and $Si(OC_2H_5)_4$ in a molar ratio of 1:1 with constant stirring. A mixture of ethyl alcohol and hydrochloric acid was then added, and the entire solution was ultrason-icated to form a homogenous pale yellow solution. For the casting process, the alumina membrane was rinsed into the above solution for a specific time period. The resultant membrane was desic-cated at an ambient temperature followed by heating it under a muffle furnace/oven up to 4,000°C for a couple of hours. A modified photocatalytic membrane exhibited a high rejection rate of sodium dodecylbenzene sulfonate (SDBS) up to 89% due to its excellent permeability and high-water flux. In addition to the complete degradation of SDBS, the membrane shows excellent anti-fouling capac-ity due to the incorporation of nanoparticles into the membrane. The mechanically-stable mem-brane can be a promising candidate for eliminating SDBS from water bodies (Zhang et al. 2006).

A multifaceted silver nanocomposite polyurethane membrane was applied for the efficient removal of certain lethal refractory contaminants including organic dyes, carcinogenic metals like arsenic, and certain bacteria. A uniform solution was prepared by dispersing polyurethane crystals along with fly ash particles into a mixture of N,N-dimethylformamide (DMF) and methy-lethylketone (MEK)/2-butanone (1:1) with constant agitation with ultrasonication. Shortly after, $AgNO_3$ (300 micrograms) was dispersed in N,N-dimethylformamide (DMF) and was added to the above prepared solution and agitated for different time periods (1 hour and 12 hours). The two membranous sheets were prepared by electrospinning at room temperature with a voltage of 15 kV. The resultant Ag-FAPs decorated nanocomposite membrane was desiccated overnight under vacuum. The morphological characteristics of the membrane were evaluated using FESEM.

The membrane was crack free with high permeability and stability. The anti-fouling membrane, with an effective filtration area of 13.4 cm^2, eliminated carcinogenic arsenic and other organic dyes. The membrane also proved its anti-bacterial nature by removing a significant number of bacteria, specifically *E.coli*, from the given water sample. The use of this membrane on a commercial scale presents as a cost-effective method and is a green approach towards water purification processes (Pant et al. 2014).

A chemically-stable organic solvent nanofiltration membrane impregnated with polybenzimidazole (PBI) was employed for water decontamination. Celazole S26 polybenzimidazole (PBI) was dissolved in a solution containing dimethylacetamide (DMAc) solvent with constant agitation and was set aside overnight to obtain a uniform solution. The solution was casted on a polypropylene membranous sheet with uniform thickness. The membrane was then transferred to a water bath at a temperature slightly lower than room temperature for a day. The resultant membrane was rinsed with propan-2-ol (IPA) to remove excess solution. The membranes were polymerized in a mixture of α,α-dibromo-p-xylene (DBX) or 1,4-dibromobutane (DBB), an acetonitrile, at 80°C for a day. The membrane was then washed with a propan-2-ol (IPA) solvent. Finally, the membrane was fabricated with a polyethylene glycol and propan-2-ol mixture for a couple of hours followed by desiccation to obtain a polymer-based nanofiltration membrane. The chemical resistance and stability of the membrane were studied by exposing the membrane to different environmental tensions, such as acidic and basics stresses. The membrane survived all the conditions without any change in permeability and filtration performance. The membrane showed excellent separation of organic contaminants/solvents from pharmaceutical and petrochemical wastewater with an effective filtration area of 14 cm^2. The application of this nano-filtration membrane on an industrial scale can be a profitable way of water purification (Valtcheva et al. 2014).

A photocatalytic nanofiltration membrane fabricated with tungsten trioxide (WO$_3$) was tailored for oil-water separation and the photocatalytic degradation of organic contaminants. A spray coating technique was used to impregnate a WO$_3$-tetrahydrofuran (THF) dispersion on a stainless-steel supportive network. The physical characteristics of the membrane were evaluated using SEM. The pore size of the membrane was kept large at 150 μm to avoid fouling problems. About 99.9% photocatalytic degradation of organic dyes, including methylene blue and direct red dye, was observed under UV light. The membrane also exhibited considerable oil-water separation up to 99% due to its superior hydrophilicity and high wettability. This simple, scalable, and cost-effective filtration process can be used on an industrial scale where oil is produced as a major effluent (Gondal et al. 2017).

A multifunctional two-dimensional molybdenum disulfide (MoS$_2$) nanofiltration membrane was introduced for the retention of organic pollutants and desalination of the given water sample on a large scale. MoS$_2$ nanosheets were formed due to the exfoliation of large amounts of MoS$_2$ powder, followed by cavitation using a high velocity of liquid jet to form a porous nano disk-like structure. SEM was used to determine the structural aspects of the modified MoS$_2$ membrane. The membrane possessed an average thickness of 1 μm with an average pore size of 0.02 μm. The membrane successfully ejected 99% of major sea salts, like sodium chloride and magnesium chloride due, to its high permeability via forward osmosis. Almost 100% of organic dyes, including methyl red (MR), methyl orange, methylene blue (MnB), and rhodamine B (RhB), were photo-degraded under UV light. The free-standing membrane was successfully tested for its anti-fouling potential and chemical stability (Sapkota et al. 2020).

19.2.4 Carbon-Based Membranes

A carbon-based nano-fibrous membrane was applied for the successful elimination of certain organic dyes and heavy metal ions. Hydrothermal carbonization (HTC) was used to manufacture carbon-based nanofiltration membranes. A mechanically and chemically stable membrane was obtained by spreading a solution of CNF on a Teflon support whilst maintaining a constant thickness. The resultant membrane was desiccated at room temperature for proper cross-linking to attain a mechanically self-sufficient membrane. The morphological characteristics of the membrane were determined via SEM. The three membranes were studied with a uniform thickness of 50 µm and varied membrane diameters, namely CNF-50, CNF-100, CNF-280, where the number corresponds to the respective diameter of the membrane. The carbon-based nanofibrous membranes adsorbed almost all the methylene blue, lead (Pb^{2+}), and certain amounts of chromium ion [Cr(IV)] from the given contaminated water sample with a high water flux of 1,580 L/m^{-2}h^{-1}. The membrane was easily regenerated and used over 6 cycles of water purification without any alteration to the filtration efficiency of the membrane. The use of this carbonaceous membrane is a cost0effective and reliable method for water purification processes because of its high rejection rate of organic dyes and heavy metals at a large scale (Liang et al. 2011).

A facile hydrothermal technique was used for preparing a nitrogen-doped reduced graphene oxide (N-rGO) membrane for the elimination of specific phenolic compounds from a contaminated water sample. A defined quantity of graphene oxide was dissolved in deionized water using ultra-sonication. Shortly after, ammonia was added was added to the GO solution with constant stirring. The resultant solution was heated to obtain a viscous hydrogel-like substance. The hydrogel was then impregnated on a basic support of cellulose-ester membranous sheet via a vacuum filtration process. The resultant nitrogen-doped reduced graphene oxide membrane was thoroughly rinsed with distilled water to eliminate excess solution. The membrane with a high flux of 0.036 ± 0.002 mmolh^{-1} successfully oxidized certain phenolic compounds such as tetracycline. The membrane was used over multiple cycles without any change in catalytic oxidation filtration performance of the membrane. The environmentally-friendly and highly stable membrane can be a powerful tool for contaminated water treatment containing certain refractory organic pollutants (Liu et al. 2016).

Carbon-based nanotubes were employed for effective salt removal from wastewater to make it potable. A vertically aligned and mixed matrix CNT membrane was used for the desalination process in this experiment. Vertically aligned CNTs were processed by arranging CNTs membranous sheets on the supportive polymeric layer with grafted silicon nitride via chemical vapor deposition. On the other hand, the mixed matrix carbon nano-tubular membrane was horizontally placed for the filtration process, which consisted of several polymeric supportive membranes. The physical characteristics of the membrane were studied using SEM. The CNTs membrane exhibited superior permeability of 7×10^{-7} mPa^{-1}s^{-1}, which is much higher than other commercially-used membrane technologies for water purification processes. The membrane possessed a pore size of 0.1–2 nm with a thickness of 2–6 µm. The membrane showed excellent desalination potential with outstanding anti-microbial properties. Due to the presence of silver nitride, the membrane interrupts the oxidative pathways of the bacterial species leading to their death. The membrane was designed with remarkable characteristics like self-cleaning mechanisms and fouling resistance. The filtration mechanism through the CNT membrane requires lower energy consumption as external pressure is not involved for completing the filtration process. These environmentally-friendly CNT membranes are probably the best alternatives to reverse osmosis, microfiltration, and nanofiltration membranes due to their high salt rejection potential and lower cost (Das et al. 2014).

Vacuum filtration along with drop-casting was applied for designing a dopamine impregnated graphene oxide membrane for water decontamination. A uniform solution of graphene oxide was obtained by dispersing graphene oxide into deionized water via ultrasonication. An alkaline solution was obtained by the addition of small concentrations of dopamine solvent into a GO solution. The resultant alkaline solution was then transferred to fabric material followed by desiccation to obtain a cross-linked colloidal gel. The resultant graphene-oxide PDA gel was mixed with a β-cyclodextrin (CD) solution along with glutaraldehyde. The desired dopamine grafted graphene oxide polymeric ultrafiltration membrane was attained via vacuum filtration. The membrane with a high permeability of 12 L m^{-2}h^{-1} successfully rejected 99.2% of methylene blue and absorbed 99.9% of lead ions (Pb^{2+}). The membrane was used for longer periods without any change in its filtration performance. The hydrophilic GO membrane was easily regenerated using HCl. It was reusable and highly stable and was easily cleaned with ethanol. The graphene oxide membrane is a promising membrane technology for many industries where lead is produced as a major contaminant (Wang et al. 2018).

An activated carbon (AC) fabricated graphene oxide membrane was employed for the decontamination of water containing antibiotics like tetracycline. A homogenous solution was prepared by adding activated carbon in a graphene oxide solvent with constant agitation using ultrasonication. The resultant solution was transferred onto a polyvinylidene fluoride sheet to mold it into a stable and robust GO/AC hybrid membrane. The structural characteristics of the GO/AC membrane were studied by SEM. The membrane has a thickness of 15 μm and pore size of 3–10 nm and exhibited a high water permeability rate. The membrane adsorbed 98.9% of tetracycline antibiotic and organic dyes like rhodamine B and methylene blue from the given contaminated water. The use of this crack-free and mechanically-stable membrane is environment friendly and is a cost-effective way to remove organic contaminants from polluted wastewater from the pharmaceutical and chemical industries. Activated carbon GO membranes can be used on an industrial scale to make water free from organic pollutants (Liu et al. 2017).

A hybrid silver nanoparticle grafted CNTs polymer-based membrane was constructed to remove refractory pollutants like oil in order to decrease the COD of a water body. A solution containing CNTs and acrylic acid was ultrasonicated in an acetone solvent for 10 minutes. The resultant solution was set aside for a few hours at a high temperature of 75^0C after the addition of benzoyl peroxide (BPO). The polymeric CNTs membrane was separated from the solution and dehydrolysed overnight. On the other hand, tollens reagent was prepared by sonicating a mixture of silver nitrate and ammonia to obtain a brownish ppt. Sodium dodecyl sulfate was introduced into a poly-acrylic acid solvent followed by the addition of tollens reagent with constant agitation. The resultant mixture was centrifuged by adding formaldehyde. The pellet form silver impregnated polymer-based CNT was rinsed with ethyl alcohol before filtering it through a polyvinylidene fluoride membrane to obtain a final silver fabricated PAA/CNT membrane. Transmission electron microscopy was used to evaluate the morphology of the modified CNTs. The average diameter of the membrane was found to be approximately 40 nm with a uniform thickness of 1 μm. The membrane was highly efficient with a high water flux of 3,000 L/m^{-2} h^{-1} bar^{-1} due to its higher adsorption of bacteria, specifically *E.coli*, and excellent oil-water separation potential. This hydrophobic, multifunctional membrane can be applied on an industrial scale where oil is produced as a major pollutant (Gu et al. 2016).

A robust graphene nanofiltration membrane was employed for the effective elimination of organic dyes and salt using the electrostatic processes. Chemically converted graphene (CCG) was used to prepare a precursor of an ultrathin graphene nanofiltration membrane, i.e., a base refluxing reduced graphene oxide solvent. Sodium hydroxide was used for the cleaning of brGO obtained in the above step. The brGO solution was then transferred onto a polyvinylidene fluoride (PVDF) support membrane to obtain an ultrathin graphene membrane. The thickness of the membrane was maintained between 22–53 nm. The membrane with a high water flux of 21.8 L/m^{-2}h^{-1} bar^{-1} offers more permeability facilitating oxidative retention of most of the organic and inorganic pollutants. The modified graphene membrane exhibited high retention of organic dyes including methylene blue (99.8%), direct red 81 (99.9%), and rhodamine B (78%). A high rejection rate of salt like Na_2SO_4 (<60%) and NaCl, $MgSO_4$, and $MgCl_2$ was observed using this membrane. The membrane was mechanically and chemically stable and easy to handle. The use of this reusable and cost-effective modified graphene membrane on a commercial level would be a green approach towards water purification processes (Han et al. 2013).

A mesoporous modified-graphene oxide nanofiltration membrane was designed for the retention of diverse macromolecules including organic and some ionic compounds. A two-dimensional membrane was oxidized using potassium permanganate ($KMnO_4$) to obtain a mesoporous graphene oxide membrane. The membrane with an average thickness of 1 μm exhibited great water permeability of 191 L/m^{-2}h^{-1}bar^{-1}. The modified membrane successfully removed 100% of cytochrome C, 90% of Evan's blue (EB), and ~40% of $Fe(CN_6)_3$ from the contaminated water sample. The membrane can be used for large-scale industries due to its low cost and long-term filtration potential. The membrane presents a good alternative to many other commercially-available membranes due to its high mechanical stability, flexibility, and remarkable separation capability (Ying et al. 2014).

In another example, a robust polyamide membrane fabricated with various nanoparticles was introduced for the effective removal organic dyes from the given water sample. An appropriate amount of polyamide 6 was dispersed into a solution of formic acid and acetic acid (3:2) using a magnetic stirrer. The resultant polymeric solution was subjected to electrospinning at a 25 kV voltage to obtain an upper and lower fibrous membrane. The central graphene oxide membrane was manufactured by using the electrospraying method. A graphene oxide solution was ultrasonicated overnight with respective nanoparticles including TiO_2, SiO_2, and Si_3N_4. The modified nanoparticles incorporated into the GO membrane were also prepared via electrospinning. The customized GO membrane was fabricated on a lower polyamide membrane by electrospraying for 2 hours. Similarly, the upper polyamide layer was electro-sprayed on a central GO membrane. The membrane was desiccated to remove excess solution for a couple of hours before using it for filtration. FESEM and X-ray fluorescence analysis were used to determine the morphological characteristics of the membrane. The membrane, with a pore size of 0.45 μm, rejected 99.36% of molybdenum oxide (MO), 99.85% of Evans blue (EB), and 92.38% of methylene blue (MB) from the contaminated water sample, maintaining a high water flux up to 13.77 $L/m^{-2}h^{-1}bar^{-1}$. The GO-modified membrane is preferred by many industries due to its high rejection rate for organic dyes, cost-effectiveness, and higher mechanical stability (Chen et al. 2018).

A cross-linked graphene oxide isophoronediisocyanate (GO/IPDI) membrane was employed for the retention of organic dyes and heavy metal ions from contaminated water bodies. A dehydrated graphene oxide powder was dissolved into an N,N-dimethylformamide (DMF) solvent via ultrasonication. Briefly, a graphene oxide isophoronediisocyanate mixture was obtained by the addition of isophoronediisocyanate (IPDI) into a GO solution with constant stirring. The solution was cast upon a commercially-available polyvinylidene fluoride (PVDF) membrane and ultimately rinsed with an N-Methy-2-pyrrolidone solvent to remove the residual solution. The membrane, with a thickness of 1μm and pore size of 0.22 μm, maintained a high water flux up to 100 L/m^{-2} h^{-1} bar^{-1}. It exhibited excellent water decontamination potential by eliminating organic dyes like MB (97.6%), RB (96.2%), MO (96.9%), and CR (98.24%). It also removed certain amounts of specific metal ions like Cu^{2+} (46.2%), Pb^{2+} (66.4%), Cr^{3+}(71.1%), and Cd^{2+}(52.8%). The membrane was relatively more stable during the filtration process with higher permeability and pollutant retention rates. Due to its cost effectiveness and lower energy consumption the membrane can be used on a large scale (Zhang et al. 2017).

A reduced graphene oxide nano-composite membrane was introduced for the efficient removal of organic pollutants from water bodies. A melamine powder was calcinated for a couple of hours to form a bulk-g-C_3N_4, which was further ultrasonicated with ethyl alcohol to obtain g-C_3N_4 nanosheets. The g-C_3N_4 nanosheets were transformed into nanotubes by vacuum desiccation overnight. The resultant g-C_3N_4 nanotubes were dissolved into a reduced graphene oxide solvent to obtain a g-C_3N_4/rGO composite. A cellulose acetate (CA) membrane was treated with polydopamine in a tris buffer for a day for the formation of apolymeric membranous layer on the CA support surface. Another solution containing g-C_3N_4/rGO along with ethylenediamine (1%) was cast on the modified CA membrane at 27°C. SEM analysis was used to evaluate the physiological characteristics of the membrane. The thickness of the membrane was maintained at 700 nm with an effective filtration area of 11.3 cm^2 and high water permeability up to 4.87 $L/m^{-2}.h^{-1}.bar^{-1}$. The membrane successfully eliminated 98.9% of Rhodamine B (Rh B) dye due to its remarkable photocatalytic properties. The membrane is self-cleanable and possesses higher chemical and mechanical stability and durability relative to other commercially-available membranes (Wei et al. 2019).

Vacuum filtration was used to prepare a two-dimensional graphene oxide membrane for dye water filtration process. Graphene oxide powder was prepared by adding graphene crystal and $KMnO_4$ (1:6) into a mixture of sulfuric acid and phosphoric acid with constant stirring. The resultant solution was mixed with hydrogen peroxide before subjecting it to centrifugation. Shortly after, polyethylene imine (PEI) was mixed with graphene oxide and the resultant mixture was centrifugated at 5,000 rpm and washed with distilled water to form a PEI grafted GO suspension. The resultant suspension was poured on a nylon microfiltration layer via vacuum filtration to form the PEI grafted graphene oxide membranes (PGOMs). FESEM was used to study the structural aspects of the membrane. The thickness of the membrane was 9.75 μm and the diameter was 4.10 cm. The membrane exhibited a high water flux of 450.2 $L/m^{-2} \cdot h^{-1} \cdot bar^{-1}$ (PGOM-10000). The membrane illustrated a high rejection rate of <99% for organic dyes including direct red (DR) and methylene blue. The membrane can be a promising candidate for water purification processes due to its remarkably high permeability and stability (Lu et al. 2019).

A halloysite nanotubes-grafted graphene oxide membrane was constructed for water purification for the removal of organic dyes and certain amounts of specific salts from a given water sample. Poly-sodium-p-styrene sulfonate (PSS) was dispersed into the distilled water followed by exposing the calcinated halloysite nanotubes to the PSS suspension for an hour for proper electrostatic interaction. The resultant modified HNTs were centrifugated and then rinsed with distilled water and dehydrated under vacuum. Graphene oxide was obtained by oxidation of natural graphene. Further, the graphene oxide was treated with sodium hydroxide followed by hydrochloric acid to get reduced GO. The mixture was prepared by mixing the three-solutions containing PRGO, polyvinyl alcohol (PVA), and HNTs. The resultant mixture was transferred onto the polyacrylonitrile (PAN) micro-filtration membrane for a specific time period followed by air dying it at a very high temperature to form a PRGO/HNTs-PSS membrane, as demonstrated in **Figure 19.4**. Transmission electron microscopy was used to estimate the structural characteristic of the membrane. The filtration area was 100 μm with a high permeability of water up to 8.8 $L/m^2 h$ bar. The membrane exhibited a low rejection of salts: 4.7% for $MgSO_4$, 4.7% for $MgCl_2$, 6.8% for NaCl, and 14.3% for Na_2SO_4, but displayed a high rejection of reactive black 5 dye (up to 97.9%). This anti-fouling and stable membrane is a remarkable tool for water purification processes in coming years (Liping Zhu et al. 2017).

A silver-nanoparticles incorporated graphene oxide membrane was employed for the removal of bacteria and other organic pollutants from water bodies. Primarily, graphene oxide membranes were dissolved in deionized water to obtain a uniform mixture. Silver nitrate was then added to the mixture and ultrasonicated for 30 minutes. The resultant mixture was then placed at a very high temperature after the addition of sodium citrate. The graphene oxide-Ag mixture was then subsequently filtered through a cellulose acetate membrane of 0.45 μm pore size to get a resultant GO-Ag hybrid membrane. SEM was used to evaluate the structural characteristics of the membrane. The thickness

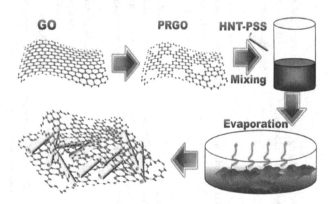

FIGURE 19.4 Fabrication of PRGO/HNTs-PSS hybrid membranes. Reprinted with permission from Zhu, Liping, Huixian Wang, Jing Bai, Jindun Liu, and Yatao Zhang. 2017. "A Porous Graphene Composite Membrane Intercalated by Halloysite Nanotubes for Efficient Dye Desalination." Desalination 420 (June). Elsevier: 145–57. doi:10.1016/j.desal.2017.07.008.

TABLE 19.1

Examples of Potential Membrane Technologies Applied to Remove Organic Pollutants From Wastewater

No.	Nature of Membrane	Type of Pollutants	Parameters	Important Properties	Efficiency	Reference
1.	$Al_2O_3/CoFe_2O_4$ catalytic ceramic membranes	Organic micro-pollutants	Pore size of 267.5 nm with a thickness of approximately 420 ± 40 mm	Reusable and low fouling	High removal rate of MB (~98%) and ibuprofen (~99.5%)	(Wang et al. 2020b)
2.	PAN/PVDF chelating membrane with amidoxime groups	Heavy metal and organic pollutants from mimic industrial wastewater	-	Mechanically stable	High efficiency Cu^{2+} (99.77%), Pb^{2+} (95.25%), Cd^{2+} (79.57%), lysozyme (94.36%), and BSA (97.67%)	(Xu et al. 2020)
3.	GO/PVDF membrane	Seafood wastewater treatment (high salinity organic wastewater) (ammonia removal)	-	Antifouling ability and cost effective	Ammonia removal efficiency (98%) and reduction in COD (95%)	(Chang Li et al. 2019a)
4.	Photocatalytic and antimicrobial multifunctional nanocomposite membranes (Ag-TiO2/PVDF-HFP)	Degrading a resilient pollutant (norfloxacin) under UV and visible radiation as well as microbes removal	Pore diameter from 1–8 μ	Reusable membrane	Maximum efficiency 80.7%	(Salazar et al. 2020)
5.	Cross-linked biocatalytic CNTs (PVDF) membrane	micro-pollutants were studied, namely bisphenol-A (BPA), carbamazepine (CBZ), diclofenac (DCF), clofibric acid (CA), and ibuprofen (IBF)	Effective membrane area was observed to be 8.4 cm² and the average pore size was 20 nm	Stable, reusable, regenerative, and negligible membrane fouling	Average removal rate of 95%	(Ji et al. 2016).
6.	Copper oxide modified ceramic hollow fiber membrane with insitu peroxymonosulfate activation (NC)	Organic pollutants removal	0.22 μm pore size	Superior stability and reusability	Eliminating methylene blue (MB), bisphenol A (BPA), phenol, 4-chlorophenol (4-CP), and sulfamerazine (SMZ) with removal rates of 93.2%, 94.5%, 91.8%, 95.3%, and 91.6%, respectively	(Wang et al. 2020a)

(Continued)

TABLE 19.1 (CONTINUED)

Examples of Potential Membrane Technologies Applied to Remove Organic Pollutants From Wastewater

No.	Nature of Membrane	Type of Pollutants	Parameters	Important Properties	Efficiency	Reference
7.	A poly-ceramic TiO_2 nanofibrous membrane	Organic matter, humic acids, bacteria, herbicides (isoproturon), and highly acidic effluents	-	Antifouling property	eliminated ~99% of isoproturon and methylene blue along with other micropollutants	(Geltmeyer et al. 2017)
8.	Tailor-made polyelectrolyte multilayeredMembrane	Obidoxime residues (OBD)	Pore size of ~0.45 μm	-	Rejected 99.26% of obidoxime (OBD)	(Nikhil Chandra and Mothi Krishna Mohan 2020)
9.	Polyethersulfone (PES) ultrafiltration membranes were casted and modified by dispersing nano-sized alumina (Al_2O_3) particles uniformly in a PES solution	Effluent includes starch and casein, $(NH_4)_2SO_4$, and KH_2PO_4	Pore sizes of 0.01–0.1 μm	Anti-fouling performance	High efficiency for 0.05 Al_2O_3/PES ratio	(Maximous et al. 2010)
10.	PTFE (polytetrafluoroethylene) membranes	Wastewaters containing volatile organic compounds such as acetone and ethanol	Pore diameter of 0.2 mm and membrane thickness of 165 mm	Mechanically stable and reusable	Separation efficiency is strictly related to the fluid-dynamic conditions existing in the phases external to the membrane	(Boi et al. 2005)
11.	Polymeric (PVDF) hydrophobic and hydrophilic membranes	Filtration for primary effluents and for the removal of bacteria including fecal-coli from bacteria like *Escherichia coli* and *Enterococci*	Optimum pore size of 0.2 μm	Stable, highly permeable, durable, and reusable	Hydrophobic PF-0.3 membrane or hydrophilic AC+0.8 membranes are best	(Modise et al. 2006)
12.	Carbonaceous nanofiber membranes	Removal of organic dyes and toxic metal ions	Uniform thickness of 50 μm	Reusable and can be easily regenerated	Membranes adsorbed almost all the methylene blue, lead (Pb^{2+}), and certain amounts of chromium ion [Cr(IV)]	(Liang et al. 2011)

(Continued)

TABLE 19.1 (CONTINUED)
Examples of Potential Membrane Technologies Applied to Remove Organic Pollutants From Wastewater

No.	Nature of Membrane	Type of Pollutants	Parameters	Important Properties	Efficiency	Reference
13.	Nitrogen-doped graphene nanosheet membrane	Degradation of selected phenolic compounds (tetracycline)	-	Environmentally friendly and reusable	Fairly efficient	(Liu et al. 2016)
14.	Vermiculite nanoparticles incorporated polyvinylidene fluoride (PVDF) flat sheet membrane	Different types of dyes and humic acid	Uniform thickness of 200 μm	Antifouling properties	Removed 100% of some harmful dyes like Malachite green (MG), Methylene blue (MB), Congo red (CR), and Safranin O (SO) and 94.56% of humic acid	(Isawi 2019)
15.	Synthetic polyethersulfone nanofiltration membrane modified by magnetic graphene oxide/metformin hybrid	Removal of dyes and heavy metal ions	Effective filtration area of 12.56 cm^2	Reusable and anti-fouling property	Copper ion removal (~92%) and dye removal (~99%)	(Abdi et al. 2018)
16.	Polymer-based dynamic membrane	Organic pollutants and heavy metals	Effectual filtration region of 0.042 m^2	Fouling problems (disadvantage)	High rejection efficiency by eliminating pollutants like ammonia (NH$_3$-H), total nitrogen (TN), total phosphorous (TP), and some volatile suspended solids	(Chu et al. 2014).
17	Nanofiltration (NF) polyester thin-film composite (TFC) membranes	Removal of natural organic matter substances (humic acid)	-	Cost effective and environmentallyfriendly	Eliminating ~75% of humic acid	(Jalanni et al. 2012)
18	UF and RO membrane technology coupled with coagulation/flocculation	Purification of palm oil mill effluent (POME)	Effectual filtration area of 0.9 m^2	Reversible membrane fouling and reusable	Removing ammonia (99%) and reduction in COD up to 78%	(Ahmad et al. 2006)

(Continued)

TABLE 19.1 (CONTINUED)

Examples of Potential Membrane Technologies Applied to Remove Organic Pollutants From Wastewater

No.	Nature of Membrane	Type of Pollutants	Parameters	Important Properties	Efficiency	Reference
19.	Ag–SiO2 nanocomposite polyethersulfone (PES) membranes	Bacteria removal	Membrane effective area 13.4 cm^2 And thickness of 200 µm	Anti-biofouling properties	Eliminated *Escherichia coli* and *Pseudomonas* sp.	(Huang et al. 2014)
20.	Nanofilteration membrane (thin-film composite polyamide)	Effluent from a textile plant [black dye (Cibacron Black B) and reactive red dye (Cibacron Red RB)]	Effective filtration area of 119.6 cm^2	Highly stable	Removing 94% Cibacron Black B and 92% Cibacron Red RB	(Chakraborty et al. 2003)
21.	Novel thin-film composite (TFC) membrane [organoclay/chitosan nanocomposite coated on the commercial polyvinylidene fluoride (PVDF) microfiltration membrane]	Dye removal (methylene blue and acid orange 7)	Pore size of 0.22 µm	Highly stable	Highly efficient	(Daraei et al. 2013).
22.	Titania-based ultrafiltration membrane	Organic dye (methylene blue) and phenolic compounds	Average thickness of about 3 µm	Anti-biofouling properties	Moderately efficient	(Djafer et al. 2010)
23.	Ag-decorated TiO2 (Ag–TiO2) photocatalytic membranes	Removal of bacteria (*E.coli*) and dyes like Rhodamine B (RdB)	-	Excellent chemical and thermal stability and anti-biofouling property	Highly efficient	(Goei and Lim 2014)
24.	Meldrum's acid modified cellulose nanofiber-based polyvinylidene fluoride microfiltration membrane	Dye removal (crystal violet) and nanoparticle removal (Fe$_2$O$_3$)	Thickness of membrane 170 ± 2 µm.	Cost effective	Effectively eliminated over 99% of CV dye and Fe$_2$O$_3$	(Gopakumar et al. 2017)
25.	Dopamine cross-linked graphene oxide membrane [β-cyclodextrin (CD)]	Removal of organic pollutants and trace heavy metals (Pb2+) from aqueous solution	Thickness of the membrane is about 4 µm	Regenerated easily and highly Stable	Rejected 99.2% of methylene blue and absorbed 99.9% of lead ions (Pb^{2+})	(Wang et al. 2018)

(Continued)

TABLE 19.1 (CONTINUED)
Examples of Potential Membrane Technologies Applied to Remove Organic Pollutants From Wastewater

No.	Nature of Membrane	Type of Pollutants	Parameters	Important Properties	Efficiency	Reference
26.	Ceramic membranes impregnated with cross-linked silylated dendritic and cyclodextrin polymers	Organic pollutants	-	Stable and regenerative capacity	Remove of polycyclic aromatic hydrocarbons (up to 99%), monocyclic aromatic hydrocarbons (up to 93%), trihalogen methanes (up to 81%), pesticides (up to 43%), and methyl-tert-butyl ether (up to 46%)	(Allabashi et al. 2007)
27.	Polyamide thin-film composite membranes containing ZIF-8	For the separation of pharmaceutical compounds from aqueous streams	Thickness ZIF-8 (0.4–0.5 μm) on the PA separation layer (2–3 μm)	Anti-fouling property, cost effective, and stabile	55% acetaminophen retention and 91% salt retention	(Basu and Balakrishnan 2017)
28.	Silica/titania nanorods/ nanotubes composite membrane	Removal of sodium dodecylbenzene sulfonate (SDBS)	Mesopores of diameters ranging from 1.4–10 nm	Stable	Removal of SDBS (89%)	(Zhang et al. 2006)
29.	Silver-doped fly ash/ polyurathene (Ag-FA/PU) nanocomposite multifunctional membrane	To remove carcinogenic arsenic (As) and toxic organic dyes and pathogenic bacteria	Effective filtration area of 13.4 cm^2	Antifouling effect and cost effective	Highly efficient	(Pant et al. 2014)
30.	Organic solvent nanofiltration (OSN) membranes fabricated from polybenzimidazole (PBI)	Organic solvent	Effective filtration area of 14 cm^2	Superior chemical stability	Highly efficient	(Valtcheva et al. 2014)
31.	Sulfonated homo and co-polyimides incorporated polysulfone ultrafiltration blend membranes	Heavy metals and proteins	Effective filtration area of 0.25 cm^2	Stable and cost-effective	Rejection of metal ions (≈98% of Pb^{2+}; ≈92% of Cd^{2+}) and proteins (>98% of BSA; >86% of Pepsin)	(Jafar Mazumder et al. 2020)

(Continued)

TABLE 19.1 (CONTINUED)
Examples of Potential Membrane Technologies Applied to Remove Organic Pollutants From Wastewater

No.	Nature of Membrane	Type of Pollutants	Parameters	Important Properties	Efficiency	Reference
32.	Graphene oxide (GO) and activated carbon (AC) hybrid membrane	Antibiotics [tetracycline hydrochloride (TCH)]	Thickness of 15 μm and pore size of 3–10 nm	Stable	Effectively removed 98.9% of (TCH)	(Liu et al. 2017)
33.	Polyetherimide membrane with grafted bentonite clay	Removal of metal ions and humic acids	-	Fouling resistance	Humic acid rejection of 87.6% and hazardous heavy metal ions rejection of 80%	(Hebbar et al. 2018)
34.	Sub-nanometre sieve composite MoS$_2$ membranes	Salt and organic pollutants removal	0.02 μm pore size	Highly stable	Rejected more than 99% of salts and nearly 100% rejection for organic dyes(MB, MR, RhB) and pollutants	(Sapkota et al. 2020)

of the membrane was found to be uniform at 1.3 μm with an effective filtration area of 17.34 cm². The membrane exhibited superior antibacterial potential by inactivating 86% of *E.coli* from the prepared water sample at neutral pH. The anti-fouling property of the membrane was maintained even after many cycles of filtration. The membrane possesses the potential for large-scale application due to its low cost and effectiveness in the membrane filtration process (Sun et al. 2015).

A cross-linked, three-dimensional graphene oxide hydrogel membrane was used for water purification. Graphene oxide was dispersed into distilled water followed by the addition of ferrous chloride at an acidic pH. The resultant solution was ultrasonicated for an hour to get a homogenous solution. Briefly, the resultant solution was transferred to a nitrocellulose membrane to obtain a graphene oxide hydrogel membrane. This highly stable membrane showed high permeability of 111.5 L/m⁻² h⁻¹ bar⁻¹. The membrane retained more than 99% of methylene blue dye with an effective filtration area of 10.17 cm². The membrane is a good candidate for water purification processes in the industries where organic pollutants are produced on a large scale due to its higher stability, cost effectiveness, and outstanding dye rejection potential (Chen et al. 2019).

19.3   CONCLUSION

In the modern era, due to the rapid increase in industrialization and urbanization, large amounts of pollutants, especially refractory pollutants, are released into the aquatic environment. Most of these pollutants are life threatening as they cannot be degraded easily, leading to problems like bioaccumulation which results in negative effects for human beings and, ultimately, the ecosystem. Membrane technologies have emerged as a superior alternative option to other conventional processes of water purification. A large number of membrane techniques like reverse osmosis, nanofiltration, ultrafiltration, and microfiltration are used for the removal of refractory pollutants like inorganic heavy metals, organic dyes, oils, pesticides, pharmaceutical products, and other effluents released from various industries. Various types of membranes made from different materials like polymer-based membranes, ceramic membranes, carbon-based membranes, or nanoparticle-incorporated membranes are employed for this purpose. Most of these processes are supplemented with electrochemical interactions, oxidation processes, or photocatalytic reactions for the absolute removal of these harmful refractory pollutants intended for the decontamination of water bodies. With the optimization of filtration performance and further research being undertaken to overcome the limitations of their applications, membrane techniques can be used as a resourceful and green approach towards water filtration processes. The potential membrane technologies applied to remove organic pollutants are described in **Table 19.1**.

REFERENCES

Abdi, Gisya, Abdolhamid Alizadeh, Sirus Zinadini, and Golshan Moradi. 2018. "Removal of Dye and Heavy Metal Ion Using a Novel Synthetic Polyethersulfone Nanofiltration Membrane Modified by Magnetic Graphene Oxide/Metformin Hybrid." *Journal of Membrane Science* 552: 326–35. doi:10.1016/j.memsci.2018.02.018.

Achiou, B., H. Elomari, A. Bouazizi, A. Karim, M. Ouammou, A. Albizane, J. Bennazha, S. Alami Younssi, and I. E. El Amrani. 2017. "Manufacturing of Tubular Ceramic Microfiltration Membrane Based on Natural Pozzolan for Pretreatment of Seawater Desalination." *Desalination* 419: 181–87. doi:10.1016/j.desal.2017.06.014.

Ahmad, A. L., M. F. Chong, S. Bhatia, and S. Ismail. 2006. "Drinking Water Reclamation from Palm Oil Mill Effluent (POME) Using Membrane Technology." *Desalination* 191 (1–3): 35–44. doi:10.1016/j.desal.2005.06.033.

Aljundi, Isam H. 2017. "Desalination Characteristics of TFN-RO Membrane Incorporated with ZIF-8 Nanoparticles." *Desalination* 420: 12–20. doi:10.1016/j.desal.2017.06.020.

Allabashi, Roza, Michael Arkas, Gerold Hörmann, and Dimitris Tsiourvas. 2007. "Removal of Some Organic Pollutants in Water Employing Ceramic Membranes Impregnated with Cross-Linked Silylated Dendritic and Cyclodextrin Polymers." *Water Research* 41 (2): 476–86. doi:10.1016/j.watres.2006.10.011.

Arora, M., R. C. Maheshwari, S. K. Jain, and A. Gupta. 2004. "Use of Membrane Technology for Potable Water Production." *Desalination* 170 (2): 105–12. doi:10.1016/j.desal.2004.02.096.

Barakat, M.A. 2008. "Removal of Cu (II), Ni (II) and Cr (III) Ions from Wastewater Using Complexation-Ultrafiltration Technique." *Journal of Environmental Science and Technology* 1 (3): 151–56. doi:10.3923/jest.2008.151.156.

Basu, Subhankar, and Malini Balakrishnan. 2017. "Polyamide Thin Film Composite Membranes Containing ZIF-8 for the Separation of Pharmaceutical Compounds from Aqueous Streams." *Separation and Purification Technology* 179: 118–25. doi:10.1016/j.seppur.2017.01.061.

Bladergroen, B. J., and V. M. Linkov. 2001. "Electrosorption Ceramic Based Membranes for Water Treatment." *Separation and Purification Technology* 25 (1–3): 347–54. doi:10.1016/S1383-5866(01)00062-4.

Boi, Cristiana, Serena Bandini, and Giulio Cesare Sarti. 2005. "Pollutants Removal from Wastewaters through Membrane Distillation." *Desalination* 183 (1–3): 383–94. doi:10.1016/j.desal.2005.03.041.

Chakraborty, S., M. K. Purkait, S. DasGupta, S. De, and J. K. Basu. 2003. "Nanofiltration of Textile Plant Effluent for Color Removal and Reduction in COD." *Separation and Purification Technology* 31 (2): 141–51. doi:10.1016/S1383-5866(02)00177-6.

Chandra P. Nikhil, and Mothi Krishna Mohan. 2020. "Tailor-Made Polyelectrolyte Multilayers for the Removal of Obidoxime from Water in Microfiltration Process." *Membranes and Membrane Technologies* 2 (2): 132–47. doi:10.1134/s2517751620020031.

Chen, Long, Na Li, Ziyan Wen, Lin Zhang, Qiong Chen, Lina Chen, Pengchao Si, et al. 2018. "Graphene Oxide Based Membrane Intercalated by Nanoparticles for High Performance Nanofiltration Application." *Chemical Engineering Journal* 347: 12–18. doi:10.1016/j.cej.2018.04.069.

Chen, Zhangjingzhi, Jun Wang, Xiaoguang Duan, Yuanyuan Chu, Xiaoyao Tan, Shaomin Liu, and Shaobin Wang. 2019. "Facile Fabrication of 3D Ferrous Ion Crosslinked Graphene Oxide Hydrogel Membranes for Excellent Water Purification." *Environmental Science: Nano* 6 (10): 3060–71. doi:10.1039/c9en00638a.

Chu, Huaqiang, Yalei Zhang, Xuefei Zhou, Yangying Zhao, Bingzhi Dong, and Hai Zhang. 2014. "Dynamic Membrane Bioreactor for Wastewater Treatment: Operation, Critical Flux, and Dynamic Membrane Structure." *Journal of Membrane Science* 450: 265–71. doi:10.1016/j.memsci.2013.08.045.

Daraei, Parisa, Sayed Siavash Madaeni, Ehsan Salehi, Negin Ghaemi, Hedayatolah Sadeghi Ghari, Mohammad Ali Khadivi, and Elham Rostami. 2013. "Novel Thin Film Composite Membrane Fabricated by Mixed Matrix Nanoclay/Chitosan on PVDF Microfiltration Support: Preparation, Characterization and Performance in Dye Removal." *Journal of Membrane Science* 436: 97–108. doi:10.1016/j.memsci.2013.02.031.

Das, Rasel, Md Eaqub Ali, Sharifah Bee Abd Hamid, Seeram Ramakrishna, and Zaira Zaman Chowdhury. 2014. "Carbon Nanotube Membranes for Water Purification: A Bright Future in Water Desalination." *Desalination* 336 (1): 97–109. doi:10.1016/j.desal.2013.12.026.

Djafer, Lahcène, André Ayral, and Abdallah Ouagued. 2010. "Robust Synthesis and Performance of a Titania-Based Ultrafiltration Membrane with Photocatalytic Properties." *Separation and Purification Technology* 75 (2): 198–203. doi:10.1016/j.seppur.2010.08.001.

EL-Bourawi, M. S., M. Khayet, R. Ma, Z. Ding, Z. Li, and X. Zhang. 2007. "Application of Vacuum Membrane Distillation for Ammonia Removal." *Journal of Membrane Science* 301 (1–2): 200–209. doi:10.1016/j.memsci.2007.06.021.

Ge, Yuanyuan, Yuan Yuan, Kaituo Wang, Yan He, and Xuemin Cui. 2015. "Preparation of Geopolymer-Based Inorganic Membrane for Removing Ni2+ from Wastewater." *Journal of Hazardous Materials* 299: 711–18. doi:10.1016/j.jhazmat.2015.08.006.

Geltmeyer, Jozefien, Helena Teixido, Mieke Meire, Thibaut Van Acker, Koen Deventer, Frank Vanhaecke, Stijn Van Hulle, Klaartje De Buysser, and Karen De Clerck. 2017. "TiO2 Functionalized Nanofibrous Membranes for Removal of Organic (Micro)Pollutants from Water." *Separation and Purification Technology* 179: 533–41. doi:10.1016/j.seppur.2017.02.037.

Goei, Ronn, and Teik Thye Lim. 2014. "Ag-Decorated TiO2 Photocatalytic Membrane with Hierarchical Architecture: Photocatalytic and Anti-Bacterial Activities." *Water Research* 59: 207–18. doi:10.1016/j.watres.2014.04.025.

Gondal, Mohammed A., Muhammad S. Sadullah, Talal F. Qahtan, Mohamed A. Dastageer, Umair Baig, and Gareth H. McKinley. 2017. "Fabrication and Wettability Study of WO3 Coated Photocatalytic Membrane for Oil-Water Separation: A Comparative Study with ZnO Coated Membrane." *Scientific Reports* 7 (1): 1–10. doi:10.1038/s41598-017-01959-y.

Gopakumar, Deepu A., Daniel Pasquini, Mariana Alves Henrique, Luis Carlos De Morais, Yves Grohens, and Sabu Thomas. 2017. "Meldrum's Acid Modified Cellulose Nanofiber-Based Polyvinylidene Fluoride Microfiltration Membrane for Dye Water Treatment and Nanoparticle Removal." *ACS Sustainable Chemistry and Engineering* 5 (2): 2026–33. doi:10.1021/acssuschemeng.6b02952.

Gopal, Renuga, Satinderpal Kaur, Chao Yang Feng, Casey Chan, Seeram Ramakrishna, Shahram Tabe, and Takeshi Matsuura. 2007. "Electrospun Nanofibrous Polysulfone Membranes as Pre-Filters: Particulate Removal." *Journal of Membrane Science* 289 (1–2): 210–19. doi:10.1016/j.memsci.2006.11.056.

Gu, Jincui, Peng Xiao, Lei Zhang, Wei Lu, Ganggang Zhang, Youju Huang, Jiawei Zhang and, Tao Chen. 2016. "Construction of Superhydrophyllic and Under-Water Superoleophobic Carbon-Based Membranes for Water Purification." *RSC Advances* 6 (77): 6.

Han, Yi, Zhen Xu, and Chao Gao. 2013. "Ultrathin Graphene Nanofiltration Membrane for Water Purification." *Advanced Functional Materials* 23 (29): 3693–3700. doi:10.1002/adfm.201202601.

He, Yingran, Yu Pan Tang, Dangchen Ma, and Tai Shung Chung. 2017. "UiO-66 Incorporated Thin-Film Nanocomposite Membranes for Efficient Selenium and Arsenic Removal." *Journal of Membrane Science* 541: 262–70. doi:10.1016/j.memsci.2017.06.061.

Hebbar, Raghavendra S., Arun M. Isloor, Balakrishna Prabhu, Abdullah M. Asiri, and A. F. Ismail. 2018. "Removal of Metal Ions and Humic Acids through Polyetherimide Membrane with Grafted Bentonite Clay." *Scientific Reports* 8 (1). doi:10.1038/s41598-018-22837-1.

Huang, Jian, Huanting Wang, and Kaisong Zhang. 2014. "Modification of PES Membrane with Ag-SiO2: Reduction of Biofouling and Improvement of Filtration Performance." *Desalination* 336 (1): 8–17. doi:10.1016/j.desal.2013.12.032.

Hubadillah, Siti Khadijah, Mohd Hafiz Dzarfan Othman, Zhong Sheng Tai, Mohd Riduan Jamalludin, Nur Kamilah Yusuf, A. Ahmad, Mukhlis A. Rahman, Juhana Jaafar, Siti Hamimah Sheikh Abdul Kadir, and Zawati Harun. 2020. "Novel Hydroxyapatite-Based Bio-Ceramic Hollow Fiber Membrane Derived from Waste Cow Bone for Textile Wastewater Treatment." *Chemical Engineering Journal* 379 (March 2019). doi:10.1016/j.cej.2019.122396.

Isawi, Heba. 2019. "Evaluating the Performance of Different Nano-Enhanced Ultrafiltration Membranes for the Removal of Organic Pollutants from Wastewater." *Journal of Water Process Engineering* 31 (April): 100833. doi:10.1016/j.jwpe.2019.100833.

Jalanni, N. A., M. N. Abu Seman, and C. K.M. Faizal. 2012. "Synthesis and Performance of Thin Film Composite Nanofiltration Polyester Membrane for Removal of Natural Organic Matter Substances." *ASEAN Journal of Chemical Engineering* 12 (1): 73–80. doi:10.22146/ajche.49757.

Ji, Chao, Jingwei Hou, and Vicki Chen. 2016. "Cross-Linked Carbon Nanotubes-Based Biocatalytic Membranes for Micro-Pollutants Degradation: Performance, Stability, and Regeneration." *Journal of Membrane Science* 520: 869–80. doi:10.1016/j.memsci.2016.08.056.

Li, Chang, Xiong Li, Lei Qin, Wei Wu, Qin Meng, Chong Shen, and Guoliang Zhang. 2019a. "Membrane Photo-Bioreactor Coupled with Heterogeneous Fenton Fluidized Bed for High Salinity Wastewater Treatment: Pollutant Removal, Photosynthetic Bacteria Harvest and Membrane Anti-Fouling Analysis." *Science of the Total Environment* 696: 133953. doi:10.1016/j.scitotenv.2019.133953.

Li, Chen, Wenjun Sun, Zedong Lu, Xiuwei Ao, Chao Yang, and Simiao Li. 2019b. "Systematic Evaluation of TiO2-GO-Modified Ceramic Membranes for Water Treatment: Retention Properties and Fouling Mechanisms." *Chemical Engineering Journal* 378 (July): 122138. doi:10.1016/j.cej.2019.122138.

Liang, Hai Wei, Xiang Cao, Wen Jun Zhang, Hong Tao Lin, Fei Zhou, Li Feng Chen, and Shu Hong Yu. 2011. "Robust and Highly Efficient Free-Standing Carbonaceous Nanofiber Membranes for Water Purification." *Advanced Functional Materials* 21 (20): 3851–58. doi:10.1002/adfm.201100983.

Liu, Ming Kai, Ying Ya Liu, Dan Dan Bao, Gen Zhu, Guo Hai Yang, Jun Feng Geng, and Hai Tao Li. 2017. "Effective Removal of Tetracycline Antibiotics from Water Using Hybrid Carbon Membranes." *Scientific Reports* 7: 1–8. doi:10.1038/srep43717.

Liu, Yanbiao, Ling Yu, Choon Nam Ong, and Jianping Xie. 2016. "Nitrogen-Doped Graphene Nanosheets as Reactive Water Purification Membranes." *Nano Research* 9 (7): 1983–93. doi:10.1007/s12274-016-1089-7.

Lu, Jia Jie, Yi Hang Gu, Yan Chen, Xi Yan, Ya Jun Guo, and Wan Zhong Lang. 2019. "Ultrahigh Permeability of Graphene-Based Membranes by Adjusting D-Spacing with Poly (Ethylene Imine) for the Separation of Dye Wastewater." *Separation and Purification Technology* 210: 737–45. doi:10.1016/j.seppur.2018.08.065.

Maximous, Nermen, G. Nakhla, K. Wong, and W. Wan. 2010. "Optimization of Al2O3/PES Membranes for Wastewater Filtration." *Separation and Purification Technology* 73 (2): 294–301. doi:10.1016/j.seppur.2010.04.016.

Mazumder M.A.J., Panchami H. Raja, Arun M. Isloor, Muhammad Usman, Shakhawat H. Chowdhury, Shaikh A. Ali, Inamuddin, and Amir Al-Ahmed. 2020. "Assessment of Sulfonated Homo and Co-Polyimides Incorporated Polysulfone Ultrafiltration Blend Membranes for Effective Removal of Heavy Metals and Proteins." *Scientific Reports* 10 (1): 1–13. doi:10.1038/s41598-020-63736-8.

Modise, Claude M., John A. Bendick, C. J. Miller, Ronald D. Neufeld, and Radisav D. Vidic. 2006. "Use of Hydrophilic and Hydrophobic Microfiltration Membranes to Remove Microorganisms and Organic Pollutants from Primary Effluents." *Water Environment Research* 78 (6): 557–64. doi:10.2175/106143006x99777.

Mouiya, M., A. Abourriche, A. Bouazizi, A. Benhammou, Y. El Hafiane, Y. Abouliatim, L. Nibou, et al. 2018. "Flat Ceramic Microfiltration Membrane Based on Natural Clay and Moroccan Phosphate for Desalination and Industrial Wastewater Treatment." *Desalination* 427 (June 2017): 42–50. doi:10.1016/j. desal.2017.11.005.

Mustereţ, Corina Petronela, and Carmen Teodosiu. 2007. "Removal of Persistent Organic Pollutants from Textile Wastewater by Membrane Processes." *Environmental Engineering and Management Journal* 6 (3): 175–87. doi:10.30638/eemj.2007.022.

Oun, Abdallah, Nouha Tahri, Samia Mahouche-Chergui, Benjamin Carbonnier, Swachchha Majumdar, Sandeep Sarkar, Ganesh C. Sahoo, and Raja Ben Amar. 2017. "Tubular Ultrafiltration Ceramic Membrane Based on Titania Nanoparticles Immobilized on Macroporous Clay-Alumina Support: Elaboration, Characterization and Application to Dye Removal." *Separation and Purification Technology* 188: 126–33. doi:10.1016/j.seppur.2017.07.005.

Pant, Hem Raj, Han Joo Kim, Mahesh Kumar Joshi, Bishweshwar Pant, Chan Hee Park, Jeong In Kim, K. S. Hui, and Cheol Sang Kim. 2014. "One-Step Fabrication of Multifunctional Composite Polyurethane Spider-Web-like Nanofibrous Membrane for Water Purification." *Journal of Hazardous Materials* 264: 25–33. doi:10.1016/j.jhazmat.2013.10.066.

Rivas, Bernabé L., Julio Sánchez, and Manuel Palencia. 2016. "Organic Membranes and Polymers for the Removal of Pollutants." *Nanostructured Polymer Membranes* 1 (November): 203–35. doi:10.1002/9781118831779.ch6.

Salazar, H., P. M. Martins, Bruno Santos, M. M. Fernandes, Ander Reizabal, Víctor Sebastián, G. Botelho, Carlos J. Tavares, José L. Vilas-Vilela, and S. Lanceros-Mendez. 2020. "Photocatalytic and Antimicrobial Multifunctional Nanocomposite Membranes for Emerging Pollutants Water Treatment Applications." *Chemosphere* 250: 126299. doi:10.1016/j.chemosphere.2020.126299.

Sapkota, Bedanga, Wentao Liang, Armin VahidMohammadi, Rohit Karnik, Aleksandr Noy, and Meni Wanunu. 2020. "High Permeability Sub-Nanometre Sieve Composite MoS2 Membranes." *Nature Communications* 11 (1): 1–9. doi:10.1038/s41467-020-16577-y.

Saucedo-Rivalcoba, V., A. L. Martínez-Hernández, G. Martínez-Barrera, C. Velasco-Santos, J. L. Rivera-Armenta, and V. M. Castaño. 2011. "Removal of Hexavalent Chromium from Water by Polyurethane-Keratin Hybrid Membranes." *Water Air and Soil Pollution* 218 (1–4): 557–71. doi:10.1007/s11270-010-0668-6.

Sun, Xue Fei, Jing Qin, Peng Fei Xia, Bei Bei Guo, Chun Miao Yang, Chao Song, and Shu Guang Wang. 2015. "Graphene Oxide-Silver Nanoparticle Membrane for Biofouling Control and Water Purification." *Chemical Engineering Journal* 281: 53–59. doi:10.1016/j.cej.2015.06.059.

Valtcheva, Irina B., Santosh C. Kumbharkar, Jeong F. Kim, Yogesh Bhole, and Andrew G. Livingston. 2014. "Beyond Polyimide: Crosslinked Polybenzimidazole Membranes for Organic Solvent Nanofiltration (OSN) in Harsh Environments." *Journal of Membrane Science* 457: 62–72. doi:10.1016/j. memsci.2013.12.069.

Velizarov, Svetlozar, João G. Crespo, and Maria A. Reis. 2004. "Removal of Inorganic Anions from Drinking Water Supplies by Membrane Bio/Processes." *Reviews in Environmental Science and Bio/technology* 3 (4): 361–80. doi:10.1007/s11157-004-4627-9.

Waeger, F., T. Delhaye, and W. Fuchs. 2010. "The Use of Ceramic Microfiltration and Ultrafiltration Membranes for Particle Removal from Anaerobic Digester Effluents." *Separation and Purification Technology* 73 (2): 271–78. doi:10.1016/j.seppur.2010.04.013.

Wang, Jing, Tiefan Huang, Lin Zhang, Qiming Jimmy Yu, and Li'an Hou. 2018. "Dopamine Crosslinked Graphene Oxide Membrane for Simultaneous Removal of Organic Pollutants and Trace Heavy Metals from Aqueous Solution." *Environmental Technology* 39 (23): 3055–65. doi:10.1080/09593330.2017.1371797.

Wang, Songxue, Jiayu Tian, Zhihui Wang, Qiao Wang, Jialin Jia, Xiujuan Hao, Shanshan Gao, and Fuyi Cui. 2020a. "Integrated Process for Membrane Fouling Mitigation and Organic Pollutants Removal Using Copper Oxide Modified Ceramic Hollow Fiber Membrane with In-Situ Peroxymonosulfate Activation." *Chemical Engineering Journal* 396: 125289. doi:10.1016/j.cej.2020.125289.

Wang, Xueling, Youling Li, Hongtao Yu, Fenglin Yang, Chuyang Y. Tang, Xie Quan, and Yingchao Dong. 2020b. "High-Flux Robust Ceramic Membranes Functionally Decorated with Nano-Catalyst for Emerging Micro-Pollutant Removal from Water." *Journal of Membrane Science* 611: 118281. doi:10.1016/j.memsci.2020.118281.

Wei, Yibin, Yuxiang Zhu, and Yijiao Jiang. 2019. "Photocatalytic Self-Cleaning Carbon Nitride Nanotube Intercalated Reduced Graphene Oxide Membranes for Enhanced Water Purification." *Chemical Engineering Journal* 356 (September 2018): 915–25. doi:10.1016/j.cej.2018.09.108.

Xu, Shugang, Yuanfa Liu, Yue Yu, Xitong Zhang, Jun Zhang, and Yanfen Li. 2020. "PAN/PVDF Chelating Membrane for Simultaneous Removal of Heavy Metal and Organic Pollutants from Mimic Industrial Wastewater." *Separation and Purification Technology* 235 (August 2019): 116185. doi:10.1016/j. seppur.2019.116185.

Yin, Jun, Eun Sik Kim, John Yang, and Baolin Deng. 2012. "Fabrication of a Novel Thin-Film Nanocomposite (TFN) Membrane Containing MCM-41 Silica Nanoparticles (NPs) for Water Purification." *Journal of Membrane Science* 423–424: 238–46. doi:10.1016/j.memsci.2012.08.020.

Ying, Yulong, Luwei Sun, Qian Wang, Zhuangjun Fan, and Xinsheng Peng. 2014. "In-Plane Mesoporous Graphene Oxide Nanosheet Assembled Membranes for Molecular Separation." *RSC Advances* 4 (41): 21425–28. doi:10.1039/c4ra01495b.

Zhang, Haimin, Xie Quan, Shuo Chen, Huimin Zhao, and Yazhi Zhao. 2006. "The Removal of Sodium Dodecylbenzene Sulfonate Surfactant from Water Using Silica/Titania Nanorods/Nanotubes Composite Membrane with Photocatalytic Capability." *Applied Surface Science* 252 (24): 8598–8604. doi:10.1016/j. apsusc.2005.11.090.

Zhang, Peng, Ji Lai Gong, Guang Ming Zeng, Can Hui Deng, Hu Cheng Yang, Hong Yu Liu, and Shuang Yan Huan. 2017. "Cross-Linking to Prepare Composite Graphene Oxide-Framework Membranes with High-Flux for Dyes and Heavy Metal Ions Removal." *Chemical Engineering Journal* 322: 657–66. doi:10.1016/j.cej.2017.04.068.

Zhang, Zewen, Yueping Bao, Xun Sun, Ke Chen, Mingjiong Zhou, Liu He, Qing Huang, Zhengren Huang, Zhifang Chai, and Yujie Song. 2020. "Mesoporous Polymer-Derived Ceramic Membranes for Water Purification via a Self-Sacrificed Template." *ACS Omega* 5 (19): 11100–105. doi:10.1021/ acsomega.0c01021.

Zhu, Li, Kadalipura Puttaswamy Rakesh, Man Xu, and Yingchao Dong. 2019. "Ceramic-Based Composite Membrane with a Porous Network Surface Featuring a Highly Stable Flux for Drinkingwater Purification." *Membranes* 9 (1). doi:10.3390/membranes9010005.

Zhu, Liping, Huixian Wang, Jing Bai, Jindun Liu, and Yatao Zhang. 2017. "A Porous Graphene Composite Membrane Intercalated by Halloysite Nanotubes for Efficient Dye Desalination." *Desalination* 420 (June): 145–57. doi:10.1016/j.desal.2017.07.008.

Zou, Dong, Minghui Qiu, Xianfu Chen, Enrico Drioli, and Yiqun Fan. 2019. "One Step Co-Sintering Process for Low-Cost Fly Ash Based Ceramic Microfiltration Membrane in Oil-in-Water Emulsion Treatment." *Separation and Purification Technology* 210: 511–20. doi:10.1016/j.seppur.2018.08.040.

20 Microalgae-Based Bioremediation of Refractory Pollutants in Wastewater

Ayesha Algade Amadu, Shuang Qiu, Abdul-Wahab Abbew,
Mohammed Zeeshan Qasim, Lingfeng Wang, Zhipeng Chen,
Yeting Shen, Zhengshuai Wu, and Shijian Ge

CONTENTS

DOI: 10.1201/9781003204442-20

20.1   INTRODUCTION

Advances in drug research, the increased use of pesticides and fertilizers, and industrial growth have led to great improvements in the quality and standard of living. The massive and indiscriminate application of these products and activities have, however, resulted in adverse effects on the environment due to the inevitable generation of certain environmental pollutants such as refractory pollutants. Refractory pollutants originate from industrial wastes either directly or indirectly from municipal wastewaters which receive industrial effluents (Iloms et al. 2020). There also exists naturally-occurring refractory pollutants which can be converted into harmful compounds through processes such as chlorination. This leads to disinfection by-products (DBPs) such as trihalomethanes (THMs) and haloacetic acids (HAAs) (Yang et al. 2018) which is why the bioremediation of wastewater is of great importance in recent times.

In the past, emphasis was placed on wastewater characteristic such as oxygen depletion, odor, and general aesthetic properties. Presently, however, more attention is being paid to biopersistent organic contaminants (i.e., refractory pollutants) that bioaccumulate in the environment through the food chain as a result of biological amplification. For instance, *Cladophora glomerata* can exert a bioenrichment effect on nonylphenol so that the concentration in the organism is much higher than the concentration in the environment (Wang et al. 2019). The five-day biochemical oxygen demand (BOD_5) has been a standard of wastewater quality for decades. Thus, industries with high BOD values for wastewater, such as paper manufacturing industries, have tried to lower the BOD values of their discharge waters through effective treatment processes. Others also employ a shortcut method of lowering BOD by switching from readily biodegradable organics to refractory organics that do not immediately create a high BOD because they resist degradation. This shortcut approach is undertaken to meet discharge standards of regulatory agencies that rely solely on BOD_5 criterion. Such a practice can be far more polluting to the environment due to the toxicity effects on flora and fauna, and the persistence and long-term oxygen depletion in receiving waters. Other examples of refractory organic pollutants are halogenated organics including pesticides, dioxins, polychlorinated biphenyl (PCBs), polycyclic aromatic hydrocarbons (PAHs), nitroaromatics, certain organic acids, aldehydes, certain pharmaceutical active ingredients (PAIs), and detergents.

Several conventional technologies, such as the advanced oxidation process (AOP), are currently being used to remove these toxic refractory pollutants from wastewater. The AOP removes refractory pollutants from wastewater due to the low oxidation selectivity and high reactivity of the radicals (Srivastav et al. 2018). However, most of the methods involved in AOPs are effective at acidic and neutral pH levels, like the Fenton process, but in many instances, industrial pollutants are formed at basic pH conditions and can thus be effectively treated through biological methods, like microalgae bioremediation. Microalgae bioremediation involves metabolic and passive processes such as biouptake, bioaccumulation, biosorption, ion exchange, surface complexation, and precipitation.

In the 1950s, Oswald and Gotaas (Oswald et al. 1957) were the pioneer researchers to propose the use of microalgae in wastewater bioremediation. Presently, microalgae wastewater bioremediation is regarded as a sustainable biotechnological approach to wastewater treatment (WWT). Several studies have exploited this technique by both dealing with the removal of wastewater pollutants and generating multiple bioproducts from wastewater-cultivated biomass. This ties in well with the sustainable development goals (SDGs) which aim to achieve a win–win situation in terms of environmental and economic goals that require the use of more sustainable means of controlling environmental pollution, such as wastewater bioremediation. Microalgal species growing in WWT systems are especially tolerant to pollution. It has been reported that *Chlorella*, *Nitzschia*, and *Scenedesmus* sp. are the most tolerant microalgal genera with high populations in wastewater systems (Chu and Phang 2019). Microalgae productivity rates have been found to be strongly linked to their strains (e.g., marine or freshwater species), climatic conditions such as light intensity and temperature, and the operational setup of the production systems such as reactor design (Ge et al. 2017). These individual factors contribute greatly towards the success of the microalgae-based wastewater bioremediation system.

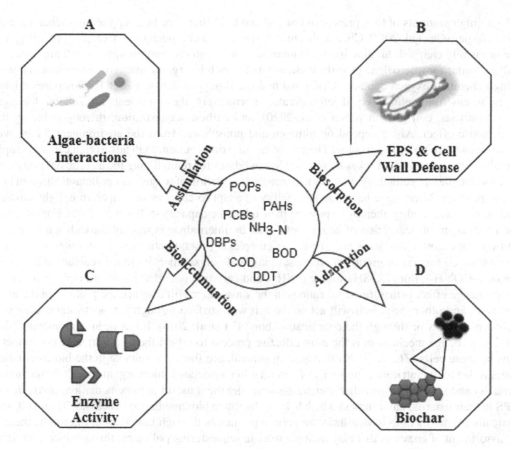

FIGURE 20.1 Microalgae mechanisms for the removal of refractory pollutants by (A) assimilation through algae–bacteria interactions, (B) biosorption through extracellular polymeric substances (EPS) and cell wall defense, (C) bioaccumulation and subsequent biodegradation through enzyme activity, and (D) adsorption through biochar.

The feasibility of utilizing microalgae in the removal of refractory pollutants such as pharmaceutical residues, radionuclides, and ammoniacal nitrogen has been demonstrated by several studies. This chapter will discuss the various types and sources of refractory pollutants, the influence of wastewater characteristics on bioremediation processes, as well as the interaction mechanisms of algae for the removal of refractory pollutants. *Figure 20.1 illustrates these mechanisms.*

20.2 ALGAE INTERACTION MECHANISMS IN WASTEWATER BIOREMEDIATION

The word microalgae is generally considered as a generic term that includes cyanobacteria (prokaryotic blue-green algae) and microalgae (eukaryotic). Microalgae are unicellular microorganisms, usually ranging in size from 1–400 μm (Yu et al. 2017b). They can be photosynthetic or non-photosynthetic and are usually subdivided into autotrophic, heterotrophic, mixotrophic, and photoheterotrophic based on their specific energy source and carbon requirements (organic or inorganic). Temperatures of 20–30°C and a basic pH level are optimal for their growth, although some species can thrive in extreme environments, thus making them ubiquitous in nature. Microalgae can also mitigate the amount of CO_2 in the environment by assimilating it into their biomass. It is not surprising that about half of their biomass is made up of carbon. Certain microalgae species can tolerate high amounts of CO_2 present in flue gas and have therefore been applied in carbon capture technologies (Ge et al. 2017). Chemically, microalgae have a composition of $CO_{0.48}H_{1.83}N_{0.11}P_{0.01}$ and are negatively charged. Judging by this composition, nitrogen (N) and phosphorus (P) are undoubtedly essential for microalgae growth. Wastewaters are rich in organic and inorganic nutrients, with a high chemical oxygen demand (COD) and biological oxygen demand (BOD). This poses a huge threat to environmental safety if left untreated. Fortunately, these nutrients are suitable for algal photosynthesis, cell growth (Chen et al. 2020), and carbon sequestration, thereby reducing the greenhouse effect. Algae depend on nitrogen and phosphorus from the environment for growth because they are non-diazotrophic. Their wide ranging distribution, high adsorption capacity, rapid metabolism, and ubiquity makes them suitable candidates for remediating the aquatic environment.

Bioremediation technology provides a thorough degradation of wastewater pollutants through biodecomposition. Microalgae have a unique ability to adapt to changes in temperature, light salinity, and nutrient availability, thereby improving their tolerance capacity (Silva et al. 2019). For instance, the mechanism of adsorption of heavy metal ions by microalgae is mainly through ion exchange and complexation. Complexes are formed by complexing metal cations in wastewater with negatively charged functional groups in proteins, lipids, and polysaccharides in microalgae cells such as amide (–$CONH_2$), thiol (R–SH), amine (–NH_2), and carboxyl (–COOH). The principle of the main ion exchange effect is that the metal cations in the wastewater will displace the protons on the algal cell wall, and other metal ions will act on the cell wall surface through the electrostatic attraction between the ions or through the coordination bond (Li et al. 2016). It has been demonstrated that the ion exchange mechanism is the most effective process to reflect the biosorption of heavy metal ions by algae cells (Zhu et al. 2019). Algae, in general, are therefore suitable in the bioremediation process due to the intricate community of several other associated microorganisms. Their metabolic diversity and specific growth characteristics also render them useful in bioremediation, partly due to EPS in their external environment which helps to facilitate bioremediation processes. Their cell wall structure allows them to have affinity for certain pollutants through biosorption. Internally, there is an assortment of enzymes that play multiple roles in sequestering pollutants through biomineralization, bioaccumulation, biodegradation, and volatilization (stripping) (Silva et al. 2019).

20.2.1 EXTRACELLULAR MECHANISMS (EPS PRODUCTION AND CELL WALL DEFENSE)

EPS play a critical role in the wastewater bioremediation by algae. EPS is a macromolecule secreted by microorganisms with a complex structure, numerous functional groups, and multiple binding sites that are important for the biosorption process (Pierre et al. 2019). EPS shows many functions in microbial communities, including the formation of a compact protective layer to guard cells against external toxic substances, the formation of a structural network with activated sludge in order to absorb and degrade pollutants, and acting as a source of carbon to provide energy in the absence of nutrients. EPS is made up of polysaccharides, proteins, humic-like substances, DNA/nucleic acids, lipids, uronic acids, and other micromolecules (Pierre et al. 2019). These constituents contribute to the unique properties of EPS. The amphipathic nature of EPS is largely dependent on hydrophilicity which is primarily determined by the polysaccharide component and hydrophobicity that is controlled by the protein component (Cao 2017). These components have implications in the WWT process. Polysaccharides are composed of a variety of homopolysaccharides and heteropolysaccharides and these are known to be associated with the interaction between the cell surface and external environment. The network structure of polysaccharide molecules is conducive to sludge bioflocculation in a wastewater treatment plant (WWTP). Within this network are α- and β-polysaccharides which maintain the structural stability of flocs and improve sludge bioflocculation (Tian et al. 2019). Through the protective layer of EPS, microalgae are able to withstand harsh external conditions in wastewater. Thus, EPS plays a major protective role, such as the protection of sludge aggregates in WWTPs when exposed to toxic substances. This is possible due to the presence of functional groups like hydroxyl, carboxylic, and phosphate groups (Tian et al. 2019). There is evidence on the role of EPS in WWT with regard to sludge flocculation, settling, oxidation-reduction properties, and bio degradability (Singh et al. 2018). When live cells come into contact with refractory pollutants in wastewater, microalgae produce EPS to protect themselves against the toxicity. Thus, EPS production is partly a defense mechanism that microalgae employ to deal with the toxic external environment. Likewise, EPS production in algae–bacteria consortia is a symbiotic relationship of great importance in the WWT industry (Perera et al. 2019) with the promise of cost-reduction due to improved biomass harvesting processes.

The cell wall barrier comes after the EPS matrix. The algal cell wall has functional groups and polymer groups that are very similar to cellulose, hemicellulose, and proteins (Xie et al. 2020). Electrostatic neutralization and hydrophobic interactions between the algal cell wall and refractory pollutants play a crucial role in the biosorption process (Wang et al. 2019). Hence, the hydrophobicity, structural properties, and ionization behavior of the pollutant are of considerable importance. In addition to the algal cell wall, EPS also offers active binding sites for further adsorption. It is important to note that biosorption is crucial for further degradation mechanisms, such as bioaccumulation and intracellular biodegradation (Xie et al. 2020). There is a positive correlation between biosorption and biodegradation rates of pollutant removal because biosorption facilitates the subsequent transport of refractory pollutants into the algal cell (Silva et al. 2019). For effective biosorption to take place, the pollutant should have an opposite charge to the algal cell wall; thus, the pollutant is expected to be hydrophobic rather than hydrophilic (Xie et al. 2020) knowing that microalgae are negatively charged. The cell wall is involved in the protective response of microalgae to radiation (Vojvodić et al. 2020) because the cell wall is the region of contact between the algal cell and the external environment, and the first line of chemical and physical defense (Baudelet et al. 2017). It also appears that microalgae contribute significantly towards the control of metal movement in aquatic ecosystems through biosorption by the cell wall matrix.

20.2.2 Intracellular Mechanisms (Enzymes)

The intracellular mechanism involves two processes: bioaccumulation, which serves as a preliminary process, and a subsequent biodegradation process. Firstly, bioaccumulation is an intracellular process, unlike the extracellular biosorption process. It is another excellent mechanism by which algal species uptake and assimilate refractory pollutants which cross algal cell membranes and find their way inside the cellular matrix (Wang et al. 2019). In this process, enzymes play a crucial role in pollutant removal due to the presence of active sites which emphasizes their narrow or broad range of specificity. They can either have chemo-, region-, or stereo-selectivity that enables them to function intracellularly or extracellularly. The active sites are the regions of the enzyme that are directly involved in the catalytic process. The bioaccumulation of refractory pollutants induces the production of reactive oxygen species (ROS) within the cell. Typically, alkoxy radicals (RO•), hydroxyl radicals (OH•), peroxide radicals (H_2O_2•), superoxide radicals (O_2•), and the non-radical forms, such as hydrogen peroxide (H_2O_2) and singlet oxygen[1].[O_2•], are produced depending on the type of pollutant and its concentration. Hydroxyl radicals are the most cytotoxic ROS, and no known enzymatic system exists for their detoxification (Koskimäki et al. 2016). At normal levels, ROS regulate essential cellular metabolic activities, act as a pathogenic defense, and are involved in programmed cell death. However, elevated levels of ROS result in severe oxidative damage to major structural components such as carbohydrates, lipids, and proteins. Therefore, the bioaccumulated pollutants must subsequently undergo biodegradation, otherwise they could cause cellular toxicity and hinder growth of the algal cells. Bioaccumulation serves as a preliminary process for the subsequent biodegradation of the refractory pollutants in a systematic order. In essence, the biodegradation of pollutants by algal enzyme systems is influenced by numerous mechanisms such as chemical, environmental, and physical factors. These include algal strains, pH levels, temperatures, and the concentration and hydrophobicity or hydrophilicity of the pollutants.

20.2.3 LIVING BIOMASS (ALGAE–BACTERIA INTERACTIONS)

Apart from the extracellular and intracellular mechanisms aforementioned, another mechanism is related to algae–bacteria interactions. Both microalgae and bacteria play complementary or competitive roles in consortia, such that the interactions between them enhance the efficient removal of refractory wastewater pollutants such as organic and inorganic nutrients and heavy metals. Organic matter mineralization by aerobic bacteria produces the inorganic carbon needed by the microalgae. In return, the O_2 required for bacterial degradation is photosynthetically produced by the microalgae (Chu and Phang 2019). Through the provision of phytohormones or macro- and micronutrients, heterotrophic bacteria influence physiological changes within algal cells. The most notable among these changes is enhanced growth rates and survival. It has been reported that up to 70% of microalgae growth is influenced by bacterial associations (Ramanan et al. 2016; Lee et al. 2019). Apart from influencing microalgal growth, bacteria have also been noted to impact flocculation of microalgae. In WWTPs, bacteria increases the floc size of microalgae, thus enhancing settlement (Ramanan et al. 2016). To achieve desirable results in algae–bacteria-based WWT systems, the microbial diversity of the influent must be well monitored and controlled. Microalgae also present competitive relationships with bacteria. From a metabolic perspective, the photosynthetic activity of microalgae results in an increase in the dissolved oxygen (DO) and pH of the cultivation medium. High DO concentrations promote photo-oxidative damage of pathogenic cells and increased pH reduces pathogen survival, thereby resulting in efficient pathogen removal. The microalgal excretion of inhibitory metabolites such as polyphenols, terpenes, alkaloids, and sterols also compete with bacteria leading to pathogen removal in microalgae-based WWT systems (Bhowmick et al. 2020). The complementary role of algal–bacterial consortia in the degradation of organophosphate insecticides such as monocrotophos, quinalphos, and methyl parathion has been successfully demonstrated along with countless toxic pesticides such as DDT, atrazine, α-endosulfan, phenol, naphthalene, benzopyrene, dibenzofuran, and azo compounds (Mahdavi et al. 2015). The degradation of toxic thiocyanate (SCN^-) in wastewater through algal–bacterial interactions was also demonstrated by Ryu et al. (2015). In the study, the consortium of SCN^--degrading bacteria with microalgae effectively removed not only SCN^- but also oxidized nitrogen. These case studies of algae–bacteria interactions demonstrate the shift from single species-based approaches to a community-based approaches in microbial bioremediation. In essence, algae–bacteria interactions are necessary in the entire valorization of microalgae, from growth to the final stages of biomass harvesting, ultimately reducing costs. Moreover, the interactions closely resemble what occurs in the natural ecosystem, therefore ensuring sustainable production and development.

20.2.4 NON-LIVING BIOMASS (BIOCHAR AND DEAD BIOMASS)

Live microalgae biomass utilized in WWT or residual biomass, which has been used in the extraction of lipids for biofuel production for instance, can be further converted into secondary by-products for further removal of refractory pollutants. Biochar is an important by-product derived from such conversion processes. It is a solid carbon-rich, porous substance produced by the thermal decomposition of the organic biomass under limited O_2 supply at relatively low temperatures (300–700°C) via thermochemical processes, such as hydrothermal liquefaction (HTL) and hydrothermal carbonization (HTC) or torrefaction (Chen et al. 2015). These processes can be fed with wet microalgal biomass (Yu et al. 2017a), eliminating the need for drying, which is an energy intensive process. Biochar properties such as high surface area, degree of porosity, and stable carbon matrix largely depend on the type of feedstock (Li et al. 2016). Microalgae and activated sludge are among several sources of biochar utilized as adsorbents (Othman et al. 2018). Microalgae-derived biochar consists of large aggregates with a lower surface area and carbon content, but a higher cation exchange capacity compared to lignocellulose biochar (Yu et al. 2017a). The sorption property of pollutants such as heavy metal onto biochar is influenced by operational parameters such as initial metal concentration, contact time, temperature, and pH (Qian et al. 2016). Kołodyńska et al. (2017) compared the sorption performance of Cu(II), Zn(II), Cd(II), Co(II), and Pb(II) using commercial activated carbon and biochar and concluded that heavy metal ions were more efficiently removed by biochar than activated carbon. Similarly, effluent organic matter enhanced Cu(II) removal efficiency onto a sludge-derived biochar, and the sorption kinetics data were better fitted with a pseudo-second order model and Freundlich equation. The highly porous adsorptive structure, low cost, and fast adsorption kinetics of biochar enhances water treatment processes, with evidence of removal of organic pollutants, volatile organic compounds, dyes, and heavy metals (Othman et al. 2018). Biochar derived from microalgae *Scenedesmus dimorphus* demonstrated an effective adsorption capacity for the removal of Co(II) ion in aqueous solutions with a correlation coefficient value of 0.9981 from the Freundlich model (Bordoloi et al. 2017). The dead biomass of *Chlorella vulgaris* was employed in the reduction of a heavy metal [Cd (II)]. The dead biomass was not transformed into biochar, but it had a high adsorption capacity for Cd, with 96.8% of the total Cd being removed by the dead algae, compared to 95.2% by the live biomass (Cheng et al. 2017). Although there was no significant difference in the adsorption capacities, this still proves the efficiency and feasibility of dead algal biomass in WWT.

20.3 REFRACTORY INDEX AND CLASSIFICATION OF WASTEWATER REFRACTORY POLLUTANTS

The inability of refractory pollutants to biodegrade leads to their build up in the environment by bioaccumulation. This can inhibit certain life forms through biomagnification or bioamplification (Dodds and Whiles 2020). Their persistence in the environment occurs at varying degrees as a result of their unique characteristics. Thus, laboratory protocols have been established to quantitatively classify these refractory pollutants into an index based on their level of persistence. Helfgott et al. (1977) firstly proposed the refractory index (RI). Values close to 1.0 indicate pollutants that are readily biodegradable. Intermediate RI values between 0.3 and 0.7 are pollutants with partial biodegradability. Pollutants with RI values close to zero (0) are refractory, while negative RI values indicate inhibitors. These are compounds with high toxicity and extremely low degradability and are inhibitory to organisms. These include compounds like phenols, antibiotics, and polycyclic aromatic hydrocarbons (PAHs), among others. In the past, the use of toxic refractory pollutants was widespread. These compounds have a high lipid solubility and thus accumulate in the fatty tissues of organisms subjected to repeated exposure (Alewu and Nosiri 2011). Toxic organic compounds decrease microbial activity; nonetheless, robust microalgae species like *Chlorella*, *Nitzschia*, and *Chlamydomonas* seem to have a high tolerance for refractory pollutants. This explains their predominance in microalgal-based WWT systems (Xiong et al. 2017). RI values have implications for practical application in the WWT industry as they serve as a guide for industries discharging wastewater into the environment. For instance, glucose has an RI value of 1.0, which means it is readily biodegradable, while a compound such as DDT has a negative RI value. The presence of such a compound (with a negative RI value) in an effluent will ultimately interfere with the operations of the WWTP. These industries must endeavor to meet the RI criteria for the proper functioning of a WWTP. The ability of microbes to thrive well in the presence of these refractory pollutants largely depends on the type of pollutant and the community structure surrounding the microorganism. Typical contaminants in wastewater include nitrogen, phosphorus, heavy metals, and dyes with high COD and BOD levels. Other contaminants of emerging concern (CEC) such as pharmaceutical and personal care products (PPCPs), persistent organic pollutants (POPs), endocrine disrupters, disinfection by-products (DBPs), and many other refractory pollutants are still being discovered by the scientific community. These wastewater pollutants have varying degrees of toxicity to organisms and to the environment at large. Microalgae were able to remove some of these important refractory pollutants by bioremediation.

20.3.1 Ammoniacal Nitrogen

The utilization of algae as a new technology was proposed due to its low operational cost and the reuse of nitrogen resource in high value-added products (Ge et al. 2018). Ammoniacal nitrogen (NH_3–N) exists in almost all kinds of wastewater, such as agricultural, anaerobic-digesters, and industrial and municipal wastewaters. Chemical, physical, and biological methods of NH_3-N removal exist, but most of these technologies are confronted with the problem of high costs. Even the most economical activated sludge process still faces the cost of aeration and waste nitrogen in the environment (Acién et al. 2016).

Studies have demonstrated that the mechanism of ammonium removal from wastewater in microalgae-based systems is through assimilation (Gonçalves et al. 2017). Ammoniacal nitrogen can be assimilated directly and converted into amino acids. Nitrites and nitrates are also reduced to ammonium, followed by conversion into amino acids. Although algae usually prefer ammoniacal nitrogen as a growth substrate, high ammoniacal nitrogen concentrations produce free ammonia to dissipate transmembrane proton gradients in algae. This can easily be found in anaerobically digested effluents of many WWTPs. Therefore, the ammoniacal nitrogen level should be carefully considered in wastewater microalgae cultivation (Shen et al. 2020). Different modes of microalgal cultivation for ammoniacal nitrogen treatment have been explored for optimization. The monoculture of microalgae has been extensively researched in ammoniacal nitrogen removal from different wastewaters. For example, *Chlorella vulgaris* was feasible for ammoniacal nitrogen removal with over 85% efficiency in the treatment of domestic secondary effluent and centrate wastewater (Gao et al. 2016; Ge et al. 2018). Due to the complex and changeable wastewater environment, microalgal consortia have been considered for ammoniacal nitrogen removal. For instance, more than 83% of nitrogen removal was achieved by Qiu et al. (2020b) by alternating free nitrous acid and free ammonia (FNA/FA) sidestream sludge treatments, which ultimately led to a 20% cost reduction. To date, various techniques are employed in ammonium removal. For instance, the algal granulation caused by EPS and the supplementation of natural zeolite enhanced the removal efficiency by 93% and 16%, respectively (Liu et al. 2017; Tao et al. 2020). However, optimizing conditions (such as the ratio of constituent parts of consortia and the concentration of ammoniacal nitrogen) are required for further exploration for the application of pilot or full-scale tests.

20.3.2 COD

Microalgae have demonstrated impeccable capabilities for COD reduction. COD is a measure of the amount of organic pollutants in water. It is usually expressed in milligrams per liter (mg/L) or parts per million (ppm). The most common application of COD is in determining the amount of oxidizable pollutants found in water (Arvia n.d.). Refractory COD is a mixture of organic compounds that are not readily biodegradable in nature but are being chemically oxidized in a COD test (Srivastav et al. 2018). COD concentration is an important parameter in the WWT industry. Textile, pharmaceutical, and paper and pulp industries have larger amounts of refractory COD in their wastewater (Wu 2017) compared to sugar, fatty acids, and alcohol-producing industries. Primary and secondary WWTs can remove between 75% and 85% of biodegradable COD, but the refractory COD still remains. In the E.U., the laws require COD discharge to be below 120 mg/L and, in China, the value should be below 50 mg/L (Arvia n.d.). In order to meet the standards of regulatory bodies, additional tertiary treatment is often required, but this process is costly. Microalgae presents superior removal efficiency of COD due to their photosynthetic nature. Microalgae can utilize bicarbonate ions ($HCO3^-$) or CO_2 as carbon sources for photosynthesis with carbonic anhydrites enzymes, thereby enhancing COD removal (82.3%) (Choi and Lee 2012). Varied removal rates were observed in different wastewaters. By optimizing sludge inoculation timing, Lee et al. (2019) also enhanced COD removal by the formation of a desirable algae–bacteria consortium achieved through the supplementation of activated sludge with the wastewater-cultivated microalgae. The cultivation of microalgae in a high rate algal pond (HRAP) achieved 62.85% and 92% of total chemical oxygen demand (TCOD) removal during the treatment of piggery effluents, potato processing waste, and slaughterhouse wastewater, respectively (Hernández et al. 2016). A large-scale bioactive ponds–wetland system, in operation for 8 years treating petrochemical industrial wastewater, achieved 75% of COD removal (Liu et al. 2014). About 78% COD reduction was also observed by Santhosh et al. (2020) during the treatment of tannery effluents using *Chlorella* sp. SRD3 and *Oscillatoria* sp. SDR2. The differences in COD removal rates were likely attributed to variations in the biodegradability of the different wastewaters (Hernández et al. 2016). This reiterates the point of the effects of WW composition on the performance of microalgae, which is discussed in more details in Section 20.4.1. Overall, the results obtained from these studies all had a high pollutant removal efficiency by algae treatment and have proven to be applicable on both a laboratory scale and large scale.

20.3.3 Heavy Metals

To meet their micronutrient requirements, microalgae require small quantities of several metals for their normal growth and metabolism. On a small scale, metals are quenched by biouptake through biosorption into a cell's intracellular matrix. These metals are bioaccumulated in a cell's vacuole by a metabolically active biological process of diffusion (Tripathi and Kumar 2017). Key heavy metals in wastewater include Cu, Cd, Cr, Hg, Zn, Pb, and Ni. Toxic metal pollutants like Cr, Pb, and Cd are especially abundant in wastewaters, such as those from the leather industry (Balaji et al. 2016). The treatment of such wastewaters is usually done via cost-intensive physicochemical processes that cannot be adopted by industries in developing countries. The concept of metal removal from wastewater through bioremediation is mainly based on the relationship between heavy metals and negatively charged groups in the carbohydrates and EPS of microalgal and bacterial cell surfaces that involves both adsorption and biouptake (Cheng et al. 2017). Metals are chemically altered once taken up, and stored or transported to various parts of the cell. Extensive studies have focused on the removal of heavy metals in wastewater by means of EPS, and the interaction mechanisms mainly include biosorption, adsorption, complexation or chelation, and ion exchange (Vimalnath and Subramanian 2018). The complex stereoscopic structure of EPS and functional groups such as hydroxyl, carboxylic, sulfhydryl, and phosphate amine groups influences the interactions between EPS and heavy metals. Important factors such as species of metal ions, valence state, and environmental conditions, critically affect the interaction of EPS with heavy metals (Tian et al. 2019). Santhosh et al. (2020) achieved 60–80% of heavy metal degradation in effluents containing Cr and Mg. Chen et al. (2020) noted that absorption was the main mechanism behind the removal of Mg^{2+} from Mg^{2+} from enriched nickel laterite ore wastewater (NLOWW). However, high metal levels also lead to toxicity, which is usually exhibited through photosynthesis. This is probably because some metals are able to substitute metal atoms in the prosthetic groups for specific photosynthetic enzymes (Chu and Phang 2019). For instance, Chen et al. (2020) observed a 1.89-fold higher biomass yield, a 3.77-fold enhanced photosynthetic activity, and improved nutrient removal with the culture mixed with 0.13% NLOWW. Conversely, excessive Mg^{2+} at 100% NLOWW produced the highest ROS suppressing microalgal growth and photosynthesis. In such instances, algal–bacterial interactions can detoxify and assimilate metals from metal-rich environments through mutualism (Ramanan et al. 2016). Microalgal growth leads to the release of metal chelators and an increase in pH (due to CO_2 release) which usually precipitate heavy metals that are subsequently taken-up by associated bacterial communities. These accumulation processes occur through physical adsorption, covalent bonding, ion exchange, surface precipitation, redox reactions, or crystallization on the cell surface (Ramanan et al. 2016). The mutualistic interactions between microalgae and bacteria is discussed in Section 20.2.3. Based on the capacity of microalgae to accumulate heavy metals, several biofilms have been developed and currently commercialized. These include ALGASORB™, which is manufactured by Bio-Recovery System, Inc. (USA). It consists of *Chlorella vulgaris* (microalgae) immobilized in a silica gel polymer matrix. It can be used for a range of heavy metal concentrations between 1 and 100 mg g/L (AlgaSORB n.d.). The process takes advantage of the natural affinity of algal cell structures for heavy metal ions. BV-SORBEX™ is another biofilm on the market, produced by BV Sorbex, Inc. (Canada). It comprises *Sphaerotilus natans* (bacteria), *Ascophyllum nodosum* (macroalgae), *Halimeda opuntia* (macroalgae), *Palmaria palmata* (macroalgae), *Chondrus crispus* (macroalgae), and *Chlorella vulgaris* (microalgae). This adsorbent is able to recover up to 99% of metals in solution (Chu and Phang 2019; BV Sorbex Inc. n.d.). The major advantage of this type of biosorbent is its ability to be regenerated for multiple reuse, similar to ion exchange resins but at a lower cost. The enhanced metal removal rates could be due to the association between algae (both macroalgae and microalgae) and bacteria.

20.3.4 PHARMACEUTICAL RESIDUES

There is an abundance of pharmaceutical residues in the aquatic environment from numerous sources, such as agricultural activities, hospital effluents, and industrial and domestic wastes. PPCPs leave residues in the environment long after their production and use. They are therefore classified as toxic/inhibitory refractory pollutants because of their prolonged persistence. The removal of these pollutants from wastewater is not a straightforward approach as it involves different transformation pathways. In some cases, the compound is either completely biodegraded into H_2O and CO_2, or it is biotransformed into various metabolites which are often deemed to be less toxic to the environment than the parent drugs. Microalgae have been reported to be feasible in the removal of these PPCPs. This is evident in their ability to remove small amounts of antibiotics through absorption and biodegradation mechanisms. Although microalgae have a much higher tolerance for antibiotics than bacteria, their growth may still be inhibited at higher concentrations (Bai and Acharya 2016). This has positive impacts in a WWT facility where low concentrations of pharmaceuticals are used to eliminate pathogenic bacteria in order to achieve axenic cultures of microalgae. Although several advanced treatment technologies are available for removing PPCPs, there are limitations, such as low removal efficiencies, since conventional activated-sludge processes in current WWTPs were not specifically designed for handling and removing PPCPs (Iloms et al. 2020). These limitations have motivated research into better treatment strategies, such as microalgae technology, for the sustainable removal of these refractory contaminants in terms of environmental friendliness and energy consumption. There have been several successful studies with the application of algae for the remediation of various PPCPs including *Arthrospira, Botryococcus, Chlaymydomonas, Chlorella, Cyanothece, Desmodesmus, Nodularia, Phormidium, Oscillatoria, Scenedesmus*, and *Spirulina* (Xie et al. 2020; Xiong et al. 2017) . For instance, Xie et al. (2020) reported the biodegradation of two common antibiotics, i.e., ciprofloxacin and sulfadiazine, by *Chlamydomonas* sp. Tai-03. This green algae effectively biotransformed both drugs into various metabolites, such as sulfo-ciprofloxacin, oxo-ciprofloxacin, and N-acetyl sulfadiazine, that have lower environmental toxicity. Thus, it can be inferred that microalgae have the ability to biotransform a toxic refractory pollutant with a negative RI (< 0) into a less toxic form, such as intermediary (RI <0.5) or biodegradable (RI <1.0) pollutants. Similarly, an enhanced removal of levofloxacin and sulfamethazine by S. *obliquus* through intracellular biodegradation was reported in synthetic saline wastewater (Xiong et al. 2017). Microalgae biodegradation of PPCPs occurs via the natural activity of the sophisticated phase I and phase II enzymatic systems. Typical metabolic biodegradation pathways include hydroxylation, decarboxylation, demethylation, side chain breakdown, and ring cleavage (Xiong et al. 2017). These mechanisms have been reported to be superior to photo-degradation in terms of the metabolic fate of PPCPs. The majority of the concentrations of PPCPs in literature are reported at ng/L to µg/L levels, which is practical for employing algal-based technologies in WWT (Villar-Navarro et al. 2018). The rationale behind this is that algae can efficiently remove these lower environmental concentrations of pharmaceuticals through bioaccumulation and subsequent biodegradation without much toxicity to the algal cells. These PPCPs are a good carbon source for growing algal biomass. For example, a recent study has reported the removal of around 64 diverse PPCPs (223 mg/L) from municipal wastewater with a removal efficiency ranging between 5% and 50% (Villar-Navarro et al. 2018), which was incorporated in the biomass through nutrient removal. It is important to note that different algal species have different removal rates of PPCPs. The genus *Chlorella* has shown adequate effectiveness for removing several antibiotics. For instance, various β-lactam and cephalosporins, such as amoxicillin, cephalexin, cefixime, cefradine, and ceftazidime, were effectively removed by *C. pyrenoidosa*. However, some antibiotics may also cause toxicity in algal cells, even at low concentrations, resulting in reduced removal rates. Several algal species such as *Chlamydomonas mexicana, C. pitschmannii, C. vulgaris,* and *O. multisporus* have only removed 0–25% of ciprofloxacin and enrofloxacin (Silva et al. 2019). The level of algal toxicity can be evaluated by analyzing the generation of ROS within the cells. Usually, higher toxicity levels lead to more ROS production (Shen et al. 2020; Chen et al. 2020; Gomes et al. 2017). Due to their abundance in nature, suitable algal strains with high PPCP tolerance and degradation performance levels can be used in conjunction with mixed microbial cultures. Such conditions can yield remarkable removal efficiencies coupled with the stability of microalgae-based WWT systems (Silva et al. 2019).

20.3.5 RADIONUCLIDES

Currently, the shift from fossil-based energy resources to renewable and sustainable sources have led to an increased awareness of nuclear technology. This technology is also applied in the medical and agriculture industries and has therefore led to an increment in radioactive material or radionuclides in wastewater. These radionuclides, from both man-made and natural sources, enter the aquatic environment and cause harm to plants and animals due to their radiation upon decay. At high doses and prolonged exposure, radionuclides pose a great threat to biomolecules (Vojvodić et al. 2020). The exposure of microalgae to radiation may cause a decrease in photosynthetic efficiency and growth as well as oxidative damage of lipids and DNA (Gomes et al. 2017). However, to a large extent, microalgal cells have some level of resistance to radiation. This exhibition of radioresistance is evidenced by the upregulation of antioxidative defense and photoprotection, and alterations in carbohydrate and general metabolic profiles. Firstly, microalgae have the ability to treat spent nuclear fuel. Their resilience to radiation stress and the ability to thrive in such ecosystems is best illustrated by seasonal algal blooms that occur in spent nuclear fuel storage at Sellafield, U.K. (Foster et al. 2020; MeGraw et al. 2018). For instance, *Haematococcus pluvialis*, freshwater proteobacteria, and *Pseudoanabaena catenata* are examples of microbial species known to colonize spent nuclear fuel storage pools and uranium tailings pounds with high levels of radiation and pollution from heavy metals (MeGraw et al. 2018). Secondly, they are also resistance to electromagnetic radiation. In aquatic systems, High energy electromagnetic radiation (e.g., gamma or X-ray) has the ability to cause water radiolysis due to its high penetrating power. Vojvodić et al. (2020) demonstrated how microalgae *C. sorokiniana* cells responded to ionizing radiation (X-rays) through cell wall defense mechanisms. Within 24 hours after irradiation with doses of 1-5 Gy, the fibrilar layer of the cell wall became thicker and the capacity to remove the main reactive product of water radiolysis increased. In addition, microalgae show significant binding capacities for metals. For instance, isolated cell wall fractions showed significant binding capacity for Cu^{2+}, Mn^{2+}, and Cr^{3+}. Vanhoudt et al. (2018) investigated the potential of higher plants, algae, and cyanobacteria for the remediation of radioactively contaminated waters and noted that there was a distinction between bioaccumulation- and biosorption-driven element removal. The pollutant removal process for higher plants or algae is usually slow and measured in days, whereas for cyanobacteria and biosorbents derived from dead biomass the removal process is much faster and is measured in hours or even minutes. This means that employing live microalgae in WWT requires a longer contact time for the effective removal of refractory pollutants.

20.4 INFLUENCING PARAMETERS FOR MICROALGAE-BASED BIOREMEDIATION PROCESSES

20.4.1 WASTEWATER CHARACTERISTICS

The characteristics of wastewater influence microalgae growth and treatment efficiencies in WWT systems. The levels and types of pollutants in wastewater, pH, color, turbidity, and reactor design employed all play a synergistic role in the overall performance of the system. During the complicated bioremediation process, including biouptake and biosorption, natural microorganism such as algae present several benefits that transcend both economic and environmental boundaries. These benefits include, but are not limited to, (1) excellence in retention capability, (2) the requirement of infinitesimal preparatory steps, (3) natural renewability, recyclability, and ease of availability all year round, (4) diversified multifunctional groups on their surface, (5) a relatively small and uniform distribution of surface binding sites, and (6) the absence or minimal use of harsh chemicals (Bilal et al. 2018). The main factors responsible for microalgae performance are strongly related to wastewater characteristics, species interaction, variations in the environmental conditions, reactor setup, and operational parameters. These are contributing factors to pollutant bioremediation in wastewater. Thus, the algal bioremediation process is influenced by several factors which could be of biotic or abiotic origin.

20.4.1.1 Nitrogenous Compounds

Nitrogen influences the performance of microalgae in biological WWT. Certain nitrogenous compounds produced during WWT or originally present in the wastewater could cause inhibitory effects in microalgae, such as free ammonia (FA) (Shen et al. 2020) and free nitrous acid (FNA) (Duan et al. 2020). Inhibitory concentrations of such toxic pollutants are usually species dependent. For instance FA concentrations of up to 51 g/m^3 have proven to reduce photosynthetic rates by 90% in dense cultures of *Scenedesmus obliquus*, *Phaeodactylum tricornutum*, and *Dunaliella tertiolecta* (Sutherland et al. 2015). Nitrogen (N) in wastewater, for instance, varies widely depending on the wastewater source. Other forms of nitrogen exist in wastewater, including inorganic molecules such as NH_4^+, NO_3^{-1}, and NO_2^{-1}, and organic molecules such as polypeptides or amino acids (De Lourdes et al. 2017). The assimilation of NH_4^+ is faster than that of other nitrogen sources, and it is typically consumed first by microalgae, because it does not require modification for its use (Chu and Phang 2019). However, the amount of NH_4^+ in some types of wastewaters can reach inhibitory values. Although the NH_4^+ tolerance is strain-dependent, for many genera, the upper limit is between 100 and 200 mg NH_3/L (Qiu et al. 2020b; De Lourdes et al. 2017).

20.4.1.2 Phosphorus Compounds

Phosphorus (P) is another important nutrient found in wastewater that is essential for microalgae energy metabolism. The primary P sources are $H_2PO_4^-$ and HPO_4^{-2}, both of which are directly incorporated into algal biomass and utilized in several metabolic processes. Phosphorus is used for the production of nucleic acids, lipids, proteins, and energetic molecules such as ATP, GTP, and ADP (Gonçalves et al. 2017). In conventional systems, P is chemically removed through precipitation which is known to enhance autoflocculation in WWTPs.

20.4.1.3 Oxygen and Other Macro and Micro Nutrients

Dissolved oxygen (O_2) in wastewater is usually characterized with increases during the day and decreases at night. High O_2 contents (above 20 mg/L) exert toxicity to the algae through photorespiration and O_2 radical formation (de Godos et al. 2017) which inhibit their biosorption capacity and efficiency. In addition to nitrogen and phosphorus, microalgae also require carbon (mainly in the form of CO_2), S, and other micronutrients like Mg, Mn, Ca, Si, Co, Mo, K, and Na which are present in some wastewaters. A recent study confirmed that carbon deficiency and suspended solids are two major problems compromising microalgae wastewater remediation (Xu et al. 2020). The overall wastewater composition affects nutrient uptake due to the sole fact that the optimal C:N:P ratio differs between microalgae species. The ratio of C:N:P in wastewater is critical in determining the efficiency of the algae biosorption process. C:N ratios of various wastewaters below the optimum 100:18 (Posadas et al. 2017) correspond to reduced biodegradability which reduces biomass generation and, by extension, reduces the pollutant removal capacity due to biosorption inefficiencies. Nutrient uptake in microalgae WWT systems is also dependent on operational parameters such as pH, temperature, light intensity, turbidity, and water color, among others, which influence microbial growth and performance.

20.4.1.4 Color

Effluents from paper, plastic, textile, and tanning industries are major sources of dyes that pollute the aquatic environment. Up to 200,000 tons of these dyes are lost in the textile industry effluents through inefficient dyeing processes (Chu and Phang 2019). Apart from the discoloration of receiving water bodies, these industries use substantial amounts of water and chemicals. Treating these wastewaters is extremely difficult due to the presence of toxic refractory compounds such as chromophoric azo groups (-N¼N-) contained in the dyes. Such azo dyes are resistant to aerobic digestion. Synthetic dyes have adverse effects on the growth and metabolic activities of microalgae, especially photosynthetic activities for the obvious reason of light deflection due to the coloration of the water. For instance, exposure to Congo red (a dye formerly used to dye cotton) reduced the growth rate and adversely affected the photosynthetic efficiency of *Chlorella vulgaris* (Hernández-Zamora et al. 2014; Tao et al. 2020). This translates to reduced nutrient removal.

20.4.2 Operational Parameters

The design of algal treatment facilities is also another factor that influences the bioremediation process. Several methodologies of algal cultivation have been introduced at laboratory scale under controlled conditions by using open or closed photo-bioreactors and open raceway systems at field scale (Villar-Navarro et al. 2018). Another engineered system includes the algal turf scrubber (ATS), whereby pulsed wastewater flows over sloping surfaces with attached naturally-seeded filamentous macroalgae. In principle, this design is similar to ALGASORB™ commercial biofilms used for heavy metal remediation that was discussed in Section 20.3.3.

20.4.2.1 pH

pH is a significant factor in the biosorption process because the range under which biosorption occurs largely influences the capacity of the biosorption process. Extreme pH values might destroy the sorbent's structure, thus optimal pH ranges are ideal. These optimal pH ranges are usually different for various biosorption systems. Generally, an increase in pH leads to an increase in biosorption until the process reaches optimal levels where the maximum biosorption capacity can be observed. For example, an increase in pH up to 5 facilitated increased biosorption capacity (98%); however, a continued increase in pH reduced the biosorption capacity (Ali Redha 2020; Bilal et al. 2018). This can be explained by high pH conditions ('9) which reduce the CO_2 absorption capacity and the maintenance of RuBisCO activity (Sutherland et al. 2015). Secondly, it stimulates ammonium (NH_4^+) dissociation to release more ammonia (NH_3) which inhibits microalgae growth and activity, thereby impacting their biosorption capacity (Molinuevo-Salces et al. 2019). The pH of the medium also determines the protonation and deprotonation of functional groups, thus affecting the capacity of biosorption of heavy metals. When the pH is low, a reduced biosorption capacity occurs due to the repulsive forces of the protonated states of carboxylic acid groups (due to surplus H^+ and H_3O^+) with heavy metal ions carrying a positive charge (Ahmad et al. 2018). During pH increases, deprotonation exposes functional groups such as amine, carboxyl, and hydroxyl groups which promotes electrostatic attraction with heavy metal ions due to their negative charges. An increased pH also results in the formation of hydroxide anionic complexes and the onset of precipitation. Changes in pH also influences the population of microorganisms (Ryu et al. 2015).

20.4.2.2 Temperature

Temperature is another important factor in the bioremediation process due to the influences on the biosorbent's surface activity. Again, similar to the pH scenario, the bioremediation capacity is increased with an increase in temperature until its optimum value is reached, after which further temperature increases significantly reduce bioremediation efficiencies (Molinuevo-Salces et al. 2019). Even though the optimum temperature value is genus and strain dependent, generally temperature variations of 20–35°C do not significantly influence the bioremediation process (Ali Redha 2020). It could therefore be inferred that temperatures of 20–35°C are generally acceptable for a wide array of microorganisms. Lower temperatures decrease algae metabolism, affecting the nutrient removal efficiency while higher temperatures facilitate oxidative stress, reduced photosynthetic activities, and biomass structure destruction (Ahmad et al. 2018; Posadas et al. 2017; Molinuevo-Salces et al. 2019); hence, an overall reduced bioremediation capacity. Variations in temperature bring about alterations in the sorption capacity due to changes in the thermodynamic parameters. The influence of temperature on the bioremediation process depends on the nature of the process, such that in an endothermic sorption process an increase in temperature consequently results in an increase in bioremediation, whereas in an exothermic sorption process an increase in temperature reduces biosorption (Ali Redha 2020). For example, it was observed that algae biosorption of Pb(II) increased as temperature increased (Brouers and Al-Musawi 2015). Conversely, the biosorption of Fe(II), Mn(II), and Zn(II) by freely-suspended and Ca–alginate immobilized *Chlorella vulgaris* was observed to reduce with an increase in temperature (from 25–45°C) since the biosorption process was exothermic.

20.4.2.3 Light

Light significantly influences photosynthesis and the metabolic pathways of microalgae. The light intensity and photoperiod are the two essential determining factors with respect to light (Zhu et al. 2019). Intensities lower than the light saturation level inhibit microalgae growth and activities (Thawechai et al. 2016). However, the biosorption capacity is increased with an increase in light till its optimum value, after which further increase significantly reduces biosorption due to the onset of photoinhibition, which occurs when photosynthetic activity is saturated at an irradiance of 100–200 µE/m^2 per day (Molinuevo-Salces et al. 2019). Higher culture densities with a suitable mixing regime can overcome this type of inhibition (de Godos et al. 2017). Thus, mixing speeds are positively correlated with biosorption capacity or efficiency (Salam 2019). Nonetheless, too much mixing leads to shear stress and ruptured cells (Molinuevo-Salces et al. 2019), which inevitably affects the biosorption capacity and efficiency. A light/dark cycle leads to increased organic carbon removal rates in comparison to continual light exposure, while the reverse is true for nitrite removal. Thus, based on the wastewater composition and the desired effluent quality, light/dark cycles should be strictly monitored.

20.4.2.4 Contact Time

Total biosorption is not directly affected by the contact time of the biosorbent, although it can be a limiting factor. There is a corresponding increase in the biosorption process as the contact time increases up to optimum contact time. When the contact time exceeds the optimum, the process becomes relatively constant. An equilibrium state occurs due to biomass saturation when all the active sites have been occupied (Ali Redha 2020). In practice, rapid sorption is preferred since it decreases the size of the biosorption column and renders the process more cost effective (Salam 2019). The optimal contact time varies between the various biosorbent types, from 300 min for immobilized algal biomass and 240 min for free suspended mass (Ahmad et al. 2018). The intracellular absorption, extracellular adsorption, and biodegradation of the hormone nonylphenol by four species of marine microalgae were observed. The amount of nonylphenol absorbed and adsorbed by all four microalgae decreased with increasing time in culture, and the intracellular absorption was greater than the extracellular adsorption, with efficiencies ranging from 43.43–90.94% (Wang et al. 2019).

20.4.2.5 HRT

Hydraulic retention time (HRT) is an important parameter in microalgae-based WWT systems because its duration influences the efficiency of the system. HRT is the volume of the photo-bioreactor divided by the flow rate of the wastewater. Longer HRTs promote increased rates of organic and inorganic contaminant removal from wastewaters in natural treatment systems due to the improved biodegradation, photodegradation, and sorption processes (Matamoros et al. 2015). HRTs adopted when cells are in the exponential phase of growth facilitate biosorption efficiency. In microalgae-based systems, HRT usually ranges between 2 and 10 days (Posadas et al. 2017) and this is regarded as the optimal range for an efficient biosorption process. However, longer HRTs are necessary during cold temperatures due to reduced metabolic activity and minimal algal growth rates (Molinuevo-Salces et al. 2019) and this negatively affects WWT procedures. Nitrifying bacteria [ammonium-oxidizing bacteria (AOB) and nitrite oxidizing bacteria (NOB)] grow slower than microalgae and other aerobic bacteria present in the wastewater. Therefore, they require longer HRTs than typical operational conditions. A short HRT washes out most AOB and NOB before they are able to perform the nitrification process. Thus, in order to enhance AOB and NOB growth and help aerobic bacteria to avoid NH_3 stripping, the microalgae-bacteria sludge retention time (SRT) should be manipulated for improved efficiency.

20.4.2.6 Microbial Community

The composition of the microbial culture has a huge influence on the performance of the WWT system. In microalgae-based WWT systems, the culture composition could be either a monoculture of microalgae–microalgae, or a co-culture of microalgae–bacteria or microalgae–fungi. Biomass concentration is another important factor affecting the biosorption efficiency and equilibrium sorption capacity. The initial algal density is crucial because higher cell densities correspond with better growth and a higher nutrient removal efficiency. On the contrary, high algal density also brings about self-shading, an accumulation of auto inhibitors, and a reduced efficiency of photosynthesis (Franco-Morgado et al. 2017). Thus, adjusting to an optimum level may be required to maximize the amount of pollutant removal. An increase in biomass concentration enhances the removal efficiency till the equilibrium sorption capacity, above which could reduce the biosorption capacity as well as the removal efficiency (Salam 2019).

Microalgae-based WWT systems present several advantages compared to traditional WWT technologies. In conventional activated sludge treatment plants, carbon is oxidized to CO_2, nitrogen (N) is stripped to the atmosphere in the form of gaseous N_2, and phosphorus (P) is usually precipitated. However, in microalgae-based systems, pollutants, pathogens, and CO_2 emissions are effectively reduced and nutrients are recovered through valuable biomass. Algae biomass grown in wastewater have been established as a powerhouse for protein and antioxidant content and have been successfully utilized as a protein source for insect feed (Qiu et al. 2020a) to enhance valorization.

20.5 CONCLUSION

As we continue to explore more bioremediation technologies, we have to consciously reduce or eliminate the use of certain products that become refractory in the environment, and also use environmentally-friendly alternatives, like Azolla biofertilizer, and incorporate soil conditioners like algae biochar which will inevitably reduce eutrophication caused by run-off. Pesticide residues in soil and their subsequent movements in the soil-water system are key aspects of their environmental behavior. The process of biosorption by EPS from algae–bacteria consortia seems to be more effective in remediation as compared to other mechanisms. The performance of the process is determined by the contact time, the alkalinity of the water, the concentration of the pollutant, and the combined effort of the algae and bacteria systems. Moreover, bacteria not only enhance algal growth but also help in bioflocculation, two essential processes in algal biotechnology. Microbial aggregates, formed through the production of EPS, have higher protection from nanotoxicity compared to planktonic cultures. These aggregates could be used for WWT for removing toxic nanoparticles. With regard to reactor conditions, temperature, HRT, pH, and mixing are important parameters to consider in a microalgal-based system. Mixing ensures that microalgae are intermittently exposed to light by providing turbulence. Optimal temperatures promote biomass productivity and efficient nutrient removal by microalgae while adequate HRTs enhance biosorption processes. Microalgae photosynthesis increases wastewater pH, and a rise in pH shifts the NH_4^+/NH_3 equilibrium toward NH_3 formation, which subsequently increases the rate of N removal via ammonia. Overall, microalgae bioremediation provides compelling evidence of an emerging nature-friendly technology that has come to stay in this era of environmental consciousness and the desire to achieve sustainable development.

ACKNOWLEDGMENT

Dr. Shijian Ge specially acknowledges the support of Distinguished Professorship of Jiangsu Province, China (Special Talents).

REFERENCES

Acién, F. Gabriel, C. Gómez-Serrano, M. M. Morales-Amaral, J. M. Fernández-Sevilla, and E. Molina-Grima. 2016. "Wastewater Treatment Using Microalgae: How Realistic a Contribution Might It Be to Significant Urban Wastewater Treatment?" *Applied Microbiology and Biotechnology* 100 (21): 9013–22. https://doi.org/10.1007/s00253-016-7835-7.

Ahmad, Ashfaq, A.H. Bhat, and Azizul Buang. 2018. "Biosorption of Transition Metals by Freely Suspended and Ca-Alginate Immobilised with Chlorella Vulgaris: Kinetic and Equilibrium Modeling." *Journal of Cleaner Production* 171 (January): 1361–75. https://doi.org/10.1016/j.jclepro.2017.09.252.

Alewu, B, and C Nosiri. 2011. "Pesticides and Human Health." In *Pesticides in the Modern World: Effects of Pesticides Exposure*, Margarita Stoytcheva, IntechOpen, 72: 231–50. https://doi.org/10.1136/oemed-2014-102454.

AlgaSORB. n.d. "Resource Management & Recovery (AlgaSORB© Biological Sorption)." Accessed October 9, 2020. https://clu-in.org/products/site/complete/resource.htm.

Ali Redha, Ali. 2020. "Removal of Heavy Metals from Aqueous Media by Biosorption." *Arab Journal of Basic and Applied Sciences* 27 (1): 183–93. https://doi.org/10.1080/25765299.2020.1756177.

Arvia. n.d. "COD Reduction and Removal in Wastewater."

Bai, Xuelian, and Kumud Acharya. 2016. "Removal of Trimethoprim, Sulfamethoxazole, and Triclosan by the Green Alga Nannochloris Sp." *Journal of Hazardous Materials* 315: 70–75. https://doi.org/10.1016/j.jhazmat.2016.04.067.

Balaji, S., T. Kalaivani, B. Sushma, C. Varneetha Pillai, M. Shalini, and C. Rajasekaran. 2016. "Characterization of Sorption Sites and Differential Stress Response of Microalgae Isolates against Tannery Effluents from Ranipet Industrial Area: An Application towards Phycoremediation." *International Journal of Phytoremediation* 18 (8): 747–53. https://doi.org/10.1080/15226514.2015.1115960.

Baudelet, Paul Hubert, Guillaume Ricochon, Michel Linder, and Lionel Muniglia. 2017. "A New Insight into Cell Walls of Chlorophyta." *Algal Research* 25 (October 2016): 333–71. https://doi.org/10.1016/j.algal.2017.04.008.

Bhowmick, Sukanya, Aninda Mazumdar, Amitava Moulick, and Vojtech Adam. 2020. "Algal Metabolites: An Inevitable Substitute for Antibiotics." *Biotechnology Advances* 43, 107571 (June). https://doi.org/10.1016/j.biotechadv.2020.107571.

Bilal, Muhammad, Tahir Rasheed, Juan Eduardo Sosa-Hernández, Ali Raza, Faran Nabeel, and Hafiz M.N. Iqbal. 2018. "Biosorption: An Interplay between Marine Algae and Potentially Toxic Elements: A Review." *Marine Drugs* 16 (2): 1–16. https://doi.org/10.3390/md16020065.

Bordoloi, Neonjyoti, Ritusmita Goswami, Manish Kumar, and Rupam Kataki. 2017. "Biosorption of Co (II) from Aqueous Solution Using Algal Biochar: Kinetics and Isotherm Studies." *Bioresource Technology* 244: 1465–69. https://doi.org/10.1016/j.biortech.2017.05.139.

Brouers, F., and Tariq J. Al-Musawi. 2015. "On the Optimal Use of Isotherm Models for the Characterization of Biosorption of Lead onto Algae." *Journal of Molecular Liquids* 212 (December): 46–51. https://doi.org/10.1016/j.molliq.2015.08.054.

BV Sorbex Inc. n.d. "BV Sorbex." Accessed October 9, 2020. http://www.bvsorbex.net/sx.htm.

Cao, F. 2017. "Hydrophobic Features of Extracellular Polymeric Substances (EPS) Extracted from Biofilms: An Investigation Based on DAX-8 Resin Technique." https://www.theses.fr/2017PESC1048.

Chen, Wei Hsin, Bo Jhih Lin, Ming Yueh Huang, and Jo Shu Chang. 2015. "Thermochemical Conversion of Microalgal Biomass into Biofuels: A Review." *Bioresource Technology* 184: 314–27. https://doi.org/10.1016/j.biortech.2014.11.050.

Chen, Zhipeng, Shuang Qiu, Ayesha Algade Amadu, Yeting Shen, Lingfeng Wang, Zhengshuai Wu, and Shijian Ge. 2020. "Simultaneous Improvements on Nutrient and Mg Recoveries of Microalgal Bioremediation for Municipal Wastewater and Nickel Laterite Ore Wastewater." *Bioresource Technology* 297 (September 2019): 122517. https://doi.org/10.1016/j.biortech.2019.122517.

Cheng, Jinfeng, Wenke Yin, Zhaoyang Chang, Nina Lundholm, and Zaimin Jiang. 2017. "Biosorption Capacity and Kinetics of Cadmium(II) on Live and Dead Chlorella Vulgaris." *Journal of Applied Phycology* 29 (1): 211–21. https://doi.org/10.1007/s10811-016-0916-2.

Choi, Hee Jeong, and Seung Mok Lee. 2012. "Effects of Microalgae on the Removal of Nutrients from Wastewater: Various Concentrations of Chlorella Vulgaris." *Environmental Engineering Research* 17 (S1): 3–8. https://doi.org/10.4491/eer.2012.17.S1.S3.

Chu, Wan Loy, and Siew Moi Phang. 2019. *Microalgae Biotechnology for Development of Biofuel and Wastewater Treatment.* https://doi.org/10.1007/978-981-13-2264-8_23.

Dodds, Walter K., and Matt R. Whiles. 2020. "Responses to Stress, Toxic Chemicals, and Other Pollutants in Aquatic Ecosystems." *Journal of Freshwater Ecology.* 2020. pp. 453–502. https://doi.org/10.1016/b978-0-12-813255-5.00016-8.

Duan, Haoran, Shuhong Gao, Xuan Li, Nur Hafizah Ab Hamid, Guangming Jiang, Min Zheng, Xue Bai, et al. 2020. "Improving Wastewater Management Using Free Nitrous Acid (FNA)." *Water Research* 15;171:115382. https://doi.org/10.1016/j.watres.2019.115382.

Foster, Lynn, Howbeer Muhamadali, Christopher Boothman, David Sigee, Jon K. Pittman, Royston Goodacre, Katherine Morris, and Jonathan R. Lloyd. 2020. "Radiation Tolerance of Pseudanabaena Catenata, a Cyanobacterium Relevant to the First Generation Magnox Storage Pond." *Frontiers in Microbiology* 11 (April): 1–13. https://doi.org/10.3389/fmicb.2020.00515.

Franco-Morgado, Mariana, Cynthia Alcántara, Adalberto Noyola, Raúl Muñoz, and Armando González-Sánchez. 2017. "A Study of Photosynthetic Biogas Upgrading Based on a High Rate Algal Pond under Alkaline Conditions: Influence of the Illumination Regime." *Science of The Total Environment* 592 (August): 419–25. https://doi.org/10.1016/j.scitotenv.2017.03.077.

Gao, Feng, Chen Li, Zhao Hui Yang, Guang Ming Zeng, Jun Mu, Mei Liu, and Wei Cui. 2016. "Removal of Nutrients, Organic Matter, and Metal from Domestic Secondary Effluent through Microalgae Cultivation in a Membrane Photobioreactor." *Journal of Chemical Technology and Biotechnology* 91 (10): 2713–19. https://doi.org/10.1002/jctb.4879.

Ge, Shijian, Pascal Champagne, William C. Plaxton, Gustavo B. Leite, and Francesca Marazzi. 2017. "Microalgal Cultivation with Waste Streams and Metabolic Constraints to Triacylglycerides Accumulation for Biofuel Production." *Biofuels, Bioproducts and Biorefining* 11 (4): 325–43. https://doi.org/10.1002/bbb.

Ge, Shijian, Shuang Qiu, Danielle Tremblay, Kelsey Viner, Pascale Champagne, and Philip G. Jessop. 2018. "Centrate Wastewater Treatment with Chlorella Vulgaris: Simultaneous Enhancement of Nutrient Removal, Biomass and Lipid Production." *Chemical Engineering Journal* 342 (February): 310–20. https://doi.org/10.1016/j.cej.2018.02.058.

de Godos, Ignacio, Zouhayr Arbib, Enrique Lara, Raúl Cano, Raúl Muñoz, and Frank Rogalla. 2017. "Wastewater Treatment in Algal Systems." In *Innovative Wastewater Treatment & Resource Recovery Technologies: Impacts on Energy, Economy and Environment,* 76–95. International Water Association. https://doi.org/10.2166/9781780407876_0076.

Gomes, Tânia, Li Xie, Dag Brede, Ole Christian Lind, Knut Asbjørn Solhaug, Brit Salbu, and Knut Erik Tollefsen. 2017. "Sensitivity of the Green Algae Chlamydomonas Reinhardtii to Gamma Radiation: Photosynthetic Performance and ROS Formation." *Aquatic Toxicology* 183: 1–10. https://doi.org/10.1016/j.aquatox.2016.12.001.

Gonçalves, Ana L., José C.M. Pires, and Manuel Simões. 2017. "A Review on the Use of Microalgal Consortia for Wastewater Treatment." *Algal Research* 24: 403–15. https://doi.org/10.1016/j.algal.2016.11.008.

Helfgott, T., Hart F., Bedard R. 1977. *An Index of Refractory Organics.* EPA Environmental Protection Technology Series. EPA-600/2-77-174.

Hernández, D., B. Riaño, M. Coca, M. Solana, A. Bertucco, and M. C. García-González. 2016. "Microalgae Cultivation in High Rate Algal Ponds Using Slaughterhouse Wastewater for Biofuel Applications." *Chemical Engineering Journal* 285: 449–58. https://doi.org/10.1016/j.cej.2015.09.072.

Hernández-Zamora, Miriam, Hugo Virgilio Perales-Vela, César Mateo Flores-Ortíz, and Rosa Olivia Cañizares-Villanueva. 2014. "Physiological and Biochemical Responses of Chlorella Vulgaris to Congo Red." *Ecotoxicology and Environmental Safety* 108: 72–77. https://doi.org/10.1016/j.ecoenv.2014.05.030.

Iloms, Eunice, Olusolao O. Ololade, Henry J.O. Ogola, and Ramganesh Selvarajan. 2020. "Investigating Industrial Effluent Impact on Municipal Wastewater Treatment Plant in Vaal, South Africa." *International Journal of Environmental Research and Public Health* 17 (3): 1–18. https://doi.org/10.3390/ijerph17031096.

Kołodyńska, D., J. Krukowska, and P. Thomas. 2017. "Comparison of Sorption and Desorption Studies of Heavy Metal Ions from Biochar and Commercial Active Carbon." *Chemical Engineering Journal* 307: 353–63. https://doi.org/10.1016/j.cej.2016.08.088.

Koskimäki, Janne J., Marena Kajula, Juho Hokkanen, Emmi Leena Ihantola, Jong H. Kim, Heidi Hautajärvi, Elina Hankala, et al. 2016. "Methyl-Esterified 3-Hydroxybutyrate Oligomers Protect Bacteria from Hydroxyl Radicals." *Nature Chemical Biology* 12 (5): 332–38. https://doi.org/10.1038/nchembio.2043.

Lee, Sang Ah, Nakyeong Lee, Hee Mock Oh, and Chi Yong Ahn. 2019. "Enhanced and Balanced Microalgal Wastewater Treatment (COD, N, and P) by Interval Inoculation of Activated Sludge." *Journal of Microbiology and Biotechnology* 29 (9): 1434–43. https://doi.org/10.4014/jmb.1905.05034.

Li, Ronghua, Jim J. Wang, Baoyue Zhou, Mukesh Kumar Awasthi, Amjad Ali, Zengqiang Zhang, Lewis A. Gaston, Altaf Hussain Lahori, and Amanullah Mahar. 2016. "Enhancing Phosphate Adsorption by Mg/Al Layered Double Hydroxide Functionalized Biochar with Different Mg/Al Ratios." *Science of the Total Environment* 559: 121–29. https://doi.org/10.1016/j.scitotenv.2016.03.151.

Liu, Lin, Hongyong Fan, Yuhong Liu, Chaoxiang Liu, and Xu Huang. 2017. "Development of Algae-Bacteria Granular Consortia in Photo-Sequencing Batch Reactor." *Bioresource Technology* 232 (February): 64–71. https://doi.org/10.1016/j.biortech.2017.02.025.

Liu, Shuo, Qiusha Ma, Baozhen Wang, Jifu Wang, and Ying Zhang. 2014. "Advanced Treatment of Refractory Organic Pollutants in Petrochemical Industrial Wastewater by Bioactive Enhanced Ponds and Wetland System." *Ecotoxicology* 23 (4): 689–98. https://doi.org/10.1007/s10646-014-1215-9.

De Lourdes, Franco Martínez María, Rodríguez Rosales María Dolores Josefina, Moreno Medina Cuauhtémoc Ulises, and Martínez Roldán Alfredo De Jesús. 2017. "Tolerance and Nutrients Consumption of Chlorella Vulgaris Growing in Mineral Medium and Real Wastewater under Laboratory Conditions." *Open Agriculture* 2 (1): 394–400. https://doi.org/10.1515/opag-2017-0042.

Mahdavi, Hamed, Vinay Prasad, Yang Liu, and Ania C. Ulrich. 2015. "In Situ Biodegradation of Naphthenic Acids in Oil Sands Tailings Pond Water Using Indigenous Algae-Bacteria Consortium." *Bioresource Technology* 187: 97–105. https://doi.org/10.1016/j.biortech.2015.03.091.

Matamoros, Víctor, Raquel Gutiérrez, Ivet Ferrer, Joan García, and Josep M Bayona. 2015. "Capability of Microalgae-Based Wastewater Treatment Systems to Remove Emerging Organic Contaminants: A Pilot-Scale Study." *Journal of Hazardous Materials* 288 (May): 34–42. https://doi.org/10.1016/j.jhazmat.2015.02.002.

MeGraw, Victoria E., Ashley R. Brown, Christopher Boothman, Royston Goodacre, Katherine Morris, David Sigee, Lizzie Anderson, and Jonathan R. Lloyd. 2018. "A Novel Adaptation Mechanism Underpinning Algal Colonization of a Nuclear Fuel Storage Pond." *mBio* 9 (3): 1–15. https://doi.org/10.1128/mBio.02395-17.

Molinuevo-Salces, Beatriz, Berta Riaño, David Hernández, and M. Cruz García-González. 2019. "Microalgae and Wastewater Treatment: Advantages and Disadvantages." In *Microalgae Biotechnology for Development of Biofuel and Wastewater Treatment*, 505–33. Springer. https://doi.org/10.1007/978-98 1-13-2264-8_20.

Oswald, W J, H B Gotaas, C G Golueke, W R Kellen, and E F Gloyna. 1957. "Algae in Waste Treatment [with Discussion] All Use Subject to JSTOR Terms and Conditions IN WASTE." *Sewage and Industrial Wastes* 29 (4): 437–57.

Othman, Ali, Eduard Dumitrescu, Daniel Andreescu, and Silvana Andreescu. 2018. "Nanoporous Sorbents for the Removal and Recovery of Phosphorus from Eutrophic Waters: Sustainability Challenges and Solutions." *ACS Sustainable Chemistry and Engineering* 6 (10): 12542–61. https://doi.org/10.1021/a cssuschemeng.8b01809.

Perera, Isiri Adhiwarie, Sudharsanam Abinandan, Suresh R. Subashchandrabose, Kadiyala Venkateswarlu, Ravi Naidu, and Mallavarapu Megharaj. 2019. "Advances in the Technologies for Studying Consortia of Bacteria and Cyanobacteria/Microalgae in Wastewaters." *Critical Reviews in Biotechnology* 39 (5): 709–31. https://doi.org/10.1080/07388551.2019.1597828.

Pierre, Guillaume, Cédric Delattre, Pascal Dubessay, Sébastien Jubeau, Carole Vialleix, Jean Paul Cadoret, Ian Probert, and Philippe Michaud. 2019. "What Is in Store for EPS Microalgae in the next Decade?" *Molecules* 24 (23): 1–25. https://doi.org/10.3390/molecules24234296.

Posadas, E., C. Alcántara, P.A. García-Encina, L. Gouveia, B. Guieysse, Z. Norvill, F.G. Acién, et al. 2017. "Microalgae Cultivation in Wastewater." In *Microalgae-Based Biofuels and Bioproducts*, 67–91. Elsevier. https://doi.org/10.1016/B978-0-08-101023-5.00003-0.

Qian, Linbo, Wenying Zhang, Jingchun Yan, Lu Han, Weiguo Gao, Rongqin Liu, and Mengfang Chen. 2016. "Effective Removal of Heavy Metal by Biochar Colloids under Different Pyrolysis Temperatures." *Bioresource Technology* 206: 217–24. https://doi.org/10.1016/j.biortech.2016.01.065.

Qiu, Shuang, Yeting Shen, Liang Zhang, Bin Ma, Ayesha A. Amadu, and Shijian Ge. 2020a. "Antioxidant Assessment of Wastewater-Cultivated Chlorella Sorokiniana in Drosophila Melanogaster." *Algal Research* 46 (January): 101795. https://doi.org/10.1016/j.algal.2020.101795.

Qiu, Shuang, Lingfeng Wang, Zhipeng Chen, Mingzhu Yang, Ziwei Yu, and Shijian Ge. 2020b. "An Integrated Mainstream and Sidestream Strategy for Overcoming Nitrite Oxidizing Bacteria Adaptation in a Continuous Plug-Flow Nutrient Removal Process." *Bioresource Technology* 319: 124133. https://doi.org/ 10.1016/j.biortech.2020.124133.

Ramanan, Rishiram, Byung Hyuk Kim, Dae Hyun Cho, Hee Mock Oh, and Hee Sik Kim. 2016. "Algae-Bacteria Interactions: Evolution, Ecology and Emerging Applications." *Biotechnology Advances* 34 (1): 14–29. https://doi.org/10.1016/j.biotechadv.2015.12.003.

Ryu, Byung Gon, Woong Kim, Kibok Nam, Sungwhan Kim, Bongsoo Lee, Min S. Park, and Ji Won Yang. 2015. "A Comprehensive Study on Algal-Bacterial Communities Shift during Thiocyanate Degradation in a Microalga-Mediated Process." *Bioresource Technology* 191: 496–504. https://doi.org/10.1016/j.bior tech.2015.03.136.

Salam, Kamoru A. 2019. "Towards Sustainable Development of Microalgal Biosorption for Treating Effluents Containing Heavy Metals." *Biofuel Research Journal* 6 (2): 948–61. https://doi.org/10.18331/BRJ20 19.6.2.2.

Santhosh, S., A. M. Rajalakshmi, M. Navaneethakrishnan, S. Jenny Angel, and R. Dhandapani. 2020. "Lab-Scale Degradation of Leather Industry Effluent and Its Reduction by Chlorella Sp. SRD3 and Oscillatoria Sp. SRD2: A Bioremediation Approach." *Applied Water Science* 10 (5). https://doi.org/10.1 007/s13201-020-01197-0.

Shen, Yeting, Shuang Qiu, Zhipeng Chen, Yaping Zhang, Jonathan Trent, and Shijian Ge. 2020. "Free Ammonia Is the Primary Stress Factor Rather than Total Ammonium to Chlorella Sorokiniana in Simulated Sludge Fermentation Liquor." *Chemical Engineering Journal* 397 (May): 125490. https://do i.org/10.1016/j.cej.2020.125490.

Silva, Andreia, Cristina Delerue-Matos, Sónia A. Figueiredo, and Olga M. Freitas. 2019. "The Use of Algae and Fungi for Removal of Pharmaceuticals by Bioremediation and Biosorption Processes: A Review." *Water* 11 (8). https://doi.org/10.3390/w11081555.

Singh, Nitin Kumar, Siddhartha Pandey, Rana Pratap Singh, Swati Dahiya, Sneha Gautam, and Absar Ahmad Kazmi. 2018. "Effect of Intermittent Aeration Cycles on EPS Production and Sludge Characteristics in a Field Scale IFAS Reactor." *Journal of Water Process Engineering* 23 (December 2017): 230–38. https:// doi.org/10.1016/j.jwpe.2018.03.012.

Srivastav, Manjari, Meenal Gupta, Sushil K. Agrahari, and Pawan Detwal. 2018. "Removal of Refractory Organic Compounds from Wastewater by Various Advanced Oxidation Process - A Review." *Current Environmental Engineering* 6 (1): 8–16. https://doi.org/10.2174/2212717806666181212125216.

Sutherland, Donna L., Clive Howard-Williams, Matthew H. Turnbull, Paul A. Broady, and Rupert J. Craggs. 2015. "Enhancing Microalgal Photosynthesis and Productivity in Wastewater Treatment High Rate Algal Ponds for Biofuel Production." *Bioresource Technology* 184: 222–29. https://doi.org/10.1016/j.bior tech.2014.10.074.

Tao, Ran, Robert Bair, Melanie Pickett, Jorge L. Calabria, Aino Maija Lakaniemi, Eric D. van Hullebusch, Jukka A. Rintala, and Daniel H. Yeh. 2020. "Low Concentration of Zeolite to Enhance Microalgal Growth and Ammonium Removal Efficiency in a Membrane Photobioreactor." *Environmental Technology* 1–45. https://doi.org/10.1080/09593330.2020.1752813.

Thawechai, Tipawan, Benjamas Cheirsilp, Yasmi Louhasakul, Piyarat Boonsawang, and Poonsuk Prasertsan. 2016. "Mitigation of Carbon Dioxide by Oleaginous Microalgae for Lipids and Pigments Production: Effect of Light Illumination and Carbon Dioxide Feeding Strategies." *Bioresource Technology* 219 (November): 139–49. https://doi.org/10.1016/j.biortech.2016.07.109.

Tian, Xiangmiao, Zhiqiang Shen, Zhenfeng Han, and Yuexi Zhou. 2019. "The Effect of Extracellular Polymeric Substances on Exogenous Highly Toxic Compounds in Biological Wastewater Treatment: An Overview." *Bioresource Technology Reports* 5 (November 2018): 28–42. https://doi.org/10.1016/j.biteb.2018.11.009.

Tripathi, Bhumi Nath, and Dhananjay Kumar. 2017. *Prospects and Challenges in Algal Biotechnology*, 1–326. https://doi.org/10.1007/978-981-10-1950-0.

Vanhoudt, Nathalie, Hildegarde Vandenhove, Natalie Leys, and Paul Janssen. 2018. "Potential of Higher Plants, Algae, and Cyanobacteria for Remediation of Radioactively Contaminated Waters." *Chemosphere* 207: 239–54. https://doi.org/10.1016/j.chemosphere.2018.05.034.

Villar-Navarro, Elena, Rosa M. Baena-Nogueras, Maria Paniw, José A. Perales, and Pablo A. Lara-Martín. 2018. "Removal of Pharmaceuticals in Urban Wastewater: High Rate Algae Pond (HRAP) Based Technologies as an Alternative to Activated Sludge Based Processes." *Water Research* 139: 19–29. https://doi.org/10.1016/j.watres.2018.03.072.

Vimalnath, S., and S. Subramanian. 2018. "Studies on the Biosorption of Pb(II) Ions from Aqueous Solution Using Extracellular Polymeric Substances (EPS) of Pseudomonas Aeruginosa." *Colloids and Surfaces, part B: Biointerfaces* 172 (Ii): 60–67. https://doi.org/10.1016/j.colsurfb.2018.08.024.

Vojvodić, Snežana, Jelena Danilović Luković, Bernd Zechmann, Mima Jevtović, Jelena Bogdanović Pristov, Marina Stanić, Alessandro Marco Lizzul, Jon K. Pittman, and Ivan Spasojević. 2020. "The Effects of Ionizing Radiation on the Structure and Antioxidative and Metal-Binding Capacity of the Cell Wall of Microalga Chlorella Sorokiniana." *Chemosphere* 260, 127553. https://doi.org/10.1016/j.chemosphere.2020.127553.

Wang, Luyun, Han Xiao, Ning He, Dong Sun, and Shunshan Duan. 2019. "Biosorption and Biodegradation of the Environmental Hormone Nonylphenol By Four Marine Microalgae." *Scientific Reports* 9 (1): 1–11. https://doi.org/10.1038/s41598-019-41808-8.

Wu, Y. 2017. "The Removal of Methyl Orange by Periphytic Biofilms: Equilibrium and Kinetic Modeling." In *Periphyton*, 367–87. https://doi.org/10.1016/B978-0-12-801077-8.00016-8.

Xie, Peng, Chuan Chen, Chaofan Zhang, Guanyong Su, Nanqi Ren, and Shih Hsin Ho. 2020. "Revealing the Role of Adsorption in Ciprofloxacin and Sulfadiazine Elimination Routes in Microalgae." *Water Research* 172: 115475. https://doi.org/10.1016/j.watres.2020.115475.

Xiong, Jiu Qiang, Mayur B. Kurade, Dilip V. Patil, Min Jang, Ki Jung Paeng, and Byong Hun Jeon. 2017. "Biodegradation and Metabolic Fate of Levofloxacin via a Freshwater Green Alga, Scenedesmus Obliquus in Synthetic Saline Wastewater." *Algal Research* 25 (March): 54–61. https://doi.org/10.1016/j.algal.2017.04.012.

Xu, Mingyu, Qingqing Zeng, Huankai Li, Yuming Zhong, Linying Tong, Roger Ruan, and Hui Liu. 2020. "Contribution of Glycerol Addition and Algal–Bacterial Cooperation to Nutrients Recovery: A Study on the Mechanisms of Microalgae-Based Wastewater Remediation." *Journal of Chemical Technology and Biotechnology* 95 (6): 1717–28. https://doi.org/10.1002/jctb.6369.

Yang, Linyan, Xueming Chen, Qianhong She, Guomin Cao, Yongdi Liu, Victor W.C. Chang, and Chuyang Y. Tang. 2018. "Regulation, Formation, Exposure, and Treatment of Disinfection by-Products (DBPs) in Swimming Pool Waters: A Critical Review." *Environment International* 121 (November): 1039–57. https://doi.org/10.1016/j.envint.2018.10.024.

Yu, Kai Ling, Beng Fye Lau, Pau Loke Show, Hwai Chyuan Ong, Tau Chuan Ling, Wei Hsin Chen, Eng Poh Ng, and Jo Shu Chang. 2017a. "Recent Developments on Algal Biochar Production and Characterization." *Bioresource Technology* 246 (June): 2–11. https://doi.org/10.1016/j.biortech.2017.08.009.

Yu, Kai Ling, Pau Loke Show, Hwai Chyuan Ong, Tau Chuan Ling, John Chi-Wei Lan, Wei Hsin Chen, and Jo Shu Chang. 2017b. "Microalgae from Wastewater Treatment to Biochar – Feedstock Preparation and Conversion Technologies." *Energy Conversion and Management* 150 (October): 1–13. https://doi.org/10.1016/j.enconman.2017.07.060.

Zhu, Shunni, Shuhao Huo, and Pingzhong Feng. 2019. "Developing Designer Microalgal Consortia: A Suitable Approach to Sustainable Wastewater Treatment." In *Microalgae Biotechnology for Development of Biofuel and Wastewater Treatment*, 569–98. Springer. https://doi.org/10.1007/978-981-13-2264-8_22.

21 Photo-Assisted Fenton Decomposition of Organic Contaminants Under Visible-Light Illumination

Titikshya Mohapatra, Parmesh Kumar Chaudhari, and Prabir Ghosh

CONTENTS

21.1 INTRODUCTION

Water pollution is a serious environmental issue in many countries as a result of growing industrial activities. The discharge of wastewater containing toxic organic pollutants from various industries into water bodies affects the environment as well as human health (Bafana et al., 2011). Moreover, for sustainable development and to protect the environment from further damage, persistent organic pollutants (POPs) should be removed from wastewater before their release to the environment. Several wastewater treatment technologies, such as physical, chemical, and biological processes, have been studied to remove POPs from the wastewater. The physical processes like adsorption and coagulation/flocculation only alter the phase of the POPs (from solution to solid phase) without destroying the pollutant (Soltani and Lee, 2020). As most of the POPs are complex in structure, biological processes are inefficient in completely degrading the pollutants (Torrades et al., 2004). The perfect wastewater treatment is employed for the complete mineralization of the concerned POPs without leaving behind any harmful residues. Consequently, advanced oxidation technologies (AOTs) have emerged as a potential solution for the complete degradation of POPs present in wastewater streams. AOTs are based on the in-situ generation of hydroxyl radicals ($^{\bullet}$OH), react with

DOI: 10.1201/9781003204442-21

the POPs, and convert them into innocuous derivatives, i.e., carbon dioxide and water. Based on the •OH formation by different processes, AOTs are categorized as Fenton, photocatalytic oxidation, wet air oxidation, etc. (Yao et al., 2014). Among the different AOTs, the photo-assisted Fenton process has attracted attention among environmental researchers due to its higher reaction rates, efficiency, and process simplicity (Arslan-Alaton and Olmez-Hanci, 2012).

21.1.1 BASICS OF THE PHOTO-FENTON PROCESSES

Over the past years, the photo-Fenton process has been studied for the destruction of POPs present in wastewater. The introduction of light irradiation in the Fenton reaction is commonly known as the photo-Fenton process (Palter et al., 2013). In this process, oxidizing species, i.e., •OH radicals, are formed by the following means (represented in Equations 21.1–21.3): (i) the catalytic decomposition of hydrogen peroxide, (ii) the photolysis of hydrogen peroxide under light irradiation, and (iii) the generation of •OH radicals plus the regeneration of ferrous iron by the photo-reduction of ferric hydroxide complexes (Arslan-Alaton and Olmez-Hanci, 2012).

$$Fe^{2+} + H_2O_2 \rightarrow Fe^{3+} + HO^• + OH^- \tag{21.1}$$

$$H_2O_2 + h\text{Å} \rightarrow HO^• + HO^• \tag{21.2}$$

$$Fe^{3+} + H_2O_2 \rightarrow Fe(OH)^{2+} + H^+ + h\text{Å} \rightarrow HO^• + Fe^{2+} + H^+ \tag{21.3}$$

These produced •OH radicals further react with the POPs and destroy them completely, as shown in Equation (21.4).

$$HO^• + POPs \rightarrow CO_2 + H_2O \tag{21.4}$$

The main advantage of the photo-Fenton process is the higher POP degradation rate owing to more •OH radical formation than the standard Fenton process (Navalon et al., 2010). Also, the utilization of less amounts of Fenton's reagent in this process makes it suitable for large-scale application (Yang et al., 2019). The process shows good efficiency at a low pH because of the generation of more soluble and photoactive hydroxyl-Fe(III) complexes (Herney-Ramirez et al., 2010). However, this process is unsuitable for industrial application due to the use of harmful ultraviolet rays, i.e., UV-C, as a light supply which increases the handling risk, and the high amounts of electrical energy required which make it expensive. To overcome these drawbacks, environmental researchers have been focusing on visible light supplies like fluorescent light bulbs, xenon lamps, black light bulbs, and sunlight as light sources for the photo-Fenton process (Ikehata and El-Din, 2006). A few scientific reports have been published that relate to visible light supplies, such as "visible-light assisted photo-Fenton" and "visible-light driven photo-Fenton" in the early 1990s. After 2009, this research field intensified enormously, as shown in Figure 21.1. However, the development of Fenton catalysts is required for the utilization of a wider range of visible light to trigger the Fenton process at broader pH ranges for the photodegradation of POPs.

This book chapter discusses the modification in the visible-light assisted photocatalyst materials and their utilization in the photo-Fenton process for the reclamation of POPs in wastewater. This chapter also enlightens readers about the influence of certain operating parameters like pH, catalyst loading, oxidant (H_2O_2) concentration, and temperature on the process's efficiency.

21.2 DEVELOPMENTS IN VISIBLE LIGHT-INDUCED PHOTOCATALYTIC MATERIAL

There are many research groups working on the development or modification of photocatalytic materials to improve the economical and technical feasibility of the photo-Fenton processes for wastewater

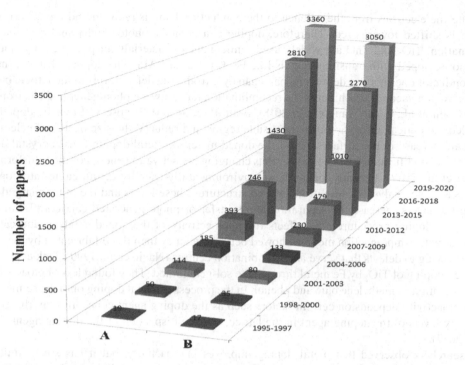

FIGURE 21.1 Results of a literature search based on Google Scholar: (A) by using keywords "visible light assisted photo-Fenton", and (B) by using keywords "visible-light driven photo-Fenton."

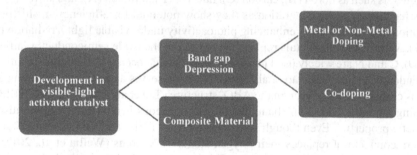

FIGURE 21.2 Methods to develop the visible light-active heterogeneous catalyst.

treatment under visible light irradiation. In the photo-assisted process, the band gap of the catalyst or semiconductor plays a significant role, as the catalytic activity depends on this energy. Pure iron has a wide band gap that exhibits photoactivity under shorter wavelengths of UV light which only accounts for 3–5% of solar energy absorbed to form photogenerated charge carriers. Therefore, pure iron is inadequate to absorb incident sunlight. To achieve the visible light-driven photoactivity, the band gap of the semiconductor needs to be narrowed. This is responsible for shifting the optical absorption boundary from the UV region to the visible region (44–47% of solar spectrum) (Soltani and Lee, 2020). Figure 21.2 represents various ways to reduce the band gap as well as to modify the surface of semiconductors, such as doping with metal or non-metal mono atoms, doping with multi atoms (known as co-doping), and combining with other metals to form new composite materials.

21.2.1 METAL OR NON-METAL DOPED AND CO-DOPED FENTON CATALYST

The inclusion of doping agents to the semiconductor will change the band gap: it may form a new state under the conduction band or over the valence band. As a result, the energy requirement to excite the electrons from the valence to the conduction band is reduced and the absorption of photon is shifted towards red. Therefore, doping can enhance photoactivity under visible light illumination. TiO_2 and ZnO are widely used semiconductor materials for photoactivity, and they can also be doped with transition metals like Fe, Cu, Co, and Mn. The doping of metal into the semiconductor does not just depress its band gap by introducing defect levels in the lattice, but also creates oxygen vacancy that hinders the recombination rate between photogenerated electron–hole pairs (Chen et al., 2001; Wang et al., 2018). Asilturk et al. (2009) observed that Fe-doped TiO_2 extended the optical absorption into the visible region and reduced the size of the particle, resulting in an increase in the surface area of the doped material. Metal doping causes crystal defects and surface modifications to the catalyst, thus changing the activation energy of the transformation (Wellia et al., 2017). Recently, bismuth, an environmentally-friendly metal, came into consideration as a photocatalyst because of its layered structure. These layers and the inner charged fields among the layers increase the separation and transfer of photo-generated carriers (Tekin et al., 2018). The doping procedure also affects the photoactivity of the doped catalyst, and catalysts prepared by the impregnation method showed better efficiency than those fabricated by the sol-gel method owing to defects that serve as recombination centres (Navio et al., 1999). Jack et al. (2015) reviewed doping of TiO_2 by Fe metal through the sol-gel method. They found less photoactivity for the degradation of methylene blue and phenol. In the process of metal doping onto the semiconductor, photoactivity depends on certain factors such as the doping agent concentration, the position of energy levels of the doping agent in the lattice, and the dispersion of the doping agent (Wellia et al., 2017).

Researchers observed that metal doping improves photoactivity, but it has some limitations like potential toxicity and the leaching of metal after consecutive runs because the metal particles mostly get embedded on the surface of the semiconductor (Jia et al., 2018). In view of this, the doping of non-metals such as boron (B), carbon (C), fluorine (F), nitrogen (N), and sulfur (S) with semiconductors has come into consideration as they show potential for efficiency in shifting the band gap absorption and subsequently enhancing photoactivity under visible light irradiation (Saravanan et al., 2017). Mainly, the non-metals replace the oxygen in the oxide semiconductor lattice (Wu and Hung, 2009). C and N are widely used non-metal dopants that occupy the cationic or anionic sites in the semiconductor lattice (Sullivan et al., 2014). Nitrogen serves as an effective doping agent as its anion size is close to that of oxygen in the ABO_3 structure (Tan et al., 2011). Jia et al. (2018) observed that N doping not only reduces the band gap of heterogeneous Fenton catalysts, but also increases their magnetic properties. Even though sulfur has a bigger ionic radius than N and C, after doping with the semiconductor it replaces their oxygen atoms by S atoms (Wellia et al., 2017). However, various reports have concluded that non-metal doping red-shifted the optical absorption range by depressing the band gap. But the depressed band gap causes an increase in the recombination rate of the photoexcited electron–hole pair, resulting in a decrease in the efficiency of photoactivity (Jing et al., 2005). Doping with transition metals enhances the photoactivity, but it significantly reduces the photoexcited carrier mobility due to its strong localized state in the band gap (Mu et al., 1989). Researchers found that co-doping can solve these issues, where the addition of metal on the non-metal doped semiconductor can inhibit the recombination of the photoexcited pair, since the metal element acts as a mediator for interfacial charge transfer (Jia et al., 2011). Multi-element doping also provides better photoactivity than single-element doping (Zhang and Song, 2009). They synthesized multi-metal co-doped magnesium ferrite from abundant saprolite laterite ore through the hydrothermal method. They observed that the high surface area of the catalyst and the combined reaction between photocatalytic oxidation and Fenton-like oxidation resulted in a higher photodegradation of Rhodamine B dye under visible light illumination (Diao et al., 2018). TiO_2 co-doping with N and Fe causes lattice distortion which modifies the dipole moments and simplifies the separation of photo-generated electron–hole pairs. Hence, it can significantly improve the efficiency of photoactivity (Jia et al., 2011).

21.2.2 COMPOSITE MATERIAL

Composite materials combine catalyst or support material with other metals, semiconductors with lower band gaps, or polymeric materials, which depresses the band gap and improves the visible light-driven photo-degradation process. Recently, nano-composite materials are gaining wider attention as the electrical conductivity increases with a reduction in particle size. Pal et al. (2020) synthesized a goethite/akageneite nanocomposites by using a "green" co-precipitation method. Furthermore, they functionalized the nanocomposites with a starch solution which converted their shape into rice grain-like structures. This also makes the surface of the nanocomposites more hydrophilic and modified their band structure to provide better performance under visible light conditions. Various carbon-based materials such as graphite, carbon black, graphene, carbon dots, etc., used as a support for metals have been studied as catalysts. Yao et al. (2014) observed that combining $ZnFe_2O_4$ nanoparticles with reduced graphene oxide resulted better photoactivity than pure $ZnFe_2O_4$. The graphene prevented the $ZnFe_2O_4$ agglomeration, enhanced the charge separation, and at the same time acted as a catalyst for the activation of oxidant to form free radicals. Di et al. (2019) combined two semiconductors, i.e., Ag_2S and $BiFeO_3$, to form a heterojunction composite with enhanced photoactivity. Additionally, they observed that the heterostructure reduced the photocorrosion of Ag_2S, and due to the Z-scheme electron transfer, separation of the photoexcited electron–hole pairs occurred efficiently.

21.3 EFFECT OF OPERATING PARAMETERS IN THE PHOTODEGRADATION PROCESS

The rate of photodegradation of POPs by the photo-Fenton process under visible light irradiation depends on various parameters such as light intensity, pH, oxidant concentration, and catalyst loading.

21.3.1 LIGHT INTENSITY

Light irradiation is a key element of the photo-Fenton process. The degradation rate of the photo-Fenton process depends on the intensity of irradiated light. The quanta of light absorbance by any catalyst or reactant are represented by quantum yields. Quantum yield (Φ_λ) is defined as the ratio of the rate of reactant consumption to the absorption of photon flux by highly reactive species, as show in Equation 21.5 (Einschlag et al., 2013).

$$\Phi_{M,\lambda} = \frac{d[M]/dt}{P_{p,a,\lambda}} \qquad (21.5)$$

Where:
d[M]/dt = rate of transformation of M
$P_{p,a,\lambda}$ = photon flux absorbed by M under irradiation of wavelength λ

The efficiency of the photo-Fenton process depends on the photochemical reduction of Fe^{3+}. The quantum yield for photochemical Fe^{2+} regeneration varies under different wavelengths of light irradiation (Saravanan et al., 2017; Einschlag et al., 2013). Chen et al. (2001) observed that irradiation under a light source produced active species which improved the Fenton reaction. Tekin et al. (2018) compared the effect of different powers of metal halogen bulbs (100 W, 500 W, and 1,000 W) for the degradation of dye wastewater through the photo-Fenton process. They noticed that changing the power of the light from 100 W to 500 W was not very effective for photodegradation, but the degradation was decreased with 1,000 W, which might be attributed to the self-scavenging effect because of the recombination of photo-generated radicals with increased energy of light. Hence, the visible light intensity should be optimized to get better photodegradation efficiency.

21.3.2 INITIAL pH OF THE SOLUTION

In the Fenton reaction, the initial pH value plays a key role, as it changes the charges on the surface of the catalyst which affects the reaction between POPs, the catalyst, and the oxidants. The homogenous Fenton process shows better degradation efficiency at lower pH levels near 3. But, heterogeneous Fenton catalysis can be utilized across a wide range of pH levels, which cuts down the need of adding acidifying reagents into the wastewater solution as well as provides eco-friendly catalysis. Tekin et al. (2018) observed that adjusting the pH to 3 or 9 can provide less degradation efficiency because at low pH intermediates formed which further decreased the pH to 2–2.5. This reduced the amount of metal complex needed for the catalytic reaction. Also, at low pH, •OH radicals get scavenged by H^+ ions. On the other hand, at high pH levels the decomposition of •OH into water and oxygen, and the precipitation of metal hydroxide reduced the photodegradation (Phan et al., 2018). Most of the heterogeneous catalysts have shown good photodegradation of POPs at neutral pH levels. This can save the extra cost of adding acidifying reagents and diminish the leaching of transition metal from the catalyst surface (Gao et al., 2020).

21.3.3 H₂O₂ CONCENTRATION

Hydrogen peroxide is a commonly-used oxidizing agent in the Fenton-related process. With an increase in H_2O_2 concentration, the rate of degradation also increases due to the increasing formation of •OH radicals that further attack the POPs (Phan et al., 2018). The addition of H_2O_2 in excess reduces the POP degradation efficiency, as H_2O_2 can react with •OH radicals to form hydroperoxyl radicals, as shown in Equation (21.6). Hydroperoxyl radicals are a weaker reactive species than •OH radicals, which slows down the POP degradation process; this phenomenon is known as the scavenging effect (Wang et al., 2018).

$$HO^\bullet + H_2O_2 \rightarrow HO_2^\bullet + H_2O \qquad (21.6)$$

21.3.4 CATALYST LOADING

Catalysts are one of the major elements of the Fenton process. The degradation efficiency improves with an increase in catalyst loading, as more Fe^{3+}/Fe^{2+} engagement in the reaction produces more •OH radicals (Wang et al., 2018). But the excess loading of catalysts decreases the POP removal efficiency, which is attributed to the turbid reaction medium which inhibits the transmission of light and reduces the formation of reactive oxidation species by the photo-driven Fenton process (Tekin et al., 2018; Phan et al., 2018). Too many Fe^{3+}/Fe^{2+} ions can scavenge •OH radicals by Fe^{2+}, as represented in Equation (21.7) (Wang et al., 2018).

$$Fe^{2+} + HO^\bullet \rightarrow Fe^{3+} + OH^- \qquad (21.7)$$

21.4 DEGRADATION OF ORGANIC CONTAMINANTS VIA VISIBLE LIGHT-DRIVEN PHOTO-FENTON PROCESS

Du et al. (2018) examined the efficiency of prepared Fe/Si co-doped TiO_2 Fenton catalyst for the degradation of metronidazole, an antibiotic, with and without light illumination. Through the xenon lamp (220W) driven photo-Fenton process 93% of metronidazole degradation was achieved within 50 min at neutral pH. The photo-Fenton process provided better degradation efficiency than the photocatalysis and only the Fenton process. The better performance of photo-Fenton process attributed to the more •OH formation simultaneously by both the photocatalysis and Fenton process.

Deng et al. (2017) studied the degradation of methyl orange (MO) dye and phenol (both having 10 mg/L initial concentrations) using a triple heterojunction $TiO_2/Fe_2TiO_5/Fe_2O_3$ composite as a catalyst through the visible light-driven photo-Fenton process. They used a 300 W xenon lamp enabled with a wavelength cut-off filter ($\lambda \geq 420$ nm) as a visible light source. They found that the three-phase composite offers good phase interfaces for the movement of photogenerated electrons among different components, which enhanced the Fe^{3+}/Fe^{2+} cycle reaction. Consequently, it improved the photo-Fenton process and attained 95% MO degradation at a neutral pH within 60 min by using 20 wt% of the Fe/Ti-loaded triple heterojunction composite. Additionally, complete phenol degradation was achieved and the related total organic carbon (TOC) removal rate was 85% within 60 min and 120 min, respectively. Di et al. (2019) also used the same visible light radiation source to degrade the same MO pollutant through photo-Fenton and photocatalytic processes by using a different heterojunction composite material, i.e., Z-scheme $Ag_2S/BiFeO_3$. The photo-Fenton process offered superior degradation efficiency of MO. As in this process, both the photocatalytic and Fenton reactions occurred simultaneously under visible light irradiation, which helped in the formation of more $^{\bullet}OH$ radicals and the regeneration of Fe^{2+}. The degradation of MO was also examined in the photo-Fenton process using Cu-doped $LaFe_2O_3$ as a catalyst under the lighting of a xenon lamp with 400 nm cut-off (Phan et al., 2018). Doping with 15 mol% Cu onto $LaFe_2O_3$ resulted in better degradation efficiency than $LaFe_2O_3$, and it also enhanced the MO degradation rate by 60% as it enhanced the decomposition of H_2O_2 under visible light illumination to produce more $^{\bullet}OH$ radicals than $LaFe_2O_3$. Under visible light irradiation, small amounts of catalysts (0.8 g/L) yielded 92.9% MO removal within 60 min by using 0.3 g/L of H_2O_2 at pH 6. The study also revealed the stability and reusability of the prepared catalyst to treat wastewater containing other cationic dyes.

However, $BiFeO_3$ magnetic nanoparticles are capable of the catalytic activation of H_2O_2 for the degradation of POPs in the dark Fenton process, but they have weak catalytic activity for the decomposition of highly stable POPs. Sultani and Lee (2016) doped barium (Ba) into $BiFeO_3$ magnetic nanoparticles which enhanced the photo-Fenton deterioration of toluene to 98%. It also reduced the COD by 94% and TOC by 85% after 40 min of irradiation under a 55 W fluorescent lamp, having an emission peak at 550 nm at a weak acidic pH of 5.5. As Ba doping onto the catalyst increased the oxygen vacancies and defects on its surface, this helped in improving photo-Fenton degradation by generating more $^{\bullet}OH$ radicals and acting as an active center to capture photo-generated electrons. Jia et al. (2017) examined the catalytic activity of N-doped $BiFeO_3$ for the degradation of bisphenol A (BPA) through the visible light-driven photo-Fenton process with and without the addition of a chelating agent (L-cysteine). The addition of 0.25 mmol/L L-cysteine amplified the degradation efficiency from 60% to 94% because L-cysteine helped the H_2O_2 to adsorb onto the catalyst surface and to produce more free radicals for the quick and complete degradation of BPA.

Degradation of methylene blue (MB) through solar light-driven photo-Fenton was investigated by using a metalloporphyrine-based porous organic polymer as a catalyst (Gao et al., 2020). The catalyst accelerated the photo-Fenton process by producing more $^{\bullet}OH$ and it also acts as photosensitizer to produce highly active O_2. The combined attack of these two reactive organic species on 50 ml of synthetic wastewater of 70 ppm MB offered complete degradation within 80 min at a neutral pH by utilizing only 4 mg of the catalyst. They also monitored this process by replacing the visible light source with a 500 W xenon lamp with an enabled 450–800 nm cut-off filter. The MB degradation rate, constantly under xenon lamp irradiation, was lower than that of solar light irradiation. This was attributed to the higher light intensity and the wide optical adsorption range (200–1200 nm) of the catalyst which is able to utilize the UV, near-infrared light, along with visible light from the solar light. Both resulted in more electron–hole pairs to enhance the MB degradation rate.

Cho et al. (2006) utilized a solar light with an average intensity of 1.7 mW/cm^2 at 365 nm for the degradation of benzene, toluene, ethylbenzene, and xylene (BTEX) compounds along with total petroleum hydrocarbons (TPHs) through the photo-Fenton process. Almost total degradation of BTEX and TPHs was achieved within 2 and 4 hr, respectively. Most of the studies have been conducted on laboratory scale. Table 21.1 summarizes the catalyst, operational parameters, and results

TABLE 21.1

Summary of Operational Parameters with Degradation Efficiency

Pollutant	Catalyst	Light Source	Operational Parameters	Degradation Efficiency (%)	Reference
Methyl orange (MO) and phenol	$TiO_2/Fe_2TiO_5/Fe_2O_3$ composite	Xe Lamp ($\lambda > 420$ nm)	MO: 10 mg/L Cat.: 1 mg/L H_2O_2: 130 µL Initial Ph: 4.0 Time: 60 min	MO: 95 Phenol: 100 TOC: 85	(Deng et al., 2017)
Metronidazole (MNZ)	Fe/Si co-doped TiO_2	200 W Xe lamp ($\lambda = 420$ nm)	MNZ: 6 ppm H_2O_2: 10 Mm pH: 7.0 Catalysts: 0.3 g Time: 50 min	MNZ: 93	(Du et al., 2018)
Methylene blue (MB)	Metalloporphyrin-based porous organic polymer	Sunlight	MB: 70 ppm pH: 7.0 Catalysts: 4 mg Time: 80 min	MB: 100	(Gao et al., 2020)
Rhodamine B (RhB)	Fe-supported bentonite	Light-emitting diodes (LED)	MNZ: 80 mg/L H_2O_2: 12 Mm pH: 4.2 Catalysts: 0.25 g/L Time: 300 min	COD: 80	(Gao et al., 2015)
P-nitrophenol (PNP)	Fe-doped ZnS	Xe lamp	PNP: 80 mg/L H_2O_2: 3 mmol/L pH: 4.2 Catalysts: 0.8 g/L Time: 180 min	PNP: 83.8	(Wang et al., 2018)

of earlier studies and confirms the synthesized heterogeneous catalysts having excellent POP degradation under visible light illumination.

21.5 CONCLUSION

One of the major issues in recent decades is the reclamation of highly toxic POPs in wastewater as these are damaging to the environment. The Fenton-related process provides the complete mineralization of these POPs. The modification of the heterogeneous Fenton catalyst has shown superior catalytic activity for the degradation of POPs under visible light irradiation. But few studies have been reported on the photo-Fenton process for the degradation of POPs under solar light irradiation. Though the use of inexpensive lamps as visible light sources for the photodegradation of POPs reduced the material and operational cost, more focus should be given to the solar-induced photo-Fenton process as it can reduce power consumption along with maintenance costs.

21.6 FUTURE PERSPECTIVE

The efficiency of the photo-Fenton process can be further improved by combining it with the fluidized-bed reactor as it provides more contact frequency between the pollutant and reactants. The utilization of heterogeneous catalysts beds can enhance the efficacy of the photo-Fenton process under solar as well as visible light irradiation.

REFERENCES

Arslan-Alaton, I., & Olmez-Hanci, T. (2012). Advanced oxidation of endocrine disrupting compounds: review on photo-fenton treatment of alkylphenols and bisphenol A. In *Green Technologies for Wastewater Treatment* (pp. 59–90). Springer, Dordrecht.

Asiltürk, M., Sayılkan, F., & Arpaç, E. (2009). Effect of Fe3+ ion doping to TiO2 on the photocatalytic degradation of Malachite Green dye under UV and vis-irradiation. *Journal of Photochemistry and Photobiology A: Chemistry, 203*(1), 64–71.

Bafana, A., Devi, S. S., & Chakrabarti, T. (2011). Azo dyes: past, present and the future. *Environmental Reviews, 19*, 350–371.

Chen, F., Xie, Y., He, J., & Zhao, J. (2001). Photo-Fenton degradation of dye in methanolic solution under both UV and visible irradiation. *Journal of Photochemistry and Photobiology A: Chemistry, 138*(2), 139–146.

Cho, I. H., Kim, Y. G., Yang, J. K., Lee, N. H., & Lee, S. M. (2006). Solar-chemical treatment of groundwater contaminated with petroleum at gas station sites: ex situ remediation using solar/TiO2 photocatalysis and solar photo-Fenton. *Journal of Environmental Science and Health: A, 41*(3), 457–473.

Deng, Y., Xing, M., & Zhang, J. (2017). An advanced TiO2/Fe2TiO5/Fe2O3 triple-heterojunction with enhanced and stable visible-light-driven fenton reaction for the removal of organic pollutants. *Applied Catalysis B: Environmental, 211*, 157–166.

Di, L., Yang, H., Xian, T., Liu, X., & Chen, X. (2019). Photocatalytic and photo-Fenton catalytic degradation activities of Z-scheme Ag2S/BiFeO3 heterojunction composites under visible-light irradiation. *Nanomaterials, 9*(3), 399.

Diao, Y., Yan, Z., Guo, M., & Wang, X. (2018). Magnetic multi-metal co-doped magnesium ferrite nanoparticles: an efficient visible light-assisted heterogeneous Fenton-like catalyst synthesized from saprolite laterite ore. *Journal of Hazardous Materials, 344*, 829–838.

Du, W., Xu, Q., Jin, D., Wang, X., Shu, Y., Kong, L., & Hu, X. (2018). Visible-light-induced photo-Fenton process for the facile degradation of metronidazole by Fe/Si codoped TiO 2. *RSC Advances, 8*(70), 40022–40034.

Einschlag, F. S. G., Braun, A. M., & Oliveros, E. (2013). Fundamentals and applications of the photo-Fenton process to water treatment. In *Environmental Photochemistry Part III* (pp. 301–342). Springer, Berlin, Heidelberg.

Gao, W., Tian, J., Fang, Y., Liu, T., Zhang, X., Xu, X., & Zhang, X. (2020). Visible-light-driven photo-Fenton degradation of organic pollutants by a novel porphyrin-based porous organic polymer at neutral pH. *Chemosphere, 243*, 125334.

Gao, Y., Wang, Y., & Zhang, H. (2015). Removal of Rhodamine B with Fe-supported bentonite as heterogeneous photo-Fenton catalyst under visible irradiation. *Applied Catalysis B: Environmental, 178*, 29–36.

Herney-Ramirez, J., Vicente, M. A., & Madeira, L. M. (2010). Heterogeneous photo-Fenton oxidation with pillared clay-based catalysts for wastewater treatment: a review. *Applied Catalysis B: Environmental, 98*(1–2), 10–26.

Ikehata, K., & El-Din, M. G. (2006). Aqueous pesticide degradation by hydrogen peroxide/ultraviolet irradiation and Fenton-type advanced oxidation processes: a review. *Journal of Environmental Engineering and Science, 5*(2), 81–135.

Jack, R. S., Ayoko, G. A., Adebajo, M. O., & Frost, R. L. (2015). A review of iron species for visible-light photocatalytic water purification. *Environmental Science and Pollution Research, 22*(10), 7439–7449.

Jia, L., Wu, C., Han, S., Yao, N., Li, Y., Li, Z., ... & Jian, L. (2011). Theoretical study on the electronic and optical properties of (N, Fe)-codoped anatase TiO2 photocatalyst. *Journal of Alloys and Compounds, 509*(20), 6067–6071.

Jia, Y., Wu, C., Kim, D. H., Lee, B. W., Rhee, S. J., Park, Y. C., ... & Liu, C. (2018). Nitrogen doped BiFeO3 with enhanced magnetic properties and photo-Fenton catalytic activity for degradation of bisphenol A under visible light. *Chemical Engineering Journal, 337*, 709–721.

Jing, D., Zhang, Y., & Guo, L. (2005). Study on the synthesis of Ni doped mesoporous TiO2 and its photocatalytic activity for hydrogen evolution in aqueous methanol solution. *Chemical Physics Letters, 415*(1–3), 74–78.

Mu, W., Herrmann, J. M., & Pichat, P. (1989). Room temperature photocatalytic oxidation of liquid cyclohexane into cyclohexanone over neat and modified TiO 2. *Catalysis Letters, 3*(1), 73–84.

Navalon, S., Martin, R., Alvaro, M., & Garcia, H. (2010). Gold on diamond nanoparticles as a highly efficient Fenton catalyst. *Angewandte Chemie International Edition, 49*(45), 8403–8407.

Navío, J. A., Testa, J. J., Djedjeian, P., Padrón, J. R., Rodríguez, D., & Litter, M. I. (1999). Iron-doped titania powders prepared by a Sol–gel method: part II: photocatalytic properties. *Applied Catalysis A: General*, *178*(2), 191–203.

Pal, S., Singh, P. N., Verma, A., Kumar, A., Tiwary, D., Prakash, R., & Sinha, I. (2020). Visible light photo-Fenton catalytic properties of starch functionalized iron oxyhydroxide nanocomposites. *Environmental Nanotechnology Monitoring and Management*, 100311.

Palter, J. B., Marinov, I., Sarmiento. J. L., & Gruber, N. (2013). Large-scale, persistent nutrient fronts of the world ocean: Impacts on biogeochemistry. In *The Handbook of Environmental Chemistry*. Berlin, Heidelberg: Springer. https://doi.org/10.1007/698_2013_241

Phan, T. T. N., Nikoloski, A. N., Bahri, P. A., & Li, D. (2018). Heterogeneous photo-Fenton degradation of organics using highly efficient Cu-doped LaFeO3 under visible light. *Journal of Industrial and Engineering Chemistry*, *61*, 53–64.

Saravanan, R., Gracia, F., & Stephen, A. (2017). Basic principles, mechanism, and challenges of photocatalysis. In *Nanocomposites for Visible Light-Induced Photocatalysis* (pp. 19–40). Springer, Cham.

Soltani, T., & Lee, B. K. (2016). Novel and facile synthesis of Ba-doped BiFeO3 nanoparticles and enhancement of their magnetic and photocatalytic activities for complete degradation of benzene in aqueous solution. *Journal of Hazardous Materials*, *316*, 122–133.

Soltani, T., & Lee, B. K. (2020). Photocatalytic and photo-fenton catalytic degradation of organic pollutants by non-TiO2 photocatalysts under visible light irradiation. In *Current Developments in Photocatalysis and Photocatalytic Materials* (pp. 267–284). Elsevier, Amsterdam.

Sullivan, J. A., Neville, E. M., Herron, R., Thampi, K. R., & MacElroy, J. D. (2014). Routes to visible light active C-doped TiO2 photocatalysts using carbon atoms from the Ti precursors. *Journal of Photochemistry and Photobiology A: Chemistry*, *289*, 60–65.

Tan, X., Chen, C., Jin, K., & Luo, B. (2011). Room-temperature ferromagnetism in nitrogen-doped BaTiO3. *Journal of Alloys and Compounds*, *509*(34), L311–L313.

Tekin, G., Ersöz, G., & Atalay, S. (2018). Visible light assisted Fenton oxidation of tartrazine using metal doped bismuth oxyhalides as novel photocatalysts. *Journal of Environmental Management*, *228*, 441–450.

Torrades, F., Montano, J. G., Hortal, J. A. G., Domènech, X., & Peral, J. (2004). Decolorization and mineralization of commercial reactive dyes under solar light assisted photo-Fenton conditions. *Solar Energy*, *77*(5), 573–581.

Wang, Q., Xu, P., Zhang, G., Zhang, W., Hu, L., & Wang, P. (2018). Characterization of visible-light photo-Fenton reactions using Fe-doped ZnS (Fe x-ZnS) mesoporous microspheres. *Physical Chemistry Chemical Physics*, *20*(27), 18601–18609.

Wellia, D. V., Kusumawati, Y., Diguna, L. J., & Amal, M. I. (2017). Introduction of nanomaterials for photocatalysis. In *Nanocomposites for Visible Light-induced Photocatalysis* (pp. 1–17). Springer, Cham.

Wu, K. R., & Hung, C. H. (2009). Characterization of N, C-codoped TiO2 films prepared by reactive DC magnetron sputtering. *Applied Surface Science*, *256*(5), 1595–1603.

Yang, X., Cheng, X., Elzatahry, A. A., Chen, J., Alghamdi, A., & Deng, Y. (2019). Recyclable Fenton-like catalyst based on zeolite Y supported ultrafine, highly-dispersed Fe2O3 nanoparticles for removal of organics under mild conditions. *Chinese Chemical Letters*, *30*(2), 324–330.

Yao, Y., Qin, J., Cai, Y., Wei, F., Lu, F., & Wang, S. (2014). Facile synthesis of magnetic ZnFe$_2$O$_4$-reduced graphene oxide hybrid and its photo-Fenton-like behavior under visible iradiation. *Environmental Science and Pollution Research*, *21*(12), 7296–7306.

Zhang, S., & Song, L. (2009). Preparation of visible-light-active carbon and nitrogen codoped titanium dioxide photocatalysts with the assistance of aniline. *Catalysis Communications*, *10*(13), 1725–1729.

22 Microbial Fuel Cells for Simultaneous Wastewater Treatment and Energy Generation
A Comprehensive Review

Sudheer Kumar Shukla and Shatha Al Mandhari

CONTENTS

DOI: 10.1201/9781003204442-22

22.1 INTRODUCTION

Wastewater management is a challenge faced by modern society. Effective wastewater treatment is essential to comply with drainage regulations and make water resources available to society. Treated wastewater can be utilized for commercial and domestic activities (Sonawane *et al.* 2017). Wastewater disposal makes up around 3–4% of the electricity demand in the United States, approximately 110 TWh/year, or equivalent to 9.6 million households' annual electricity consumption (McCarty *et al.* 2011). Wastewater treatment is an energy-intensive activity. However, energy can be extracted from wastewater using modern wastewater treatment technologies, unlike current energy-exhaustive processes. Microbial fuel cell (MFC), as an emerging technology, is applicable in wastewater treatment because of its unique capability of converting the chemical energy of organic waste. It can convert low-strength wastewaters and lignocellulosic biomass into electricity, hydrogen, or chemical products. Activities in such electrical and chemical cells are either anodic or cathodic (Akshay 2016). The electrons produced during oxidation are moved to the electrode immediately. The electron flux is passed to the cathode. The balance of charge in the body is regulated by the ion's movement within the cell, usually via the ion exchange membrane. Microbial fuel cells may also be incorporated into the existing treatment of wastewater systems to improve water quality and nutrient recovery (Akshay 2016). One of the microbial fuel cells' uses, apart from wastewater treatment, is power production (Altenergymag 2010).

22.2 PRINCIPLES OF MICROBIAL FUEL CELLS

Electrochemical reactions are exergonic, i.e., the reaction has negative free reaction energy (Gibb's free energy), and it proceeds spontaneously with the release of energy (electric or electron release). Standard free energy can easily be converted into a standard cell voltage [or electromotive force (emf)] at $\Delta E°$, as shown in Equation (22.1).

$$\Delta E° - \left[\pounds v_i \Delta G°_{i, \text{products}} - \pounds v_i \Delta G°_{i, \text{reactants}} \right] / n_F = -\frac{\Delta G}{nF} \qquad (22.1)$$

Those quantities reflect the free energies of the production of the corresponding compounds and reactants (J/mol), n (moles) reflect the stoichiometry variables of the redox reaction, and F Faraday's constant (96,485.3 C/mol). The Gibbs free energy of the reaction measures the maximum amount of useful work obtained from the thermodynamic system response. The actual cell voltage or emf of the total reaction (the disparity between the anode and the cathode potential) defines whether the device is capable of producing electricity (see Equation (22.2)).

$$\Delta Ecell° = \Delta E_{\text{cathode}} - \Delta E_{\text{anode}} \qquad (22.2)$$

As seen in Equation (22.3), negative free-reaction energy corresponds to a higher average cell voltage. This distinguishes a galvanic cell from an electrolysis cell since the latter is associated with positive free reaction energy and therefore with a negative cell voltage, which requires electrical energy input. Generic cell voltage may also be extracted from the respective redox couples' biological standard redox potentials, as seen below (Harnisch and Schroder 2010).

Within the MFC, the free energy of the Gibbs reaction is negative. The emf is thus optimistic, suggesting capacity for the spontaneous production of electricity from the reaction. For example, when acetate is used as an organic substrate [$(CH_3COO^-) = (HCO_3_-) = 10$ mM, pH 7, 298.15 K, $pO_2 = 0.2$ bar] for the loss of oxygen, a simultaneous redox reaction should have resulted (Rozendal *et al.* 2008), as shown in Equations (22.3–22.5):

$$\text{Anode: } CH3COO^- + 4H_2O \; 2HCO3^- + 9H^+ + 8e^- \; (E = -0.289 \text{ V vs. SHE}) \qquad (22.3)$$

$$\text{Cathode: } 2O_2 + 8H^+ + 8e^- \; 4H_2O \; (E = 0.805 \text{ V vs. SHE}) \tag{22.4}$$

$$\text{Total: } CH3COO^- + 2O_2 \; 2HCO_3^- + H^+ (\Delta G = -847.60 \text{ kJ/mol; emf} = 1.094 \text{ V}) \tag{22.5}$$

22.2.1 OXIDATION-REDUCTION REACTIONS (ORR) IN MFCS

Wastewater contaminants such as biological compounds and other minerals and metals may generate safe and direct energy. Electricity generation in MFCs is the product of oxidation-reduction reactions in electron release, transition, and acceptance via biochemical or electrochemical reactions in anode and cathode chambers. One acts as an electron donor, while the other essentially acts as an electron acceptor. Chemical substances responsible for the acceptance of electrons are called terminal electron acceptors (TEAs).

The following oxidation-reduction reactions (Equations (22.6)–(22.18)) describe potential bioelectrochemical reactions in microbial fuel cells producing electricity using wastewater as a substrate (electron donor) and other contaminants such as nitrates, phosphates, and others as electron receptors.

22.2.1.1 Oxidation Reaction (anode)

Glucose:

$$C_6H_{12}O_6 + 12H_2O \; 6HCO_3 \; + 30H^+ + 24e^- E^\circ = -0.429 \text{ V vs. SHE} \tag{22.6}$$

Glycerol:

$$C_3H_8O_3 + 6H_2O \rightarrow 3HCO^-_3 + 17H^+ + 14e^- E^\circ = -0.289 \text{ V vs. SHE} \tag{22.7}$$

Malate:

$$C_4H_5O_5^- + 7H_2O \rightarrow 4H_2CO_3 + 11H^+ + 12e^- E^\circ = -0.289 \text{ V vs. SHE} \tag{22.8}$$

Sulfur:

$$HS^- \rightarrow S^0 + H^+ + 2e^- E^\circ = -0.230 \text{ V vs. SHE} \tag{22.9}$$

22.2.1.2 Reduction Reaction (cathode)

$$O_2 + 4H^+ + 4e^- \rightarrow 2H_2O \; E^\circ = +1.230 \text{ V vs. SHE} \tag{22.10}$$

$$O_2 + 2H^+ + 2e^- \rightarrow H_2O_2 \; E^\circ = +0.269 \text{ V vs. SHE} \tag{22.11}$$

$$NO_3^- + 2e \; + 2H^+ \rightarrow NO_2^- + H_2O \; E^\circ = +0.433 \text{ V vs. SHE} \tag{22.12}$$

$$NO_2^- + e^- + 2H^+ \rightarrow NO + H_2O \; E^\circ = +0.350 \text{ V vs. SHE} \tag{22.13}$$

$$NO + e^- + H+ \rightarrow 1/2N_2O + 1/2H_2O \; E^\circ = +1.175 \text{ V vs. SHE} \tag{22.14}$$

$$1/2N_2O + e^- + H^+ \rightarrow 1/2N_2 + 1/2H_2O \; E^\circ = +1.355 \text{ V vs. SHE} \tag{22.15}$$

$$2NO_3^- + 12H^+ + 10 \; e^- \rightarrow N_2 + 6H_2O E^\circ = +0.734 \text{ V vs. SHE} \tag{22.16}$$

$$Fe^{3+} + e^- + H^+ \rightarrow Fe^{2+} + 1/2H_2O \ E^\circ = +0.773 \text{ V vs. SHE} \tag{22.17}$$

$$MnO_2 + 4H^+ + 3e^- \rightarrow Mn^{2+} + 2H_2O \ E^\circ = +0.602 \text{ V vs. SHE} \tag{22.18}$$

22.3 CONFIGURATION OF MICROBIAL FUEL CELLS

MFCs are a form of novel bioelectrochemical systems (BESs) that can directly transform chemical energy into electrical energy utilizing microbial biocatalysts and are among the primary renewable energy generation systems. Figure 22.1 displays an MFC unit's schematic layout attached to a data logger that records a multimeter signal to be transmitted to a personal computer via a data acquisition network and fitted with a cathode aeration air pump. Usually, the MFC consists of anodic and cathodic chambers divided by a proton exchange membrane (PEM) (Du *et al.* 2007). The microbial fuel cell, composed of plexiglass, glass, or polycarbonate, comprises carbon-cloth electrodes, graphite, carbon fiber, graphite felt, platinum (Pt), black platinum, or reticulated vitreous carbon (RVC) (Flimban *et al.* 2019). Such electrodes are linked by an external circuit (Kumar *et al.* 2016). The anode is packed with wastewater where electrochemically-active bacteria, produced from anaerobic sludge, absorb organic substrates and emit protons and electrons. Electrons are passed into the outer circuit when protons are moved to the cathode chamber to ensure electroneutrality (Flimban *et al.* 2019; Palanisamy *et al.* 2019).

Such parallel waves of electrons and protons from the anode to the cathode are accompanied by interactions with electron receptors such as oxygen, ferric chloride, potassium dichromate, etc., for the processing of water (Figure 22.1) (Kumar *et al.* 2017). However, for specific experiments, membranes with a greater capacity to carry only protons are of considerable value. The electronic data logger tracks the passage of electrons through the circuit for one microsecond. The data logger is attached to a personal computer to make data saving simpler and to view polarization curves with PD, CD, and voltage values. However, the use of low-cost and safe anode and cathode separator products to make MFCs have drawn considerable interest and attracted much publicity. The different parts of an MFC are discussed below.

22.3.1 ELECTRODES

Features of the electrode are widely known to be essential factors for enhancing the efficiency of MFCs. Electrodes are the place where bacterial populations grow and electrochemical reactions

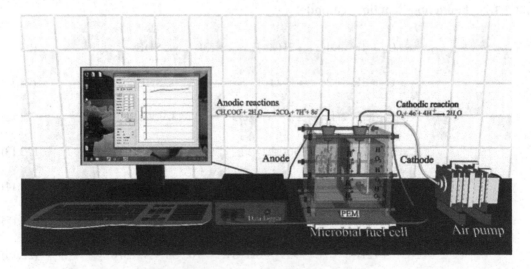

FIGURE 22.1 Diagram of the MFC system. Used with permission from Shabani et al. 2020.

take place. The electrode's design influences the efficiency of MFCs in electrical and mechanical areas such as internal resistance, inoculum size, and fluid dynamics. In particular, the electrode must be compliant with strong mechanical power, biocompatibility, high conductivity, low resistance, and low-cost MFCs. The electrode's morphology affects the redox reaction rate and biocompatibility (Tang *et al.* 2020). The electrode system configuration can be classified as macroscopic and microscopic (Tang *et al.* 2020).

22.3.1.1 Anodes Used in Microbial Fuel Cells

New materials for MFCs are being developed to increase their economic viability and efficiency. In this respect, a broad range of products and designs have been examined for the anode, such as carbonaceous, metal and metal oxides, and composite materials (Gnana Kumar *et al.* 2013). Carbonaceous materials demonstrate excellent biocompatibility, chemical stability, and conductivity; they are also relatively inexpensive and have also been the most-commonly used materials for MFC anodes. There are three different kinds of anode arrangement: air, packed, and brushed (Wei *et al.* 2011).

22.3.1.2 Cathodes Used in Microbial Fuel Cells

Air cathodes and aqueous cathodes are the most popular structures used by MFCs. The air cathodes are in direct contact with oxygen (exposed to the atmosphere). This cathode form is the most utilized because of its simple nature, which does not involve aeration and provides a high density of electricity (Hernández-Fernández *et al.* 2015). It usually consists of three layers: a diffusion layer exposed to the atmosphere, a conductive-supporting substance that often serves as a diffusion layer, and a mixture of catalyst and binder in contact with moisture. Nevertheless, in the case of an aqueous cathode, the electrode frequently consists of conductive-supporting material, such as carbon fabric or platinum wire, covered with a binder/catalyst film. It is submerged in an aqueous environment with minimal oxygen content (Logan *et al.* 2005). The most popular component for air cathodes is carbon cloth (Gude 2016), and the binder being used, which attaches the catalyst to the electrode, is often perfluorosulfonic acid (Nafion) or polytetrafluoroethylene (PTFE).

22.3.2 Membranes

Anodes and cathodes in MFCs are separated by a membrane that allows the transfer of protons from anode to cathode. The membrane, which provides a contactor or physical divider, is placed between the anode and cathode dual chambers (Venkata Mohan *et al.* 2014). Another essential function of the membrane is to avoid transferring oxygen from the aerated cathode to the anaerobic anode chamber (Sadhasivam *et al.* 2017). In certain instances, the dividers or membranes used in MFCs contain polymers (Bakonyi *et al.* 2018; Koók *et al.* 2019). Certain features of a membrane are essential such as low internal resistance, mass transfer between cathodic and anodic chambers, fast proton conductivity, and high ion conductivity (Daud *et al.* 2015; Yang *et al.* 2019). Ion exchange membranes are an essential class of polymeric membranes typically containing workable sets in their form that enable the exchange of cations and anions. Membranes selectively enable the passage of certain opposite charge ions (counter ions) while blocking related charge ions (co-ions) (Luo *et al.* 2018). Ion exchange membranes are classified into five groups: cation exchange membranes (CEMs), anion exchange membranes (also recognized as monopolar membranes) (Daud *et al.* 2015), mosaic ion exchange membranes, bipolar membranes (BPMs), and amphoteric ion exchange membranes (Strathmann 2004). CEMs, often referred to as PEMs, are built to facilitate the passage of protons from other cations via the membrane to the cathode chamber. Hence, the membrane's charge is a negative networkable group (Harnisch and Schröder 2010) (Figure 22.2). Thus, positive charges, such as phosphate or carbonate, connected to the AEM membranes can facilitate proton transfer by adding proton carriers (pH buffers) (Li *et al.* 2011).

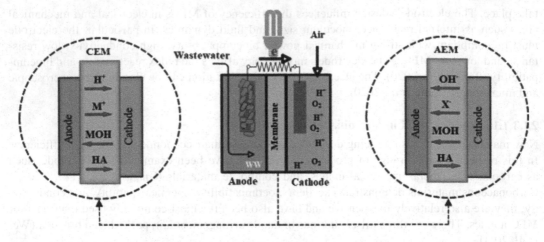

FIGURE 22.2   Schematic design of a CEM and an AEM. Used with permission from Shabani et al. 2020.

22.3.3   SUBSTRATES

MFCs can use numerous biodegradable organic compounds derived from the waste streams produced by farms, livestock, domestic (municipal wastewater), poultry, industrial and landfill leachate, and many others. Earlier research examined the efficiency of MFCs with artificial (synthetic) wastewater to understand the viability of the working theory and pathways for enhancing energy recovery and organic removal efficiencies. Current experiments have concentrated on using natural wastewater from the multiple waste streams mentioned above to assess the functional viability of MFCs since the structure of real wastewater is somewhat different from that of conventional wastewater.

22.3.3.1   MFCs with Synthetic Wastewater as Substrates

MFCs have demonstrated increased levels of carbon sequestration (>90%) from wastewater. Synthetic wastewater utilized in MFCs contains acetate, creatine, sucrose, and xylose, and several other organic substrates for microbial degradation in the anode chamber (Pant *et al.* 2010). Acetate is the primary and most widely utilized substrate in MFCs as a carbon source for exoelectrogenic bacteria. Acetate is easily biodegradable by these bacteria because it provides a simple metabolism. Acetate is also the outcome of many metabolic pathways for higher-grade carbon sources (Biffinger *et al.* 2008), such as in the anaerobic digestion of wastewater sludge, the carbonaceous matter is turned into smaller chain organic acids like acetic acid. MFCs with acetate substrates generated a higher energy density compared to glucose substrates and domestic wastewater.

Liu *et al.* (2005b) recorded that the power created with acetate (506 mW/m², 800 mg/l) was up to 66% higher than that of butyrate (305 mW/m², 1,000 mg/l). Chae *et al.* (2009) recently contrasted four specific substrates' performance in terms of coulombic efficiency (CE) and power production. Acetate-fed MFCs achieved the largest CE (72.3%), then butyrate (43.0%), propionate (36.0%), and glucose (15.0%), respectively. Once acetate was applied to protein-rich wastewater as a substrate in the MFCs, MFCs based on acetate-induced consortia obtained more than two-fold average electrical strength and one-half maximal external load resistance relative to MFCs based on protein-rich wastewater consortia (Liu *et al.* 2009). Glucose is another widely studied substrate in MFC science. A total power density of 216 W/m³ was generated from glucose fed by batch MFC using 100 mM of ferric cyanide as a cathode oxidant (Rabaey *et al.* 2003). The findings of a variety of empirical experiments indicate that the power levels for dynamic wastewater are factor 5 or lower relative to separate substrates. Although substrate removal levels for real wastewater vary from 0.5–2.99 kg COD/m³ reactor/day, the removal levels for artificial wastewater can be as high as 8.9 kg COD per m³ reactor/day, demonstrating the capacity for high-strength substrates (Aelterman *et al.* 2006).

22.3.3.2 MFCs with Actual Wastewater as Substrates

Municipal wastewaters have reduced BOD concentrations of <300 mg/l and are classified as low-energy carriers or feedstocks for microbial fuel cells. MFCs can also treat high-strength wastewater, which means high energetic density with BOD levels ranging from 2,000 mg/l due to anaerobic situations in the anodic chamber. These high-powered wastewater sources are created by the food processing industry, brewing plants, dairy farms and animal feed, and other industrial waste streams. Feed wastewater is abundant in readily-biodegradable carbohydrates and organic acids with reasonably small amounts of organic nitrogen (proteins) (Fuchs *et al.* 2003; Lepisto and Rintala 1997). MFCs may generate energy in the scope of 2–260 kWh/ton of the product from the wastewater utilized to process food products, depending on the BOD and the volume of water used in the process. A total of 46 MW of energy can potentially be generated from limited BOD wastewater from dairy farms in the U.S. At the same time, a maximum of 1,960 volts of power can be created from high BOD wastewater from the dairy industry (Borole and Hamilton 2010). The livestock industry's animal wastewater is also exceptionally high in organic substance content (approximately 100,000 mg/l COD for animal waste). It may include large amounts of nitrogen-containing materials, such as proteins, and harder-to-degrade organic materials, such as cellulose (Hoffmann *et al.* 2008).

22.4 TYPES OF MICROBIAL FUEL CELLS

To improve the performance of microbial fuel cells, scientists have carried out a variety of studies to achieve a microbial fuel cell that is efficient in the production of electric power and the treatment of wastewater. Various types of microbial fuel cells are discussed below.

22.4.1 DOUBLE-CHAMBER MFC

The double-chamber or dual-chamber microbial fuel cell (DCMFC) is the most commonly researched and utilized type of MFC, as seen in Figure 22.3(A) (Palanisamy *et al.* 2019).

The two-chamber configuration consists of one anodic and one cathodic chamber in which oxidation and degradation of each of the reactions occurs, divided by a membrane (Jafary *et al.* 2018). Inserting a specific proton membrane between two chambers decreases the diffusion of oxygen to the anode and, as a consequence, improves columbic efficiency (CE). Nevertheless, the utilization of a membrane induces the accumulation of protons in the anodized space, which decreases the pH of the anolyte and eventually deteriorates the efficiency of the microorganisms and the production of electricity (Yang *et al.* 2016). In addition, the membrane imposes high internal resistance and expense (62.5% of the capital cost) on MFCs, allowing exterior input energy for cathodic aeration (Li *et al.* 2018; Palanisamy *et al.* 2019).

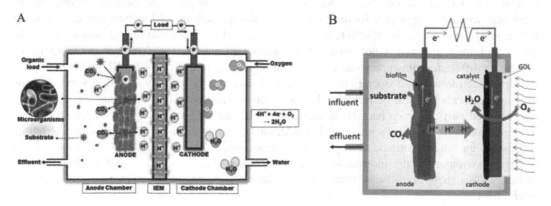

FIGURE 22.3 A) Dual-chamber microbial fuel cell (Palanisamy et al. 2019), and B) air cathode single-chamber membraneless microbial fuel cell (Massaglia et al. 2019). Used with permission.

22.4.2 SINGLE-CHAMBER MFC

In order to overcome the difficulties faced due to the application of membranes in DCMFCs, the air cathode single-chamber microbial fuel cell (SCMFC) was introduced, as seen in Figure 22.3(B) (Massaglia *et al.* 2019). Throughout the SCMFC, the membrane is extracted. The air cathode is equipped with a gas diffusion layer (GDL) to regulate oxygen diffusion from the air to the cathode, reducing the additional aeration requirement. Lower capital costs, simplicity, and fast power production have rendered the SCMFC a viable concept for large-scale deployment (Liu *et al.* 2015). Logan *et al.* recorded an 80% improvement in power production (28 W/m2 in a membrane-dependent MFC vs. 146 W/m^2 in a membraneless MFC) in an SCMFC working on domestic wastewater (Liu and Logan 2004). For SCMFCs, the cathode interacts closely with the liquid on one side and the electrolyte on the other side (Vogl *et al.* 2016). The elimination of the SCMFC architecture membrane significantly decreases its capital expense and the internal resistance (Vicari *et al.* 2016). However, it improves the diffusion of organic matter to the cathode layer, which ultimately contributes to aerobic biofilm creation. This biofilm formation obstructs the active catalytic layer sites coated on the air cathode and, as a result, degrades the efficiency of MFCs over long-term activity (Chen *et al.* 2019). Over the last decade, many review studies have been conducted dealing with different aspects of MFC architecture, the development of modern electrode materials, and the enhancement of a device output for high power recovery and the remediation of emissions. However, none of the previous research analyzed the air cathode biofouling pathways in SCMFC or outlined the biofouling mitigation strategies.

22.5 MICROBIOLOGY OF MFC

MFCs are systems built to transform waste material into electric power utilizing microorganisms. Microorganisms, such as bacteria, can produce energy using organic material and recyclable substrates like wastewater. They can also aid in the biodegradation/treatment of biodegradable goods, such as wastewater.

Many MFCs utilize wastewater as a source of electricity. The selection of electrogenic populations happens in open sediment-based environments. Consortia in the anodic chamber have similar roles as methanogens in the anaerobic process, except that the microorganisms can move electrons to the electrode substitute methanogens. This microbiocenosis is called androphilic (anodophilic consortia) (Konovalova *et al.* 2018). G-Proteobacteria is found in sediment systems from 50–90% of anode microorganisms. To lesser degrees, Cytophagal (to 33%), Firmicutes (11.6%), and J-Proteobacteria (9–10%) (Konovalova *et al.* 2018). This means that the quality of the community of accessible structures is reliant on the form of sediment. For example, freshwater microorganisms of Geobacteraceae are responsible for the creation of freshwater sediments. On the other hand, a strong presence of Geobacteraceae (G-Proteobacteria) in acidophilous populations of various substrate forms (bottom sediments) is not unexpected. Geobacteraceae are the prevalent microorganisms present in different sedimentary environments in which the reduction of Fe(III) oxide is the key terminal electron-acceptor operation. It has been shown that the composition of microorganisms in open platforms can also be influenced by a variety of mediator substances capable of transferring charged particles from bacteria to electrodes. The utilization of exogenous electron carriers in open systems is not always justified. The addition of certain extracellular quinones, like humic substances and their synthetic analogs [e.g., anthraquinone 2,6-disulfonate (AQDS)] recognized in the improvement of the movement of electrons among Geobacteraceae and insoluble Fe(III) oxides, causes changes in the structure of microbial anode-associated communities (Konovalova *et al.* 2018).

22.6 OPERATION OF MICROBIAL FUEL CELLS

First, inoculate the reactor with the bacteria's origins and the medium, then close the anode chamber to avoid oxygen leakage. Put the resistor as the "load" for the MFC power source in the electrodes' connection. It will not be feasible to utilize a particular unit, such as a light bulb or a tiny generator, unless the researcher constructs a massive reactor. Moreover, the voltage would be weak, which would not make it practical to light up a tiny LED form of light (which requires little power) unless a few MFCs are hooked in a sequence to pump up the voltage. Start utilizing a resistor in the range of 500–1,000 ohm to get the machine running since it appears to fit well in the experiments. Low resistance (1–10 ohm) may seem to make sense intuitively since it would encourage more current to pass, but the researchers note that this does not appear to be the case. If the device is in operation, the maximal strength can be calculated by generating a polarization curve using a set of resistors (Logan *et al.* 2006). A basic voltmeter may be used to test the voltage decrease around the resistor, enabling the researcher to determine the current ($I = V/R$) and then the strength ($P = IV$). A multimeter is utilized to calculate the voltage at fixed periods (20–30 minutes) from 20–40 separate units, but these meters are costly and need a data storage device. On the other hand, if the researcher has access to a device with some sort of data collection capacity, this could possibly be utilized for the MFC. The data acquisition machine attached to a computer in a laboratory typically only calculates the signal in terms of voltage (analog input), typically from 0–1 V or 10 V. A basic analog-to-digital conversion board may be acquired from various electronic supply stores (to store the value as an amount), enabling the construction of a data acquisition device as part of the experimental equipment.

The cathode chamber of the MFC was loaded with ferricyanide [50 mM $K_3Fe(CN)_6$ in a phosphate buffer (50 mM K_2HPO_4)] as a catholyte mediator reported (Venkata Mohan *et al.* 2008). The catholyte solution was continuously stirred (50 rpm) using magnetic beads to maintain successful interaction with the proton, electron, and mediator. In contrast, the MFC cathode chamber was loaded with a phosphate buffer (50 mM K_2HPO_4) as a catholyte. The catholyte was actively sparged with air by using an air pump connected to the sparger network. Under these situations, the dissolved oxygen (DO) concentration in the MFC_{AC} cathode chamber was between 4 and 5 mg/l. The pH of both catholytes was preserved at 7.5. Both fuel cells' anode chambers are identical to the anaerobic suspended touch bioreactor system, which is usually used for wastewater treatment. The anodic chamber before start-up was injected with specifically enriched H_2 containing mixed microflora (2 g VSS/l) employing synthetic feed (0.7 l) (Venkata Mohan *et al.* 2008). The mixture of anolytes was continually recirculated at 100 ml/min, utilizing a peristaltic pump to minimize the substrate's gradient and maintain successful interaction with the substrate and the mixed consortia. The pH of anolyte was maintained at 6.0 ± 0.1 to ensure the viability of the acidogenic bacteria (AB) while at the same time suppressing the development of MB. Both fuel cells' efficiency was evaluated under a constant substrate load of 4.316 g COD/l and performed in batch mode at room (mesophilic) temperature (29 ± 2°C) and atmospheric pressure. The constant COD elimination and steady voltage production were deemed measures of healthy operating conditions before the feed transition. Fresh synthetic feed was filled until the voltage began to decrease. Before adjusting the feed, the inoculum was permitted to settle (30 minutes), and the drained feed (650 ml) was supplemented with fresh feed under anaerobic conditions. A set biomass (50 ml by volume) was utilized for subsequent experiments. The anode chamber was stored with oxygen-free N_2 gas for 4 minutes to preserve the anaerobic microenvironment after each feed event (Venkata Mohan *et al.* 2008). Nutrition, decanting, and recirculation processes were carried out using electronic timer-operated peristaltic pumps. Furthermore, there are two modes of operation for a microbial fuel cell: batch mode and continuous mode.

22.7 APPLICATION

The MFC is an outstanding technology with unique characteristics of energy, environmental, and economic productivity. It is a platform technology that is diverse.

22.7.1 ELECTRICITY GENERATION

Chemical energy (organic waste matter) is converted to electrical power with only a limited energy loss compared to other wastewater treatment technologies, such as anaerobic digestion. In the conventional treatment operation, energy-containing organic compounds are extracted from wastewater. In the MFC, electricity is produced by substrate oxidation in an anode chamber, which results in the production and transport of protons and electrons (He *et al.* 2005). Simultaneously, the electron is transferred from the anode to the cathode via an external circuit, and the proton is relocated through a polymer electrolyte membrane (Rabaey and Verstraete 2005). Eventually, electrons and protons merge with oxygen to create water molecules in the cathode chamber (Sharma and Li 2010). In the MFC, microorganisms' catalytic operations result in the transformation of chemical forms of energy to electric power (Kim *et al.* 2002; Rahimnejad *et al.* 2011). This energy is captured immediately by the MFC in the form of electricity. For instance, effective reductions in chemical oxygen demand (COD) (79 ± 7%), total nitrogen removal (71 ± 8%), and electricity production (0.47 Wm^{-3}) were achieved with a low pilot operating energy (250 L horizontal MFC stackable) compared to conventional treatments (Feng *et al.* 2014). MFCs can be termed as energy-saving technologies because they reduce the energy utilized for aeration and produce less sludge than most other technologies (Oh *et al.* 2010; He 2013). Moreover, the MFC's environmental and economic advantages also include the elimination of pollutants, the recovery of nutrients, and the production of electrical energy from wastewater. MFCs are produced and grouped into various types to achieve greater efficiency, depending on alignment (membrane and electrode assembly), electrolyte nature, etc. The main types are single-chambered MFCs (SCMFCs), double-chambered MFCs (DCMFCs), stacked MFCs, and upstream MFCs (Ou *et al.* 2016; Wu *et al.* 2016).

22.7.2 REMOVAL OF ORGANIC MATTER

The largest number of carbon removal (>90%) from wastewater was noted in MFCs. As researchers explained, MFC technology is utilized to treat wastewater and produce electricity. The oxidation of organic material in wastewater by exoelectrogens generates electrons and protons in the anodic compartment. The membrane and the electrodes convert the electrons and protons to the cathode chamber. These procedures lead to the generation of clean water and energy. Wastewater from food (including fermented apple juice), vegetable oil, refineries and distilleries, seafood, dairy, cassava, swine, slaughterhouse, surgical cotton, livestock, and the petrochemical industry have been utilized for the creation of power. Dairy wastewater may also be utilized as an MFC substrate because it includes biodegradable organic compounds and nutrients (Mansoorian *et al.* 2016; Faria *et al.* 2017). Lu *et al.* (2017) produced a 20-L MFC system for the treatment of brewery effluent. The optimum COD removal efficiency was 94.6 ± 1%, which resulted in a flow rate of 1 ml/min and a hydraulic retention time of 313 hr. Firdous *et al.* (2018) noted that the performance of MFCs was increased by increasing temperature and time. A high COD removal (80–90%) with optimum voltage (5839 mV) was achieved in vegetable oil sewage treated in a double-chamber MFC at 35°C. Kook *et al.* (2016) used a fluid portion of municipal solid waste pressed in double-chambered MFCs to generate waste electrical energy. As a result, the maximum power (8–9 J^{-1} COD d^{-1}) was generated at the lowest possible COD concentrations. The potential applications of MFCs in treated wastewater and bioenergy manufacturing depend on a range of factors, like substrate choice (wastewater category), the design of the reactor, and physiological and biocatalysts parameters. Therefore, it is essential to maintain a suitable environment for operating the treatment of wastewater in MFCs to achieve enhanced results.

22.7.3 Removal of Nutrients

Wastewater exiting in the anodic chamber is rich in nitrogen and phosphorus compounds. Even so, these nutrient compounds can be eliminated in MFCs, particularly in biocathode chambers, to improve the quality of effluent water or can be returned as ammonia or magnesium ammonium phosphate ($MgNH_4PO_4.6H_2O$), recognized as struvite. Struvite is a magnesium ammonium phosphate (MAP) crystal with equivalent molar concentrations of Mg, ammonium (NH_4), and P combined with six water molecules. The purity of the struvite mineral varies depending on the elementary composition of Mg: N: P 1:1:1 (Le Corre et al. 2009; Etter 2011).

22.7.3.1 Nitrogen Removal in MFCs

Conventional biological nitrogen removal through well-known nitrification–denitrification reactions is energy intensive, carbon intensive, and cost intensive. This process includes the aerobic oxidation of nitrite ammonia (NO_2^-) and nitrate (NO_3^-) and denitrification, which reduces nitrate to nitrogen gas (Ahn 2006). The latter necessitates an electron donor, usually organic compounds, to allow the bacteria to gain power from the nitrate/nitrite reduction process. Nitrate may share electrons from organic materials decreased to nitrogen gas (e.g., conventional denitrification processes). Even an electron transfer method makes it possible to utilize nitrate as a terminal electron acceptor in a BES. Nitrate reduction can create a positive electrical potential of 0.98 V when utilizing organic compounds (e.g., acetate) as a source of the electron (Madigan et al. 2010). In the case of an MFC, the electricity needed to provide a sufficient reduction in energy for denitrification can be reduced significantly if the bacteria immediately utilize the cathodic electrode as an electron donor.

The major factors impacting the nitrogen extraction efficiency of MFCs are the concentration of dissolved oxygen (DO), the pH, the ratio of carbon to nitrogen (C/N), and the electricity generation by the anode process. Large amounts of DO inhibit denitrification as well as high pH situations. Neutral pH situations are appropriate for nitrification–denitrification procedures in MFCs. As the number of electrons released by the anode process is affected by the cathode procedure, the C/N ratio is an essential parameter for procedure quality (Kelly and He 2014). A high C/N ratio appears to be better for denitrification. However, higher C/N ratios are not entirely advantageous due to heterotrophic denitrification impacts on the performance of MFCs. The recovery of ammonia in a cathode with high pH and aeration, and the subsequent absorption by an acid solution are also achievable (Kuntke et al. 2012).

22.7.3.2 Phosphorous Removal in MFCs

Phosphorous is generally extracted from wastewater as a struvite. Controlled wastewater struvite recovery can be achieved through some chemical additions or carbon dioxide stripping, or electrolysis strategies. Most struvite recovery researchers have aimed at increasing the pH solution through the addition of a chemical base, like NaOH, Mg (OH)$_2$, and Ca(OH)$_2$ or carbon dioxide stripping via aeration (Cusick and Logan 2012). Some researchers analyzed the disposal or recovery of phosphorous from wastewater in microbial fuel cells. In earlier findings, digested sewage sludge had been used to release orthophosphate from iron phosphate. This was followed by adding magnesium and ammonium and pH adjustments, resulting in struvite formation (Fischer et al. 2011). Orthophosphate was recovered in yields of 48% and 82% of pure ferric phosphate hydrate and digested sludge, respectively. Phosphorous removal from MFCs appears to be incidental in recent studies. Large amounts of precipitate containing highly concentrated phosphorous on the surface of the liquid side of the air cathode were utilized in pig-water-fed MFCs (Ichihashi and Hirooka 2012). Since precipitate also contained a high amount of magnesium (Mg), it was assumed that phosphate was removed as a struvite because the pH near the cathode was more significant than that of the anodic side. 70–82% of phosphorus was extracted, and struvite precipitation occurred only on the cathode's surface in this study. Although the electrolyte pH was not very high (~8), it was assumed that the reduction of oxygen to the cathode increased the localized pH, which facilitated the formation of struvite.

22.7.4 HYDROGEN PRODUCTION

Biohydrogen production is another attractive feature of MFC technology (Varanasi *et al.* 2015). Exoelectrogenic bacteria oxidize the material (organic matter) in the MEC, which leads to the generation of electrons and protons (Kadier *et al.* 2016). Electrons and protons are responded to in the cathode to create hydrogen gas (H_2) in the presence of a catalyst (Kadier *et al.* 2016). Electrons are shifted from anode to cathode via an electrical circuit, and the hydrogen ions (proton) pass through the membrane to the cathode.

In the cathode segment, the electrons in the cathode are combined with the protons to create hydrogen gas. This hydrogen production process occurs by applying external power output, and the hydrogen produced can be used to meet the demand for hydrogen (Liu *et al.* 2005a). Equations (22.19) and (22.20) illustrate the generation of hydrogen in the MEC:

In the anode chamber:

$$C_2H_4O_2 + 2H2O \rightarrow 2CO_2 + 8e^- + 8H^+ \qquad (22.19)$$

In the cathode chamber:

$$8H^+ + 8e^- \rightarrow 4H_2 \qquad (22.20)$$

For instance, an integrated approach involving an acid-rich effluent and mixed vegetable waste in the integrated MFC (Mohanakrishna *et al.* 2010) was used to produce biohydrogen and to harvest the electrical current produced. Also, OPTIMASH BG® (enzyme) and Tween-80 (surfactant) pretreated palm oil effluents have been used to enhance biohydrogen production along with electricity generation (Leano and Babel 2012).

22.8 CHALLENGES

Several obstacles to MFCs, including high operating costs and low power output, must be considered before the commercial success of MFC technology, whether as a single procedure or in combination with other techniques. The initial cost of the MFC is, on average, 30 times greater than that of the standard activated-sludge treatment system for domestic sewage, mainly due to its layout and treatment capability. Generally, the high initial cost is mainly due to utilizing costly materials of the electrode, like current collectors, catalysts, and separators (He *et al.* 2017). Also, bacteria can release electrons to the anode and protons into the general MFC solution, resulting in a negative anode potential (around −0.2 V). Oxygen in the air is considered to be the most promising cathode oxidant, with a theoretical maximum potential of 0.805 V. However, the estimated maximum acquired in MFCs is +0.3 V even with Pt-catalyzed cathodes. The current energy density produced from wastewater is often less than 0.5 W/m² under optimum conditions. The temperature is almost 30°C; the solution's conductive ability is approximately 20 mS/cm, and the well-buffered solutions are neutral or slightly alkaline. The low power density is fundamentally due to high internal resistance, solution situations, low substrate degradation, and biofilm kinetics (He *et al.* 2017). The usage of stacked MFCs in series or parallel would be crucially important for a safe increase in voltage and current. Parallel stacked MFCs can primarily raise currents and power densities, and parallel coulombic efficiency seems much higher than in sequence. However, the low voltage would still limit its implementation. Stacked MFCs in the direct sequence can increase the voltages that can hardly be performed by a chemical fuel cell due to the external circuit impacts on the microbial consortium. The study should pay much attention to the matter and occurrence of the reversing of voltage and improve efficiency in direct series of stacked MFCs. The greater voltage can be gained by wiring arrays of MFCs to charge condensers in parallel and then discharging condensers in series, but this technique would certainly raise the power consumption and cost (He *et al.* 2017).

TABLE 22.1

Cost Analysis of MFC at Different Hydraulic Retention Times (Abourached et al. 2016)

Material	Cost ($/m²)	Cost ($/m³ reactor)	Cost ($/MFC)			
HRT (hours)			1.5	2	4	6
Anode	10	1,000	596,200	794,940	1,589,880	2,384,810
Cathode	50	5,000	2,981,000	3,974,700	7,949,400	11,924,050
Separator	1	100	59,620	79,494	158,988	238,481
Others (plates, wires, etc.)		1,000	597,000	795,000	1,590,000	2,385,000
Total capital cost (million $)			4.2	5.6	11.3	16.9
P_0 operational cost (million $)			2.2	2.9	5.7	8.6
Total MFC cost (million $)			**6.4**	**8.5**	**17.0**	**25.5**

In addition, the cost of the electrode (84.5% for both electrodes) and membrane (1.408%) holds a significant part of the overhead costs. Table 22.1 shows the cost analysis of the MFC reactor depending on the number of cathodes, anodes, separators, and other components at variable hydraulic retention times (HRTs). With a rise in HRT time, the cost of MFCs increases significantly (Palanisamy et al. 2019).

On the other hand, the established MFC system has been used for wastewater treatment. Still, there have been several challenges: stability, organic loading, reduced production of real wastewater, and membrane fouling (He et al. 2017). Also, MFC's increase in the wastewater treatment experience has reduced the level of power generation. For example, for the MFC scale-up, 400 W/m³ of power output should be achieved (Clauwaert et al. 2008). Several approaches have been made to achieve higher power output, such as improving electrode performance and separator performance. Carbon structured materials such as graphite rod, graphite felt, carbon cloth (CC), etc., have generally been used as electrode materials for the past few years. Several modifications have been made to overcome difficulties, such as a high surface area for bacterial colonization, mass transfer problems, for the long-term stability and easy maintenance of electrode materials.

Alternatively, the distance between the negative and positive electrodes must be reduced for the efficient transfer of ions between the electrodes that reduce the microbial fuel cell's internal resistance. However, less distance between electrodes may lead to the substrate's permeability to the cathodic and oxygen transfer to the anodic chambers. An appropriate divider/membrane for lesser distance electrodes will be needed to overcome this problem. The following challenges were identified in membranes: the enhancement of internal resistance, the effort to select a membrane with a low oxygen diffusion coefficient, and membrane charge (Li et al. 2018). Therefore, these limitations with the membrane must be corrected to improve the microbial fuel cell's efficiency. In addition, biofouling in the membrane/separator is also considered a fundamental challenge for the long-term operation of MFCs in the treatment of wastewater. As a result, development in membrane materials has been widely researched with some upcoming principles for improved MFC productivity. In addition, some issues have arisen throughout electron transport techniques between microorganisms and electrodes (Kim et al. 2015). Moreover, a deeper understanding of the molecular mechanisms of microbe–electrode relationships, the composition of biofilms, and the synergistic impacts of microorganisms in the biofilm are needed to enhance electricity generation throughout the treatment of wastewater. Investigators are therefore making substantial efforts to develop the performance of MFCs by changing the format of new structures, microorganisms, materials of the electrode, and membrane.

22.9 CONCLUSION

This review covers broad areas of MFCs, including principles, structures, operation, applications, and challenges. The MFC is a potential technology of the future where energy-positive, or at least energy-neutral, wastewater treatment processes have been envisaged. It has various applications, like energy generation, organic matter removal, nutrient removal, and hydrogen production. There are some barriers in the commercial application of the technology, such as high capital and operational cost, which is 30 times greater than that of the standard activated-sludge treatment system for domestic sewage. Generally, the high initial cost is mainly due to the electrode's costly materials, like current collectors, catalysts, and separators. Low power density is another issue. There are several other challenges like stability, organic loading, reduced production of real wastewater, and membrane fouling. Technology scaling for higher electricity generation is also another problem. The development of new electrode materials and better configurations of the system is required. A deeper understanding of the molecular mechanisms of microbe–electrode relationships, the composition of biofilms, and the synergistic impacts of microorganisms in the biofilm are needed to enhance electricity generation throughout the treatment of wastewater. MFC is a futuristic wastewater technology; however, technological and economic bottlenecks need to be overcome before the commercial implementation of the technology.

REFERENCES

Abourached, Carole, Marshall J English, and Hong Liu. 2016. Wastewater treatment by microbial fuel cell (MFC) prior irrigation water reuse. *J. Cleaner Prod.* 137, 144–49. doi:https://doi.org/10.1016/j.jclepro.2016.07.048.

Aelterman, P., K. Rabaey, P. Clauwaert, and W. Verstraete 2006. Microbial fuel cells for wastewater treatment. *Water Sci. Technol.* 54 (8), 9–15.

Ahn, YH. 2006. Sustainable nitrogen elimination biotechnologies: A review. *Process Biochem.* 41 (8), 1709–1721.

Akshay D, Namrata Sain, and W Jabez Osborne. 2016. Microbial fuel cells in bioelectricity production. *Front. Life Sci.* 9 (4), 252–66. doi:10.1080/21553769.2016.1230787.

Altenergymag. 2010. Applications of the microbial fuel cell. https://www.altenergymag.com/article/2009/12/microbial-fuel-cellsprinciples-andapplications/587#:~:text=Another%20potential%20application%20of%20the,organic%20ma tter%2C%20such%20as%20sewage.

Bakonyi, Péter, László Koók, Gopalakrishnan Kumar, Gábor Tóth, Tamás Rózsenberszki, Dinh Duc Nguyen, Soon Woong Chang, Guangyin Zhen, Katalin Bélafi-Bakó, and Nándor Nemestóthy. 2018. Architectural engineering of bioelectrochemical systems from the perspective of polymeric membrane separators: A comprehensive update on recent progress and future prospects. *J. Membr. Sci.* 564, 508–22. doi:https://doi.org/10.1016/j.memsci.2018.07.051.

Biffinger, J.C., J.N. Byrd, B.L. Dudley, and B.R. Ringeisen 2008. Oxygen exposure pro-motes fuel diversity for Shewanella oneidensis microbial fuel cells. *Biosens. Bioelectron.* 23 (6), 820–826.

Borole, A.P., and C.Y. Hamilton 2010. Energy production from food industry wastewa-ters using bioelectrochemical cells. In: *Emerging Environmental Technologies*, vol. II. Springer, Netherlands, pp. 97–113.

Carty, P.L., Bae, J., and J. Kim 2011. 'Domestic Wastewater Treatment as a Net Energy Producer–Can This be Achieved?'. *Environmental Science & Technology. American Chemical Society*, 45 (17), 7100–7106. doi: 10.1021/es2014264.

Chae, K.J., M.J. Choi, J.W. Lee, K.Y. Kim, I.S. Kim 2009. Effect of different substrates on the performance, bacterial diversity, and bacterial viability in microbial fuel cells. *Bioresour. Technol.* 100 (14), 3518–3525.

Chen, Shuiliang, Sunil A Patil, Robert Keith Brown, and Uwe Schröder. 2019. Strategies for optimizing the power output of microbial fuel cells: Transitioning from fundamental studies to practical implementation. *Appl. Energy* 233–234, 15–28. doi:https://doi.org/10.1016/j.apenergy.2018.10.015.

Clauwaert, P., P. Kim Aelterman, L. De Schamphelaire, M. Carballa, K. Rabaey, W. Verstraete 2008. Minimizing losses in bio-electrochemical systems: The road to applications. *Appl. Microbiol. Biotechnol.* 79, 901–913.

Cusick, R.D., and B.E. Logan 2012. 'Phosphate recovery as struvite within a single chamber microbial electrolysis cell', *Bioresource Technology,* 107, 110–115. doi: https://doi.org/10.1016/j.biortech.2011.12.038.

Daud, Siti Mariam, Byung Hong Kim, Mostafa Ghasemi, and Wan Ramli Wan Daud. 2015. Separators used in microbial electrochemical technologies: Current status and future prospects. *Bioresour. Technol.* 195, 170–79. doi:https://doi.org/10.1016/j.biortech.2015.06.105.

Du, Z., Li, H., and T. Gu 2007. 'A state of the art review on microbial fuel cells: A promising technology for wastewater treatment and bioenergy.' *Biotechnology advances*. England, 25 (5), 464–482. doi: 10.1016/j.biotechadv.2007.05.004.

Etter, B., E. Tilley, R. Khadka, and K.M. Udert 2011. Low-cost struvite production using source-separated urine in Nepal. *Water Res.* 45 (2), 852–862.

Faria, A., L. Gonçalves, J.M. Peixoto, L. Peixoto, A.G. Brito, G. Martins 2017. Resources recovery in the dairy industry: Bioelectricity production using a continuous microbial fuel cell. *J. Cleaner Prod.* 140, 971–976.

Feng, Yujie, Weihua He, Jia Liu, Xin Wang, Youpeng Qu, and Nanqi Ren. 2014. A horizontal plug flow and stackable pilot microbial fuel cell for municipal wastewater treatment. *Bioresour. Technol.* 156, 132–38. doi:https://doi.org/10.1016/j.biortech.2013.12.104.

Firdous, S., W. Jin, N. Shahid, Z.A. Bhatti, A. Iqbal, U. Abbasi, and A. Ali 2018. The performance of microbial fuel cells treating vegetable oil industrial wastewater. *Environ. Technol. Innovation* 10, 143–151.

Fischer, F., C. Bastian, M. Happe, E. Mabillard, and N. Schmidt 2011. Microbial fuel cell enables phosphate recovery from digested sewage sludge as struvite. *Bioresour. Technol.* 102 (10), 5824–5830.

Flimban, S.G.A., I.M.I. Ismail, T. Kim, S.-E. Oh 2019. Overview of recent advancements in the microbial fuel cell from fundamentals to applications: Design, major elements, and scalability. *Energies* 12, 3390. https://doi.org/10.3390/en12173390.

Fuchs, W., H. Binder, G. Mavrias, R. Braun 2003. Anaerobic treatment of wastewater with high organic content using a stirred tank reactor coupled with a membrane filtration unit. *Water Res.* 37 (4), 902–908.

Gnana Kumar, G, V G Sathiya Sarathi, and Kee Suk Nahm. 2013. Recent advances and challenges in the anode architecture and their modifications for the applications of microbial fuel cells. *Biosens. Bioelectron.* 43, 461–75. doi:https://doi.org/10.1016/j.bios.2012.12.048.

Gude, V.G. 2016. 'Wastewater treatment in microbial fuel cells - An overview', *Journal of Cleaner Production*. Elsevier Ltd, 122, 287–307. doi:10.1016/j.jclepro.2016.02.022.

Harnisch, F., U. Schroder 2010. From MFC to MXC: Chemical and biological cathodes and their potential for microbial bioelectrochemical systems. *Chem. Soc. Rev.* 39 (11), 4433–4448.

He, Li, Peng Du, Yizhong Chen, Hongwei Lu, Xi Cheng, Bei Chang, and Zheng Wang. 2017. Advances in microbial fuel cells for wastewater treatment. *Renewable Sustainable Energy Rev.* 71, 388–403. doi:https://doi.org/10.1016/j.rser.2016.12.069.

He, Zhen. 2013. Microbial fuel cells: Now let us talk about energy. *Environ. Sci. Technol.* 47 (1), 332–33. doi:10.1021/es304937e.

He, Zhen, Shelley D Minteer, and Largus T Angenent. 2005. Electricity generation from artificial wastewater using an upflow microbial fuel cell. *Environ. Sci. Technol.* 39 (14), 5262–67. doi:10.1021/es0502876.

Hernández-Fernández, F J, A Pérez de los Ríos, M J Salar-García, V M Ortiz-Martínez, L J Lozano-Blanco, C Godínez, F Tomás-Alonso, and J Quesada-Medina. 2015. Recent progress and perspectives in microbial fuel cells for bioenergy generation and wastewater treatment. *Fuel Process. Technol.* 138, 284–97. doi:https://doi.org/10.1016/j.fuproc.2015.05.022.

Hoffmann, R.A., M.L. Garcia, M. Veskivar, K. Karim, M.H. Al-Dahhan, and L.T. Angenent 2008. Effect of shear on performance and microbial ecology of continuously stirred anaerobic digesters treating animal manure. *Environ. Sci. Technol.* 100 (1), 38e48. https://doi.org/10.1016/j.jclepro.2019.02.172.

Ichihashi, O., and K. Hirooka 2012. Removal and recovery of phosphorus as struvite from swine wastewater using microbial fuel cell. *Bioresour. Technol.* 114, 303–307.

Jafary, T., W.R.W. Daud, M. Ghasemi, M.H. Abu Bakar, M. Sedighi, B.H. Kim, A.A. Carmona-Martínez, J.M. Jahim, and M. Ismail 2018. Clean hydrogen production in a full biological microbial electrolysis cell. *Int. J. Hydr. Energy* 44(58), 1–8. https://doi.org/10.1016/j.ijhydene.2018.01.010.

Kadier, Abudukeremu, Yibadatihan Simayi, Peyman Abdeshahian, Nadia Farhana Azman, K Chandrasekhar, and Mohd Sahaid Kalil. 2016. A comprehensive review of microbial electrolysis cells (MEC) reactor designs and configurations for sustainable hydrogen gas production. *Alexandria Eng. J.* 55 (1), 427–43. doi:https://doi.org/10.1016/j.aej.2015.10.008.

Kelly, P.T., Z. He 2014. Nutrients removal and recovery in bioelectrochemical sys-tems: A review. *Bioresour. Technol.* 153, 351–360.

Kim, Bongkyu, Junyeong An, Deby Fapyane, and In Seop Chang. 2015. Bioelectronic platforms for optimal bio-anode of bio-electrochemical systems: From nano- to macro scopes. *Bioresour. Technol.* 195, 2–13. doi:https://doi.org/10.1016/j.biortech.2015.06.061.

Kim, H.J., H.S. Park, M.S. Hyun, I.S. Chang, M. Kim, and B.H. Kim 2002. A mediator-less microbial fuel cell using a metal reducing bacterium, Shewanella putrefaciens. *Enzym. Microb. Technol.* 30, 145–152.

Konovalova, E. Y. et al. 2018. 'The microorganisms used for working in microbial fuel cells', AIP Conference Proceedings. *American Institute of Physics*, 1952(1), 20017. doi: 10.1063/1.5031979.

Koók, László, Péter Bakonyi, Falk Harnisch, Jörg Kretzschmar, Kyu-Jung Chae, Guangyin Zhen, Gopalakrishnan Kumar, et al. 2019. Biofouling of membranes in microbial electrochemical technologies: Causes, characterization methods and mitigation strategies. *Bioresour. Technol.* 279, 327–38. doi:https://doi.org/10.1016/j.biortech.2019.02.001.

Kook, L., T. Rózsenberszki, N. Nemestóthy, K. Bélafi-Bakó, and P. Bakonyi 2016. Bio- electrochemical treatment of municipal waste liquor in microbial fuel cells for energy valorization. *J. Cleaner Prod.* 112, 4406–4412.

Kumar, S.S, S.K. Malyan, and N.R. Bishnoi. 2017. Performance of buffered ferric chloride as terminal electron acceptor in dual chamber microbial fuel cell. *J. Environ. Chem. Eng.* 5 (1), 1238–43. doi:https://doi.org/10.1016/j.jece.2017.02.010.

Kumar, S.S., S.K. Malyan, A. Kumar, S. Basu, and N.R. Bishnoi 2016. Microbial fuel cells technology: Food to energy conversion by anode respiring bacteria. In: *Environmental Concerns of 21st Century: Indian and Global Context.* 139–154.

Kuntke, P., K.M. Smiech, H. Bruning, G. Zeeman, M. Saakes, T.H.J.A. Sleutels, et al. 2012. Ammonium recovery and energy production from urine by a microbial fuel cell. *Water Res.* 46 (8), 2627–2636.

Logan, B.E., et al. 2005. 'Electricity generation from cysteine in a microbial fuel cell', *Water Research*. Pergamon, 39 (5), 942–952. doi: 10.1016/J.WATRES.2004.11.019.

Le Corre, K.S., E. Valsami-Jones, P. Hobbs, and S.A. Parsons 2009. Phosphorus recovery from wastewater by struvite crystallization: A review. *Crit. Rev. Environ. Sci. Technol.* 39 (6), 433–477.

Leaño, Emmanuel Pacheco, and Sandhya Babel. 2012. The influence of enzyme and surfactant on biohydrogen production and electricity generation using palm oil mill effluent. *J. Cleaner Prod.* 31, 91–99. doi:https://doi.org/10.1016/j.jclepro.2012.02.026.

Lepisto, S.S. and J.A. Rintala 1997. Start-up and operation of laboratory-scale thermo-philic upflow anaerobic sludge blanket reactors treating vegetable processing wastewaters. *J. Chem. Technol. Biotechnol.* 68 (3), 331–339.

Li, Wen-Wei, Guo-Ping Sheng, Xian-Wei Liu, and Han-Qing Yu. 2011. Recent advances in the separators for microbial fuel cells. *Bioresour. Technol.* 102 (1), 244–52. doi:https://doi.org/10.1016/j.biortech.2010.03.090.

Li, X., G. Liu, S. Sun, F. Ma, S. Zhou, J.K. Lee, and H. Yao 2018. Power generation in dual chamber microbial fuel cells using dynamic membranes as separators. *Energy Convers. Manag.* 165, 488–494. https://doi.org/10.1016/j.enconman.2018.03.074.

Liu, H., and B.E. Logan 2004. Electricity generation using an air-cathode single chamber microbial fuel cell in the presence and absence of a proton exchange membrane. *Environ. Sci. Technol.* 38, 4040–4046.

Liu, H., S. Cheng, and B.E. Logan 2005a. Power generation in fed-batch microbial fuel cells as a function of ionic strength, temperature, and reactor configuration. *Environ. Sci. Technol.* 39 (14), 5488–5493.

Liu, H., S. Grot, and B.E. Logan 2005b. Electrochemically assisted microbial production of hydrogen from acetate. *Environ. Sci. Technol.* 39, 4317–4320.

Liu, Weifeng, Shaoan Cheng, Dan Sun, Haobin Huang, Jie Chen, and Kefa Cen. 2015. Inhibition of microbial growth on air cathodes of single chamber microbial fuel cells by incorporating enrofloxacin into the catalyst layer. *Biosens. Bioelectron.* 72: 44–50. doi:https://doi.org/10.1016/j.bios.2015.04.082.

Liu, Z., J. Liu, S. Zhang, and Z. Su 2009. Study of operational performance and electrical response on mediatorless microbial fuel cells fed with carbon-and protein-rich substrates. *Biochem. Eng. J.* 45 (3), 185–191.

Logan, B.E., P. Aelterman, B. Hamelers, R. Rozendal, U. Schroder, J. Keller, S. Freguiac, W. Verstraete, and K. Rabaey 2006. Microbial fuel cells: Methodology and technology. *Environ. Sci. Technol.* 40 (17), 5181–5192.

Lu, Mengqian, Shing Chen, Sofia Babanova, Sujal Phadke, Michael Salvacion, Auvid Mirhosseini, Shirley Chan, Kayla Carpenter, Rachel Cortese, and Orianna Bretschger. 2017. Long-term performance of a 20-L continuous flow microbial fuel cell for treatment of brewery wastewater. *J. Power Sources* 356, 274–87. doi:https://doi.org/10.1016/j.jpowsour.2017.03.132.

Luo, Tao, Said Abdu, and Matthias Wessling. 2018. Selectivity of ion exchange membranes: A review. *J. Membr. Sci.* 555, 429–54. doi:https://doi.org/10.1016/j.memsci.2018.03.051.

Madigan, M.T., J.M. Martinko, D. Stahl, and J. Parker. 2010. *Brock Biology of Microorganisms*, 13th ed. Benjamin Cummings, San Francisco, CA.

Mansoorian, Hossein Jafari, Amir Hossein Mahvi, Ahmad Jonidi Jafari, and Narges Khanjani. 2016. Evaluation of dairy industry wastewater treatment and simultaneous bioelectricity generation in a cata-lyst-less and mediator-less membrane microbial fuel cell. *J. Saudi Chem. Soc.* 20 (1), 88–100. doi:https ://doi.org/10.1016/j.jscs.2014.08.002.

Massaglia, Giulia, Valentina Margaria, Adriano Sacco, Micaela Castellino, Angelica Chiodoni, Fabrizio C Pirri, and Marzia Quaglio. 2019. N-doped carbon nanofibers as catalyst layer at cathode in single cham-ber microbial fuel cells. *Int. J. Hydrogen Energy* 44 (9), 4442–49. doi:https://doi.org/10.1016/j.ijhydene. 2018.10.008.

Mohanakrishna, G., S.V. Mohan, and P.N. Sarma 2010. Utilizing acid-rich effluents of fermentative hydrogen production process as substrate for harnessing bioelectricity: An integrative approach. *Int. J. Hydrog. Energy* 35, 3440–3449.

Oh, S.T., J.R. Kim, G.C. Premier, T.H. Lee, C. Kim, and W.T. Sloan 2010. Sustainable wastewater treatment: How might microbial fuel cells contribute. *Biotechnol. Adv.* 28, 871–881.

Ou, S., Y. Zhao, D.S. Aaron, J.M. Regan, and M.M. Mench 2016. Modeling and validation of single-chamber microbial fuel cell cathode biofilm growth and response to oxidant gas composition. *J. Power Sources* 328, 385–396.

Palanisamy, G., H.-Y. Jung, T. Sadhasivam, M.D. Kurkuri, S.C. Kim, and S.-H. Roh 2019. A comprehensive review on microbial fuel cell technologies: Processes, utilization, and advanced developments in elec-trodes and membranes. *J. Cleaner Prod.* 221, 598–621.

Pant, D., G. Van Bogaert, L. Diels, and K. Vanbroekhoven 2010. A review of the sub-strates used in microbial fuel cells (MFCs) for sustainable energy production. *Bioresour. Technol.* 101 (6), 1533–1543.

Rabaey, K., and W. Verstraete 2005. Microbial fuel cells: Novel biotechnology for energy generation. *Trends Biotechnol.* 23, 291–298.

Rabaey, K., G. Lissens, S.D. Siciliano, and W. Verstraete 2003. A microbial fuel cell capable of converting glucose to electricity at high rate and efficiency. *Bio-technol. Lett.* 25 (18), 1531–1535.

Rahimnejad, M., A.A. Ghoreyshi, G. Najafpour, and T. Jafary 2011. Power generation from organic substrate in batch and continuous flow microbial fuel cell operations. *Appl. Energy* 88, 3999–4004.

Rozendal, R.A., H.V. Hamelers, K. Rabaey, J. Keller, and C.J. Buisman 2008. Towards practical implementa-tion of bioelectrochemical wastewater treatment. *Trends Biotechnol.* 26 (8), 450–459.

Sadhasivam, T, K Dhanabalan, Sung-Hee Roh, Tae-Ho Kim, Kyung-Won Park, Seunghun Jung, Mahaveer D Kurkuri, and Ho-Young Jung. 2017. A comprehensive review on unitized regenerative fuel cells: Crucial challenges and developments. *Int. J. Hydrogen Energy* 42 (7), 4415–33. doi:https://doi.org/10.1016/j.ijhy dene.2016.10.140.

Shabani, Mehri, Habibollah Younesi, Maxime Pontié, Ahmad Rahimpour, Mostafa Rahimnejad, and Ali Akbar Zinatizadeh. 2020. A critical review on recent proton exchange membranes applied in microbial fuel cells for renewable energy recovery. *J. Cleaner Prod.* 264, 121446. doi:https://doi.org/10.1016/j.jcle pro.2020.121446.

Sharma, Yogesh, and Baikun Li. 2010. The variation of power generation with organic substrates in single-chamber microbial fuel cells (SCMFCs). *Bioresour. Technol.* 101 (6), 1844–50. doi:https://doi.org/10.1 016/j.biortech.2009.10.040.

Sonawane, Jayesh M, Samuel B Adeloju, and Prakash C Ghosh. 2017. Landfill leachate: A promising substrate for microbial fuel cells. *Int. J. Hydrogen Energy* 42 (37): 23794–98. doi:https://doi.org/10.1016/j.ijhy dene.2017.03.137.

Strathmann, Heiner B T and Membrane Science and Technology, ed. 2004. Chapter 3: Preparation and charac-terization of ion-exchange membranes. In *Ion-Exchange Membrane Separation Processes*, 9. Elsevier, Amsterdam, 89–146. doi:https://doi.org/10.1016/S0927-5193(04)80034-2.

Tang, R.C.O. et al. 2020. 'Review on design factors of microbial fuel cells using Buckingham's Pi Theorem', *Renewable and Sustainable Energy Reviews* 130, 109878. doi:https://doi.org/10.1016/j.rser.2020.109878.

Varanasi, Jhansi L, Shantonu Roy, Soumya Pandit, and D Das. 2015. Improvement of energy recovery from cellobiose by thermophillic dark fermentative hydrogen production followed by microbial fuel cell. *Int. J. Hydrogen Energy* 40, 8311–21.

Venkata Mohan, S, R Saravanan, S Veer Raghavulu, G Mohanakrishna, and P N Sarma. 2008. Bioelectricity production from wastewater treatment in dual chambered microbial fuel cell (MFC) using selectively enriched mixed microflora: Effect of catholyte. *Bioresour. Technol.* 99 (3), 596–603. doi:10.1016/j. biortech.2006.12.026.

Venkata Mohan, S, G Velvizhi, J Annie Modestra, and S Srikanth. 2014. Microbial fuel cell: Critical factors regulating bio-catalyzed electrochemical process and recent advancements. *Renewable Sustainable Energy Rev.* 40, 779–97. doi:https://doi.org/10.1016/j.rser.2014.07.109.

Vicari, Fabrizio, Adriana D'Angelo, Alessandro Galia, Paola Quatrini, and Onofrio Scialdone. 2016. A single-chamber membraneless microbial fuel cell exposed to air using shewanella putrefaciens. *J. Electroanal. Chem.* 783, 268–73. doi:https://doi.org/10.1016/j.jelechem.2016.11.010.

Vogl, Andreas, Franz Bischof, and Marc Wichern. 2016. Increase life time and performance of microbial fuel cells by limiting excess oxygen to the cathodes. *Biochem. Eng. J.* 106, 139–46. doi:https://doi.org/10.1016/j.bej.2015.11.015.

Wei, Jincheng, Peng Liang, and Xia Huang. 2011. Recent progress in electrodes for microbial fuel cells. *Bioresour. Technol.* 102 (20): 9335–44. doi:https://doi.org/10.1016/j.biortech.2011.07.019.

Wu, Dan, Liping Huang, Xie Quan, and Gianluca Puma. 2016. Electricity generation and bivalent copper reduction as a function of operation time and cathode electrode material in microbial fuel cells. *J. Power Sources* 307 (March), 705–14. doi:10.1016/j.jpowsour.2016.01.022.

Yang, Euntae, Kyu-Jung Chae, Mi-Jin Choi, Zhen He, and In S Kim. 2019. Critical review of bioelectrochemical systems integrated with membrane-based technologies for desalination, energy self-sufficiency, and high-efficiency water and wastewater treatment. *Desalination* 452, 40–67. doi:https://doi.org/10.1016/j.desal.2018.11.007.

Yang, Wei, Jun Li, Dingding Ye, Liang Zhang, Xun Zhu, and Qiang Liao. 2016. A hybrid microbial fuel cell stack based on single and double chamber microbial fuel cells for self-sustaining pH control. *J. Power Sources* 306, 685–91. doi:https://doi.org/10.1016/j.jpowsour.2015.12.073.

23 Phytoremediation
An Eco-Friendly and Sustainable Approach for the Removal of Toxic Heavy Metals

Sonal Nigam and Surbhi Sinha

CONTENTS

23.1 INTRODUCTION

In this modern era of urbanization, the environment is continuously being exposed to increasing concentration of organic, inorganic, and metallic contaminants. Heavy metals are the major contributors of soil and groundwater pollution and are reported to have high density and toxicity even at low concentrations (Duruibe et al. 2007). These heavy metals occur in the environment

DOI: 10.1201/9781003204442-23

due to numerous natural and anthropogenic activities. Volcanic eruptions, forest fires, and wind erosion are some of the natural activities contributing to heavy metal toxification in the environment. Furthermore, mine tailing, electroplating, energy-fuel production, agricultural practices, gas exhausts, and sludge dumping are some of the major anthropogenic activities contributing to the heavy metal contamination of soil and water (Cutright and van Keulen 2008). Heavy metals are more likely to accumulate in soil and water due to their ability of being non-biodegradable and soluble in aqueous solution, therefore posing risk to human health and the environment (Liu et al. 2012). In addition, they can undergo changes in their oxidation state and have a half-life of 20 years. These characteristics contribute to their persistence in the environment for longer periods (Ahmed 2018). Heavy metals can be categorized as essential and non-essential on the basis of their participation in performing the function of living systems. Manganese (Mn), copper (Cu), nickel (Ni), zinc (Zn), and iron (Fe) are some of the essential heavy metals needed in minor quantities for performing various biochemical and physiological functions of the body (Chaffai and Koyama 2011). As these essential heavy metals are demanded by the body in very small quantities they are also known as trace elements. Whereas non-essential amino acids such as lead (Pb), cadmium (Cd), mercury (Hg), arsenic (As), and chromium (Cr) are not required by living systems to perform their functions (Sarwar et al. 2017). Though these metals play an important role in various physiological and biochemical functions of plants and animals, when present in concentrations higher than the threshold value they can interfere with other biological functions, thereby posing adverse effects on health. High accumulations can lead to the production of reactive oxygen species (ROS) (Ashraf et al. 2019), which can negatively affect the activities of microbes when present in soil (Khan et al. 2010) and can lead to bioaccumulation, thereby posing a great threat to human health due to the potential risk of their entry into food chain. In plants, heavy metals can alter the chlorophyll content and photosynthetic rate, thereby reducing their growth (Sarwar et al. 2015). Also, they can lead to water stress in plants as result of cell enlargement and decrease in the number and size of xylem vessels and chloroplasts. Therefore, it becomes important to search for an effective approach to maintain and remove excessive amounts of heavy metals in order to ensure food safety and a healthy environment. Although various physicochemical methods have been proposed for the remediation of heavy metals from the environment these methods are not preferable as they are expensive, time consuming, and may have detrimental effects on soil fertility. Microbial remediation is an effective technique; however, it cannot ensure effective remediation because not all microbes can adapt according to the environment (Chen et al. 2015). Phytoremediation is an efficient, cost-effective (Singh and Santal 2015), simple, widely accepted, and long-term method for restoring the natural composition of heavy metal contaminated soil and water. This approach can exploit both native and genetically modified plants for the purpose of remediation (Parmar and Singh 2015). Furthermore, the process is generally solar driven and is carried out in situ which contributes to its low cost and minimal exposure of toxicants to wildlife, humans, and the environment (Pilon-Smits 2005). This approach is effective because plants are the primary recipients of the heavy metals and therefore can serve as important bioremediators. A recent advancement has been done in the field of phytoremediation by using nanoparticles along with plants in order to remove environmental pollutants. Recently, nano-phytoremediation has attracted considerable attention for improving the soil quality due to its robustness and high efficiency. Furthermore, nanoparticles are reported to have several positive effects on plant growth (Zhu et al. 2019a). However, when used in high concentrations they can have toxic effects on plants, such as leaf toxicity reactions and reduced growth (Zhu et al. 2019a). Therefore, further research needs to be done to explore this field in order to use nano-phytoremediation on a large scale. This chapter describes the basics of phytoremediation along with its different types, their mechanisms, and the various factors influencing the entire process of phytoremediation. The chapter also presents the possible modifications of phytoremediation techniques to enhance its efficiency in the removal of heavy metals, offering a wide spectrum of potential applications of this green technology.

23.2 PHYTOREMEDIATION AND ITS NEEDS

Phytoremediation refers to the process of restoring environmental damage by using plants and their associated microorganisms. The contaminants present in groundwater, surface water, sludge, and soil can all be removed or degraded by the direct use of plants or vegetable/fruit waste material. Phytoremediation is a solar energy-driven, eco-friendly, and cost-effective strategy for environmental cleanup. Furthermore, it demands low maintenance and has less installation costs. Figure 23.1 discusses the various advantages and disadvantages of phytoremediation. The process of phytoremediation exploits the natural ability of plants to concentrate, degrade, and accumulate toxicants from water, soil, and air. Heavy metals such as As, Cu, Zn, Cd, and Pb are highly poisonous for biological systems (Muthusaravanan et al. 2018), and when present in excessive amounts may have toxic effects on plants. Therefore, it becomes a necessity to take this problem into consideration and determine an efficient strategy for remediating heavy metals. Phytoremediation is one such approach which can help in solving this problem; however, it is known to have limited efficiency because of the non-biodegradable nature of heavy metals. Heavy metal remediation is difficult to carry out by a single approach. However, using combinations of different approaches along with some modifications can help in overcoming this problem. For instance, photocatalysis after coagulation followed by phytoextraction can increase the efficiency and reduce the time of the process (Kumar Yadav et al. 2018). Traditional methods of phytoremediation are known to produce relatively less biomass, have low growth rates, and adapt less under different environment conditions. Therefore, they demand an increased number of harvest cycles, have variable bioavailability, and cannot be used for highly contaminated sites. Furthermore, the soil fertility, inert species competition, change in pH, and plant species also determine bioavailability, thus affecting the rate of environment cleanup. The efficiency of traditional methods can, however, be enhanced by using different strategies such as by using chelating agents, biochar, nanomaterials, and microbes; by performing genetic manipulation of plants; or by modifying the entire process. Ethylene diamine tetraacetic acid (EDTA), ethylene diamine disuccinate (EDDS), and nitrilotriacetic acid (NTA) are some of the frequently used chelating agents (Muthusaravanan et al. 2018). Chelating agents work by increasing the mobility of the metal ions in the medium. Other organic acids with low molecular weight, such as citric acid, malic acid, acetic acid and oxalic acid, can also be used as chelating agents on a large scale and with low cost. Using biochar along with the non-hyperaccumulator plants can enhance their performance to a large extent (Sarwar et al. 2017). Mycorrhizal fungi also enhance the rate of phytoremediation by modifying the soil pH and exudates from the roots which in turn increases the bioavailability of metal ions (Kumar Yadav et al. 2018). Plant-growth-promoting bacteria (PGPR), when associated with plants, can lead to an increase in the rate of N-fixation and phosphate absorption and reduce

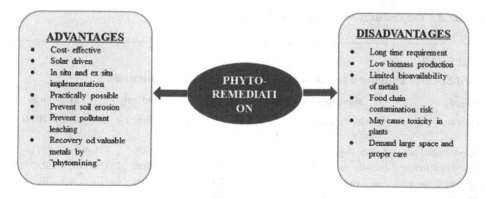

FIGURE 23.1   The advantages and disadvantages of phytoremediation.

ethylene production under stress conditions, thereby assisting in the process of phytoremediation (Kumar Yadav et al. 2018). Using various techniques with different influencing factors in combination can help to achieve higher efficiency and better cleanup strategies.

23.3 TYPES OF PHYTOREMEDIATION

Phytoremediation is divided into various categories on the basis of its process and application. Techniques of phytoremediation include phytoextraction, phytostabilization, phytovolatilization, phytodegradation, phytofiltration, and rhizofiltration. Figure 23.2 illustrates various types of phytoremediation processes. Table 23.1 depicts the various plant species utilized in different techniques of phytoremediation.

23.3.1 PHYTOEXTRACTION

The natural tendency of plants to take up metals from their surrounding makes them ideal for the process of phytoextraction. Phytoextraction, also known as phytosequestration, phytoabsorption, or

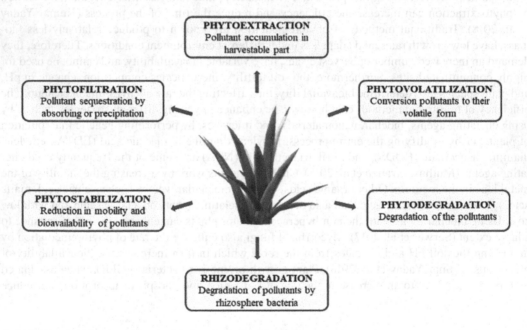

FIGURE 23.2 The various types of phytoremediation processes.

TABLE 23.1
Plant Species Employed for Different Phytoremediation Processes

Method	Plant Species	Metals	Medium	References
Phytoextraction	*Solenum nigrum L.*, *Spinacia oleracea L.*, and	Pb, Cu, Cd, and Cr	Soil	(Dinesh et al. 2014)
Phytodegradation	*Arundo donax L.*	As	Soil	(Mirza et al. 2011)
Phytovolatilization	*Brassica* sp. (wild type)	Se	Soil	(Bañuelos et al. 2005)
Phytostabilization	*Vossia cuspidata*	Cu	Water	(Galal et al. 2017)
Phytofiltration	*Phalari arundinacea*	Cr	Water	(Vymazal 2016)
Rhizodegradation	*Lactuca sativa*	Se	Water	(Hawrylak-Nowak 2013)

phytoaccumulation, is a method which utilizes plant roots to uptake contaminants from water and soil along with other essential nutrients and water required for their growth. The metals sequestered by the plants are translocated and stored in the leaves, shoots, and other aerial parts of the plant rather than undergoing the process of degradation (Rashid et al. 2014). Finally, the plant biomass is harvested for safe disposal. This method is used in order to produce long-term effects. Plants used for phytoextraction should have high growth rates, be easy to cultivate, have profusive root systems, and be able to produce high biomass (Sytar et al. 2015). This is a highly effective commercial method which can be further improved by soil property modification, increased metal bioavailability, and the enhanced sink capacity of the concerned plant species (Sarwar et al. 2017).

23.3.2 PHYTOSTABILIZATION

Phytostabilization or phytoimmobilization is a process which works by transforming toxic compounds to lesser or non-toxic forms by utilizing plants. This method reduces the mobility and thus the bioavailability of toxicants in the environment which is done by preventing the leaching and percolation of contaminants into the soil and groundwater. In immobilization, various physical and chemical processes, such as the absorption by roots, change in valency, the precipitation of heavy metal, or the formation of metal complexes facilitated by plants, can contribute to the stabilization of heavy metals in the environment (Ali et al. 2013). Furthermore, some plants are known to secrete redox enzymes in their rhizosphere which can also convert the heavy metals into less toxic forms (Ali et al. 2013). However, phytostabilization is not a long-term remediation method and is just a management strategy as it only stabilizes and decreases the toxicity levels of the heavy metals and does not completely eradicate them from soil and water (Ali et al. 2013). Some of the ideal characteristics of plants used for phytostabilization include an extended root system, the ability to tolerate high concentrations of heavy metals and reduce soil erosion, and the ability to play a role in hydraulic control (Kumar Yadav et al. 2018).

23.3.3 PHYTOVOLATILIZATION

Phytovolatilization is a method which exploits the ability of plants to convert pollutants from the environment to volatile forms which is subsequently followed by their release into the atmosphere. The process of phytovolatilization starts with the transformation of the contaminants to less toxic and water soluble forms during their journey from the root to leaves and compartmentalization into the vacuoles, followed by the release of water-soluble modified metals through the stomata by the process of transpiration. Phytovolatilzation is divided into two types: direct and indirect methods. In the direct method the plants assimilate the organic compounds; however, in the indirect method the contaminants are fluxed into the plants from the water and soil due to plant root activities (Kumar Yadav et al. 2018). The major setback of phytovolatilization is that in this method the pollutants are just transferred from the soil or water to the air rather than being completely eliminated from the environment. This method is particularly useful for organic pollutants and certain heavy metals such as Hg, As, and Se (Ali et al. 2013).

23.3.4 PHYTODEGRADATION

The process of employing plants for performing the degradation of organic contaminants present in the soil is called phytodegradation. This method is also known as phytotransformation and is not dependent on rhizospheric microorganisms (Ali et al. 2013). The plants in this method may use their metabolic activities or may exudate enzymes such as oxygenases and dehalogenases to degrade the contaminants present in soil and water. The enzymes help by catalyzing and accelerating the rate of degradation. Some major enzymes include dehalogenase, nitrilase, nitroreductase, phosphatise, and peroxidase which transform chlorinate, cyanated aromatic, nitrated, organophosphate pesticides, and phenolic compounds, respectively (Kumar Yadav et al. 2018). One major limitation of this method is that it is only applicable for organic pollutants (synthetic pesticides, insecticides, and herbicides) as heavy metals are resistant to degradation. The various factors influencing this process include the efficiency of pollutant uptake, the concentration of contaminants in the environment, and the phytochemical properties of the plant (Muthusaravanan et al. 2018). Further studies are focusing on using genetically modified plants for phytodegradation. Indian mustard (*Brassica juncea*) and yellow poplar (*Liriodendron tulipifera*) are examples of genetically modified plants with phytoremediation qualities (Kärenlampi et al. 2000). Since the breakdown of organic pollutants cannot completely convert these molecules to basic molecules (water, carbon dioxide, etc.), this process only changes the contaminants to non-toxic forms rather than eliminating them from the environment.

23.3.5 PHYTOFILTRATION

The practice of employing plants to draw out the organic and inorganic compounds present in soil and water by either absorbing, adsorbing, or precipitating the contaminants is called phytofiltration. The plants act by changing the pH of the rhizosphere which thus precipitates the toxicants present in it (Ashraf et al. 2019). This process results in the minimized movement of contaminants in the soil and water, thus reducing contamination levels. Phytofiltration based on the specific plant parts utilized for the process can be divided into rhizofiltration (roots), caulofiltration (shoots), and blastofiltration (seedlings) (Ali et al. 2013). Some of the requisite features of plants for phytofiltration include metal tolerance, a large surface area for increased absorption, and hypoxia tolerance (Cristaldi et al. 2017). It has been reported that terrestrial plants are preferred due to their extended fibrous networks when compared to aquatic plants for phytofiltration (Cristaldi et al. 2017). This process helps in remediating C, Ni, Cr, Cu, V, and Pb and also certain radionucleotides such as caesium (Cs), strontium (Sr), and uranium (U) (Rezania et al. 2016). The process of phytofiltration starts with the acclimatization of ideal plant species by exposing them to the supplied contaminated water. Followed by this, the plants are transported to the site of contamination for performing the task of remediation. Finally, the plants are harvested upon reaching the maximum absorption capacity (Cristaldi et al. 2017). Some of the major limitation of this method are the demand for pH adjustment and greenhouses (Sharma and Pandey 2014). However, it is an eco-friendly and low-cost method for the remediation of contaminants from the environment.

23.3.6 PHYTOSTIMULATION

Phytostimulation is also known as rhizodegradation. Rhizodegradation employs the microorganisms present in the rhizosphere for degrading the contaminants present in soil. The microbes applicable for this purpose may include fungi, bacteria, and yeast (Awa and Hadibarata 2020). The plant secretes various exudates including carbohydrates, amino acids, and flavonoids. This increases the metabolic activities of these microbes by 10–100 fold by providing them with enough carbon and nitrogen sources for their growth (Ali et al. 2013). The extended root systems of plants provide large surface areas which help with efficient microbial growth due to high oxygen availability. Furthermore, the enzymes released by plants help in the degradation of organic contaminants and stimulate the microbial growth. However, the potency of rhizodegradation decreases with an increase in the depth of the soil. Symbiotic mycorrhizal fungi present in association with roots are reported to be efficient in degrading the compounds which are difficult to be degraded by bacteria (Kumar Yadav et al. 2018). The efficiency of phytostimulation is primarily dependent on the heavy metal composition and concentration, the association between plant and microbes, the soil properties, optimum environmental variables, and the metabolic rate of the microorganisms (Kumar Yadav et al. 2018).

23.4 FACTORS INFLUENCING PHYTOREMEDIATION

A large number of factors play an essential role in influencing the process of phytoremediation. Some of these factors include environmental conditions, plant species, root zones, the bioavailability of metals, and the physico-chemical properties of media.

23.4.1 PLANT SPECIES

The phytoremediation capabilities of plants for the same metal, in the same environmental conditions, and on the same land can widely differ among different species (Kumar Yadav et al. 2018). Therefore, choosing a toxicant-specific species for phytoremediation determines the success of the phytoremediation process. The identification of specific hyperaccumulator species by plant screening can help in the production of large amounts of plant biomass by using standard crop production and management strategies. The brake fern (*Pteris vittata*) is a hyperaccumulator plant which is specific to arsenic. Its mechanism for hyperaccumulation includes the rapid transfer of arsenic to the aerial parts of the plant with minimum concentration of arsenic in roots (Zhang et al. 2002).

23.4.2 PHYSICO-CHEMICAL PROPERTIES OF MEDIA

The bioavailability of metal ions, which is a crucial factor for phytoremediation, is dependent on various physico-chemical properties of the media. Texture, root exudates, temperature, and nutrient content have a great effect on the process of phytoremediation. pH is considered important for governing the availability of metal ions to be taken up by plants. The metal concentration in the medium increases at low pH due to the decreased absorption by the tissues of the plant. Some metals like Ni, Zn, and Cd are reported to have comparatively high mobility while others like Cr, Pb, and Cu show less mobility at normal pH levels (Kim et al. 2015). For instance, the phytoremediation of Cd from wastewater by Spirulina (*Arthrospira* platensis) showed optimum results at pH 7 (Rangsayatorn et al. 2002). The increasing concentration of organic matter increases the mobilization of heavy metals and thus increases the availability of nutrients, promotes the formation of chelate, and enhances metal availability for plants by enhancing cation exchange capacity. Several other parameters including temperature, cation competition, and microbial activity are also known to influence metal availability but have a minor role in context to soil components and pH. With the application of genetically engineered pants, chelating agents, and fertilizers and by practicing soil and crop management strategies, the efficiency of phytoremediation can be increased.

23.4.3 BIOAVAILABILITY OF METALS

The term "bioavailability" refers to the availability of toxicants present in the environment for their uptake by exposed living systems. Plants can only act on a specific fraction of heavy metals that are readily available to them. Bioavailability is one of the major rate-limiting factors of phytoremediation at the site of contamination. Redox potential, pH, oxide minerals, and organic matter present in the media affect the bioavailability of metals. The pH of the soil can reduce the metal concentration in the soil by forming complexes with the functional groups of oxides and organic matter (Awa and Hadibarata 2020). The organic matter and oxide in the media have high exchange capacities and therefore can react with metal cations and can thus reduce the bioavailability of metals. The redox potential of soil is also an important consideration as it affects the solubility, transformation, and uptake of metals by plants. It has been reported that metals such as manganese and iron with high oxidation states have low solubility (Awa and Hadibarata 2020).

23.4.4 ROOT ZONES

A plant's root zone is of much importance in the process of phytoremediation. It is the main route of entry of heavy metals into the plant, from where they may be translocated to other aerial parts for metabolization, absorption, or storage. The root also exudates some enzymes in the rhizosphere which assist in the degradation of heavy metals from complex toxic to simpler non-toxic forms. Local environmental conditions, such as precipitation, temperature, soil moisture, and drought, immensely affect the properties of roots by influencing root growth and length (Kumar Yadav et al. 2018). The speciation of metal takes place by redox-potential modification, organic ligands, acidification/alkalinization, and chelating agent exudation by the plant roots, which also potentially raise the solubility of metals (Ma et al. 2016). Thus, it becomes important to analyze the exudate composition and the properties of the root zone to achieve the maximum accumulation of trace elements in plants.

23.4.5 ENVIRONMENTAL CONDITIONS

Climatic conditions at the site of contamination play a major role in the process of phytoremediation. The rate of metal removal by plants increases with an increase in temperature. Temperature directly affects the uptake and elimination of pollutants by controlling plant growth, plant metabolism, transpiration, and water chemistry (Bhargava et al. 2012). The metabolic pathways of plants, such as autotrophic respiration and photosynthesis, are influenced by soil heat fluxes, the moisture content of the soil, differences in soil and air temperature, and the relative humidity of air (Kumar Yadav et al. 2018). At high moisture levels a greater biomass is produced by the plant which in turn increases the phytoremediation capacity of the plants. Moreover, an increasing rate of pollution can also decrease the metal uptake by plants as plants absorb some atmospheric aerosols through the surface of the leaves. For instance, Cd, Zn, and Cu are absorbed by the plants by penetrating the leaves, while Pb remains on the surface as a precipitate (Bhargava et al. 2012).

23.4.6 CHELATING AGENT ADDITION

The addition of chelating agents or micronutrients to the media not only enhances the rate of phytoaccumulation but also stimulates the microbial population in the rhizosphere to enhance their heavy metal uptake capacity. Because an alkaline pH generally decreases the bioavailability of heavy metals, chelating agents are required by soils having oh above 5.5 or 6 (Awa and Hadibarata 2020). The chelating reagent increases the rate of phytoremediation which thus helps in reducing the time and cost of the process. EDTA, when exposed to plants for a long period, can improve the phytoremediation process by improving the translocation of metal in plant tissues (Roy et al. 2005).

23.5 MECHANISM OF METAL UPTAKE BY PLANTS

Plants are highly efficient in sequestering the heavy metals present in soil and water. The efficiency of phytoremediation depends on the soil properties, the bioavailability of toxicants, and the type of toxicant (Kumar Yadav et al. 2018). Remediation mainly occurs through the vastly extended root system of plants. The roots have a very large surface area that helps in the absorption of water and nutrients required for their growth along with non-essential heavy metals. Sometimes, the microorganisms that reside in the rhizosphere are in close connection with the plants and help in the mobilization of metal ions, thereby enhancing their availability. Sinha et al. (2007) denote plants as both "accumulators" and "excluders." Accumulators are the plants that accumulate trace elements in their aerial parts and cope with high metal concentrations by biotransforming them into inert forms. Excluders are those in which the metal ions accumulate in the roots only and are not translocated to the aerial part of the plant. Accumulator plants are reported to concentrate the heavy metals up to

1,000 times more when compared to excluder plants (Kumar Yadav et al. 2018). The mechanism of metal uptake by plants begins when they encounter trace elements in their rhizosphere. The rhizosphere allows the uptake of these metals from soil and water into the plant via the roots. Following this, the accumulated metal ions may get stored in the roots or may be translocated to the aerial parts of the plant through the xylem vessels. The translocation may occur through symplastic or apoplastic pathways (Abdel-Shafy 2018). Upon reaching the aerial parts of the plant the heavy metals are compartmentalized into the plant vacuoles which have low metabolic activities. The storing of these metal ions in the vacuoles limit their interaction with the plant's cellular metabolic process and decreases their concentration in the cytosol, thereby serving as a tolerance mechanism in hyperaccumulator plants species towards heavy metals (Ali et al. 2013). For metal accumulation and phytoremediation, the tolerance of heavy metals by plants is a key requirement. The high tolerance of heavy metals by plants provides them with a high efficiency for performing phytoremediation. Furthermore, it will also help in maintaining the health of the plant by reducing the adverse effects of heavy metals. The major mechanisms that govern the metal tolerance potential of a plant include the use of proteins and peptides as chelators of metal ions (because of donor atoms such as N, O, and S), metal binding with the cell wall, complex formation, and the translocation of metals to the vacuoles (Ali et al. 2013). Various molecules contribute to the transportation and regulation of metals from the root to the aerial parts of the plant. Special transport proteins present in the root plasma membrane act as carrier molecules for the uptake of heavy metals by the roots. For instance, the uptake of Zn^{2+} and Fe^{2+} occurs through zinc–iron permease (ZIP) family transporters. Some metals with similar chemical structures, properties, ionic radii, and oxidation states can enter the root system by transporters for other metals. For example, the uptake of As(V) in plants occurs through phosphate transport channels (Tripathi et al. 2007).

23.6 HEAVY METAL DETOXIFICATION AND METAL TOLERANCE BY PLANTS

Heavy metals present in the environment are divided into essential and non-essential forms. For instance, Cu and Zn are essential heavy metals as they serve as the major components of several enzymes and proteins and thus help in determining the growth and development of a plant. However, when present in excessive amount these essential heavy metals can also have adverse effects on the growth and development of a plant by interfering with its cellular and metabolic activities. Heavy metals exhibit their toxic effects by binding with the sulphydryl groups of cellular proteins which, as a repercussion, leads to the disruption of protein structures or inhibition of their activity (Hall 2002). Another route of toxicity by heavy metals includes the generation of oxidative stress by forming reactive oxidant species (ROS). However, plants somehow manage to endure these high concentrations of heavy metals by getting assistance from the potential cellular mechanism that plays an essential role in detoxifying the heavy metals. A broad range of strategies and techniques constitutes the heavy metal tolerance mechanisms in plants. Figure 23.3 discusses the heavy metal tolerance mechanisms of plants. The extracellular mechanisms for heavy metal detoxification include the role of mycorrhiza, extracellular exudates, and the cell wall (Hall 2002). Mycorrhiza are reported to be effective for improving the heavy metal detoxification efficiency of plants. The mechanism behind this is not yet completely understood but mostly it has been proposed that it does so by absorbing metals on the hyphal sheath, by adsorbing onto the external mycelium, or by exudating chelating agents and therefore restricting the heavy metal uptake by plants (Hall 2002). Metal detoxification differs among various species and also depends on the type of metal ion species. A study reported considerable amounts of zinc retention by the mycelium of *Glomus* species in association with ryegrass and clover (Joner et al. 2000). The role of the cell wall in detoxification is yet to be fully explored. Metal accumulation generally takes place in the apoplast. The divalent and trivalent metal ions generally under stress conditions bind to the carboxyl group of the pectin present in the plant cell wall (Kushwaha et al. 2015). The plasma membrane can also assist in detoxification either by reducing the heavy metal uptake or by pumping out the metals

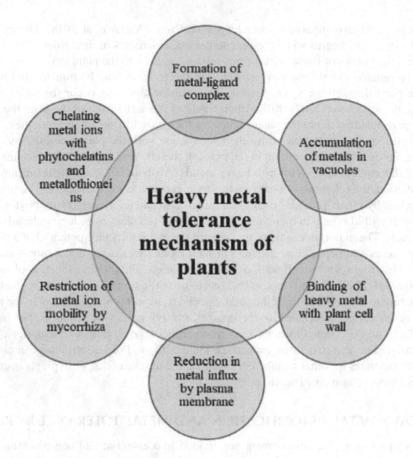

FIGURE 23.3 The heavy metal tolerance mechanisms of plants.

ions which have already entered the cytoplasm. Furthermore, the protoplast also possesses some useful mechanisms such as repairing of stress-damaged proteins. For instance, heat-shock proteins (HSPs) are the main molecules which assist in protein repairing. The production of these proteins is induced due to a variety of stress conditions such as high temperature and high metal concentration. HSPs help in protecting and repairing protein by acting as molecular chaperons in the protein folding process (Kumar Yadav et al. 2018). Furthermore, when heavy metal stress conditions occur in the surroundings, the cell leads to the activation of a complex network for detoxification and storage mechanisms. This includes the chelation of metal ions by phytochelatins and metallothioneins which is followed by their compartmentalization in the vacuoles. Phytochelatins (PC) are cysteine rich, specific, and high-affinity ligands. PCs are most widely studied in plants and are known to be highly specific for cadmium. They help in metal detoxification by playing a role in heavy metal homeostasis. PCs are synthesized from glutathione (GSH) by catalyzing the reaction with PC synthase. PCs are induced by the presence of Cu, Ag, Pb, Cd, Zn, Hg, and Au, with Cd as the strongest inducer. The complex formation between Cd and PCs takes place due to the thiol groups of cysteine (-SH). After this complex formation they are translocated to the vacuoles by ABC transporters (Kushwaha et al. 2015). A study reported the initiation of PC synthesis by *Rubia tinctorum* after exposure to Cd, Zn, and Cu metals (Maitani et al. 1996). Another family of cysteine-rich chelating agents is Metallothionenin (MT). They are low-molecular-weight molecules and help in detoxification by sequestering heavy metals. They do so by forming complexes with metal ions through multiple thiol groups of cysteine. MTs also help in protecting against the damage caused by oxidative stress. They are classified into class 1 and class 2 MTs. MT 1 can align with the mammalian renal MT due to the presence of cysteine residues, while MT 2, in spite of the presence of similar cysteine residues, cannot be aligned easily with MT 1 (Hall 2002). MTs are expressed during abiotic stress conditions and during the development of the plant. Natural senescence, heat shock proteins, nutrient starvation, hormones like ABA, UV light, cold stress, salt stress, and infection by virus are some of the stimuli which can help in upregulating the MT gene expression (Kushwaha et al. 2015). Nowadays, these genes can also be genetically modified in order to develop enhanced tolerance in plants against heavy metals. A study reported an increased Cd tolerance in tobacco plants by the over-expression of metallothionein gene PaMT3-1 from *Phytolacca americana* (Zhi et al. 2020). In addition to phytochelatins and metallothioneins, metabolic pathways also provide amino acids, peptides, or organic acids for chelation (Kumar Yadav et al. 2018). Following chelation, the complex molecules are translocated to the vacuoles for compartmentalization. This ensures minimum contact of heavy metals with the metabolic processes of the cell. The transportation of metals takes place with the help of some major transporters such as ABC transporters, ZIP family proteins, the CDF transporter family, the natural resistance-associated macrophage protein (NRAMP) family, P-type metal ATPases, the CAX family, copper transporter (COPT) family proteins, and pleiotropic drug resistance 537 (PDR) transporters (Kushwaha et al. 2015).

23.7 MODIFICATIONS OF THE PHYTOREMEDIATION PROCESS FOR ENHANCED EFFICIENCY

Some of the major drawbacks of traditional methods, such as low biomass production, long time requirements, low bioavailability of metals, and reduced efficiency due to environmental conditions, can limit their application on a large-scale level. Applying modified methods such as using genetically manipulated plants, chemical agent nanomaterials, and biochar and providing microbial assistance can help to overcome these limitations.

23.7.1 USE OF GENETICALLY MANIPULATED PLANTS

The genetic manipulation of plants can be done by using recombinant DNA technology. These plants can increase the level of phytoremediation to a great extent. By altering the genetic structure of hyperaccumulator plants the extraction potential and biomass production of these plants can be increased. The major principle behind this technique is the insertion of specific genes involved in translocation, tolerance, uptake, and sequestration into the ideal plant species (Sarwar et al. 2017). The main aim of producing transgenic plants is to enhance the tolerance of these plants towards metal toxicity and to increase the immobilization and translocation of heavy metals to the aerial parts of the plant (Kumar Yadav et al. 2018). Genetic manipulation of *Nicotiana tabacum*, *Populus angustifolia*, and *Silene cucubalus* enhanced their metal accumulation potential due to over-expression of glutamyl cysteine synthetase (Fulekar et al. 2009). One of the major drawbacks associated with the use of transgenic plants is that most of these plants are grown under controlled conditions rather than in fields. Therefore, thorough risk studies are needed before using them on a large scale. Furthermore, they are also reported to pose possible risks to wildlife and humans.

23.7.2 CHEMICAL-ASSISTED PHYTOREMEDIATION

The bioavailability of heavy metals enhances the process of phytoremediation. Employing chemicals such as chelating agents or surfactants can increase the mobility of ions in the medium, therefore increasing the accumulating potential of non-hyperaccumulating plants. 1,2-cyclohexylenedinitrilo tetraacetic acid (CDTA), ethylene diamine tetra-acetic acid (EDTA), nitrilotriacetic acid (NTA), diethylenetriaminepentaacetic acid (DTPA), ethylene diamine disuccinate (EDDS), and ethylene glycol-bis (β-aminoethyl ether)-N,N,N′,N′-tetraacetic acid (EGTA) are some of the most commonly-used chelating agents (Muthusaravanan et al. 2018). Chelating agents enhance the process of phytoremediation by increasing the concentration of heavy metals in soil, destroying the root barriers by high levels of resultant complexes, increasing the metal-EDTA complex movement towards the roots, and enhancing the metal translocation from root to shoot by increasing the complex mobility in comparison to free ions (Kumar Yadav et al. 2018). One serious concern associated with use of chelating agents is that they can lead to the leaching of metals into groundwater, therefore contaminating it (Muthusaravanan et al. 2018). Furthermore, some chelating agents and their complexes, such as that of EDTA (non-biodegradable), can cause harm to the plants and microorganisms present in the soil. EDDS is produced by microorganisms and can increase the bioavailability of Ni, Cu, and Zn to a greater extent when compared to EDTA. NTA is another chelating agent which is easy to biodegrade and has no phytotoxic effects (Kumar Yadav et al. 2018).

23.7.3 NANO-PHYTOREMEDIATION

Nano-phytoremediation uses a combination of nanotechnology and phytoremediation for removing heavy metals and inorganic and organic toxicants from the environment. Nanomaterials are small particles ranging in size from 1–100 nm. Recently, the use of nanotechnology has gained much interest in the field of phytoremediation due to its high efficiency of absorbing or adsorbing the toxicants present in the environment, and when these two processes are used in combination greater results are achieved. For instance, the degradation and removal of TNT (2,4,6-trinitrotoluene)-contaminated soil by nanophytoremediation is more effective than either nano-remediation or phytoremediation alone (Srivastav et al. 2019). Various types of nanomaterials have been explored to date, some of which include nano Fe_3O_4, nano hydroxyapatite (NHAP), nano TiO_2, nano Ni, nano Zn, and nano carbon black (NCB) (Zhu et al. 2019b). Nano-phytoremediation can be effectively applied for enhancing the remediation of Cr, Pb, Zn, Ni, and Cd contaminated sites. Despite being a promising, highly-efficient, and eco-friendly method, the field of nano-phytoremediation still needs to be explored in more detail to uncover its advantages, applications, and other implications.

23.7.4 MICROBE-ASSISTED PHYTOREMEDIATION

Microbe-assisted phytoremediation processes can enhance the efficiency of remediation by reducing soil pH; changing redox potential; releasing chelators; and increasing metal transportation, solubility, or bioavailability of the metals (Kumar Yadav et al. 2018). Mycorrhiza are the plant associated fungal organization that mediate successful remediation of pollutants. The hyphal network of mycorrhiza is vastly extended which plays a crucial role in extracting heavy metals such as Co, S, Zn, Ca, K, Cu, P, and N from the soil (Sarwar et al. 2017). They also modify the pH of the soil and chemical composition of root exudates, which in turn increase the bioavailability of the heavy metals. It has also been observed that some mycorrhiza also prevent their associated plants from the toxic effects of heavy metals. *Eucalyptus rostrata* in association with *Glomus deserticola* (arbuscular mycorrhizae fungi) showed enhanced absorption and accumulation of Pb. The mycorrhizal association also played a role in preventing the plant from excessive lead toxicity (Bafeel 2008). Another important community of microorganisms for phytoremediation is PGPR. They can be grouped into two classes: free-living rhizobacteria and symbiotic bacteria. They assist in enhancing plant growth by showing specific enzyme activity, performing nitrogen fixation, and reducing ethylene production in stress conditions. Marques et al. (2013) reported that the inoculation of PGPR strains *Ralstonia eutropha* (B1) and *Chrysiobacterium humi* (B2) enhanced the growth of plants but reduced the Zn and Cd bioaccumulation in the roots and shoots of *Helianthus annuus* (sunflower).

23.7.5 BIOCHAR-ASSISTED PHYTOREMEDIATION

Biochar is a porous carbonaceous substance released as a product after the pyrolysis of organic substances such as sludges, organic manure, and plant material. The physico-chemical properties of biochar, such as high surface area, carbon content, cation exchange capacity, high pH, and high potential for heavy metal immobilization, make it a potential candidate for being used in the field of phytoremediation (Sarwar et al. 2017). In a study, four phytoremediation plants, *Thalia dealbata*, *Phragmites* sp., *Vetiveria zizanioides*, and *Salix rosthornii Seemen*, were used to study the potential of biochar for removing N and P from aqueous solutiona. Among these, *Thalia dealbata* was reported to show the best results (Zeng et al. 2013). When used in combination with traditional methods biochar is reported to increase the efficiency of the process due to its ability to enhance the growth and biomass production of the plant. Furthermore, the chemical present in biochars may also promote the growth of beneficial microorganisms while suppressing the growth of pathogens.

23.8 LIMITATIONS AND FUTURE PROSPECTS

In spite of being a highly-efficient and cost-effective technique for removing toxicants present in the environment, phytoremediation possesses some drawbacks which limit the large-scale application of this approach. The long time period requirement, low biomass production, low bioavailability of heavy metals, and low efficiency towards highly-contaminated soil are some of the major drawbacks. Phytoremediation can sometimes also lead to food chain contamination. Furthermore, in comparison to other environmental cleanup techniques, phytoremediation is relatively new. To date, only limited studies have been carried out at the actual site of contamination, and most of the studies are performed at laboratory or greenhouse scale. Laboratories and greenhouses offer highly-controlled environmental conditions, whereas the conditions at the actual site are much more varied as multiple factors are responsible for determining the actual efficiency of the method employed (Ali et al. 2013). In addition, the efficiency of different plant species in the remediation of specific metals needs to be tested in order to obtain optimum results. After choosing the ideal plant species with desirable traits, they can be modified by genetic engineering or conventional breeding to obtain much higher efficiencies, as genetically modified plants can be manipulated to tolerate and remediate high levels of heavy metals in the environment. Genetic engineering of plants for phytoremediation has already been quoted in many studies; however, there is still a lot of investigating to be done in this particular field. Currently, research is being performed on finding the hyperaccumulator genes in plants which can also be genetically engineered in order to produce "superbug" plants (Ali et al. 2013). Therefore, it becomes important to learn more about gene expressions and interactions to attain optimal results without causing unknown or unintended harm to the environment. Another way of advancing our knowledge in this field is to make collaborative efforts by involving different fields of biology, such as biochemistry, microbiology, nanotechnology, and plant physiology in the process of phytoremediation. Understanding the coordination chemistry of metals in the tissues of the plants is also a requirement for performing successful phytoremediation. Future studies in the field of phytoremediation can be based on identifying the plants species with herbivore-repelling substances to prevent the entry of heavy metals into the food chain, generating genetically manipulated plants showing high plant-microbe interaction or releasing chelating ligands which may assist in promoting the bioavailability of metals, transforming metal transporters and their site-specific targeting for safe compartmentalization of metals, and treating multi-contaminated areas with genetically modified organisms using a multigene transfer approach (Kumar Yadav et al. 2018). Thus, in spite of its many limitations, phytoremediation as a green approach is considered as one of the best remediation techniques with future prospects.

23.9 CONCLUSION

The increasing rate of heavy metal contamination from various natural and anthropogenic sources in the environment is a major cause for concern as these metal ions may enter the food chain and affect various life forms on the Earth. To date, many environmental cleanup strategies have been developed; however, due to their high cost, the release of secondary pollutants, and harm caused to soil microorganisms they are not highly supported. Phytoremediation, also known as "green technology," is an eco-friendly, cost-effective, solar-driven technique which can serve as an effective approach for solving this issue. It is a technique which exploits the native ability of plants to extract metal ions from their surrounding media. Using hyperaccumulator plants can help in the efficient sequestration of large concentrations of metals and organic or inorganic pollutants present in the environment. However, certain limitations can restrict the large-scale application of this approach, such as low biomass production, the extended time requirements, and the need for favorable conditions. Various modifications can be done to this technique by using microorganisms, genetically modified plants, chelating agents, nanomaterials, and biochar which can enhance the performance of phytoremediation to a great extent. Future studies will need to be conducted in this field to understand the mechanism of genes associated with phytoremediation for using genetically manipulated plants and microorganisms to improve the efficiency of phytoremediation.

REFERENCES

Abdel-Shafy, Hussein. 2018. "Phytoremediation for the Elimination of Metals, Pesticides, PAHs, and Other Pollutants from Wastewater and Soil Impact of Organic Co-Compost Application on Soil and Different Plant Sp. Grown on Sandy Soils View Project Sustainable Development For Wastewater Treatment and Reuse via Constructed Wetlands In Sinai (SWWTR) View Project." In: Phytobiont and *Ecosystem Restitution*, 101–36, December. Springer. doi:10.1007/978-981-13-1187-1_5.

Ahmed, Ambreen. 2018. "Heavy Metal Pollution: A Mini Review." *Journal of Bacteriology & Mycology: Open Access* 6(3): 179–81. doi:10.15406/jbmoa.2018.06.00199.

Ali, Hazrat, Ezzat Khan, and Muhammad Anwar Sajad. 2013. "Phytoremediation of Heavy Metals-Concepts and Applications." *Chemosphere* 91(7): 869–81. Elsevier Ltd. doi:10.1016/j.chemosphere.2013.01.075.

Ashraf, Sana, Qasim Ali, Zahir Ahmad Zahir, Sobia Ashraf, and Hafiz Naeem Asghar. 2019. "Phytoremediation: Environmentally Sustainable Way for Reclamation of Heavy Metal Polluted Soils." *Ecotoxicology and Environmental Safety* 174(February): 714–27. Elsevier Inc. doi:10.1016/j.ecoenv.2019.02.068.

Awa, Soo Hui, and Tony Hadibarata. 2020. "Removal of Heavy Metals in Contaminated Soil by Phytoremediation Mechanism: A Review." *Water Air and Soil Pollution*. Springer. 231(2), 1–15. doi:10.1007/s11270-020-4426-0.

Bafeel, Sameera O. 2008. "Contribution of Mycorrhizae in Phytoremediation of Lead Contaminated Soils by Eucalyptus rostrata Plants. " *World Applied Sciences Journal* 5(4): 490–98. https://www.researchgate.net/publication/242327347.

Bañuelos, G. S., Z. Q. Lin, I. Arroyo, and N. Terry. 2005. "Selenium Volatilization in Vegetated Agricultural Drainage Sediment from the San Luis Drain, Central California." *Chemosphere* 60(9): 1203–13. Elsevier Ltd. doi:10.1016/j.chemosphere.2005.02.033.

Bhargava, Atul, Francisco F. Carmona, Meenakshi Bhargava, and Shilpi Srivastava. 2012. "Approaches for Enhanced Phytoextraction of Heavy Metals." *Journal of Environmental Management*. 105, 103–20. doi:10.1016/j.jenvman.2012.04.002.

Chaffai, Radhouane, and Hiroyuki Koyama. 2011. "Heavy Metal Tolerance in Arabidopsis Thaliana." *Advances in Botanical Research* 60: 1–49. Academic Press Inc. doi:10.1016/B978-0-12-385851-1.00001-9.

Chen, Ming, Piao Xu, Guangming Zeng, Chunping Yang, Danlian Huang, and Jiachao Zhang. 2015. "Bioremediation of Soils Contaminated with Polycyclic Aromatic Hydrocarbons, Petroleum, Pesticides, Chlorophenols and Heavy Metals by Composting: Applications, Microbes and Future Research Needs." *Biotechnology Advances*. 33(6), 745–755. doi:10.1016/j.biotechadv.2015.05.003.

Cristaldi, Antonio, Gea Oliveri Conti, Eun Hea Jho, Pietro Zuccarello, Alfina Grasso, Chiara Copat, and Margherita Ferrante. 2017. "Phytoremediation of Contaminated Soils by Heavy Metals and PAHs. A Brief Review." *Environmental Technology and Innovation*. Elsevier B.V. 8, 309–26. doi:10.1016/j.eti.2017.08.002.

Cutright, Teresa J., and Harry van Keulen. 2008. "Phytoremediation of Heavy Metal Contaminated Media." *Heavy Metal Pollution*: 283–302.

Dinesh, Mani, Mourya Vishv Kumar, Pathak Neeraj, and Balak Shiv. 2014. "Phytoaccumulation of Heavy Metals in Contaminated Soil Using Makoy (Solenum Nigrum L.) and Spinach (Spinacia oleracea L.) Plant." *Sciences* 2(4): 350–54. www.ijlsci.in.

Duruibe, J. O., M. O. C. Ogwuegbu, and J. N. Egwurugwu. 2007. "Heavy Metal Pollution and Human Biotoxic Effects." *International Journal of Physical Sciences* 2. http://www.academicjournals.org/IJPS.

Fulekar, M. H., Anamika Singh, and Anwesha M. Bhaduri. 2009. "Genetic Engineering Strategies for Enhancing Phytoremediation of Heavy Metals." *African Journal of Biotechnology* 8(4): 529–35. http://www.academicjournals.org/AJB.

Galal, Tarek M., Fatma A. Gharib, Safia M. Ghazi, and Khalid H. Mansour. 2017. "Phytostabilization of Heavy Metals by the Emergent Macrophyte Vossia cuspidata (Roxb.) Griff.: A Phytoremediation Approach." *International Journal of Phytoremediation* 19(11): 992–99. Taylor and Francis Inc. doi:10.1080/15226 514.2017.1303816.

Hall, J. L. 2002. "Cellular Mechanisms for Heavy Metal Detoxification and Tolerance." *Journal of Experimental Botany.* 53, 1–11. doi:10.1093/jxb/53.366.1.

Hawrylak-Nowak, Barbara. 2013. "Comparative Effects of Selenite and Selenate on Growth and Selenium Accumulation in Lettuce Plants under Hydroponic Conditions." *Plant Growth Regulation* 70(2): 149–57. doi:10.1007/s10725-013-9788-5.

Joner, Erik J., Roberto Briones, and Corinne Leyval. 2000. "Metal-Binding Capacity of Arbuscular Mycorrhizal Mycelium." *Plant and Soil* 226(2): 227–34. Springer. doi:10.1023/A:1026565701391.

Kärenlampi, S., H. Schat, J. Vangronsveld, J. A. C. Verkleij, D. Van Der Lelie, M. Mergeay, and A. I. Tervahauta. 2000. "Genetic Engineering in the Improvement of Plants for Phytoremediation of Metal Polluted Soils." *Environmental Pollution* 107(2): 225–31. doi:10.1016/S0269-7491(99)00141-4.

Khan, Sardar, Abd El Latif Hesham, Min Qiao, Shafiqur Rehman, and Ji Zheng He. 2010. "Effects of Cd and Pb on Soil Microbial Community Structure and Activities." *Environmental Science and Pollution Research International* 17(2): 288–96. Springer. doi:10.1007/s11356-009-0134-4.

Kim, Rog Young, Jeong Ki Yoon, Tae Seung Kim, Jae E. Yang, Gary Owens, and Kwon Rae Kim. 2015. "Bioavailability of Heavy Metals in Soils: Definitions and Practical Implementation: A Critical Review." *Environmental Geochemistry and Health* 37(6): 1041–61. Kluwer Academic Publishers. doi:10.1007/s10653-015-9695-y.

Kumar Yadav, Krishna, Neha Gupta, Amit Kumar, Lisa M. Reece, Neeraja Singh, Shahabaldin Rezania, and Shakeel Ahmad Khan. 2018. "Mechanistic Understanding and Holistic Approach of Phytoremediation: A Review on Application and Future Prospects." *Ecological Engineering.* 120, 274–98 doi:10.1016/j.ecoleng.2018.05.039.

Kushwaha, Anamika, Radha Rani, Sanjay Kumar, and Aishvarya Gautam. 2015. "Heavy Metal Detoxification and Tolerance Mechanisms in Plants: Implications for Phytoremediation." *Environmental Reviews.* Canadian Science Publishing. 24, 39–51. doi:10.1139/er-2015-0010.

Liu, Zhaoming, Yan Wu, Chengfeng Lei, Pengming Liu, and Meiying Gao. 2012. "Chromate Reduction by a Chromate-Resistant Bacterium, Microbacterium sp." *World Journal of Microbiology and Biotechnology* 28(4): 1585–92. doi:10.1007/s11274-011-0962-5.

Ma, Ying, Rui S. Oliveira, Helena Freitas, and Chang Zhang. 2016. "Biochemical and Molecular Mechanisms of Plant-Microbe-Metal Interactions: Relevance for Phytoremediation." *Frontiers in Plant Science* 7(June). Frontiers Research Foundation. 7, 918. doi:10.3389/fpls.2016.00918.

Maitani, Tamio, Hiroki Kubota, Kyoko Sato, and Takashi Yamada. 1996. "The Composition of Metals Bound to Class III Metallothionein (Phytochelatin and Its Desglycyl Peptide) Induced by Various Metals in Root Cultures of Rubia Tinctorum." *Plant Physiology* 110(4): 1145–50. doi:10.1104/pp.110.4.1145.

Marques, Ana P. G. C., Helena Moreira, Albina R. Franco, António O. S. S. Rangel, and Paula M. L. Castro. 2013. "Inoculating Helianthus annuus (Sunflower) Grown in Zinc and Cadmium Contaminated Soils with Plant Growth Promoting Bacteria: Effects on Phytoremediation Strategies." *Chemosphere* 92(1): 74–83. Elsevier Ltd. doi:10.1016/j.chemosphere.2013.02.055.

Mirza, Nosheen, Arshid Pervez, Qaisar Mahmood, Mohammad Maroof Shah, and Mustafa Nawaz Shafqat. 2011. "Ecological Restoration of Arsenic Contaminated Soil by Arundo donax L." *Ecological Engineering* 37(12): 1949–56. Elsevier. doi:10.1016/j.ecoleng.2011.07.006.

Muthusaravanan, S., N. Sivarajasekar, J. S. Vivek, T. Paramasivan, Mu Naushad, J. Prakashmaran, V. Gayathri, and Omar K. Al-Duaij. 2018. "Phytoremediation of Heavy Metals: Mechanisms, Methods and Enhancements." *Environmental Chemistry Letters* 16(4): 1339–59. Springer. doi:10.1007/s10311-018-0762-3.

Parmar, Shobhika, and Vir Singh. 2015. "Phytoremediation Approaches for Heavy Metal Pollution: A Review Assessment of Heavy Metal Pollution in Yamuna Riverine Ecosystem. View Project Heavy Metals View Project." *Researchgate.Net.* https://www.researchgate.net/publication/286440107.

Pilon-Smits, Elizabeth. 2005. "Phytoremediation." *Annual Review of Plant Biology* 56: 15–39. doi:10.1146/annurev.arplant.56.032604.144214.

Rangsayatorn, N., E. S. Upatham, M. Kruatrachue, P. Pokethitiyook, and G. R. Lanza. 2002. "Phytoremediation Potential of Spirulina (Arthrospira) Platensis: Biosorption and Toxicity Studies of Cadmium." *Environmental Pollution* 119(1): 45–53. doi:10.1016/S0269-7491(01)00324-4.

Rashid, Audil, Tariq Mahmood, Faisal Mehmood, Azeem Khalid, Beenish Saba, Aniqa Batool, and Ammara Riaz. 2014. "Phytoaccumulation, Competitive Adsorption and Evaluation of Chelators-Metal Interaction in Lettuce Plant." *Environmental Engineering and Management Journal* 13(10): 2583–92. http://eemj.eu/index.php/EEMJ/article/view/2096.

Rezania, Shahabaldin, Shazwin Mat Taib, and Mohd Fadhil, F. A. Dahalan, and H. Kamyab. 2016. "Comprehensive Review on Phytotechnology: Heavy Metals Removal by Diverse Aquatic Plants Species from Wastewater." *Journal of Hazardous Materials*. Elsevier B.V. Md Din, Farrah Aini Dahalan, and Hesam Kamyab. 318, 587–99. doi:10.1016/j.jhazmat.2016.07.053.

Roy, Sébastien, Suzanne Labelle, Punita Mehta, Anca Mihoc, Nathalie Fortin, Claude Masson, René Leblanc, et al. 2005. "Phytoremediation of Heavy Metal and PAH-Contaminated Brownfield Sites." *Plant and Soil* 272(1–2): 277–90. doi:10.1007/s11104-004-5295-9.

Sarwar, Nadeem, Muhammad Imran, Muhammad Rashid Shaheen, Wajid Ishaque, Muhammad Asif Kamran, Amar Matloob, Abdur Rehim, and Saddam Hussain. 2017. "Phytoremediation Strategies for Soils Contaminated with Heavy Metals: Modifications and Future Perspectives." *Chemosphere* 171: 710–21. Elsevier Ltd. doi:10.1016/j.chemosphere.2016.12.116.

Sarwar, Nadeem, Wajid Ishaq, Ghulam Farid, Muhammad Rashid Shaheen, Muhammad Imran, Mingjian Geng, and Saddam Hussain. 2015. "Zinc-Cadmium Interactions: Impact on Wheat Physiology and Mineral Acquisition." *Ecotoxicology and Environmental Safety* 122(December): 528–36. Academic Press. doi:10.1016/j.ecoenv.2015.09.011.

Sharma, Parul, and Sonali Pandey. 2014. "Status of Phytoremediation in World Scenario." *International Journal of Environmental Bioremediation and Biodegradation* 2(4): 178–91. doi:10.12691/ijebb-2-4-5.

Singh, N. P., and Anita Rani Santal. 2015. "Phytoremediation of Heavy Metals: The Use of Green Approaches to Clean the Environment." In: *Phytoremediation: Management of Environmental Contaminants*, Volume 2, 115–29. Springer. doi:10.1007/978-3-319-10969-5_10.

Sinha, Rajiv K., Sunil Herat, and P. K. Tandon. 2007. "Phytoremediation: Role of Plants in Contaminated Site Management." In: *Environmental Bioremediation Technologies*, 315–30. Springer. doi:10.1007/978-3-540-34793-4_14.

Srivastav, Akansha, Krishna Kumar Yadav, Sunita Yadav, Neha Gupta, Jitendra Kumar Singh, Ravi Katiyar, and Vinit Kumar. 2019. "Nano-Phytoremediation of Pollutants from Contaminated Soil Environment: Current Scenario and Future Prospects." In: *Phytoremediation: Management of Environmental Contaminants*, Volume 6, 383–401. Springer. doi:10.1007/978-3-319-99651-6_16.

Sytar, Oksana, Marian Brestic, Nataliya Taran, and Marek Zivcak. 2015. "Plants Used for Biomonitoring and Phytoremediation of Trace Elements in Soil and Water." In: *Plant Metal Interaction: Emerging Remediation Techniques*, 361–84. Elsevier Inc. doi:10.1016/B978-0-12-803158-2.00014-X.

Tripathi, Rudra D., Sudhakar Srivastava, Seema Mishra, Nandita Singh, Rakesh Tuli, Dharmendra K. Gupta, and Frans J. M. Maathuis. 2007. "Arsenic Hazards: Strategies for Tolerance and Remediation by Plants." *Trends in Biotechnology*. doi:10.1016/j.tibtech.2007.02.003.

Vymazal, R. January 2016. "Concentration Is Not Enough to Evaluate Accumulation of Heavy Metals and Nutrients in Plants." *Science of the Total Environment* 544(February): 495–98. Elsevier B.V. doi:10.1016/j.scitotenv.2015.12.011.

Zeng, Zheng, Song-da Zhang, Feng-liang Zhao, He-ping Zhao, Yang Xiao-e, Hai-long Wang, Jing Zhao, and RAFIQ Muhammad Tariq. 2013. "Sorption of Ammonium and Phosphate from Aqueous Solution by Biochar Derived from Phytoremediation Plants *." *Journal of Zhejiang University. Science. Part B (Biomedicine and Biotechnology)* 14(12): 1152–61. doi:10.1631/jzus.B1300102.

Zhang, Weihua, Yong Cai, Cong Tu, and Lena Q. Ma. 2002. "Arsenic Speciation and Distribution in an Arsenic Hyperaccumulating Plant." *Science of the Total Environment* 300(1–3): 167–77. Elsevier. doi:10.1016/S0048-9697(02)00165-1.

Zhi, Junkai, Xiao Liu, Peng Yin, Ruixia Yang, Jiafu Liu, and Jichen Xu. 2020. "Overexpression of the Metallothionein Gene PaMT3-1 from Phytolacca americana Enhances Plant Tolerance to Cadmium." *Plant Cell, Tissue and Organ Culture* 143(1): 211–18. Springer. doi:10.1007/s11240-020-01914-2.

Zhu, Yi, Fang Xu, Qin Liu, Ming Chen, Xianli Liu, Yanyan Wang, Yan Sun, and Lili Zhang. 2019. "Nanomaterials and Plants: Positive Effects, Toxicity and the Remediation of Metal and Metalloid Pollution in Soil." *Science of the Total Environment* 662: 414–21. Elsevier B.V. 662:414–21. doi:10.1016/j.scitotenv.2019.01.234.

24 Biomembrane-Based Technology for the Efficient Removal of Industrial Effluents

Subhasish Dutta and Debanko Das

CONTENTS

24.1 INTRODUCTION

The Earth is covered with 326 cubic million miles of water, which makes up 71% of its total surface. Out of the total water content only 3% of Earth's water is fresh; the rest is in saline form locked in the oceans. Consumable water makes up only 0.5% while the rest is made up of glaciers and ice-capped mountains. According to a report by the WHO, more than a billion people do not have access to fresh potable water. More than ~3,000 minors die every day due to the consumption of polluted water (United Nations International Children's Emergency Fund 2015). People across the globe are facing severe health crises due to environmental hazards caused by human activities and the discharge of industrial effluents.

The lack of proper water treatment techniques will result in a rise in health issues if the discharge of potential emerging pollutants (PEPs) such as heavy metals, toxic chemicals, agricultural chemical waste, and sludge is not treated effectively (Deblonde et al. 2011, Stuart et al. 2012, Nasseri et al. 2018). Thus, a proper technique needs to be devised for the supply of potable water and maintaining proper health and hygiene. The most commonly-used industrial technique for water purification is distillation. The need for an economic, easy-to-handle, and energy-efficient method has led to the popularity of membrane-based separation (MBS) techniques (Sairam et al. 2006). Over the past few

DOI: 10.1201/9781003204442-24

years, various membrane-based separation techniques have evolved but only a handful of them have proven to be successful in large-scale industrial use (Ortego et al. 1995).

Waste originating from various industrial effluents that are discharged into water sources need to be identified, segregated, and disposed of in a proper way so that they do not pose a threat to the environment (Thomaidis et al. 2012). PEP-contaminated water gets mixed with various water streams that are used for human consumption. The handling of PEPs has been a major problem for various industries as the techniques used are usually inefficient in properly treating industrial waste-water as they fail to eliminate various toxic elements from the waste (Tijani et al. 2013). The use of membrane-based separation techniques is much more energy intensive than other conventional techniques such as bio-oxidation, coagulation, filtration, and sedimentation. Membrane-based separation techniques are expensive when it comes to initial installation; on the other hand they provide a far better recovery of chemical/toxic metal components from wastewater and have low maintenance costs in the long run (Mohammad et al. 2015).

The main objective of this chapter is to present an overview of the emerging biomembranes and how they can be used as a better option in segregating potential emerging pollutants from industry effluents over other membrane-based separation techniques.

24.2 POTENTIAL ENVIRONMENTAL POLLUTANTS (PEPs)

Potential environmental pollutants (PEPs) are defined as the presence of synthetic or naturally occurring chemical toxic elements that enter and pose a serious threat to the environment as well as human health (Geissen et al. 2015). More than 60% of PEPs are hazardous and toxic in nature. A study shows that the amount of PEPs produced over the years is constant and no decrease in PEP production has been observed due to the lack of proper separation techniques (Wang and Wang 2016). Presently, more than 700 new potential emerging pollutants are listed in the environment. In some cases, the presence of potential emerging pollutants remains unmonitored due to the lack of proper detection methods. On the other hand, the synthesis of chemicals or the use of toxic metals with their improper disposal causes new source of emerging pollutants. Emerging pollutants can be released from point-pollution sources like wastewater treatment plants from industries, or are diffused through atmospheric deposition in various practices such as animal husbandry and agriculture. Emerging pollutants are categorized on the basis of their origin (Geissen et al. 2015).

The major categories of PEPs are as follows:

- Agriculture (pesticides)
- Industry (industrial chemicals)
- Disinfection by-products and pesticides
- Pharmaceutical waste

24.3 CHALLENGES IN MONITORING PROGRAMS

Quantitative analyses, the detection of sources, and the identification of emerging pollutants are essential for gaining knowledge in the process of devising an effective method. There are many challenges involved in this process.

Firstly, the number of known potential emerging pollutants is very high. The calculated value rises to more than 700 in number and they transform over a period of time with changes in the chemical composition of the raw materials used in various industries and agricultural units (von der Ohe et al. 2011).

Secondly, the range of emerging pollutants is vast and with varied physical and chemical properties, such as organic biowaste which can be divided into persistent bioaccumulative (PBTs) and persistent organic pollutants (POPs). Particulate matters such as nanoparticles and microplastics also fall under emerging pollutants.

24.4 CONVENTIONAL MEMBRANE-BASED SEPARATION TECHNIQUES FOR THE SEGREGATION OF POTENTIAL EMERGING POLLUTANTS

The membrane-based separation process is one of the best methods for the separation of particulate metals as they provide a variety of applications in various fields (Lee et al. 2016). These membranes are usually made up of polymers, zeolites, or ceramic. They have various filtering features which depend on membrane morphology, hydrophobicity or hydrophilicity, pore size, and surface charge. The mode of separation depends on either solution diffusion, molecular diffusion, or the size exclusion principle (Munk and Aminabhavi 2002) (Table 24.1).

24.4.1 MICROFILTRATION

Microfiltration basically involves the separation of colloidal particles, organic biowaste, and high molecular weight from pollutant-contaminated sources. Microfiltration is applied along with other conventional methods for the removal of emerging pollutants. Domestic wastewater contains endocrine disrupters such as 17α-ethynylestradiol (EE2), estrone (E1), 17β-estradiol (E2), and BPA which, even in small traces, can severely damage the human endocrine system (Vandenberg et al. 2007).

In a research done by Han et al. (2013), a method was devised for cross-flow microfiltration by using a polythene sulfone (PES), nitrocellulose, regenerated cellulose, polyester, cellulose acetate (CA), and polyamide-66 (PA) (Han et al. 2013). The surface absorption characteristics of some PEPs at high concentrations severely affect the membrane performance which in turn causes membrane fouling. Membrane fouling takes place due to the presence of organic waste remains in the channel which affects the flux of polyamide-66 membranes.

24.4.2 ULTRAFILTRATION

Ultrafiltration has a solute rejection system of above 2 kilo Dalton molecular weight, considering the fact that most of the potential emerging pollutants fall under the range of macromolecular particles. Ultrafiltration can successfully filter out a large number of PEPs from wastewater outlets (Acero et al. 2010). In industrial cases, ultrafiltration is used along with nano-filter (NF) membranes to achieve a certain level of purity for treating the industrial effluents.

24.4.3 NANOFILTRATION

Nanofiltration has a much tighter pore setup than ultrafiltration which helps it to filter industrial effluents from dye industries and divalent salts from various chemical industries. They have a pore size of 1–10 nanometres which are made up of thin polymer films. The materials that are commonly used in nanofiltration are polyethylene terephthalate or light metals such as aluminium (Akin 2007). Pore dimensions are controlled by temperature and pH.

TABLE 24.1

Membrane Filtration Spectrum by Various Filtration Techniques

	Micrometre μm	Particulate Matters	Filtration Techniques	Reference
Macro particle	100~1,000 μm	Coal particles, sand	Particle filtration	(Dharupaneedi et al. 2019)
Micro particle	1.0~100 μm	Hair, pollen	Microfiltration	(Lee et al. 2016)
Macro molecular	0.1~1.0 μm	Yeast, bacteria, red blood cell	Ultrafiltration	(Dharupaneedi et al. 2019)
Molecular	0.01~0.1 μm	Virus	Nanofiltration	(Dharupaneedi et al. 2019)
Ionic	0.0001~0.01 μm	Aqueous salt, metal ions	Reverse osmosis	(Lee et al. 2016)

24.4.4 REVERSE OSMOSIS

Reverse osmosis has gained popularity in the recent years for the purification of drinking water. Reverse osmosis is combined with nanofiltration for the treatment of potential effluent pollutants from industrial wastewater treatment (Wang et al. 2018). N-nitroso dimethylamine (NDMA) and N-nitrosamine are potential carcinogens which are usually found in wastewater (Mitch et al. 2003). The toxic level of NDMA in potable water is estimated at 10 ng/L. In a study it was observed that the solid phase extraction (SPE) method was successfully used for the detection of NDMA. The use of reverse osmosis combined with thin-film composite membranes showed a rejection rate of ~50–60% (Plumlee et al. 2008).

24.4.5 FORWARD OSMOSIS

Forward osmosis method is based on osmotic gradient which is created by the high concentration gradient. Forward osmosis is not as prone to fouling as any other membrane-based separation techniques which helps in using it in a wider field of practises(Cath et al. 2006). Forward Osmosis can operate at very low hydrostatic pressure making it more sustainable method for water treatment(Dharupaneedi et al. 2019).

24.5 COMPLEX TECHNOLOGIES FOR THE SEPARATION OF POTENTIAL EMERGING POLLUTANTS

24.5.1 PHOTOCATALYTIC MEMBRANES/REACTORS (PMs/PMRs)

Photolytic membranes or photolytic membranes have been used in recent years because they have been proven to provide more synergistic advantages for potential emerging pollutants. PMs in combination with other membrane-based separation techniques prove to be more effective in the removal of potential emerging effluents. The photocatalysts present in suspended form can mineralize the organic components in the sample, thus reducing the fouling effect (Grzechulska and Morawski 2003).

24.5.2 MEMBRANE BIOREACTORS

Membrane bioreactors separate themselves from conventional activated sludge treatment techniques by the inclusion of membrane modules and aeration steps. One of the benefits of using membrane bioreactors is that it can run the biodegradation and separation processes at the same time. In a study, both membrane bioreactors and conventional activated sludge treatment was used to remove waste efficiently under different operating conditions (Radjenovic et al. 2007). Membrane bioreactors show ~80% more efficiency than conventional activated sludge treatment methods. The membrane part of a membrane bioreactor consists of microfiber regimes (0.4 μm). In some cases, it was found that low molecular weight cut-off membranes have been used to increase the efficiency of membrane bioreactors (Zaviska et al. 2013).

24.6 THE USE OF CELLULOSE-BASED HYBRID MEMBRANES FOR THE REMOVAL OF INDUSTRIAL EFFLUENTS

Conventional methods were able to separate oil–water mixtures but they failed in the separation of oil–water emulsions (Cheryan and Rajagopalan 1998). They not only lack efficiency but also have a high setup cost which make them unsuitable for use by large-scale industries. Materials that can be super-wetted with variable affinities can be used for the separation of water and oil as they are intrinsically immiscible (Yuan et al. 2008).

Super oleophilic and ultra-hydrophobic materials such as polytetrapolyethylene-coated mesh, polysiloxane-based gel, and so on have been developed for the separation of oil–water mixtures either by affinity absorption or membrane filtration (Lu et al. 2009, Tian et al. 2011, Deng et al. 2013). In recent years, materials which are graphene based have been used as sorbents as they exhibit efficient uptake capabilities. The major disadvantages in this method are that it is not economical and the availability of raw materials for practical application is scarce (Novoselov et al. 2004, Allen et al. 2010, Hu et al. 2014). The use of natural sorbents can solve this problem to a great extent as they are easily available, thus cutting the cost of raw material extraction, and do not harm the environment. Some examples of natural sorbents such as wool fibers (Radetić et al. 2003), activated carbon (Olsson et al. 2010), and zeolites (Gui et al. 2010) have shown promising results in this field. These natural sorbents are biodegradable which makes them perfect for proper disposal after use. It is observed that by increasing the hydrophilic characteristics of a membrane, the fouling can be reduced to a large extent (Jönsson and Jönsson 1995, Cornelissen et al. 1998, Hang et al. 2003). Many studies have been done on the hydrophilization of the ultrafiltration membrane, which are referred to in this chapter (Xu et al. 1999a, b, Pieracci et al. 2002, Ochoa et al. 2003, Shen et al. 2003, Ying et al. 2003). In some studies, a hydrophilic polymer is often blended to obtain the desired result. An example of a commonly-used hydrophilic polymer is polyvinylpyrrolidone, also known as PVP (Xu et al. 1999b). In another study, sulfonated polyetherimide was used to obtain a hydrophilic polymer for ultrafiltration membranes (Shen et al. 2003).

Cellulose shows hydrophilic properties as it has three active hydroxyl groups in each repeating unit of cellulose molecule, represented in **Figure 24.1**. This property of cellulose makes it an ideal choice for being used as a raw material. Cellulose being abundantly available in nature is another cause for being the choice of raw material (Fink et al. 2001). Due to the strong hydrogen bonds between the cellulose chains, they do not dissolve or melt in any ordinary solvent. Cellulose from various sources that are abundantly available in nature is used to produce cellulose acetate (CA) which can be regenerated to form cellulose membranes. These can be widely used in cellulose membrane-based separation (CMBS) techniques such as ultrafiltration, micro filtration, and dialysis (Sivakumar et al. 1998, Zhou et al. 2002, Chen et al. 2004).

In a recent study by Li et al. (2006), cellulose was used with a high concentration of alpha-cellulose and a high degree of polymerization. The raw material was dried for a prolonged period of time until no weight loss was observed. Highly hydrated N-methylmorpholine-N-oxide was used in the preparation of the dope along with other chemical components. Scanning electron microscope (SEM) images of the cellulose-based membrane showed that the outer skin dominated the transport resistance of the fiber. In this experiment, it was demonstrated that the cellulose-based membrane was able to hold a retention factor of over 99%, and a concentration of oil as low as 10 mgL^{-1} was obtained in the filtrate. This experiment proves that the cellulose-based membrane technique is not resistant to fouling but it has a wide pH tolerance. In a study by Kollarigowda et al. (2017), a cellulose membrane was changed from super hydrophilic to super hydrophobic using the reversible addition fragmentation chain transfer technique (RAFT). They used a block co-polymer of Poly[(3-(trimethoxysilyl)propyl acrylate)-block-myrcene]. The absorption test showed superior levels of extraction and a high level of reusability.

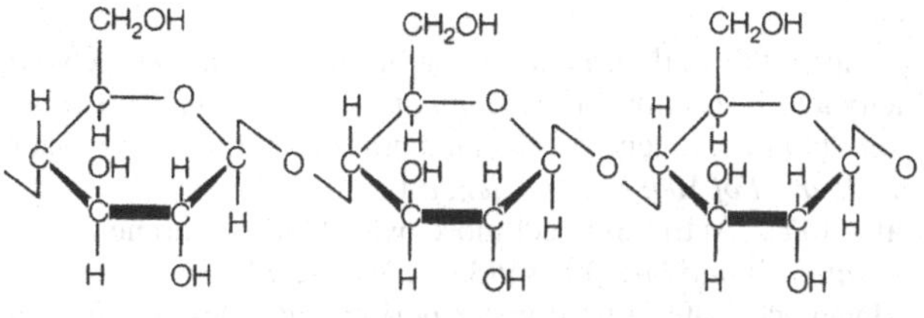

FIGURE 24.1 The structure of cellulose which shows three active hydroxyl groups.

24.7 CONCLUSION

The major reasons for the use of cellulose-based membrane separation techniques are its natural abundance, lack of toxicity, ease of processing, and ease of disposal. The fact that cellulose-based membrane separation techniques are biodegradable in nature plays a vital role in its selection over any other conventional materials and methods. Studies conducted by Kollarigowda et al. (2017) on cellulose-based membrane separation techniques prove that cellulose membranes not only display high levels of extractive properties but can also be reused for a long period of time. When cellulose are chemically treated they act as an anti-fouling agent in the treated water.

Overall, cellulose-based membrane separation techniques are versatile, promising in the long run, and an eco-friendly alternative to conventional separation techniques (Zhou et al. 2002, Ochoa et al. 2003). Li et al. (2006) have done extensive research which proves the fact that cellulose-based separation techniques are a better than conventional techniques. These studies have proven the technique to be more sustainable and eco-friendly for the environment as a whole.

REFERENCES

Acero, J. L., et al. (2010). "Retention of emerging micropollutants from UP water and a municipal secondary effluent by ultrafiltration and nanofiltration." *Chemical Engineering Journal* 163(3): 264–272.

Akin, D. (2007). "Nanotechnology in biology and medicine: Methods, devices, and applications. Edited by Tuan Vo-Dinh." *ChemMedChem: Chemistry Enabling Drug Discovery* 2(10): 1534–1535.

Allen, M. J., et al. (2010). "Honeycomb carbon: A review of graphene." *Chemical Reviews* 110(1): 132–145.

Cath, T. Y., et al. (2006). "Forward osmosis: Principles, applications, and recent developments." *Journal of Membrane Science* 281(1–2): 70–87.

Chen, Y. (2004). "Physical properties of microporous membranes prepared by hydrolyzing cellulose/soy protein blends." *Journal of Membrane Science* 241(2): 393–402.

Cheryan, M. and N. Rajagopalan. (1998). "Membrane processing of oily streams. Wastewater treatment and waste reduction." *Journal of Membrane Science* 151(1): 13–28.

Cornelissen, E., et al. (1998). "Physicochemical aspects of polymer selection for ultrafiltration and microfiltration membranes." *Colloids and Surfaces A: Physicochemical and Engineering Aspects* 138(2–3): 283–289.

Deblonde, T., et al. (2011). "Emerging pollutants in wastewater: A review of the literature." *International Journal of Hygiene and Environmental Health* 214(6): 442–448.

Deng, D., et al. (2013). "Hydrophobic meshes for oil spill recovery devices." *ACS Applied Materials* 5(3): 774–781.

Dharupaneedi, S. P., et al. (2019). "Membrane-based separation of potential emerging pollutants" Separation and *Purification Technology* 210: 850–866.

Fink, H.-P., et al. (2001). "Structure formation of regenerated cellulose materials from NMMO-solutions." *Progress in Polymer Science* 26(9): 1473–1524.

Geissen, V., et al. (2015). "Emerging pollutants in the environment: A challenge for water resource management." *International Soil and Water Conservation Research* 3(1): 57–65.

Grzechulska, J. and A. W. Morawski. (2003). "Photocatalytic labyrinth flow reactor with immobilized P25 TiO2 bed for removal of phenol from water." *Applied Catalysis B: Environmental* 46(2): 415–419.

Gui, X., et al. (2010). "Carbon nanotube sponges." *Advanced Materials* 22(5): 617–621.

Han, J., et al. (2013). "Capturing hormones and bisphenol A from water via sustained hydrogen bond driven sorption in polyamide microfiltration membranes." *Water Research* 47(1): 197–208.

Hang, S. (2003). "Oil adsorption measurements during membrane filtration." *Journal of Membrane Science* 214(1): 93–99.

Hu, H., et al. (2014). "Compressible carbon nanotube–graphene hybrid aerogels with superhydrophobicity and superoleophilicity for oil sorption." *Environmental Science and Technology Letters* 1(3): 214–220.

Jönsson, C. and A. Jönsson. (1995). "Influence of the membrane material on the adsorptive fouling of ultrafiltration membranes." *Journal of Membrane Science* 108(1–2): 79–87.

Kollarigowda, R. H., et al. (2017). "Antifouling cellulose hybrid biomembrane for effective oil/water separation" ACS *Applied Materials and Interfaces* 9(35): 29812–29819.

Lee, A., et al. (2016). "Membrane materials for water purification: Design, development, and application." *Environmental Science: Water Research and Technology* 2(1): 17–42.

Li, H., et al. (2006). "Development and characterization of anti-fouling cellulose hollow fiber UF membranes for oil–water separation." *Journal of Membrane Science* 279(1–2): 328–335.

Lu, P., et al. (2009). "Macroporous silicon oxycarbide fibers with luffa-like superhydrophobic shells." *Journal of the American Chemical Society* 131(30): 10346–10347.

Mitch, W. A., et al. (2003). "N-nitrosodimethylamine (NDMA) as a drinking water contaminant: A review." *Environmental Engineering Science* 20(5): 389–404.

Mohammad, A. W., et al. (2015). "Nanofiltration membranes review: Recent advances and future prospects." *Desalination* 356: 226–254.

Munk, P. and T. M. Aminabhavi. (2002). *Introduction to Macromolecular Science*. Wiley, New York. Copyright © 1993 WILEY-VCH Verlag GmbH & Co. KGaA, Weinheim.

Nasseri, S., et al. (2018). "Synthesis and characterization of polysulfone/graphene oxide nano-composite membranes for removal of bisphenol A from water." *Journal of Environmental Management* 205: 174–182.

Novoselov, K. S., et al. (2004). "Electric field effect in atomically thin carbon films" 306(5696): 666–669.

Ochoa, N. (2003). "Effect of hydrophilicity on fouling of an emulsified oil wastewater with PVDF/PMMA membranes." *Journal of Membrane Science* 226(1–2): 203–211.

Olsson, R. T., et al. (2010). "Making flexible magnetic aerogels and stiff magnetic nanopaper using cellulose nanofibrils as templates." *Nature Nanotechnology* 5(8): 584–588.

Ortego, J. D., et al. (1995). "A review of polymeric geosynthetics used in hazardous waste facilities." *Journal of Hazardous Materials* 42(2): 115–156.

Pieracci, J. (2002). "Increasing membrane permeability of UV-modified poly (ether sulfone) ultrafiltration membranes." *Journal of Membrane Science* 202(1–2): 1–16.

Plumlee, M. H., et al. (2008). "N-nitrosodimethylamine (NDMA) removal by reverse osmosis and UV treatment and analysis via LC–MS/MS." *Water Research* 42(1–2): 347–355.

Radetić, M. M., et al. (2003). "Recycled wool-based nonwoven material as an oil sorbent." *Environmental Science and Technology* 37(5): 1008–1012.

Radjenovic, J., et al. (2007). "Analysis of pharmaceuticals in wastewater and removal using a membrane bioreactor." *Analytical and Bioanalytical Chemistry* 387(4): 1365–1377.

Sairam, M., et al. (2006). "Polyaniline membranes for separation and purification of gases, liquids, and electrolyte solutions." *Separation and Purification Reviews* 35(4): 249–283.

Shen, L., et al. (2003). "Ultrafiltration hollow fiber membranes of sulfonated polyetherimide/polyetherimide blends: Preparation, morphologies and anti-fouling properties." *Journal of Membrane Science* 218(1–2): 279–293.

Sivakumar, M., et al. (1998). "Modification of cellulose acetate: Its characterization and application as an ultrafiltration membrane." *Journal of Applied Polymer Science* 67(11): 1939–1946.

Stuart, M., et al. (2012). "Review of risk from potential emerging contaminants in UK groundwater." *Science of the Total Environment* 416: 1–21.

Thomaidis, N. S., et al. (2012). "Emerging contaminants: A tutorial mini-review." *Global Nest Journal* 14(1): 72–79.

Tian, D., et al. (2011). "Micro/nanoscale hierarchical structured ZnO mesh film for separation of water and oil." *Physical Chemistry Chemical Physics* 13(32): 14606–14610.

Tijani, J. O., et al. (2013). "A review of pharmaceuticals and endocrine-disrupting compounds: Sources, effects, removal, and detections." *Water Air and Soil Pollution* 224(11): 1770.

United Nations International Children's Emergency Fund, WHO. (2015). "Progress on sanitation and drinking water: 2015 update and MDG assessment." from https://www.unicef.org/publications/index_82419.html.

Vandenberg, L. N., et al. (2007). "Use in food contact application." *US FDA*.

von der Ohe, P. C., et al. (2011). "A new risk assessment approach for the prioritization of 500 classical and emerging organic microcontaminants as potential river basin specific pollutants under the European water Framework Directive." *Science of the Total Environment* 409(11): 2064–2077.

Wang, J. and S. Wang (2016). "Removal of pharmaceuticals and personal care products (PPCPs) from wastewater: A review." *Journal of Environmental Management* 182: 620–640.

Wang, Y., et al. (2018). "Influence of wastewater precoagulation on adsorptive filtration of pharmaceutical and personal care products by carbon nanotube membranes." *Chemical Engineering Journal* 333: 66–75.

Xu, Z., et al. (1999a). "Effect of polyvinylpyrrolidone molecular weights on morphology, oil/water separation, mechanical and thermal properties of polyetherimide/polyvinylpyrrolidone hollow fiber membranes." *Journal of Applied Polymer Science* 74(9): 2220–2233.

Xu, Z., et al. (1999b). "Polymeric asymmetric membranes made from polyetherimide/polybenzimidazole/poly (ethylene glycol)(Pei/PBI/PEG) for oil–surfactant–water separation." *Journal of Membrane Science* 158(1–2): 41–53.

Ying, L., et al. (2003). "pH effect of coagulation bath on the characteristics of poly (acrylic acid)-grafted and poly (4-vinylpyridine)-grafted poly (vinylidene fluoride) microfiltration membranes." *Journal of Colloid and Interface Science* 265(2): 396–403.

Yuan, J., et al. (2008). "Superwetting nanowire membranes for selective absorption." *Nature Nanotechnology* 3(6): 332–336.

Zaviska, F., et al. (2013). "Nanofiltration membrane bioreactor for removing pharmaceutical compounds." *Journal of Membrane Science* 429: 121–129.

Zhou, J., et al. (2002). "Cellulose microporous membranes prepared from NaOH/urea aqueous solution." *Journal of Membrane Science* 210(1): 77–90.

25 Advances in the Photo-Oxidation of Nitro-Organic Explosives Present in the Aqueous Phase

Pallvi Bhanot, Anchita Kalsi, S. Mary Celin,
Sandeep Kumar Sahai, and Rajesh Kumar Tanwar

CONTENTS

DOI: 10.1201/9781003204442-25

441

25.1   INTRODUCTION

Explosives are reactive substances that decompose rapidly with the sudden release of large amounts of gases and energy. From a chemical point of view, most modern explosives are based on organic compounds containing nitrogen (N_2). They primarily find application in the defense sector. Explosives find their way into the environment via improper waste disposal during manufacturing; demilitarization activities; and the loading, assembly, and packaging (LAP) of munitions. Unexploded ordnance (UXO) also poses a serious threat (Pichtel 2012). These explosives are deposited onto the surface of soil and compounds are transferred into the soil layers, which leads to the contamination of groundwater resources. Commonly-used explosives like RDX (hexa hydro-1,3,5-trinitro-1,3,5-triazine) and HMX (octahydro-1,3,5,7-tetranitro-1,3,5,7-tetrazocine) are characterized by a low adsorption coefficient, leading to a higher risk of migration to groundwater (Pennington and Brannon 2002). The effluents released from the manufacturing industries can also contribute to the contamination of surface waters. The explosive compounds are undesirable in the environment due to their toxicity to both plant and animal life. Prolonged exposure to various explosives compounds can result in acute and chronic poisoning in human beings, viz., neurological, liver, and kidney damage. Moreover, the presence of explosives can also damage both the physical and chemical characteristics of soil. Hence, to mitigate these effects, the environmental remediation of nitro-organic explosives is of utmost importance.

In the last few decades, the rising demands of society for the decontamination and reclamation of severely polluted surface and groundwater resources has led to very stringent legislative guidelines for treatment technologies. This has led to the development of new, advanced, and more effective wastewater treatment technologies. Various treatment technologies such as physical (adsorption, incineration, flocculation, and filtration), chemical (photo-oxidation, hydrolysis, volatilization, and potassium permanganate), and biological (microbial and phytoremediation) are available which are being effectively used for treating anthropogenically-contaminated water resources. Owing to the complexity of the wastewaters contaminated with explosives, most of the above-mentioned conventional treatments systems fail to meet the requisite quality standards of the treated water. Hence, more effective wastewater treatment technologies are required to overcome the prevailing issues. The ultimate aim of any treatment technology should be the complete removal of the pollutants. Numerous treatment systems have been projected, tried, and practically applied to meet the expected treatment standards, but they suffer from major shortcomings which make them inefficient. Physical treatment technologies simply transfer the contaminants from one phase to another; whereas, biological treatment technologies could take a long time to obtain effective results. An efficient treatment system should provide the guaranteed removal of contaminants in the least amount of time. The chemical oxidation process could serve as an effective solution to the existing issue. Chemical oxidation processes basically use oxidants to treat explosive-contaminated effluents. They can be categorized into two types: conventional chemical processes and photo-oxidation processes. In the past few decades, a lot of work has been done to explore the effectiveness of photo-oxidation treatment systems for treating a diverse range of hazardous effluents that comprise recalcitrant and toxic waste products.

Photo-oxidation processes are based on the production of $^{\bullet}OH$, having extremely high electrochemical oxidation potential. Due to their high electrochemical oxidation potential, these strong oxidizing free radicals can easily and freely degrade all types of refractory organic complexes to CO_2, H_2O, and inorganic ions by hydroxylation or dehydrogenation. The most commonly studied photo-oxidation processes includes ozone-based treatment systems, Fenton processes, photo-Fenton processes, heterogeneous photocatalytic oxidation processes, and electrochemical oxidation processes. The classification of various photo-oxidation processes on the basis of $^{\bullet}OH$ source is shown in Figure 25.1. This chapter highlights the most recent advances and prospects of various photo-oxidation processes for the treatment of different types of simulated and actual nitro-organic explosive wastewater. We also describe the characteristics of explosive wastewaters and the limitations

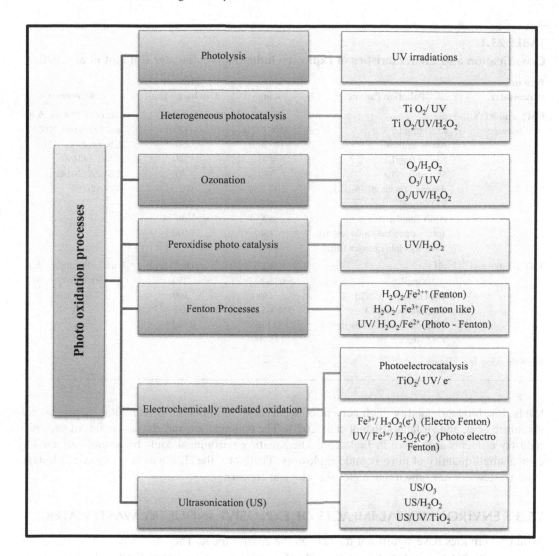

FIGURE 25.1 The classification of various photo-oxidation processes on the basis of •OH radicals as a source.

of the existing treatment techniques to efficiently remove them. In addition, it also explains the various applications of homogenous photo-oxidation processes for explosive effluent treatment, the mechanism of photo-oxidation reactions, and the major factors that affect the efficiency of the treatment processes. This chapter also provides comprehensive data on the degradation of various nitro-organic explosives by a wide variety of heterogeneous catalysts and adsorbents. Further, shortcomings of conventional photo-oxidation processes for employment at industrial level and the proposed ways to curb these drawbacks are also explained.

25.2 CHARACTERIZATION OF EXPLOSIVE INDUSTRY WASTEWATERS

Recalcitrant explosive industry effluents typically comprise very high concentrations of organic loads/compounds which is expressed as chemical oxygen demand (COD), hazardous explosive compounds [RDX, HMX, TNT (2,4,6, Trinitrotoluene), DNT (2,4, Dinitrotoluene)], colors (red water, pink water, and yellow water) based upon the type of explosive manufactured, very low pH (acidic)

TABLE 25.1

Classification and Characteristics of Explosive Industry Wastewater (Bhanot et al. 2020)

Type of Wastewater	Pollution Parameter	Range	Effluent Discharge Limit	Reference
HMX and RDX wastewater	pH	1–2.32	6.0–9.0	Raut et al. (2004), Zoh
	COD (mg/L)	350–3,420	125	and Stenstrom (2002),
	Acetic acid (%)	16–30%	50	Chen et al. (2011),
	Nitrate (mg/L)	400	200	USEPA (1997),
	Sulfate (mg/L)	350	50	USEPA (2014a),
	Ammonium nitrate (mg/L)	200–250	0.01*	EPA (2003)
	RDX (mg/L)	500	0.04*	
	HMX (mg/L)	500	50	
	Total suspended solids (mg/L)	260	100	
	Total Kjeldahl nitrogen (mg/L)	172		
TNT wastewater	pH	1–2.6	6.0–9.0	Barreto-Rodrigues et al.
	COD(mg/L)	638–673	250	(2009), USEPA (2014b)
	Total nitrogen (mg/L)	730	100	
	Total solids (mg/L)	1160–13260	50	
	Total suspended solids(mg/L)	1110	0.002*	
	TNT (mg/L)	145–160		

*EPA drinking water limits

levels, low biodegradability, high content of acetic acid, ammonium nitrate, total suspended solids, nitric acid, and sulfates (Bhanot et al. 2020). The composition and characteristics of explosive industry effluents are shown in Table 25.1. The aquatic environment could be severely affected by even a small quantity of nitro-organic explosives. Therefore, the elimination of organic pollutants from wastewater has grown into a global ecological concern.

25.3   ENVIRONMENTAL IMPACTS OF EXPLOSIVE INDUSTRY WASTEWATER

Military activities have significant impacts on the environment. They not only destroy the socio-environment, but also generate huge amounts of greenhouse gases that support anthropogenic climate change, increase terrestrial and aquatic pollution levels, and deplete resources. Explosives, their related energetic materials, and their accompanying degradation products persist in the environment as the inheritance of previous ammunition manufacturing, training, testing, and LAP processes (Talmage et al. 1999).

Many nitro-organic polluted locations hold explosives and propellants in soil, sediment, and groundwater or freshwater resources at concentrations that span several orders of magnitude (Jenkins et al. 2006). These toxic nitro-organic explosive compounds can have direct as well as indirect antagonistic consequence for human well-being and the environment. They are highly persistent in nature and can be easily transportable in the soil ecosystem. Explosive pollutants in soil can undergo metabolic transformations, catalytic degradation, and biodegradation procedures, for instance oxidation, dehydrogenation, reduction, hydrolysis, and exchange reactions (Davis et al. 2004). The migration of organic molecules by soil can happen via leaching, overspill, or volatilization (Pennington and Brannon 2002) and the contaminated soil acts as a cause of pollution to adjoining surface and groundwater resources.

Most of the nitro-organics enter the ecosystem and accumulate in soils, sediments, and water-beds. From there, these hazardous chemicals enter into the food chain of living beings and act similar to a poison (Pennington and Brannon 2002, Travis et al. 2008). Groundwater and freshwater systems are possibly the most susceptible to the effects of explosive industry wastewater, which is a concern as they are responsible for so many vital environmental resources and services to the world. The preservation of munitions in casings for long periods can also lead to the release of explosives as well as heavy metals into the environment. Heavy metals are well-known water contaminants (Harmon 2009), and several other nitro-organic explosives have shown numerous effects on marine creatures. Military activities and explosive pollution cause disturbances to the soil and vegetation in surrounding areas, leading to soil erosion and compaction. Soil erosion increases the loads of both mineral and organic sediments in the freshwater ecosystem, causing alterations in stream metabolic rates (Houser et al. 2005), decreases in macro invertebrate populations and community transferals, amplified riverbed flux, reduced coarse wood inputs (Maloney et al. 2005), and diminished diatom and fish concentrations. Soil compaction has the potential to disturb indigenous hydrology and cause flooding.

Generally, in many army installation contamination of groundwater and surface soil has been found due to RDX, HMX, and TNT. Specifically, RDX is of major environmental concern as it is easily soluble and resistant to microbial transformation in aerobic environments (Davis et al. 2004). RDX is not considerably taken up by most soils and can percolate to underground resources from soil due to its low soil sorption coefficient (K_{OC}) values (ATSDR 2012, EPA 2005). It may travel via the vadose zone and pollute fundamental groundwater resources, specifically at source regions having a superficial groundwater table and ample rainfall (USACE CRREL 2006, EPA 2012). It may also bio-accumulate in plants and might be a possible way for the exposure of herbivorous biota. RDX has been assigned under the category of a possible human carcinogen C (ATSDR 2012). When inhaled in large amounts, it targets the central nervous system and can cause seizures in human beings and in other creatures. Indications of excess exposure would be eye and skin irritations, headache, irritability, fatigue, tremor, nausea, dizziness, vomiting, insomnia, and convulsions (HSDB 2016). HMX is another nitramine explosive used commonly. It is a superior explosive to RDX, but has a very low solubility and adsorption coefficient. It is also highly resistant to hydrolysis and degradation processes. HMX may possibly be found mixed with other explosives, mainly TNT and RDX. Its effects are similar to those of RDX. TNT, another extensively-used military high explosive, has also been classified as a possible human carcinogen C (EPA IRIS 1993), and research on animals shows that the ingestion of large amounts of TNT can cause damage to the liver, blood, immune, and reproductive systems (EPA 2005). Potential exposure to explosives might take place by dermal contact, and inhalation exposure to a lesser extent. The most anticipated form of exposure to these hazardous compounds at/near explosive waste sites is the consumption of polluted potable water or crops watered with explosives-polluted water (ATSDR 2012).

The degradation rate of explosives in the aquatic ecosystem is more rapid as compared to the terrestrial biome, which make them significant for preservation and sustainable management. Upcoming studies in this field varies from clean-up information for explosives manufacturing and disposal, to sustaining military inclination by aptly managing training and testing ranges in an ecologically-liable way. Further investigation will be required to improve environmental and human health threat valuation approaches, to develop tools for the operative management of essential armed drills, and to decrease the antagonistic effects of hazardous explosive compounds on human health and the environment.

24.4 MECHANISM OF PHOTO-OXIDATION OF EXPLOSIVES

The idea of photo-oxidation/advanced oxidation processes was first recognized by Glaze et al. in 1987. This treatment system is based upon the production of highly reactive $^{\bullet}$OH radicals. The photo-oxidation treatment technology has grown and gained attention as an imperious system for

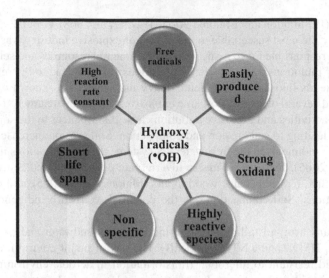

FIGURE 25.2   The characteristics of hydroxyl radicals (•OH).

the oxidation and degradation of recalcitrant pollutants present in aqueous solutions. While select-ing an oxidant for oxidation processes, the oxidant's oxidizing power should be considered. The relative oxidation power of •OH is 2.8 eV and it is the second strongest oxidizing species. •OH are primary oxidants, non-selective in nature, and are efficient to degrade refractory organic contami-nants by transforming them into less toxic by-products.

There are three possible ways through which •OH targets the organic compounds present in wastewater: (i) removal of a hydrogen atom to form H_2O, (ii) hydroxylation of a non-saturated bond, (iii) redox reaction. The characteristics of hydroxyl radicals which have established oxidation pro-cesses as influential systems for degrading hazardous organic pollutants are shown in Figure 25.2. Hydroxyl radical generation can be accomplished using different methods such as ozonation, Fenton-based processes, photochemical oxidation processes, electrochemical processes, photo-peroxidation processes, sonolysis, and photocatalyst oxidation processes. Oxidation processes are favorable for the reduction of toxic and biodegradation resilient explosive compounds present in explosive industry effluents. Much research has explored the application of these processes for explosive industry effluent treatment.

25.5   CONVENTIONAL PHOTO-OXIDATION TECHNIQUES

25.5.1   Photolysis

Photolysis processes for degrading contaminants present in wastewater are established on the basis of energy given to the contaminants in the light/radiation form, which is captivated by various chemical compounds to reach excited states, when chemical reactions are taking place. Radiant energy emitted by the light source is absorbed by the molecules from the quantized units known as photons. Energy provided by the photons excites the specific e^- and generates free radicals which go through a sequence of chain reactions to provide the reaction end products. The free radicals are produced by the transfer of electrons from the excited state to the molecular oxygen state of the organic contaminants and by the homolysis of weak bonds. Photolysis processes use ultraviolet (UV) rays due to photons having high energy as shown by the Planck's equation (Equation 25.1), given below. Therefore, in direct photolysis, UV radiation interacts with contaminants/molecules which lead to their breakdown into simpler molecules. The intensity and wavelength of the UV radiation affect the efficiency of the treatment system. Generally, mercury vapor lamps are used as a source of UV radiation. Humic acid, fulvic acid, and nitrate ions present in wastewater act as photosensitizers that intervene the oxidation process.

$$E = hc / \lambda \qquad (25.1)$$

$E\lambda$ = Photon's energy
h = Planck's constant
c = Speed of light

The photodegradation of explosive compounds present in wastewater is a genuine treatment technology that engages the photon energy from different sources leading to the division of bonds in the explosives to form simpler compounds. The chief driving force behind photolysis of explosive compounds is the high photon energy absorption capacity of NO_2 groups present in nitro-organic explosive compounds. The photodegradation ability of explosive compounds is influenced by the photon absorption capacity of the explosives at specific wavelengths. Dubroca et al. (2013) explored the absorption spectra of various explosives such as RDX, HMX, TNT, and tetryl in UV and visible ranges, and explained that the intermolecular hydrogen bonding of these explosive molecules was the main factor which affected the absorption capacity of the photons. There are many cases where investigators applied different types of radiation sources without using any catalysts, photosensitizers, or reagents in the treatment process. A summary of photolysis studies of explosives is given in Table 25.2.

25.5.2 UV/H$_2$O$_2$ System

UV radiation has been extensively applied for treating water and wastewater around the globe. Various studies have shown that the photolytic treatment is beneficial for the removal of nitro-organic explosive compounds present in wastewater. Though, this treatment system is only appropriate for wastewaters that have a low chemical oxygen demand (COD) and the presence of light-sensitive organic compounds. Generally, effluents released from explosive manufacturing sites have organic loads and CODs which might obstruct the treatment process. In such circumstances, the UV/H$_2$O$_2$ system is a good substitute. This process benefit from the dual actions of (i) the photolysis capability of the UV rays, and (ii) the reaction of the contaminants present, with the $^{\bullet}$OH produced by the homolytic breaking of the O–O bond in hydrogen peroxide. The reactions defining the UV/H$_2$O$_2$ system are given below (Equations 25.2–25.7) (Buxton et al. 1988). The UV/H$_2$O$_2$ system is a firm treatment process for mineralization of organic waste products. One of the drawbacks of this system is that it does not use solar radiation as the source of UV light because ultraviolet energy needed for oxidation does not exist in the solar spectrum (Niaounakis and Halvadakis 2006). The main parameters that might affect the treatment process are the initial organic load, the concentration of H$_2$O$_2$, the pH of the wastewater, the presence of bicarbonate ions, and time of reaction. A summary of photo-peroxidation studies of nitro-organic explosives is given in Table 25.2.

$$H_2O_2 + h\nu \rightarrow 2\,HO \qquad (25.2)$$

$$H_2O_2 + HO^{\bullet} \rightarrow HO_2^{\bullet} + H_2O \qquad (25.3)$$

$$H_2O_2 + HO_2^{\bullet} \rightarrow HO^{\bullet} + H_2O + O_2 \qquad (25.4)$$

$$2HO^{\bullet} \rightarrow H_2O_2 \qquad (25.5)$$

$$2\,HO_2^{\bullet} \rightarrow H_2O_2 + O_2 \qquad (25.6)$$

$$HO\bullet + HO_2^{\bullet} \rightarrow H_2O + O_2 \qquad (25.7)$$

TABLE 25.2

Various Conventional Photo-Oxidation Processes for Treating Nitro-Organic Explosives

Sl. No.	Treatment Process	Contaminant	Treatment Time (min)	Mode	Experimental Details	Findings	Reference
	Photolysis	RDX	8,400	Batch	Simulated sunlight from 1000 W metal halide lamp	• Photodegradation products such as N_2O, NO_2, and 4-nitro-2,4-diazabutanal were formed • 99% degradation of 8.45 mM RDX in aqueous solution	Just and Schnoor (2004)
		2,4-DNT, 2,6-DNT	300	–	Simulated solar radiations using 1.5 kW Xe lamp	• 85% and 27% removal of 55 mM 2,6 DNT and 2,4-DNT, respectively, was achieved in seawater • Dinitrobenzaldehydes and dinitrobenzoic acid intermediates were formed as degradation products	O'Sullivan et al. (2010)
1	UV/H_2O_2 system	TNT, RDX, HMX NB 1,3-DNB	25 5	Batch Batch	Low pressure mercury lamp of 15 W Hg intensity used as irradiation source H_2O_2 concentration used: 0.38–65 mM 125 W medium pressure Hg lamp was used as light source 1 kW medium-pressure UV lamp used as irradiation source	• In aqueous solution, near 100% degradation of 23–55 mg/L RDX, 4 mg/L HMX, and 29–80 mg/L TNT was observed • 100% degradation of 0.25 mM NB in presence of 10 mM H_2O_2 was observed • 97% elimination of 0.11 mM 1,3-DNB in presence of 9.3 mM H_2O_2 in 30 min was observed • 100% mineralization was achieved in 45 min • Intermediates such as resorcinol (RS), catechol (CC), hydroquinone (HQ), benzoquinone, nitrohydroquinone (NHQ), 4-nitrocatechol (4-NC), 2,4-dinitrophenol, phenol, 3-nitrophenol. and mineralization products such as NO_3^-, NO_2^- and CO_2 were formed	Alnaizy & Akgerma (1999) García Einschlag et al. (2002) Chen et al. (2004)

(Continued)

TABLE 25.2 (CONTINUED)

Various Conventional Photo-Oxidation Processes for Treating Nitro-Organic Explosives

Sl. No.	Treatment Process	Contaminant	Treatment Time (min)	Mode	Experimental Details	Findings	Reference
2	Ozone-based processes O$_3$/UV process O$_3$/H$_2$O$_2$	Groundwater polluted with high concentrations of explosive DNT isomers and 2,4,6-TNT in spent acid DDNP (Dinitrodiazophenol)	40–15	Continuous flow Batch Batch	• Low pressure UV of 12 W with continuous ozonation • 12 low-pressure Hg vapor lamps (8 W each) used as light source (254 nm) • Ozone dosage: 0–3.8 g h^{-1} • O$_3$ dosage: 18.92 mg/min • H$_2$O$_2$ dose: 18 mM • pH: 10	• Parent explosive compounds degraded below analytical detection limits (0.005 mg/l) within 40 min of treatment • 100% mineralization of TOC was achieved using 3.8 g/h ozone dosage • 82.29% COD reduction and 93.81% removal of color number	Zappi et al. in (2016) Chen et al. (2007) Gu et al. in (2019)
5	Fenton process	HMX 2, 4, 6-trinitrophenol (PA), ammonium picrate (AP), DNT, and TNT	80	Batch Batch	• pH: 3 • Temperature: 25°C • H$_2$O$_2$ dosage: 0.2 mL • FeSO$_4$.7H$_2$O dosage: 8.3 mL • Fe^{2+} concentration: 0.72 mM • H$_2$O$_2$ concentration: 0.29 M • pH: 3.0 • Temperature: 25°C	• 81.4% removal of HMX with reduction in COD concentration from 214 to 155 mg/L was observed • 80% of 2×10−4 M PA was degraded in 50 min • 90.4% of 2×10−4 M AP and 86% 2×10−4 M DNT removed in 20 min • 61% removal of TNT was observed under similar conditions	Cao and Li (2018) Liou and Lu (2007)

(*Continued*)

TABLE 25.2 (CONTINUED)
Various Conventional Photo-Oxidation Processes for Treating Nitro-Organic Explosives

Sl. No.	Treatment Process	Contaminant	Treatment Time (min)	Mode	Experimental Details	Findings	Reference
6	Photo-Fenton process	MNT, DNT, TNT 2,4-DNT Nitrophenols	15–60 60 120	Batch Batch Batch	• UV intensity: 21 W • H_2O_2 concentration: 0.15 M (0.5% v/v) • Fe^{2+} dosage: 0.72 mM (40 mg L^{-1}) • UV intensity: 125 W • Fenton's reagent: 10 ml/l having H_2O_2 and $FeSO_4 \cdot 7H_2O$ in 3:1 ratio • Natural sunlight as well as 150 W medium-pressure Hg UV used with Fenton's reagent • pH: 3	• 95% degradation of TNT and DNT in aqueous solution was achieved in 60 min and for MNT similar degradation was observed in 15 min at pH 3 • Oxalic acid, CO^2, NO^{3-}, and H_2O produced as by-products • 100% removal of 100 ppm DNT in 1 hr. • 96% and 57% removal of TOC and TN in 2 hr reaction time • >92% of 0.87 mM TNP was mineralized in solar Fenton as well as UV Fenton processes	Li et al. (1998) Celin et al. (2003) Kavitha and Palanivelu (2004)
6	Electrochemical oxidation	2,4-dinitrophenol (DNP) P-nitrophenol (PNP)	60 300	– Continuous flow	• Hydrodynamic cavitation coupled with chemical oxidants such as H_2O_2, ferrous activated persulfate, CuO, and Fenton's reagent • Orifice plate used as cavitating device • Cu/GAC used as catalyst • Microwave-assisted catalytic oxidation	• 100% degradation of DNP occurred in 60 min • 91.8% removal of PNP and 88% TOC removal observed	Bagal and Gogate (2014) Bo et al. (2008)

25.5.3 Ozone-Based Oxidation Processes

The ozonation-based oxidation method has been effectively applied for treating explosive effluents for the last two decades. This method of wastewater treatment is dual phase, comprising gas–liquid reactions. The ozonation process consists of O_3 and hydroxyl radicals. It involves O_3 in liquid form, which disintegrates via a chain of chemical reactions leading to the production of hydroxyl radicals.

O_3 targets the contaminants present in wastewater generally by two means: (i) under acidic conditions, molecular ozone is dominant and organic pollutants are exposed to electrophilic interactions leading to their decomposition, and (ii) at higher pH values, ozone disintegrates into free •OH which in turn react with the organic contaminants rapidly. O_3 is one of the strongest oxidizing agents with relative oxidation power (E_0) of 2.07 V. Therefore, it can easily and efficiently degrade the nitro-aromatic compounds present in explosive wastewater. The conditions that promote the production of molecular ozone and •OH fluctuate with the features of the contaminants, such as the water quality, concentration of pollutants, pH, temperature, and level of ozonation.

25.5.3.1 O_3/UV Process

Ozonation integrated with UV irradiation is more efficient in degrading organic contaminants in comparison to ozonation treatment alone (Fu et al. 2015). In the O_3/UV system, UV photons are used to fast track the decomposition of O_3 molecules to generate strong hydroxyl radicals (Glaze et al. 1987). On exposure to UV radiation at a wavelength of 254 nm, ozone absorbs the UV rays and undergoes decomposition. Throughout the course of the reaction, atomic oxygen with a strong oxidizing capacity is produced. Combining UV radiation with ozonation lessens the concentration of ozone required and also constrains the formation of bromate. Ozone/UV-based treatment technology transforms organic compounds into volatile substances such as H_2O, CO_2, and N_2. The combined effects of the ozone and UV systems (as described in Equations 25.8 and 25.9) result in the improved oxidation of organic compounds. The application of sequential ozonation and ultraviolet radiation that considerably accelerates the degradation and mineralization process of the nitro-organic compounds in wastewater has also been studied, and is summarized in Table 25.2.

$$O_3 + H_2O + h\nu \leftrightarrow O_2 + H_2O_2 \tag{25.8}$$

$$H_2O_2 + h\nu \leftrightarrow 2\,^\bullet OH \tag{25.9}$$

25.5.3.2 O_3 combined with H_2O_2

The addition of other oxidants to the ozonation treatment can enhance its efficacy of degradation of organic pollutants. The O_3/H_2O_2 treatment method is known as an efficient chemical oxidation process to degrade refractory organics present in effluents. When H_2O_2 is mixed with ozone, it generates •OH which in turn upsurges the decomposition rate of O_3. H_2O_2 reacts with ozone once it is partly disassociated into HO_2^- (hydroperoxide anion). Earlier research has revealed that superior effects might be attained by treating wastewater with O_3/H_2O_2 in comparison to ozonation alone. The reaction mechanism involved in the O_3/H_2O_2 process is given below in Equations 25.10–25.13 (Rosenfeldt et al. 2006). The combined reactions of O_3/H_2O_2 demonstrate that two ozone molecules lead to the generation of two •OH, which can hasten the process as well as increase its efficiency. Much work has been reported, documenting the use of O_3/H_2O_2 for the treatment of explosives in water (Table 25.2).

$$H_2O_2 \rightarrow HO_2^- + H^+ \tag{25.10}$$

$$HO_2^- + O_3 \rightarrow HO_2 + O_3^- \tag{25.11}$$

$$H_2O_2 + O_3 \rightarrow H_2O + 3O_2 \tag{25.12}$$

$$2O_3 + H_2O \rightarrow 3OH_2 + 3O_2 \tag{25.13}$$

25.5.4 FENTON-BASED PROCESSES

The Fenton system is one of the most significant photo-oxidation processes, and the chemistry of reactions associated with this process consists of reactions of peroxides with Fe^{2+} to produce reactive oxygen species that can reduce the organic and inorganic pollutants present in aqueous solutions. Fenton chemistry came into being in 1894 when the activation of hydrogen peroxide by Fe^{2+} salts was studied by Henry J. Fenton for oxidizing tartaric acid. In 1934, Haber and Weiss (1934) suggested that Fenton's reaction generated •OH with an oxidation potential of 2.73 V and therefore could be employed for degrading organic contaminants present in wastewaters. There are two types of Fenton reactions, namely homogenous and photo-Fenton, which have been explained below.

25.5.4.1 Homogeneous Fenton Reactions

The Fenton system is a cost-effective and easy process where highly reactive oxygen species are produced to degrade pollutants. Ferrous salt used in the Fenton's reagent is economical and safe to handle. Also, hydrogen peroxide used is economical, easy to use, and can decompose very easily to H_2O and O_2. The reaction mechanism of the decomposition of H_2O_2 to generated •OH is represented below in Equations (25.14)–(25.20).

$$Fe^{2+} + H_2O_2 \rightarrow Fe^{3+} + OH^- + {}^{\bullet}OH \tag{25.14}$$

$$Fe^{3+} + H_2O_2 \rightarrow Fe^{2+} + HO_2^{\bullet} + H^+ \tag{25.15}$$

$${}^{\bullet}OH + H_2O_2 \rightarrow HO_2^{\bullet} + H_2O \tag{25.16}$$

$${}^{\bullet}OH + Fe^{2+} \rightarrow Fe^{3+} + OH^- \tag{25.17}$$

$$Fe^{3+} + HO_2^{\bullet} \rightarrow Fe^{2+} + O_2H^+ \tag{25.18}$$

$$Fe^{2+} + HO_2^{\bullet} + H^+ \rightarrow Fe^{3+} + H_2O_2 \tag{25.19}$$

$$HO_2^{\bullet} + HO_2^{\bullet} \rightarrow H_2O_2 + O_2 \tag{25.20}$$

•OH radicals which are necessary for the treatment of organic pollutants are generated by reaction, shown in Equations (25.21) and (25.22). It is a rate-limiting reaction which is slow by numerous orders of magnitude. The organic substances (RH/R) may be reacted by one or amalgamations, •OH radicals, abstraction of hydrogen (•R), and the addition of hydroxyl radicals (•ROH) (Walling and Kato 1971).

$$RH + {}^{\bullet}OH \rightarrow H_2O + {}^{\bullet}R \rightarrow \text{advance oxidation} \tag{25.21}$$

$$R + {}^{\bullet}OH \rightarrow {}^{\bullet}ROH \rightarrow \text{advance oxidation} \tag{25.22}$$

Either Fe^{2+} or H_2O_2 can forage •OH, therefore it is essential to optimize the concentration of Fe^{2+}/H_2O_2 to decrease their scavenging action. The main elements that disturb the Fenton system are the pH of the aqueous solution, the concentration of Fe^{2+} and H_2O_2, the load of toxins, and the existence of additional ions. The optimal pH range required for effective treatment by Fenton's reagent is in the range of 2–4. At higher pH (greater than 4) the ferrous ions become unsteady and simply get converted to ferric ions, making complexes with $^{\bullet}$OH. Also, in high pH environments, H_2O_2 decomposes into oxygen and water and loses its oxidation energy (Niaounakis and Halvadakis 2006). Therefore, the pH adjustment of wastewater is very much essential before treating it with Fenton's process. Increasing H_2O_2 and ferrous ion concentrations marks the increment in the degradation rate of the contaminants, though high concentration of H_2O_2 will possibly affect the overall degradation effectiveness of the treatment process. The presence of various ions in wastewater, such as phosphate, sulfate, fluoride, bromide, and chloride ions, may inhibit the Fenton oxidation process. Literature supporting the use of the homogenous Fenton oxidation process for the treatment of explosives has been given in Table 25.2.

25.5.5 PHOTO-FENTON PROCESSES

The photo-Fenton process is a combination of the Fenton process and photolysis. Fenton's reagent, in grouping with UV radiation, accelerates the rate of reaction and effectiveness for degrading organic compounds in wastewater. The increase in treatment efficacy is associated with the photochemistry of $[Fe^{3+} (OH)-]^+$ and $[Fe^{3+}(RCO_2)-]^{2+}$ which disassociate into Fe^{2+}. The photochemistry of Fe^{3+} gives advantage to the Fenton treatment method as the abridged Fe^{2+} reacts with H_2O_2, thus making $^{\bullet}$OH. Figure 25.3 exhibits the mechanism of treatment with the photo-Fenton process. The best results given by photo-Fenton reaction are under acidic conditions (pH 2.8). As the pH rises (from the optimal value), the Fe^{3+} precipitates as oxyhydroxides. Whereas, as the pH drops below the optimal, the amount of $Fe(OH)_2^+$ may drop (Khataee et al. 2015). $Fe(OH)_2^+$ acts as a source of $^{\bullet}$OH. The reaction of Fe $Fe(OH)_2^+$ with UV regenerates Fe^{2+} and $^{\bullet}$OH. The photodegradation of H_2O_2 hydrogen takes place under UV radiation, as shown in Equation 23 given below. Liou et al. in (2003) described the oxidative degradation of TNP, AP, 2,4-DNT, tetryl, 2,4,6-TNT, RDX, and HMX with Fenton and photo-Fenton treatment systems. With all the explosive compounds, they observed that the oxidation rates considerably improved by rising the dosage of Fe^{2+} in the Fenton process, and increasing irradiation with UV rays in the photo-Fenton process (Equation 25.23).

$$H_2O_2 + h\nu \rightarrow 2HO\bullet \tag{25.23}$$

Similar to the Fenton system, the photo-Fenton process also has some shortcomings like higher sludge development and processing costs, and a very narrow range of optimal pH for effective

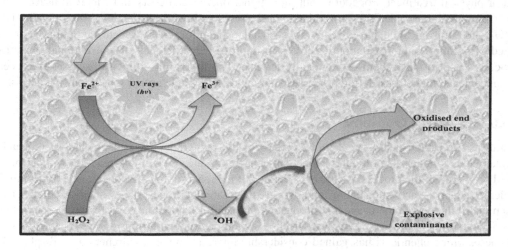

FIGURE 25.3 Mechanisms of the Photo-Fenton process.

functioning. These drawbacks confine its employment in wastewater treatment. Literature supporting the use of the photo-Fenton oxidation procedure for the management of nitro-organic explosives has been given in Table 25.2.

25.5.6 ELECTROCHEMICAL OXIDATION

The current industrial emphasis on ecological growth has gained increasing interest in the use of electrical energy for wastewater management. It is an eco-friendly treatment technology as it involves the application of electrons, which are free from chemical reagents, for contaminant removal. The regularly-used electrochemical processes are electro reduction, electro deposition, electro coagulation, electro flotation, electro photo-oxidation, electro disinfection, and electro oxidation. Amongst the various electrochemical treatment processes, electrochemical oxidation has been extensively studied owing to its ease. This process relies on the direct oxidation at the anode or indirect oxidation by appropriate anodically developed oxidants. Electrochemical oxidation processes are recognized as a favorable type of oxidation process for treating wastewater since they are proficient in handling a varied collection of toxic and refractory organic pollutants

Electrochemical oxidation involves the in-situ generation of $^{\bullet}OH$, which reacts with organic pollutants and mineralize them to carbon dioxide, water, and inorganic molecules. This treatment process is widely used by investigators for treating wastewater because it is compacted, simple, energy effective, versatile, flexible, cost effective, and environmentally friendly. Electro-Fenton (EF) and photo-EF (PEF) are two developing electrochemical-based oxidation treatment methods for the treatment of wastewater generated by industries (Zhang et al. 2007). In EF, hydrogen peroxide is generated at the cathode of the cell from due to a decrease in O_2 gas. This indirect electro-oxidation process projects benefits over Fenton process like the on-site generation of H_2O_2, greater removal proficiency, constant production of Fe^{2+} at the cathode, and lower cost. On the other hand, PEF is a mixture of electrochemical and photochemical procedures with the Fenton process. This grouping produces large amounts of $^{\bullet}OH$ in comparison to the Fenton treatment process. The UV radiations boost the EF treatment system through the photochemical redevelopment of Fe^{2+} by the photo reduction of Fe^{3+}. Applications of the photo-Fenton oxidation process for the treatment of nitro-organic explosives have been given in Table 25.2.

25.6 RECENT ADVANCES IN PHOTO-OXIDATION PROCESSES

Photo-oxidation processes have various benefits in comparison to the existing conventional biological or physical treatment procedures, but their higher operational costs and chemical usage limits their uses at a large scale. Various oxidation processes, like Fenton, ozonation, photo-Fenton, UV/H_2O_2, and ultrasonication, have been established at lab and pilot scales. Recent studies have emphasized on refining photo-oxidation processes to improve the efficiency of conventional oxidation processes and to make them economically feasible. The following segment discusses the recent advancements in photo-oxidation processes, making them a viable option for various industries.

25.6.1 CATALYTIC OXIDATION

Catalysts are established as vital tools/sources for the reduction of waste. A catalyst is usually explained as a material/compound which can change the rate of chemical reaction without being used or altered in the process. It enhances the rate of the reaction towards chemical equilibrium in order to reduce the activation energy and improve the treatment process. They are also called "green chemicals" that lessen the usage and production of harmful compounds. To achieve zero waste in chemical treatment processes, catalyst selection should be done with utmost care. The application of theses green chemicals has gained considerable interest among researchers as it displays the largest prospective of progression. Literature displays that numerous catalytic agents can be applied for various conventional photo-oxidation processes to minimalize the operational costs. To date, various catalytic agents have been recognized to be beneficial for oxidation processes in wastewater management.

Catalytic compounds can be broadly classified into two categories: homogenous catalysts and heterogeneous catalysts. Homogeneous catalysts get equivalently circulated within the reaction solution and have similar phases as the reactants. In this, the catalytic agent is dissolved in the reaction mixture and reaction occurs inside the liquid. In the case of heterogeneous catalysts, a catalytic agent is applied in a stage which is dissimilar from the reactants, and reaction happens at gas-solid or liquid-solid interfaces (Spivey et al. 2005). Heterogeneous catalysts are also known as surface catalysts because the reactions occur on the surface of the catalysts, superficially, or in the holes of the catalytic agent.

25.6.1.1 Homogeneous Catalytic Oxidation

Homogeneous catalysts are well studied and discovered. Their catalytic action is simple and implicit. Transition metals have been extensively applied as a homogenous catalytic agent in many industrial treatment processes. The utmost extensively used transition metal as a homogeneous catalyst in Fenton- and ozone-based oxidation processes are Fe^{2+}, Fe^{3+}, Zn^{2+}, Mn^{2+}, Mn^{3+}, Mn^{4+}, Ti^{2+}, Cr^{3+}, Cu^{2+}, Co^{2+}, Ni^{2+}, Cd^{2+}, and Pb^{2+} (Zhu et al. 2011). Iron salts are mostly employed as a catalytic agent to produce hydroxyl radicals in a typical Fenton-based system. A lot of work has been reported based on the Fenton oxidation of explosives using catalysts, and a typical Fenton process is usually combined with ozone, UV, ultrasonication, electrochemical oxidation, and Fe-free catalysts to increase the removal rate of refractory explosives. Homogenous Fenton reactions use an excessive quantity of iron salts and produce Fe sludge that requires secondary treatment. This restricts the usage of the Fenton method at large industrial scale. Fe^{2+} is the most commonly used transition element for homogeneous catalytic oxidation by O_3 process. In a catalytic O_3/Fe^{2+} procedure, Fe^{2+} reacts directly with O_3 to generate OH radicals, as shown in Equations 25.24–25.26. The $O_3/UV/Fe^{2+}$ system's treatment effectiveness can be upgraded with iron salts, however, the reduction/removal procedure may be repressed by the development of Fe^{3+} complexes with transitional compounds which could be photo decomposed.

$$Fe^{2+} + O_3 \rightarrow FeO_2 + O_2 \tag{25.24}$$

$$FeO_2 + H_2O \rightarrow Fe^{3+} + OH + OH^- \tag{25.25}$$

$$FeO_2 + Fe^{2+} + 2H^+ \rightarrow 2Fe^{3+} + 2H_2O \tag{25.26}$$

Wu et al. (2006) carried out the degradation of TNT with catalytic and non-catalytic ozonation treatment. They used Mn^{2+} as the catalyst. They observed that the Mn^{2+}-catalyzed ozone process displayed significantly better degradation of TNT and COD in comparison to the non-catalyzed process. The main drawbacks of the homogeneous catalytic oxidation process include the narrow pH range, Fe sludge formation, the short life span of catalysts, the difficulty in separating the catalytic agent from the treated medium, catalyst regeneration, intermediate production, expensive posttreatment, and the need for high concentrations of oxidants. So, keeping in mind all the above flaws, it is essential to develop superior catalyst-based photo-oxidation processes having less constraint for the efficient treatment of wastewater.

25.6.1.2 Heterogeneous Catalytic Oxidation

Heterogeneous catalytic agents can overcome the limitations of a homogeneous catalytic oxidation process because they are proficient in preventing Fe sludge formation. These types of catalytic agents are recognized to be significantly efficient in treating industrial effluents that have non-biodegradable pollutants due to their high catalytic action. Under this system, Fe-containing solids replace soluble Fe^{2+} thorough the interaction of organic contaminants present in the solution. The benefits of heterogeneous Fenton processes over conventional Fenton methods include a high removal efficiency under a wide range of pH levels, economic feasibility, the absence of Fe sludge, the easy removal of catalysts from treated solutions, great steadiness, and enhanced adsorption capability. In the Fenton system, the •OH formed degrade quickly and the rate of oxidation is slow owing to the Fe^{2+}-to-Fe^{3+} regeneration. Heterogeneous catalytic agents are experts in regulating the generation and usage of Fe ions throughout the oxidation process. Some of the heterogeneous catalysts provide high degradation rates owing to their large surface areas and high dispersal activities. An excellent heterogeneous catalytic agent would have high catalytic activity applicable in varied ranges of pH, and would be economical, photo sensitive, and at the same time have less catalyst leaching. A number of FeO and transition metals have been effectively applied as heterogeneous catalytic agents for degrading refractory organic contaminants present in wastewater. Heterogeneous catalysts that have been successfully used to date in photo-oxidation processes consist of TiO_2 (photocatalysts), ferrihydrite, α-Fe_2O_3, α-FeOOH, lepidocrocite (γ-FeOOH), magnetite, pyrite, and activated carbon (AC) (Matta et al. 2007).

The amalgamation of transition metals on various support systems, for instance Nafion, metals, AC, graphene, carbon aerogel, carbon nanotube (CNT), clays, polymers, alumina, fly ashes, and zeolite, are recognized to have enhanced the catalytic action in heterogeneous oxidation processes for treating various types of wastewaters (Dantas et al. 2006). In heterogeneous systems, porous matrices act as the active phase. In the course of the preparation of heterogeneous catalytic agents, the choice of catalyst supports should be appropriate. The immobilization of catalytic agents on various supports facilitates their use in an extensive pH range. The types of support and their characteristics act as significant factors for controlling the actions occurring in the catalytic positions.

i) FeO Minerals

The successful application of FeO as a heterogeneous catalytic agent to reduce an extensive range of contaminants is presented in the literature (Dantas et al. 2006, Liou and Lu 2007, Hanna et al. 2008). FeO has been applied as an effective catalytic agent with a narrow band gap of 2.0–2.3 eV. The most largely used FeO minerals to degrade contaminants which are reported in literature include αFeOOH, α-Fe_2O_3, ferrihydrite, and magnetite (Hanna et al. 2008). The factors which might affect the degradation rate of the contaminants by FeO includes the categories and surface areas of FeO, H_2O_2, and Fe, the salt concentration, the pH value, and the characteristics of the contaminants. FeO minerals deliver a novel favorable category of heterogeneous catalytic agents that can successfully degrade refractory organic pollutants by photo-oxidation processes. Transition metals (FeO) are preferred because of their low cost, simple magnetic split-up, great strength, and adsorption capabilities. In addition, the combination and association of semiconductors by various band-gap energy intensities roots a substantial rise in the photocatalytic productivity of FeO catalyst (Aleksić et al. 2010).

α-FeOOH (goethite) is a commonly available oxy-hydroxide in the surroundings and is present together with α-Fe_2O_3 (Auerbach et al. 2003). Goethite is the most abundant and steady of all oxy-hydroxides and it has the ability to combine a varied range of ecologically vital oxy-anions and cations in its composite matrix. It is generally applied for degrading organic contaminants as it is effective at a varied range of pH, comparatively inexpensive, eco-friendly, thermodynamically constant, and has a greater productivity under UV rays (Li et al. 2007). Liou and Lu (2008) used granular-size goethite (-FeOOH) particles as an adsorbent for the removal of picric acid (PA) and ammonium picrate (AP) present in a solution under acidic conditions. Experimental results demonstrated that target explosives (PA and AP) were adsorbed superficially on goethite whilst oxidation was taking place. Optimal amounts of goethite for the treatment were observed to be 0.4 g/50 mL. It was also reported that the further addition of goethite repressed the removal rate but boosted the adsorption rate of explosive compounds.

Fe$_3$O$_4$ (magnetite) nanoparticles are commonly claimed as appropriate heterogeneous catalytic agents. This is because of their inherent peroxidase-like action that speeds up the breakdown of hydrogen peroxide and convenient parting from the reaction solution by peripheral magnetic field (Chen et al. 2010). The catalytic activity of Fe$_3$O$_4$ can be enhanced by replacing Fe with other transition metals (cations), or by trapping/arresting metals on the exterior of the Fe$_3$O$_4$. A study was carried out by Matta et al. (2008) to examine the oxidation of TNT in the presence of different iron-bearing minerals at a neutral pH in an aqueous suspension. In their research, the TNT mineralization rate considerably increased when magnetite was used at pH 7. It was also reported that the addition of non-toxic iron chelatant, carboxy-methyl-cyclodextrin (CMCD), and magnetite increased the degree of TNT degradation by threefold in comparison to without CMCD treatment. Magnetite is commonly employed in Fenton treatment processes due to its magnetic and redox features. The chief difficulty related to the usage of magnetite is the generation of an Fe^{3+} oxide layer throughout the oxidation process, which decreases the catalytic action of Fe$_3$O$_4$ to generate $^{\bullet}$OH for the oxidation process. So, it is essential to make constant and effective magnetite catalytic agents to oxidize H$_2$O$_2$ to produce reactive free radicals for the oxidation. Fe$_3$O$_4$ is extremely sensitive to high temperatures and can be oxidized to α-Fe$_2$O$_3$ when heated beyond 600°C. The activity of hematite can also be upgraded by adding various types of metals into its arrangement, as they improve the catalytic action and strength of the catalytic agent.

ii) Zero-Valent Iron

Iron is the fourth most abundant component present in the Earth's crust. Over the last ten years, studies have been conducted on the elimination of refractory pollutants using zero-valent iron (ZVI) as it is safe, abundant, affordable, simple to produce, and its removal procedure is hassle free. It is a reactive metal with a standard redox potential of E0=−0.44 V. The degradation mechanism of pollutants by ZVI is based upon the direct transfer of e$^-$ from ZVI to the pollutants which are then converted to non-hazardous species. Figure 25.4 represents the major applications of ZVI and its target contaminants for environmental remediation. To improve the activity of ZVI, bimetallic elements with Fe as the primary metal and the deposition of a thin coating of transition metals such as Pd, Cu, Ni, and Pt on the exterior of Fe have been used. The transition metals trapped on the ZVI surface are supposed to boost the declining rates by functioning as hydrogen catalysts. Bimetallic

FIGURE 25.4 Applications of ZVI and its target contaminants.

particles have numerous latent benefits above ZVI, like rapid reaction kinetics and delayed oxidization. To increase the degradation rates, ZVI joined with the Fenton process was evaluated and applied for treating nitro-aromatic compounds. The integration of ZVI with the Fenton system was studied for pre-treatment of 2,4-dinitroanisole containing wastewater (Shen et al. 2013). After pre-treatment by the ZVI under optimum conditions, the degradation of nitro-aromatic compounds was found to be 77.2%. Barreto-Rodrigues et al. (2009) reported the application of the integrated ZVI and Fenton process for the treatment of wastewater from the TNT manufacturing industry. The treatment scheme removed 100% of TNT, 100% of organic nitrogen, and 95.4% of the COD. Jiang et al. (2011) combined a reductive pre-treatment through the combined ZVI–Fenton process to remove nitrobenzene present in aqueous solutions and industrial wastewater. They observed that the combined treatment process amplified the biodegradation ability and reduced the severity of toxicity of the effluent. It was observed to be significantly better in comparison to conventional Fenton treatment systems.

The application of ZVI for treating hazardous pollutants in wastewater has received extensive attention, but several challenges still exist. ZVI has been used to treat a variety of impurities present in effluents. The mechanism of reaction of ZVI with pollutants differs on the basis of the type or nature of the pollutants. It is quite complex as oxidation, reduction, adsorption, surface precipitation, and co-precipitation are included in the remediation process. More research has to be done to disclose the reaction mechanism between ZVI and pollutants. The degradation efficiency of the pollutants by ZVI depends upon the category of pollutants, the interactions between the pollutant and ZVI, the ZVI size, the initial pH, the temperature, the firmness, the dissolved oxygen, and the occurrence of natural organic complexes. So a more thorough study on these relations is required to gain a better understanding. Further investigation is still required regarding how to use the dissolve oxygen more efficiently in ZVI-based processes.

iii) *Zeolite-Based Catalysts*

Zeolites are commonly known as molecular sieves. They are microporous crystalline alumina silicates solid structures built of silicon and aluminum, which are connected to one another through oxygen ions that form a framework by cavities and channels where cations, water, and other small molecules may be located. The unification of aluminum into a silica structure creates a negatively charged frame that is possibly adjusted by using either additional organic or inorganic cations. Chemical compounds like boron, phosphorus, zinc, germanium, and transition metals like cobalt, iron, and manganese can be fused into zeolite to modify its structure/frame. The external and internal surface areas of zeolite is 300–700 m^2/g, which brands zeolite as an appropriate adsorbent and soil transformer meant for the industrial sector (Auerbach et al. 2003). Factors like porosity and the specific surface area of the zeolite affect its adsorption capacity. Zeolites are environmentally friendly, economic, and have a simple production procedure – factors that draw the attention of investigators worldwide. Natural zeolites can be modified to enhance their porosity and surface activity using cationic surfactants. Therefore, the adsorption, electrostatic attraction, exchange capacity, and catalytic activity are all improved, and the usefulness of natural zeolite is also amended (Chen et al. 2006). Surfactants cannot move inside the cavities of zeolites owing to their large size; thus, reactions simply take place on its surface. Due to its ion exchange capacity, zeolites can interchange heavy metals in effluents (Chen 2006). Zeolites to date have been considered and efficiently used to eliminate explosives, ethylbenzene, phenols, xylenes, dyes, benzene, heavy metals, insecticides, and humic acid from natural or industrial effluents (Chen et al. 2006, Aleksić et al. 2010). Zeolite comprising Fe is employed to generate •OH along with H_2O_2. Iron salts get superficially immersed in zeolite, repressing the reaction between iron and H_2O_2, thus producing complexes of Fe^{2+}/Fe^{3+} on the zeolites surface. Fe-modified zeolite increases the speed and efficacy of Fenton and photo-Fenton oxidation systems.

Nezamzadeh-Ejhieh and Khorsandi 2014 investigated the removal of 4-nitrophenol using ZnO/ nanoclinoptilolite zeolite under UV radiation According to their study, the degradation in COD from 784 to 120 mg/L was accomplished in 150 minutes at pH 8, 0.25 g/l of catalyst loading, and radiation with a medium-pressure UV lamp of 75 W. The stability and reusability of the catalytic agent was validated in three continuous tests. Reungoat et al. 2007 studied the degradation of nitrobenzene (NB) by joining zeolite adsorption and ozone restoration. In a successive treatment process combining adsorption and oxidation by ozone, NB was 100% degraded from the aqueous solution and the preliminary adsorption capability of the zeolite was also completely restored, thus proving the recyclability of zeolite. Actually, this study established the viability of a three-stage system for the treatment of wastewater by integrating the adsorption of recalcitrant organic compounds on hydrophobic zeolites and the in-situ revival of the zeolites by O_3 which will oxidize the desorbed particles. Qin et al. (2007) studied the adsorption of nitrobenzene on top of mesoporous molecular sieves (MCM-41) from an aqueous medium in batch mode. The results exhibited that NB was adsorbed onto MCM-41 rapidly. The effects of temperature, pH, ionic strength, and humic acid, and the existence of solvent in the adsorption procedures were also studied. The monolayer adsorption capability for NB reduced from 3.705 to 1.841 μmol/g with a rise in temperature (from 278 to 308 K), ionic strength (from 0.001 to 0.1 mol/L), and pH (from 1.0 to 11.0). The maximum adsorption capacity of nitrobenzene was found to be 54.3% at pH 1.0 which gradually decreases to 18.1% with an increase in pH to 11.0. On the other hand, the quantity of NB adsorbed onto MCM-41 did not show distinguished change in the presence of humic acid. On the basis of these finding, they suggested that the adsorption is principally carried out by hydrophobic collaboration between NB and the MCM-41 surface.

iv) *Nanostructured Materials*

In recent times, interest in nanostructured materials has grown for their use in wastewater treatment through photo-oxidation processes. Minimum diffusion confrontation, easy availability to reactants, and accessibility to numerous active sites are amongst the benefits in comparison to micro- and macro-sized molecules in chemical processes specifically for treating wastewater. In most circumstances like in Fenton-based processes, nanoparticles act as heterogeneous catalytic agents. TiO_2 has been one of the most widely studied semiconductors for the photocatalytic oxidation of explosive wastewater (Lee et al. 2002). The heterogeneous photocatalysis treatment of explosives-contaminated water using a circular photocatalytic reactor using TiO_2 as a catalytic agent was studied by Lee et al. (2002). The studies exhibited that the photocatalytic process was an efficient treatment method for the removal of TNT, RDX, and HMX. RDX and HMX removal rates were superior at a neutral pH in comparison to acidic and basic environment, and complete degradation was observed using 1.0 g/L TiO_2 at pH 7 in 150 minutes. Whereas, in case of TNT, the removal rate was found to highest at pH 11. Almost 82% TOC (Total organic carbon) reduction was reached after 150 minutes in TNT, whereas 24% and 59% TOC reduction in RDX and HMX, respectively, was observed. Nitrate and ammonium ions were also identified as the nitrogen by-products from the photocatalytic process and more than 50% of the total nitrogen was recuperated as nitrate ions in each explosive compound. The $UV/Fe/H_2O_2/TiO_2$ and $UV/ferrioxalate/H_2O_2/TiO_2$ -P25 processes are found to be more efficient and superior in comparison to the specific photo-Fenton and UV/TiO_2 -P25 processes. In another study carried out by Kong et al. (2018), the photodegradation of 2, 4-DNT, and 2,4,6-TNT with UV light-powered TiO_2/Ptjanusmicromotors was demonstrated. Favorable outcomes were witnessed with a reduction in DNT and TNT concentration by light-powered TiO_2/Pt Janus micromotors. Photodegradation proficiency of 1 mM DNT by TiO_2/Pt Janus micromotors irradiated by using a xenon lamp of 40 mW/cm^2 in 1 hour enhanced the degradation efficiency of DNT from 2.4% to 27.4%, whereas TNT degradation efficiency rises from 2.6% to 16.8% under similar conditions. The redox reactions, made by photo-generated holes and electrons on the TiO_2/Pt Janus micromotor exteriors, produced a native electric arena that propelled the micromotors as well as oxidative species that can photodegrade 2,4-DNT and 2,4,6-TNT.

Nanoscale zero-valent iron (nZVI) can also be used in photo oxidation processes. In the last decade, the addition of nZVI to eliminate various kinds of pollutants has gained much attention due to its greater exterior area and more advanced reactivity when compared to ZVI. The published literature on nZVI comprises 15.8% of the total ZVI. NZVI has a solid affinity to get oxidized.

The application of NZVI for remediating contaminated surface and groundwater resources is restricted because of its absence of constancy, easy accumulation, and effort in the removal of nZVI from the treated solutions. nZVI immobilized on solid porous materials like carbon, resin, bentonite, kaolinite, and zeolite have been applied for the treatment of various pollutants. Several researchers have effectively used nZVI for treating explosives. For example, Zhang et al. (2010) examined the of TNT wastewater degradation by 5 g/L nZVI and reported that >99% TNT was removed in 3 hours with the original TNT concentration at pH 4. A new process for producing subnano-sized ZVI by smectite clay layers as simulation was established (Gu et al. 2010) and verified for the elimination of NB. It was observed that a reaction efficacy of 83% was attained at a 1:3 molar ratio of NB/non-structural Fe and 80% of the NB was degraded in 1 minute. Extremely reactive nZVI immobilized in a nylon membrane was produced and assessed for the elimination of NB in groundwater. It was reported that 68.9% of NB was quickly reduced during first 20 minutes.

nZVI particles have greater reactivity due to their tiny particle size, the capability to ascribe on huge particles and the ability to restrained in or on solid provisions. Owing to their nanoparticle size, high superficial area, and great in-situ reactivity, nZVI particles have ecological application in the treatment of polluted soil, sediment, and groundwater. Moreover, they cannot migrate to groundwater, therefore can be inserted as a sub-colloidal metal element into polluted lands and aquifers which made nZVI particles a better treatment option in comparison to ZVI.

The use of nanoparticles is a developing technology in the photo-oxidation treatment of aqueous solutions and presents a good prospect for the treatment of refractory contaminants. The generation of more composite nanomaterials is of utmost importance as these materials can improve treatment efficiencies. Also, nanoparticles can also be triggered by energy-dissipating means comprising UV radiations and ultrasonication. The arrest of nanoparticles on the support can diminish the ecological issues as post-management is not required.

25.7 ADVANTAGES AND DISADVANTAGES OF PHOTO-OXIDATION PROCESSES

Photo-oxidation processes have been around for several years. Consequently, these treatment processes have proved their effectiveness and usefulness; however, these processes are still being explored and improved according to the contaminant nature or type. Photo-oxidation processes are based upon the production of strong, non-selective OH radicals that are capable of degrading generally all types of hazardous organic contaminants and toxic by-products from the parent compound that do not reach complete oxidation. Oxidation processes lead to the degradation of organic contaminants to CO_2, H_2O, and inorganic ions such as chlorides and nitrates. Like other treatment processes, oxidation processes do not cause secondary pollution because they do not tranfer the pollutants from one phase to another. A powerful treatment process like the photo-oxidation process has many benefits, but it also has its share of shortcomings. The advantages and disadvantages of various advanced oxidation processes are described in Table 25.3.

25.8 CHALLENGES FOR THE FUTURE

25.8.1 H_2O_2

The principal shortcoming of photo-oxidation processes is the high operational rate correlated by the usage of expensive chemical compounds, specifically H_2O_2. The in-situ generation of H_2O_2 has drawn much attention from investigators in recent times, but further studies and research are required

TABLE 25.3

Advantages and Disadvantages of Various Photo-Oxidation Processes

Photo-Oxidation Treatment Process	Advantages	Disadvantages	Reference
Photolysis	• No consumption of chemicals. • Low maintenance and operational costs. • In-situ production of OH• radicals. • Fast process. • Ability to stimulate cleavage of the chemical bonds of various refractory compounds.	• Some organic molecules behave as photosensitizers and increase the turbidity of the aqueous solution, thus reducing efficiency of the process. • High energy requirement. • Fouling of UV lamps. • High maintenance cost	Cuerda-Correa et al. (2019)
Fenton's based processes	• Fenton's reagent combined with ZVI/other catalysts enhances the treatment efficiency in comparison with individual processes. • Fenton's reagent reaction is effective at low pH which is the main characteristic of ammunition effluents. • Reagents used in Fenton and photo-Fenton processes are safe to handle and non-threatening to the environment. • Do not require any complicated apparatus, thus the transition from laboratory scale to a large scale is easy.	• Strong reliance on the pH of aqueous solutions. • Treatment efficiency depends on the concentrations of H_2O_2 and ferric/ferrous ion. • Formation of iron sludge might cause waste disposal problems. • pH adjustment increases the treatment costs.	Kavitha and Palanivelu (2004)
Peroxidation photocatalysis	• Widely used for dye removal. • No potential formation of bromated compounds. • Full-scale drinking water treatment system exists. • Larger amounts of OH• radicals are generated as compared to photolysis.	• Potential bromated by-product. • UV light penetration can be obstructed by turbidity. • Compounds such as nitrate can interfere with absorbance of UV light.	Brienza and Katsoyiannis (2017)
Ozonation	• Installations are relatively simple and require only a little space. • Sludge is not formed. • Good pre-treatment to a biological treatment, as complex organics are converted into simple biodegradable complexes. • Remove color and odor. • On-site production of ozone. • Simple, rapid, and efficient process. • UV/O_3 system is more efficient in generating OH• radicals as compared to the UV/H_2O_2 system.	• Relatively expensive treatment process. • Energy consumption is very high. • In some cases, does not lead to complete mineralization. • Short half life. • No effect on salinity.	Cuerda-Correa et al. (2019)

(Continued)

TABLE 25.3 (CONTINUED)
Advantages and Disadvantages of Various Photo-Oxidation Processes

Photo-Oxidation Treatment Process	Advantages	Disadvantages	Reference
Electrochemically mediated oxidation	• Increase efficiency of removal of organic pollutants. • Rapid process. • No addition of chemical reagents except a catalytic quantity of ferrous ions is required. • Complete degradation of the pollutant.	• High initial and maintenance costs of the equipment. • Anode passivation and deposition of sludge on electrode that hampers continuous operation of the process.	Crini and Lichtfouse (2018).
Ultrasonication	• Produces mass transfer effects at macroscopic and microscopic levels. • Initiates reactions without any external reagent. • Less heat generation in comparison to the UV system. • No bromated formation.	• Full-scale application does not exist. • In order to increase the efficiency of the system, chemicals might be required, thus increasing the treatment cost.	Ashok Kumar (2016)
Heterogeneous photocatalysis	• Use of sunlight or near UV light proved to be economical especially for large-scale operations. • More efficient than conventional photo-oxidation processes. • No sludge formation. • Work at very low concentrations. • Very good reduction in COD and TOC. • Recycling of catalysts.	• No full-scale application exists. • Separation step is required when catalyst is added as slurry. • Economically non-viable for small and medium industries.	Augugliaro et al. (2019).

on the effective in-situ generation of H_2O_2. H_2O_2 usage can also be minimized by continuous feeding during the course of the reaction and appropriate optimization of photo-oxidation processes.

25.8.2 TOXICITY ASSESSMENT

The ecological influence of the treatment/management process on the basis of harmfulness is an essential characteristic to be taken into account while evaluating the efficiency the photo-oxidation systems. The treated wastewater would be less contaminated or harmful in comparison to the original wastewater. This is due to the fact that some by-products that are produced in the course of the process might amplify the detrimental impacts of a pollutant, for instance long-existing intermediate products could be more hazardous than the parent explosive compounds.

25.8.3 REACTOR DESIGN

Designing and developing an effectual pilot-scale reactor is a major challenge, therefore photo-oxidation process reactors are hardly covered in the literature. There are some factors that need to be studied carefully while designing reactors to ensure constant and even radiations on the solutions in the system at the incident radiation strength. Moreover, mass transfer difficulties happen in ozone-based processes.

25.8.4 EXISTENCE OF IONS IN WASTEWATER

The anions and cations present in the system, which might disturb the treatment processes by reaction with •OH and the absorption of UV radiations, affect the treatment process drastically. CO_3^{-}, HCO_3^{-}, CI^{-}, SO_4^{2-}, PO_4^{3-}, and NO_3^{-} are examples of ions that exist in wastewaters and which have a significant impact on the pollutant removal processes. Hence, more work needs to be done to evaluate the effect of anions and cations on oxidation processes.

25.8.5 RESTRICTION ON CATALYSTS

It is essential to produce catalysts that exhibit high activity, recoverability, and cost effectiveness. Costly catalytic agents cannot be employed in industries. Also, in the case of heterogeneous catalysts, overcoming the percolation of catalytically energetic species by intermediate products is imoprtant for their application in actual effluents. Another important factor to consider is the deactivation of the catalyst due to its specific exterior area and poisoning of by-intermediates designed, superficial deposition, and strong adsorption on a polymeric carbon layer.

25.8.6 FORMATION OF BY-PRODUCTS

By-products produced by the photo-oxidation system may have high polarity and water solubility as associated to the parental complexes. The by-products formed could be more toxic to the aquatic environment and to humans as well. And so, further research should be carried out to scrutinize the productivity of combining a diversity of catalysts with photo-oxidation processes.

25.8.7 INDUSTRIAL-SCALE APPLICATIONS

Even though photo-oxidation processes have been successfully used at lab scale, more and more investigations are required to appraise their applicability at industrial level as well. Cost-effectiveness and the efficiency of oxidation processes are the most important aspects that constrain their application at industrial scale. Amongst photo-oxidation processes, the ozonation and UV/H_2O_2 systems have been extensively engaged at industrial level for drinking and groundwater treatment, but they are hardly used for treating original wastewaters. It should be noted that the industrial application of photo-oxidation processes, mainly Fenton-based treatment procedures for actual wastewater treatment, has not been fully investigated as most of the treatment studies have been done to find out the reduction in performance by simulated wastewater. Because the reduction mechanisms vary with impurities, dissimilarity amongst different photo-oxidation treatment processes should be made in terms of elimination proficiency, operation safety, manageability, applicability, technical efficiency for actual wastewater, prevention of secondary pollution, reduction in toxicity, and economic sustainability. Also, studies on the cost valuation of various photo-oxidation processes are inadequate. The cost of various photo-oxidation processes comprises investment, operational, and maintenance overheads. An overall kinetic model is also essential to design these oxidation processes and to define the best effectual or cost-effective operative areas. Alongside the full-scale kinetic models, cost is reliant on functional/operational factors like H_2O_2 concentration, contaminant concentration, intensity of irradiated light, and catalyst concentration. Generally, extra efforts are required to implement this effective and economically-feasible treatment system for industrial-scale systems.

25.9 CONCLUSION

Water pollution, particularly the treatment of recalcitrant organics, has long been considered a significant environmental problem. This chapter provides information about the recent advances in various photo-oxidation processes for explosive wastewater treatment. The improvement of conventional photo-oxidation processes by heterogeneous catalytic agents and energy-dissipating agents outweighs the restrictions of conventional oxidation systems. Heterogeneous materials like FeO minerals, zeolite-based catalytic agents, ZVI, and nano-sized materials display high catalytic activity in the oxidation reactions in comparison to homogeneous catalysts.

The combination of photocatalysts and advanced oxidation reactions can effectively reduce the consumption of energy and materials for the remediation of wastewater. Due to the abundance and low cost of iron, using iron as a harmless catalytic agent has a broad application prospect. The usage of heterogeneous catalytic agents is encouraging, for the reason that the thorough mineralization of organic contaminants can generally be attained with the relaxed parting of the catalytic agents and there is no production/formation of secondary contaminants. Advanced research is requisite to promote the potential of various types of oxidation processes to improve the proficiency of contaminant elimination with nominal treatment budgets.

Hybrid oxidation systems might offer improved degradation efficacy of recalcitrant compounds when matched to specific oxidation processes. Typically, amalgamations of various oxidation processes boost the formation of free radicals, which raise the reaction rate. In general, it can be established that photo-oxidation processes can possibly become an effective method if the difficulties of scheming and optimizing the system are addressed and resolved. In addition, all the parameters that support the improvement of removal productivity must be carefully uncovered and estimated.

ACKNOWLEDGMENTS

The support and encouragement provided by Sh. R. K. Tanwar, Associate Director and Sh. Rajiv Narang OS and Director and Officers and Staff of ESRG, Centre for Fire Explosive and Environment Safety (CFEES), DRDO, India, is greatly acknowledged.

REFERENCES

Agency for Toxic Substances and Disease Registry (ATSDR). (2012). Toxicological profile for RDX. www. atsdr.cdc.gov/toxprofiles/tp78.pdf.
Aleksić, M., Kušić, H., Koprivanac, N., Leszczynska, D., & Božić, A. L. (2010). Heterogeneous Fenton type processes for the degradation of organic dye pollutant in water: The application of zeolite assisted AOPs. *Desalination*, 257(1–3), 22–29. doi:10.1016/j.desal.2010.03.016.

Alnaizy, R., & Akgerman, A. (1999). Oxidative treatment of high explosives contaminated wastewater. *Water Research*, 33(9), 2021–2030. doi:10.1016/s0043-1354(98)00424-2 .

Ashok kumar, M. (2016). Advantages, disadvantages and challenges of ultrasonic technology. In *Ultrasonic Synthesis of Functional Materials*, 41–42. Springer. doi:10.1007/978-3-319-28974-8_3.

Auerbach, S. M., Carrado, K. A., & Dutta, P. K. 2003. *Handbook of Zeolite Science and Technology*. Marcel Dekker, Inc.

Augugliaro, V., Palmisano, G., Palmisano, L., & Soria, J. (2019). Heterogeneous photocatalysis and catalysis. *Heterogeneous Photocatalysis*, 1–24. doi:10.1016/b978-0-444-64015-4.00001-8.

Bagal, M. V., &Gogate, P. R. (2014). Wastewater treatment using hybrid treatment schemes based on cavitation and Fenton chemistry: A review. *Ultrasonics Sonochemistry*, 21(1), 1–14. doi:10.1016/j.ultsonch.2013.07.009.

Barreto-Rodrigues, M., Silva, F. T., & Paiva, T. C. B. (2009). Optimization of Brazilian TNT industry wastewater treatment using combined zero-valent iron and Fenton processes. *Journal of Hazardous Materials*, 168(2–3), 1065–1069. doi:10.1016/j. jhazmat.2009.02.172.

Bhanot, P., Celin, S. M., Sreekrishnan, T. R., Kalsi, A., Sahai, S. K., & Sharma, P. (2020). Application of integrated treatment strategies for explosive industry wastewater: A critical review. *Journal of Water Process Engineering*, 35, 1–14. doi:10.1016/j.jwpe.2020.101232. 101232.

Bo, L., Zhang, Y., Quan, X., & Zhao, B. (2008). Microwave assisted catalytic oxidation of p-nitrophenol in aqueous solution using carbon-supported copper catalyst. *Journal of Hazardous Materials*, 153(3), 1201–1206. doi:10.1016/j.jhazmat.2007.09.082.

Brienza, M., & Katsoyiannis, I. (2017). Sulfate radical technologies as tertiary treatment for the removal of emerging contaminants from wastewater. *Sustainability*, 9(9), 1604. doi:10.3390/su9091604.

Buxton, G. V., Greenstock, C. L., Helman, W. P., & Ross, A. B. (1988). Critical review of rate constants for reactions of hydrated electrons, hydrogen atoms and hydroxyl radicals (OH/O: In aqueous solution. *Journal of Physical and Chemical Reference Data*, 17(2), 513–886. doi:10.1063/1.555805.

Cao, T., & Li, J. (2018). Experimental study on the treatment of HMX explosive wastewater by Fenton process. *IOP Conference Series: Earth and Environmental Science*, 170, 032115. doi:10.1088/1755-1315/170/3/032115.

Chen, F., Li, Y., Cai, W., & Zhang, J. (2010). Preparation and sono-Fenton performance of 4A-zeolite supported α-Fe$_2$O$_3$. *Journal of Hazardous Materials*, 177(1–3), 743–749.

Chen, Q. M., Yang, C., Goh, N. K., Teo, K. C., & Chen, B. (2004). Photochemical degradation of 1,3-dinitrobenzene in aqueous solution in the presence of hydrogen peroxide. *Chemosphere*, 55(3), 339–344. doi:10.1016/j.chemosphere.2003.11.028.

Chen, W., Juan, C., & Wei, K. (2007). Decomposition of dinitrotoluene isomers and 2,4,6-trinitrotoluene in spent acid from toluene nitration process by ozonation and photo-ozonation. *Journal of Hazardous Materials*, 147(1–2), 97–104. doi:10.1016/j.jhazmat.2006.12.052.

Chen, X. L. 2006. *Technique Research on Surface-Active Agent Modified Zeolite Treating DDNP Wastewater*. Taiyuan: North University of China.

Chen, X. L., Yuan, F. Y., & Guo, F. B. (2006). Experimental investigation on CPB and HDTMA modified zeolite for the treatment of DDNP wastewater. *Mechanical Management and Development*, 1, 11–14.

Chen, Y., Hong, L., Han, W., Wang, L., Sun, X., & Li, J. (2011). Treatment of high explosive production wastewater containing RDX by combined electrocatalytic reaction and anoxic–oxic biodegradation. *Chemical Engineering Journal*, 168(3), 1256–1262. doi:10.1016/j.cej.2011.02.032.

Crini, G., & Lichtfouse, E. (2018). Advantages and disadvantages of techniques used for wastewater treatment. *Environmental Chemistry Letters*, 17, 145–155. doi:10.1007/s10311-018-0785-9.

Cuerda-Correa, E. M., Alexandre-Franco, M. F., & Fernández-González, C. (2019). Advanced oxidation processes for the removal of antibiotics from water. An Overview. *Water*, 12(1), 102. doi:10.3390/w12010102.

Dantas, T. L. P., Mendonça, V. P., José, H. J., Rodrigues, A. E., & Moreira, R. F. P. M. (2006). Treatment of textile wastewater by heterogeneous Fenton process using a new composite Fe2O3/carbon. *Chemical Engineering Journal*, 118(1–2), 77–82. doi:10.1016/j.cej.2006.01.016 .

Davis, J. L., Wani, A. H., O'Neal, B. R., & Hansen, L. D. (2004). RDX biodegradation column study: Comparison of electron donors for biologically induced reductive transformation in groundwater. *Journal of Hazardous Materials*, 112(1–2), 45–54. doi:10.1016/j.jhazmat.2004.03.020.

Dubroca, T., Moyant, K., & Hummel, R. E. (2013). Ultra-violet and visible absorption characterization of explosives by differential reflectometry. *Spectrochimica Acta. Part A: Molecular and Biomolecular Spectroscopy*, 105, 149–155. doi:10.1016/j.saa.2012.11.090.

EPA 822-R-03-007, (2003). Drinking water advisory: Consumer acceptability advice and health effects analysis on sulfate, February. https://www.epa.gov/sites/ production/files/201409/documents/support_ccl_sulfate_healtheffects.pdf.

EPA. (2005). EPA handbook on the management of munitions response actions. EPA 505-B-01-001. http://nepis.epa.gov/Exe/ZyPURL.cgi? Dockey=P100304J.txt.

EPA. (2012). Site characterization for munitions constituents. *EPA Federal Facilities Forum Issue Paper.* EPA.

EPA, & Integrated Risk Information System (IRIS). (1993). 2,4,6-Trinitrotoluene (TNT) (CASRN 118–96-7). www.epa.gov/iris/subst/0269.htm.

Fenton, H. J. H. (1894). LXXIII: Oxidation of tartaric acid in presence of iron. *Journal of the Chemical Society, Transactions,* 65, 899–910. doi:10.1039/ct8946500899.

Fu, P., Feng, J., Yang, T., & Yang, H. (2015). Comparison of alkyl xanthates degradation in aqueous solution by the O_3 and UV/O_3 processes: Efficiency, mineralization and ozone utilization. *Minerals Engineering,* 81, 128–134. doi:10.1016/j.mineng.2015.08.001.

García Einschlag, F. S., Lopez, J., Carlos, L., Capparelli, A. L., Braun, A. M., & Oliveros, E. (2002). Evaluation of the efficiency of photodegradation of nitroaromatics applying the UV/H_2O_2 technique. *Environmental Science and Technology,* 36(18), 3936–3944. doi:10.1021/es0103039.

Glaze, W. H., Kang, J.-W., & Chapin, D. H. (1987). The chemistry of water treatment processes involving ozone, hydrogen peroxide and ultraviolet radiation. *Ozone: Science and Engineering,* 9(4), 335–352. doi:10.1080/01919518708552148.

Gu, C., Jia, H., Li, H., Teppen, B. J., & Boyd, S. A. (2010). Synthesis of highly reactive subnano-sized zero-valent iron using smectite clay templates. *Environmental Science and Technology,* 44(11), 4258–4263. doi:10.1021/es903801r.

Gu,Z.,Wang, Y., Feng, K., &Aiping Zhang, A. (2019). A comparative study of dinitrodiazophenol industrial wastewater treatment: Ozone/hydrogen peroxide versus microwave/persulfate. *Process Safety and Environmental Protection,* 130, 39–47. doi:10.1016/j.psep.2019.07.019.

Haber, F., & Weiss, J. (1934). The catalytic decomposition of hydrogen peroxide by iron salts. *Proceedings of the Royal Society A: Mathematical, Physical and Engineering Sciences,* 147(861), 332–351. doi:10.1098/rspa.1934.0221.

Hanna, K., Kone, T., &Medjahdi, G. (2008). Synthesis of the mixed oxides of iron and quartz and their catalytic activities for the Fenton-like oxidation. *Catalysis Communications,* 9(5), 955–959. doi:10.1016/j.catcom.2007.09.035.

Harmon, S. M. (2009). Effects of pollution on freshwater organisms. *Water Environment Research,* 21, 2030–2069.

Hazardous Substance Data Bank (HSDB). (2016). Cyclonite. toxnet.nlm.nih.gov/cgi-bin/ sis/htmlgen?HSDB.

Houser, J. N., Mulholland, P. J., & Maloney, K. O. (2005). Catchment disturbance and stream metabolism: Patterns in ecosystem respiration and gross primary production along a gradient of upland soil and vegetation disturbance. *Journal of the North American Benthological Society,* 24(3), 538–552. doi:10.1899/04-034.

Jenkins, T. F., Hewitt, A. D., Grant, C. L., Thiboutot, S., Ampleman, G., Walsh, M. E., Ranney, T. A., Ramsey, C. A., Palazzo, A. J., & Pennington, J. C. (2006). Identity and distribution of residues of energetic compounds at army live-fire training ranges. *Chemosphere,* 63(8), 1280–1290. doi:10.1016/j.chemosphere.2005.09.066.

Jiang, B.-C., Lu, Z.-Y., Liu, F.-Q., Li, A.-M., Dai, J.-J., Xu, L., & Chu, L.-M. (2011). Inhibiting 1,3-dinitrobenzene formation in Fenton oxidation of nitrobenzene through a controllable reductive pretreatment with zero-valent iron. *Chemical Engineering Journal,* 174(1), 258–265. doi:10.1016/j.cej.2011.09.014.

Just, C. L., &Schnoor, J. L. (2004). Phytophotolysis of hexahydro-1,3,5-trinitro-1,3,5-triazine (RDX) in leaves of reed canary grass. *Environmental Science and Technology,* 38(1), 290–295. doi:10.1021/es034744z .

Kavitha, V., &Palanivelu, K. (2004). The role of ferrous ion in Fenton and photo-Fenton processes for the degradation of phenol. *Chemosphere,* 55(9), 1235–1243. doi:10.1016/j.chemosphere.2003.12.022.

Khataee, A., Salahpour, F., Fathinia, M., Seyyedi, B., &Vahid, B. (2015). Iron rich laterite soil with mesoporous structure for heterogeneous Fenton-like degradation of an azo dye under visible light. *Journal of Industrial and Engineering Chemistry,* 26, 129–135. doi:10.1016/j.jiec.2014.11.024.

Kong, L., Mayorga-Martinez, C. C., Guan, J., & Pumera, M. (2018). Fuel-free light-powered TiO_2/Pt Janus micromotors for enhanced nitroaromatic explosives degradation. *ACS Applied Materials and Interfaces,* 10(26), 22427–22434. doi:10.1021/acsami.8b05776.

Lee, S. J., Son, H. S., Lee, H. K., &Zoh, K. D. (2002). Photocatalytic degradation of explosives contaminated water. *Water Science and Technology,* 46(11–12), 139–145. doi:10.2166/wst.2002.0729.

Li, W., Zhang, S., & Shan, X. (2007). Surface modification of goethite by phosphate for enhancement of Cu and Cd adsorption. *Colloids and Surfaces. Part A: Physicochemical and Engineering Aspects,* 293(1–3), 13–19. doi:10.1016/j.colsurfa.2006.07.002.

Li, Z. M., Shea, P. J., & Comfort, S. D. (1998). Nitrotoluene destruction by UV-catalyzed fenton oxidation. *Chemosphere*, 36(8), 1849–1865. doi:10.1016/s0045-6535(97)10073-x.

Liou, M.-J., Lu, M.-C., & Chen, J.-N. (2003). Oxidation of explosives by Fenton and photo-Fenton processes. *Water Research*, 37(13), 3172–3179. doi:10.1016/s0043-1354(03)00158-1.

Liou, M.-J., & Lu, M.-C. (2007). Catalytic degradation of nitroaromatic explosives with Fenton's reagent. *Journal of Molecular Catalysis A: Chemical*, 277(1–2), 155–163. doi:10.1016/j.molcata.2007.07.030.

Liou, M. J., & Lu, M. C. (2008). Catalytic degradation of explosives with goethite and hydrogen peroxide. *Journal of Hazardous Materials*, 151(2–3), 540–546. doi:10.1016/j.jhazmat.2007.06.016.

Maloney, K. O., Mulholland, P. J., &Feminella, J. W. (2005). Influence of catchment-scale military land use on stream physical and organic matter variables in small Southeastern Plains catchments (USA). *Environmental Management*, 35(5), 677–691. doi:10.1007/s00267-004-4212-6.

Mary Celin, S., Pandit, M., Kapoor, J. C., & Sharma, R. K. (2003). Studies on photo-degradation of 2,4-dinitro toluene in aqueous phase. *Chemosphere*, 53(1), 63–69. doi:10.1016/s0045-6535(03)00358-8.

Matta, R., Hanna, K., & Chiron, S. (2007). Fenton-like oxidation of 2,4,6-trinitrotoluene using different iron minerals. *Science of the Total Environment*, 385(1–3), 242–251. doi:10.1016/j.scitotenv.2007.06.030.

Matta, R., Hanna, K., Kone, T., & Chiron, S. (2008). Oxidation of 2,4,6-trinitrotoluene in the presence of different iron-bearing minerals at neutral pH. *Chemical Engineering Journal*, 144(3), 453–458. doi:10.1016/j.cej.2008.07.013.

Nezamzadeh-Ejhieh, A., &Khorsandi, S. (2014). Photocatalytic degradation of 4-nitrophenol with ZnO supported Nano-clinoptilolite zeolite. *Journal of Industrial and Engineering Chemistry*, 20(3), 937–946. doi:10.1016/j.jiec.2013.06.026 .

Niaounakis, M., & Halvadakis, C. P. (2006). *Olive Processing Waste Management: Literature Review and Patent Survey*, 2nd ed. Elsevier.

O'Sullivan, D. W., Denzel, J. R., &LuningPrak, D. J. (2010). Photolysis of 2,4-dinitrotoluene and 2,6-dinitrotoluene in seawater. *Aquatic Geochemistry*, 16(3), 491–505. doi:10.1007/s10498-010-9089-9.

Pennington, J. C., & Brannon, J. M. (2002). Environ fate of explosives. *Thermochim*, 38, 163–172.

Pichtel, J. (2012). Distribution and fate of military explosives and propellants in soil: A review. *Applied and Environmental Soil Science*, 2012, 1–33. doi:10.1155/2012/617236.

Qin, Q., Ma, J., & Liu, K. (2007). Adsorption of nitrobenzene from aqueous solution by MCM-41. *Journal of Colloid and Interface Science*, 315(1), 80–86. doi:10.1016/j.jcis.2007.06.060.

Raut, V. D., Khopade, R. S., Rajopadhye, M. V., & Narasimhan, V. L. (2004). Recovery of ammonium nitrate and reusable acetic acid from effluent generated during HMX production. *Defence Science Journal*, 54(2), 161–167.

Reungoat, J., Pic, J. S., Manéro, M. H., &Debellefontaine, H. (2007). Adsorption of nitrobenzene from water onto high silica zeolites and regeneration by ozone. *Separation Science and Technology*, 42(7), 1447–1463. doi:10.1080/01496390701289948 .

Rosenfeldt, E. J., Linden, K. G., Canonica, S., & von Gunten, U. (2006). Comparison of the efficiency of OH radical formation during ozonation and the advanced oxidation processes O3/H_2O_2 and UV/H_2O_2. *Water Research*, 40(20), 3695–3704. doi:10.1016/j.watres.2006.09.008.

Shen, J., Ou, C., Zhou, Z., Chen, J., Fang, K., Sun, X., Li, J., Zhou, L., & Wang, L. (2013). Pretreatment of 2,4-dinitroanisole (DNAN) producing wastewater using a combined zero-valent iron (ZVI) reduction and Fenton oxidation process. *Journal of Hazardous Materials*, 260, 993–1000. doi:10.1016/j.jhazmat.2013.07.003.

Spivey, J. J., Centi, G. , Nam, I. S., &Perathoner, S. 2005. *Catalysis*. Cambridge, UK: Royal Society of Chemistry.

Talmage, S. S., Opresko, D. M., Maxwell, C. J., Welsh, C. J. E., Cretella, F. M., Reno, P. H., & Daniel, F. B. (1999). NitroaromaticMunition compounds: Environmental effects and screening values. *Reviews of Environmental Contamination and Toxicology*, 161, 1–156. doi:10.1007/978-1-4757-6427-7_1.

Travis, E. R., Bruce, N. C., & Rosser, S. J. (2008). Microbial and plant ecology of a long-term TNT-contaminated site. *Environmental Pollution*, 153(1), 119–126. doi:10.1016/j.envpol.2007.07.015.

USACE Cold Regions Research and Engineering Laboratory (CRREL). (2006). Conceptual model for the transport of energetic residues from surface soil to groundwater by range activities. TR-06-18 ERDC/CRREL. www.dtic.mil/cgi-bin/ GetTRDoc?Location=U2&doc=GetTRDoc.pdf&AD =ADA472270.

U.S. Department of Health and Human Services. (1997). Public Health Service Agency for toxic substances and disease registry. Toxicological profile for HMX. https://www.atsdr.cdc.gov/ToxProfiles/tp98.pdf.

U.S. Environmental Protection Agency (EPA). (2014a). Technical fact sheet: Hexahydro1,3,5-Trinitro1,3,5-triazine (RDX). https://www.epa.gov/sites/production/files/2014-03/documents/ffrrofactsheet_cont aminant_rdx_january2014_final.pdf.

U.S. Environmental Protection Agency (EPA). (2014b). Technical fact sheet: 2,4,6-Trinitrotoluene (TNT). https://www.epa.gov/sites/production/files/ 201403/documents/ffrrofactsheet_contaminant_tnt_january 2014_final.pdf.

Walling, C., & Kato, S. (1971). Oxidation of alcohols by Fenton's reagent. Effect of copper ion. *Journal of the American Chemical Society*, 93(17), 4275–4281. doi:10.1021/ja00746a031.

Wu, Y.-G., Hui, L., Zhao, C.-H., Wang, Q.-H., & Feng, W.-L. (2006). Mechanism of 2,4,6-trinitrotoluene-removal by Mn(II)-catalyzed ozonation. *Chinese Journal of Explosives and Propellants*, 2006–2005.

Zappi, M. E., Hernandez, R., Gang, D., Bajpai, R., Kuo, C. H., & Hill, D. O. (2016). Treatment of groundwater contaminated with high levels of explosives using advanced oxidation processes. *International Journal of Environmental Science and Technology*, 13(12), 2767–2778. doi:10.1007/s13762-016-1109-x.

Zhang, H., Fei, C., Zhang, D., & Tang, F. (2007). Degradation of 4-nitrophenol in aqueous medium by electro-Fenton method. *Journal of Hazardous Materials*, 145(1–2), 227–232. doi:10.1016/j.jhazmat.2006.

Zhang, X., Lin, Y., Shan, X., & Chen, Z. (2010). Degradation of 2,4,6-trinitrotoluene (TNT) from explosive wastewater using nanoscale zero-valent iron. *Chemical Engineering Journal*, 158(3), 566–570. doi:10.1016/j.cej.2010.01.054.

Zhu, X., Tian, J., Liu, R., & Chen, L. (2011). Optimization of Fenton and electro-Fenton oxidation of biologically treated coking wastewater using response surface methodology. *Separation and Purification Technology*, 81(3), 444–450. doi:10.1016/j.seppur.2011.08.023.

Zoh, K. D., & Stenstrom, M. K. (2002). Application of a membrane bioreactor for treating explosives process wastewater. *Water Research*, 36(4), 1018–1024. doi:10. 1016/S0043-1354(01)00284-6.

26 Cellulose-Based Nanoadsorbents for Wastewater Remediation

Garima Kumari, Eder Lima, Kulvinder Singh, Nitesh Kumar, Anupam Guleria, Dinesh Kumar and Ashish Guleria

CONTENTS

26.1 INTRODUCTION

Over the last decade, the pollution of clean and drinking water resources has become a serious issue which the potential to lead the world into crisis. Water resources have been continuously contaminated with various pollutants that are released into the environment mostly by human activities, which includes industrial, municipal, and agricultural activities. The pollutants that are released from various activities include agrochemicals, dyes, pigments, pharmaceuticals, synthetic detergents, and toxic inorganic cations and anions. Most of these pollutants have a high toxicity, carcinogenicity, and even at trace levels cause harm to aquatic life and will also cause a variety of diseases and disorders in humans [1, 2]. In to order remove these pollutants from water, various techniques have been employed which includes chemical precipitation, flocculation, membrane separation, ion exchange, adsorption, evaporation, electrolysis, sedimentation, distillation, filtration, reverse osmosis, adsorption, electronic purification, and treatment with UV radiation [3–5].

Among the various methods mentioned above, adsorption is considered the most convenient for operational use [6]. It is a surface phenomenon that involves the accumulation of toxic pollutants

DOI: 10.1201/9781003204442-26

on the surface of a solid adsorbent and hence it is widely used for wastewater remediation. For the adsorption of toxic pollutants in wastewater remediation, various adsorbents such as inorganic adsorbents (metal hydroxides, silica gels, metal oxides, and zeolites), carbon-based adsorbents (activated carbon, mesocarbon, and carbon nanotubes), polymeric adsorbents (ion exchange resins, and polymeric membranes), and biomass-based adsorbents (lignocellulosic biomass, cellulose, lignin, and nanocellulose) have been widely used across the globe [7–10]. Among the various adsorbents mentioned above, the use of biomass-based adsorbents is attracting research interest due to their availability, renewability, and sustainability. Further, these biomass adsorbents have the ability to get modified and hence the removal efficiency of toxic pollutants from wastewater increases.

Wastewater remediation by nanoadsorbents has recently gained much attention due to advancements in the field of nanotechnology. However, nanomaterials from biomass have gained more attention for their role in the removal of toxic pollutants from wastewaters. In this regard, cellulose nanomaterials obtained from renewable resources have high surface-area-to-volume ratios, and the unique structural and mechanical properties have driven their applications as a new generation of biomass-based nanoadsorbents [11]. Cellulose nanomaterials are generally biodegradable, nontoxic, have no negative impacts on the environment, and their use eliminates the safety concerns associated with nanomaterials of carbon and inorganic nanoparticles [10]. Cellulose nanoadsorbents have many advantages as compared to carbon and inorganic nanoadsorbents, which includes low-cost materials, a high aspect ratio, sustainable, high removal efficiency for pollutants, and selective adsorption possibility via appropriate chemical modification [11, 12].

This book chapter focuses on the recent progress made related to the application of nanocellulose-based adsorbents for the removal of toxic pollutants from wastewater. Nanocellulose-based adsorbents are making significant advances in terms of the high removal efficiency for toxic inorganic pollutants. The isolation and properties of different kinds of nanocellulosic materials and their major adsorption mechanisms are described briefly, followed by a detailed discussion about the different kinds of chemical modifications of cellulose and their role in the removal of contaminants from water.

26.2 CELLULOSE NANOMATERIALS

Cellulose $(C_6H_{10}O_5)_n$, which is the most abundant organic biopolymer obtained from the primary cell wall of green plants, has become a subject of intense research due to its utility in various fields such as in textiles, biorefining, wastewater remediation, packaging, and biomedical applications [13]. Nowadays, nanocellulose obtained from different biomass attracts more attention in various fields due to its large surface area, high strength, chemical inertness, and versatile surface chemistry [14]. The degree of polymerization and chain length of cellulose-based nanomaterials depend on the composition of the cellulosic material [15]. The degree of polymerization in wood cellulose is nearly 10,000 glucose units, whereas cotton cellulose has 15,000 glucose units [16, 17]. Figure 26.1 shows the structure of cellulose that represent intra- and inter-molecular H-bonds in the ring [18]. Cellulose nanomaterials like nanocellulose, nanocrystalline cellulose (NCC), cellulose nanofibers (CNFs), cellulose nanocrystals (CNCs), and spherical nanocellulose can be obtained from several cellulosic sources such as wood pulp, bamboo, okra, corncob, wheat straw, cotton, rice husk, and pine leaves [14, 19, 20]. Different forms of nanostructured cellulose mentioned above can be obtained from different biomass which depend upon the processing and pre-treatments processes being followed. Various mechanical, physical, and chemical treatments have been used to obtain different forms of nanocellulose [14, 19, 21]. Due to its abundance, surface chemistry, and eco-friendly nature, cellulose nanomaterials have been a topic of great research interest in wastewater remediation. In the next section we discuss the use of cellulose nanomaterials for the removal of toxic heavy metals from wastewaters.

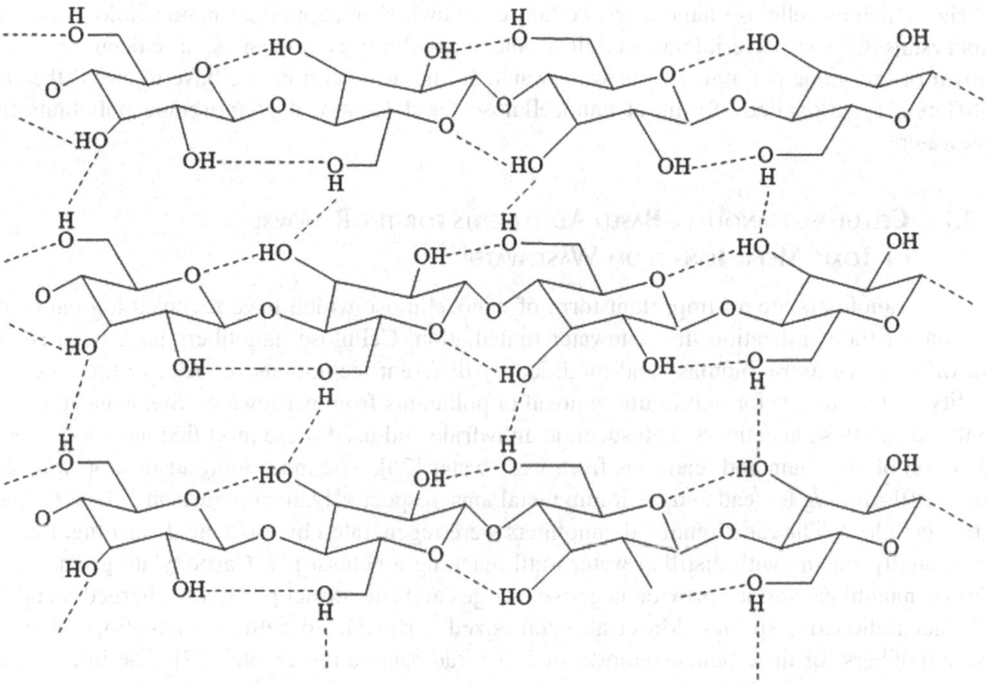

FIGURE 26.1 The chemical structure of cellulose [18].

26.3 CELLULOSE-BASED NANOADSORBENTS FOR THE REMOVAL OF TOXIC METAL IONS FROM WASTEWATER

Toxic metal ions in water bodies originate from either natural sources or industrial sources. Thus, in order to make water bodies free from toxic metal ions, various methods are used. However, most of the methods are expensive, not eco-friendly, and some of the methods generate high quantities of sludge during the removal of these pollutants. Advances in nanotechnology resulted in new adsorbents from cellulose which are effective for the removal of pollutants from wastewater. Nanocellulose-based adsorbents obtained from cellulose are economical and eco-friendly materials for wastewaters remediation. Nanocellulose has a low adsorption capacity of metals and other pollutants, despite its high specific surface area, and therefore the functionalization of nanocellulose materials has been carried out for the efficient removal of pollutants. The functionalization of nanocellulose by different functional groups is one of the main parameters for the removal of a specific class of toxic pollutant and it also increases the removal efficiency of pollutants from wastewater. As nanocellulose has an abundance of hydroxyl groups, the functionalization of cellulose nanomaterials can be done through different techniques, such as amidation, carboxylation, etherification, esterification, phosphorylation, silyation, sulfonation, (2,2,6,6-tetramethylpiperidin-1-yl) oxyl (TEMPO)-mediated oxidation, graft copolymerization, and various other techniques which have been reported in the literature. These modified products of nanocellulose have been used for wastewater remediation [14, 19, 22–25]. Functionalized nanocellulose adsorbents exhibit promising application in the removal of inorganic pollutants from wastewater, which might be due to the presence of various functional groups such as carboxylic, amine, thiol, and sulfate on the surface of the nanocellulose. The efficient removal of pollutants from wastewaters by nanocellulose adsorbents are dependent on the behavior of pollutants in an aqueous solution. Another limiting factor is the complexing or ionic sites on the surface of the nanocellulose. Various types of nanocellulose materials, such as cellulose nanofibers, cellulose nanowhiskers, spherical nanocellulose, cellulose nanocrystals (CNCs), bacterial nanocellulose, and nanocellulose composites, have been used for the removal of inorganic pollutants from wastewater. In the next section, we have reported the study of different functionalized forms of nanocellulose for the removal of inorganic pollutants from wastewater.

26.3.1 Cellulose Nanofiber-Based Adsorbents for the Removal of Toxic Metal Ions from Wastewater

Cellulose nanofibers are an important form of nanocellulose which have received a great deal of attention for their utilization in wastewater remediation. Cellulose nanofibers have been isolated from different cellulose biomass and modified by different techniques in order to increase functionality on the surface for maximum removal of pollutants from wastewater. Stephena et al. functionalized cellulose nanofibers with succinic anhydride and used these modified nanofibers for the adsorption of cadmium and lead ions from wastewater [26]. The maximum adsorption was about 1.0 and 2.91 mmol/g for lead and cadmium metal ions, respectively, in comparison to 0.002 mmol/g for raw cellulose. The functionalized nanofibers were regenerated by nitric acid washing, followed by repeatedly rinsing with distilled water until reaching a neutral pH. Carboxylate groups on the cellulose nanofibers surface provide negative charges and can attract positively charged metal ions and other radioactive species. Ma et al. synthesized carboxylated cellulose nanofibers and used these nanofibers for the efficient removal of UO_2^{2+} radioactive metal ions [27]. The interaction of carboxylate groups with UO_2^{2+} led to a maximum adsorption capacity of 167 mg/g, which was about three times higher than that of typical adsorbents such as montmorillonite, modified silica particles, fibrous membranes, ion imprinted polymer particles, and hydrogels. Hokkanen et al. modified cellulose nanofibers with 3-aminopropyltriethoxysilane and used these modified nanofibers for the adsorption of Ni^{2+}, Cu^{2+}, and Cd^{2+} metal ions from solution [28]. Metal ion adsorption was found to be pH dependent and in an acidic medium, a relatively high concentration of protons competed with cationic metal ions on free amine sites, thus decreasing the adsorption of metal ions. The competition of protons with metal ions for the amine groups became less significant with an increase in pH. The interaction of amino groups with divalent metal ions occurred through ion exchange and complexation processes. The maximum adsorption capacities reached 4 mmol g^{-1} and the regeneration of adsorbent was done by an alkaline treatment.

Bansal et al. extracted cellulose nanofibers from bagasse biomass by using the ultrasonication technique and functionalized these nanofibers with L-cysteine [29]. The modification with cysteine introduced thiol and amine functional groups in the cellulose nanofibers, which was characterized by SEM, TEM, FT-IR, EDS, and XRD techniques. The L-cysteine functionalized nanofibers were used for the selective removal of Hg^{2+} ions from the wastewater. The maximum adsorption capacity of functionalized nanofibers was found to be 116.82 mg/g. Cellulose nanofibrils (CNFs) isolated from rice straw by using a soda cooking method and isolated CNF were functionalized using L-methionine [30]. Functionalization with L-methionine introduced sulfides and amino functional groups in the CNF, which was confirmed by FTIR, XRD, TEM, and elemental analysis techniques. Modified CNFs were used for the removal of Hg(II) ions and an adsorption capacity of 131.86 mg/g for Hg (II)ions was noted. Adsorption parameters, such as the effect of pH and time, was studied for the removal of Hg(II) ions. Adsorption kinetics follows the pseudo-second-order model and hence chemisorption was involved for the removal of Hg(II) ions from wastewater.

26.3.2 Cellulose Nanocrystal-Based adsorbents for the Removal of Toxic Metal Ions from Wastewater

Cellulose nanocrystals are another emerging green candidate and are utilized for the removal of toxic pollutants from water bodies due to their sustainability, abundant availability, and high surface area. Acid hydrolysis and steam explosion methods have been reported in the literature for the preparation of cellulose nanocrystals from cellulose biomass. Kardam et al. used the acid hydrolysis technique for the preparation of cellulose nanocrystals from rice straw and used nanocrystals for the removal of Pb^{2+}, Ni^{2+}, and Cd^{2+} heavy metal ions [31]. The adsorption uptake of heavy metal ions by the cellulose nanocrystals increases with pH and remains constant at pH 6.5. Cellulose nanocrystals were regenerated by using a nitric acid solution. For 25 mg/L of metal ion solution, CNCs adsorbed about 9.7 mg/g of Cd(II), 9.42 mg/g of Pb(II), and 8.55 mg/g of Ni(II) ions. Liu et al. prepared cellulose nanocrystals by H_2SO_4 acid hydrolysis and cellulose nanofibrils were prepared by a process of mechanical grinding [32]. Both cellulose nanocrystals and cellulose nanofibrils were used for the adsorption of Ag(I) ions. It was found that cellulose nanocrystals uptake 34 mg/g of Ag(I), whereas cellulose nanofibrils uptake 14 mg/g of Ag(I). The higher uptake of Ag(I) by cellulose nanocrystals as compared to cellulose nanofibrils was due to the presence of sulfate functional groups on the surface of CNCs. The uptake of Ag(I) was pH dependent and best results were obtained near neutral pH levels. The adsorption uptake was found low under acidic pH conditions, as H^+ competes with Ag^+ for the $-SO_3^-$ functional groups on the surface of cellulose nanocrystals.

Liu et al. prepared cellulose nanocrystals via sulfuric acid hydrolysis and studied the adsorption capacity of nanocrystals for Ag(I), Cu(II), and Fe(III) metal cations [33]. Adsorption capacities of 56, 20, and 6.5 mg/g were reported for Ag(I), Cu(II) and Fe(III) metal cations, respectively. In order to enhance uptake capacity, cellulose nanocrystals were functionalized with phosphate groups. It was found that the adsorption capacity of modified cellulose nanocrystals with phosphate groups was enhanced and the adsorption capacities of modified cellulose nanocrystals were 136, 117, and 115 mg/g for Ag(I), Cu(II), and Fe(III) metal cations, respectively. It was found that the kinetics of adsorption increase with modified CNCs and it was higher than the cellulose nanocrystals.

Yu et al. studied the surface modification of cellulose nanocrystals with succinic anhydride, and used modified cellulose nanocrystals for the removal of Pb(II) and Cd(II) toxic metal ions [34]. It was reported that succinic anhydride modified CNCs exhibited a higher adsorption capacity than cellulose nanocrystals. Modified CNCs showed a ten times higher adsorption capacity than unmodified CNCs. The regeneration of CNC adsorbent was carried out by using saturated sodium chloride solution and it was found that the adsorbent showed no loss of adsorption capacity even after two recycles. Ion exchange and complexation mechanisms were involved in the adsorption process with predominance of the former at a pH over 7.

The surface modification of cellulose materials by graft copolymerization is an exciting and promising technique and many studies for cellulose nanofibers have been reported in the literature [35]. Rani et al. prepared CNCs from banana fiber using the steam explosion method and graft copolymerized CNCs with butyl acrylate monomer using ceric ammonium nitrate as an initiator [36]. The graft copolymer of cellulose nanocrystals was characterized by using FT-IR, X-RD, SEM, and EDAX techniques. From FT-IR analysis, it was found that methylene linkages were effective in the formation of grafted copolymer, and X-RD results elucidated that the crystalline nature of the cellulose nanocrystals changes by grafting with butyl acrylate. Butyl acrylate graft copolymerized cellulose nanocrystals was used as adsorbent for the removal of lead ions from an aqueous solution. Adsorption parameters such as the effect of pH, contact time, initial metal ion concentration, and adsorbent dosage were investigated. From adsorption studies it was observed that the maximum removal of lead ions occurred at pH 5 with six hours as the optimum time. The adsorption isotherm models of lead ions were best described by use of the Freundlich model, and the kinetic studies indicated that pseudo-second-order kinetics showed a better fit. Desorption studies of the butyl acrylate grafted cellulose nanocrystals were carried out with 0.1 N HCl and 0.1 M EDTA to regenerate ion adsorption capacity. The initial removal rate with 0.1 N HCl was 95.89% and with 0.1 M EDTA it was 94.52%. The percentage desorption was found to be higher in HCl as compared to EDTA for metal ions and adsorbents, which was due to fact that EDTA desorbed metal ions by complexation and HCl by electrostatic interaction. After the second cycle, the removal rate became stable and adsorption–desorption behavior of grafted cellulose nanocrystals remained at a high level for

60 minutes and then reached equilibrium. Thus, cellulose nanocrystals grafted with butyl acrylate has excellent reusability and great potential to be used as an economical adsorbent for the removal of metal ions from aqueous solutions.

26.3.3 BACTERIAL CELLULOSE AND NANOCELLULOSE-BASED ADSORBENTS FOR THE REMOVAL OF TOXIC METAL IONS FROM WASTEWATER

Bacterial cellulose (BC) and nanocellulose have been utilized as adsorbents for the removal of metal ions from water bodies. Shen et al. functionalized bacterial cellulose nanofibrils with diethylenetriamine to generate amine functions on the surface, and functionalized bacterial cellulose was used for the adsorption of copper and lead metal ions [37]. The uptake capacity of functionalized bacterial cellulose was found to be 50 and 40 mg g^{-1} for Cu^{2+} and Pb^{2+}, respectively, at pH 5 within 2 hours. The functionalized bacterial cellulose was regenerated successfully by treatment with acid solution to strip off the adsorbed metal. Chen et al. prepared carboxymethylated bacterial cellulose in situ with *Acetobacter xylinum* by adding water-soluble carboxymethyl cellulose to the culture medium [38]. The generation of carboxylic groups was shown to increase the adsorption capacity by more than twofold compared to non-modified bacterial cellulose, with an adsorption capacity of about 20 mg g^{-1} (copper) and 65 mg g^{-1} (lead), compared to 11 mg g^{-1} (copper) and 25 mg g^{-1} (lead) for bacterial cellulose.

Hokkanen et al. prepared succinic anhydride-modified mercerized nanocellulose and used it for the adsorption of Ni^{2+}, Cu^{2+}, Co^{2+}, and Cd^{2+} from aqueous solutions [39]. The rate of adsorption was very fast for all the metals and it was found that 50% of the adsorption capacity was achieved within 5 minutes, and equilibrium was reached within 60 minutes, depending on the metal. The adsorption was dependent on the difference in the availability of the adsorption sites and the time that a metal takes to diffuse inside the pores of the cellulosic material to reach the hidden sites. The maximum adsorption capacities were about 1.3, 0.8, 2, 1.6, and 2 mmol g^{-1} for cobalt, nickel, copper, zinc, and cadmium metal ions, respectively. Adsorbent regeneration was done with acid washing and a sonication treatment for at least 15 seconds to remove the adsorbed metal.

Vadakkekara et al. prepared nanocellulose from jute fiber and esterified this nanocellulose with itaconic anhydride followed by treatment with sodium bicarbonate for the introduction of sodium ions by replacing protons [40]. The modified nanocellulose was characterized by FTIR, SEM, EDX, TEM, AFM, and BET N_2 adsorption isotherm techniques for its surface morphology and modification. The modified nanocellulose was used for the removal of lead ions from a solution, and lead ion concentration was determined by AAS and ICP-MS methods. The various adsorption parameters such as time, temperature, ion concentration, and pH on the adsorption capability were evaluated. From adsorption kinetic studies, it was found that the intraparticle diffusion and pseudo-second-order models were best fit for the removal of lead ions. From adsorption isotherm data, it was found that the Freundlich model was suitable for describing the isotherm equilibrium. The maximum chemisorption capacity of sodium itaconate nanocellulose for lead ions from an aqueous solution was found to be 85 mg/g at pH 5.5. From the regeneration of sodium itaconate nanocellulose adsorbent, sodium chloride solution was used and it was found that the regenerated adsorbent could be reused for four consecutive cycles with no loss of removal efficiency. Prepared adsorbent was also found effective for the removal of other metal ions and it could be an economical alternative to the conventional cation exchange resins.

26.3.4 SPHERICAL NANOCELLULOSE-BASED ADSORBENTS FOR THE REMOVAL OF TOXIC METAL IONS FROM WASTEWATERS

Spherical nanocellulose are smart multifunctional materials from sustainable biomass which have a low cost and no environmental impact. They have extensive applications in various fields. In recent years, spherical nanocellulose materials have gained attention as nanoadsorbents and even chemo-sensors for the removal of toxic pollutants from wastewater.

Bhagat et al. prepared spherical nanocellulose (SNC) multifunctional material by reacting SNC with 2-aminoethane-1-sulfonic acid and mercaptopropionic acid (MPA). The SNC multifunctional materials prepared showed both naked-eye sensing and adsorption properties for Cr(VI) ions [41]. At very low concentrations of Cr(VI) ions, a high fluorescent intensity was observed and the detection limits of Cr(VI) ions was observed to be less than 30 ppb from 100 ppm. The SNC chemosensor showed rapid adsorption for Cr(VI) ions even at pH 7.0 and it showed mild biocidal activity against microorganisms Gram (+), *S. aureus,* Gram (−), *E. coli,* and *A. niger.* 0.1 N sodium hydroxide was used as a desorbing agent for the regeneration of spherical nanocellulose chemosensor. Bhagat and Chauhan further used spherical nanocellulose functional materials for the adsorption of Hg^{2+} from aqueous solutions [42]. They prepared it from the cotton cellulose via acid hydrolysis and further functionalized it with lipase catalyzed esterification using 3-mercaptopropionic acid. Thiol functional groups were introduced in the spherical nanocellulose which were employed for use as Hg^{2+} ion adsorbents. It was found that adsorption was rapid, efficient, and selective for mercury ions. Figure 26.2 shows the synthetic route for thiol functionalized SNC and the adsorption mechanism for the removal of mercury ions from aqueous solutions.

It was found that 98.6% of mercury ions was removed within 20 minutes from a 100 ppm solution. The adsorption follows the pseudo-second-kinetic model and Freundlich isotherm model. Adsorption process was spontaneous and endothermic in nature. Regeneration and reusability of thiol functionalized SNC adsorbent was carried out for nine cycles with a cumulative adsorption capacity of 404.95 mg/g. Regeneration studies were carried out by using 0.1 N HCl, HNO_3, NaCl,

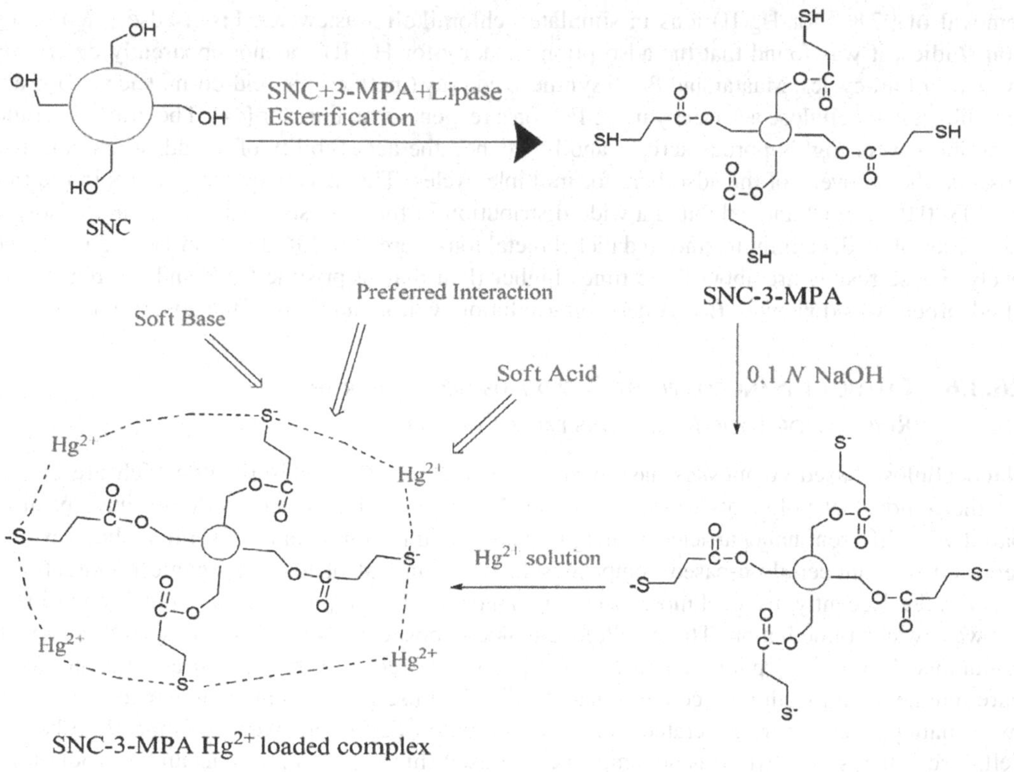

FIGURE 26.2 Synthetic route for SNC-3-MPA and the mechanism of its Hg^{2+} ions adsorption [42]. Adapted with permission.

NaNO$_3$, and CH$_3$COONa; among these 0.1 N HCl returned the best results. The maximum adsorptive removal of 88% was obtained from tap water and in the presence of other ions it was found to be 78%. Thus, spherical nanocellulose functional materials are cost effective and eco-friendly and exhibit a high capacity in the targeted application.

26.3.5 NANOCELLULOSE AEROGEL-BASED ADSORBENTS FOR THE REMOVAL OF TOXIC METAL IONS FROM WASTEWATER

Nanocellulose aerogel adsorbents have been utilized by researchers for wastewater remediation. Nanocellulose aerogels are mesoporous materials having a low density, high surface area, and porosity between 95–99%. These materials are prepared by gel in which a liquid solvent is replaced by air using supercritical drying, freeze drying, and room temperature drying techniques. By using these techniques, no change was noticed in the network structure and gel volume of aerogel. A facile freeze-drying technique was used by Geng et al. for the fabrication of recyclable and reusable bamboo nanocellulose aerogel biosorbents for the highly efficient removal of very toxic Hg(II) ions from water bodies [43]. Bamboo-derived nanofibrillated cellulose was oxidized with 2,2,6,6-tetramethylpiperidine-1-oxyl (TEMPO). The oxidized samples were then suspended in hydrolyzed 3-mercaptopropyltrimethoxysilane (MPTs) sols. Thiol and carboxyl groups were introduced in the nanocellulose aerogel by TEMPO and MPTs treatment. Nanocellulose aerogel prepared was used for the removal of Hg(II) ions in concentration ranges of 0.01–85 mg/L, either exclusively or coexisting with other heavy metal solutions. The maximal removal of Hg(II) ions by nanocellulose aerogel was found to be 718.5 mg/g and 92% of Hg(II) ions was removed from the solution. Adsorption isotherm and kinetics studies suggested monolayer adsorption of mercury ions and a chemisorptive mechanism, respectively. Nanocellulose aerogel exhibited an adsorptive removal of 97.8% for Hg(II) ions in simulated chloralkali wastewater. From adsorption/desorption studies, it was found that the adsorption capacity for Hg(II) was not apparently deteriorated even after four cycles. Maatar and Boufi synthesized poly (methacrylic acid-co-maleic acid)-grafted nanofibrillated cellulose aerogel by using Fenton's reagent as an initiator [44]. The grafted cellulose nanofibers were highly porous aerogel and broadened the accessibility of the adsorbent and made possible the recovery of the adsorbent for multiple cycles. The density of aerogel was in the range of 0.03–0.06 g cm^{-3} and exhibited a wide distribution in the pore size. The maximum adsorption capacities of lead, cadmium, zinc, and nickel metal ions were 166, 134, 136, and 115 mg g^{-1}, respectively. These results are about three times higher than that of pristine CNF and the regeneration of adsorbent was done with EDTA desorbing solution without any loss of the adsorption capacity.

26.3.6 CELLULOSE NANOCOMPOSITE-BASED ADSORBENTS FOR THE REMOVAL OF TOXIC METAL IONS FROM WASTEWATER

Nanocellulose-based composites have been a major class of nanoadsorbents which are effective for the removal of pollutants from wastewater. Nanocomposites of nanocellulose have been prepared with different nanoparticles of inorganic and organic compounds. Several studies have been reported on nanocellulose-based composites for the removal of toxic inorganic pollutants from wastewater. Recently, nanocellulose-magnetic nanoparticle composites have gained much interest for wastewater remediation. These cellulose nanocomposites enhanced the maximum removal of pollutants during adsorption and further facilitated the recovery process in adsorption. To prepare magnetic nanocellulose, cellulose nanofibrils were used as a template on which the magnetic oxide nanoparticles were generated. Zhu et al. prepared Fe$_3$O$_4$-nanoparticle-embedding bacterial cellulose spheres as a hybrid nanocomposite and used this nanocomposite as an adsorbent for the removal of lead (II), manganese (II), and chromium (III) metal ions [45]. Magnetic nanocomposites had adsorption capacities of 65, 33, and 25 mg g^{-1} for lead (II), manganese (II), and chromium (III) metal ions, respectively.

Anirudhan et al. prepared poly(itaconicAcid)-poly(methacrylicacid)-grafted-nanocellulose/na nobentonite composite (P(IA/MAA)-g-NC/NB) for the effective removal of uranium (VI) from aqueous solutions [46]. The absorbent was synthesized by a graft copolymerization reaction of methacrylic acid (MAA) and itaconic acid (IA) on nanocellulose/nanobentonite (NC/NB) composite using ethylene glycol dimethacrylate (EGDMA) as a cross-linker and potassium peroxydisulphate (KPS) as an initiator. The co-polymerization of IA and MAA was reported for the effective removal of the metals from the aqueous solution which could give three carboxyl functionalities that enhance the adsorption of uranium. The adsorbent was characterized using FTIR, XRD, SEMEDS, TG, and potentiometric titrations. The carboxylic groups from two monomers increased the effective adsorptive removal of U(VI) and the optimum pH was found to be 5.5. The adsorbent dose is found to be 2.0 g/L for the removal of U(VI) from 100 mg/L. The kinetic data followed the pseudo-second-order model and equilibrium was attained at 120 minutes and the Sips isotherm model fits well. The adsorption–desorption studies conducted over 6 cycles and adsorbed U(VI) was desorbed using 0.1 N HCl. It was found that in 10 mg/L of U(VI), only 0.20 g/L of adsorbent is required for complete removal. The decrease in the adsorption capacity of the adsorbent with liquid nuclear 20 active waste is due to the competition matrix ions with binding sites/screening effects by other metal ions. The results reveal that the P(IA/MAA)-g-NC/NB composite was 2.44 times more effective than the NC/NB composite for the removal of U(VI).

Anirudhan et al. furthermore used the above prepared composite [P(IA/MAA)-g-NC/NB] for the effective removal of Cobalt (II) from aqueous solutions [47]. For Cobalt (II) removal from the solution, pH 6.0 was found to be optimum. 2.0 g/L of adsorbent dose was found to be sufficient for the complete removal of Co(II) from 100 mg/L at room temperature. Figure 26.3 shows the synthetic route for preparation of [P(IA/MAA)-g-NC/NB] nanocomposite and the mechanism of its Co^{2+} ions adsorption. Kinetic data of adsorption was studied using kinetic models, and the adsorption of Co(II) follows the pseudo-second-order model. The equilibrium time for adsorption was obtained at 120 minutes. Langmuir, Freundlich, and Sips isotherm models were used for equilibrium isotherm studies, and the Sips model was found to be the best fit. From thermodynamic studies, adsorption was found endothermic and physical nature of adsorption of the Co(II) onto the adsorbent. Desorption experiments done with 0.1 M HCl proved that without significant loss in performance, the adsorbent could be reused for six cycles. Anirudhan et al. prepared a magnetite nanocellulose composite [(MB-IA)-g-MNCC] by the graft co-polymerization of itaconic acid onto magnetite nanocellulose (MNCC) using potassium persulfate as a free radical initiator [48]. For the preparation of the composite [(MB-IA)-g-MNCC], various steps were involved. In the first step, cellulose is converted to nanocellulose and then nanocellulose is converted to magnetite nanocellulose composite (MNCC). Thereafter, MNCC was graft copolymerized with itaconic acid using ethylene glycol dimethyl methacrylate (EGDMA) as a cross-linking agent and potassium persulfate as a free radical initiator. In the final step, modification was done with 2-mercaptobenzamide. Prepared functionalized magnetite nanocellulose composite was further used for the removal of Co(II) metal ions. Adsorption experiments were conducted by agitating 0.05 g of adsorbent with 25 mL of Co(II) solution of desired initial concentrations and pH at 30°C in a thermostatic water bath shaker with a shaking speed of 200 rpm using 100 mL Erlenmeyer flasks. Different adsorption parameters such as time, pH, dose, concentration, and temperature were studied. Maximum adsorption was obtained at pH 6.5.

From kinetic studies, it was found that the pseudo-second-order model was best fitted to data, hence the mechanism of Co(II) adsorption onto adsorbent follows ion exchange followed by complexation. From equilibrium isotherm studies, it was observed that the Langmuir model was the best suited for the adsorption of Co(II) onto the (MB-IA)-g- MNCC adsorbent. The (MB-IA)-g-MNCC composite adsorbent was further used for simulated nuclear power plant coolant water samples for the removal of Co(II) in the presence of other metal ions and this adsorbent showed good results with the removal of cobalt metal ions. For the regeneration of

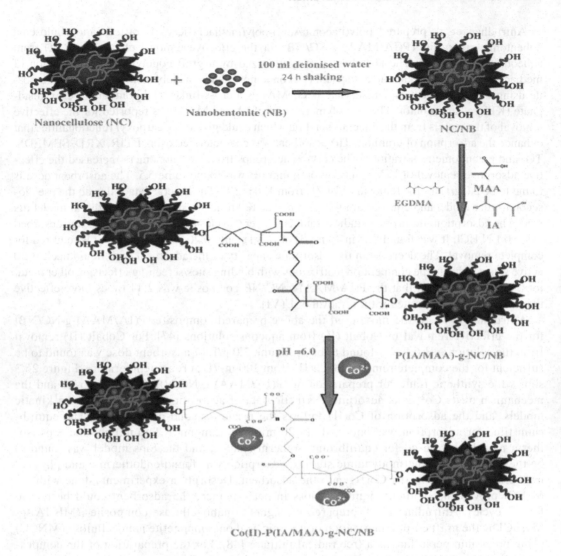

FIGURE 26.3 Synthetic route for the preparation of [P(IA/MAA)-g-NC/NB] nanocomposites and the mechanism of their Co^{2+} ion adsorption [47]. Adapted with permission.

adsorbents after adsorption, 0.1 M HCl was used as a suitable desorbing agent where six cycles of adsorption–desorption experiments were conducted. It was found that the adsorption capacity of adsorbent was decreased from 97.5% in the first cycle to 84.7% in the sixth cycle and the recovery of Co(II) using 0.1 M HCl decreased from 93.2% in the first cycle to 79.3% in the sixth cycle.

Ionic liquids have been used for the preparation of magnetic cellulose nanocomposites and these nanocomposites have been utilized for wastewater remediation. Xiong et al. synthesized magnetic nanocomposites of cellulose–iron oxide nanoparticles by co-precipitation using 1-butyl-3-methylimidazolium chloride ionic liquid as a co-solvent for cellulose and iron salt [49]. It was found that synthesized magnetic nanocomposites exhibited a superparamagnetic behavior, were sensitive under an external magnetic field, and used a nanoadsorbent in water treatment. Magnetic nanocomposites were used for the removal of Pb(II) and methylene blue from water bodies and displayed excellent adsorption efficiency as compared with other reported magnetic materials. The adsorption capacity of the magnetic nanocomposites for the removal of Pb(II) metal ions were 21.5 mg g^{-1}. Furthermore, the prepared magnetic nanoadsorbent could be efficiently recycled and reused by applying an external magnetic field. Ionic liquids have the ability for green and size controlled synthesis of cellulose nanopartilces.

Dong et al. prepared an iron oxide–cellulose nanocomposite adsorbent for the removal of As(III) and As(V) from an aqueous solution [50]. Cellulose nanocrystals (CNCs) by acid catalyzed hydrolysis of microcrystalline cellulose. 6 g of microcrystalline cellulose was mixed with a sulfuric acid solution (90 mL) and was stirred vigorously for 2 hours at 40°C. After 2 hours, the reaction was stopped by diluting the suspension ten times and the suspension was repeatedly centrifuged, washed with water, and dialyzed against water. For the preparation of the iron oxide–cellulose nanocomposite adsorbent, 0.4 g of hydrated ferric chloride and 2 g of urea were mixed with 40 mL of 1 wt% cellulose nanocrystals suspension in a reactor which was reacted at 110°C for 2–24 hours. After the temperature decreased to room temperature, the brick-red-colored product was centrifuged and washed with distilled water and ethanol. A brick-red-colored powder was obtained by freeze-drying and was used for the removal of arsenic from an aqueous solution. The adsorption behaviour was studied for the removal of As(III) and As(V) pollutants and the maximum adsorption of synthesized adsorbent was 13.866 mg/g and 15.712 mg/g which occurred at pH levels of 7 and 3, respectively. The adsorption process followed the quasi-second-order kinetic and Langmuir adsorption isotherm models, indicating that the adsorption process consists of a chemical adsorption of monolayers. The results indicate that this composite can be used as a potential adsorbent for the treatment of water containing harmful substances. Table 26.1 summarizes the removal of toxic metal ions by various cellulose family nanoadsorbents.

26.4 CONCLUSIONS AND FUTURE PERSPECTIVES

In this chapter, we have summarized the recent developments carried out in synthesizing a new class of cellulose-based nanoadsorbents and the applications of these nanoadsorbents for the removal of heavy metal pollutants from aqueous solutions. We have given a detailed overview regarding the utilization of cellulose nanofibers, cellulose nanocrystals, bacterial cellulose, spherical nanocellulose, nanocellulose aerogel, and nanocellulose nanocomposites as adsorbents in the removal of toxic metal ions from wastewater. Unmodified nanostructured cellulose adsorbents possess less adsorption capacities. Hence, nanocellulose can be tailored by appropriate surface modifications techniques for the generation of new functional or binding sites to enhance the adsorption of a specific class of pollutant, and various modification techniques are reported in this chapter. Graft copolymerization of nanocellulose and magnetic nanocellulose composites led to the introduction of different functional groups on the surface of nanocellulose which are effective for the removal of heavy metal pollutants from aqueous solutions. Functionalized nanocellulose adsorbents display adsorptive removal capacities for toxic metal ions which are comparable or even better than conventional adsorbents, e.g., activated carbon and ion exchange resins. Although the cellulose nanoadsorbents are new bio-based adsorbents used for different metal ions, which is demonstrated through different studies, more input is required to further enhance the metal ion uptake capacities. However, there is a need to address and investigate selectivity in the presence of more complex water systems; hence, more research is required to generate nanoscale hybrid structures on the nanocellulose surface that must be capable of adsorbing multiple chemical species at one time. Batch mode adsorption should be transferred to adsorption columns filled with nanocellulose so that continuous filtration can be considered in real water treatment. Furthermore, very few adsorption experiments were carried out for the treatment of industrial wastewater and therefore more research work is required in this area.

TABLE 26.1

Removal of Toxic Metal Ions by Cellulose Nanoadsorbents

Nanocellulose Family	Functionalization Agent	Toxic Metal Ion	Maximum Removal Capacity	pH	Contact Time (min.)	References
Cellulose nanofibers	oxolane-2,5-dione	Pb(II)	1.0 mmol/g	5.8	120	[26]
	oxolane-2,5-dione	Cd(II)	2.91 mmol/g	5.8	60	[26]
	Amino-propyl-triethoxy-silane	Ni(II)	159.5 mg/g	6.0	5	[28]
	Amino-propyl-triethoxy-silane	Cu(II)	200.2 mg/g	6.0	5	[28]
	Amino-propyl-triethoxy-silane	Cd(II)	471.5 mg/g	6.0	5	[28]
	L-cysteine	Hg(II)	116.82 mg/g	7.0	30	[29]
	L-methionine	Hg(II)	131.86 mg/g	7.8	40	[30]
Cellulose nanocrystals	Acid hydrolysis	Pb(II)	9.42 mg/g	6.5	40	[31]
	Acid hydrolysis	Cd(II)	9.7 mg/g	6.5	40	[31]
	Acid hydrolysis	Ni(II)	8.55 mg/g	6.5	40	[31]
	Acid hydrolysis	Ag(I)	34 mg/g	6.39	120	[32]
	Acid hydrolysis and phosphylration	Ag(I)	136 mg/g	6.5	40	[33]
	Acid hydrolysis and phosphylration	Cu(II)	117 mg/g	6.5	40	[33]
	Acid hydrolysis and phosphylration	Fe(III)	115 mg/g	6.5	40	[33]
	butyl acrylate	Pb(II)	140.95 mg/g	5.0	360	[36]
Bacterial cellulose nanofibrils	Diethylenetriamine	Cu(II)	50 mg/g	5.0	120	[37]
	Diethylenetriamine	Pb(II)	50 mg/g	5.0	120	[37]
	Carboxylation	Cu(II)	20 mg g^{-1}	6.0	60	[38]
	Carboxylation	Pb(II)	65 mg g^{-1}	6.0	60	[38]
Nanocellulose	Succinic anhydride	Co(II)	1.3 mmol g^{-1}	5.0	60	[39]
	Succinic anhydride	Ni(II)	0.8 mmol g^{-1}	5.0	60	[39]
	Succinic anhydride	Cu(II)	2.0 mmol g^{-1}	5.0	60	[39]
	Succinic anhydride	Zn(II)	1.6 mmol g^{-1}	5.0	60	[39]
	Succinic anhydride	Cd(II)	2.0 mmol g^{-1}	5.0	60	[39]
	Itaconic anhydride	Pb(II)	85 mg/g	5.5	60	[40]
Spherical nanocellulose	Lipase and mercaptopropionic acid	Hg(II)	98.6 mg/g removal from 100 ppm	5.6	20	[42]

(Continued)

TABLE 26.1 (CONTINUED)
Removal of Toxic Metal Ions by Cellulose Nanoadsorbents

Nanocellulose Family	Functionalization Agent	Toxic Metal Ion	Maximum Removal Capacity	pH	Contact Time (min.)	References
Nanocellulose aerogel	2,2,6,6-tetramethylpiperidine-1-oxyl and 3-mercaptopropyltrimethoxysilane	Hg(II)	718.5 mg/g	7.0	1,440	[43]
	Methacrylic acid and maleic acid	Pb(II)	165 mg g^{-1}	5.0	60	[44]
	Methacrylic acid and maleic acid	Cd(II)	134 mg g^{-1}	5.0	60	[44]
	Methacrylic acid and maleic acid	Zn(II)	134 mg g^{-1}	5.0	60	[44]
	Methacrylic acid and maleic acid	Ni(II)	115 mg g^{-1}	5.0	60	[44]
Cellulose-Fe$_3$O$_4$ nanocomposites	Fe$_3$O$_4$-NPs bacterial cellulose nanocomposites	Pb(II)	65 mg g^{-1}	4.0	120	[45]
	Fe$_3$O$_4$ NPs bacterial cellulose nanocomposites	Mn(II)	33 mg g^{-1}	4.0	120	[45]
	Fe$_3$O$_4$ NPs bacterial cellulose nanocomposites	Cr(III)	25 mg g^{-1}	4.0	120	[45]
Cellulose-iron oxide magnetic nanocomposites	Iron oxide-nanocellulose composites	Pb(II)	21.5 mg g^{-1}	6.0	1,440	[49]
	Iron oxide nanocellulose composites	As(III)	13.866 mg/g	7.0	120	[50]
	Iron oxide nanocellulose composites	As(V)	15.712 mg/g	3.0	120	[50]

REFERENCES

1. Martins, R. J. E., Pardo, R., and Boaventura, R. A. R. 2004. Cadmium(II) and zinc(II) adsorption by the aquatic moss *Fontinalis antipyretica*: Effect of temperature, pH and water hardness. *Water Res* 38(3):693–699.
2. Iqbal, M., Saeed, A., and Akhtar, N. 2002. Petiolar felts heath of palm: A new biosorbent for the removal of heavy metals from contaminated water. *Bioresour Technol* 81(2):151–153.
3. Namasivayam, C., and Ranganathan, K. 1995. Removal of Pb(II), Cd(II) and Ni(II) and mixture of metal ions by adsorption onto waste Fe(III)/ Cr(III) hydroxide and fixed bed studies. *Environ Technol* 16:851–860.
4. O'Connell, D. W., Birkinshaw, C., and O'Dwyer, T. F. 2008. Heavy metal adsorbents prepared from the modification of cellulose: A review. *Bioresour Technol* 99(15):6709–6724.
5. Singha, A. S., Guleria, A., and Rana, R. K. 2013. Adsorption and equilibrium isotherms study on removal of copper (II) ions from aqueous solution by chemically modified *Abelmoschus esculentus* fibers. *Int J Polym Anal Charact* 18(6):451–463.
6. Ali, I., and Gupta, V. K. 2007. Advances in water treatment by adsorption technology. *Nat Protoc* 1(6):2661–2667.
7. Adeleye, A. S., Conway, J. R., Garner, K., Huang, Y., Su, Y., and Keller, A. A.. 2016. Engineered nanomaterials for water treatment and remediation: Costs, benefits, and applicability. *Chem Eng J* 286:640–662.
8. Rengaraj, S., Yeon, J. W., Kim, J. W. Y., Jung, Y., Ha, Y. K., and Kim, W. H. 2007. Adsorption characteristics of Cu(II) onto ion exchange resins 252H and 1500H: Kinetics, isotherms and error analysis. *J Hazard Mater* 143(1–2):469–477.
9. Bhatnagar, A., and Sillanpa, M. 2010. Utilization of agro-industrial and municipal waste materials as potential adsorbents for water treatment: A review. *Chem Eng J* 157(2–3):277–296.
10. Mahfoudhi, N., and Sami, B. 2017. Nanocellulose as a novel nanostructured adsorbent for environmental remediation: A review. *Cellulose* 24(3):1171–1197.
11. Carpenter, A. W., Lannoy, C. F. D., and Wiesner, M. R. 2015. Cellulose nanomaterials in water treatment technologies. *Environ Sci Technol* 49(9):5277–5287.
12. Voisin, H., Bergstrom, L., Liu, P., and Mathew, A. P. 2017. Nanocellulose based materials for water purification. *Nanomaterials* 7(3):57–74.
13. Klemm, D., Heublein, B., Fink, H. P., and Bohn, A. 2005. Cellulose: Fascinating biopolymer and sustainable raw material. *Angew Chem Int Ed Engl* 44(22):3358–3393.
14. Klemm, D., Kramer, F., Moritz, S., Lindstrom, T., Ankerfors, M., Gray, D., and Dorris, A. 2011. Nanocelluloses: A new family of nature based materials. *Angew Chem Int Ed Engl* 50(24):5438–5466.
15. Borjesson, M., and Westman, G. 2015. Crystalline nanocellulose: Preparation, modification, and properties. In: *Cellulose: Fundamental Aspects and Current Trends*, InTech.
16. Singha, A. S., and Guleria, A. 2014. Chemical modification of cellulosic biopolymer and its use in removal of heavy metal ions from wastewater. *Int J Biol Macromol* 67:409–417.
17. Williamson, R. E., Burn, J. E., and Hocart, C. H. 2002. Towards the mechanism of cellulose synthesis. *Trends Plant Sci* 7(10):461–467.
18. Guleria, A., Kumari, G., and Saravanamurugan, S. 2020 *Cellulose Valorization to Potential Platform Chemicals. Biomass, Biofuels, Biochemicals. Recent Advances in Development of Platform Chemicals.* pp. 433–457.
19. Thomas, B., Raj, M. C., Athira, K. B., Rubiyah, M. H., Joy, J., Moores, A., et al. 2018. Nanocellulose, a versatile green platform: From biosources to materials and their applications. *Chem Rev* 118(24):11575–11625.
20. Singla, R., Guliani, A., Kumari, A., and Yadav, S. K. 2016. *Nanoscale Material Target. Drug Deliv. Theragnosis Tissue Regen*, Springer. pp. 103–125.
21. Khalil, H. P. S. A., Davoudpour, Y., Islam, M. N., Mustapha, A., Sudesh, K., Dungani, R., and Jawaid, M. 2014. Production and modification of nanofibrillated cellulose using various mechanical processes: A review. *Carbohydr Polym* 99:649–665.
22. Bozic, M., Liu, P., Mathew, A. P., and Kokol, V. 2014. Enzymatic phosphorylation of cellulose nanofibers to new highly-ions adsorbing, flame-retardant and hydroxyapatite-growth induced natural nanoparticles. *Cellulose* 21(4):2713–2726.
23. Isogai, A., Saito, T., and Fukuzumi, H. 2011. TEMPO-oxidized cellulose nanofibers. *Nanoscale* 3(1):71–85.

24. Chin, K. M., Ting, S. S., Ong, H. L., and Omar, M. 2018. Surface functionalized nanocellulose as a veritable inclusionary material in contemporary bioinspired applications: A review. *J Appl Polym Sci* 135(13):46065–46083.

25. Bagheri, S., and Julkapli, N. M. 2018. *Biopolymer Grafting: Synthesis and Properties*, Elsevier Inc. pp. 521–549.

26. Stephena, M., Catherine, N., Brendaa, M., Andrew, K., Leslie, P., and Corrinec, G. 2011. Oxolane-2,5-dione modified electrospun cellulose nanofibers for heavy metals adsorption. *J Hazard Mater* 192(2):922–927.

27. Ma, H., Hsiao, B. S., and Chu, B. 2012. Ultrafine cellulose nanofibers as efficient adsorbents for removal of UO_2^{2+} in water. *ACS Macro Lett* 1(1):213–216.

28. Hokkanen, S., Repo, E., Suopajarvi, T., Liimatainen, H., Niinimaa, J., and Sillanpää, M. 2014. Adsorption of Ni(II), Cu(II) and Cd(II) from aqueous solutions by amino modified nanostructured microfibrillated cellulose. *Cellulose* 21(3):1471–1487.

29. Bansal, M., Ram, B., Chauhan, G. S., and Kaushik, A. 2018. l-cysteine functionalized bagasse cellulose nanofibers for mercury(II) ions adsorption. *Int J Biol Macromol* 112:728–736.

30. Bisla, V., Rattan, G., Singhal, S., and Kaushik, A. 2020. Green and novel adsorbent from rice straw extracted cellulose for efficient adsorption of Hg (II) ions in an aqueous medium. *Int J Biol Macromol* 161:194–203.

31. Kardam, A., Raj, K. R., Srivastava, S., and Srivastava, M. M. 2014. Nanocellulose fibers for biosorption of cadmium, nickel, and lead ions from aqueous solution. *Clean Technol Environ Policy* 16(2): 385–393.

32. Liu, P., Sehaqui, H., Tingaut, P., Wichser, A., Oksman, K., and Mathew, A. P. 2014. Cellulose and chitin nanomaterials for capturing silver ions (Ag?) from water via surface adsorption. *Cellulose* 21(1):449–461.

33. Liu, P., Borrell, P. F., Bozi, M., Kokol, V., Oksman, K., and Mathew, A. P. 2015. Nanocelluloses and their phosphorylated derivatives for selective adsorption of Ag^+, Cu^{2+} and Fe^{3+} from industrial effluents. *J Hazard Mater* 294:177–185.

34. Yu, X., Tong, S., Ge, M., Wu, L., Zuo, J., Cao, C., and Song, W. 2013. Adsorption of heavy metal ions from aqueous solution by carboxylated cellulose nanocrystals. *J Environ Sci* 25(5):933–943.

35. Guleria, A., Kumari, G., and Lima, E. C. 2020. Cellulose-g-poly-(acrylamide-co-acrylic acid) polymeric bioadsorbent for the removal of toxic inorganic pollutants from wastewaters. *Carbohydr Polym* 228:115396.

36. Rani, K., Gomathi, T., Vijayalakshmi, K., Saranya, M., and Sudha, P. N. 2019. Banana fiber cellulose Nano Crystals graftedwith butyl acrylate for heavy metal lead (II) *Removal. Int J Biol Macromol* 131:461–472.

37. Shen, W., Chen, S., Shi, S., Li, X., Zhang, X., Hu, W., and Wang, H. 2009. Adsorption of Cu(II) and Pb(II) onto diethylenetriamine bacterial cellulose. *Carbohydr Polym* 75(1):110–114.

38. Chen, S., Zou, Y., Yan, Z., Shen, W., Shi, S., Zhang, X., and Wang, H. 2009. Carboxymethylated-bacterial cellulose for copper and lead ion removal. *J Hazard Mater* 161(2–3):1355–1359.

39. Hokkanen, S., Repo, E., and Sillanpa, M. 2013. Removal of heavy metals from aqueous solutions by succinic anhydride modified mercerized nanocellulose. *Chem Eng J* 223:40–47.

40. Vadakkekara, G. J., Thomas, S., and Nair, C. P. R. 2020. Sodium itaconate grafted nanocellulose for facile elimination of lead ion from water. *Cellulose* 27(6):3233–3248.

41. Ram, B., Chauhan, G. S., Mehta, A., Gupta, R., and Chauhan, K. 2018. Spherical nanocellulose-based highly efficient and rapid multifunctional naked eye Cr(VI) ion chemosensor and adsorbent with mild antimicrobial properties. *Chem Eng J* 349:146–155.

42. Ram, B., and Chauhan, G. S. 2018. New spherical nanocellulose and thiol-based adsorbent for rapid and selective removal of mercuric ions. *Chem Eng J* 331:587–596.

43. Geng, B., Wang, H., Wu, S., Ru, J., Tong, C., Chen, Y., et al. 2017. Surface-tailored nanocellulose aerogels with thiol-functional moieties for highly efficient and selective removal of Hg(II) ions from water. *ACS Sustain Chem Eng* 5(12):11715–11726.

44. Maatar, W., and Boufi, S. 2015. Poly (methacylic acid-co-maleic acid) grafted nanofibrillated cellulose as a reusable novel heavy metal ions adsorbent. *Carbohydr Polym* 126:199–207.

45. Zhu, H., Jia, S., Wan, T., Jia, Y., Yang, H., Li, J., et al. 2011. Biosynthesis of spherical Fe3O4/bacterial cellulose nanocomposites as adsorbents for heavy metal ions. *Carbohydr Polym* 86(4): 1558–1564.

46. Anirudhan, T. S., Deepa, J. R., and Jayan, B. 2015. Synthesis and characterization of multi-carboxyl-functionalized nanocellulose/nanobentonite composite for the adsorption of uranium(VI) from aqueous solutions: Kinetic and equilibrium profiles. *Chem Eng J* 273:390–400.

47. Anirudhan, T. S., Deepa, J. R., and Christa, J. 2016. Nanocellulose/nanobentonite composite anchored with multi-carboxyl functional groups as an adsorbent for the effective removal of cobalt(II) from nuclear industry wastewater samples. *J Colloid Interface Sci* 467:307–320.

48. Anirudhan, T. S., Shainy, F., and Deepa, J. R. 2019. Effective removal of cobalt(II) ions from aqueous solutions and nuclear industry wastewater using sulfhydryl and carboxyl functionalised magnetite nanocellulose composite: Batch adsorption studies. *Chem Ecol* 35(3):235–255.

49. Xiong, R., Wang, Y., Zhang, X., and Lu, C. 2014. Facile synthesis of magnetic nanocomposites of cellulose@ultrasmall iron oxide nanoparticles for water treatment. *RSC Adv* 4(43):22632–22641.

50. Dong, F., Xu, X., Shaghaleh, H., Guo, J., Guo, L., Qian, Y., et al. 2020. Factors influencing the morphology and adsorption performance of cellulose nanocrystal/iron oxide nanorod composites for the removal of arsenic during water treatment. *Int J Biol Macromol* 156:1418–1424.

27 Application of Green Bioremediation Technology for Refractory and Inorganic Pollutants Treatment

Eman A. Mahmoud, Hekmat R. Madian, and Mostafa M. El-Sheekh

CONTENTS

27.1 INTRODUCTION

Water is one of the most essential substances on the Earth. If there was no water on Earth, there would be no life. Industrialization and urbanization release huge amounts of wastewater, primarily used as a vital irrigation resource in agriculture. This urbanization promotes large-scale economic activities, promotes the livelihoods of innumerable farmers, and significantly modifies the water quality of natural water systems. The consumption of this contaminated water is increasing, and its sanitation problem in most developing countries is growing day by day. This growing water scarcity crisis has a major negative effect on human livelihoods, global economic growth, and the quality of the environment. Therefore, the protection of water from contamination and the establishment of cost-effective solutions are crucial processes (Rajasulochana and Preethy 2016). Different chemical pollutants exist in wastewater, like heavy metals and pesticides, as well as phosphorus

DOI: 10.1201/9781003204442-27

and nitrogen which are the most common limiting nutrients in eutrophication (Cheng *et al.* 2019). Water is used as a solvent or transport medium for a large number of industrial processes, so a great number of attempts have been made in the last two decades to eliminate contaminants from industrial waste streams (Luan *et al.* 2017). Wastewater from industrial operations frequently contains harmful and unsuitable organic molecules, which can be divided into two categories: easily-extracted and refractory chemicals. Ethers, alcohols, and aldehydes were the most easily extracted chemicals, whereas heterocyclic and benzene-derived compounds were the most difficult to extract. The refractory compounds are classified as organic (halogenated organics, aromatic and aliphatic hydrocarbons, pharmaceutical products, and pesticides) and inorganic (heavy metals and nanoparticles). In order to remove the settled materials and oxidize the organic materials present in the wastewater, standard primary and secondary processes are applied (Rajasulochana and Preethy 2016). There are non-conventional and conventional techniques that have been demonstrated and observed to be effective for wastewater treatment. Different conventional methods are available, such as activated charcoal, ion exchange, flocculation, and evaporation (Crini and Lichtfouse 2019). Conventional techniques have a somewhat high level of mechanization as these techniques require pumping and power supplies and involve trained labor for the processing and preservation of the system (Fahad *et al.* 2019). Furthermore, these conventional methods are very expensive and ineffective in reducing harmful heavy metals, nitrogen, and phosphorus, particularly when concentrations are very high. Currently, researchers have been focusing on biological processes and offer preference to green technology because of the difficulties and disadvantages of traditional methods. Green technology offers an environmentally-friendly and cost-effective method toward synthesizing nonstructural material with adaptable structures, morphologies, and distributions of particles (Khan *et al.* 2017). Green bioremediation technology can use plants, microbes, or algae to get rid of pollutants. Microalgae play an important role in processing nutrients, heavy metals, and antibiotics and have the greatest abundance of plant biomass in aquatic environments and a higher tolerance than bacteria for pollutants (Liu *et al.* 2012). Microalgae culture systems have shown potential for the cost-effective removal of various nutrients from wastewater. They also play a crucial role in the production of oxygen in aquatic environments and have been widely used in the treatment of wastewater with an intense focus on biomass production and nutrient removal (Ramanan *et al.* 2016). Microalgae biomass comprises abundant energy-rich components that can be converted into various biofuels and high-value components. The cultivation of microalgae with wastewater is considered to be an effective development for the production of microalgae feedstock or bioenergy. It can also enhance the cost of microalgal cultivation in industrial production (Li *et al.* 2018). While advanced treatment methods like microalgae removal, photocatalysis, or adsorption have been used to extract pollutants from wastewater, it is also a challenge to treat antibiotics, heavy metals, and high concentrations of ammonia nitrogen generated by many kinds of wastewaters in an effective manner to ensure biosafety. Intense rises in ammonium, chemical oxygen demand (COD), and large quantities of antibiotics from the release of wastewater into the environment are major concerns in many areas (Lou *et al.* 2018). Abou-Shanab *et al.* (2013) found that *Chlamydomonas Mexicana* was capable of removing 62% of nitrogen, 28% of phosphorus, and 29% of inorganic carbon from piggery wastewater. However, there are few records of combined treatments of refractory ammonia nitrogen, Cu(II), and pig wastewater antibiotics cultivated with microalgae. More importantly, the isolation of microalgae from their growth medium is difficult and expensive because of their small size and negatively charged algal cells. Traditional microalgae growing systems, such as open ponds or closed photobioreactor have high harvesting costs, and are bulky. For the treatment of wastewater, a novel cultivation system called the attached biofilm has been applied. Biofilm-attached cultivation is an additional technique in which high-density algal cells immobilize and settle on artificial supporting materials. Microalgal cells attach to a supporting structure to form an artificial leaf. In this system, several of these leaves are inserted vertically into a glass chamber (Cheng *et al.* 2017). Biofilm-attached microalgae cultivation in wastewater isolate algae cells from wastewater instead of concentrating or filtrating them, and the treated wastewater can be effectively recycled.

Nano-bioremediation is the latest theory that combines the use of nanoparticles and bioremediation in polluted matrices for the sustainable remediation of environmental pollutants (Cecchin *et al.* 2017). Plant and marine algal biomass can be used for making nanoparticles that have important potential in wastewater treatment. These green nanoparticles have the ability to eliminate different pollutants such as phosphate, dyes, and ammonia nitrogen (Devatha *et al.* 2016; Roy 2019).

Finally, it seems clear that existing wastewater treatment methods are based on obsolete ideas developed in the early 20th century. It seems unavoidable that if we are to cope with population growth and improve living standards, we need to develop new technology for wastewater treatment (Kehrein *et al.* 2020). The present chapter discusses the different new green technical methods available and the role of the various algal species for wastewater treatment.

27.2   WATER CRISIS

Water is the foundation for all life and is a vital source for almost all the economic and social activities involving humans, such as food production, human health, the production of electricity, and the development of many merchandise and services. In addition, water has been described as a basic human right and a vital element for human survival. Until now, more than one billion people do not have reasonable access to clean drinking water. The assumption that 70% of the Earth is covered in sufficient water resources is incorrect. Almost all the water, either saltwater or water preserved in glaciers and ice caps, is not suitable or unavailable for human use (Jackson *et al.* 2001). Freshwater sources such as springs, rivers, reservoirs, and wetlands make up less than 0.05% of the land's water (Postel 2000). More than 40% of the world population will live on river reservoirs, which are severely water-stressed. By 2050, the demand for water will rise by 400% from development and 130% from household consumption (OECD 2012). It is expected that securing clean water for human needs will be challenged by 2030, as there is tremendous pressure on limited natural sources of freshwater worldwide (Wichelns *et al.* 2015).

Sixty percent of the total surface freshwater comes from river basins shared internationally. Transboundary waters pose huge challenges when it comes to addressing water insecurity. There are about 276 large cross-border watersheds throughout the world, spanning almost half of the Earth's surface and overlapping a territory of 145 states (MacQuarrie and Wolf 2013). There have also been more than 300 cross-border aquifers, the majority being in two or more countries. In order to ensure that the water is provided for human, economic, and environmental needs, continued cooperation and coordination among nations is crucial. Despite the hundreds of international water agreements signed, the way countries cooperate to manage increasing pressures on water resources to avoid more water disputes is often not clear. Transboundary water management encompasses a broad variety of industries and disciplines, including international water law, the management of water supplies and conservation of habitats, food and energy security, international peace and stability, foreign relations, and local growth. Unilateral development policies such as the development of hydropower and water mining can have significant environmental impacts on surrounding states in the same basin without ongoing dialog and co-operation (Wolf 2007). Such effects can result in river fragmentation, disruptions to the health of aquatic ecosystems, and other adverse effects on downstream communities. The growing scarcity of water and instability will lead to further drought and waterborne illness deaths, political conflict about limited resources, and the depletion of freshwater. Access to adequate amounts of water is a high priority for many parts of the world and is exacerbating tensions between and within shared sources of water (Gleick 2009). Droughts can aggravate water insecurity and affect more people than any other form of crisis. A total of 411 million people were affected by crises in 2016, and 94% were affected by drought. Droughts are also one of the costliest hazards with major impacts, particularly on agriculture where droughts cause an estimated annual loss of US$ 6–8 billion in U.S. agriculture (Zhang *et al.* 2015). Additionally, droughts in China have created a loss of over 27 million tons of annual grain production over the last two decades (Chen *et al.* 2014). Findings show that improving water safety may help to stabilize the production and prices of food crops. The probability of world wheat production in a water-secure scenario has decreased from 83% to 38% (Sadoff *et al.* 2015).

For the alleviation of poverty, water safety is essential, and the management of water resources impacts almost every aspect of economic activity such as industry, food production and safety, transport, and energy production. These human activities may destroy water supplies. Every day, two million tons of man-made wastes are deposited in watercourses, and 15–18 billion m³ of freshwater resources are polluted every year by fossil fuel generation (WWAP 2017). Furthermore, in high-income and low-income countries, the food industry contributes 40% and 54% to the production of organic water pollutants, respectively. Approximately one-third of all rivers are infected by extreme pathogenic pollution; about a seventh of rivers are seriously contaminated by severe organic pollution; and about a tenth of all Latin American, African, and Asian rivers are polluted by severe and moderate salt pollution (UNEP 2016).

The quantity and quality of water supplies have been dramatically compromised in recent years, which is an indication of rapid industrialization and unsustainable growth. Water levels are dropping. Furthermore, industrial effluents and human waste cause significant water quality degradation in rivers and streams (Postel 2000). Owing to the indiscriminate dumping of wastewater into water supplies, droughts in several parts of the country, and the wasteful usage of water resources, water quality has been exacerbated by the pollution of water bodies at both groundwater and surface levels. Additionally, increasing urbanization, industrialization, and pollution are placing more pressure on the amount of available water; these challenges raise concerns that safe water will run out. Developed countries have decided in recent decades to reduce pollution, particularly in water, by amending environmental policies. In this respect, governments have developed several regulations that have restricted industrial activity that releases heavy metal pollution into the environment (Sud *et al.* 2008). Technology has played a critical role in water management. Desalination has provided water-scarce countries with a reliable source of drinking water, while water treatment plants allow water to be reused. Technological advances are creating new and creative ways to efficiently manage water use and overcome some of the water sector's challenges. As technology and industrial development progress, freshwater supplies are threatened all over the world. One-sixth of the world's population lacks access to freshwater (Elimelech 2006). It is seen that developed countries suffer most from chemical discharge problems, while developing countries suffer most from agricultural pollutants. Polluted water causes health problems and leads to waterborne diseases, although these can be avoided even at the household level by taking necessary steps, but supplying clean water is a challenging task (Sharma and Bhattacharya 2017). The contamination of water is a common worldwide epidemic; it may be anthropogenic (made by man) or geological. Natural contaminate forms and concentrations depend on the quality of the geological structures from which the groundwater flows and the consistency of the recharge water. Groundwater movement through sedimentary rocks and soils can have a wide variety of contaminants, for example calcium, magnesium, chloride, nitrate, and iron. The impact of these natural contaminations depends on the nature and concentration of the pollutants. Water may also be polluted by natural elements that occur at undesirable levels (Ghrefat *et al.* 2014).

Currently, the treatment of wastewater and reuse of processed water is difficult since there is a broad variety of pollutants that vary from place to place and may change over time, each requiring different treatment methods. Managing wastewater plays a significant role in sustainable urban planning. The aim of wastewater treatment has traditionally been to protect downstream users from health risks. Protecting biodiversity by the avoidance of nutrient contamination in surface waters has become a focus area in recent decades (Kehrein *et al.* 2020).

27.3   CONVENTIONAL WASTEWATER TREATMENT TECHNIQUES

Primary and secondary treatments and discarding or using the solids, as well as treated water, constitute conventional wastewater treatment methods. A primary treatment can be as easy as a septic tank to eliminate settling solids that can be found in low soil and high groundwater areas. In spite of these techniques being cheap and needing minimal maintenance, they are susceptible to failure and can still leave a pathogen-rich waste stream even if running effectively (Rajasulochana and Preethy 2016). Although secondary methods of treatment depended on sand filters and have successful pathogen removal in areas with deep permeable soils, they can be unsuccessful in other locations. The content of residual toxic metals in wastewater treatment plants affects the option of removal technique to be used. Traditional wastewater treatment requires different physical, chemical, and/or biological methods for the elimination of solids from effluents, nutrients, organic matter, colloids, and soluble pollutants (Akpor *et al.* 2014).

27.3.1 PHYSICAL METHODS

While physical treatment processes play vital roles in wastewater remediation, they have their flaws and, in some cases, involve the application of a combination of remediation processes. Physical treatments cause changes in the physical structure of a pollutant but do not induce any difference in the pollutant chemistry. Therefore, some physical remediation methods are always used in conjunction with other treatment methods. These physical methods include adsorption, flocculation and coagulation, sedimentation, ion exchange, and membrane processes (Yogalakshmi *et al.* 2020).

Adsorption by activated carbon or other appropriate sorbents has been applied to eliminate dissolution and final treatment of heavy metal (Liu *et al.* 2020). Adsorption is a balancing technique in which the pollutant is adsorbed into the adsorbent in the solution. Activated carbon, due to its porosity, wide surface area by unit mass, and improved efficiency in wastewater decolorization, is the most commonly used adsorbent. However, sludge output containing saturated dye materials present a major problem associated with this technique. Another significant issue with adsorbents is requirement for the additional regeneration of the adsorbent (Crini and Lichtfouse 2019).

The coagulation/flocculation method is performed by a blend of coagulants and suspended polluted particles for the production of gelatinous clusters. These clusters are big enough to settle down and constrained in a filter. The drawbacks associated with this method are problems of sludge resulting from the use of coagulants and the necessity of required chemicals in order to maintain the best pH for removing the pollutant particles. Some conventional approaches, such as precipitation, flocculation, and adsorption, require a post-treatment process to remove the pollutants from the newly polluted environment (Danis *et al.* 1998).

The ion exchange method is used to remove unwanted charged ions from the effluent. There are organic and inorganic ionic exchangers, which are produced naturally or synthetically, respectively. In this procedure, the incoming ionic solution is exchanged electrostatically with the ions attached to the efficient groups. When all ion exchanging active sites are saturated, they get washed back to be cleaned and retrofitted. This process is not successful for lower pH solutions or when heavy metal concentrations are high. Furthermore, the waste produced is very concentrated, and organic solvents are important for the regeneration of sorbents and raise their operating costs (Crini and Lichtfouse 2019).

Filtration with membranes is another wastewater treatment method; it is commonly used as an effective technique for dye and heavy metal removal from the polluted wastewater. The water obtained from membrane processes can be used indirectly for potable water purposes. Nanofiltration, microfiltration, ultrafiltration, and reverse osmosis are all commonly-used membrane methods. This system utilizes costly membranes that have a shorter membrane life because the membrane pores are often obstructed (Yogalakshmi *et al.* 2020).

27.3.2 CHEMICAL METHODS

Chemical remediation processes include the use of chemical compounds in the treatment of pollutants. It involves radiation from an electron beam, chemical extraction, radiocolloids, and oxidation processes. The chemical oxidation process and advanced oxidation process (AOP) are known as oxidation methods; however, the effectiveness of chemical oxidation is less than that of advanced oxidation. AOP is used to remove harsh contaminants by using hydroxyl-radicals as potent oxidants. Luan *et al.* (2017) demonstrated that wet air oxidation (WAO) is one of the most economical and environmentally-friendly advanced oxidation processes for removing and treating organic refractory pollutants. The AOP also involves photocatalytic oxidation, using sunlight to stimulate the catalyst for semiconductors and the reaction of Fenton. Ferrous ion and hydrogen peroxide produce OH radicals in Fenton oxidation which removes a wide variety of dyes within a short period of reaction (Ay *et al.* 2008). Due to the flocculation of the reagent and dye molecules, the Fenton method leads to enormous quantities of iron sludge. Ozone is also one of the most active oxidants that react either directly or by creating secondary hydroxyl-free radicals with a great number of organic compounds (Baig and Liechti 2001). Either diffuser tubes or turbine mixers supply ozone to the wastewater. It is used for the oxidation of organic and inorganic compounds and the elimination of dyes and other pollutants. Pretreatment is necessary prior to ozonation to prevent the excessive ingestion of ozone by the contaminants existing in the wastewater. Another example of a chemical oxidation method is the electrochemical anodic oxidation process (Al-Raad *et al.* 2019). During this process, the contaminants are either oxidized directly to the electrode's surface or oxidized using electrochemically-induced oxidizing species.

27.3.3 BIOLOGICAL METHODS

Another form of wastewater treatment is biological remediation. It includes the use of microorganisms (bacteria, fungi, protozoa, and algae) to break down organic compounds and toxins. Microorganisms degrade pollutants through the aerobic or anaerobic process. The aerobic biological process is reported to be insufficient in the degradation of certain contaminants such as textile dyes (Ibrahim *et al.* 2009). It also needs more space and higher retention times and generates an unknown oxidation compound. Furthermore, anaerobic treatment generates more toxic aromatic amines. Biological remediation processes by microorganisms are dependent on different environmental conditions, such as molecular oxygen, pH, and nutrient conditions. For instance, the degradation of halogenated compounds requires anaerobic conditions, while the degradation of petroleum products requires aerobic conditions (Akpor *et al.* 2014).

Activated sludge systems, trickling filters, and rotating biological contactors are examples of conventional biological treatment techniques. Rotating biological contactors and trickling filters are sensitive to temperature and remove only a small amount of biochemical oxygen demand, and trickling filters are costly to install compared to activated sludge systems. Even though the conventional activated sludge (CAS) method is competitive in achieving legal effluent quality requirements, it is considered unsustainable due to its low resource recovery capacity. It is also highly expensive because blowers and pumps require more energy (Fahad *et al.* 2019). During the different stages of the manufacturing process, factories produce saline effluents due to the addition of various chemicals. These complicated saline metal ions cannot be absorbed by activated sludge in large amounts, and they have adverse effects on the action of activated sludge (Quesnel and Nakhla 2005).

Traditional techniques are not effective in removing certain pollutants such as nitrogen and phosphorous. Furthermore, conventional techniques need pumping and power supplies and require skilled work for processing and device protection (Fahad *et al.* 2019). Additionally, these methods are inefficient in the treatment of heavy metals, principally at high concentrations (Rajasulochana and Preethy 2016). Each conventional treatment has its own restrictions, not only relating to cost but also to proficiency, poisonous by-products, sludge production, operational complexity, specifications for pretreatment, and environmental effects. However, for technical and economic reasons, only a few of the multiple treatment processes presently mentioned for the treatment of wastewater are widely utilized by the industry sector. Generally, the physico-chemical or biological removal of contaminants in wastewater treatment processes is carried out by work based on cheaper, efficient system combinations or new options (Crini and Lichtfouse 2019). The choice of a specific treatment method depends on a number of factors like the type and concentration of the waste, the variability of the effluent, the required clean-up levels, and economic factors (Rajasulochana and Preethy 2016).

The restrictions and disadvantages of different conventional treatment methods are described in Table 27.1. These restrictions on different conventional methods described above have motivated researchers to develop a more effective and eco-friendly wastewater treatment system (Luan *et al.* 2017).

27.4 NOVEL BIOREMEDIATION STRATEGIES FOR WASTEWATER TREATMENT

27.4.1 PHYCOREMEDIATION TECHNOLOGIES

The Phycoremediation method is economically efficient, environmentally sustainable, and relatively safe. It can significantly minimize the nutrient load of wastewater and thereby reduce the overall dissolved solids. Algae are used with good results, as long as the biomass produced is reused in an environmentally-friendly way without creating secondary pollution (El-Sheekh et al. 2016, 2017, 2020; Baghour 2017). The biomass produced during the bioremediation process has great potential as feedstock for the production of biofuels. Singh and Olsen (2011) indicated that the algae used are photoautotrophic microorganisms and are among the most important bioresources currently gaining tremendous popularity due to their ability to grow at a faster pace, the possibility of cultivation on non-arable land, and less water absorption and land requirements. *Scenedesmus*, *Chlorella*, *Chlamydomonas*, *Spirulina*, *Oscillatoria*, and *Arthrospira* are among the microalgal organisms used for bioremediation (Dubey *et al.* 2013). Additionally, other macroalgae such as *Kappaphycus alvarezii* and *Ulva lactuca* were studied for wastewater treatment. During normal growth and development, macroalgae will assimilate large quantities of nutrients into their tissues. Phycoremediation is theoretically applicable to a range of contaminants such as heavy metals, dyes, organic refractory pollutants, ammonia, pharmaceutical toxins, and antibiotics.

27.4.1.1 Removal of Heavy Metals by Algae

"Heavy metals" is a term commonly used for metals and metalloids with a density greater than 5 g/cm3, including arsenic (As), cadmium (Cd), chromium (Cr), copper (Cu), iron (Fe), lead (Pb), mercury (Hg), silver (Ag), zinc (Zn), and others. They are widely involved in human activities, such as the combustion of fossil fuels, mining, electroplating, the manufacturing of dyes and pigments, fertilizers, and other industrial activities, which are then released daily in large quantities into the environment through wastewater or other routes (Zakhama *et al.* 2011). Because of their non-biodegradable properties, heavy metals appear to remain in nature, leading to bioaccumulation in the food chain and causing significant environmental and health problems (Yang *et al.* 2015). These heavy metals and their derivatives are highly toxic, carcinogenic, mutagenic, and teratogenic, even at extremely low concentrations. The direct interaction, inhalation, and ingestion of these heavy metals pose significant threats to human health, causing mutations and genetic harm, weakening the central nervous system, and increasing the risk of cancer (Jaafari and Yaghmaeian 2019). Therefore, the remediation of these hazardous heavy metals in wastewater effluent prior to discharge is essential. Because of their eco-friendliness and cost efficiency at low concentrations of heavy metals, the bioremediation of heavy metals using various microorganisms like bacteria, microalgae, yeasts, and fungi has been widely implemented as an alternative to traditional methods. Microalgae with outstanding biological characteristics, such as high photosynthetic efficiency and simple structure, have the potential to develop well under severe environmental conditions such as the presence of heavy metals, high salinity, nutrient stress, and extreme temperatures among all microorganisms. Thus, due to its high binding affinity, the abundance of binding sites, and large surface area there is a growing trend in the use of microalgae in the phycoremediation of toxic heavy metals (Cameron *et al.* 2018). In addition, both living cells and non-living microalgae

TABLE 27.1

Major Restrictions of Conventional Wastewater Treatment Techniques (modified from Crini and Lichtfouse 2019)

Conventional Methods	Restrictions
Flotation	• High initial capital and energy costs
	• High operational and maintenance costs
	• Requires chemicals
	• Selectivity depends on pH
Coagulation	• Introduce non-reusable chemicals
	• Selectivity depends on pH
	• Generation of increased volumes of sludge
	• Low arsenic removal
Ion exchange	• Needs a physico-chemical pretreatment
	• High volumes need large columns
	• The cationic exchanger is saturated before the anionic resin
	• Blocking of the reactor duo to precipitation of the metals
	• Degradation of the matrix with time
	• Inefficient for some contaminants
	• Conventional resins not selective
Membrane filtration	• High cost and energy necessities
	• Rapid membrane clogging
	• Low quantity and incomplete flow rates
	• Specific processes
Adsorption	• Non-selective methods and need expensive materials
	• Various types of adsorbents are required
	• Fast saturation and reactor clogging
	• Renewal is costly and leads to material loss
	• Not effective with certain types of metals and dyestuffs
	• Removal of the adsorbent
Chemical oxidation	• Chemicals required
	• Pretreatment necessary
	• Quality is highly affected by oxidant form
	• Generation of unidentified intermediates
	• No reduction in the values of chemical oxygen demand
	• Volatile compounds and aromatic amines released
	• Generation of enlarged volumes of sludge
Chemical precipitation	• Unsuccessful in low-level metal ion removal
	• Chemical consumption
	• Generation of high volumes of sludge
	• A step of oxidation is necessary when the metal is complex
Electrochemical	• High start-up and maintenance costs of the equipment
	• Requires the addition of chemicals
	• Post-treatment required
	• Creation of sludge and the sludge treatment process is very costly
Biodegradation	• Slow process
	• Poor biodegradability of certain molecules
	• During the decomposition process the composition of mixed cultures will change

biomass can be used as biosorbent substances. Apart from exceptional removal capacities and being environmentally friendly, the bioremediation of heavy metals using microalgae has the advantages of a robust and easy procedure, lack of toxicity constraints, rapid growth compared to higher plants, and the creation of value-added products such as biofuels and fertilizers (Abinandan *et al.* 2019). Moreover, heavy-metal-induced oxidative stress increases the lipid content of microalgae. Microalgae can also be used to recover valuable metal ions, such as thallium, silver, and gold (Jaafari and Yaghmaeian 2019).

Mechanisms of Heavy Metal Removal by Microalgae

Microalgae absorb heavy metals such as boron (B), cobalt (Co), copper (Cu), iron (Fe), molybdenum (Mo), manganese (Mn), and zinc (Zn) as trace elements for the enzymatic process and cell metabolism, while other heavy metals such as Cd, Cr, Pb, and Hg are toxic to microalgae. Low-toxic heavy metal concentrations can stimulate the growth and metabolism of microalgae due to the hormesis phenomenon (Sun *et al.* 2015). Many cyanobacterial species such as *Anabaena*, *Oscillatoria*, *Phormidium*, and *Spirogyra* can grow naturally in heavy metal polluted water because of their resistance to heavy metal stress. In addition to heavy metals, microalgae have reactive groups with active binding sites that can form pollutant complexes in wastewater. This results in flocculation and consequently decreases both total dissolved solids (TDS) and total suspended solids (TSS) content (Balaji *et al.* 2016). Microalgae organisms have several strategies with specific mechanisms of self-protection against heavy metal toxicity such as heavy metal immobilization, gene control, exclusion, and chelation as well as antioxidants or the reduction of enzymes that absorb heavy metals through redox reactions. Microalgae can form complexes of cellular-protein heavy metals without affecting their own activity (Priatni *et al.* 2018). The organometallic complexes are further separated within the vacuoles to help control the concentration of heavy metal ions in the cytoplasm, which then reduce their toxic effects. In addition, heavy metals enable the biosynthesis of phytochelatins (PCs), which are thiol-rich peptides, and proteins that, by interacting with them, reduce heavy metal stress. Microalgae synthesize antioxidant enzymes such as catalase, glutathione reductase, ascorbate peroxidase, and superoxide dismutase (SOD), as well as non-enzymatic antioxidants such as carotenoids, cysteine, ascorbic acid (ASC), glutathione (GSH), and prolongation to combat the free radicals generated during adsorption by heavy metals (Upadhyay *et al.* 2016). By breaking it down into oxygen molecules and hydrogen peroxide, SOD serves as the first line of protection against superoxide anions. The hydrogen peroxide in water and oxygen molecules is further depleted by the catalase. Cysteine indirectly or specifically acts as the precursor for compounds containing PCs, GSH, metallothionein, and other sulpher, thereby acting as the source for the synthesis of various antioxidants. GSH and ASC are essential endogenous microalgae-synthesized antioxidants, and play a key role in reducing reactive oxygen species (ROS) and free radicals. In addition to preserving the ROS development and removal equilibrium, ASC protects microalgal cells by controlling metal-containing enzyme activity and ascorbic acid-glutathione (ASC-GSH) pathways, as well as sustaining the dissipation of excess excitation energy and scavenging ROS. In addition, microalgae secrete a high degree of ASC as a hydrophilic redox buffer, which is responsible for shielding cytosol and other cellular components from oxidative threats. On the other hand, high GSH levels protect the microalgae by providing resistance, scavenging free radicals, promoting PC and ASC synthesis, and restoring the substrate for other antioxidants (Upadhyay *et al.* 2016). The removal of heavy metals by microalgae is accomplished by means of a two-stage process. The first stage is rapid passive extracellular adsorption (biosorption), while the second stage is slow positive intracellular diffusion and accumulation (bioaccumulation). In addition to cell polymeric substances, such as peptides and exopolysaccharides with uronic groups, microalgae cell walls are composed primarily of polysaccharides (cellulose and alginate), lipids, and organic proteins, including various functional groups (such as amino, carboxyl, hydroxyl, imidazole, phosphate, sulfonate, thiol, and others) capable of binding heavy metals (Priatni *et al.* 2018). In addition, they consist of huge amounts of monomeric alcohols, laminar, deprotonated sulphate, and carboxyl groups that attract both anionic and cationic species of various heavy metals. Microalgae biomass also has numerous functional groups that contribute to biosorption, such as amide, carbonyl, carboxylic acid, ether, and hydroxyls (Pradhan *et al.* 2019). The adsorption of heavy metals on the surface of microalgae is a rapid process and can occur through various pathways: the formation of a covalent bond between ionized cell wall with heavy metals, the ionic exchange of heavy metal ions with cell wall cation, and the binding of heavy metal cations with microalgae exopolysaccharides' negative uronic acids. However, the process of aggregation of heavy metal within the cell is much slower. Heavy metals are deliberately transported across the cell membrane and into the cytoplasm, accompanied by diffusion and ultimate interaction with internal binding sites of proteins and peptides, such as GSH, metal transporters, oxidative stress reduction agents, and phytochelatins (Pradhan *et al.* 2019). There are only a few variables that should be considered. Firstly, the screening and selection of necessary microalgae strains is an essential step in the bioremediation of heavy metal wastewater. It is beneficial if the microalgae have the following attributes: (i) capable of accumulating high lipid content and other useful co-products, (ii) high capacity for CO_2 sequestration and low nutrient requirements, (iii) resistant to predation by grazers and robustness towards the presence of other microorganisms, and (iv) self-flocculating capability for cost-effective cell harvesting. Therefore, genetic, metabolic, and molecular engineering to improve the adaptive ability, specificity, and robustness of microalgae strains is a potential research field. Similarly, ongoing research is also crucial to better understand the underlying mechanisms as well as to establish a more detailed equilibrium and kinetic model for the biosorption of heavy metals and microalgae bioaccumulation. Recently, techniques such as whole-cell immobilization, pelletization, and microalgal biofilm for the removal and recovery of heavy metals have gained significant attention due to their potential in industrial applications. Strategies such as surface and chemical alteration techniques and the combination of other heavy metal removal techniques for microalgae biomass can also help increase the performance of heavy metals.

27.4.1.2 Dyes Removal and Recovery Using Algal Membrane

Dyes and dyestuffs are used for the most important sectors of textile production such as feed, medicine, cosmetics, and leather. A big economic field of industry is the textile industry. Textile manufacturing absorbs large quantities of water and discharges large volumes of contaminated effluents. Strong color, high salinity, high temperature, variable pH, and high chemical oxygen demand (COD) characterize the textile industry effluents. Also the presence of complex aromatic molecules in the synthetic dyes makes it difficult to biodegraded. (Aravindhan *et al.* 2007). The toxicity of the specific dyes is therefore very significant as they have a cancerous effect on human health because of their various effects on the environment and living organisms. Many traditional methods are employed for the isolation of these dyes from industrial effluents, such as membrane filtration, biological oxidation, coagulation, and adsorption (Mansor *et al.* 2020). These methods, however, have drawbacks, such as their high energy demand, high costs, slow dye removal process, large quantities of chemical requirements, and hazardous by-products. Many researchers were therefore interested in the need for more environmentally-appropriate and economical techniques, such as biosorption, for industrial and domestic wastewater treatment. Biosorption is a method that uses biosorbents for dyes in the form of biomasses such as algae, bacteria, fungi, agricultural waste, and polysaccharide materials. A significant number of researchers were concerned about the use of marine algae (seaweed) as a biosorbent material (El Atouani *et al.* 2019). Seaweed is characterized by the most important constituent of having alginate gel in their cell wall. It can be inferred in this regard that biosorbent materials are strong biosorbents due to their binding sites, such as carboxyl, sulfonate, alginate gel-induced amine, and hydroxyl groups. Where harvesting facilities have been detected after the treatment process, restrictions on the use of such dried algae in the biosorption process still exist. Furthermore, no additional surface area is feasible and will not be simply reused.

Therefore, the concept of integrating dried algae into polymer substances with a regulated porosity, which can be shaped into beads, membranes, and fiber filaments, has gained popularity. As a result, a simpler harvesting process, higher surface-area-to-volume ratio, further pore site development, and economic value increase were observed by a successful regeneration application. Polymers are characterized by hydrophobicity and are free of functional groups. To use these polymers in the adsorption process, several functional groups are generated as grafting by chemical modification (El-Shahat et al. 2020).

Two key factors hinder the industrial implementation of biotechnology strategies: the lack of long-term operating stability and the challenge of cell recovery and reuse. Cell immobilization that constitutes the chemical or physical restriction of cell mobility may therefore resolve these disadvantages. The application of cell immobilization technology in the field of bioremediation is becoming increasingly relevant because it offers many advantages such as high biomass, cell reuse, high mechanical power, high tolerance to toxic chemicals, enhanced genetic stability and the elimination of cell washout problems (Kadimpati et al. 2013). The selection of carrier or support material is a very important decision to take in an immobilization process, since it affects the viability of immobilized microorganisms and thus affects the efficiency of the wastewater treatment process in which the immobilized device is to be used. Microbial cell carriers should be insoluble, biomass non-biodegradable, non-toxic, cheap, and easy to handle and regenerate. In addition, they must provide high cell mass loading capabilities; high mechanical, biological, and chemical stability; as well as optimum nutrient diffusion from flowing material to the carrier center. Membranes for treatment procedures provide a modern form of cell immobilization. Membranes are used to promote the creation of a biofilm for bioremediation as an immobilization medium, forming a buffer between the effluent and the nutrient source.

27.4.1.3 Recovery of Inorganic Pollutants Using Algae

Wastewater is an excellent platform for microalgal growth, which is low-cost and readily available. It contains macro- and micronutrients, which sustain microalgae growth (Ajayan et al. 2015). The main nutrients found in wastewater are nitrate, ammonia, phosphate, urea, and trace minerals. Carbon (C), nitrogen (N), and phosphorus (P) are the three most considered nutrients when determining a wastewater source for microalgal growth enhancement (Ji Kabra et al. 2014). Their molar ratio should be adequate to the stoichiometric ratio of microalgal biomass in the water/growth medium to avoid growth limitations. The N-to-P ratio should approximate the Redfield ratio (16:1) which is not a global optimum microalgal biofilm technology for wastewater treatment; however, it means that an average of species-specific N:P ratios range from 8–45, suggesting that if wastewater is used to provide nutrients, N or P may need to be added to meet the ratio that will provide adequate nutrients to sustain it. The key advantages of combining wastewater treatment with the cultivation of microalgae are: (i) low-cost biomass production for biofuel generation, (ii) recovery of essential nutrients, and (iii) advanced wastewater treatment. Phosphorus recovery from waste has become an emerging topic since the main source of phosphorus in the world is phosphate rock, which is predicted to become depleted in this century. Concentrations of phosphorus in domestic and industrial wastewater can be high, highlighting the ability to recover it as struvite ($MgNH_4PO_4 \cdot 6H_2O$). Recovering P and N via struvite precipitation requires the addition of magnesium and a high pH solution (>8), resulting in a minimal recovery of N. A promising alternative to restoring both P and N is the cultivation of microalgae. The high efficiency of microalgal photosynthesis offers an effective method to recover nutrients in biomass or precipitate form. The recovery of N and P from wastewater during the active cultivation of microalgae has been investigated (Caporgno et al. 2015). The growth of the microalga *Chlorella vulgaris* using wastewater in a photo-microbial fuel cell resulted in the removal of 70% of P with simultaneous organic removal (99.6%) and N (87.6%). Wastewater from various sources, including municipal wastewater treatment plants, restricted animal feeding operations, and other industrial activities, can be effectively treated by microalgae due to their composition (mainly N and P). Microalgae is able to develop in wastewater because it provides an appropriate environment, such as pH, CO_2 dissolved, and HCO_3^-. Thus, the use of wastewater as the growth medium for microalgal biomass will increase the economic viability of large-scale biofuel production (Pires *et al.* 2013). Microalgae cultivation using wastewater has the double benefit of reducing the cost of biofuels production and enhancing the management of wastewater to produce freshwater for reuse. Nitrogen, for example, can be obtained cheaply from different wastewaters, a crucial element essential for the growth of all plants. Nitrogen is an important part of the biological macromolecules, including proteins, peptides, enzymes, fragments of energy transfer (ATP/ADP), chlorophylls, and genetic constituents (DNA/RNA). Organic N originates from inorganic sources, including nitrite (NO_2^- and NO_3^-), nitric acid (HNO_3), ammonium (NH_4^+), nitrogen gas (N_2), and ammonia (NH_3). Algae play a vital role in the transformation of inorganic N (including NO_3^-, NO_2^-, and NH_4^+) to its organic form using a method called assimilation, carried out by all eukaryotic microalgae. NO_2^- and NO_3^- are both reduced via nitrite reductase. Nitrate reductase catalyzes the transfer of two electrons from the reduced form of adenine nicotinamide dinucleotide (NADH) to nitrate, leading to nitrite formation. Nitrite is further reduced by nitrite reductase and ferredoxin (Fd) to NH_4^+ which transfers a total of 6 electrons in the reaction. Lastly, all sources of inorganic N are reduced to NH_4^+ before combining into the intracellular liquefied amino acids. In the final step, glutamine synthase makes the synthesis of NH_4^+ with glutamate (Glu) and adenosine triphosphate (ATP) into the amino acid glutamine. The addition of glutamate was reported to cause an additional 70% NH_4^+ reduction in each cell through the growth of *Chlorella vulgaris* in wastewater (Khan and Yoshida 2008).

Phosphorus is an essential part of basic life molecules in microalgae such as DNA, RNA, ATP, protein/amino acid, lipids/fatty acid, and carbohydrate/sugar metabolism intermediates and cell membrane materials. Phosphorus is present regularly as inorganic anionic types in wastewater, including $H_2PO_4^-/HPO_4^{2-}$. Inorganic PO_4^{2-} plays a vital role in the growth and metabolism of microalgae cells, and their absence or depletion can affect the photosynthetic process considerably (Suganya *et al.* 2016). During microalgae metabolism, P is ideally incorporated in the form of $HPO_4^{2-}/H_2PO_4^-$ into organic compounds during phosphorylation, abundantly involving the synthesis of ATP from adenosine diphosphate (ADP), with a type of energy input. Apart from the inorganic forms of phosphorus, some microalgae species use the P originated in organic esters for their growth. Although in freshwater systems ortho-phosphate is widely known as the limiting nutrient, several cases of eutrophication are produced by superfluous P, which may occur from excess wastewater. Similar to the elimination of N, P removal from wastewater is controlled not only by consumption in the microalgal cells but also by environmental conditions (including dissolved oxygen and pH) (Ji Kabra *et al.* 2014).

27.4.1.4   Microalgal Biofilm Applications for Wastewater Treatment

As mentioned earlier, microalgae can grow efficiently in wastewater and remove the primary nutrients and micropollutants. In addition, microalgae do not compete with arable land crops. However, there are still some fundamental barriers to the industrial application of microalgae for biofuel production, and they include high growth, harvesting, and freshwater requirements. The main techniques for microalgae harvesting and concentration (centrifugation, filtration, flocculation, gravity sedimentation, and flotation) are still not economically viable for large-scale microalgae industries (Barros *et al.* 2015). Algal biofilms are becoming increasingly popular as a new strategy for the concentration of microalgae, making harvesting/dewatering easier and cheaper. Biofilm is defined as microorganisms attached to the biological or non-biological surfaces embedded in an extracellular polymeric matrix from microbial layers. Biofilms are layers composed of multi-species microbial communities, extracellular polymeric substances (EPSs), inorganic material, and water. Algal biofilms represent three-dimensional, multilayered, and multi-species structures involving consortia of prokaryotic and eukaryotic heterotrophic and photoautotrophic organisms (Berner *et al.* 2015). Photosynthetic organisms include macro- and microalgae, and cyanobacterial species, both filamentous and unicellular. These species are distinguished by their usual content of pigments, which is used to absorb solar energy and protect the cells against radiation. In this very diverse environment, heterotrophic organisms can include protozoa, bacterial and fungal cells, and flagellates. Typically, organisms colonize various areas within the biofilm structure that are better suited for their growth (Gross *et al.* 2015). Within this complex and highly productive environment, heterotrophic and photoautotrophic species can be kept together by an EPS matrix whose functions include adhesion, aggregation, water retention and nutrients, toxins and heavy metal diffusion barriers, cell motility, grazer safety barriers and harmful chemicals, or environmental conditions. EPS, which can account for 90% of the dry biofilm mass, is usually composed of high molecular weight heteropolysaccharides comprising linear or branched repeating units consisting of 2–10 monosaccharides such as hexoses, pentoses, glucose, mannose, arabinose, uronic acids, and deoxy-sugars. Low intracellular polymers and proteins infected with EPS indicate low rates of rupture of the cells inside biofilms. Wastewater treatment is one of the favored applications of algal biofilm systems because it offers a simple, energy-efficient technology for the absorption of key nutrients, nitrogen, and phosphorus accompanied by an easy and robust separation of algal biomass from the bulk of wastewater (Li *et al.* 2015). Despite the obvious benefits of low-cost nutrient removal and low-energy biomass development of biofilm-based technology, algal biofilm-based treatment methods have not been widely used until recently, as sources of urban, commercial, and agricultural waste have been limited (Kesaano and Sims 2014). As a result, the use of biofilms in the production of biodiesel from microalgae can reduce harvesting and re-inoculation expenses. Following harvest, the leftover adherent cells, together with the biofilms, can be utilized as a fresh inoculum. This stress-induced mode enhances the total lipid productivity of the desired algae.

27.4.1.5   Removal of Antibiotics from Wastewater

Antibiotics are widely used to control and prevent bacterial infections. Antibiotics such as penicillins, sulphonamides, cephalosporins, macrolides, and quinolones were commonly used particularly in developing countries with a rise of 36% between 2000 and 2010. However, the prevalence of bacterial resistance is a major concern due to the resulting increased treatment costs, extended hospital stays, and increased mortality rates. Wastewater treatment plants (WWTPs) are significant sources of antibiotic release into the environment. In WWTPs, several antibiotics cannot be effectively eliminated by activated sewage sludge, which consists of a community of bacteria, due to the antibacterial natures (Villar-Navarro *et al.* 2018). Many processes have been examined for antibiotic elimination, such as adsorption, advanced oxidation, and photocatalysis. The adsorption process' antibiotic removal efficiency is highly adsorbent dependent, and the adsorbent is commonly expensive. Advanced oxidation and photocatalysis can generally be efficient, but they need intensive, costly chemical agents or catalysts and are likely to produce secondary pollutants (Jiang *et al.* 2018). On the other hand, the method of microalgae-based wastewater treatment is a biological process and involves few chemical agents and can be designed to effectively eliminate the emerging contaminants such as antibiotics (Wang *et al.* 2017).

Microalgae-based technology has recently been widely documented as an efficient method for treating urban and industrial effluents, with benefits such as CO_2 fixation, the elimination of pollutants, saving nutrient inputs, and the potential for developing algae-derived products (Bai and Acharya 2019). Microalgae-based technologies for treating water or wastewater that contains pharmaceutical and personal care products (PPCPs) were also introduced. Studies have shown that microalgae can effectively remove PPCPs from wastewaters, such as antibiotics (Villar-Navarro *et al.* 2018). For example, a system dominated by *Coelastrum* sp. with high-rate algal pond (HRAP) removed 64 PPCPs including 33 antibiotics (with an average concentration of 223 mg/L) from urban wastewater with an average antibiotic removal rate of 5–50% higher than the traditional activated sludge process over a six-month period (Villar-Navarro *et al.* 2018). Wang *et al.* (2017) summarized the feasibility of eliminating PPCPs using microalgae. Over the past three years, the number of publications relating to microalgae-based antibiotic removal processes has rapidly increased.

27.4.1.5.1   Inhibition Challenge of Antibiotics to Microalgae

The inhibition of antibiotic-induced microalgae is an important factor to consider when developing antibiotic-based removal processes based on algae. Antibiotics can affect algal growth by the inhibition of chemical synthesis, including chlorophyll-a, pigments, and enzyme activities such as catalase and superoxide dismutase. The weakened photosynthetic activity of *Chlorella vulgaris* was observed when streptomycin was introduced to the algae. Species like *Pseudokirchneriella subcapitata*, *Chlorella vulgaris*, *Scenedesmus Vacuolatus*, and *Desmodesmus subspicatus* are antibiotic-sensitive and are therefore commonly used as biomarkers for tracking levels of environmental contamination with antibiotics (Valitalo *et al.* 2017). The inhibitory effect of antibiotics on microalgae is generally quantified by the half-maximum effective EC50 concentration, which is characterized as the antibiotic concentration inhibiting 50% of algal growth. Many antibiotics have EC50 values of many orders of magnitudes higher than wastewater or surface/groundwater concentrations, suggesting that microalgae usually have a strong resistance to these antibiotics. For instance, the concentrations of ciprofloxacin, tetracycline, and sulfamethoxazole in municipal wastewater are approximately 1 mg/L whereas the EC50 values for these antibiotics are 65 mg/L (for *C. mexicana*), 3.31 mg/L (for *P. subcapitata*), and 0.146 mg/L (for *P. subcapitata*) (Wang *et al.* 2017). In general, the EC50 value of an antibiotic depends on the particular species of antibiotics and microalgae, e.g., sulfamonomethoxine has an EC50 value of 5.9 for *C. vulgaris* and 9.7 mg/L for *Isochrysis galbana* (Huang *et al.* 2014). Since the concentrations of antibiotics in WWTP effluents are several orders of magnitudes below the EC50 value, the use of the microalgae for antibiotics treatment may be feasible after process development.

27.4.1.5.2   Mechanisms for Removing Antibiotics

Microalgae show a series of responses when exposed to antibiotics to survive and eliminate the harmful antibiotics. During the removal process, antibiotics are either adsorbed, collected, biodegraded, or photodegraded.

27.4.1.5.2.1 Adsorption   Adsorption is the mechanism by which contaminants are eliminated by passive binding to solids (Bai and Acharya 2016). There have been detailed studies of the degradation of antibiotics by adsorbents such as biochar, activated carbon, and nano-materials. Microalgae can be an important adsorbent for the elimination of antibiotics. Functional groups and polymer assemblages (similar to cellulose, hemicelluloses, and proteins, etc.) may achieve adsorption on the algal cell walls, and this is an extracellular method. The efficacy of the adsorption process can be evaluated by using dead algal biomass to adsorb antibiotics, and the efficiency of adsorption differs greatly depending on particular antibiotics and microalgae due to their varying hydrophilicity, functionality, and structures. In general, it is beneficial when the antibiotic is more hydrophobic than hydrophilic, and when it bears microalgae with an opposite charge (Xiong *et al.* 2018). Microalgae biomass adsorption of antibiotics is achieved mostly through hydrogen bonds, electrostatic attraction, partition, and hydrophobic effect. Adsorption was not a dominant antibiotic removal mechanism in some other studies, but it may serve as the first step for subsequent antibiotic removal mechanisms, such as accumulation and biodegradation, based on microalgae (Yu *et al.* 2017).

27.4.1.5.2.2 Accumulation   Unlike extracellular adsorption, accumulation is an intracellular mechanism for removing pollutants from water. Many antibiotics may cross algal cell membranes, and the cells can then likely assimilate. Sonication coupled with a dichloromethane/methanol mixture (v: v = 1:2) can extract the intracellular accumulated antibiotics. Accumulation of algae has been stated to have played an important role in the elimination of antibiotics such as trimethoprim, sulfamethoxazole, and doxycycline. Some accumulated antibiotics can induce the development of reactive oxygen species which are necessary for controlling cellular metabolism at normal concentrations but result in severe cell damage or, eventually, death if excessive (Xiong *et al.* 2019).

The algal cell, on the other hand, will resist metabolism to deplete the antibiotics. In this case, accumulation becomes a pre-step for biodegradation, and the combination of accumulation and biodegradation in algal cells significantly leads to the completion of certain antibiotics assimilation. Sulfamethazine aggregations in *Chlorella pyrenoidosa* were observed and followed biodegradation (Sun *et al.* 2017).

27.4.1.5.2.3 Biodegradation Biodegradation describes the process of breaking down antibiotics by algae inside or outside the cells, with the algal cells degrading certain broken-down derivatives further (Naghdi *et al.* 2018). The intracellular degradation of ceftazidime and its basic parent structure 7-ACA per C is an example of this process. In this process, antibiotic removal is carried out in three steps: ceftazidime is first adsorbed to algae, then slowly transmitted to the algal cell wall, and finally broken down by algal enzymes (Yu *et al.* 2017). Other cases of the elimination of antibiotics based on intracellular degradation include the elimination of levofloxacin by *Scenedesmus Oblquus* and *C. vulgaris*. For the extracellular degradation process, algal metabolites such as extracellular enzymes break down the antibiotic, and algal cells can further metabolize the biodegraded intermediates/end products (Naghdi *et al.* 2018). The contributions of intracellular and extracellular antibiotic removal pathways can be distinguished by studying the intermediate/end-product antibiotics in the medium and biomass, or by using extra- and intracellular enzymes derived from algae to degrade antibiotics. Regardless of the various antibiotic removal mechanisms mentioned above, hydrolysis, side-chain breakdown, hydroxylation, ring cleavage, demethylation, decarboxylation, and antibiotic dehydroxylation can occur during the biodegradation of antibiotics such as the levofloxacin biodegradation process (Xiong *et al.* 2019).

27.4.1.5.2.4 Photodegradation Photodegradation of antibiotics involves direct photolysis of the light-contributing antibiotics and their indirect photodegradation, caused in the presence of light by the reactive components formed by algae. Many antibiotics can be eliminated in algae-free environments by direct photolysis when light is present (Jiang *et al.* 2018). By comparing the rate of removal between daytime and night, 40% of tetracycline in water can be removed by direct sunlight. The removal of tetracycline, ciprofloxacin, cefazolin, and cephapirin in various wastewater treatment systems based on microalgae was also confirmed to be a direct consequence of photolysis (Bai and Acharya 2016). When algae are present in the environment, the removal of such antibiotics can be facilitated by indirect photodegradation. Reactive oxygen species, such as hydroxyl radicals, can be produced from algal components during indirect photodegradation, leading to the breakdown of the antibiotics. Some antibiotics, however, are resistant to light and the algal reactive species caused by light. Photodegradation is inadequate for antibiotic removal in these situations.

27.4.1.6 Enhancement of Refractory Organic Matter Removal

Despite recent advancements in remediation technology, processes of electrochemical oxidation have limitations so it is important to establish efficient and economical strategies. Wang *et al.* (2016) successfully cultivated two strains of microalgae in reverse osmosis (RO) concentrate with low salinity (<900 mg Cl$^-$ L^{-1}), resulting in nutrient and hardness elimination at the same time. The treatment of algae-mediated wastewater effectively eliminates nutrients and heavy metals, decreases the need for chemical and biochemical oxygen, and removes and/or degrades xenobiotic compounds and other pollutants (Rawat *et al.* 2016). Despite previous observations of algae as useful mediators for wastewater remediation, there has, to date, been limited research in the treatment of reverse osmosis (RO) concentrate. RO is increasingly used in advanced wastewater treatment, especially for water reuse, but prior work with RO membranes focused primarily on the efficiency and quality of the treated wastewater. While numerous studies have concentrated on characterizing and reducing fouling, which is the most significant limitation for the operation of membrane systems, it is more urgent to treat the concentrate resulting from the filtration processes. RO concentrate typically contains high concentrations of dissolved salts and recalcitrant organics, and both the characteristics of the raw sewage and the efficiency of the reclamation processes will influence the levels of those components. In particular, free chlorine is often used as a biocide to prevent RO membranes from biofouling, and its use results in high salinity. High salinity, which prevents microbial development, may thus inhibit traditional biological treatments or may be inefficient due to recalcitrance of organic constituents in the RO concentrate. As such, given the expense and efficiency of treating the RO concentrate, various innovative treatment approaches have been proposed but these are still in their infancy. Lee *et al.* (2009) explored the impact of biological activated carbon (BAC) pretreatment on deionization efficiency and also combined ozonation with BAC to increase the biodegradability of organic matter in the RO concentrate. Some studies have reported that electrochemical oxidation can improve the degradation of dissolved organic matter in the RO concentrate produced from municipal wastewater, and the combination of photochemical and electrochemical processes based on UV has also been investigated (Hurwitz *et al.* 2014). Most previous studies emphasize the processing of algal biomass and the subsequent removal of inorganic pollutants from a wide range of wastewater, and insufficient research has been carried out on the biodegradation of refractory organic compounds other than the removal of nutrients. Eukaryotic algae have been shown to be capable of biotransforming and biodegrading aromatic contaminants typically found in natural and wastewaters, although the extent of the process for degradation is not yet fully understood. As an example, *Scenedesmus quadricauda* is a green algae that could be used to degrade refractory organic matter and then use assailable fractions of organics with inorganic nutrients in highly saline RO concentrate. *Scenedesmus* is a dominant algae genus typically found in wastewater ponds, and it was previously found that *S. quadricauda* is highly efficient in extracting inorganic nutrients from piggery wastewater at a high salinity of up to 4.8% Cl^- (Kim *et al.* 2016). Investigations revealed that *S. quadricauda* tolerates a wide range of salinities and can lead to heterotrophic growth, in addition to the typical autotrophic growth when CO_2 is used as the sole source of carbon. No studies have addressed the biodegradation/assimilation of refractory organic matter and the concurrent use of mixotrophic microalgae for inorganic nutrients from highly saline RO concentrate. In particular, there is a lack of information about the impact of algae on the biodegradability of refractory organics in RO concentrate. Different phycoremediation modes were evaluated to recognize the biodegradation potential of *S. quadricauda*, and the mechanisms were investigated by measuring reactive oxygen species. The reactive oxygen species are produced in an algal cell because of the excessive light under photosynthetic reactions and can be excreted out of the external water environment in the form of H_2O_2. Hydrogen peroxide is the only molecule among those ROS members that is membrane-permeable and can be readily released from algal cells. ROS are continually developed as by-products in photosynthetic organisms via different metabolic pathways located in the mitochondria, peroxisome, chloroplast, and even on the plasma membrane (Pérez-Pérez *et al.* 2012). Direct oxidation has historically been regarded not to play a major role in the degradation of organic compounds by hydrogen peroxide, but for most applications it eventually requires activation of some form or another. Hydrogen peroxide dissociates in hydroperoxide anion (HO^{2-}) in an alkaline solution, which is a strong nucleophile that readily attacks electron-deficient substrates. It can also be used by mixing with electron-deficient compounds to produce more strong oxidants, and is also of use in traditional peroxide bleaching reactions. Oxidative peroxide bleaching then reduces the absorption of light due to a break in the aromatic rings. Degradation caused by algae alters the fluorescent properties of organic matter. In addition to the substantial decrease in fluorescence strength, the humic-like fluorophore change to a shorter wavelength occurrs as a result of a decrease in the π-electron system (e.g., decrease in the number of aromatic rings or conjugated bonds in a chain structure, or conversion of a linear to a non-linear ring system) (Kim *et al.* 2006). The decrease in fluorescence intensity of organic matter indicates the degradation or, at times, mineralization of the corresponding fluorophores or functional groups present in the fluorescent organic matter's chemical structure (Mostofa *et al.* 2013). The chemistry of the hydrogen peroxide reaction is complex and it can potentially degrade a wide range of organic compounds, depending on the conditions. pH has a significant effect on the chemistry and efficacy of hydrogen peroxide. A high pH is thought to catalyze hydrogen peroxide activation, but oxidative peroxide bleaching has been observed even at a lower pH. Considering the exoelectrogenic behaviour of eukaryotic photoautotrophs (McCormick *et al.* 2011), hydrogen peroxide activation with a higher reduction potential than oxygen can also be catalyzed by electrons transferred from the cell surface through the plasma membrane. Hydrogen peroxide can produce a wide range of free radicals and other reactive species that can transform or decompose organic chemicals. In conclusion, processes of algae-mediated oxidation are correlated with increased COD removal for pretreated RO concentrate during subsequent mixotrophic algae treatment as a contrast with raw RO concentrate treatment. Biological treatment after ozonation is a well-established technique to efficiently extract dissolved organic matter by converting organic refractory compounds into easily biodegradable fractions, rather than by oxidizing them completely (Kim *et al.* 2014). Nevertheless, our results show that ozonation in highly saline RO concentrate was ineffective not only in reducing COD, but also in enhancing the biodegradability of organic matter. This is not surprising because certain factors may deteriorate the ability of ozonation to degrade and/or mineralize organic compounds completely. With *S. quadricauda*, phycoremediation, using two separate cultivation methods, was found to be highly feasible as a robust treatment for highly saline RO concentrate, although a minor nitrogen inhibition was observed, possibly due to rapid phosphate depletion in high pH wastewater. In particular, algae-induced refractory organic degradation facilitated the removal of COD in the subsequent mixotrophic algae treatment, which is more successful than the application of ozone before biological therapy in the pretreatment of highly saline RO. Ozone decomposition in an aqueous phase can result in the development of hydroxyl radicals which are known to effectively destroy organic matter, but in the presence of high-strength hydroxyl radical predators (HCO_3^- and CO_3^{2-}), which could be consumed by algal cells, it was not efficient in converting refractory organics into biodegradable ones.

27.4.2 GREEN NANOTECHNOLOGY

27.4.2.1 Green Synthesis of Nanoparticles

The synthesis of nanoparticles by microorganisms or plant extracts, called green nanotechnology, has attracted much interest all over the world as it is a much simpler process than physical and chemical methods. Additionally, green nanoparticle synthesis methods are less harmful, non-toxic in nature, and environmentally friendly. The utilization of algae for the biosynthesis of nanometals is known as phytonanotechnology. There are various procedures for the synthesis of different metallic nanoparticles from their corresponding aqueous salt solutions using micro and macroalgae extracts as reducing agents. Live and dead (dried) biomasses can be used to synthesize metal nanoparticles. Algal extracts containing sulfated polysaccharides as a major component, which has sulphate, hydroxyl, and aldehyde groups, are used for the stabilization of the nanoparticles. Silver nanoparticles have been synthesized by using different marine microalgal organisms (Merin *et al.* 2010). Microalgae are able to produce nanoparticles of other metal ions such as gold, platinum, and cadmium. Additionally, other nanoparticles, such as iron oxide nanoparticles (Fe_3O_4-NPs) have been synthesized using seaweed aqueous extracts, such as brown (*Colpomenia sinuosa*) and red (*Pterocladia capillacea*) marine algae (Salem *et al.* 2019). Plants are also used for nanoparticles synthesis. The extracts from different plant parts (leaves, stems, and seeds) contain proteins, amino acids, enzymes, vitamins, and organic polymers which act as reducing and stabilizing agents in the production of nanoparticles (Iravani 2011). For example, Aromal and Philip (2012) used *Trigonella foenum-graecum* extract as a reducing agent for gold nanoparticles production. The formation of nanoparticles using biological extracts depends on several parameters, including temperature, pH, and extract concentration (Iravani 2011). The nucleating rate for metal ions and the shape of synthesized nanoparticles depend on pH and temperature. During nanoparticle synthesis, the production rate changes when the reaction temperature increases. For example, gold nanoparticles exhibit various color effects (yellow-brown, purplish-pink, and pink-brown) during synthesis at different

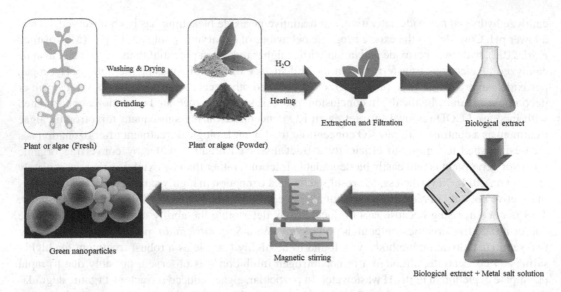

FIGURE 27.1 Design process for green nanoparticle synthesis using plant and algal extracts.

temperatures (Bankar *et al.* 2010). The mechanism of green synthesis of nanoparticles using plant and algal extracts is shown in Figure 27.1.

27.4.2.2 Application of Green Nanotechnology in Wastewater Treatment

Green nanotechnology has a range of uses in different fields. However, green synthesized nanoparticles have been commonly used in the treatment of various wastewater types. For example, silver nanoparticles were synthesized via *Piliostigma thonningii* leaf extract, then used to effectively remove heavy metals from wastewater such as copper, magnesium, lead, and iron (Shittu and Ihebunna 2017). Treated wastewater can be effectively reused for industrial and household needs. Furthermore, El-Kassas *et al.* (2016) used two seaweeds' (*Padina pavonica* and *Sargassum acinarium*) water extracts for the production of Fe_3O_4-NPs. In addition, the algal extract was used as a reduction agent for $FeCl_3$ enabling the photosynthesis of Fe_3O_4-NPs. The biosynthesized Fe_3O_4-NPs were taken in beads of calcium alginates and used in the removal of Pb. Green synthesized Fe_3O_4-NPs alginate beads by *P. pavonica* had a high capacity (91%) for the removal of Pb whereas that of *S. acinarium* had a 78% removal capacity after 75 min (El-Kassas *et al.* 2016). The silver nanoparticles synthesized by using *Sargassum longifolium* extract show high antifungal activity that can be used therapeutically in biomedical applications (Rajeshkumar *et al.* 2014). Moreover, *Padina tetrastromatica*, a marine brown macroalgae, was used for the biosynthesis of silver nanoparticles, which degrade brown and Congo red dye and are used in water purification (Roy 2019).

Rosales *et al.* (2017) found that the reactivity of synthesized zero-valent iron nanoparticles is greater when using rooibos (*Aspalathus linearis*) extract. Additionally, green zero-valent iron nanoparticles were used to dissolve textile dyes and demonstrated improved performance in the treatment of wastewater. Choudhary *et al.* (2017) reported a rapid approach to green chemistry for the synthesis of gold nanoparticles (AuNPs) using leaf extract from *Lagerstroemia speciosa*. AuNPs showed good photocatalytic activity in reducing dyes such as bromophenol blue, bromo-cresol green, methylene blue, and methyl orange. Devatha *et al.* (2016) demonstrated that the iron nanoparticles were formed by different plant extracts and used for domestic wastewater treatment. For example, *Azadirachta indica*, in particular, demonstrated good results in the elimination of phosphate (98.08%), ammonia nitrogen (84.32%), and COD (82.35%). Accordingly, the green synthesis of NPs is a very promising approach to green nanotechnology, and green-synthesized NPs can be easily isolated and reused due to their unique character and their high stability. In addition, the use of green synthesized NPs in water and wastewater treatment is not only an environmentally-friendly choice but also a promising technology for low-income countries, as it also fulfills the principle of zero effluent discharge after wastewater treatment using low cost and energy.

27.5 CHALLENGES OF THE BIOREMEDIATION PROCESS

Algal bioremediation has emerged as a low-cost alternative to traditional remediation methods that are environmentally harmful, expensive, and generate secondary emissions, thereby impacting the ecosystem negatively. However, it may be limited by several factors, such as pollutant distribution among microorganisms, which often require the pretreatment of wastewater. Also, the internal shading resulting from algal growth and other factors is a limiting factor in the bioremediation process.

a. **Wastewater Pretreatment**

Raw wastewater contains high levels of contaminants like bacteria, protozoa, fungi, and solid particles which inhibit microalgae growth. These species compete with algae for nutrients and other resources in the wastewater. Prior to the introduction of microalgae, wastewater requires a pretreatment to kill all species. Different pretreatment methods have recently been introduced to manage large volumes of wastewater. Methods of pretreatment, including filtration and autoclaving, are commonly employed. Even so, findings indicated that autoclaving was noted as the most effective method of pretreatment for microbial removal, but the authors also reported in the same study that autoclaving may interfere with the wastewater nutrient content (Ramsundar et al. 2017). Many researchers tested and proved that hypothesis. The authors observed high concentrations of biomass for the filtration method as opposed to autoclaving when microalgae were grown in municipal effluent. This indicates that the particulates present in the autoclaved pretreated effluent may have blocked complete access to light utilization for photosynthetic activities by the microalgae. Commercial implementation of these pretreatment procedures may not be feasible because of the high energy costs involved. Other alternative methods of pretreatment with wastewater have been published. Ultraviolet (UV) irradiation and chlorination, for example, were noted for the pretreatment of wastewater and used effectively by Qin et al. (2014).

b. **Selection of Suitable Microalgae Strains for Wastewater Treatment**

The selection of microalgae species is very important for treating wastewater. Microalgae should be adaptable enough to cope with changes in environmental conditions due to the physical and chemical composition of wastewater. The species should also have the ability to exchange metabolites to withstand tension, overcome any unwanted species attack, and restrict nutrient limitations (Qin et al. 2016).

c. **Inner Shading**

Inner shading restricts microalgal photosynthetic operations. Wastewater is rich in nutrients, and microalgae in this effluent may rapidly multiply within a day (24 h), but its multiplication within the log phase may be limited (3.5 h). The rapid increase in the number of cells can decrease the amount of light accessed by a portion of the effluent, as the dense culture in the upper part limits the intensity of light that penetrates into the water (Amaro et al. 2011). This problem can be solved by using raceway ponds or photobioreactors (Figure 27.2). In raceway ponds, the bottom portion of the microalgae is circulated near the surface for the microalgae to absorb light energy in the culture media by rotating the shift paddle. In a photobioreactor device, light is placed close to the upper portion of the photobioreactor, and air is ignited into the device to rotate the lower portion of the microalgae close to the surface to absorb light energy from the microalgae. Wastewater also contains a

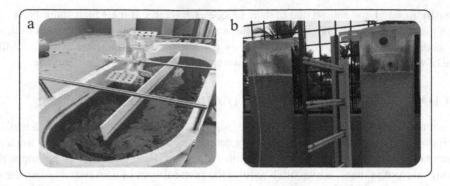

FIGURE 27.2 Example of (a) a raceway pond and (b) tubular photobioreactors for wastewater treatment (Mohd *et al.* 2017).

significant amount of solids which are suspended. That can interfere with the microalgae's growth process. Also, high wastewater turbidity restricts the penetration of light through the wastewater, which affects the microalgae photosynthesis process (Larsdotter 2006). Pretreatment methods of wastewater, such as flocculation, can also be used to eliminate suspended solids in the wastewater. Adding agitation to the effluent can also be used to resolve this issue by exposing the microalgae in the effluent to light for a short period of time, thus improving the efficiency of wastewater treatment based on microalgae.

27.6 CONCLUSION

Environmental pollution is a significant concern for public health because of the adverse effects of toxins on humans and other living organisms. Chemical and physical remediation strategies are costly and do not result in complete pollutant elimination. In addition, both methods could lead to increased contamination and disturbance of the site, thereby adversely affecting humans and other biota in the immediate vicinity of the contaminated site. The chemical and physical remediation approaches are also not considered eco-sustainable. In comparison to these approaches, bioremediation, which is based on biological processes (mediated by various classes of living organisms), contributes to the permanent removal of contaminants. It is a low-cost biological treatment method of wastewater that requires less capital investment and minimal operational attention. Phycoremediation technology is highly efficient in the treatment of a wide range of refractory pollutants. It can significantly minimize the nutrient load of wastewater and thereby reduce the overall dissolved solids. Coupling microalgae to eliminate pollutants from wastewater and biofuel generation may be an effective techno-economic strategy to reduce the cost of wastewater treatment and biofuel production. While some progress has been made in neutralizing contaminants from the water via algae, several challenges remain that need to be addressed. This is a newly-emerging field of applied technology, where the main emphasis is on the development of environmentally-sustainable technologies with economic viability.

REFERENCES

Abinandan, S., Subashchandrabose, S.R., Venkateswarlu, K., Perera, I.A., and Megharaj, M. 2019. Acid-tolerant microalgae can withstand higher concentrations of invasive cadmium and produce sustainable biomass and biodiesel at pH 3.5. *Bioresour Technol* 281: 469–473.

Abou-Shanab, R., Ji, M., Kim, H., Paeng, K., and Jeon, B. 2013. Microalgal species growing on piggery wastewater as a valuable candidate for nutrient removal and biodiesel production. *J Environ Manag* 115: 257–264.

Ajayan, K.V., Selvaraju, M., Unnikannan, P., and Sruthi, P. 2015. Phycoremediation of tannery wastewater using microalgae Scenedesmus species. *Int J Phytorem* 17(10): 907–916.

Akpor, O.B., Otohinoyi, D.A., Olaolu, T.D., and Aderiye, B.I. 2014. Pollutants in wastewater effluents: Impacts and remediation processes. *Int J Environ Res Earth Sci* 3(3): 050–059.

Al-Raad, A., Hanafiah, M.M., Naje, A.S., et al. 2019. Treatment of saline water using electrocoagulation with combined electrical connection of electrodes. *Processes* 7(5): 242.

Amaro, H.M., Guedes, A.C., and Malcata, F.X. 2011. Advances and perspectives in using microalgae to produce biodiesel. *Appl Energy* 88(10): 3402–3410.

Aravindhan, R., Rao, J.R., and Nair, B.U. 2007. Removal of basic yellow dye from aqueous solution by sorption on green alga Caulerpa scalp elliformis. *J Hazard Mater* 142(1–2): 68–76.

Aromal, S.A., and Philip, D. 2012. Green synthesis of gold nanoparticles using Trigonella foenum-Graecum and its size dependent catalytic activity. *Spectrochim Acta Mol Biol Spectrosc* 97: 1–5.

Ay, F., Catalkaya, E.C., and Kargi, F. 2008. Advanced oxidation of Direct Red (DR 28) by Fenton treatment. *Environ Eng Sci* 25(10): 1455–1462.

Baghour, M. 2017. Effect of seaweeds in phyto-remediation. In: *Biotechnological Applications of Seaweeds*, ed. E. Nabti, 47–83. Nova Science Publishers, New York.

Bai, X., and Acharya, K. 2016. Removal of trimethoprim, sulfamethoxazole, and triclosan by the green alga *Nannochloris* sp. *J Hazard Mater* 315: 70 –75.

Bai, X., and Acharya, K. 2019. Removal of seven endocrine disrupting chemicals (EDCs) from municipal wastewater effluents by a freshwater green alga. *Environ Pollut* 247: 534–540.

Baig, S., and Liechti 2001. Ozone treatment for biorefractory COD removal. *Water Sci Technol* 43(2): 197–204.

Balaji, S., Kalaivani, T., Shalini, M., Gopalakrishnan, M., Rashith Muhammad, M.A., and Rajasekaran, C. 2016. Sorption sites of microalgae possess metal binding ability towards Cr (VI) from tannery effluents: A kinetic and characterization study. *Desalin Water Treat* 57(31): 14518–14529.

Bankar, A., Joshi, B., Ravi Kumar, A., and Zinjarde, S. 2010. Banana peel extract mediated synthesis of gold nanoparticles. *Colloids Surf B Biointerfaces* 80(1): 45–50.

Barros, A.I., Goncalves, A.L., Simoes, M., and Pires, J.C.M. 2015. Harvesting techniques applied to microalgae: A review. *Renew Sustain Energy Rev* 41: 1489–1500.

Berner, F., Heimann, K., and Sheehan, M. 2015. Microalgal biofilms for biomass production. *J Appl Phycol* 27(5): 1793–1804.

Cameron, H., Mata, M.T., and Riquelme, C. 2018. The effect of heavy metals on the viability of *Tetraselmis marina* AC16-meso and an evaluation of the potential use of this microalga in bioremediation. *PeerJ* 6: e5295.

Caporgno, M.P., Taleb, A., Olkiewicz, M., et al. 2015. Microalgae cultivation in urban wastewater: Nutrient removal and biomass production for biodiesel and methane. *Algal Res* 10: 232–239.

Cecchin, I., Krishna, R., Thom, A., Tessaro, E.F., and Schnaid, F. 2017. Nanobioremediation: Integration of nanoparticles and bioremediation for sustainable remediation of chlorinated organic contaminants in soils. *Int Biodeterior Biodegrad* 119: 419–428.

Chen, H., Wang, J., and Huang, J. 2014. Policy support, social capital, and farmers' adaptation to drought in China. *Glob Environ Change* 24: 193–202.

Cheng, D., Ngo, H., Guo, W., Chang, S., Nguyen, D., and Kumar, S. 2019. Microalgae biomass from swine wastewater and its conversion to bioenergy. *Bioresour Technol* 275: 109–122.

Cheng, P., Wang, Y., Liu, T., and Liu, D. 2017. Biofilm attached cultivation of Chlorella pyrenoidosa is a developed system for swine wastewater treatment and lipid production. *Front Plant Sci* 8: 1594.

Choudhary, B.C., Paul, D., Gupta, T., et al. 2017. Photocatalytic reduction of organic pollutant under visible light by green route synthesized gold nanoparticles. *J Environ Sci (China)* 55: 236–246.

Crini, G., and Lichtfouse, E. 2019. Advantages and disadvantages of techniques used for wastewater treatment. *Environ Chem Lett* 17(1): 145–155.

Danis, T.G., Albanis, T.A., Petrakis, D.E., and Pomonis, P.J. 1998. Removal of chlorinated phenols from aqueous solutions by adsorption on alumina pillared clays and mesoporous alumina aluminum phosphates. *Water Res* 32(2): 295–302.

Devatha, C.P., Thalla, A.K., and Katte, S.Y. 2016. Green synthesis of iron nanoparticles using different leaf extracts for treatment of domestic waste water. *J Clean Prod* 139: 1425–1435.

Dubey, S.K., Dubey, J., Mehra, S., Tiwari, P., and Bishwas, A. 2013. Potential use of cyanobacterial species in bioremediation of industrial effluents. *Afr J Biotechnol* 10: 1125–1132.

El Atouani, S., Belattmania, Z., Reani, A., et al. 2019. Brown seaweed *Sargassum muticumas* low-cost biosorbent of methylene blue. *Int J Environ Resour* 13(1): 131–142.

Elimelech, M. 2006. The global challenge for adequate and safe water. *J Water Supply Res Technol Aqua* 55(1): 3–10.

El-Kassas, H.Y., Aly-Eldeen, M.A., and Gharib, S.M. 2016. Green synthesis of iron oxide (Fe3O4) nanoparticles using two selected brown seaweeds: Characterization and application for lead bioremediation. *Acta Oceanol Sin* 35(8): 89–98.

El-Shahat, M., Abdelhamid, A.E., and Abdelhameed, R.M. 2020. Capture of iodide from wastewater by effective adsorptive membrane synthesized from MIL-125-NH$_2$ and crosslinked chitosan. *Carbohydr Polym* 231: 115742.

El-Sheekh, M.M., Farghl, A.A., Galal, H.R., and Bayoumi, H.S. 2016. Bioremediation of different types of polluted water using microalgae. *Rendiconti Lincei* 27(2): 401–410.

El-Sheekh, M.M., Metwally, A.M., Allam, N.G., and Hemdan, H.E. 2017. Effect of algal cell immobilization technique on sequencing batch reactors for sewage wastewater treatment. *Int J Environ Res* 11(5–6): 603–611.

El-Sheekh, M.M., Metwally, A.M., Allam, N.G., and Hemdan, H. 2020. Simulation treatment of industrial wastewater using microbiological cell immobilization technique. *Iran J Sci Technol Trans A Sci* 44(3): 595–604.

Fahad, R.M.S., Mohamed, B., Radhi, M., and Al-Sahari 2019. Wastewater and its treatment techniques: An ample review. *Indian J Sci Technol* 12(25): 1–13.

Ghrefat, H., Nazzal, Y., Batayneh, A., et al. 2014. Geochemical assessment of ground water contamination with special emphasizes on fluoride, a case study from Midyan Basin, north western Saudi Arabia. *Environ Earth Sci* 71(4): 1495–1505.

Gleick, P.H. 2009. Facing down the hydro-crisis. *World Policy J* 26(4): 17–23.

Gross, M., Jarboe, D., and Wen, Z. 2015. Biofilm-based algal cultivation systems. *Appl Microbiol Biotechnol* 99(14): 5781–5789.

Huang, D.J., Hou, J.H., Kuo, T.F., and Lai, H.T. 2014. Toxicity of the veterinary sulfonamide antibiotic sulfamonomethoxine to five aquatic organisms. *Environ Toxicol Pharmacol* 38(3): 874–880.

Hurwitz, G., Hoek, E.M.V., Liu, K., Fan, L., and Roddick, F.A. 2014. Photo-assisted electrochemical treatment of municipal wastewater reverse osmosis concentrate. *Chem Eng J* 249: 180–188.

Ibrahim, Z., Amin, M.F.M., Yahya, A., et al. 2009. Characterisation of microbial flocs formed from raw textile wastewater in aerobic biofilm reactor (ABR). *Water Sci Technol* 60(3): 683–688.

Iravani, S. 2011. Green synthesis of metal nanoparticles using plants. *Green Chem* 13(10): 2638.

Jaafari, J., and Yaghmaeian, K. 2019. Optimization of heavy metal biosorption onto freshwater algae (*Chlorella coloniales*) using response surface methodology (RSM). *Chemosphere* 217: 447–455.

Jackson, R.B., Carpenter, S.R., Dahm, C.N., et al. 2001. Water in a changing world. *Ecol Appl* 11(4): 1027–1045.

Ji Kabra, A.N., Salama, S., Roh, H.S. et al. 2014. Effect of mine wastewater on nutrient removal and lipid production by a green microalga *Micratinium reisseri* from concentrated municipal wastewater. *Bioresour Technol* 157: 84–90.

Jiang, L., Yuan, X., Zeng, G. et al. 2018. Metal-free efficient photocatalyst for stable visible-light photocatalytic degradation of refractory pollutant. *Appl Catal B* 221: 715–725.

Kadimpati, K.K., Mondithoka, K.P., Bheemaraju, S., and Challa, V.R.M. 2013. Entrapment of marine microalga, *Isochrysis galbana*, for biosorption of Cr (III) from aqueous solution: Isotherms and spectroscopic characterization. *Appl Water Sci* 3(1): 85–92.

Kehrein, P., Van Loosdrecht, M., Osseweijer, P., Garfi, M., Dewulf, J., and Posada, J. 2020. A critical review of resource recovery from municipal wastewater treatment plants: Market supply potentials, technologies and bottlenecks. *Environ Sci Water Res Technol* 6(4): 877–910.

Kesaano, M., and Sims, R.C. 2014. Algal biofilm based technology for wastewater treatment. *Algal Res* 5: 231–240.

Khan, I., Saeed, K., and Khan, I. 2017. Nanoparticles: Properties, applications and toxicities. *Arab J Chem* 12(7): 908–931.

Khan, M., and Yoshida, N. 2008. Effect of L-glutamic acid on the growth and ammonium removal from ammonium solution and natural wastewater by *Chlorella vulgaris* NTM06. *Bioresour Technol* 99(3): 575–582.

Kim, H.C., Choi, W.J., Chae, A.N., Park, J., Kim, H.J., and Song, K.G. 2016. Evaluating integrated strategies for robust treatment of high saline piggery wastewater. *Water Res* 89: 222–231.

Kim, H.C., Choi, W.J., Maeng, S.K., Kim, H.J., Kim, H.S., and Song, K.G. 2014. Ozonation of piggery wastewater for enhanced removal of contaminants by *S. quadricauda* and the impact on organic characteristics. *Bioresour Technol* 159: 128–135.

Kim, H.C., Yu, M.J., and Han, I. 2006. Multi-method study of the characteristic chemical nature of aquatic humic substances isolated from the Han River, Korea. *Appl Geochem* 21(7): 1226–1239.

Larsdotter, K. 2006. Wastewater Treatment with Microalgae—A literature review. *Sol Energy* 62: 31–38.

Lee, L.Y., Ng, H.Y., Ong, S.L., et al. 2009. Ozone-biological activated carbon as a pretreatment process for reverse osmosis brine treatment and recovery. *Water Res* 43(16): 3948–3955.

Li, T., Lin, G.Y., Podola, B., and Melkonian, M. 2015. Continuous removal of zinc from wastewater and mine dump leachate by a microalgal biofilm PSBR. *J Hazard Mater* 297: 112–118.

Li, X., Yang, W.L., He, H. et al. 2018. Responses of microalgae *Coelastrella* sp. to stress of cupric ions in treatment of anaerobically digested swine wastewater. *Bioresour Technol* 251: 274–279.

Liu, Y., Cheng, H., and He, Y. 2020. Application and mechanism of sludge-based activated carbon for phenol and cyanide removal from bio-treated eluent of coking wastewater. *Processes* 8(1): 82.

Liu, Y., Guan, Y., Gao, B., and Yue, Q. 2012. Antioxidant responses and degradation of two antibiotic contaminants in Microcystis aeruginosa. *Ecotoxicol Environ Saf* 86: 23–30.

Lou, Y., Ye, X., Ye, Z., Chiang, P., and Chen, S. 2018. Occurrence and ecological risks of veterinary antibiotics in struvite recovered from swine wastewater. *J Cleaner Prod* 201: 678–685.

Luan, M., Jing, G., Piao, Y., Liu, D., and Jin, L. 2017. Treatment of refractory organic pollutants in industrial wastewater by wet air oxidation. *Arab J Chem* 10: S769–S776.

MacQuarrie, P., and Wolf, A.T. 2013. Understanding water security. In: *Environmental Security: Approaches and Issues*, ed. R. Floyd, and R.A. Matthew, 169–186. Routledge, London.

Mansor, E.S., Abdallah, H., and Shaban, A.M. 2020. Fabrication of high selectivity blend membranes based on poly vinyl alcohol for crystal violet dye removal. *J Environ Chem Eng* 8(3): 103706.

McCormick, A.J., Bombelli, P., Scott, A.M., et al. 2011. Photosynthetic biofilms in pure culture harness solar energy in a mediator less bio-photovoltaic cell (BPV) system. *Energy Environ Sci* 4(11): 4699–4709.

Merin, D., Prakash, S., and Bhimba, B.V. 2010. Antibacterial screening of silver nanoparticles synthesized by marine micro algae. *Asian Pac J Trop Med* 3(10): 797–799.

Mohd Udaiyappan, A.F., Abu Hasan, H., Takriff, M.S., and Sheikh Abdullah, S.R. 2017. A review of the potentials, challenges and current status of microalgae biomass applications in industrial wastewater treatment. *J Water Process Eng* 20: 8–21.

Mostofa, K.M.G., Liu, C.q., Vione, D., et al. 2013. Colored and chromophoric dissolved organic matter in natural waters. In: *Photobiogeochemistry of Organic Matter: Principles and Practices in Water Environments*, ed. M.G.K. Mostofa, T. Yoshioka, A. Mottaleb, and D. Vione, 365–428. Springer, Berlin Heidelberg.

Naghdi, M., Taheran, M., Brar, S.K., Kermanshahi-pour, A., Verma, M., and Surampalli, R.Y. 2018. Removal of pharmaceutical compounds in water and wastewater using fungal oxidoreductase enzymes. *Environ Pollut* 234: 190–213.

OECD 2012. Environmental Outlook to 2050: The consequences of inaction. OECD, March, 1–8. http://www. oecd.org/environment/outlookto2050.

Pérez-Pérez, M.E., Lemaire, S.D., and Crespo, J.L. 2012. Reactive oxygen species and autophagy in plants and algae. *Plant Physiol* 160(1): 156–164.

Pires, J.C., Alvim-Ferraz, M.C., Martins, F.G., and Simoes, M. 2013. Wastewater treatment to enhance the economic viability of microalgae culture. *Environ Sci Pollut Res Int* 20(8): 5096–5105.

Postel, S. 2000. Entering an era of water scarcity: The challenges ahead. *Ecol Appl* 10(4): 941–948.

Pradhan, D., Sukla, L.B., Mishra, B.B., and Devi, N. 2019. Biosorption for removal of hexavalent chromium using microalgae *Scenedesmus* sp. *J Cleaner Prod* 209: 617–629.

Priatni, S., Ratnaningrum, D., Warya, S., and Audina, E. 2018. Phycobiliproteins production and heavy metals reduction ability of *Porphyridium* sp. *IOP Conf Ser Earth Environ Sci* 160(1): 012006.

Qin, L., Shu, Q., Wang, Z., et al. 2014. Cultivation of *chlorella vulgaris* in dairy wastewater pretreated by UV irradiation and sodium hypochlorite. *Appl Biochem Biotechnol* 172(2): 1121–1130.

Qin, L., Wang, Z., Sun, Y., et al. 2016. Microalgae consortia cultivation in dairy wastewater to improve the potential of nutrient removal and biodiesel feedstock production. *Environ Sci Pollut Res Int* 23(9): 8379–8387.

Quesnel, D., and Nakhla, G. 2005. Optimization of the aerobic biological treatment of thermophilically treated refractory wastewater. *J Hazard Mater* 125(1–3): 221.

Rajasulochana, P., and Preethy, V. 2016. Comparison on efficiency of various techniques in treatment of waste and sewage water: A comprehensive review. *Resour Effic Technol* 2(4): 175–184.

Rajeshkumar, S., Malarkodi, C., Paulkumar, K., Vanaja, M., Gnanajobitha, G., and Annadurai, G. 2014. Algae mediated green fabrication of silver nanoparticles and examination of its antifungal activity against clinical pathogens. *Int J Met* 2014: 1–8.

Ramanan, R., Kim, B., Cho, D.H., Oh, H.M., and Kim, H.S. 2016. Algae-bacteria interactions: Evolution, ecology and emerging applications. *Biotechnol Adv* 34(1): 14–29.

Ramsundar, P., Guldhe, A., Singh, P., and Bux, F. 2017. Assessment of municipal wastewaters at various stages of treatment process as potential growth media for *Chlorella sorokiniana* under different modes of cultivation. *Bioresour Technol* 227: 82–92.

Rawat, I., Gupta, S.K., Shriwastav, A., Singh, P., Kumari, S., and Bux, F. 2016. Microalgae applications in wastewater treatment. In: *Algae Biotechnology: Products and Processes*, ed. F. Bux, and Y. Chisti, 249–268. Springer International Publishing, Cham.

Rosales, E., Meijide, J., Pazos, M., and Sanromán, M.A. 2017. Challenges and recent advances in biochar as low-cost biosorbent: From batch assays to continuous-flow systems. *Bioresour Technol* 246: 176–192.

Roy, S. 2019. A review: Green synthesis of nanoparticles from seaweeds and its some applications. *Austin J Nanomed Nanotechnol* 7(1): 105.

Sadoff, C.W., Hall, J.W., Grey, D., et al. 2015. *Securing Water, Sustaining Growth: Report of the GWP/OECD Task Force on Water Security and Sustainable Growth*, 180. University of Oxford, Oxford, UK.

Salem, D.M., Ismail, M.M., and Aly-Eldeen, M.A. 2019. Biogenic synthesis and antimicrobial potency of iron oxide (Fe_3O_4) nanoparticles using algae harvested from the Mediterranean Sea, Egypt. *Egypt J Aquat Res* 45(3): 197–204.

Sharma, S., and Bhattacharya, A. 2017. Drinking water contamination and treatment techniques. *Appl Water Sci* 7(3): 1043–1067.

Shittu, K.O., and Ihebunna, O. 2017. Purification of simulated waste water using green synthesized silver nanoparticles of *Piliostigma thonningii* aqueous leave extract. *Adv Nat Sci Nanosci Nanotechnol* 8(4): 1–9.

Singh, A., and Olsen, S.I. 2011. A critical review of biochemical conversion, sustainability and life cycle assessment of algal biofuels. *Appl Energy* 88(10): 3548–3555.

Sud, D., Mahajan, G., and Kaur, M.P. 2008. Agricultural waste material as potential adsorbent for sequestering heavy metal ions from aqueous solutions: A review. *Bioresour Technol* 99(14): 6017–6027.

Suganya, T., Varman, M., Masjuki, H.H., and Renganathan, S. 2016. Macroalgae and microalgae as a potential source for commercial applications along with biofuels production: A biorefinery approach. *Renew Sustain Energy Rev* 55: 909–941.

Sun, J., Cheng, J., Yang, Z., Li, K., Zhou, J., and Cen, K. 2015. Microstructures and functional groups of *Nannochloropsis* sp. cells with arsenic adsorption and lipid accumulation. *Bioresour Technol* 194: 305–311.

Sun, M., Lin, H., Guo, W., Zhao, F., and Li, J. 2017. Bioaccumulation and biodegradation of sulfamethazine in *Chlorella pyrenoidosa*. *J Ocean Univ China* 16(6): 1167–1174.

United Nations Environment Programme (UNEP) 2016. *A Snapshot of the World's Water Quality: Towards a Global Assessment*, 162pp. United Nations Environment Programme, Nairobi, Kenya.

United Nations World Water Assessment Programme (WWAP). 2017. The United Nations world water development report 2017. Wastewater: The untapped resource. UNESCO, Paris. http://unesdoc.unesco.org/images/0024/002471/247153e.pdf.

Upadhyay, A.K., Mandotra, S.K., Kumar, N., Singh, N.K., Singh, L., and Rai, U.N. 2016. Augmentation of arsenic enhances lipid yield and defense responses in alga *Nannochloropsis* sp. *Bioresour Technol* 221: 430–437.

Valitalo, P., Kruglova, A., Mikola, A., and Vahala, R. 2017. Toxicological impacts of antibiotics on aquatic micro-organisms: A mini-review. *Int J Hyg Environ Health* 220(3): 558–569.

Villar-Navarro, E., Baena-Nogueras, R.M., Paniw, M., Perales, J.A., and Lara-Martín, P.A. 2018. Removal of pharmaceuticals in urban wastewater: High rate algae pond (HRAP) based technologies as an alternative to activated sludge based processes. *Water Res* 139: 19–29.

Wang, X.X., Wu, Y.H., Zhang, T.Y., Xu, X.Q., Dao, G.H., and Hu, H.Y. 2016. Simultaneous nitrogen, phosphorous, and hardness removal from reverse osmosis concentrate by microalgae cultivation. *Water Res* 94: 215–224.

Wang, Y., Liu, J., Kang, D., Wu, C., and Wu, Y. 2017. Removal of pharmaceuticals and personal care products from wastewater using algae-based technologies: A review. *Rev Environ Sci Bio Technol* 16(4): 717–735.

Wichelns, D., Drechsel, P., and Qadir, M. 2015. Wastewater: Economic asset in an urbanizing world. In: *Wastewater*, 3–14. Springer, Netherlands.

Wolf, A.T. 2007. Shared waters: Conflict and cooperation. *Annu Rev Environ Resour* 32(3): 3.1–3.29.

Xiong, J.Q., Kim, S.J., Kurade, M.B., et al. 2019. Combined effects of sulfamethazine and sulfamethoxazole on a freshwater microalga, *Scenedesmus obliquus*: Toxicity, biodegradation, and metabolic fate. *J Hazard Mater* 370: 138–146.

Xiong, J.Q., Kurade, M.B., and Jeon, B.H. 2018. Can microalgae remove pharmaceutical contaminants from water? *Trends Biotechnol* 36(1): 30–44.

Yang, J., Cao, J., Xing, G., and Yuan, H. 2015. Lipid production combined with biosorption and bioaccumulation of cadmium, copper, manganese and zinc by oleaginous microalgae *Chlorella minutissima* UTEX2341. *Bioresour Technol* 175: 537–544.

Yogalakshmi, K.N., Das, A., Rani, G., Jaswal, V., and Randhawa, J.S. 2020. Nano-bioremediation: A new age technology for the treatment of dyes in textile effluents. In: *Bioremediation of Industrial Waste for Environmental Safety*, ed. G. Saxena, and R. Bharagava. Springer, Singapore.

Yu, Y., Zhou, Y., Wang, Z., Torres, O.L., Guo, R., and Chen, J. 2017. Investigation of the removal mechanism of antibiotic ceftazidime by green algae and subsequent microbic impact assessment. *Sci Rep* 7(1): 1–11.

Zakhama, S., Dhaouadi, H., and M'Henni, F. 2011. Nonlinear modelisation of heavy metal removal from aqueous solution using *Ulva lactuca* algae. *Bioresour Technol* 102(2): 786–796.

Zhang, D., Yan, D., Lu, F., Wang, Y., and Feng, J. 2015. Copula-based risk assessment of drought in Yunnan Province, China. *Nat Hazards* 75(3): 2199–2220.

28 Role of Enzymes in the Bioremediation of Refractory Pollutants

Viresh R. Thamke, Ashvini U. Chaudhari,
Kisan M. Kodam, and Jyoti P. Jadhav

CONTENTS

28.1 INTRODUCTION

With the advent of the industrial revolution in the 18th century, there has been a huge dependency on machinery, chemicals, and advanced technology for the production of various commodities. Industrialization has played a key role in reducing unemployment issues, even with an increasing global population, and with the advances in science and technology, medicine has improved the standard of living of the common man. Several agricultural and industrial activities, in recent years, simultaneously increased the usage of organic chemicals. Although these activities have resulted in the improvement in the economic growth of various countries across the globe, rapid industrial growth, commercialization, and modern agricultural practices have disturbed environmental conditions. Due to anthropogenic interference, the utilization of synthetic chemicals has been a major

cause of environmental pollution. The chemicals used in processes have a relatively complex structure which is difficult to metabolize by biotic life. The challenges associated with these chemicals
is the toxicity of metabolized products, recalcitrant behavior, insolubility, and volatile nature. A
well-known phenomenon called the "grasshopper" effect explains the persistence of these xenobiotic compounds in the natural food chain, affecting even the remote places of the globe. These
compounds thus cause irreversible damage to the natural environment and biotic life.

Industries mainly release untreated effluents in rivers and nearby water bodies, resulting in
adverse effects on the associated ecosystem. Based upon their chemical nature, effluents can be
broadly classified as organic and inorganic industrial effluents. Wastes from electroplating, coal,
and metal industries constitute the inorganic waste effluents, while wastes generated due to the
usage of organic chemicals in chemical industries, with pesticide as the main ingredient, constitute
organic industrial effluents. Pollutants in the environment not only harms the environment, it also
affects the health of humans and animals. Numerous diseases may appear due to their exposure:
pollutants may inhibit respiration, may provoke reduced reproduction of fish-eating birds, as well as
contribute to the birth of premature babies or babies with genetic defects such as Down's syndrome,
anencephaly, and spina bifida. Pollution continues to be a major cause of illness and long-term
health damage (Rao et al. 2010). The increase in contaminated sites around the world that impact
environmental and human health makes it imperative to decontaminate the environment by imple-
menting effective bioremediation strategies. Previously, the wastes generated by various industries
were treated by dumping in landfills, by incineration, and by using UV rays. However, these methods are ineffective, complex, costly, and generate recalcitrant by-products (Vidali 2001). Recently,
bioremediation has proved to be an efficient way of breaking down these chemicals (Dzionek et al.
2016). The major goals of bioremediation should be the rejuvenation of soil health and fertility, the
detoxification of water sources, and effective water treatment and its reutilization.

The utilization of microbes to curb the pollution caused by industrial wastewater has been implemented all around the world. Bioremediation is a microorganism mediated transformation or degradation of contaminants into non-hazardous or less-hazardous substances. The ability of various
organisms like bacteria, fungi, algae, and plants for the efficient bioremediation of pollutants has
been reported. The process of bioremediation mainly depends on microorganisms which enzymatically attack the pollutants and convert them to innocuous products. As remediation can be effective
only where environmental conditions permit microbial growth and activity, its application often
involves the manipulation of environmental parameters to allow microbial growth and degradation to proceed at a faster rate. The microbes tend to be highly adaptable and tolerant, and have
a robust physiology and highly efficient enzyme system that can metabolize a plethora of organic
compounds with varied functional groups. Because they are easy to culture in laboratories and are
ideal for in-situ applications, these tiny bugs provide a cost-effective and eco-friendly solution to
organic pollutants. The most favored biomolecules for bioremediation are enzymes since they show
chemio-, regio-, and stereospecificity (Rao et al. 2010). Enzymes have been widely used in various
bioremediation strategies. Apart from specificity, they are easy to maintain and handle and require
mild conditions for operation. *Curvularia haloperoxidase* acts as an enzymatic disinfection which
is useful in wastewater treatment, quickly killing microbes by oxidizing halides such as iodine and
chloride. This mechanism resulted in 6 order of magnitude reductions in the *E. coli* population in 10
min, 5 order of magnitude reductions in the *S. epidermidis* population in 20 min, and also caused a
large reduction in other microbial populations (Hansen et al. 2003). Apart from this, the enzyme has
the capability to break large sludge particles, creating more surface area for microbe attacks; this
allows for the more efficient and complete degradation of sludge particles in wastewater treatment
(Whiteley et al. 2003).

Enzymes enhance the rate of reactions which otherwise would take a lot of time to biotransform. Therefore, microbial enzymes are powerful tools that help to maintain a clean environment
in several ways. They are utilized for environmental purposes in a number of industries. Enzymes
also help to maintain an unpolluted environment through their use in waste management. With this

background, we have attempted to provide an account of various enzymes that are harnessed in the biotransformation of industrial wastes and have various biotechnological implications. For a better understanding of the functional diversity of enzymes, we have attempted to provide a broad description of the enzymes based on the classification suggested by the Enzyme Commission.

28.2 MICROBIAL OXIDOREDUCTASES

Microbes extract energy via energy-yielding biochemical reactions mediated by enzymes to cleave chemical bonds and catalyze the transfer of electrons and protons from a donor to an acceptor. The reactions involve electron transfer, proton abstraction, hydrogen extraction, hydride transfer, oxygen insertion, and other key steps. During such oxidation-reduction reactions, the contaminants are finally oxidized to harmless compounds. Oxidoreductases detoxify various toxic xenobiotics, such as phenolic or anilinic compounds, through polymerization, copolymerization with other substrates, or binding to humic substances. Based on the type of reaction catalyzed, oxidoreductases have been divided into 22 different EC-subclasses. Mono- or di-oxygenases, reductases, dehalogenases, cytochrome P450 monooxygenases, polyphenol oxidases, and peroxidases (lignin and manganese peroxidases) are the main subclasseses of oxidoreductases (Rao et al. 2010). The oxidoreductase has the highest potential in the production of polymer building within lignocellulose biorefineries. While the chemical industries, especially in the bulk treatment/production enzymatic processes, do not work effectively. They are used in the degradation of several natural and man-made pollutants. The oxidoreductase enzymes from many bacterial species are involved in the reduction of radioactive metals as a result of redox reactions. Apart from the microbes, some plant families like *Fabaceae*, *Gramineae*, and *Solanaceae* secrete extracellular oxidoreductase for the degradation of soil contaminants such as chlorinated compounds and petroleum-containing hydrocarbons.

28.2.1 Oxygenase

Oxygenases have a broad substrate range and are active against a wide range of compounds. The introduction of O_2 atoms into the organic molecule by oxygenase results in cleavage of the aromatic rings. Oxygenases act as biocatalysts in the bioremediation process and synthetic chemistry due to their high regioselectivity, stereoselectivity, and enantiospecificity of a wide range of substrates (Arora et al. 2010). The variety of pollutants that persists in the environment, including chlorinated compounds, halogen-containing herbicides, fungicides, and pesticides, are degraded by oxygenase (Sharma et al. 2018). Glyphosate oxidase isolated from bacterium *Pseudomonas* sp. LBr is involved in the bioremediation of pesticides; similarly, oxygenase produced by certain marine bacteria degrades organic pollutants (Scott et al. 2008; Sivaperumal et al. 2017). On the basis of oxygen molecules involved in the enzymatic reaction, oxygenases are classified into two subclasses: monooxygenase (catalyze the insertion of a single oxygen atom) and dioxygenase (catalyze the insertion of both oxygen atoms).

28.2.1.1 Monooxygenases

Monooxygenases inset one oxygen atom and catalyze the degradation of aromatic compounds and enhance their reactivity and solubility [Figure 28.1(A)]. Some monooxygenases require a cofactor for their catalytic activity, while some reports show the independence of cofactors in an enzyme system, i.e., the enzymes require only molecular oxygen for their activities and utilize the substrate as a reducing agent (Karigar and Rao 2011). The biotransformation and biodegradation of various aromatic, heterocyclic hydrocarbons and aliphatic compounds are catalyzed by monooxygenases like desulfurization, dehalogenation, denitrification, ammonification, and hydroxylation (Grosse et al. 1999). This enzyme is also involved in the degradation of hydrocarbon such as substituted methane, alkanes, cycloalkanes, alkenes, haloalkenes, ethers, and aromatic ethers [Figure 28.1(B)]. Monooxygenases are further classified into two groups based on the cofactor used: Flavin-dependent monooxygenases and P450 monooxygenases.

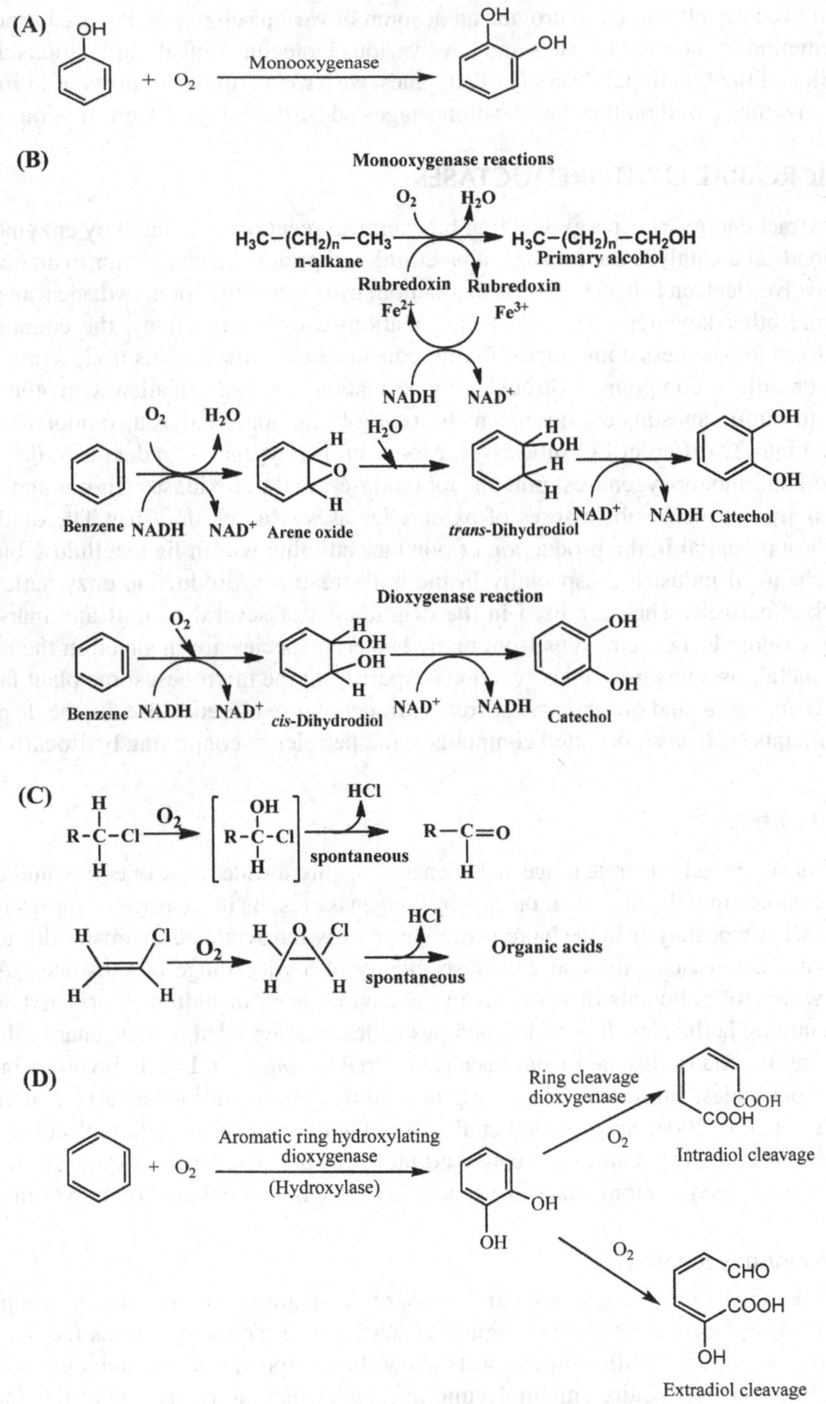

FIGURE 28.1 (A) The degradation of an aromatic compound by monooxygenase (Arora et al. 2010), (B) The role of monooxygenase and dioxygenase in the bacterial degradation of aromatic compounds (Fritsche and Hofrichter 1999), (C) The mechanism of oxygenase-mediated dehalogenation reaction of a chlorinated compound (Field and Sierra-Alvarez 2004), (D) The degradation of an aromatic compound by dioxygenase (Agrawal and Dixit 2015).

Flavin-dependent monooxygenases consist of flavin as a prosthetic group and require NADP or NADPH as the coenzyme. Endosulfan diol and Endosulfan ether are the members of the two-component flavin diffusible monooxygenase family used for the degradation of chlorine-containing pesticides such as endosulfan (Bajaj et al. 2010).

Halogenated organic compounds comprise the largest groups of environmental pollutants as a result of their widespread use as herbicides, insecticides, fungicides, hydraulic and heat transfer fluids, plasticizers, and intermediates for chemical synthesis. Oxidative dehalogenation of these halogenated aromatics is catalyzed by monooxygenase, dioxygenases, and peroxidase (Wang et al. 2010). Oxygenases mediate dehalogenation reactions of halogenated methanes, ethanes, and ethylenes in association with multifunctional enzymes. If oxygen is inserted into the same carbon as a chloro group, the resulting molecule is chemically unstable and will spontaneously convert to an aldehyde, releasing the chloro group as HCl. A monooxygenase attack of chlorinated ethenes results in the formation of unstable epoxides that spontaneously decompose to organic acids, including chloroacetic acids (Field and Sierra-Alvarez 2004) [Figure 28.1(C)].

28.2.1.2 Cytochrome P450 Monooxygenases

Cytochrome P450 was targeted by directed evolution for enhancing the degradation of recalcitrant aromatic and aliphatic pollutants. Cytochrome P450 is a superfamily of ubiquitous heme proteins that perform a number of difficult oxidative reactions, such as N-, O-, and S-dealkylations; hydroxylations; sulfoxidations; epoxidations; deaminations; desulfurations; dehalogenations; peroxidations; and N-oxide reductions of numerous endogenous and exogenous compounds (Das and Chandran 2011). Due to its extremely diverse reaction mechanism, Cytochrome P450 plays an important role in the biotransformation and degradation of diverse industrial waste. All members of P450 oxidoreductase have iron-containing porphyrin groups, and to recycle their redox center use a non-covalently bound cofactor [most frequently NAD(P)H is used]. The *E. coli* expressing rat NADPH-cytochrome P450 oxidoreductase was created by scientists using ice-nucleation proteins from *Pseudomonas syringae* for the selective synthesis of new chemicals and pharmaceuticals, bioconversion, bioremediation, and bio-chip development (Yim et al. 2006).

28.2.1.3 Dioxygenases

Dioxygenase degrades aromatic pollutants by adding two atoms of an oxygen molecule in the ring [Figure 1(D)]. The aromatic dioxygenases are classified according to their mode of action as aromatic ring hydroxylation dioxygenase (ARHDs) and aromatic ring cleavage dioxygenase (ARCDs). The ARHDs degrade aromatic compounds by adding two molecules of oxygen into the ring, whereas ARCDs break aromatic rings of compounds. The catechol dioxygenases that are found in the soil bacteria cause the biotransformation of aromatic precursors into aliphatic products (Muthukamalam et al. 2017). Toluene dioxygenase (TOD) catalyzes the first reaction in the degradation of toluene. This multi-component enzyme system acts on the broad substrate and behaves as monooxygenase and dioxygenase. TOD also has the ability to catalyze sulfoxidation reactions and convert ethyl phenyl sulfide, methyl phenyl sulfide, methyl p-nitrophenyl sulfide, and p-methoxymethyl sulfide into their respective sulfoxides. TOD also detoxifies polychlorinated hydrocarbons, chlorotoluenes, and BTEX residues very effectively (Scott et al. 2008).

28.2.1.4 Laccases

Laccases couple the electron reduction of dioxygen into two molecules of water with the oxidation of a vast variety of aromatic compounds besides lignin. The potential applications of laccases include bleaching, decolorization, and detoxification of dye-containing textile effluents and effluents containing lignin-related or phenolic compounds, and the biobleaching of pulp for the paper industries (Sharma et al. 2018). The enzymes turn phenolic compounds into less toxic end products via degradation or polymerization reactions and/or cross-coupling of pollutant phenols with naturally-occurring phenols (Abadulla et al. 2000). Laccases are also efficient on various inorganic effluents such as sulfides, sulfates, chlorides, carbonates, peroxides, chlorine bleach compounds, and heavy metals. Laccase isolated from fungus having better properties such as high salt and heavy metal tolerance and thermostability with potential for industrial waste treatment. Laccases purified from various organisms show different properties depending on the organism. Laccase can also be used for the management of environmental waste produced in the agro-industries (Ruqayyah et al. 2013). Laccase can help in the oxidation of the hydrocarbon backbone of polyethylene by reducing the average molecular weight and average molecular number of polyethylene by 20% and 15%, respectively (Sivan 2011).

These are perhaps the most widely distributed oxidases in white-rot fungi. These four-copper metalloenzymes catalyze the O_2-dependent oxidation of a variety of phenolic compounds and do not require H_2O_2 or Mn(II) for activity. Similar to peroxidases, laccases catalyze the subtraction of one electron from phenolic hydroxyl groups of phenolic compounds to form phenoxy radicals as intermediates. They also oxidize non-phenolic substrates in the presence of mediators which are oxidized to reactive radical or cation substrates by laccase and undergo further oxidation of non-phenolic targets (Rodgers et al. 2010). Dodor et al. (2004) examined the ability of immobilized and free laccases from *Trametes versicolor* to degrade anthracene and benzo[a]pyrene. Both forms of this enzyme were able to degrade between 80% and 85% for each of these compounds when in the presence of the mediator 2,2-azino-bis (3-ethylbenzothiazoline-6-sulfonic acid), and between 5% and 17% in the absence of a mediator.

Before implementation into an industrial sector, enzyme catalytic properties are to be improved for their repeated cycle and high yield. The immobilization provides a stable catalyst for practical application in industries; also it is convenient in handling easy isolation from complex reaction mixture ultimately impacts on the process economy. It is of great value to evaluate the toxicity status of untreated and treated dye/effluent samples as an additional bio-treatment efficiency tool, since these dye pollutants, as well as their degradation metabolites, are recognized to be a potential threat to human health and aquatic biota. Different immobilized laccases from different sources have been utilized to treat different azo dyes; it is more likely that the detoxification differences arise from the sensitivity of the toxicity assays to be employed (Bilal and Asgher 2015). The *T. versicolor* laccase immobilized on CPC-silica beads efficiently detoxified sulfathiazole (STZ) and sulfamethoxazole (SMZ), as confirmed by a micro-toxicity assay (Rahmani et al. 2015). The treatment of anthraquinonic dyes with the alumina-immobilized *T. hirsuta* laccase reduced their toxicities by up to 80% (Abadulla et al. 2000).

28.2.1.5 Peroxidases

Peroxidases are the enzymes that catalyze the oxidation of lignin and other phenolic compounds with the cost of hydrogen peroxide. They have many important functions in living systems, such as the immune system and hormone regulation in humans, and auxin metabolism, lignin and suberin formation, cell elongation, and defense against pathogens in plants. Although peroxidases have a very important role in plant and animal cells, they have also been used in the treatment of industrial waste (Karigar and Rao 2011).

28.2.1.5.1 Classification of Peroxidases

The peroxidase enzyme can be divided into haem and non-haem proteins. The haem peroxidase is classified into two groups as found only in animals and found in plants: fungi, and prokaryotes. The second group is subdivided into three classes: Class I includes yeast cytochrome c peroxidase, ascorbate peroxidase from the plant, and bacterial gene duplicated catalase peroxidase. Class II includes lignin peroxidase and manganese peroxidase which are mainly involved in the degradation of lignin in wood. Class III contains secondary plant peroxidases that are categorized as biosynthetic enzymes mainly involved in plant cell wall formation and lignification. The non-haem peroxidase forms five different families: thiol peroxidase, alkylhydroperoxidase, non-haem halo-peroxidase, manganese catalase, and NADH peroxidase. Among these thiols, peroxidases are the largest and include two subfamilies: glutathione peroxidase and peroxy redoxins.

Out of these different peroxidases, lignin peroxidase, manganese peroxidase, and versatile peroxidases are studied the most because of their high degradability of industrial waste and toxic substances in nature.

28.2.1.5.2 Lignin Peroxidase

Lignin peroxidase (LiP), first discovered in *Phanerochaete chrysosporium*, is a glycosylated haem protein which catalyzes a reaction in the presence of hydrogen peroxidase. Its natural function is the degradation of lignin, a compound found in the cell wall of the higher plants. After LiP has been oxidized by hydrogen peroxidase, oxidized LiP gains an electron from veratryl alcohol (cofactor), gets reduced back, and the veratryl aldehyde is formed. Veratryl aldehyde is then again reduced to veratryl alcohol by gaining an electron from a substrate. This results in the oxidation of halogenated phenolic compounds, polycyclic aromatic compounds, and other aromatic compounds, followed by a series of non-enzymatic reactions (Karigar and Rao 2011) [Figure 28.2(A)]. These enzymes are valuable in pulp, paper, and cardboard industrial waste because of their lignin

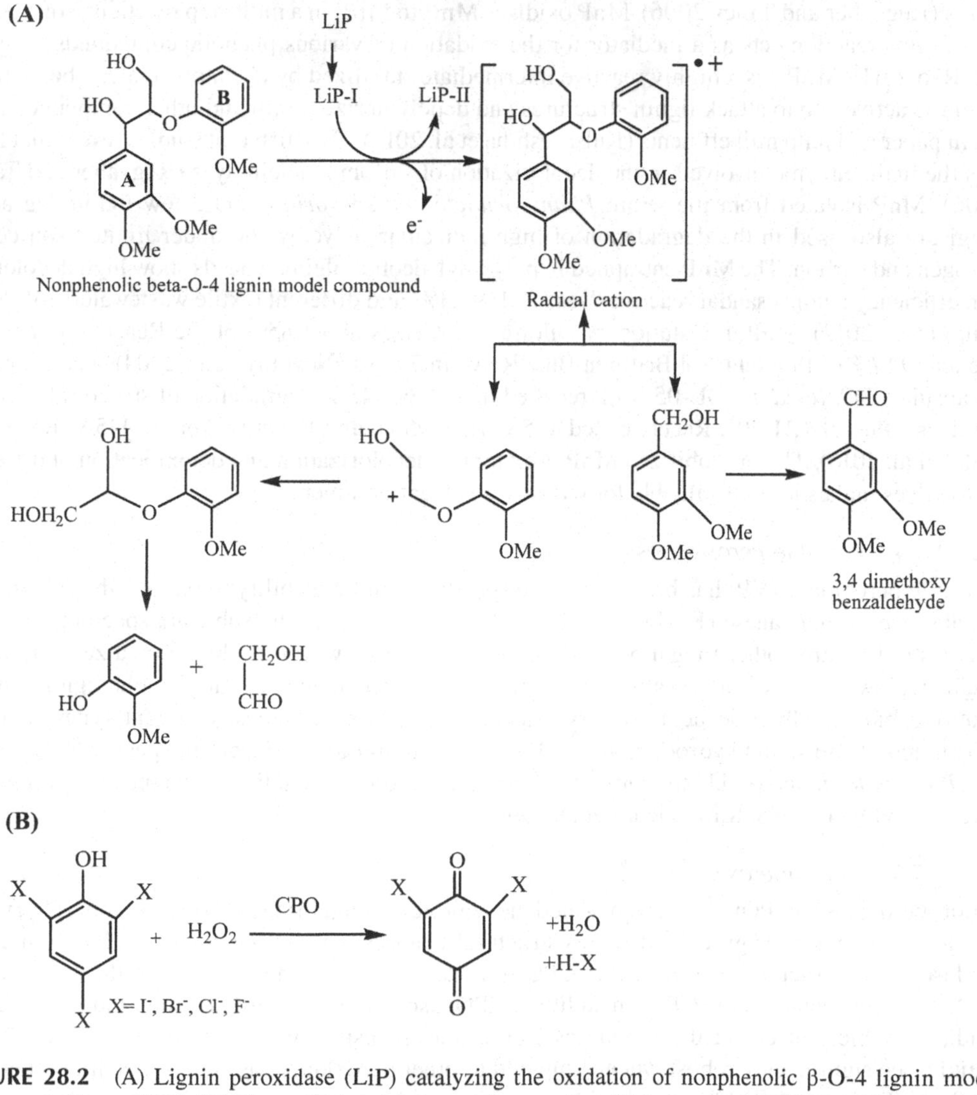

FIGURE 28.2 (A) Lignin peroxidase (LiP) catalyzing the oxidation of nonphenolic β-O-4 lignin model compound (Wong 2009), (B) The oxidative dehalogenation reaction (Osborne et al. 2006).

as well as various pollutants degrading ability. Lignin-degrading enzymes are applicable in the degradation of highly toxic environmental chemicals such as dioxins, polychlorinated biphenyls, various dyes and polyaromatic hydrocarbons, and decolorization of Kraft bleach plant effluents (Kulshreshtha et al. 2013).

The stability of LiP can be increased by immobilizing on the porous ceramic surface. It has an inconceivable potential for the biodegradation of several organic pollutants like environmentally-persistent aromatics. The removal of 70% decolorization, 55% phenol, and 15% of total organic carbon in 3 h from pulp mill effluents was achieved by the immobilization of LiP on Amberlite IRA-400 resin (Peralta-Zamora et al. 1998b). The report shows that LiP is capable of degrading benzo[a]pyrene into 52% 1,6-quinone, 25% 3,6-quinone, and 23% 6,12-quinone; the ratios of these products were similar to those found after chemical and electrochemical degradation of benzo[a] pyrene (Fiechter et al. 1986). Also, nanoporous gold immobilized LiP was used for the effective removal of fuschine, rhodamine B, and pyrogallol red (Qiu et al. 2009)

28.2.1.5.3 Manganese Peroxidases

Manganese peroxidases (MnP) are glycoproteins with an iron porphyrin IX secreted in multiple iso-forms (Ruggaber and Talley 2006). MnP oxidizes Mn^{2+} to Mn^{3+} in a multistep reaction; Mn^{3+} generated in this reaction acts as a mediator for the oxidation of various phenolic compounds (Karigar and Rao 2011). MnP^{3+} is a highly reactive intermediate stabilized by chelators and can be diffused from the active site to attack lignin structures and depolymerize lignin and other compounds present in paper and pulp mill effluents (Kulshreshtha et al. 2013). Two different studies reveal that MnP was the main enzyme involved in the decolorization of sulfonphthalein dyes (Ruggaber and Talley 2006). MnP isolated from the strain *Phanerochaete chrysosporium* and a few lignin-degrading fungi are also used in the degradation of high molecular polyethylene under limited sources of nitrogen and carbon. The MnP entrapped in polyvinyl alcohol-alginate beads show high decolorization efficiency against sandal reactive dyes (78.1–92.3%) and different textile wastewaters (61–80%) (Bilal et al. 2017). MnP in a stationary culture decolorizes about 85% of the Reactive Orange 16 dye and 99.7% of the Remazol Brilliant Blue R dye in 7 days (Novotný et al. 2004). The agar-agar encapsulated *G. lucidum* IBL-05 MnP resulted in an 8.6–84.7% degradation of structurally different dyes, Blue 21 (RB 21), Reactive Red 195A (RR 195A), and Reactive Yellow 145A (RY 145A) (Bilal et al. 2016). The immobilized MnP used for the decolorization and detoxification of different textile dyes makes it more valuable for industrial effluent treatment.

28.2.1.5.4 Versatile Peroxidases

Versatile peroxidase (VP) has broad substrate specificity and the ability to oxidize the substrate in the absence of manganese (Karigar and Rao 2011). VP combines the substrate specificity characteristics of the three other fungal peroxidase families. In this way, it is able to oxidize a variety of (high and low redox potential) substrates including Mn^{2+}, phenolic and non-phenolic lignin dimers, a-keto-g-thiomethylbutyric acid, veratryl alcohol, dimethoxybenzenes, different types of dyes, substituted phenols, and hydroquinones. VP has been found to be present in species of *Pleurotus* and *Bjerkandera*, and is able to oxidize both phenolic and nonphenolic compounds including dyes (Mester and Field 1998; Ruiz-dueñas et al. 1999).

28.2.1.5.5 Chloroperoxidase

Chloroperoxidase (CPO) is a glycosylated hemoprotein containing an iron protoporphyrin IX (heme) as its prosthetic group and shares structural features with both cytochrome P450s and peroxidases. This structural feature makes CPO function as cytochrome P450 due to the activation of the ferric heme center in the CPO. In addition, CPO also undergoes peroxidations and two-electron oxidation typical of classical peroxidases and catalases, respectively [Figure 28.2(B)]. CPO was found to be significantly robust when compared to other peroxidases as well as harsh to other bio-catalysts (Osborne et al. 2006).

CPO efficiently degrades monofluorophenols as well as oxidizes 10 different organophosphorus pesticides. Organophosphorus pesticides containing the phosphorothioate group, azinphosmethyl, chlorpyrifos, dichlorofenthion, dimethoate, parathion, phosmet, and terbufos enzymes from *Caldariomyces fumago* (Hernandez et al. 1998). These reports show that the CPO enzyme is versatile in nature and applicable in the removal of complex industrial waste.

28.2.2 REDUCTASE

Reductase is the most important subclass of an oxidoreductase which catalyzed different organic and inorganic waste containing azo dye, chromate, nitrate, etc. Wastewater from textile and dyestuff industries is difficult to treat because these effluents contain a variety of chemicals of a complex structure, like dyes having the azo group, aromatic and heterocyclic groups, surfactants, heavy metals, and salts. There are various reports available on flavin reductase which is catalyzed by various reductase activity. Flavin reductases from *Citrobacter freundii* A1 and *Sphingomonas* sp. strain BN6 are involved in the biotransformation of azo dyes from effluents (Abdul-Wahab et al. 2013). Azoreductases are involved in bioremediation and detoxification of azo dyes to eliminate the color effluents found in wastewaters. The mechanism involves the reductive cleavage of azo bonds (-N=N-) with the help of azoreductase under anaerobic conditions. The biotransformation of different complex dyes has been reported by several researchers from NADH-dependent azoreductase isolated from *Alishewanella* sp. strain KMK6, *Lysinibacillus* sp. KMK-A, and *Bacillus* sp. (Ooi et al. 2007). FMN-dependent aerobic azoreductase from *Enterococcus faecalis* is catalyzed by different dyes as well as sulfonated azo dyes (Chen et al. 2004). The azoreductase of *Clostridium perfringens* was able to reduce azo dye in strictly anaerobic conditions. While Božič et al. (2010) have reported an NADH-dependent reduction of redox dyes isolated from *Bacillus subtilis* (Figure 28.3).

Nitroreductase, responsible for the nitroreduction of RDX both under aerobic and anaerobic conditions, involves a reduction of the nitro-group of RDX or TNT by the enteric bacterium *Morganella morganii* and *Enterobacter cloacae*. However, nitroreductases do not reduce inorganic nitrate or nitrite like nitrate/nitrite reductase and they do not provide a source of reduced nitrogen for cell growth.

Heavy metal pollutants like Cr, As, Hg, Cd, etc., cause major environmental concern because of their higher toxicity. Chromate is a serious environmental toxicant due to the wide use of chromium compounds in industries such as tanning, corrosion control, plating, pigment manufacturing, and nuclear weapons production. The enzymatic detoxification of Cr(VI) to Cr(III) has been reported by chromate reductase isolated from *Escherichia coli* ATCC 33456, *Lysinibacillus* sp., *Pseudomonas putida*, and *Lysinibacillus sphaericus* G1 (Bae et al. 2005; Chaudhari et al. 2013). Chromate reductase converts the highly toxic and soluble hexavalent chromium to its less toxic trivalent form having much lower solubility; thereby the reduction by the enzymes affords a means of chromate bioremediation. Chromate reductase from *Pseudomonas putida* and YieF from *Escherichia coli* both reduced Cr(VI) ($kcat/Km = \sim 2 \times 10^4$ M^{-1} • s^{-1}) as well as reduced quinones, potassium

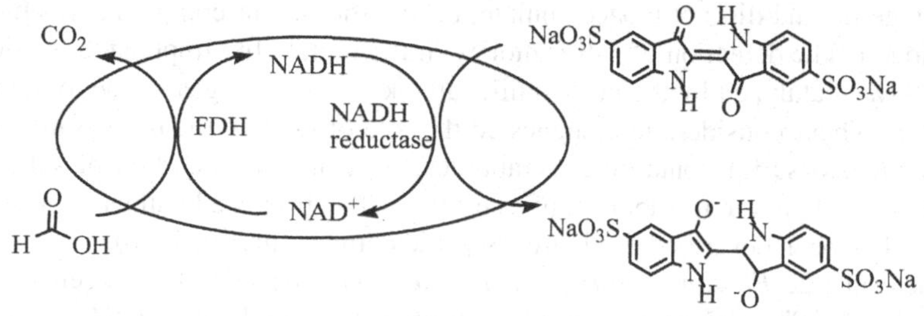

FIGURE 28.3 The regeneration of NADH during the NADH-dependent reductase reduction of CI Acid Blue 74 using formate dehydrogenase (FDH) and the formic acid regeneration system (Božič et al. 2010).

ferricyanide, and 2,6-dichloroindophenol. YieF protein effectively reduces high-valence metals like V(V), Mo(VI), and methylene blue. Thus, chromate reductases from *Pseudomonas putida* and YieF from *Escherichia coli* appears to have a broader substrate range (Ackerley et al. 2004). Nitroreductases from *Escherichia coli* DH5 and *Vibrio harveyi* KCTC 2720 show nitrofurazone and 2,4,6-trinitrotoluene as well as chromate reductase activity (Kwak et al. 2003). The involvement of Cytochrome C_7 in the reduction of Cr(VI) to Cr(III) from *Desulfuromonas acetoxidans* has also been demonstrated. Mercury can exist in three forms in the environment: metallic Hg(0), inorganic Hg(I)/Hg(II), and organic Hg. Mercuric reductase is an important enzyme involved in the detoxification of mercury by the reduction of Hg(II) to metallic Hg in effluents. Mercuric reductases from *Lysinibacillus sphaericus*, *Pseudomonas aeruginosa* PA09501, and *Bacillus* sp. strain RC607 have been reported (Schiering et al. 1991; Bafana et al. 2013).

28.3 TRANSFERASES

Transferase is a class of enzyme that transfers specific functional groups from one molecule (donor) to another (acceptor). This enzyme is also involved in the biodegradation of various industrial wastes. Hydrogen cyanide (HCN) is a major worldwide environmental pollutant produced from the metallurgic industry. HCN contains cyanide which has a strong binding ability to cytochrome oxidase and causes toxicity to aerobic life. Even though it is highly toxic, it can be converted to less toxic forms by an enzyme called rhodanase (thiosulfate: cyanide sulfurtransferases), which is one of the mechanisms employed for cyanide detoxification (Raybuck 1992). Similarly, chloromethane produced in industries has wide use in the production of silicon polymers, the manufacturing of butyl rubber, petroleum refining, etc. Although it has valuable use it is very toxic to the environment and has harmful effects on living systems. An anaerobic acetogenic bacterial strain *Acetobacterium dehalogenans* isolated from sewage sludge utilizes chloromethane as a sole source of energy (Stanlake and Finn 1982). The aerobic facultative methylotrophic bacteria carry out the degradation of dichloromethane by glutathione-S-transferases to S-chloromethyl glutathione which is an unstable intermediate goes to abiotic hydrolysis to form formaldehyde (Gisi et al. 1998). The *cis*-dichloroethene and vinyl chloride are the organic pollutants which are degraded by aerobic bacterial monooxygenase resulting in the formation of epoxides which are further degraded by epoxy alkane: coenzyme M transferase (Field and Sierra-Alvarez 2004).

Glutathione-S-transferase (GST) is one of the more widely-studied enzymes due to its relevance in the detoxification system; However, its significance with respect to bioremediation has been less illustrated. GST plays a key role in maintaining the redox status of a cell. Glutathione (GSH) plays a key role by mediating a conjugation reaction with nonpolar compounds that contain an electrophilic carbon, nitrogen, or sulfur atom to form less-toxic, water-soluble, and excretable products (Oakley 2005). The role of GST in detoxification is well illustrated with respect to herbicide alachlor (Field and Thurman 1996). A study reported that a novel sulfonated metabolite of herbicide alachlor, i.e., 2-[(2,6-diethylphenyl)(methoxymethyl)amino]-2-oxoethanesulfonic acid, was detected in the groundwater of the Midwestern region of the United States. This sulfonated metabolite was found to be non-mutagenic and did not bioaccumulate, unlike the parent compound alachlor which is a known carcinogen. The detection of sulfur moiety in the metabolite implicated the dechlorination reaction which was catalyzed by the nucleophilic attack of the thiol group due to the microflora of the groundwater. These considerations suggested the role of GST in the transformation of the said herbicide. Conjugated sulfur-containing metabolites, in turn, act as carbon or sulfur sources for microbes and bring about the metabolism of a substrate. The proposed pathway for the mineralization of alachlor has been described in Figure 28.4 (Field and Thurman 1996).

Several bacteria like *Flavobacterium*, *E.coli*, *Pseudomonas,* etc., have been studied to demonstrate the role of GST in the biotransformation of organic pollutants. GSTs play a key role in the metabolism of drugs, insecticides, and pesticides. Elevated levels of GSHs or GSTs are used as biomarkers for explaining the oxidative stress in plants and microbes (Martins et al. 2011).

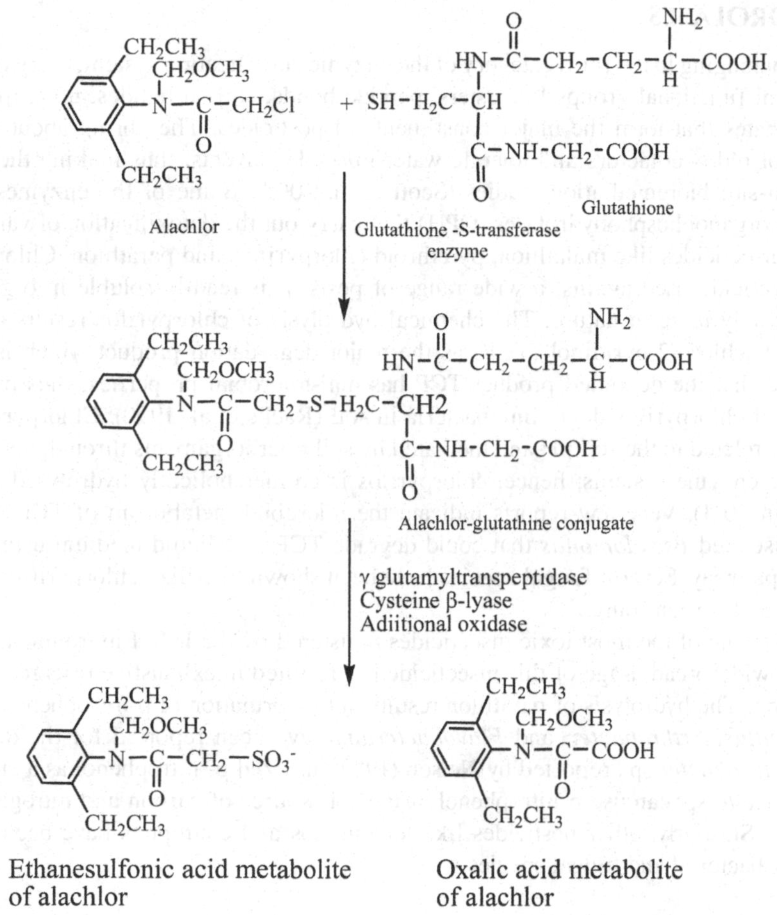

Alachlor

Glutathione

Glutathione -S-transferase enzyme

Alachlor-glutathine conjugate

γ glutamyltranspeptidase
Cysteine β-lyase
Adiitional oxidase

Ethanesulfonic acid metabolite
of alachlor

Oxalic acid metabolite
of alachlor

FIGURE 28.4 A glutathione-S-transferase enzyme-mediated reaction between glutathione and the chloro-acetanilide herbicide alachlor to form sulfonated and non-sulfonated metabolites (Field and Thurman 1996).

Kulkarni et al. (2014) reported the role of GST to combat the oxidative stress under the elevated dosage of Tri-butyl phosphate by bacterial isolate *Klebsiella pneumoniae*. GSTs play a significant role in the bioremediation of organic pollutants as they form conjugated metabolites of the substrate that are less toxic, more soluble, and are easily metabolized by the robust activity of the microbes. They maintain the cellular integrity of the microbes, thereby enhancing the tolerance level of the microbes towards the pollutant and aiding in bioremediation.

Demethylases are the enzymes which remove methyl ($-CH_3$) groups from a compound. This kind of reaction generally can be seen in DNA and protein for epigenetic modifications. However, there are few reports that suggest the role of this enzyme in the degradation of organic pollutants. The imidazolium-based ionic liquid is degraded by *Rhodococcus hoagii* VRT1 by removing methyl groups attached to the third position of the imidazolium ring (Thamke and Kodam 2016). The same demethylation mechanism is seen in the case of other imidazolium, pyridinium, phosphonium, and ammonium-based ionic liquids when degraded by aerobic bacteria granules (a bacterial consortium) (Thamke et al. 2019). The degradation of benzidine-based dye Direct Blue-6 by *Pseudomonas desmolyticum* was carried out by the demethylase enzyme in bacteria (Kalme et al. 2007). Demethylation is a major mechanism in dye degradation and includes the degradation of some common dyes like malachite green by using this enzyme (Jadhav and Govindwar 2006).

28.4 HYDROLASES

Hydrolases, belonging to the third category of the enzyme classification system, carry out the hydrolysis of several functional groups like esters, peptide bonds, carbon halides, urea, thioesters, and organophosphates that form the major constituents of pesticides. They bring about hydrolysis in the absence of redox cofactors and tolerate water miscible solvents, thus making them ideal candidates for in-situ bioremediation studies (Scott et al. 2008). Some of the enzymes included in this class are organophosphohydrolases (OPH) that carry out the detoxification of various organophosphorus insecticides like malathion, pyrethroid chlorpyrifos, and parathion. Chlorpyrifos is an effective insecticide used against a wide range of pests. It is readily soluble in organic solvents and is moderately toxic in nature. The chemical hydrolysis of chlorpyrifos results in the formation of 3,5,6-trichloro-2-pyridinol (TCP) as the major degradation product which is more toxic. It is observed that the degraded product TCP has anti-microbial properties, thus preventing the proliferation of chlorpyrifos-degrading bacteria in soil (Racke et al. 1990). Chlorpyrifos degradation is mainly related to the soil pH and mediated by soil microorganisms through non-specific and non-inducible enzyme systems; hence, chlorpyrifos is co-metabolically hydrolysed to TCP (Das and Chandran 2011). Very few reports indicate the microbial metabolism of TCP. Stanlake and Finn (1982) isolated *Pseudomonas* that could degrade TCP in a liquid medium using a reductive chlorination pathway. Several fungal species have been shown to utilize chlorpyrifos as their chief phosphorous and carbon source.

Parathion is one of the most toxic insecticides registered by the U.S. Environmental Protection Agency. The widespread usage of this insecticide has resulted in exhaustive research on microbial bioremediation. The hydrolysis of parathion results in the formation of p-nitrophenol. Several species like *Bacillus*, *Arthrobacter*, and *Flavobacterium* have been reported for the degradation of parathion. *Arthrobacter* sp., reported by Nelson (1982), utilized p- nitrophenol as a chief source of carbon. *Moraxella* sp. can use p-nitrophenol as the sole source of carbon and nitrogen (Spain and Gibson 1991). Similarly, other pesticides like fenamiphos and coumphos have been biodegraded using various bacterial and fungal species.

28.4.1 Phosphotriesterases

Another efficient enzyme belonging to the same class is phosphotriesterase. It belongs to an amidohydrolase metalloenzyme family which catalyzes the hydrolysis of organophosphorus (OP) triesters [Figure 28.5(A)]. Phosphotriesterases are divided into two categories: OpdA from *Agrobacterium radiobacter* and OPH from *Pseudomonas diminuta* and *Flavobacterium* (Harcourt et al. 2002). OpdA have been studied extensively and are used commercially for the remediation of OP-contaminated soil. It is sold under the brand name LandGuard™ from Orica Watercare (Australia) at a lower cost than the pesticides themselves (Scott et al. 2008).

Belonging to the amido hydrolases group, phosphotriesterases exhibit catalytic promiscuity which is an added advantage in the field of bioremediation. Apart from the hydrolysis of P–O bonds in organophosphorus pesticides, they are operational towards the hydrolysis of P–S, P–F, P–C–N, and C–O bonds. These attributes make phosphotriesterase an ideal candidate for the bioremediation of chemical warfare agents like tabun, sarin, toman, and VX. Several bacterial strains have been isolated, namely *Alteromonas*, *Pseudomonas diminuti*, *Burkholderia caryophilli*, and *Pseudomonas testosteronis*, and have been reported for the hydrolysis of these compounds (Singh and Walker 2006).

28.4.2 Alkaline Phosphatases

Alkaline phosphatases are known to hydrolyse various organic compounds in the nuclear industry. It is most active in alkaline environments and its physiological role is dephosphorylating compounds. Tributyl phosphate (TBP) is one of the best-known extractants used in the plutonium

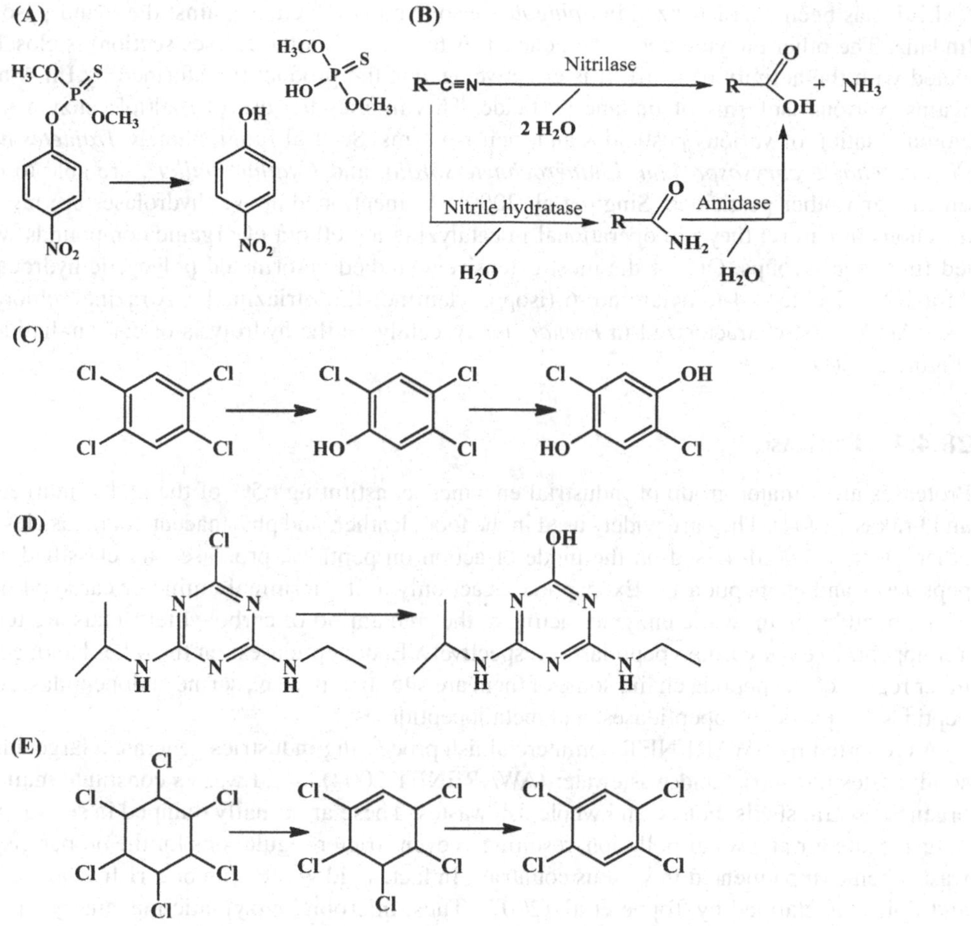

FIGURE 28.5 (A) The hydrolysis of the insecticidal phosphotriester parathion by the bacterial phosphotriesterase OpdA (Scott et al. 2008), (B) An enzymatic pathway for the hydrolysis of nitriles (Rao et al. 2010), (C) Sequential hydrolytic dechlorinations of 2,4,5,6- tetrachloro-3,6-cyclohexadiene by LinB (Scott et al. 2008), (D) Atrazine chlorohydrolysis catalyzed by AtzA and TrzN (Scott et al. 2008), (E) Chloroelimination of γ-hexachlorocyclohexane by the hexachlorocyclohexane dehydrochlorinase LinA (Scott et al. 2008).

uranium extraction process (PUREX). Several bacteria belonging to the family *Enterobacteriaceae* have been reported to tolerate and degrade high doses of TBP aided with induced alkaline phosphatases in a bacterial system. One of the more well-studied bacterial isolates is *Klebsiella pneumonia* (Kulkarni et al. 2014), which can tolerate high concentration of TBP and bring about transformation to a non-toxic metabolite di-butyl phosphate with the help of induced alkaline phosphates.

28.4.3 HALOALKANE DEHALOGENASES

Haloalkane dehalogenases are another class of enzymes that are effective against the halogenated compounds. One group of widely-studied compounds is γ-isomer of hexachlorocyclohexane, commonly known as lindane (γ-HCH). It is extensively used as an insecticide against agricultural pests. The two genes mainly responsible for the biodegradation of lindane are *LinA* and *LinB*, which have been extensively studied in bacterial systems. LinB is a haloalkane dehalogenase that catalyzes the conversion of 2,3,5,6-tetrachloro-1,4-cyclohexadiene to 3,6-dichloro-2,5-dihydroxy-1,4-cyclohexadiene [Figure 28.5(C)] (Negri et al. 2007)

LinB has been characterized in *Spingobium* sp. And is effective against the β and γ isomers of lindane. The other enzyme coded by gene LinA (discussed in the Lyases section) is closely associated with the activity of LinB. It is effective against the product transformed by LinA and acts against various isoforms of lindane pesticide. This implies the use of multiple enzymes for the bioremediation of various pesticides and their isoforms. Several fungi, namely *Trametes hirsutus*, *Phanerochaete chrysosporium*, *Phanerochaete sordia*, and *Cyathus bulleri*, are able to degrade lindane and other pesticides (Singh et al. 2000). As mentioned above, hydrolases are highly promiscuous in nature; they are operational in catalyzing a plethora of organic compounds with varied functional groups. One of the most extensively studied chlorinated polycyclic hydrocarbon is Atrazine [2-chloro-4-(ethylamino)-6-(isopropylamino)-1,3,5-triazine)]. Atrazine chlorohydrolase (AtaA), first characterized in *Pseudomonas*, catalyzes the hydrolysis of carbon–halide bonds [Figure 28.5(D)].

28.4.4 PROTEASES

Proteases are a major group of industrial enzymes, constituting 65% of the global market (Banik and Prakash 2004). They are widely used in the food, leather, and pharmaceutical industries, among others. (Singh 2003). Based on the mode of action on peptides, proteases are classified as endopeptidases and exopeptidases. Exopeptidases act only at the terminal amino or carboxyl positions of the peptide chain, while enzymes acting at the free amino or carboxyl terminals are termed as aminopeptidases or carboxypeptidases, respectively. Endopeptidases, on the other hand, act on the inner region of the peptide chain; some of them are site specific, e.g., serine endopeptidase, cysteine peptidase, aspartic endopeptidases, and metallopeptidases.

As reported by AWARENET, commercial fish processing industries generate a large number of solid wastes (20–60%) and wastewater (AWARENET 2004). Solid wastes constitute mainly body organs, viscera, shells, bones, and whole fish wastes. These are usually dumped in seawater, which in turn cause air and water pollution, resulting in even stricter regulations for the proper disposal of wastes being implemented in various countries. In fact, solid wastes can be a rich source of protein and lipids, as claimed by Toppe et al. (2007). Thus, microbial enzymatic machinery can be harnessed to degrade the wastes and obtain products which have commercial importance.

28.4.5 LIPASES

A lipase is defined as a carboxylesterase which catalyzes the hydrolysis and synthesis of long-chain acylglycerols, with trioleoylglycerol being the standard substrate. There are many biotechnological implications of lipases. Many bacteria like *Pseudomonas, Bacillus, Streptococcus, Chromobacterium*, and *Staphylococcus,* and fungi like *Candida, Rhizopus, Penicillium*, and *Aspergillus* produce lipases, which have been studied well. The operational strategy implemented by lipases is a unique feature, unlike other esterases. The activity of lipases is enhanced on insoluble substrates (such as emulsions) that form aggregates in water, compared with the same substrates in true monomeric solutions (Verger 1997). Lipases are employed in the food, leather, fishing, oil, paper making, and detergent manufacturing industries. One of the most well-studied examples of lipases is the degradation of crude oil during oil spills.

Accidental oil spills that have been extensively studied in the past are the Exxon Valdez spill in 1989 and the BP Deepwater Horizon spill in 2010 in the U.S. Although different in nature with respect to the oil spilled in the ocean, the mode of dispersion, etc., the impact due to the spill was tremendous and affected the aquatic biota. Various measures, both physical and biological, were attempted to curb the impact caused due to the oil spill. The oil spilled in the BP Deepwater Horizon spill was a leak from a well 5,000 ft. (1,500 m) below the ocean surface, while the Exxon Valdez spill was a surface phenomenon. Bioremediation strategies were extensively used in both the oil spills where hydrocarbonoclastic bacteria played a key role in reducing the pollution (Atlas and Hazen 2011).

28.4.6 NITRILASE

Nitrilase (nitrile aminohydrolase) enzymes act on nonpeptide C-N bonds produced by both bacterial species such as *Nocardia* sp. and *Rhodococcus* sp. as well as fungal species such as *Fusarium solani* and *Aspergillus niger*. Nitrilase acts on a broad range of nitrile substrates and converts them into respective carboxylic acid and ammonia [Figure 28.5(B)]. Nitrilase from filamentous fungi has a high affinity toward aromatic nitriles, aliphatic, and alicyclic nitriles. Due to their high activity, thermostability, excellent region, and enantioselectivity they are good biodegraders of nitrile contaminants (Rao et al. 2010). Based on the substrate specificity, nitrilase are classified into three classes: aliphatic nitrilase, aromatic nitrilase, and arylacetonitrilase. Aromatic nitrilase degrades nitrile in one step, aliphatic nitrilase degrades it in two steps, and aromatic nitrilase is unable to degrade aliphatic nitriles. Arylacetonitrilase, a new type of microbial nitrilase isolated from *Alcaligenes faecalis* JM3, acts on acetonitrile and acrylonitrile which are aromatic and aliphatic nitriles, respectively (Basile 2008).

28.5 LYASES

Lyases constitute the fourth category of the enzyme classification system. They initiate cleavage of the C-C bond which is energetically demanding. One of the most well-studied examples of this group is the lindane dehydrochlorinase LinA which acts against the γ-hexachlorocyclohexane [Figure 28.5(E)]. It initiates the first two dechlorination reactions, which are further acted upon by LinB. Mercury contamination is a worldwide problem and methylmercury is the most abundant form of organic mercury which is found to be more toxic than ionic mercury. The bacterial organomercurial lyase (MerB) catalyzes the carbon–mercury bond by protonolysis and releases Hg(II), a less-toxic, non-biomagnified form of mercury.

$$RHg(I) \rightarrow RH + Hg(II)$$

MerB converts organic mercury to its inorganic form which is further detoxified by mercury reductase enzymes (Benison et al. 2004). Pectinases are utilized to a large extent in food processing, jute, and paper and pulp industries. The wastewater obtained from the citrus fruit industry contain a large number of pectin fibers that are difficult to degrade. The physical treatments used for treating such wastes include spray irrigation, chemical coagulation, chemical hydrolysis, and sludge treatment that leads to the generation of toxic methane gas. Above all, these treatments are highly expensive and require a prolonged time to degrade. Thus, the usage of microbial pectinases can effectively remediate pectic wastes and are eco-friendly and cost effective. Tanabe et al. (1987) reported the *Bacillus* species as effective against industrial pectic substances.

28.6 MODERN TECHNOLOGIES ENHANCING BIOREMEDIATION

In a natural environment, microorganisms are incapable of producing sufficient quantities of enzymes. Thus, the practical application of enzymes in bioremediation faces several difficulties such as less stability, productivity, and activity. Enzymes are complex molecular structures and any physical or chemical changes result in the inactivation of the enzyme (Nigam and Shukla 2015). To resolve this problem many scientists continuously engage to isolate novel enzymes producing microbes which could potentially degrade toxic chemicals from various sources. Moreover, new genetic engineering, immobilization, and enzyme engineering techniques have been created to overcome the challenges faced by traditional bioremediation (Baweja et al. 2016; Gupta and Singh 2017). The genetic engineering tool offers an easy way to produce the desired enzyme in bulk quantities for bioremediation purposes. Recombinant DNA technology is used for the overexpression of enzymes. It is cost effective and produces a stable and active forms of enzymes. The genetic engineering approach can enhance the shelf life, substrate range, pH, and temperature stability of enzymes. Techniques use such as DNA shuffling, site-directed mutagenesis, or error-prone PCR are used to increase the range of contaminants as substrates (Dua et al. 2002). The peroxidase from *Thanatephorous cucumeris* strain Dec 1 is expressed in *Aspergillus Oryzae* RD 005 and is successfully used for the decolorization of dyes.

Another aspect of modern engineering is enzyme engineering which involves the transformation or modification of the amino acid structure of an enzyme to improvise certain properties like activity, stress tolerance, temperature, pH, etc. Similar to the expression of a particular enzyme in microbes, recombinant DNA technology is used in the same way in which enzyme engineering is performed. The amino acids in the enzyme determine the basic structure and functioning of the enzyme; therefore, alteration in the amino acid sequence can modify the characteristics of the enzyme. Nowadays, enzyme engineering has been explored for the specific and efficient bioremediation of heavy metals and radionucleotides (Dhanya 2014). The 2-nitrotoluene dioxygenase was modified at position 258 by site-directed mutagenesis which then converts nitrotoluene to 3-methyl catechol and nitrite (Singh et al. 2008).

Enzyme immobilization is another aspect through which effective bioremediation can be achieved. The technique involves the immobilization of enzymes in different ways, such as affinity-tag binding, adsorption on glass and alginate beads or on the matrix, and by covalently binding to the support-like silica gel (Sirisha et al. 2016). Immobilization on the solid support can enhance the catalytic efficiency of enzymes and can be recovered easily and reused multiple times. Immobilized LiP can remove up to 55% of the total phenol from paper mill effluents without losing activity during the process (Peralta-Zamora et al. 1998a). The immobilization of laccase increases its stability and resistance to proteases (Dodor et al. 2004). Nanozymes are nanoparticle-based enzyme mimics. They are the next generation of artificial enzymes which possess enzyme-like properties and follow the same kinetics and mechanisms of natural enzymes in physiological conditions (Gao and Yan 2016). There are many nanozymes that mimic naturally occurring enzymes like catalase, oxidase, peroxidase, phosphatase, protease, nuclease, esterase, superoxide dismutase, and peroxidase. These are used for the detection and degradation of pollutants such as dyes, lignin-containing wastes, and organic compounds. (Liang et al. 2017). The magnetic nanoparticles (Fe_3O_4 - MNPs) resemble peroxidases and have been useful in the degradation of many organic pollutants like methylene blue, phenol, and rhodamine (Wu et al. 2015). These methods are found to be cost effective and simple for the degradation and mineralization of organic dyes in an industrial process.

28.7 APPLICATIONS OF ENZYMES

Enzymes have numerous applications in different fields. They have been observed to catalyze a mixture of diverse pollutants and soluble toxic to insoluble non-toxic materials in industrial waste. The ligninolytic enzymes produced by white-rot fungi are useful for the remediation of pollutants present in the environment in high concentrations. Generally, physical and chemical treatments are not used for the remediation of textile wastewater because of the high costs associated with these treatments and the nature of waste. Conversely, the industrial effluents containing various dyes are effectively removed by microbes, and this method is very economical for the industrial sector. The relatively extensive use of organophosphorus pesticides in agriculture is a great concern as it causes pollution and toxicity in the environment. Naturally occurring bacteria containing enzyme phosphotriesterase effectively catalyze the organophosphorus compounds and reduce their toxicity. Genetic engineering techniques like DNA shuffling ans site-directed mutagenesis enhance the catalytic power of this enzyme and help to reduce pollution. Some other enzymes like 2,4–dinitrotoluene dioxygenase and the cytochrome P450 enzyme system sufficiently degrade certain explosive materials and contribute in the remediation of such toxic waste generated from ammunition factories and explosive testing areas. The application of biological catalysts in the field of immobilization and biosensors is one of the emerging trends today. Enzymes offer a major advantage over chemical treatments especially in terms of specificity, selectivity, high yield of products, low costs of production, less generation of heat and toxic by-products, and mild conditions of operation. Biosensors have been extensively used for the bioremediation of several organic pollutants such as organophosphorus pesticides and phenols. Biosensors can even be utilized to screen the contaminants in barren lands and landfills. This is of great significance as it aids to monitor the leaching of pollutants from landfills. In turn, groundwater pollution can be controlled.

28.8 CONCLUSION

This chapter provided a brief account of the functional diversity and selective catalytic potential of enzymes. Bioremediation strategies mediated using these biological catalysts provide an eco-friendly, cost-effective alternative. The robust catalytic machinery of the microbes can be fully exploited to degrade pollutants. With the advent of biosensing technology, the gap between the enzymes and their in-situ application can be bridged.

REFERENCES

Abadulla E, Tzanov T, Costa S, Robra K, Cavaco-Paulo A, Gu¨Bitz GM (2000) Decolorization and detoxification of textile dyes with a laccase from *Trametes hirsuta*. *Appl Environ Microbiol* 66(8): 3357–3362.

Abdul-Wahab MF, Chan GF, Mohd Yusoff AR, Abdul Rashid NA (2013) Reduction of azo dyes by flavin reductase from *Citrobacter freundii* A1. *J Xenobiot* 3: 9–13.

Ackerley DF, Gonzalez CF, Park CH, Ii RB, Keyhan M, Matin A (2004) Chromate-reducing properties of soluble flavoproteins from *Pseudomonas putida* and *Escherichia coli*. *Am Soc Microbiol* 70: 873–882.

Agrawal N, Dixit AK (2015) An environmental clean-up strategy - Microbial transformation of xenobiotic compounds. *Int J Curr Microbiol Appl Sci* 4: 429–461.

Arora PK, Srivastava A, Singh VP (2010) Application of monooxygenases in dehalogenation, desulphurization, denitrification and hydroxylation of aromatic compounds. *J Biorem Biodegrad* 1: 1–8.

Atlas RM, Hazen TC (2011) Oil biodegradation and bioremediation: A tale of the two worst spills in U.S. history. *Environ Sci Technol* 45(16): 6709–6715.

AWARENET (2004) *Handbook for the Prevention and Minimization of Waste and Valorization of By-Products in European Agro-Food Industries. Agro-Food Waste Minimization and Reduction Network (AWARENET)*. Grow Programme, European Commission. 1–7.

Bae W, Lee H, Choe Y, Jahng D, Lee S, Kim S, Lee J, Jeong B (2005) Purification and characterization of NADPH-dependent Cr(VI) reductase from *Escherichia coli* ATCC 33456. *J Microbiol* 43(1): 21–27.

Bafana A, Chakrabarti T, Krishnamurthi K (2013) Mercuric reductase activity of multiple heavy metal-resistant *Lysinibacillus sphaericus* G1. *J Basic Microbiol* 55(3): 285–292.

Bajaj A, Pathak A, Mudiam MR, Mayilraj S, Manickam N (2010) Isolation and characterization of a *Pseudomonas* sp. strain IITR01 capable of degrading α-endosulfan and endosulfan sulfate. *J Appl Microbiol* 109(6): 2135–2143.

Banik RM, Prakash M (2004) Laundry detergent compatibility of the alkaline protease from *Bacillus cereus*. *Microbiol Res* 159(2): 135–140.

Basile LJ (2008) Cyanide-degrading enzymes for bioremediation. PhD thesis, A&M University.

Baweja M, Nain L, Kawarabayasi Y, Shukla P (2016) Current technological improvements in enzymes toward their biotechnological applications. *Front Microbiol* 7: 965–978.

Benison GC, Di Lello P, Shokes JE, Cosper NJ, Scott RA, Legault P, Omichinski JG, Uni V, February RV, Re V, Recei M, May V (2004) A stable mercury-containing complex of the organomercurial lyase MerB : Catalysis, product release, and direct transfer to MerA: 8333–8345.

Bilal M, Asgher M (2015) Dye decolorization and detoxification potential of Ca-alginate beads immobilized manganese peroxidase. *BMC Biotechnol* 15: 111–125.

Bilal M, Asgher M, Iqbal HM (2016) Polyacrylamide gel-entrapped fungal manganese peroxidase from *Ganoderma lucidum* IBL-05 with enhanced catalytic, stability, and reusability characteristics. *Protein Pept Lett* 23(9): 812 – 818.

Bilal M, Asgher M, Parra-Saldivar R, Hu H, Wang W, Zhang X, Iqbal HMN (2017) Immobilized ligninolytic enzymes: An innovative and environmental responsive technology to tackle dye-based industrial pollutants: A review. *Sci Total Environ* 576: 646–659.

Božič M, Pricelius S, Guebitz GM, Kokol V (2010) Enzymatic reduction of complex redox dyes using NADH-dependent reductase from *Bacillus subtilis* coupled with cofactor regeneration. *Appl Microbiol Biotechnol* 85(3): 563–571.

Chaudhari AU, Tapase SR, Markad VL, Kodam KM (2013) Simultaneous decolorization of reactive Orange M2R dye and reduction of chromate by *Lysinibacillus* sp. KMK-A. *J Hazard Mater* 262: 580–588.

Chen H, Wang RF, Cerniglia CE (2004) Molecular cloning, overexpression, purification, and characterization of an aerobic FMN-dependent azoreductase from *Enterococcus faecalis*. *Protein Expr Purif* 34(2): 302–310.

Das N, Chandran P (2011) Microbial degradation of petroleum hydrocarbon contaminants: An overview. *Biotechnol Res Int* 2011: 1–13.

Dhanya MS (2014) Advances in microbial biodegradation of chlorpirifos. *J Environ Res Dev* 9: 232–240.

Dodor DE, Hwang HM, Ekunwe SIN (2004) Oxidation of anthracene and benzoaPyrene by Laccases from *Trametes versicolor*. *Enzyme Microb Technol* 35(2–3): 210–217.

Dua M, Singh A, Sethunathan N, Johri A (2002) Biotechnology and bioremediation: Successes and limitations. *Appl Microbiol Biotechnol* 59(2–3): 143–152.

Dzionek A, Wojcieszyńska D, Guzik U (2016) Natural carriers in bioremediation: A review. *Electron J Biotechnol* 23: 28–36.

Fiechter A, Leisola MS, Sanglard D, Fiechter A (1986) Oxidation of benzo(a)pyrene by extracellular ligninases of *Phanerochaete chrysosporium*. Veratryl alcohol and stability of ligninase. *J Biol Chem* 261(15): 6900–6903.

Field JA, Sierra-Alvarez R (2004) Biodegradability of chlorinated solvents and related chlorinated aliphatic compounds. *Growth*: 185–254. University of Arizona.

Field JA, Thurman EM (1996) Glutathione conjugation and contaminant transformation. *Environ Sci Technol* 30(5): 1413–1418.

Fritsche W, Hofrichter M (1999) Aerobic, degradation, by microorganisms: Principles of bacterial degradation. In: Rehm HJ, Reed G, Puhler A, Stadler A, editors, *Biotechnology, Environmental Processes II*, vol IIb. Wiley-VCH, Weinhein. 145–167.

Gao L, Yan X (2016) Nanozymes: An emerging field bridging nanotechnology and biology. *Sci China* 59(4): 400–402.

Gisi D, Willi L, Traber H, Leisinger T, Vuilleumier S (1998) Effects of bacterial host and dichloromethane dehalogenase on the competitiveness of methylotrophic bacteria growing with dichloromethane. *Appl Environ Microbiol* 64(4): 1194–1202.

Grosse S, Laramee L, Wendlandt KD, McDonald IR, Miguez CB, Kleber HP (1999) Purification and characterization of the soluble methane monooxygenase of the type II methanotrophic bacterium *Methylocystis* sp. strain WI 14. *Appl Environ Microbiol* 65(9): 3929–3935.

Gupta S, Singh D (2017) Role of genetically modified microorganisms in heavy metal bioremediation. In: Kumar R, Sharma AK, Ahluwalia SS, editors, *Advances in Environmental Biotechnology*. Springer Nature, Singapore.

Hansen EH, Albertsen L, Schäfer T, Johansen C, Frisvad JC, Molin S, Gram L (2003) *Curvularia* haloperoxidase : Antimicrobial activity and potential application as a surface disinfectant. *Appl Environ Microbiol* 69(8): 4611–4617.

Harcourt RL, Horne I, Sutherland TD, Hammock BD, Russell RJ, Oakeshott JG (2002) Development of a simple and sensitive fluorimetric method for isolation of coumaphos-hydrolysing bacteria. *Lett Appl Microbiol* 34(4): 263–268.

Hernandez J, Robledo NR, Velasco L, Quintero R, Pickard MA, Vazquez-Duhalt R (1998) Chloroperoxidase-mediated oxidation of organophosphorus pesticides. *Pestic Biochem Physiol* 61(2): 87–94.

Jadhav JP, Govindwar SP (2006) Biotransformation of malachite green by *Saccharomyces cerevisiae* MTCC 463. *Yeast* 23(4): 315–323.

Kalme SD, Parshetti GK, Jadhav SU, Govindwar SP (2007) Biodegradation of benzidine based dye Direct Blue-6 by *Pseudomonas desmolyticum* NCIM 2112. *Bioresour Technol* 98(7): 1405–1410.

Karigar CS, Rao SS (2011) Role of Microbial enzymes in the bioremediation of pollutants: A review. *Enzyme Res* 2011: Article ID 805187.

Kulkarni SV, Markad VL, Melo JS, D'Souza SF, Kodam KM (2014) Biodegradation of tributyl phosphate using *Klebsiella pneumoniae* sp. S3. *Appl Microbiol Biotechnol* 98(2): 919–929.

Kulshreshtha S, Mathur N, Bhatnagar P (2013) Mycoremediation of paper, pulp and cardboard industrial wastes and pollutants. In: Goltapeh EM, Danesh YR, Varma A, editors, *Fungi as Bioremediators*. Springer-Verlag: Berlin Heidelberg.

Kwak YH, Lee DS, Kim HB (2003) *Vibrio harveyi* nitroreductase is also a chromate reductase. *Appl Environ Microbiol* 69(8): 4390–4395.

Liang H, Lin F, Zhang Z, Liu B, Jiang S, Yuan Q, Liu J (2017) Multicopper laccase mimicking nanozymes with nucleotides as ligands. *ACS Appl Mater Interfaces* 9(2): 1352–1360.

Martins PF, Carvalho G, Gratão PL, Dourado MN, Pileggi M, Araújo WL, Azevedo RA (2011) Effects of the herbicides acetochlor and metolachlor on antioxidant enzymes in soil bacteria. *Process Biochem* 46(5): 1186–1195.

Mester T, Field JA (1998) Characterization of a novel manganese peroxidase-lignin peroxidase hybrid isozyme produced by *Bjerkandera* species strain BOS55 in the absence of manganese. *J Biol Chem* 273(25): 15412–15417.

Muthukamalam S, Sivagangavathi S, Dhrishya D, Sudha Rani S (2017) Characterization of dioxygenases and biosurfactants produced by crude oil degrading soil bacteria. *Braz J Microbiol* 48(4): 637–647.

Negri A, Marco E, Damborsky J, Gago F (2007) Stepwise dissection and visualization of the catalytic mechanism of haloalkane dehalogenase LinB using molecular dynamics simulations and computer graphics. *J Mol Graph Modell* 26(3): 643–651.

Nelson LM (1982) Biologically-induced hydrolysis of parathion in soil: Isolation of hydrolyzing bacteria. *Soil Biol Biochem* 14(3): 219–222.

Nigam VK, Shukla P (2015) Enzyme based biosensors for the detection of environmental pollutants- A review. *J Microbiol Biotechnol* 25(11): 1773–1781.

Novotný Č, Svobodová K, Erbanová P, Cajthaml T, Kasinath A, Lang E, Šašek V (2004) Ligninolytic fungi in bioremediation: Extracellular enzyme production and degradation rate. *Soil Biol Biochem* 36(10): 1545–1551.

Oakley AJ (2005) Glutathione transferases: New functions. *Curr Opin Struct Biol* 15(6): 716–723.

Ooi T, Shibata T, Sato R, Ohno H, Kinoshita S, Thuoc TL, Taguchi S (2007) An azoreductase, aerobic NADH-dependent flavoprotein discovered from *Bacillus* sp.: Functional expression and enzymatic characterization. *Appl Microbiol Biotechnol* 75(2): 377–386.

Osborne RL, Raner GM, Hager LP, Dawson JH (2006) C. fumago *Chloroperoxidase Is Also a Dehaloperoxidase: Oxidative Dehalogenation of Halophenols*. *J Am Chem Soc* 128(4): 1036–1037.

Peralta-Zamora P, de Moraes SG, Esposito E, Antunes R, Reyes J, Durán N (1998a) Decolorization of pulp mill effluents with immobilized lignin and manganese peroxidase from *Phanerochaete chrysosporium*. *Environ Technol* 19(5): 521–528.

Peralta-Zamora P, Esposito E, Pelegrini R, Groto R, Reyes J, Durán N (1998b) Effluent treatment of pulp and paper, and textile industries using immobilised horseradish peroxidase. *Environ Technol* 19(1): 55–63.

Qiu H, Li Y, Ji G, Zhou G, Huang X, Qu Y, Gao P (2009) Immobilization of lignin peroxidase on nanoporous gold: Enzymatic properties and in situ release of H_2O_2 by co-immobilized glucose oxidase. *Bioresour Technol* 100(17): 3837–3842.

Racke KD, Laskowski DA, Schultz MR (1990) Resistance of chlorpyrifos to enhanced biodegradation in soil. *J Agric Food Chem* 38(6): 1430–1436.

Rahmani K, Faramarzi MA, Mahvi AH, Gholami M, Esrafili A, Forootanfar H, Farzadkia M (2015) Elimination and detoxification of sulfathiazole and sulfamethoxazole assisted by laccase immobilized on porous silica beads. *Int Biodeterior Biodegrad* 97: 107–114.

Rao M, Scelza R, Scotti R, Gianfreda L (2010) Role of enzymes in the remediation of polluted environments. *J Soil Sci Plant Nutr* 10(3): 333–353.

Raybuck SA (1992) Microbes and microbial enzymes for cyanide degradation. *Biodegradation* 3(1): 3–18.

Rodgers CJ, Blanford CF, Giddens SR, Skamnioti P, Armstrong FA, Gurr SJ (2010) Designer laccases: A vogue for high-potential fungal enzymes? *Trends Biotechnol* 28(2): 63–72.

Ruggaber TP, Talley JW (2006) Enhancing bioremediation with enzymatic processes: A review. *Pract Period Hazard Toxic Radioact Waste Manage* 10(2): 73–85.

Ruiz-dueñas FJ, Martinez AT, Martinez MJ (1999) Molecular characterization of a novel peroxidase isolated from the ligninolytic fungus *Pleurotus eryngii*. *Mol Microbiol* 31(1): 223–235.

Ruqayyah TID, Jamal P, Alam MZ, Mirghani MES (2013) Biodegradation potential and ligninolytic enzyme activity of two locally isolated *Panus tigrinus* strains on selected agro-industrial wastes. *J Environ Manag* 118: 115–121.

Schiering N, Kabsch W, Moore MJ, Distefano MD, Walsh CT, Pai EF (1991) Structure of the detoxification catalyst mercuric ion reductase from *Bacillus* sp. strain RC607. *Nature* 352(6331): 168–172.

Scott C, Pandey G, Hartley CJ, Jackson CJ, Cheesman MJ, Taylor MC, Pandey R, Khurana JL, Teese M, Coppin CW, Weir KM, Jain RK, Lal R, Russell RJ, Oakeshott JG (2008) The enzymatic basis for pesticide bioremediation. *Indian J Microbiol* 48(1): 65–79.

Sharma B, Dangi AK, Shukla P (2018) Contemporary enzyme based technologies for bioremediation: A review. *J Environ Manag* 210: 10–22.

Singh BK, Walker A (2006) Microbial degradation of organophosphorus compounds. *FEMS Microbiol Rev* 30(3): 428–471.

Singh BK, Kuhad RC, Singh A, Tripathi KK, Ghosh PK (2000) Microbial degradation of the pesticide lindane (ghexachlorocyclohexane). *Adv Appl Microbiol* 47: 269–298.

Singh CJ (2003) Optimization of an extracellular protease of *Chrysosporium keratinophilum* and its potential in bioremediation of keratinic wastes. *Mycopathologia* 156(3): 151–156.

Singh S, Kang SH, Mulchandani A, Chen W (2008) Bioremediation: Environmental clean-up through pathway engineering. *Curr Opin Biotechnol* 19(5): 437–444.

Sirisha VL, Jain A, Jain A (2016) Enzyme immobilization: An overview on methods, support material, and applications of immobilized enzymes. In: *Advances in Food and Nutrition Research*, 1st ed. Elsevier Inc.

Sivan A (2011) New perspectives in plastic biodegradation. *Curr Opin Biotechnol* 22(3): 422–426.

Sivaperumal P, Kamala K, Rajaram R (2017). Bioremediation of industrial waste through enzyme producing marine microorganisms. In: *Advances in Food and Nutrition Research*, 1st ed. Elsevier Inc.

Spain JC, Gibson DT (1991) Pathway for biodegradation of *p*-nitrophenol in a *Moraxella* sp. *Appl Environ Microbiol* 57(3): 812–819.

Stanlake GJ, Finn RK (1982) Isolation and characterization of a pentachlorophenol- degrading bacterium. *Appl Environ Microbiol* 44(6): 1421–1427.

Tanabe H, Yoshihara K, Tamura K, Kobayashi Y, Akamatsu I, Niyomwan N, Footrakul P (1987) Pretreatment of pectic wastewater from orange canning process by an alkalophilic *Bacillus* sp. *J Ferment Technol* 65(2): 243–246.

Thamke VR, Kodam KM (2016) Toxicity study of ionic liquid, 1-butyl-3-methylimidazolium bromide on guppy fish, *Poecilia reticulata* and its biodegradation by soil bacterium *Rhodococcus hoagii* VRT1. *J Hazard Mater* 320: 408–416.

Thamke VR, Chaudhari AU, Tapase SR, Paul D, Kodam KM (2019) *In vitro* toxicological evaluation of ionic liquids and development of effective bioremediation process for their removal. *Environ Pollut* 250: 567–577.

Toppe J, Albrektsen S, Hope B, Aksnes A (2007) Chemical composition, mineral content and amino acid and lipid profiles in bones from various fish species. *Comp Biochem Physi B Biochem Mol Biol* 146(3): 395–401.

Verger R (1997) 'Interfacial activation' of lipases: Facts and artifacts. *Trends Biotechnol* 15(1): 32–38.

Vidali M (2001) Bioremediation. An overview. *Pure Appl Chem* 73(7): 1163–1172.

Wang G, Li R, Li S, Jiang J (2010) A novel hydrolytic dehalogenase for the chlorinated aromatic compound chlorothalonil. *J Bacteriol* 192(11): 2737–2745.

Whiteley CG, Melamane X, Pletschke B, Rose PD (2003) The enzymology of sludge solubilisation utilising sulphate reducing systems: The role of lipases. *Water Sci Technol* 48(8): 159–167.

Wong DWS (2009) Structure and action mechanism of ligninolytic enzymes. *Appl Biochem Biotechnol* 157(2): 174–209.

Wu J, Xiao D, Zhao H, He H, Peng J, Wang C, Zhang C, He J (2015) A nanocomposite consisting of graphene oxide and Fe_3O_4 magnetic nanoparticles for the extraction of flavonoids from tea, wine and urine samples. *Microchim Acta* 182(13–14): 2299–2306.

Yim SK, Jung HC, Pan JG, Kang HS, Ahn T, Yun CH (2006) Functional expression of mammalian NADPH-cytochrome P450 oxidoreductase on the cell surface of *Escherichia coli*. *Protein Expr Purif* 49(2): 292–298.

29 Microalgae
A Natural Tool for Water Quality Improvement

Diana Pacheco, Ana Cristina Rocha,
Tiago Verdelhos, and Leonel Pereira

CONTENTS

29.1 INTRODUCTION

Access to drinking water and sanitation are basic human rights. However, water resources continue to grow increasingly scarce and it is expected that until 2050, water demand will increase by up to 20–30% of the current use. Furthermore, the exponential growth of the population and industrialization has increased the quantity and complexity of effluents. Aquatic ecosystems are the final destination of a wide range of pollutants from anthropogenic activities, namely effluents from urban or industrial wastewater treatment plants (WWTPs) or through agriculture effluent run-off (Gauthier et al. 2020). In this context, the chemical composition of wastewater differs according to their origin, and they can be classified into three main categories: urban, industrial, or rural wastewater (Figure 29.1). Inadequate wastewater treatment can have negative repercussions on society and ecosystems, having noxious effects on human and aquatic organism health, and negatively affecting economic activities (Lipponen and Nikiforova 2017). Therefore, the development of an integrated approach to protect freshwater resources, as well as innovative and feasible wastewater treatment methods, are of outmost importance (Azoulay and Houngbo 2019).

Bearing that in mind, the regulation of wastewater discharges has been revised over the years due to the presence of contaminants of emerging concern (European Commission 2019). For instance, Europe adopted the Directive 91/271/EEC (EU 1991). In the United States of America, this matter is under the scope of the Clean Water Act (EPA 1972), while in Brazil it is regulated by the decree 18.328/97. However, the enforcement of current legislation is still a challenge in many areas (Wall 2018). In fact, more than 80% of wastewater derived from anthropogenic activities is still discharged into aquatic ecosystems without any treatment (United Nations 2017). Hence, there is a consistent

DOI: 10.1201/9781003204442-29

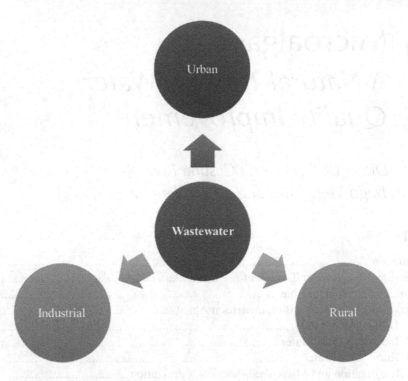

FIGURE 29.1   A schematic representation of the different wastewater sources.

determination to improve wastewater treatment and water quality, which is reflected in the 2030 Agenda, under the scope of the Sustainable Development Goal #6 (United Nations 2017). This concern is supported by highly sensitive analytical methodologies, which allow for the detection of several chemical compounds in the aquatic environment, including emerging pollutants (EPs) such as pharmaceuticals, radioactive substances, personal care products, surfactants, plasticizers, micro-plastics, pesticides, and inorganic compounds (Eerkes-Medrano et al. 2019; Muñoz et al. 2009; Menció et al. 2016). The technological evolution of highly sensitive instrumental techniques, such as gas chromatography-atmospheric pressure chemical ionization-mass spectrometry (GC-APCI-MS/MS) or liquid chromatography–mass spectrometry (LC-MS/MS), facilitates EP detection in water, sediment, and biota. Nevertheless, there is a need to standardize these methods in order to accredit EP analysis (Hernández et al. 2012). Moreover, the analytical cost of these analyses discourages the monitoring of EPs (Montes-Grajales et al. 2017).

Most of these substances are also considered refractory pollutants, which are toxic and poorly biodegraded under natural conditions. The occurrence of these compounds is a result of point or diffuse pollution and currently WWTPs are not equipped to efficiently remove these types of pollut-ants (Frimmel et al. 2008). Moreover, EPs are known to accumulate in aquatic organisms and some of these chemicals may even transfer throughout trophic chains, triggering harmful effects in higher levels of biological organization (Ruhí et al. 2016; Esperanza et al. 2020). Despite that, their occur-rence is not regularly surveyed (Geissen et al. 2015). Due to the chemical complexity of EP-rich wastewaters, there is a need to develop innovative and environmentally-friendly methodologies in order to remove these pollutants and safely discharge the effluents into the aquatic receptor bodies.

Microalgae represent a natural tool with recognized efficiency to remove several pollutants from different types of wastewater sources (Ramírez et al. 2017; Marchão et al. 2018; Baglieri et al. 2016; Bai and Acharya 2016). This process is denominated phycoremediation and is defined by the application of microalgae biomass to remove refractory pollutants from wastewaters. Afterwards, the treated effluents can be reused (e.g., irrigation and car wash), whereas the produced microalgal

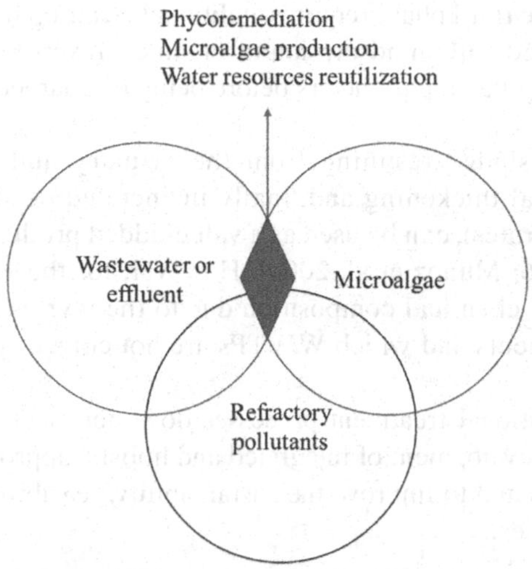

Phycoremediation
Microalgae production
Water resources reutilization

FIGURE 29.2 A schematic representation of the phycoremediation process.

biomass is a sub-product of added value, presenting a win–win approach based on a Blue and Circular Economy (Figure 29.2) (Lavrinovičs and Juhna 2018).

However, the application of microalgae for wastewater phycoremediation still faces some challenges that need to be tackled regarding technical and operational factors that hinder the scale-up of this process.

In this book chapter, the reader will find a comprehensive analysis of the feasibility of phycoremediation, as well as the current challenges and opportunities faced by the application of microalgae for wastewater treatment from different sources.

29.2 CONVENTIONAL WASTEWATER TREATMENT

The treatment of domestic or urban wastewater is mandatory so that the concentrations of pollutants, as well as physico-chemical features of effluents, comply with the maximum or recommended thresholds regulated by each country's legislation, before their discharge into aquatic systems (i.e., rivers, lakes, or oceans) (Arbib et al. 2012; FAO 1992). Conventional wastewater treatment consists of a set of physic and chemical processes to remove sediments, metals, or microorganisms from effluents. The methodology that is currently implemented includes mainly four key steps (FAO 1992):

- Pre-treatment: The first phase of wastewater treatment aims to clear its color and to remove large, solid materials and oils through a coarse screening and degreasing process (Voulvoulis 2018). Afterwards, the primary treatment undergoes a primary sediment settlement in order to reduce the total suspended solids, organic and inorganic loads, and biochemical oxygen demand. At this phase, the primary sludges are generated.
- Secondary treatment: The second step is biological treatment in which the wastewater liquid phase is directed to biological reactors, where microorganisms (bacteria) perform a nitrification/denitrification process, removing compounds such as carbon (C), nitrogen (N), and phosphorus (P). In this phase, several technologies can be applied alone or in combination, according to the needs of the WWTPs, such as activated sludge, trickling filters, biofilters, oxidation ditches, or rotating biological contactors. The mentioned technologies are based on the enhancement of bacteria fitness, allowing the increase of wastewater treatment efficacy (FAO 1992). Thereafter, the wastewater goes through another settling process in secondary settlers (Ruiz-Marin et al. 2010).

- Tertiary treatment: The third phase requires additional clean-up methods, such as coagulation, filtration, activated carbon adsorption of organics, reverse osmosis, or disinfection, in order to meet the regulated parameters before being discharged (Dadson et al. 2020).

Concurrently, the sewage sludge resulting from the primary and secondary settlement are treated through gravitational thickening and finally incinerated or, due to their high inorganic load (i.e., phosphate and nitrates), can be used as a value-added product in agriculture to fertilize crops (Panepinto et al. 2016; Muñoz et al. 2009). However, for the valorization of this residue, it is necessary to assess the chemical composition due to the myriad of emergent contaminants currently appearing in effluents and which WWTPs are not currently fully equipped to remove (Rizzo et al. 2019).

The efficiency of conventional treatment processes does not meet the growing complexity of wastewater. Therefore, the development of integrated and holistic approaches is imperative to overcome the current challenges and to improve the sustainability, feasibility, and efficiency of wastewater treatment methodologies.

29.3 MICROALGAE'S ECOLOGICAL ROLE

With improvements in microscopy techniques, it is possible to describe and classify a wide range of microalgal species. In general, algae are divided taxonomically according to their dominant pigments, in which species belonging to the phyla Chlorophyta, Rhodophyta, and Phaeophyceae are commonly characterized by green, red, and brown algae, respectively. Regarding cyanobacteria, which are also photosynthetic microorganisms, some authors classify them as blue microalgae (Fawley and Fawley 2020).

Microalgae and cyanobacteria are resilient microorganisms that can tolerate hostile conditions (i.e., temperature, salinity, and pH). Their presence in extreme habitats, such as hydrothermal vents, caves, or volcanos (Hamilton and Havig 2017; Kamenev et al. 1993; Ciniglia et al. 2005), is evidence that these microorganisms can adapt and survive in harsh conditions.

Furthermore, microalgae have a pivotal role in the ecosystem, being part of a community where they form the basis of the trophic chain as primary producers, while benefiting from the interactions among other microorganisms (Padmaperuma et al. 2018). These photosynthetic microorganisms provide numerous ecosystem services, namely carbon dioxide sequestration, the biotransformation of toxic compounds, and also as a source of food (Steudel 2014; Hope et al. 2020). But, above all, it is also important to highlight that microalgae communities are considered as the lungs of the planet, providing a major part of the oxygen produced (Jacquin et al. 2014).

Microalgae and cyanobacteria have been used by mankind for years, as food, feed, and as a pharmaceutical source. However, with the improvement of microalgae cultivation and extraction methodologies of bioactive compounds, interest in microalgae's biotechnological applications has been continuously increasing (Wikfors and Ohno 2001). Among the myriad of applications, wastewater treatment has been confirmed by researchers and by industries as a multipurpose solution for water pollution, due to their fast growth and resilience. Thus, many studies have been developed to implement microalgae as novel tools for wastewater treatment.

The occurrence of microalgae blooms is usually linked with environmental pollution, having a negative implications for society (Jacquin et al. 2014). However, algae blooms are not the cause of pollution, but a consequence of the discharge of nutrient-enriched effluents (i.e., nitrogen and phosphate) derived from rural, urban, or industrial origins (European Environment Agency 2018). For effective environmental management, it is necessary to assess the biodiversity and analyze a set of ecological indicators. In this context, microalgae are considered suitable bioindicators, revealing the environmental quality of water bodies in a certain time and locality (Kadam et al. 2020).

29.4 MICROALGAE AS WASTEWATER BIOREMEDIATION AGENTS

Algae are natural bioremediation agents, being able to metabolize or accumulate pollutants that are naturally present in marine ecosystems, for example arsenic (As) (Papry et al. 2020) or mercury (Hg) (Beauvais-Flück et al. 2018). These compounds are toxic and their potential for biomagnification through the food web is a matter of concern, having affects from the cell to the community. Thus, microalgae play an important role in the biogeochemical cycle of these metalloids. These microorganisms bioaccumulate these compounds, balancing As and Hg concentrations in the aquatic ecosystems. Moreover, many studies have already highlighted that microalgae have the potential to remove a wide range of organic and inorganic contaminants from wastewaters derived from anthropogenic activities (Deviram et al. 2020; Choudhary et al. 2016; Bai and Acharya 2016).

According to these findings, the scientific community has considered the advantages of using microalgae as an additional and sustainable tool to complement wastewater treatment (Pacheco et al. 2020; Singh et al. 2016).

Wastewater bioremediation through microalgae application is possible due to their resilience. These microorganisms have metabolic processes which allow them to grow despite the EP concentration, namely bioadsorption, biological uptake, and biodegradation (Sutherland and Ralph 2019).

Bioadsorption is a process that occurs due to microalgae cell wall exopolysaccharides and several functional groups (i.e., carboxyl, hydroxyl, and sulfate) that interact with the contaminants through ion exchange, complexation, or chelation (Norvill et al. 2016; Kumar et al. 2016; Hansda et al. 2016). Through this process, microalgae can remove pharmaceuticals from wastewaters (Gojkovic et al. 2019). A study revealed that the green microalgae *Scenedesmus obliquus* (currently, *Tetradesmus obliquus*) was able to remove sulfamethazine and sulfamethoxazole through the hydroxylation, methylation, nitrosation, and deamination of these compounds (Xiong et al. 2019).

The biological uptake of contaminants can occur through the cell membrane from three pathways: passive diffusion (Vernouillet et al. 2010), passive-facilitated diffusion (Bai and Acharya 2016), or active uptake (Wilde and Benemann 1993). Through biological uptake, researchers showed that microalgae are capable of removing antibiotics (Gentili and Fick 2017), hormones (Hom-Diaz et al. 2015), and organic synthetic compounds (Ji et al. 2014).

Furthermore, microalgae can also degrade or transform EPs through their metabolism. In this case, this strategy will tackle the need to treat the microalgal biomass after their disposal (Tiwari et al. 2017). Through this metabolic pathway, microalgae are able to remove alkaloids (Matamoros et al. 2016), hormones (Zhang et al. 2014), organic compounds (Ji et al. 2014), and a wide range of pharmaceuticals (Gojkovic et al. 2019).

Considering that effluents differ in their nature depending on their origin (Chiu et al. 2015), it is critical to determine their composition in terms of pollutants in order to select the most suitable microalgae species to be successfully applied in the bioremediation process (Khanna et al. 2012; Kadam et al. 2020). It is also necessary to consider that the feasibility and rate of these processes will vary according to abiotic parameters (i.e., pH, temperature, carbon dioxide concentration, and photoperiod) (Sutherland and Ralph 2019).

29.4.1 URBAN WASTEWATER

The chemical composition of wastewater released into water bodies differs according to the region, area, and population density, in addition to the local characteristics and land use. Urban wastewater is the main source of EPs present in aquatic ecosystems (Tran et al. 2018), with urban wastewater containing up to 200 EPs, mainly categorized into pharmaceuticals, hormones, personal care products, microplastics, and household cleaning chemicals (Petrie et al. 2015; Norvill et al. 2016; Tran et al. 2018). Urban WWTPs are still unable to remove these persistent compounds from wastewaters, so in this context, WWTPs can significantly contribute to the high concentrations of these substances in aquatic systems (Windsor et al. 2019). Therefore, more efficient methodologies of wastewater treatment are necessary in order to reduce the impact of discharged effluents on aquatic ecosystems.

Proof that microalgae are suitable remediators of urban effluents rich in EPs can be found in the literature. For instance, researchers demonstrated that the green microalgae *Selenastrum capricornutum* (formerly known as *Rhapidocelis subcapitata*) and *Chlamydomonas reinhardtii* were able to remove two hormones (β-Estadiol and α-Ethynylestradiol) from the centrate collected from an urban WWTP (Hom-Diaz et al. 2015). *Chlorella vulgaris* (Chlorophyta) was shown to remove a wide range of pharmaceuticals, such as amitriptyline, biperiden, bupropion, codeine, diclofenac, diphenhydramine, flecainide, hydroxyzine, memantine, mirtazapine, orphenadrine, trihexyphenidyl, tramadol (Gojkovic et al. 2019; Santos et al. 2017), tetracycline (de Godos et al. 2012), levofloxacin (Xiong et al. 2017), and iohexol (Akao et al. 2020).

Personal care products contain a wide range of refractory compounds usually used for household purposes, such as health and cleaning (Montes-Grajales et al. 2017). Substances such as triclosan, trimethoprim, and sulfamethoxazole are also widely used and can be found in the aquatic environment, having adverse effects on marine biota (Kolpin et al. 2002). Algae are also capable of accumulating these compounds (Coogan et al. 2007). Under controlled conditions, specimens of green algae *Nannochloris* sp. were cultivated in a glass flask and after 7 days they were able to totally remove triclosan and trimethoprim, while removing 68% of sulfamethoxazole (Bai and Acharya 2016).

With regard to surfactants and biocides, which are widely present in effluents due to the utilization of detergents, shampoos, and other cleaning agents (Jardak et al. 2016; Wieck et al. 2016), a consortia of bacteria (*Spirosoma lingua* and *Pseudomonas aeruginosa*) and microalgae (*Scenedesmus dimorphus*, currently known as *Tetradesmus dimorphus*) were shown to be effective in the removal of alkylbenzene sulfonate, a surfactant widely applied in detergents. In this study, the microalgae–bacteria consortia were cultivated on a fiber bioreactor reinforced with plastic, a volume of 325 L, a hydraulic retention time of 2.7 hours and a flow rate of 40 L/h. With this strategy, it was possible to achieve an alkylbenzene sulfonate removal of 94.6%, which is more effective than conventional wastewater treatments (81.7%) (Tu et al. 2020).

Microalgae are a natural tool for the bioremediation of refractory pollutants. Promoting microalgae cultivation in wastewater will help mitigate the environmental impacts of treated effluents since this biological method will complement conventional wastewater treatments and improve not only the removal of organic and inorganic load but also the removal of EPs.

29.4.2 Industrial Wastewater

The industrial sector is diverse including several different industries, such as pulp and paper, mining, petrochemical, energy supply, textile, and food and beverages (European Environment Agency 2018). Such diversity leads to the production of heterogenous effluents, composed of miscellaneous pollutants – inorganic, organic, and chlorinated compounds; phenol; oils; and metals (Abdelbasir and Shalan 2019). These compounds are extremely harmful to the environment, so regulations regarding industrial wastewater treatment and discharge are becoming stricter.

Industrial wastewater treatment processes also consist of physical (sedimentation and flotation), chemical (coagulation and precipitation), and/or biological processes. However, due to the complex composition of industrial effluents, these processes are more challenging than those from urban WWTPs (Crini and Lichtfouse 2019), and often are more expensive, particularly the physical and chemical treatments. Biological treatments using microalgae can be beneficial to tackle this problem and complement wastewater treatment methodologies (Wang et al. 2016). In fact, several studies evince that microalgae can remove noxious substances from industrial effluents (Table 29.1).

The paper and pulp industries, in particular, produce wastewaters rich in nutrients (e.g., phosphates, nitrates, nitrites, and carbon), but they also produce noxious substances, namely sulfates and

TABLE 29.1

Microalgae Application in Industrial Wastewaters from Different Sectors

Industrial Wastewater	Microalgae Species	Pollutant	Removal Rate (%)	Reference
Paper	Chlorella vulgaris	Phosphorus	54	(Porto et al. 2020)
	Mixed consortium	Chloride	66	(Ramírez et al. 2017)
Petrochemical	Chlorella vulgaris	Nitrogen	100	(Madadi et al. 2016)
		Phosphorus	100	
		Petroleum hydrocarbons	27	
	Tribonema sp.	Nitrogen	98.4	(Huo et al. 2018a)
Mining	Stichococcus bacillaris	Zinc	85	(Li et al. 2015)
	Chlorella vulgaris	Manganese	99.4	(Saavedra et al. 2018)
	Chlorophyceae spp.	Zinc	91	
	Chlorophyceae spp.	Copper	88	
Textile	Chlorella pyrenoidosa	Nitrate	82	(Pathak et al. 2015)
		Phosphate	87	
	Chlorella vulgaris	Ammonium	45.1	(Lim et al. 2010)
		Phosphate	33.3	
		Color	50	
	Chlorella sp.	Ammonium	80.2	(Wu et al. 2020)
		Color	77.9	
	Anabaena ambigua	Chloride	55.5	(Brar et al. 2019)
		Nitrate	52.9	
		Phosphate	63	
	Chlorella (Auxenochlorella) pyrenoidosa	Chloride	62	
		Nitrate	74	
		Phosphate	28	
	Scenedesmus abundans	Chloride	44	
		Nitrate	68	
		Phosphate	70	
Food	Chlorella (Auxenochlorella) pyreniodosa	Ammonia	100	(Yadavalli et al. 2014)
		Phosphate	99	
	Euglena gracilis	Ammonia	95	
		Phosphate	97	
	Chlorella sorokiniana	Ammonia	92	(Riaño et al. 2016)
		Phosphate	20	
	Chlorella sorokiniana	Total nitrogen	55	(Cheah et al. 2018)
		Total phosphorus	77	
Beverage	Chlorella sorokiniana	Nitrate	95	(Solovchenko
		Sulfate	35	et al. 2014)
		Phosphate	77	
	Scenedesmus (Tetradesmus) obliquus	Ammonia	99	(Marchão et al. 2018)
		Phosphate	43	
	Chlorella sp.	Total nitrogen	97	(Ganeshkumar
		Total phosphorus	35	et al. 2018)

chloride (Usha et al. 2016; Pokhrel and Viraraghavan 2004) due to paper washing and bleaching. Microalgae have been shown to efficiently remove phosphorus from a diluted secondary effluent from a paper industry. With 11 days of cultivation on borosilicate glass flasks, the green microalgae *Chlorella vulgaris* was able to remove 54% of the phosphorus (Porto et al. 2020). Another study shows evidence that a microalgae consortium (with a concentration of 10% from the initial), previously isolated from chlorinated waters, was able to significantly reduce chloride concentrations in wastewater. Moreover, wastewater enriched with 0.5 g/L of a fertilizer (NPK) enhanced microalgal chloride uptake, achieving a removal of 66% (Ramírez et al. 2017).

In petrochemical wastewaters, alkanes, aromatic compounds, and organic compounds are the main refractory and noxious constituents (Lai et al. 2012). Neutralization, flocculation, filtration, clarification, and biodegradation are some of the processes used to treat these effluents. Biodegradation can also be used as a remediation methodology. There are several procedures that could be employed, and microalgae can be a useful tool to achieve this goal (Huo et al. 2018b). *Chlorella vulgaris* (Chlorophyta) was cultivated on petrochemical wastewater diluted with BG11 growth medium, whereas the most suitable concentration of wastewater was 25%. In this context, the green microalgae totally removed the nitrogen and phosphorus, while removing 27% of the petroleum hydrocarbons present in the wastewater (Madadi et al. 2016). Huo et al. (2018a) used open tubular photobioreactors to cultivate the green filamentous microalgae *Tribonema* sp. in petrochemical wastewater, in order to reduce its inorganic load. It was observed that in 7 days practically all the nitrogen was removed from the wastewater (98.4%) (Huo et al. 2018a).

Mining produces metal-rich wastewater (e.g., zinc, manganese, iron, and copper) which have harmful impacts on aquatic ecosystems if not properly treated (Li et al. 2019). In fact, metal levels have been successfully reduced in microalgae-based treatments. Saavedra et al. (2018) cultivated the green microalgae *Scenedesmus* sp. on glass flasks and it was observed that within 1 week, removal rates above 80% for As, Zn, and Cu were achieved (Saavedra et al. 2018).

Textile industry effluents are rich in nutrients, acids, alkalis, dyes, hydrogen peroxide, starch, surfactants, or metals (El-Kassas and Mohamed 2014) due to procedures such as scouring, bleaching, and dyeing. Microalgae-based treatments have been shown to considerably reduce the loads of nitrate, phosphate, ammonium, and chloride from textile effluents (Lim et al. 2010; Wu et al. 2020; Brar et al. 2019). The removal of dyes is still a current challenge due to their synthetic and non-biodegradable nature. The application of microalgae to reduce the color of textile wastewaters has been under study (Darwesh et al. 2019; Aragaw and Asmare 2018). More recently, Wu et al. (2020) tested immobilized microalgae (*Chlorella* sp.) in alginate beads in order to remove ammonium and color from textile wastewater. The best results were achieved under a semi-continuous strategy and within 18 days, 80.2% and 77.9% of ammonium and color were removed, respectively. Furthermore, *Chlorella vulgaris* cultivated in textile effluent, in a raceway system, was shown to remove 50% of color, and 45% and 33% of ammonium and phosphate, respectively (Lim et al. 2010).

The use of microalgae as biological treatment for industrial wastewaters is promising. Furthermore, there are different systems, methods, and strategies that could be employed. The correct selection of microalgae species, their origin (native or culture), and cultivation processes will define the refractory compounds removal rate success. So, further research on this matter is of utmost importance to improve and optimize the procedures. Over the years, different cultivation techniques have been used, the most common being aerated ponds or raceways (Schade and Meier 2019). Currently, innovative methods are under research, such as bubble-column photobioreactors, built to cultivate *Scenedesmus* (*Tetradesmus*) *obliquus* and remove nutrients from brewery wastewater. The food and beverage sector is a source of nutrient-rich wastewaters (e.g., nitrite, nitrate, or phosphate). The study conducted by Marchão et al. (2018) achieved removal rates ranging between 82–99% and 43–87% of ammonia and phosphates, respectively. These removal rate differences are justified by the different strategies adopted, namely in batch and in continuous systems, respectively.

29.4.3 RURAL WASTEWATER

Rural wastewaters are considered diffuse pollution nodes (Topp et al. 2008), due to the run-off of agriculture and farming-related pollutants being widely spread in aquatic ecosystems (e.g., estuaries, rivers, lakes, and groundwaters) (Mane et al 2005). Apart from agriculture and cattle farming, this sector also includes aquaculture and other activities that are sources of wastewaters containing nutrients, pharmaceuticals, and pesticides (Díaz-Cruz and Barceló 2008; García-Galán et al. 2018; Watanabe et al. 2010).

Agricultural wastewater is primarily received via drainage or irrigation channels, but diverting them into urban WWTPs is often unfeasible due to the economic cost (Christou et al. 2017). Therefore, the excess of pesticides and inorganic fertilizers used in agricultural crops is frequently discharged in aquatic environments. Within this context, microalgae-based treatment systems could be used as sustainable tools for farmers to treat their wastewater Recently, new approaches using microalgae have been implemented to treat agricultural run-off discharges. García-Galán et al. (2018) designed a low-cost semi-closed tubular horizontal photobioreactor in which the predominant microalgal genera were *Pediastrum* sp., *Chlorella* sp., *Scenedesmus* sp. (Chlorophyta), and the cyanobacteria *Gloeothece* sp. With this strategy, it was possible to substantially reduce the content of organic compounds such as tonalide (73%), galaxolide (68%), and the pharmaceutical diclofenac (61%).

Moreover, other researchers provided evidence that green microalgae *Chlorella vulgaris* and *Scenedesmus quadricauda* were able to biodegrade some of the compounds present in commercially available pesticides applied in crops, namely metalaxyl, fenhexamid, iprodione, triclopyr, and adsorb pyrimethanil (Baglieri et al. 2016). Thus, proving that microalgae are effective in reducing the content of pesticides in agricultural wastewaters.

Aiming at a possible upscale of a microalgae-based treatment, a study regarding the removal of pharmaceuticals in rural wastewaters was performed in a lagoon-based system. While clay mineral and its nonionic and cationic organoclay derivatives adsorbed pharmaceuticals, the native microalgae–bacteria consortium contributed to the degradation of organic compounds (Oliveira et al. 2020). Hence, the disadvantage of this system was the discharge of microalgae and bacteria into water bodies. Cost-effective removal techniques of the produced microalgal biomass could be a possible solution for overcoming this problem.

Cattle farming is the main source of veterinary pharmaceuticals in rural wastewaters (e.g., sulfadimethoxine, sulfamethoxazole, trimethoprim, sulfamerazine, and sulfadiazine) (Wei et al. 2011; Tong et al. 2009; Naderi Beni et al. 2020). However, most studies conducted with effluents from farming units have focused on the recycling of nutrients from these wastewaters. For instance, a high efficiency removal of nitrate (80%) from a dairy cattle farm was obtained by using a native microalgae species (*Chroococcus* sp.) (Prajapati et al. 2014). The feasibility of wastewater treatment implemented in a poultry production was studied by Oliveira et al. (2019). Bubble-column photobioreactors were used to cultivate *Scenedesmus obliquus* and remove inorganic nutrients (ammonia and phosphate) from the wastewater, while producing a valued-added microalgal biomass. Furthermore, *C. vulgaris* and *Scenedesmus accuminatus* were demonstrated to be potential candidates to remove ammonium and phosphate from cattle production wastewater with different cultivation bioreactors, namely an Erlenmeyer flask (Ganeshkumar et al. 2018), tubular borosilicate reactors (Kim et al. 2014), and plastic bags (Kumar et al. 2010). Nevertheless, it is necessary to consider that the variation of each technic-operational parameter will define the success of wastewater treatment.

Aquaculture (e.g., fish, oysters, and shrimp) is also a source of pharmaceuticals, inorganic nutrients, and organic matter (Martinez-Porchas et al. 2014). The implementation of integrated multitrophic aquaculture (IMTA) has been proven to be a feasible approach to address this problem. Microalgae can efficiently harvest and recycle nutrients (Mane et al. 2005). Li et al. (2019) demonstrated that by coupling fish and oysters cultured in a recirculating aquaculture system with raceways for a native microalgae consortium production, it was possible to efficiently reduce the concentrations of nitrate (98.2%), nitrite (98%), ammonium (97.3%), and phosphate (96.1%) in 60 days. Kumar et al. (2016) implemented a different treatment approach; a green microalgae *Picochlorum maculatum* was isolated in alginate blocks and achieved high removal rates of nutrients, namely 46% for nitrate, 89% for nitrite, 98% for ammonia, and 57% for phosphate. Moreover, microalgae can also remediate metals from aquaculture effluents. Recently, *Arthrospira* sp. (also known as *Spirulina*) was shown to be capable of completely removing metals (Al, Fe, Ca, and Mg) from fish aquaculture effluents (Molina-Miras et al. 2020). The main advantage of using a microalgal-based IMTA approach to clean up aquaculture effluents is that the obtained biomass can be valorized.

29.5 WATER REUTILIZATION AND MICROALGAL BIOMASS VALORIZATION

Water is an essential resource that is becoming increasingly scarce. Thus, it is important to preserve it and promote water recycling processes. As it was highlighted, microalgae can remove a wide range of pollutants from wastewaters, complementing the existing wastewater treatments and reducing the load of EPs entering aquatic ecosystems.

Within a sustainability-oriented circular economy paradigm, biological treatments using microalgae may provide additional benefits, namely the reutilization of water and valorization of biomass. Once efficiently treated, water can be reintroduced in the industrial circuit, diminishing costs associated with water expenses. This is feasible for several types of industries such as textile, and paper and pulp (Aragaw and Asmare 2018). Furthermore, treated wastewater can also be recycled for crop irrigation (Mateo-Sagasta et al. 2013).

Another advantage of biological treatments is the production of microalgal biomass that, through biorefinery, can be transformed into added-value products, such as biodiesel, fertilizer, natural plastics, and protein-rich feed (Sutherland and Ralph 2019; Deviram et al. 2020).

29.6 CONCLUSION

Global human population growth and the intensification of anthropogenic activities augmented the demand for water resources and the anthropogenic pressure over aquatic ecosystems. Wastewater nutrient recycling through microalgae cultivation has been extensively studied. However, in the last decades, EP detection in water resources alarmed the scientific community and competent authorities. By mimicking natural environment conditions, researchers have been investigating the capability of microalgae to remove these refractory pollutants. Successful studies have been conducted and the ability of microalgae to remove or considerably reduce the concentration of EPs in different types of effluents has been demonstrated.

Nevertheless, it is still necessary to optimize and validate the upscaling of this sustainable and feasible wastewater treatment in order to enable the industrialization of microalgae-based bioremediation technology.

ACKNOWLEDGMENTS

This work was supported by national funds through FCT—Foundation for Science and Technology, I.P., within the scope of the project UIDB/04292/2020—MARE—Marine and Environmental Sciences Centre. Ana Cristina Rocha thanks to CENTRO2020 - SAICT – Integrated Program, Operation CENTRO-01-0145-FEDER-000006, with the reference SAICT 000006 SUSpENsE - SUSpENsE Sustainable built Environment under Natural Hazards and Extreme Events.

REFERENCES

Abdelbasir, Sabah Mohamed, and Ahmed Esmail Shalan. 2019. "An Overview of Nanomaterials for Industrial Wastewater Treatment." *Korean Journal of Chemical Engineering* 36(8): 1209–25. https://doi.org/10.1007/s11814-019-0306-y.

Akao, Patricia K., Hadas Mamane, Aviv Kaplan, Igal Gozlan, Yaron Yehoshua, Yael Kinel-Tahan, and Dror Avisar. 2020. "Iohexol Removal and Degradation-Product Formation via Biodegradation by the Microalga *Chlorella vulgaris*." *Algal Research* 51(October). https://doi.org/10.1016/j.algal.2020.102050.102050.

Aragaw, Tadele Assefa, and Abraham M. Asmare. 2018. "Phycoremediation of Textile Wastewater Using Indigenous Microalgae." *Water Practice and Technology* 13(2): 274–84. https://doi.org/10.2166/wpt.2018.037.

Arbib, Zouhayr, Jesus Ruiz, Pablo Alvarez, Carmen Garrido, Jesus Barragan, and Jose Antonio Perales. 2012. "*Chlorella stigmatophora* for Urban Wastewater Nutrient Removal and CO_2 Abatement." *International Journal of Phytoremediation* 14(7): 714–25. https://doi.org/10.1080/15226514.2011.619237.

Azoulay, A., and G. Houngbo. 2019. *The United Nations World Water Development Report 2019.* Edited by UNESCO. 1st ed. Paris: UNESCO. https://unesdoc.unesco.org/ark:/48223/pf0000367306.

Baglieri, Andrea, Sarah Sidella, Valeria Barone, Ferdinando Fragalà, Alla Silkina, Michèle Nègre, and Mara Gennari. 2016. "Cultivating *Chlorella vulgaris* and *Scenedesmus quadricauda* Microalgae to Degrade Inorganic Compounds and Pesticides in Water." *Environmental Science and Pollution Research International* 23(18): 18165–74. https://doi.org/10.1007/s11356-016-6996-3.

Bai, Xuelian, and Acharya Kumud. 2016. "Removal of Trimethoprim, Sulfamethoxazole, and Triclosan by the Green Alga *Nannochloris* sp.. " *Journal of Hazardous Materials* 315(September): 70–5. https://doi.org/10.1016/j.jhazmat.2016.04.067.

Beauvais-Flück, Rebecca, Vera I. Slaveykova, and Claudia Cosio. 2018. "Molecular Effects of Inorganic and Methyl Mercury in Aquatic Primary Producers: Comparing Impact to a Macrophyte and a Green Microalga in Controlled Conditions." *Geosciences (Switzerland)* 8(11). https://doi.org/10.3390/geosciences8110393.

Brar, A., M. Kumar, V. Vivekanand, and N. Pareek. 2019. "Phycoremediation of Textile Effluent-Contaminated Water Bodies Employing Microalgae: Nutrient Sequestration and Biomass Production Studies." *International Journal of Environmental Science and Technology* 16(12): 7757–68. https://doi.org/10.1007/s13762-018-2133-9.

Cheah, Wai Yan, Pau Loke Show, Joon Ching Juan, Jo-Shu Chang, and Tau Chuan Ling. 2018. "Waste to Energy: The Effects of *Pseudomonas* sp. on *Chlorella sorokiniana* Biomass and Lipid Productions in Palm Oil Mill Effluent." *Clean Technologies and Environmental Policy* 20(9): 2037–45. https://doi.org/10.1007/s10098-018-1505-7.

Choudhary, Poonam, Sanjeev Kumar Prajapati, and Anushree Malik. 2016. "Screening Native Microalgal Consortia for Biomass Production and Nutrient Removal from Rural Wastewaters for Bioenergy Applications." *Ecological Engineering* 91(June): 221–30. https://doi.org/10.1016/j.ecoleng.2015.11.056.

Ciniglia, C., G. M. Valentino, P. Cennamo, M. Stefano, D. Stanzione, G. Pinto, and A. Pollio. 2005. "Influences of Geochemical and Mineralogical Constraints on Algal Distribution in Acidic Hydrothermal Environments: Pisciarelli (Naples, Italy) as a Model Site." *Archiv fur Hydrobiologie* 162(1): 121–142. https://doi.org/10.1127/0003-9136/2005/0162-0121.

Chiu, Y. S., C. Y. Kao, T. Y. Chen, Y. B. Chang, C. M. Kuo, and C. S. Lin. 2015. "Cultivation of Microalgal Chlorella for Biomass and Lipid Production Using Wastewater as Nutrient Resource." *Bioresource Technology* 184: 179–189. https://doi.org.10.1016/j.biortech.2014.11.080.

Christou, Anastasis, Popi Karaolia, Evroula Hapeshi, Costas Michael, and Despo Fatta-Kassinos. 2017. "Long-Term Wastewater Irrigation of Vegetables in Real Agricultural Systems: Concentration of Pharmaceuticals in Soil, Uptake and Bioaccumulation in Tomato Fruits and Human Health Risk Assessment." *Water Research* 109(February), 24–34. https://doi.org/10.1016/j.watres.2016.11.033.

Coogan, Melinda A., Regina E. Edziyie, Thomas W. La Point, and Barney J. Venables. 2007. "Algal Bioaccumulation of Triclocarban, Triclosan, and Methyl-Triclosan in a North Texas Wastewater Treatment Plant Receiving Stream." *Chemosphere* 67(10): 1911–18. https://doi.org/10.1016/j.chemosphere.2006.12.027.

Crini, Grégorio, and Eric Lichtfouse. 2019. "Advantages and Disadvantages of Techniques Used for Wastewater Treatment." *Environmental Chemistry Letters* 17(1): 145–55. https://doi.org/10.1007/s10311-018-0785-9.

Dadson, Simon, Dustin Garrick, Edmund Penning-Rowsell, Jim Hall, Rob Hope, and Jocelyne Hughes. 2020. *Water Science, Policy and Management: A Global Challenge.* Edited by Simon Dadson, Dustin Garrick, Edmund Penning-Rowsell, Jim Hall, Rob Hope, and Jocelyne Hughes. 1st ed. Oxford: Wiley. https://books.google.pt/books?id=2ey2DwAAQBAJ&printsec=frontcover&hl=pt-PT#v=onepage&q&f=false.

Darwesh, Osama M., Ibrahim A. Matter, and Mohamed F. Eida. 2019. "Development of Peroxidase Enzyme Immobilized Magnetic Nanoparticles for Bioremediation of Textile Wastewater Dye." *Journal of Environmental Chemical Engineering* 7(1): 102805. https://doi.org/10.1016/j.jece.2018.11.049.

Deviram, Garlapati, Thangavel Mathimani, Susaimanickam Anto, Tharifkhan Shan Ahamed, Devanesan Arul Ananth, and Arivalagan Pugazhendhi. 2020. "Applications of Microalgal and Cyanobacterial Biomass on a Way to Safe, Cleaner and a Sustainable Environment." *Journal of Cleaner Production* 253: 119770. https://doi.org/10.1016/j.jclepro.2019.119770.

Díaz-Cruz, M.S., and D. Barceló. 2008. "Input of Pharmaceuticals, Pesticides and Industrial Chemicals as a Consequence of Using Conventional and Non-Conventional Sources of Water for Artificial Groundwater Recharge." In: *Emerging Contaminants from Industrial and Municipal Waste*. Edited by D. Barceló and M. Petrovick. Berlin Heidelberg: Springer, 219–38.

Dinesh Kumar, S., P. Santhanam, Min S. Park, and Mi-Kyung Kim. 2016. "Development and Application of a Novel Immobilized Marine Microalgae Biofilter System for the Treatment of Shrimp Culture Effluent." *Journal of Water Process Engineering* 13(October): 137–42. https://doi.org/10.1016/j.jwpe.2016.08.014.

Eerkes-Medrano, Dafne, Heather A. Leslie, and Brian Quinn. 2019. "Microplastics in Drinking Water: A Review and Assessment." *Current Opinion in Environmental Science & Health* 7 (February): 69–75. https://doi.org/10.1016/j.coesh.2018.12.001.

El-Kassas, Hala Yassin, and Laila Abdelfattah Mohamed. 2014. "Bioremediation of the Textile Waste Effluent by *Chlorella vulgaris*." *The Egyptian Journal of Aquatic Research* 40(3): 301–8. https://doi.org/10.1016/j.ejar.2014.08.003.

EPA, U.S. 1972. *Federal Water Pollution Control Act*.AU

Esperanza, Marta, Marta Seoane, María J. Servia, and Ángeles Cid. 2020. "Effects of Bisphenol A on the Microalga Chlamydomonas reinhardtii and the Clam *Corbicula fluminea*." *Ecotoxicology and Environmental Safety* 197(July): 110609. https://doi.org/10.1016/j.ecoenv.2020.110609.

EU. Parlamento Europeu. 1991. "Diretiva n.o 91/271/CEE: Relativa Ao Tratamento de Águas Residuais Urbanas." *Jornal Oficial Das Comunidades Europeias L* 135/40: 13. https://op.europa.eu/en/publication-detail/-/publication/34860b42-95cc-4a57-b998-2e445d5befa9.

European Commission. 2019. "Evaluation of the Urban Waste Water Treatment Directive." 186. https://ec.euro pa.eu/environment/water/water-urbanwaste/pdf/UWWTD Evaluation SWD 448–701. web.pdf.

European Environment Agency (EEA). 2018. *Industrial Waste Water Treatment Pressures on Environment*. Publications Office of the European Union, Luxembourg (2019). doi:10.2800/496223

FAO. 1992. *Wastewater Treatment and Use in Agriculture*. http://www.fao.org/3/t0551e/t0551e03.htm.

Fawley, Marvin W., and Karen P. Fawley. 2020. "Identification of Eukaryotic Microalgal Strains." *Journal of Applied Phycology*, 32, 2699–2709. https://doi.org/10.1007/s10811-020-02190-5.

Frimmel, Fritz, Gudrun Abbt-Braun, Klaus Heumann, Berthold Hock, and Hans Ludemann. 2008. *Refractory Organic Substances in the Environment | Wiley*. Wiley.

Ganeshkumar, Vimalkumar, Suresh R. Subashchandrabose, Rajarathnam Dharmarajan, Kadiyala Venkateswarlu, Ravi Naidu, and Mallavarapu Megharaj. 2018. "Use of Mixed Wastewaters from Piggery and Winery for Nutrient Removal and Lipid Production by Chlorella Sp. MM3." *Bioresource Technology* 256(May): 254–58. https://doi.org/10.1016/j.biortech.2018.02.025.

García-Galán, María Jesús, Raquel Gutiérrez, Enrica Uggetti, Víctor Matamoros, Joan García, and Ivet Ferrer. 2018. "Use of Full-Scale Hybrid Horizontal Tubular Photobioreactors to Process Agricultural Runoff." *Biosystems Engineering* 166: 138–49. https://doi.org/10.1016/j.biosystemseng.2017.11.016.

Gauthier, Léa, Juliette Tison-Rosebery, Soizic Morin, and Nicolas Mazzella. 2020. "Metabolome Response to Anthropogenic Contamination on Microalgae: A Review." In *Metabolomics*. Springer. https://doi.org/10.1007/s11306-019-1628-9.

Geissen, Violette, Hans Mol, Erwin Klumpp, Günter Umlauf, Marti Nadal, Martine van der Ploeg, Sjoerd E.A.T.M. van de Zee, and Coen J. Ritsema. 2015. "Emerging Pollutants in the Environment: A Challenge for Water Resource Management." *International Soil and Water Conservation Research* 3(1): 57–65. https://doi.org/10.1016/j.iswcr.2015.03.002.

Gentili, Francesco G., and Jerker Fick. 2017. "Algal Cultivation in Urban Wastewater: An Efficient Way to Reduce Pharmaceutical Pollutants." *Journal of Applied Phycology* 29(1): 255–62. https://doi.org/10.1007/s10811-016-0950-0.

Godos, Raúl Muñoz Ignacio de, and Benoit Guieysse. 2012. "Tetracycline Removal during Wastewater Treatment in High-Rate Algal Ponds." *Journal of Hazardous Materials* 229–230(August): 446–49. https://doi.org/10.1016/j.jhazmat.2012.05.106.

Gojkovic, Zivan, Richard H. Lindberg, Mats Tysklind, and Christiane Funk. 2019. "Northern Green Algae Have the Capacity to Remove Active Pharmaceutical Ingredients." *Ecotoxicology and Environmental Safety* 170(April): 644–56. https://doi.org/10.1016/j.ecoenv.2018.12.032.

Hamilton, T.L., and J. Havig. 2017. "Primary Productivity of Snow Algae Communities on Stratovolcanoes of the Pacific Northwest." *Geobiology* 15(2): 280–95. https://doi.org/10.1111/gbi.12219.

Hansda, Arti, Vipin Kumar, and Anshumali. 2016. "A Comparative Review Towards Potential of Microbial Cells for Heavy Metal Removal with Emphasis on Biosorption and Bioaccumulation." *World Journal of Microbiology and Biotechnology* 32(10): 1–14. https://doi.org/10.1007/s11274-016-2117-1.

Hernández, F., J.V. Sancho, M. Ibáñez, E. Abad, T. Portolés, and L. Mattioli. 2012. "Current Use of High-Resolution Mass Spectrometry in the Environmental Sciences." In *Analytical and Bioanalytical Chemistry*. Springer. https://doi.org/10.1007/s00216-012-5844-7.

Hom-Diaz, Andrea, Marta Llorca, Sara Rodríguez-Mozaz, Teresa Vicent, Damià Barceló, and Paqui Blánquez. 2015. "Microalgae Cultivation on Wastewater Digestate: β-Estradiol and 17α-Ethynylestradiol Degradation and Transformation Products Identification." *Journal of Environmental Management* 155(May): 106–13. https://doi.org/10.1016/j.jenvman.2015.03.003.

Hope, Julie A., David M. Paterson, and Simon F. Thrush. 2020. "The Role of Microphytobenthos in Soft-Sediment Ecological Networks and Their Contribution to the Delivery of Multiple Ecosystem Services." *Journal of Ecology*. Blackwell Publishing Ltd. 108(3): 815–830. https://doi.org/10.1111/1365-2745.13322.

Huo, Shuhao, Jing Chen, Xiu Chen, Feng Wang, Ling Xu, Feifei Zhu, Danzhao Guo, and Zhenjiang Li. 2018a. "Advanced Treatment of the Low Concentration Petrochemical Wastewater by *Tribonema* sp. Microalgae Grown in the Open Photobioreactors Coupled with the Traditional Anaerobic/Oxic Process." *Bioresource Technology* 270(December): 476–81. https://doi.org/10.1016/j.biortech.2018.09.024.

Huo, Shuhao, Feifei Zhu, Bin Zou, Ling Xu, Fengjie Cui, and Wenhua You. 2018b. "A Two-Stage System Coupling Hydrolytic Acidification with Algal Microcosms for Treatment of Wastewater from the Manufacture of Acrylonitrile Butadiene Styrene (ABS) Resin." *Biotechnology Letters* 40(4): 689–96. https://doi.org/10.1007/s10529-018-2513-8.

Jacquin, Anne Gaëlle, Stéphanie Brulé-Josso, M. Lynn Cornish, Alan T. Critchley, and Patrick Gardet. 2014. "Selected Comments on the Role of Algae in Sustainability." *Advances in Botanical Research* 71: 1–30. Academic Press Inc. https://doi.org/10.1016/B978-0-12-408062-1.00001-9.

Jardak, K., P. Drogui, and R. Daghrir. 2016. "Surfactants in Aquatic and Terrestrial Environment: Occurrence, Behavior, and Treatment Processes." *Environmental Science and Pollution Research International* 23(4): 3195–216. https://doi.org/10.1007/s11356-015-5803-x.

Ji, Min-Kyu, Akhil N. Kabra, Jaewon Choi, Jae-Hoon Hwang, Jung Rae Kim, Reda A.I. Abou-Shanab, You-Kwan Oh, and Byong-Hun Jeon. 2014. "Biodegradation of Bisphenol A by the Freshwater Microalgae *Chlamydomonas mexicana* and *Chlorella vulgaris*." *Ecological Engineering* 73(December): 260–69. https://doi.org/10.1016/j.ecoleng.2014.09.070.

Kadam, Abhijeet D., Garima Kishore, Deepak Kumar Mishra, and Kusum Arunachalam. 2020. "Microalgal Diversity as an Indicator of the State of the Environment of Water Bodies of Doon Valley in Western Himalaya, India." *Ecological Indicators* 112(May): 106077. https://doi.org/10.1016/j.ecolind.2020.106077.

Kamenev, G. M., V. I. Fadeev, N. I. Selin, V. G. Tarasov, and V. V. Malakhov. 1993. "Composition and Distribution of Macro- and Meiobenthos around Sublittoral Hydrothermal Vents in the Bay of Plenty, New Zealand." *New Zealand Journal of Marine and Freshwater Research* 27: 407–418.

Khanna, D.R., R. Bhutiani, Gagan Matta, Vikas Singh, and Fouzia Ishaq. 2012. "Seasonal Variation in Physico-Chemical Characteristic Status of River Yamuna in Doon Valley of Uttarakhand." *Environment Conservation Journal* 13(1&2): 119–24.

Kim, Hyun Chul, Wook Jin Choi, Jun Hee Ryu, Sung Kyu Maeng, Han Soo Kim, Byung Chan Lee, and Kyung Guen Song. 2014. "Optimizing Cultivation Strategies for Robust Algal Growth and Consequent Removal of Inorganic Nutrients in Pretreated Livestock Effluent." *Applied Biochemistry and Biotechnology* 174(4): 1668–82. https://doi.org/10.1007/s12010-014-1145-2.

Kolpin, Dana W., Edward T. Furlong, Michael T. Meyer, E. Michael Thurman, Steven D. Zaugg, Larry B. Barber, and Herbert T. Buxton. 2002. "Pharmaceuticals, Hormones, and Other Organic Wastewater Contaminants in U.S. Streams, 1999–2000: A National Reconnaissance." *Environmental Science and Technology* 36(6): 1202–11. https://doi.org/10.1021/es011055j.

Kumar, Dhananjay, Lalit K. Pandey, and J.P. Gaur. 2016. "Metal Sorption by Algal Biomass: From Batch to Continuous System." *Algal Research* 18(September): 95–109. https://doi.org/10.1016/j.algal.2016.05.026.

Kumar, Martin S., Zhihong H. Miao, and Sandy K. Wyatt. 2010. "Influence of Nutrient Loads, Feeding Frequency and Inoculum Source on Growth of *Chlorella vulgaris* in Digested Piggery Effluent Culture Medium." *Bioresource Technology* 101(15): 6012–18. https://doi.org/10.1016/j.biortech.2010.02.080.

Lai, Bo, Yuexi Zhou, Ping Yang, and Ke Wang. 2012. "Comprehensive Analysis of the Toxic and Refractory Pollutants in Acrylonitrile–Butadiene–Styrene Resin Manufacturing Wastewater by Gas Chromatography Spectrometry with a Mass or Flame Ionization Detector." *Journal of Chromatography. Part A* 1244(June): 161–67. https://doi.org/10.1016/j.chroma.2012.04.058.

Lavrinovičs, Aigars, and Tālis Juhna. 2018. "Review on Challenges and Limitations for Algae-Based Wastewater Treatment." *Construction Science* 20(1): 17–25. https://doi.org/10.2478/cons-2017-0003.

Li, Meng, Myriam D. Callier, and Jean-Paul Blancheton, Amandine Galès, Sarah Nahon, Sébastien Triplet, Thibault Geoffroy, Christophe Menniti, Eric Fouilland, and Emmanuelle Roque d'orbcastel. 2019. "Bioremediation of Fishpond Effluent and Production of Microalgae for an Oyster Farm in an Innovative Recirculating Integrated Multi-Trophic Aquaculture System." *Aquaculture* 504(April): 314–25. https://doi.org/10.1016/j.aquaculture.2019.02.013.

Li, Tong, Gengyi Lin, Björn Podola, and Michael Melkonian. 2015. "Continuous Removal of Zinc from Wastewater and Mine Dump Leachate by a Microalgal Biofilm PSBR." *Journal of Hazardous Materials* 297(October): 112–18. https://doi.org/10.1016/j.jhazmat.2015.04.080.

Li, Yongchao, Zheng Xu, Hongqing Ma, and Andrew S. Hursthouse. 2019. "Removal of Manganese(II) from Acid Mine Wastewater: A Review of the Challenges and Opportunities with Special Emphasis on Mn-Oxidizing Bacteria and Microalgae." *Water* 11(12): 2493. https://doi.org/10.3390/w11122493.

Lim, S.-L., W.-L. Chu, and S.-M. Phang. 2010a. "Bioresource Technology Use of *Chlorella vulgaris* for Bioremediation of Textile Wastewater." *Bioresource Technology* 101(19): 7314–22. https://doi.org/10.1016/j.biortech.2010.04.092.

Lipponen, Annukka, and Nataliya Nikiforova. 2017. *Wastewater: The Untapped Resource.* Edited by UNESCO. Paris: UNESCO. www.unwater.org.

Madadi, R., A.A. Pourbabaee, M. Tabatabaei, M.A. Zahed, and M.R. Naghavi. 2016. "Treatment of Petrochemical Wastewater by the Green Algae *Chlorella vulgaris.*" *International Journal of Environment and Resource* 10(4): 555–60.

Mane, V.R., A.A. Chandorkar, and R. Kumar. 2005. "Prevalence of Pollution in Surface and Ground Water Sources in the Rural Areas of Satara Region, Maharashtra, India." *Asian Journal of Water Environment and Pollution* 2(2): 81–7.

Marchão, Leonilde, Teresa Lopes da Silva, Luísa Gouveia, and Alberto Reis. 2018. "Microalgae-Mediated Brewery Wastewater Treatment: Effect of Dilution Rate on Nutrient Removal Rates, Biomass Biochemical Composition, and Cell Physiology." *Journal of Applied Phycology* 30(3): 1583–95. https://doi.org/10.1007/s10811-017-1374-1.

Martinez-Porchas, Marcel, Luis Rafael Martinez-Cordova, Jose Antonio Lopez-Elias, and Marco Antonio Porchas-Cornejo. 2014. "Bioremediation of Aquaculture Effluents." In: *Microbial Biodegradation and Bioremediation*: 539–53. Elsevier https://doi.org/10.1016/B978-0-12-800021-2.00024-8.

Matamoros, Víctor, Enrica Uggetti, Joan García, and Josep M. Bayona. 2016. "Assessment of the Mechanisms Involved in the Removal of Emerging Contaminants by Microalgae from Wastewater: A Laboratory Scale Study." *Journal of Hazardous Materials* 301(January): 197–205. https://doi.org/10.1016/j.jhazmat.2015.08.050.

Mateo-Sagasta, Javier, Kate Medlicott, Manzoor Qadir, Liqa Raschid-Sally, Pay Drechsel, and Jens Liebe. 2013. *Proceedings of the UN-Water Project on the Safe Use of Wastewater in Agriculture.* Edited by Jens Liebe and Reza Ardakanian. 1st ed. Vol. 1. Bonn, Germany: UNW-DPC. www.unwater.unu.edu

Menció, Anna, Josep Mas-Pla, Neus Otero, Oriol Regàs, Mercè Boy-Roura, Roger Puig, Joan Bach, et al. 2016. "Nitrate Pollution of Groundwater; All Right..., but Nothing Else?. " *Science of the Total Environment* 539(January): 241–51. https://doi.org/10.1016/j.scitotenv.2015.08.151.

Molina-Miras, A., L. López-Rosales, M.C. Cerón-García, A. Sánchez-Mirón, A. Olivera-Gálvez, F. García-Camacho, and E. Molina-Grima. 2020. "Acclimation of the Microalga *Amphidinium carterae* to Different Nitrogen Sources: Potential Application in the Treatment of Marine Aquaculture Effluents." *Journal of Applied Phycology* 32(2): 1075–94. https://doi.org/10.1007/s10811-020-02049-9.

Montes-Grajales, Diana, Mary Fennix-Agudelo, and Wendy Miranda-Castro. 2017. "Occurrence of Personal Care Products as Emerging Chemicals of Concern in Water Resources: A Review." *Science of the Total Environment* 595(October): 601–14. https://doi.org/10.1016/j.scitotenv.2017.03.286.

Muñoz, Ivan, María José Gómez-Ramos, Ana Agüera, Amadeo R. Fernández-Alba, Juan Francisco García-Reyes, and Antonio Molina-Díaz. 2009. "Chemical Evaluation of Contaminants in Wastewater Effluents and the Environmental Risk of Reusing Effluents in Agriculture." *TrAC Trends in Analytical Chemistry* 28(6): 676–94. https://doi.org/10.1016/j.trac.2009.03.007.

Naderi Beni, Nasrin, Daniel D. Snow, Elaine D. Berry, Aaron R. Mittelstet, Tiffany L. Messer, and Shannon Bartelt-Hunt. 2020. "Measuring the Occurrence of Antibiotics in Surface Water Adjacent to Cattle Grazing Areas Using Passive Samplers." *Science of the Total Environment* 726(July): 138296. https://doi.org/10.1016/j.scitotenv.2020.138296.

Norvill, Zane N., Andy Shilton, and Benoit Guieysse. 2016. "Emerging Contaminant Degradation and Removal in Algal Wastewater Treatment Ponds: Identifying the Research Gaps." *Journal of Hazardous Materials*. Elsevier B.V. 313: 291–309. https://doi.org/10.1016/j.jhazmat.2016.03.085.

Oliveira, Ana Cristina, Ana Barata, Ana P. Batista, and Luísa Gouveia. 2019. "*Scenedesmus obliquus* in Poultry Wastewater Bioremediation." *Environmental Technology (United Kingdom)* 40(28): 3735–44. https://doi.org/10.1080/09593330.2018.1488003.

Oliveira, Tiago, Mohammed Boussafir, Laëtitia Fougère, Emilie Destandau, Yoshiyuki Sugahara, and Régis Guégan. 2020. "Use of a Clay Mineral and Its Nonionic and Cationic Organoclay Derivatives for the Removal of Pharmaceuticals from Rural Wastewater Effluents." *Chemosphere* 259: 127480. https://doi.org/10.1016/j.chemosphere.2020.

Pacheco, Diana, Ana Cristina Rocha, Leonel Pereira, and Tiago Verdelhos. 2020. "Microalgae Water Bioremediation: Trends and Hot Topics." *Applied Sciences* 10(5): 1886. https://doi.org/10.3390/app10051886.

Padmaperuma, Gloria, Rahul Vijay Kapoore, Daniel James Gilmour, and Seetharaman Vaidyanathan. 2018. "Microbial Consortia: A Critical Look at Microalgae Co-Cultures for Enhanced Biomanufacturing." *Critical Reviews in Biotechnology* 38(5): 690–703. https://doi.org/10.1080/07388551.2017.1390728.

Panepinto, Deborah, Silvia Fiore, Giuseppe Genon, and Marco Acri. 2016. "Thermal Valorization of Sewer Sludge: Perspectives for Large Wastewater Treatment Plants." *Journal of Cleaner Production* 137: 1323–29. https://doi.org/10.1016/j.jclepro.2016.08.014.

Papry, Rimana Islam, Shogo Fujisawa, Zai Yinghan, Okviyoandra Akhyar, M. Abdullah Al Mamun, Asami S. Mashio, and Hiroshi Hasegawa. 2020. "Integrated Effects of Important Environmental Factors on Arsenic Biotransformation and Photosynthetic Efficiency by Marine Microalgae." *Ecotoxicology and Environmental Safety* 201(September): 110797. https://doi.org/10.1016/j.ecoenv.2020.

Pathak, Vinayak V., Richa Kothari, A.K. Chopra, and D.P. Singh. 2015. "Experimental and Kinetic Studies for Phycoremediation and Dye Removal by *Chlorella pyrenoidosa* from Textile Wastewater." *Journal of Environmental Management* 163(November): 270–77. https://doi.org/10.1016/j.jenvman.2015.08.041.

Petrie, Bruce, Ruth Barden, and Barbara Kasprzyk-Hordern. 2015. "A Review on Emerging Contaminants in Wastewaters and the Environment: Current Knowledge, Understudied Areas and Recommendations for Future Monitoring." *Water Research* 72(April): 3–27. https://doi.org/10.1016/j.watres.2014.08.053.

Pokhrel, D., and T. Viraraghavan. 2004. "Treatment of Pulp and Paper Mill Wastewater: A Review." *Science of the Total Environment* 333(1–3): 37–58. https://doi.org/10.1016/j.scitotenv.2004.05.017.

Porto, Bruna, Ana L. Gonçalves, Ana F. Esteves, Selene M.A. Guelli Ulson de Souza, Antônio A. Ulson de Souza, Vítor J.P. Vilar, and José C.M. Pires. 2020. "Microalgal Growth in Paper Industry Effluent: Coupling Biomass Production with Nutrients Removal." *Applied Sciences* 10(9): 3009. https://doi.org/10.3390/app10093009.

Prajapati, Sanjeev Kumar, Poonam Choudhary, Anushree Malik, and Virendra Kumar Vijay. 2014. "Algae Mediated Treatment and Bioenergy Generation Process for Handling Liquid and Solid Waste from Dairy Cattle Farm." *Bioresource Technology* 167(September): 260–68. https://doi.org/10.1016/j.biortech.2014.06.038.

Ramírez, M.E., Y.H. Vélez, L. Rendón, and E. Alzatezate. 2017. "Potential of Microalgae in the Bioremediation of Water with Chloride Content." *Brazilian Journal of Biology* 78(3): 472–76. https://doi.org/10.1590/1519-6984.169372.

Riaño, B., S. Blanco, E. Becares, and M.C. García-González. 2016. "Bioremediation and Biomass Harvesting of Anaerobic Digested Cheese Whey in Microalgal-Based Systems for Lipid Production." *Ecological Engineering* 97(December): 40–5. https://doi.org/10.1016/j.ecoleng.2016.08.002.

Rizzo, Luigi, Sixto Malato, Demet Antakyali, Vasiliki G. Beretsou, Maja B. Đolić, Wolfgang Gernjak, Ester Heath, et al. 2019. "Consolidated vs New Advanced Treatment Methods for the Removal of Contaminants of Emerging Concern from Urban Wastewater." *Science of the Total Environment* 655(March): 986–1008. https://doi.org/10.1016/j.scitotenv.2018.11.265.

Ruhí, Albert, Vicenç Acuña, Damià Barceló, Belinda Huerta, Jordi Rene Mor, Sara Rodríguez-Mozaz, and Sergi Sabater. 2016. "Bioaccumulation and Trophic Magnification of Pharmaceuticals and Endocrine Disruptors in a Mediterranean River Food Web." *Science of the Total Environment* 540(January): 250–59. https://doi.org/10.1016/j.scitotenv.2015.06.009.

Ruiz-Marin, Alejandro, Leopoldo G. Mendoza-Espinosa, and Tom Stephenson. 2010. "Growth and Nutrient Removal in Free and Immobilized Green Algae in Batch and Semi-Continuous Cultures Treating Real Wastewater." *Bioresource Technology* 101(1): 58–64. https://doi.org/10.1016/j.biortech.2009.02.076.

Saavedra, Ricardo, Raúl Muñoz, María Elisa Taboada, Marisol Vega, and Silvia Bolado. 2018. "Comparative Uptake Study of Arsenic, Boron, Copper, Manganese and Zinc from Water by Different Green Microalgae." *Bioresource Technology* 263(September): 49–57. https://doi.org/10.1016/j.biortech.2018.04.101.

Santos, Carla Escapa, Ricardo Nuno de Coimbra, Sergio Paniagua Bermejo, Ana Isabel García Pérez, and Marta Otero Cabero. 2017. "Comparative Assessment of Pharmaceutical Removal from Wastewater by the Microalgae *Chlorella sorokiniana*, *Chlorella vulgaris* and *Scenedesmus obliquus*."In *Biological Wastewater Treatment and Resource Recovery.* InTech. https://doi.org/10.5772/66772.

Schade, S., and T. Meier. 2019. "A Comparative Analysis of the Environmental Impacts of Cultivating Microalgae in Different Production Systems and Climatic Zones: A Systematic Review and Meta-Analysis." *Algal Research* 40(June): 101485. https://doi.org/10.1016/j.algal.2019.101485.

Singh, Poonam, Sheena Kumari, Abhishek Guldhe, Rohit Misra, Ismail Rawat, and Faizal Bux. 2016. "Trends and Novel Strategies for Enhancing Lipid Accumulation and Quality in Microalgae." *Renewable and Sustainable Energy Reviews* 55: 1–16. https://doi.org/10.1016/j.rser.2015.11.001.

Solovchenko, Alexei, Sergei Pogosyan, Olga Chivkunova, Irina Selyakh, Larisa Semenova, Elena Voronova, Pavel Scherbakov, et al. 2014. "Phycoremediation of Alcohol Distillery Wastewater with a Novel *Chlorella sorokiniana* Strain Cultivated in a Photobioreactor Monitored On-Line via Chlorophyll Fluorescence." *Algal Research* 6(PB): 234–41. https://doi.org/10.1016/j.algal.2014.01.002.

Steudel, Bastian. 2014. "Microalgae in Ecology: Ecosystem Functioning Experiments." https://doi.org/10.4172/2332-2632.1000122.

Sutherland, Donna L., and Peter J. Ralph. 2019. "Microalgal Bioremediation of Emerging Contaminants - Opportunities and Challenges." *Water Research* 164(November): 114921. https://doi.org/10.1016/j.watres.2019.114921.

Tiwari, Bhagyashree, Balasubramanian Sellamuthu, Yassine Ouarda, Patrick Drogui, Rajeshwar D. Tyagi, and Gerardo Buelna. 2017. "Review on Fate and Mechanism of Removal of Pharmaceutical Pollutants from Wastewater Using Biological Approach." *Bioresource Technology* 224(January): 1–12. https://doi.org/10.1016/j.biortech.2016.11.042.

Tong, Lei, Ping Li, Yanxin Wang, and Kuanzheng Zhu. 2009. "Analysis of Veterinary Antibiotic Residues in Swine Wastewater and Environmental Water Samples Using Optimized SPE-LC/MS/MS." *Chemosphere* 74(8): 1090–97. https://doi.org/10.1016/j.chemosphere.2008.10.051.

Topp, Edward, Sara C. Monteiro, Andrew Beck, Bonnie Ball Coelho, Alistair B.A. Boxall, Peter W. Duenk, Sonya Kleywegt, et al. 2008. "Runoff of Pharmaceuticals and Personal Care Products Following Application of Biosolids to an Agricultural Field." *Science of the Total Environment* 396(1): 52–9. https://doi.org/10.1016/j.scitotenv.2008.02.011.

Tran, Ngoc Han, Martin Reinhard, and Karina Yew Hoong Gin. 2018. "Occurrence and Fate of Emerging Contaminants in Municipal Wastewater Treatment Plants from Different Geographical Regions-A Review." *Water Research*. Elsevier Ltd. 133: 182–207. https://doi.org/10.1016/j.watres.2017.12.029.

Tu, Renjie, Wenbiao Jin, Song-Fang Han, Binbin Ding, Shu-hong Gao, Xu Zhou, Shao-feng Li, et al. 2020. "Treatment of Wastewater Containing Linear Alkylbenzene Sulfonate by Bacterial-Microalgal Biological Turntable." *Korean Journal of Chemical Engineering* 37(5): 827–34. https://doi.org/10.1007/s11814-020-0499-0.

United Nations. 2017. *World Water Development Report: Wastewater: The Untapped Resource.*

Usha, M.T., T. Sarat Chandra, R. Sarada, and V.S. Chauhan. 2016. "Removal of Nutrients and Organic Pollution Load from Pulp and Paper Mill Effluent by Microalgae in Outdoor Open Pond." *Bioresource Technology* 214(August): 856–60. https://doi.org/10.1016/j.biortech.2016.04.060.

Vernouillet, Gabrielle, Philippe Eullaffroy, André Lajeunesse, Christian Blaise, François Gagné, and Philippe Juneau. 2010. "Toxic Effects and Bioaccumulation of Carbamazepine Evaluated by Biomarkers Measured in Organisms of Different Trophic Levels." *Chemosphere* 80(9): 1062–68. https://doi.org/10.1016/j.chemosphere.2010.05.010.

Voulvoulis, Nikolaos. 2018. "Water Reuse from a Circular Economy Perspective and Potential Risks from an Unregulated Approach." *Current Opinion in Environmental Science and Health*. Elsevier B.V. 2: 32–45.https://doi.org/10.1016/j.coesh.2018.01.005.

Wall, Kevin. 2018. "The Evolution of South African Wastewater Effluent Parameters and Their Regulation: A Brief History of the Drivers, Institutions, Needs and Challenges." *The Journal for Transdisciplinary Research in Southern Africa* 14(1): 581. https://doi.org/10.4102/td.v14i1.581.

Wang, Yue, Shih Hsin Ho, Cheng Chieh Lun, Wan Qian Guo, Dillirani Nagarajan, Nan Qi Ren, Duu Jong Lee, and Jo Shu Chang. 2016. "Perspectives on the Feasibility of Using Microalgae for Industrial Wastewater Treatment." *Bioresource Technology.* Elsevier Ltd. 222: 485–497. https://doi.org/10.1016/j.biortech.2016.09.106.

Watanabe, Naoko, Brian A. Bergamaschi, Keith A. Loftin, Michael T. Meyer, and Thomas Harter. 2010. "Use and Environmental Occurrence of Antibiotics in Freestall Dairy Farms with Manured Forage Fields." *Environmental Science and Technology* 44(17): 6591–600. https://doi.org/10.1021/es100834s.

Wei, Ruicheng, Feng Ge, Siyu Huang, Ming Chen, and Ran Wang. 2011. "Occurrence of Veterinary Antibiotics in Animal Wastewater and Surface Water around Farms in Jiangsu Province, China." *Chemosphere* 82(10): 1408–14. https://doi.org/10.1016/j.chemosphere.2010.11.067.

Wieck, Stefanie, Oliver Olsson, and Klaus Kümmerer. 2016. "Possible Underestimations of Risks for the Environment Due to Unregulated Emissions of Biocides from Households to Wastewater." *Environment International* 94(September): 695–705. https://doi.org/10.1016/j.envint.2016.07.007.

Wikfors, Gary H., and Masao Ohno. 2001. "Impact of Algal Research in Aquaculture." *Journal of Phycology. Journal of Phycology* 37(6): 968–974. https://doi.org/10.1046/j.1529-8817.2001.01136.x.

Wilde, Edward W., and John R. Benemann. 1993. "Bioremoval of Heavy Metals by the Use of Microalgae." *Biotechnology Advances* 11(4): 781–812. https://doi.org/10.1016/0734-9750(93)90003-6.

Windsor, Fredric M., M. Glória Pereira, Charles R. Tyler, and Steve J. Ormerod. 2019. "Persistent Contaminants as Potential Constraints on the Recovery of Urban River Food Webs from Gross Pollution." *Water Research* 163(October): 114858. https://doi.org/10.1016/j.watres.2019.114858.

Wu, Jane-Yii, Chyi-How Lay, Chin-Chao Chen, Shin-Yan Wu, Dandan Zhou, and Peer Mohamed Abdula. 2020. "Textile Wastewater Bioremediation Using Immobilized Chlorella sp, April. Wu-G23 with Continuous Culture." *Clean Technologies and Environmental Policy*, 23: 153–161. https://doi.org/10.1007/s10098-020-01847-6.

Xiong, Jiu-Qiang, Sun-Joon Kim, Mayur B. Kurade, Sanjay Govindwar, Reda A.I. Abou-Shanab, Jung-Rae Kim, Hyun-Seog Roh, Moonis Ali Khan, and Byong-Hun Jeon. 2019. "Combined Effects of Sulfamethazine and Sulfamethoxazole on a Freshwater Microalga, *Scenedesmus obliquus*: Toxicity, Biodegradation, and Metabolic Fate." *Journal of Hazardous Materials* 370(May): 138–46. https://doi.org/10.1016/j.jhazmat.2018.07.049.

Xiong, Jiu-Qiang, Mayur B. Kurade, and Byong-Hun Jeon. 2017. "Biodegradation of Levofloxacin by an Acclimated Freshwater Microalga, Chlorella vulgaris." *Chemical Engineering Journal* 313(April): 1251–57. https://doi.org/10.1016/j.cej.2016.11.017.

Yadavalli, Rajasri, C.S. Rao, Ramgopal S. Rao, and Ravichandra Potumarthi. 2014. "Dairy Effluent Treatment and Lipids Production by *Chlorella pyrenoidosa* and *Euglena Gracilis* : Study on Open and Closed Systems." *Asia-Pacific Journal of Chemical Engineering* 9(3): 368–73. https://doi.org/10.1002/apj.1805.

Zhang, Dongqing, Richard M. Gersberg, Wun Jern Ng, and Soon Keat Tan. 2014. "Removal of Pharmaceuticals and Personal Care Products in Aquatic Plant-Based Systems: A Review." *Environmental Pollution* 184(January): 620–39. https://doi.org/10.1016/j.envpol.2013.09.009.

Index

A